Applications of Artificial Intelligence in Engineering VI

SIXTH INTERNATIONAL CONFERENCE
ON
ARTIFICIAL INTELLIGENCE IN ENGINEERING
AIENG/91

SCIENTIFIC COMMITTEE

Acknowledgement is made to N. Woodcock *et al.* for the use of Figure 3 on p. 907, which appears on the front cover of this book.

Applications of Artificial Intelligence in Engineering VI

Editors: G. Rzevski, The Open University, U.K.
R.A. Adey, Wessex Institute of Technology, U.K.

Computational Mechanics Publications
Southampton Boston

Co-published with

Elsevier Applied Science
London New York

G. Rzevski
Faculty of Technology
The Open University
Walton Hall
Milton Keynes
MK7 6AA
UK

R.A. Adey
Computational Mechanics Institute
Wessex Institute of Technology
Ashurst Lodge
Ashurst
Southampton
SO4 2AA, UK

Co-published by

Computational Mechanics Publications
Ashurst Lodge, Ashurst, Southampton, UK

Computational Mechanics Publications Ltd
Sole Distributor in the USA and Canada:

Computational Mechanics Inc.
25 Bridge Street, Billerica, MA 01821, USA

and

Elsevier Science Publishers Ltd
Crown House, Linton Road, Barking, Essex IG11 8JU, UK

Elsevier's Sole Distributor in the USA and Canada:

Elsevier Science Publishing Company Inc.
655 Avenue of the Americas, New York, NY 10010, USA

Cataloguing In Publication Data are available from the British Library

ISBN 1-85166-678-8 Elsevier Applied Science, London, New York
ISBN 1-85312-141-X Computational Mechanics Publications, Southampton
ISBN 1-56252-069-5 Computational Mechanics Publications, Boston, USA

Library of Congress Catalog Card Number 91-72244

CONTENTS

SECTION 1: ENGINEERING DESIGN

SECTION 6: ROBOTICS

SECTION 7: KNOWLEDGE ELICITATION AND REPRESENTATION

SECTION 8: THEORY AND METHODS FOR SYSTEM DEVELOPMENT

PREFACE

This book contains papers presented at the sixth International Conference on Application of Artificial Intelligence in Engineering held in Oxford, UK in July 1991. The first conference in this series was held in Southampton, UK in 1986, the second in Cambridge, Massachusetts, USA in 1987, the third in Palo Alto, California, USA in 1988, the fourth in Cambridge, UK in 1989 and the fifth in Boston, Massachusetts, USA in 1990. The conference series has now established itself as the unique forum for the presentation of the latest research, development and application of artificial intelligence (AI) in all fields of engineering. Consequently, books of conference proceedings provide a historical record of the application of AI in engineering design, analysis, simulation, planning, scheduling, monitoring, control, diagnosis, reliability and quality, as well as in robotics and manufacturing systems, from the early beginnings to mature applications of today.

Whilst previously the field was dominated by knowledge-based systems, in this latest volume, for the first time, a significant proportion of papers cover the paradigms of neural networks and genetic algorithms. Learning and self-organising behaviour of systems based on these paradigms are particularly important in engineering applications.

From a large number of submitted proposals over sixty papers have been selected by members of the Advisory Committee who acted as referees. Papers have been grouped under the following headings.

SECTION 1 - Engineering Design
SECTION 2 - Engineering Analysis and Simulation
SECTION 3 - Planning and Scheduling
SECTION 4 - Monitoring and Control
SECTION 5 - Diagnosis, Safety and Reliability
SECTION 6 - Robotics
SECTION 7 - Knowledge Elicitation and Representation
SECTION 8 - Theories and Methods for System Development

I am very grateful to members of the Advisory Committee and all those who have helped with the difficult task of refereeing, selecting and presenting this valuable collection of papers.

George Rzevski
The Open University, UK

SECTION 1: ENGINEERING DESIGN

Genetic Algorithms as a Computational Theory of Conceptual Design

D.E. Goldberg

Department of General Engineering, University of Illinois at Urbana-Champaign, Urbana, IL 61801, USA. E-mail: goldberg@vmd.cso.uiuc.edu

ABSTRACT

The essentially inductive processes of conceptual design have received scant attention in those portions of the design literature concerned with effective computation or mathematical rigor. This paper draws a connection between the discriminative and recombinative processes of conceptual design and *genetic algorithms* (GAs)—search procedures based on the mechanics of natural genetics and natural selection. Recent empirical and theoretical results with a type of GA called a messy genetic algorithm support the conjecture that GAs can solve all problems no harder than the functions of *bounded deception* in time that grows no more quickly than a polynomial function of the number of decision variables. These results suggest that inductive designers—far from wasting their efforts when they bet on some combination of past designs—are engaging in a computationally effective means of solving very difficult, even misleading, design problems. Although more theoretical and computational work is needed, the paper shows one path to a more rigorous theory of conceptual design, a path that should help put design on the same mathematical foundations as analysis without detracting from the joy, or the necessity, of human invention.

INTRODUCTION

The creative processes of engineering design have long been regarded a black art. While the engine of analysis steamrolls ever forward, our understanding of conceptual design seems locked in a timewarp of platitudes, vague design procedures, and problem-specific design rules. Attempts to make portions of the design process more rigorous have been most successful when they have concentrated on those processes downstream of conceptual design, processes that are more nearly deductive. For example, the design use of some of the techniques of artificial intelligence and expert systems have largely concentrated on detailed component design, configuration, and the satisfaction of well-specified design codes and constraints. On the other hand, few efforts have reasoned quantitatively about the essential inductive or synthetic processes of *conceptual design*. Put another way, few studies have examined the eureka moment and the mental computations that get us there; yet, surely it is the shroud of mystery surrounding the processes of discovery, innovation, and invention that most urgently needs to be lifted if we are to get beyond the current witches brew.

An exception to this state of affairs may be found in the literature of *genetic algorithms* (GAs) [1,2], although the connection of this body of work to design theory has been poorly understood, if recognized at all. GAs are search procedures based on the mechanics of natural genetics and natural selection, and they have been used to good effect in solving a number of difficult problems of search, optimization, and machine learning [1,3]. A detailed discussion of genetic algorithms in the light of induction thinking has been presented elsewhere [4], and a brief discussion connecting genetic algorithms to design theory has been offered [5], but for the most part the connection to design theory has languished. That a connection should exist is not surprising, as qualitative discussions viewing design as evolution [6] and evolution as design [7,8] abound. What GAs bring to the party is a refreshing dose of *computational* and *quantitative* rigor.

This paper seeks to make the connection between GAs and design more clearly by viewing genetic algorithms *as* a computational theory of design. Specifically, the paper considers an idealized framework of conceptual design and shows how a type of GA called a *messy genetic algorithm* (mGA) fits within that framework. This fit is no parlor curiosity as theoretical calculations and computational experiments support the conjecture that mGAs solve a large class of difficult conceptual design problems to global optimality in polynomial time. The paper points out that if this conjecture can be proven, conceptual design will have no less rigorous an intellectual footing than analysis has long enjoyed, and this will have been accomplished in a manner that helps explain the magic of innovation in both qualitative and quantitative terms without destroying the beauty of the eureka moment, without detracting from the designer's art.

AN IDEALIZED FRAMEWORK FOR CONCEPTUAL DESIGN

A number of approaches to the study of engineering design have been taken over the years. The oldest among these might be called the list-of-design-lists approach, where the design process is identified by lists of design procedures and methods (see [9] for a thorough, current example of this approach). These qualitative discussions and listings sometimes capture important truths about design, but somehow they lack the intellectual weight of modern engineering analysis, which relies from square one on mathematical classification and quantification. Attempts to remove some of the ambiguity of the list-of-lists approach have divided along a number of lines, including, an axiomatic approach [10] and what might be called design-as-computer-program approaches [11]. Suh [10] derives principles of design from two axioms, while the design-as-program approaches attempt to codify portions of the design process using techniques of artificial intelligence (AI) and expert systems. While one might doubt whether all of design theory may be derived from two axioms, Suh's call for mathematical rigor rings true, as does the AI camp's call for a more computational approach. Yet, so far these more rigorous approaches have failed to capture some of the truths about design, discovery, and innovation that are at least qualitatively captured by the lists of lists. Here, we try to blend the best of these worlds by first defining an idealized framework of conceptual design, thereafter defining an algorithm to do design that appeals to our own intuitive notions of innovation and invention. The attempt will not capture the full richness of human innovation; it is unclear whether this is possible. Instead, the focus is on incorporating important mechanisms and showing that an efficient and convergent computational procedure results. This will contrast starkly with the ploddingly deductive schemes that now dot the design-computations landscape.

What is the essence of conceptual design? Certainly there must be some problem to solve, there must be someone or something to solve it, there must be one or more designs, and there must be a means to comparing alternative designs or designers. We capture these entities by defining the four objects of idealized conceptual design:

1. the *design challenge* C;
2. the *design* Z;
3. the *designer* D; and
4. the *design competition* K.

The *design challenge* or *problem* C is exactly that. It is the problem that must be solved, and it is an ordered pair:

$$C = (S, R), \tag{1}$$

where S is the *solution space* (elements $s \in S$ are called *solutions*), and R is the *design relation*, a partial ordering on S (recall that a partial ordering is a relation that is antisymmetric and transitive; in this context think of the design relation as a "better-than" ordering, but remember that it need not order all pairs of elements of S). Formally, C is a partially ordered set or poset, and recognizing its definition almost immediately poses the *essential search problem of conceptual design*: that of finding the subset of linearly ordered elements of S that are maximal with respect to the design relation. Note that nothing has been said here about objective functions or utility values. In certain problems, the design relation R may be induced by the existence of scalar or vector-valued functions and an appropriate relation over the function values. In other words, R may itself be determined by an ordered pair (f, R_f), where f is an *objective function*, $f : S \rightarrow \mathbf{R}^n$, and R_f is an appropriate partial-ordering relation over the function values. When the design relation may be written in such terms we say that we have a problem in *design optimization*. When $n = 1$ the optimization is *scalar* or *simple*, and when $n > 1$ the problem is *multicriterion* or *multiobjective*. Note that under these definitions all optimization problems are search problems, but the converse is not necessarily true. Also note that the requirement for a partial ordering can be weakened by requiring a probability distribution over the ordered pairs of the relation and by requiring density function values (when they are nonzero) to favor one set element over another (probabilistic antisymmetry). In optimization problems such a probability distribution is often induced with a noisy f and a deterministic relation. The extension to probabilistic problems is important, and a number of other functions and relations in this theory may be similarly relaxed, but in the remainder we only consider the deterministic theory.

The observation that the essential task of design is to find good *subsets* of the solution space leads to a definition of a *design proposal.* A design proposal O is simply a member of the power set of the solution space S, that is $O \in \mathcal{P}(S)$. The maximal subset of linearly ordered elements of S under the design relation is called the *ideal proposal*, and individual solutions are said to be *optimal* if they are elements of the ideal proposal. In discrete solution spaces, we may define *k-enlarged* or *near-ideal proposals* as being that subset of S that includes all solutions within a chain length k of some optimal solution, and any such solution is called *k-optimal* or *near-optimal at length k.* For infinite solution spaces, an appropriate metric is required over chains to define near-optimality in an analogous manner. In those cases where the design relation is induced by a scalar or vector function f and an appropriate ordering relation over the f values, near-ideality and near-optimality may be defined in terms of some distance metric over the f values fairly directly.

Once a proposal is defined, we are close to a good definition of an idealized conceptual *design.* In this framework, a design Z is an ordered pair

$$Z = (O, t), \tag{2}$$

where O is a proposal as just defined, and $t \in \mathbf{R}$ is the design time, a measure of the computational cost of *producing* the solution (it could easily be made a vector-valued function, but the scalar definition suffices here). Note, that this time or cost has nothing to do with satisfaction of the design relation R; the cost is not a cost of manufacturing or producing the design. It simply captures the essential point that some designers work faster (or charge less) than others. We define an *ideal design* as one that produces an ideal proposal in zero time, by assigning zero time to near-ideal proposals we may likewise define *near-ideal designs.*

Designs do not emanate from thin air, so we imagine a designer D:

$$D = (\mathcal{C}, \mathcal{C}', \mathcal{Z}, a, d, b) \tag{3}$$

where $\mathcal{C}$ is the set of challenges, $\mathcal{C}'$ is the set of *perceived challenges*, a is an *abstraction function*, d is a *design function* or method, and b is a *billing function.* To follow the idealization, we imagine that the designer first examines the challenge through his abstraction function a, thereby generating an abstracted or *perceived challenge*, $a : \mathcal{C} \rightarrow \mathcal{C}'$. The perceived challenge is used as input to the design method d, $d : \mathcal{C}' \rightarrow \mathcal{P}(S')$, a mapping from the space of perceived challenges to the power set of the perceived solution space. In other words, a design method takes a problem and creates a design proposal. (Other definitions of a designer are possible, and one might argue that the designer would be better defined as an automaton, with some notion of time as well as internal design state or memory. Such an extension is straightforward and explicitly recognizes the adaptive capability of individual designers, but it is unnecessary here.) The billing function

$b : C' \rightarrow \mathbf{R}$, takes the perceived challenge as input and generates the design time or cost associated with the generated proposal. Together, d and b generate the proposal and time components of a design.

These definitions invite a number of other definitions using straightforward set theory. For example, we say that a designer is *ideally perceptive* or *perceptive* when his perception coincides with reality, that is when $C = C'$; he is *sufficiently perceptive* when despite perceptual difficulty the maximal set of the perceived challenge coincides with that of the original challenge: $\max(C') = \max(C)$. Beyond this, various degrees of perceptual difficulty (myopia=sees too little, hyperopia=sees too much, stigmatism=sees too little in some places and too much elsewhere) can be defined. This exercise is fairly straightforward and is not pursued here further.

We should, however, make a distinction between the design capacity of different designers. One straightforward measure considers the maximum cardinality of the proposals in a designer's image. Many (most?) optimization algorithms generate singleton proposals, and we call these *cardinality-one* designers. Other population-oriented schemes such as simplex methods, complex methods, and many genetic algorithms converge to some fixed-size, finite population. We might call these *cardinality-κ* or *fixed-κ* schemes, where κ is the population size. Schemes that converge to sets of infinite cardinality can be classified depending upon whether their proposals are *countable* or *uncountable*. Of course, the cardinality of a designer is only an issue if the challenges being tackled have cardinality higher than those of the designer. We can calculate the *designer-ideal-design cardinality ratio* or *DIDC ratio* as the ratio of the designers maximum proposal cardinality to the cardinality of the ideal design. When that number equals or exceeds one, the designer is said to have *sufficient capacity* to tackle the problem, and other classifications are possible.

Sufficient capacity is important, but the utilization of that capacity to get a good distribution of points is another matter. It is useful to subdivide the ideal proposal (or near-ideal proposal) into connected subsets of points. In the case of *cardinality-κ* designers on continuous problems we can't hope to define the solution exactly, but we can hope for good *representation*, a good distribution of proposal elements among the connected sets of the ideal or near-ideal proposal and a good spread of solutions within connected sets as can be measured on the basis of some meaningful distance metric.

It is also useful to classify billing functions in relation to the size of the underlying search space, but we save this discussion until later when we consider a particular space S. Note, that although the terminology implies that we are talking about a single designer, the framework places few limitations on the form of the design method used. Certainly, a method d can be envisioned as integrating the efforts of a number of cooperative design agents.

At another level, we do explicitly accommodate the notion of parallel effort, specifically competitive parallel effort. We define a *design competition* K as

$$K = (\mathcal{C}, \mathcal{G}, \mathcal{R}_k, \mathcal{I}, c), \tag{4}$$

where $\mathcal{C}$ is the set of design challenges as before, $\mathcal{G}$ is *guild set* or the set of m-tuples of designers, $\mathcal{G} = \mathcal{D}^m$, $\mathcal{R}_k$ is a set of relations among designers (call them the *K-relations*), $\mathcal{I}$ is the set of index sets over the integers between 1 and m,

$\mathcal{I} = \mathcal{P}(I)$, $I = \{1 \ldots m\}$, and c is the *choice* function, $c : \mathcal{R}_k \times \mathcal{C} \times \mathcal{G} \rightarrow \mathcal{I}$. In words, the design competition chooses a subset of a particular guild's designers as having won the competition, using knowledge of the particular challenge C and the competition relation. Although the competition relation has few restrictions placed on it, it is in the K-relation that tradeoffs between solution quality and design time are considered. Note also that the decision here is cast as one of choosing a subset of designs (designers). The choice of a particular solution from among the chosen designs is here viewed as a separate problem termed the *decision-making problem.*

Taken together, the idealized framework helps identify the important entities in the design process without restricting the type of problem to be solved, the methods used, or the means of comparing different designs or different methods. Despite the generality of the framework, its identification has helped make some important distinctions, distinctions that are often overlooked. The framework has helped clarify the relationship between search and optimization with search being the more general category, requiring only a partial-ordering relation among the elements of the search space, and optimization being somewhat more restrictive by requiring a function over the solution space and a subsidiary relation. The general setting has also helped point out that design is (or should be) about finding subsets of solutions. Although sometimes we get confused about this because we are forced to choose a single solution from the proposals generated during the design process, the fundamental problem posed by a partial ordering is the location of the maximal subset (or sufficiently close approximations to it). This has important implications for a designer's approach to design, and raises many questions regarding the adequacy of cardinality-one or low-κ designers. The framework has also encouraged us to transform a number of loose semantic distinctions into more formal set-theoretic definitions. This surface has only begun to be scratched, and these distinctions and classifications should help us better understand and compare different conceptual design algorithms and methods.

COMING BACK DOWN TO EARTH

Lofty frameworks may lift the soul (and they may help us make important theoretical distinctions clearly), but unfortunately they do little to actually solve problems. To bring us down to earth, we restrict the solution space to the binary l-tuples, $S = \mathbf{B}^l$, $\mathbf{B} = \{0, 1\}$. Moreover, we restrict the design relations to those that may be written as a scalar function $f : \mathbf{B}^l \rightarrow \mathbf{R}$ and the usual $>$ relation over the f values. In addition, we restrict ourselves to considering functions that have a unique singleton maximal set. Despite these restriction, from both a design and a computational perspective, this class of problems is relevant.

From a design perspective, each of the binary variables—call them $x_i \in \mathbf{B}$—may be thought of as some design *feature* or *attribute*. For example, the presence or absence of a particular widget may be encoded by a 1 or a 0, as can the choice between any other two mutually exclusive alternatives. Moreover, numerical parameters may be encoded with any degree of precision using mapped binary arithmetic [1]. In these ways, many conceptual design problems of practical import can be cast over $\mathbf{B}^l$.

From a computational perspective, this class of problems should not be underestimated. Although finite, these problems are far from trivial. With no restriction on the structure of f, there exist problems that can only be solved through exhaustive enumeration. For example, the needle in a haystack,

$$f(\mathbf{x}) = \begin{cases} 1, & \text{when, } \mathbf{x} = \mathbf{x}^*; \\ 0, & \text{otherwise,} \end{cases} \tag{5}$$

requires enumeration or random search to find the needle. Examining the billing function of such algorithms, we recognize that the amount of computation required grows exponentially as the number of decision variables increases: $b(l) = O(2^l)$. For example, a 100-bit problem with one function evaluation per nanosecond requires $\sim$ 40,000 billion years for solution, a time that exceeds current estimates of the age of the universe ($\sim$ 16.5 billion years). On the other hand, many other problems have fairly simple structure that can be solved by less computationally intensive algorithms.

For example, consider the bitwise linear problems determined by an equation of the form

$$f(\mathbf{x}) = \sum_{i=1}^{l} a_i x_i + b, \tag{6}$$

where $a_i, b \in \mathbf{R}$. It is clear that almost any greedy algorithm will work on this problem class. Imagine starting by evaluating a randomly generated or predetermined bit vector. Then progressing bit by bit, make each single-bit change, compare the function values of the changed solution and the current solution, and keep as the new solution only those solutions that yield improvement. With this simple problem, we expect to converge to the best solution in exactly $l + 1$ function evaluations, yielding a billing function complexity of $b = O(l)$.

These two extremes are quite telling, both from the standpoint of the problems being solved and the algorithms effective for their solution. On the one hand, the needle in a haystack has very little information or landscape structure to guide us toward a solution, and as a result, algorithms for its solution can do no better than enumerate or walk randomly. On the other hand, problems with a lot of structure can be solved more quickly if that structure is utilized, but that convergence is usually bought at the price of ineffectiveness on other problems. What would be nice is if design algorithms existed that could exploit structure when it is there, but do no worse than random search when it is not. Of course, the harder part of this is to exploit arbitrary structure when it exists and do so in a computationally effective—in a polynomial or $b(l) = O(l^k)$—manner. One can always do no worse than random or enumerative search by keeping a background component of randomness in operation in hybrid with the more structure-exploiting component. As we shall soon see, genetic algorithms use a hybrid of structure exploitation and diversity generation in their operator mix to good effect.

GENETIC ALGORITHMS AS COMPUTATIONAL INNOVATION

Of the two algorithms discussed in the previous section, which is closest to the way designers really design? Neither, seems very close, yet in one important sense, the enumerative scheme seems furthest from the way people handle difficult design tasks. Human designers haven't the patience to plow through a solution space enumeratively, and even if they did, their swashbuckling, speculative instincts would get the better of them; in some sense, humans are born to experiment, to go boldly where no one has gone before. The greedy algorithm seems closer to the mark in that it does incorporate an element of experimentation and it does decide matters quite quickly (too quickly). Yet, the greedy algorithm, too, misses some of the speculative essence of human innovation. To identify the key features, this section examines the views of several writers regarding the mental processes of invention and shows how those views combine to form a loose prescription for the design of a genetic algorithm. This connection is then made firmer with both intuitive and straightforward mathematical arguments. The end result is a view of genetic algorithms as a lower bound on the performance of human innovation.

Views of invention

The mathematician Hardamard [12] was quite explicit in identifying his view of the essential mechanics of invention:

> We shall see a little later that the possibility of imputing discovery to pure chance is already excluded... On the contrary that there is an intervention of chance but also a necessary work of unconsciousness, the latter implying and not contradicting the former, appears... when we take account not merely of the results of introspection, but of the very nature of the question.
>
> Indeed, it is obvious that invention or discovery, be it in mathematics or anywhere else, takes place by combining ideas. Now, there is an extremely great number of such combinations, most of which are devoid of interest, while on the contrary, very few of them can be fruitful. Which ones does our mind—I mean our conscious mind—perceive? Only the fruitful ones, or exceptionally, some which could be fruitful.

Related sentiments have been voiced by the poet and critic Valéry:

> It takes two to invent anything. The one makes up combinations; the other chooses, recognizes what he wishes and what is important to him in the mass of the things which the former has imparted to him.

Both writers understand that human discovery consists of at least two things:

1. the ability to discriminate, the ability to select good from bad; and
2. the ability to recombine different ideas.

In a moment will show that computational counterparts exist in genetic algorithms that correspond to these qualitative descriptions. Before we make that connection, we examine one other view of the inventive process, a view expressed by Bulwer-Lytton:

> Invention is nothing more than a fine deviation from, or enlargement on a fine model...Imitation, if noble and general, insures the best hope of originality.

Certainly this writer recognizes the importance of selection, as his object of modification is a good idea or "fine model," although the process of discovery here is viewed as one of blind, yet modest, change. This, too, connects with the theory of genetic algorithms, and labels the writer a proponent of innovation through what a biologist might call *point mutations.* As we shall soon see, the union of these viewpoints gives us a remarkably faithful, albeit somewhat vague, prescription for the design of a genetic algorithm.

Genetic algorithm essentials

Loosely stated, genetic algorithms are search procedures based on the mechanics of natural genetics and natural selection. To be a GA in the sense of Holland [2], a procedure must contain one or more representatives of each of the following types of operators:

1. selection;
2. recombination; and
3. diversity generation.

The similarity between the views of invention briefly examined in the previous subsection and this list is already quite remarkable, and we examine representatives of each class of operator to be sure we are here talking about well-posed computations.

Selection is the survival of the fittest within a GA, and there are many ways to do it effectively. Ranking, tournament, and proportionate schemes have all been used with success, but the key idea is to give more copies (more offspring) to better individuals.

There are also many ways to implement effective recombination operators. Simple crossover assumes fixed-length structures, and after random mating, a cross site along the mated strings is chosen at random, and position values are swapped between the two strings following the cross site. For example, starting with the two strings $A = 11111$ and $B = 00000$, if the random selection of a cross site generates a 3, we would obtain the two strings $A' = 11100$ and $B' = 00011$ following crossover, and these new strings would be placed in the population. This process can be performed in overlapping populations, or full nonoverlapping populations can be created in each generation through repeated selection, mating, and crossing.

Selection and recombination are surprisingly simple operators, involving nothing more complex than random number generation, string copying, and partial string exchanges, but their combined action is responsible for much of a GA's search punch. Yet this is not surprising when viewed from the standpoint of the processes of innovation or discovery discussed earlier. What is it that we are doing when we are being creative or innovative? Most often we are combining sets of features or attributes that worked well in one context (a *notion*) with other notions that worked well in some other context to form a new, possibly better *idea* of how to perform the task at hand. In a similar way, GAs juxtapose many different substrings through the combined action of reproduction and recombination to form new strings. This intuitive view of GA operation has been around for some time [5], as has a more rigorous view of GA power [2]. We will examine the mathematical argument for GAs in a moment. Before we do, we must examine the role of diversity-generating mechanisms such as mutation.

If reproduction and crossover provide much of the innovation of genetic search, what then is the role of the diversity-generating or mutation operators? There are many different mutation operators. For example, in a binary-coded GA, one commonly used operator changes a 1 to a 0 or vice versa with specified mutation probability, p_m. By itself, mutation motivates a random walk through the solution space. When used in concert with selection, it may be viewed as a form of genetic hillclimbing. When used with selection and recombination, it may be viewed as an insurance policy against the loss of needed genetic material. Taken together, the random-walking, hillclimbing, and insurance-policy roles of mutation can be important to the functioning of practical GAs. On the other hand, there has been much too much talk in the literature of natural and artificial genetics about the primacy of mutation to the exclusion of recombination and other mixing operators. To combat such mutation chauvinism, I have elsewhere [1,5] taken the contrary view that mutation is secondary to recombination. This might be viewed as taking one extreme position to combat another, but when genetic operators are viewed in the setting of human invention, the more interesting operations are those of Hardamard and Valéry versus that of Bulwer-Lytton. Moreover, when the operators are examined from a more mathematical perspective, the view supporting the primacy of recombination is the more justifiable. To be fair, however, diversity must come from somewhere (both in the beginning and after substantial convergence), and the hillclimbing role of selection+mutation is an important one in refining solutions after recombination and selection have narrowed the search.

A more rigorous view

The viewpoint of innovation as selection, recombination, and a little mutation jives with some of the literature of invention, and it is consistent with an intuitive explanation of why genetic algorithms work, but if these were the only arguments supporting the case, we would be on shaky ground indeed. Fortunately, Holland [2] placed the theory of genetic algorithms on firm foundations by shifting our attention from individual strings to *similarity subsets* or *schemata*. The loose talk of notions and ideas can be replaced by a discussion of solutions and their schemata, and qualitative arguments can be replaced by quantitative theory. The idea is threefold:

1. Selection increases the numbers of representatives of above-average schemata.
2. Recombination and mutation do not materially degrade the selection-only growth pattern, except when schemata are overly specific or insufficiently tightly coded.
3. Recombination probabilistically juxtaposes good, short, low-order schemata to form new structures.

The first two of these items are described quantitatively by the *schema theorem*, sometimes called the *fundamental theorem of genetic algorithms.* Defining a growth factor γ for a schema $\mathbf{h}$ as

$$\gamma(\mathbf{h}) = \phi(\mathbf{h})[1 - \epsilon(\mathbf{h})], \tag{7}$$

where ϕ is the growth ratio under selection alone and ϵ is an upper bound on the total expected operator disruption, the theorem says that the expected ratio of the number of individuals in $\mathbf{h}$ in the next generation to the number in the current generation is no less than γ. For good schemata under all commonly used selection schemes, $\phi > 1$. The trick in getting good growth then is to make sure the disruption due to recombination, mutation, and other operators is sufficiently small. It may easily be shown that under most recombination and mutation operators that $\epsilon \ll 1$ for short, low-order schemata. The highly fit, short, low-order schemata—call them *building blocks*—therefore grow and dominate the population. This growth is good, but it doesn't say that building blocks are likely to come together and form good strings. This is guaranteed by the *second law of genetic algorithms* [2], sometimes called the *law of positive effect.* Imagine that m building blocks are needed to form a particular solution and imagine that each building block is represented by a proportion of no less than p. Repeated crossing with simple crossover or other low-disruption recombination operators yields a stochastic fixed point of no fewer than p^m individuals with the desired building blocks. Thus we can see that selection with sufficiently nondisruptive operators allows building blocks to grow and take over the population and that recombination juxtaposes those building blocks to find optimal structures with high probability.

There is an article of faith here, however. What happens when the best, short, low-order building blocks do not combine to form the best structures. In fact, we can imagine functions so that on average, low-order building blocks lead toward the complement of the global optimum, and if taken to the extreme these *deceptive* functions [13–15] are harder than the previous needle-in-a-haystack example, because not only is there an isolated optimum but it is surrounded by points that lead away from it. As with the needle in a haystack, there is little hope of solving order-l, fully deceptive problems in polynomial time, but if the degree of maximum deception is bounded, then the theory of GAs says we should be able to solve such problems if we get the linkage right.

Unfortunately, traditional GA theory gives few answers here. Standard suggestions for using reordering operators such as inversion have been unsuccessful in autonomously finding good linkage, and there has been one suggestion that such operators may be inherently ineffective [16]. Furthermore, algorithms that depend on the user encoding good linkage require more information than a supposedly robust method should require. In response to these difficulties, a radically rearranged genetic algorithm called a *messy genetic algorithm* or *mGA* has been invented [17, 18]. As nature has formed complex organisms from simple organisms, messy GAs find fuller solutions by cutting and splicing bits and pieces of partial solutions. The details of mGAs are covered elsewhere [17, 18], but it is interesting to note that in functions of bounded deception, mGAs have found global optima, and they have done so in a time that grows only as a polynomial function of the number of decision variables: $b(l) = O(l^k)$, where k is the maximum degree of deception. Moreover, theoretical computations have shown that if a parallel computer is used (with a polynomial number of processors) mGAs operate in a time that grows only as $b(l) = O(\log l)$. Whether done serially or in parallel, it has been conjectured that mGAs converge to global optima of functions of bounded deception with a probability that can be made arbitrary close to one. A number of the details of the proof should not be underestimated, but recent variance calculations [19] suggest good pathways to fruitful bounds. Moreover, simple modifications to the mGA should enable the simultaneous solution of multimodal problems, and extensions have also been suggested that will enable mGAs to perform classification and machine learning tasks.

Without getting lost in the details, it should be clear that this is good news for those who want a theory of conceptual design that (1) has some connection with what designers actually do and (2) shows that those processes are computationally efficient and convergent. We are not there yet, but we are getting closer, and the ramifications for design theory are just beginning to be appreciated. It interesting (and important) that the connection between GAs and conceptual design is a loose one. The argument being made here is not that genetic algorithms mimic a clever designer exactly, rather GAs form a lower bound on the performance of a designer that uses recombinative and selective processes. That such a crude lower bound on the performance of such processes appears to yield polynomial performance in tough problems is a surprise. No doubt, genetic algorithmists and design theorists will spend the coming years trying to close the gap between genetic design and the human kind, but for the present, this theory helps do no less than explain in quantitatively and computationally rigorous terms why the mental processes of human innovation are as effective as we have long known they are.

CONCLUSIONS

This paper has (1) constructed an idealized framework for conceptual design, (2) considered a number of views regarding the mental processes of human innovation, and (3) shown how genetic algorithms can be thought of as a bounding model of discriminative and recombinative invention. As human designers recombine bits and pieces of previous designs to form new, possibly better proposals, GAs recombine bits and pieces of artificial chromosomes to search for globally optimal solutions. Although these connections between human innovation and genetic algorithms have been made qualitatively before, recent computational and theoretical results on a particular type of genetic algorithm called a messy

GA support the conjecture that certain GAs solve difficult deceptive problems in a computing time that grows no more quickly than a polynomial function of the number of decision variables. Although efforts should be redoubled to prove this conjecture, the connection between design theory and genetic algorithms drawn in this paper gives us more than a little hope that conceptual design will soon have the same intellectual footing that analysis has always enjoyed. That the addition of such rigor appears as though it will occur without detracting from the designer's art is a welcome, if unexpected, bonus.

ACKNOWLEDGMENTS

This work has been supported by Subcontract No. 045 of Research Activity AI.12 under the auspices of the Research Institute for Computing and Information Systems (RICIS) at the University of Houston, Clearlake under NASA Cooperative Agreement NCC9-16 and by the National Science Foundation under Grants CTS-8451610 and ECS-9022007. Continued support of mGA research under U.S. Army Contract DASG60-90-C-0153 is also acknowledged.

REFERENCES

1. Goldberg, D. E., *Genetic Algorithms in Search, Optimization, and Machine Learning*, Addison-Wesley, Reading, MA, 1989.

2. Holland, J. H., *Adaptation in Natural and Artificial Systems*, University of Michigan Press, Ann Arbor, 1975.

3. Davis, L. (Ed.), *The Handbook of Genetic Algorithms*, Van Nostrand Reinhold, New York, 1991.

4. Holland, J. H., Holyoak, K. J., Nisbett, R. E. and Thagard, P. R., *Induction: Processes of Inference, Learning, and Discovery*, The MIT Press, Cambridge, MA, 1986.

5. Goldberg, D. E., *Computer-aided Gas Pipeline Operations Using Genetic Algorithms and Rule Learning* (Ph.D. dissertation, Civil Engineering), University of Michigan, Ann Arbor, 1983.

6. French, M. J., *Invention and Evolution: Design in Nature and Engineering*, Cambridge University Press, Cambridge, 1988.

7. Thompson, D., *On Growth and Form*, Abridged Edition (J. T. Bonner, Ed.), Cambridge University Press, 1961.

8. Tributsch, H., *How Life Learned to Live* (translated by M. Varon), The MIT Press, Cambridge, MA, 1982.

9. Pahl, G. and Beitz, W., *Engineering Design* (translated by A. Pomerans and K. Wallace), Springer-Verlag, Berlin, 1988.

10. Suh, N. P., *The Principles of Design*, Oxford University Press, New York, 1990.

11. Finger, S. and Dixon, J. R., A Review of Research in Mechanical Engineering Design. Part I: Descriptive, Prescriptive and Computer-based Models of Design Processes, *Research in Engineering Design*, Vol. 1, pp. 51–67, 1989.

12. Hadamard, J., *The Psychology of Invention in the Mathematical Field*, Princeton University Press, Princeton, 1945.

13. Goldberg, D. E., Simple Genetic Algorithms and the Minimal Deceptive Problem. Chapter 6. *Genetic Algorithms and Simulated Annealing* (L. Davis, Ed.), Pitman, London, pp. 74–88, 1987.

14. Goldberg, D. E., Genetic Algorithms and Walsh Functions: Part I, a Gentle Introduction, *Complex Systems*, Vol. 3, pp. 129–152, 1989.

15. Goldberg, D. E., Genetic Algorithms and Walsh Functions: Part II, Deception and Its Analysis, *Complex Systems*, Vol. 3, pp. 153–171, 1989.

16. Goldberg, D. E. and Bridges, C. L., An Analysis of a Reordering Operator on a GA-hard Problem, *Biological Cybernetics*, Vol. 62, 397–405, 1990.

17. Goldberg, D. E., Korb, B. and Deb, K., Messy Genetic Algorithms: Motivation, Analysis, and First Results, *Complex Systems*, Vol. 3, pp. 493–530, 1989.

18. Goldberg, D. E., Deb, K. and Korb, B., Messy Genetic Algorithms Revisited: Studies in Mixed Size and Scale, *Complex Systems*, Vol. 4, pp. 415–444, 1990.

19. Goldberg, D. E. and Rudnick, M., *Genetic Algorithms and the Variance of Fitness* (IlliGAL Report No. 91001). The Illinois Genetic Algorithms Laboratory, University of Illinois at Urbana-Champaign, Urbana, IL, 1991.

Pre-schematic Electronic Circuit Designer

D.D. Harris, T. McNeill, P.H. Sydenham

Measurement & Instrumentation Systems Centre, University of South Australia, P.O. Box 1, Ingle Farm, SA 5098, Australia

ABSTRACT

The design and detailing of electronic circuitry is largely an intuitive process conducted by skilled persons. System requirements, architecture, circuit and component choice are not yet subject to automation processes. Using A.I. techniques, coupled to heuristic and formalised engineering procedures of design, the AJITA CAE system described is proving that automation can be achieved in the pre-schematic stage of electronic system design.

INTRODUCTION

There has been considerable work done in applying A.I. techniques to electronic circuit design. In various papers on the analogue circuit designer BLADES, for example, (El-Turky et al, [1],[2],[3]) the authors provide a detailed description of this system as well as considerable background information on the task. Intelligent software techniques are also available in commercial software applicable to Very Large Scale Integration (VLSI) integrated circuit design. There is, however, very little knowledge based software commercially available to the designer of circuitry that is usually realised as Printed Circuit Boards (PCBs) which comprise a very large part of the electronic design carried out in industry.

Once a circuit layout has been determined there are many tools which facilitate its conversion into a manufactured product. This paper describes the development of "Ajita", a pre-schematic electronic circuit design system which is targeted at a perceived gap in the software tools currently available. Ajita operates ahead of schematic capture, using intelligent techniques to assist the designer in moving from the product concept to the schematic layout.

Ajita is intended to complement CAD/CAM systems in current use. It presents itself to the designer in a format which, in its "look and feel", has similarities with practices which have become widely accepted in CAD/CAM systems. However, its operation includes the application of rules, both formal and heuristic, applying expertise to guide the designer in the creation of the circuit layout required to meet the operational specification.

ELECTRONIC CIRCUIT DESIGN

Circuit Design Methodology

There are several steps involved in the design of electronic systems, Fig. 1 from Mc Neill [4]. The process begins with a requirement to be satisfied. This requirement is usually identified by someone external to the design process so a specification is required. This is a very important step that is often neglected in current practice as it represents for the designer a time consuming step. M'Pherson [5] and Berlin [6] see it as one of the more important steps.

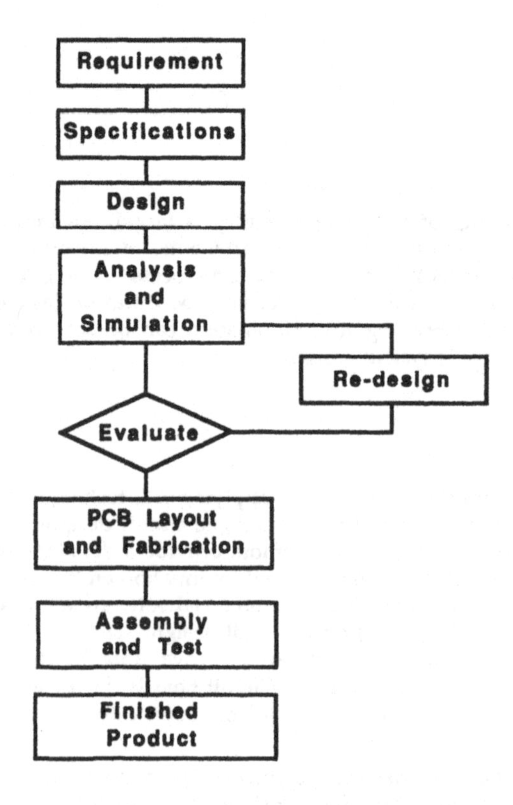

Figure 1. The Engineering of an Electronic System.

With the specification established, the design process follows several sub steps, Fig. 2 also from Mc Neill [4]. These sub steps follow the top down design strategy of formulating a solution in a general sense and then filling in the detail in subsequent steps. Rowles and Leckie [7] and Finkelstein and Finkelstein [8] see this approach as the most productive in design. In the case of electronic systems the initial design in the general sense takes the form of a block diagram where the details of circuit implementation within the blocks is not initially considered but a general functional description is used; for example, "a low pass filter".

In the next sub step the block diagram is refined with more detail being added but there is little development of actual circuit implementation. For example, "a low pass filter with cut-off at 1 kHz and passband gain of 10 dB." The circuit level design then takes place with actual circuit topologies and component values being calculated. The process until this point can be considered as pre-schematic design since there is no need until now to consider a schematic representation of the system.

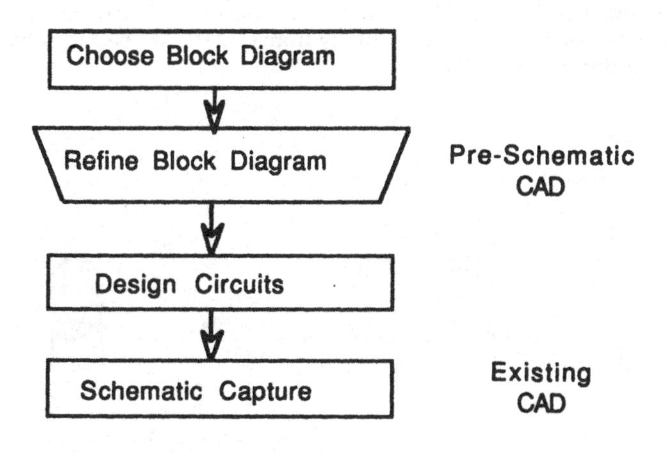

Figure 2. The Design Step.

The circuit is then captured in a schematic form, that is converted into circuit symbology. Analysis and simulation then follow to ensure that specifications are met. M'Pherson [5] suggests that the success of the evaluation step is highly dependent on the specification step. If the design fails to meet specification then a redesign step is undertaken.

The physical implementation of the system can be in a number of forms ranging from custom designed integrated circuit to printed circuit board (PCB) implementations. This paper is concerned with the PCB implementation. Layout and fabrication follow to lead on to the assembly stage where components are loaded into the board. There is a final testing stage to ensure that circuits perform as the simulation would suggest.

CAD Tools

Computer Aided Design and Manufacturing (CAD/CAM) tools are available to support many of the steps from design through to manufacture. In general, the existence and state of development of these tools reflect three factors:-

Need. The cost and time savings provided by CAD tools are critical in meeting market introduction time-scales for electronic products, while the current level of VLSI design of integrated circuits would not be economically possible without the highly developed CAD tools available for their design.

Ease of implementation. Interactive circuit design CAD tools which start with a given circuit level design involve, essentially, the application of procedural and algorithmic programming which is a mature discipline. This is not to minimise the skill required to develop a powerful, useful and user-friendly CAD system but it is in principle a straight forward programming task.

Available computer power. This is reflected, for example, in operating speed, multiple work displays and integrated circuit simulation. Computer power used in this way provides greatly increased utility from CAD/CAM packages but does not extend their domain of operation.

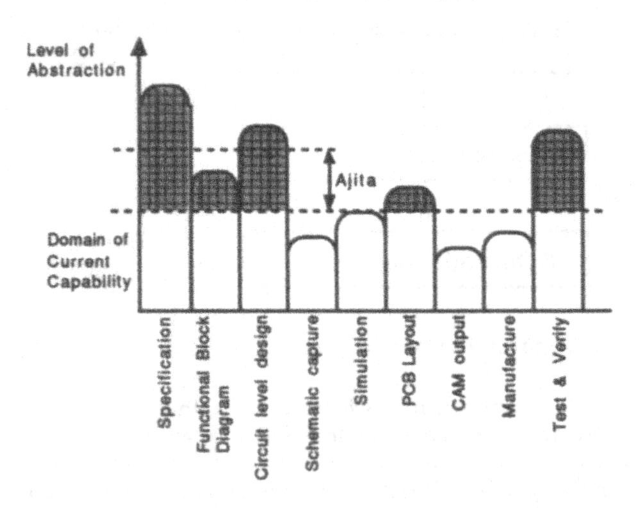

Figure 3. Levels of Knowledge Required in Engineering Design

There are few CAD/CAM tools which extend computing power into more abstract areas of the larger domain of total electronic design, in particular the steps from translation of the product concept into an operating specification through to generation of the schematic of a circuit to meet this specification. This situation is illustrated pictorially in Fig. 3. Note that in PCB layout a skilled person is able to produce a better layout than existing autoplace and autoroute software.

NEED FOR PRE-SCHEMATIC DESIGN

In moving from a product concept to a schematic layout skilled electronic designers apply many rules and techniques of which they are often unaware. This is, of course, a situation frequently encountered by knowledge engineers.

Ajita is a design tool taking up a position between the operating specification and the circuit schematic. Such a system extends the domain of operation of electronic CAD systems beyond currently available technology. Once the circuit schematic has been prepared there is a wide range of CAD/CAM tools available to progress the design through to physical realisation, so Ajita can also be regarded as a "front end" to these tools.

The essential role of a pre-schematic tool is to translate broad functional requirements (amplification, filtering etc) into circuit blocks which will achieve this function *to the needs of the specification*. It must provide circuit schematic pieces with all components selected and sized and with all inter-connections formed and validated.

If the pre-schematic package carries out this task completely then it achieves full automation of this step. However, even if it only achieves partial automation it provides major benefits in building a partial circuit schematic very rapidly, leaving its completion to the designer. In other words, such a package is a very useful design tool even if full automation cannot yet be completely achieved.

A.I. IN CIRCUIT DESIGN TO DATE

The use of A.I. in circuit design can be seen to fall into some distinct categories. There have been quite separate developments in the area of digital design and analogue design , so most of the work treats the two as separate. The categories are as follows:-

Digital Design for VLSI Custom Design

In large production run situations, advantage in custom designed integrated circuits (ICs) was recognised some time ago. The ability to produce complex Very Large Scale Integration (VLSI) ICs in large run sizes for a low cost per unit rapidly surpassed the ability to design such circuits. It was this imbalance that led to the idea of silicon compilers, (Vanhoof et al, [9]), essentially a process by which certain functional circuits are stored as library blocks and can be recalled and placed by the designer thus saving the designer the repetition of laying out identical shapes for each individual component in the circuit. This approach rose from the fact that the circuit level design and floor plan layout are inseparable in contrast to the PCB case in which the circuit level design is a precursor to the layout design.

The reasoning used to decide the appropriate combination of gates to produce a design was easy to identify and makes use of symbolic manipulation. This led directly to the use of A.I. techniques which were especially useful when the hierarchical representation was recognised. (Lang and Mc Cormick, [10]) For example a NAND gate consists of the same combination of transistors and hence the same shape on the silicon, and a flip flop consists of the same combination of NAND gates and so on. Much of the work, Fujita et al [11] is a typical example, avoided analogue design due to reasons that are discussed later.

Digital Semi-custom Design

This follows on from the fully custom design approach but makes the advantages of VLSI design accessible to the smaller, medium scale production run applications. The approach involves having a number of predesigned and masked sub circuits on the same chip. A design is implemented by making the connections on the general chip between various sub circuits.. The process can be in the form of a final masking, in the case of ASIC's (Vanhoof et al, [9]), or by programming in an EPROM fashion in the case of programmable gate arrays. The limitations of the semi-custom approach arise, firstly in the lack of analogue circuits available, and secondly in the fact that small production runs can not be economically implemented using this approach.

Software packages are used to assist the designer in selecting the appropriate connections and optimising the design. A.I. techniques become useful in this area and have been used with much success.

Analogue Circuit IC Design

A relatively recent area of work has been in the area of analogue circuit I.C. design. Digital circuits are presently easy to design with a great deal of automated assistance. The trend is to bring analogue design to a comparative level. There are two reasons for work on analogue circuits lagging work in the digital field. (El-Turky, [2]) Firstly the non-structured nature of analog circuits and the fact that analogue signals are continuous. Secondly the non-linear nature of circuit components has meant that it is necessary to incorporate simulation and accurate models to analyse the implications of each design modification. This is emphasised by work done by Stoffels [12]. A.I. plays a big part here because obviously an experienced designer does not perform such analysis and simulation in achieving optimal analogue design. Work on the BLADES (El-Turky et al, [1].[2],[3]) project recognised this and has achieved much in the understanding of the heuristic and intuitive knowledge employed by the experienced analogue designer. Harjani et al [13] have identified an hierarchical classification of analogue circuits and have built on this to make progress in automating analogue circuit design.

Component Level Design

Another area of circuit design that has been the subject of attention is the area of component level design. This work (Dincbas, [14], Hanrahan and Caetano, [15] and Di Cataldo and Nunnari, [16]) has concentrated on the very bottom of the hierarchy in attempting to design circuits using a selection of components and capturing the reasoning used by an experienced designer to arrive at a circuit design.

This can be seen as a formidable task when one considers the multitude of components in an electronic system and the reasoning associated with them. The AJITA strategy represents a significantly different approach to the problem.

THE AJITA STRATEGY

The AJITA strategy is compatible with the top down approach to electronic system design. It differs from existing approaches in that it provides partially designed blocks in an undimensioned form that are dimensioned to suit the designer's requirements. The nature of this partial design depends upon whether the block is an analogue or digital block.

Circuit level design of analogue circuits can be separated into two steps, a topology design step and a then a dimensioning step. The topology design in a majority of cases consists of selecting an appropriate topology. eg. An amplifier circuit using an operational amplifier will probably have one of a few topologies, one for an inverting amplifier, another for a non-inverting amplifier and so on. Digital circuit level design can be seen as a "sub-circuit selection" step followed by a "connection making" step. eg. In the design of a binary counter we select the type of flip flop we require and then connect the flip flops accordingly.

Upon examination of the various blocks that make up the lowest level of an electronic system, it can be seen that many systems rely on a relatively small number of different types of blocks. The AJITA strategy uses this fact by providing partially pre-designed blocks. (About 30 analogue and 15 digital) (Harpas, [17] and Phan, [18]) In the case of the analogue blocks the partial pre-design takes the form of circuits with pre-defined topologies arranged in an hierarchical menu system. The designer (user) selects the appropriate topology, eg. Inverting Amplifier, and then by answering a series of questions determines the transfer function for the block. Ajita then selects appropriate components for the block to achieve (as nearly as possible) the required transfer function. By taking this approach we have captured the circuit level design knowledge. The dimensioning process is achieved by pre-programmed algebraic equations and heuristics.

For digital blocks, (Phan, [18]), the sub-circuit is selected and then connections are made depending on the function required. This process is very similar to that used in programmable gate arrays. This "make connections" process makes use of procedural programs and heuristics.

Where as there are thousands of different types of component there are few different types of circuit block. AJITA is intended as a rapid prototyping tool, in that it will produce a circuit in very short time that will not necessarily result in a minimum parts count design. The strategy does not allow for reasoning at the component level, but it could conceivably be used as a initial designer to a circuit optimising tool in the future. It should be noted that there is a large number of situations in which a minimum parts count is not as important as a rapid solution to a design problem. AJITA still requires the user to have enough electronic knowledge to know what the appropriate overall block diagram will be.

THE CAENIC CONCEPT

The CAENIC concept is discussed in detail in Mc Neill [4] and is briefly summarised here. CAENIC is intended as an integrated electronic systems designer that derives a specification from a user, who may not be an electronic expert, and uses the specification to produce a system design to the schematic stage at which it can be picked up by commercial schematic and PCB layout packages.

CAENIC, Fig 4, will incorporate AJITA to perform the circuit level design and will contain knowledge about the appropriate use of each of the blocks in AJITA. It is responsible for deriving the specification and developing the block diagram. It will be possible to use AJITA in conjunction with the CAENIC package but AJITA should realise potential as a stand alone pre-schematic design tool.

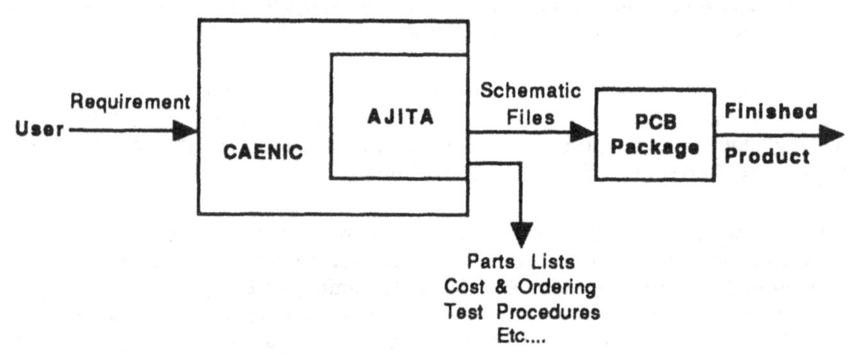

Figure 4. The Caenic Strategy.

CURRENT STATUS

The current version of Ajita (Ajita Version 2) is written in Turbo Pascal, uses the MetaWindow graphics library and is mouse-driven. The user interface reflects current electronic CAD practice, presenting the user with a work area and a menu from which functional blocks may be selected.

In the program, the blocks are held in an array of special data structures. The data structure carries the block generic name, which is used to link to other parts of the system and a number of fixed items which are specific to this generic block, such as the name of its circuit diagram file. The data structure also carries a number of items which initially are loaded with default values but which are changed when the block is dimensioned as a result of user input. For analogue blocks these values are generally component values and units while for digital blocks they are interconnection details.

To the extent that the blocks simply carry values they are basically parametric rather than "prototypes" as defined by Gero [19]. However, as one fixed (generic) item is a pointer to a design calculation procedure from which the actual values are calculated, the blocks are not simply parametised objects.

Ajita allows the user to assemble a functional block diagram of the circuit. The user may place functional blocks in the work area in any order, but would normally commence with an input block. This is a block whose input is user-defined and does not depend on the output of any block "upstream". Typically this may be a sensor or signal generator or an output from a another electronic system. The user then places other functional blocks to condition the signal from the input block to the requirement of the last, output block.

Once placed on screen the user can activate a "hot circle" in a block. This takes the user into the next level which produces a screen that is divided into four quarters. The first quarter contains input parameters required by AJITA. The second quarter contains a schematic diagram of the block. The third contains a parts list that is completed when all inputs have been entered. The last quarter is for notes and comments to the user.

The user can then return to the main screen and dimension the next block in the system. Connections can be made between blocks on the screen and are validated by the system upon entry. Checks are made for matched impedances, correct voltage levels and so on.

ANCILLARY RESEARCH

A truly useful automated electronic designer tool must provide more than just a circuit block building facility. A well engineered version will need facility for calculating the predicted system reliability and for estimating the overall cost of the assembly.

Experienced circuit designers also know that many interblock connections are routine and, therefore, can be automated to free the system user from systematic operations. This section discusses the research progress made to date in implementing these aspects.

Reliability

Methods for calculating an estimated system reliability from the expected failure rates of electronic components are well developed and probably best expressed in the military standard, MIL-HBK-217E "Reliability Prediction of Electronic Equipment", 1986. The methods explained there are used as the basis of reliability estimation in this software system, Karafoulidis [20] and Leifting [21]. An early consideration was whether to perform the calculations after the complete circuit was developed or to carry them out as each building block was completed. The latter was adopted as it allows generic block development and yields detailed information that is needed if the reliability level needs changing. Overall circuit reliability is easily calculated from combined values of each block allowing for their connection in series or parallel.

Costing

The designer, especially if not well experienced, can easily select components that are unusually costly. Facility is needed to allow the user to rapidly be provided with parts lists and cost estimates. As well as taking away a lot of the tedium of circuit development this also can be used to hone down the cost of the design as each block is developed. Costing features were also investigated on a block by block basis as the procedure for each block is similar. To allow Ajita to provide block cost information the facility has been added for it to access a component supplier's electronic data base of part numbers and prices (Nguyen, [22]).

The basic requirement for such a concept to be implemented is availability of a regularly updated cost data base that is in electronic format. Although most vendors now use electronic supply and accounting systems they are generally highly customised, costly to interface to and not accessible to this project because of their confidential nature. However one supplier, R.S. Components, was able to provide this information on disc.

Choice of Integrated Circuit

Early customer trials of the first version of AJITA often led to the suggestion that users want to use the IC they are used to using - not that supplied in the software. More investigation soon revealed that they did not always know, in objective terms, why they wanted a certain component but they did have confidence in that certain IC due to their past experience with it. It was decided that a useful CAE tool would be one that assisted the user to select, at best cost, an appropriate IC for a task, this choice being better substantiated by a mixture of mathematical and heuristic reasoning. This system was demonstrated by Temby [23] using CRYSTAL Expert System shell software.

Block Connectivity

Skilled designers readily know how to make most of the connections between blocks and what terminal conditions are needed. Similarly so do they know about the input and output port information where through the electronic system connects to the external world via sensors and actuators.

As the number of rules about interconnections between blocks - especially for analogue systems - appears to be small and well defined it should be possible to make automatically many of the necessary connections. Several projects have contributed to this aspect. The main strategy investigated was to develop a list of parameters for the input and output ports for block - such as impedance, electrical form, frequency, signal type. This generic form is then filled in by the software as a block pair are characterised by the user. Rules are then applied to see if these are allowable connections. If not, a warning is issued.

One project, relating to input and output block conditions Kieu [24], covered how to set up a suitable input block for Wheatstone bridge input stages to be well conditioned for interrogating sensors of many types. It calculates bridge sensitivity, self heating limitations on excitation level, lead effects and d.c. bridge type.

More recent work has set up an improved foundation for automated block connections, Kashef [25]. This project explored analogue system connectivity along expert systems lines. Generic rules about connections (that is, those that apply to most of the 45 plus blocks of AJITA) have been externalised and then encapsulated in an expert system manner. These deal with inputs, outputs, number of ports, types of ports, addition of new functions, editing, redimensioning if needed, need for intermediate block, a.c or d.c signals, frequency of signals, voltage and current levels, nature of impedance variables, loading effect, component availability and graphics of symbols.

This work is being carried out in close cooperation with the CAENIC project that aims for circuit automation from the operational requirement document. As such connectivity and block generation form the basic sub elements need for that next stage of automation.

CONCLUSIONS

The Ajita project has demonstrated that an A.I. approach to electronic design software, and especially the application of heuristic and logical rules, extends the domain of operation of electronic CAD beyond currently available systems. In particular, it is able to extend computer support back towards the designer, bringing it closer to a specification-based CAD tool.

In addition to the obvious need to implement more functional blocks, there are important developments to increase the usefulness and capability of the system in ways of value to the designer. Reliability and cost calculation are important in this regard, and the ability of knowledge based tools to implementt this type of functionality will be important in the future development of the system.

ACKNOWLEDGEMENTS

This work has extended over several years and contributions have been made by many final year students. The award of a National Research Fellowship, by the Australian Research Council, to partially support David Harris to manage these projects, has allowed the concept to reach a stage where viability has been proven enough to attract commercial interest.

REFERENCES

1. El-Turky, F.M. and Nordin, R.A. "BLADES: An Expert System for Analog Circuit Design." IEEE Int. Symposium on Circuits and Systems, May 1986.

2. El-Turky, F.M. "A Fully Automated Expert System Design Environment for Operational Amplifiers." Circuit Theory and Design 87, ECCTD '87, Sept 1987.

3. El-Turky, F. and Perry, E.E. "BLADES: An Artificial Intelligence Approach to Analog Circuit Design." IEEE Transactions on Computer Aided Design, 8, (6), June 1989.

4. Mc Neill, T. "An Integrated Approach to the Automation of Class Three Electronic Design Problems." Artificial Intelligence in Engineering, July 1991.

5. M'Pherson, P.K. "Systems Engineering: Approach to Whole-System Design." The Radio and Electronic Engineer, 50, (11/12), pp 545-558, Nov/Dec 1980.

6. Berlin, L.M. "ser-Centred Application Definition: A Methodology and Case Study." Hewlett-Packard Journal, October 1989.

7. Rowles, C.D. and Leckie, C. A "Design Automation System Using Explicit Models of Design." (submitted to) Int. Journal for Engineering Applications of A.I., July 1988.

8. Finkelstein, L. and Finkelstein, A.C.W. "Review of Design Methodology." IEE Proceedings, 130, Pt A, (4), June 1983.

9. Vanhoof, J., Rabaey, J., and De Man, H. A "Knowledge Based CAD System for Synthesis of Multiprocessor Digital Signal Processing Chips." Presented at VLSI 87, IFIP, 10-12 Aug 1987.

10. Lang, M.H. and Mc Cormick, P.E. "Hierachical Design Methodologies: A VLSI Necessity." Design Methodologies, S. Goto Ed. North Holland, 1986.

11. Fujita, T., Nakakuki, Y. and Goto, S. "A New Knowledge Based Approach to Circuit Design." Presented at ICCAD 87, IEEE Int. Conf. on Computer Aided Design, Nov. 1987.

12. Stoffels, J. "Automation in High-Performance Negative Feedback Amplifier Design." PhD Thesis, Delft University of Technology, The Netherlands.

13. Harjani, R., Rutenbar, R.A. and Carley, L.R. "A Prototype Framework for Knowledge Based Analog Circuit Synthesis." 24th Design Automation Conf., ACM/IEEE, June 1987.

14. Dincbas, M.A "Knowledge Based Expert System for Automatic Analysis and Synthesis." CAD, IFIP 1980.

15. Hanrahan, H.E. and Caetano, S.S. "A Knowledge Based Aid for DC Circuit Analysis." IEEE Transactions on Education, 32, (4), Nov 1989.

16. Di Cataldo, G. and Nunnari, G. "New Trends in Circuit Design via Expert System." Presented in IEEE Int Symposium on Circuits and Systems, 1988.

17. Harpas, P. "CAD of Analogue Circuits." Degree Project Report, School of Electronic Engineering, Universtiy of South Australia. 1988

18. Phan Tri, "Digital Blocks for AJITA Design System" Degree Project Report, School of Electronic Engineering, University of South Australia. 1990.

19. Gero, J.S. "Chunking Structural Design Knowledge as Prototypes." Artificial Intelligence in Engineering Design, Elsevere, 3-21. 1988.

20. Karafouldis C, "Circuit Design Goal Assessment Software". Degree Project Report, School of Electronic Engineering, University of South Australia 1989.

21. Leifting C, "AJITA Reliability Package", Internal Report, Measurement and Instrumentation Systems Centre, University of South Australia 1990.

22. Nguyen T T, "Sensor Costing and Ordering Interface", Degree Project Report, School of Electronic Engineering, University of South Australia. 1988.

23. Temby C, "Expert System for circuit block IC selection", Degree Project Report, School of Electronic Engineering, University of South Australia. 1989.

24. Kieu L D, "Circuit I/O Specification" Degree Project Report, School of Electronic Engineering, University of South Australia. 1989.

25. Kashef Kaveh, "AJITA Block Connectivity" Degree Project Report, School of Electronic Engineering, University of South Australia. 1990.

Expert Systems in Mechanical Engineering Design

J. Duhovnik, R. Žavbi

Laboratory of Computer Aided Design, Faculty of Mechanical Engineering, University of Ljubljana, Ljubljana, Slovenia

ABSTRACT

The paper presents an engineering design model for innovative design (i.e. variation of working principles) , which represents a deep knowledge of an expert system for configuring technical systems (i.e. from simple to complex assemblies). A knowledge base contains functional descriptions of building blocks (i.e. components of different level of complexity). An expert system can also synthesize functional structures which are used then as shallow knowledge to configure technical systems with equivalent models of shape. Also a flexible functional structure is used to manage models of shape properly.

INTRODUCTION

The design has been treated as an intuitive process, e.g. Begg [4], Suh [15] (even as an art), difficult to formalize and as such, a problem for an efficient expert system design.

Efficient and successfull expert system design is based on the appropriate knowledge formalization of the field of interest. Shallow knowledge was one of characteristics of first-generation expert systems and which describes personal approach of experts. But shallow knowledge is just a consequence of mechanisms which were not embedded in the first-generation expert systems. Such expert systems can solve only very narrow situations known in advance, e.g. Bratko [1], but they simply fail to operate in unpredictible situations, because they do not know the mechanisms (i.e. first principles), which rule the particular problems. Nowadays tendency in developing expert systems is incorporating deep-knowledge (i.e. first principles, mechanisms) into expert systems, which are also known as model-based expert systems.

Most of such systems are developed for diagnostic purposes in medicine, e.g. Bratko, Mozetič and Lavrač [12], electronics, e.g. Keravnou and Washbrook [6] etc., but few (comparing to those for diagnosis) in design e.g. Cremonini, Lamma and Mello [5], Mittal, Dym and Morjaria [10], Adeli [11].
We must emphasize, that models for model based expert systems are qualitative and not quantitative ,e.g. Bratko [1], Keravnou and Washbrook [6], Kunz, Stelzner and Williams [7].

A DESIGN MODEL

A designer composes elements and higher-order components (building blocks, in general) into assemblies, which fulfil a basic function (we can also say, a task) (Fig. 1). Basic functions can be simple and they require simple assemblies or complex ones and as such, they require complex assemblies, e.g. Duhovnik and Žavbi [8], Tessa and Trucco [14].
So, a designer is in a conceptual phase of design confronted to a problem, to design a technical system, which will satisfy a particular task, which also defines a physical function. A set of working principles belongs to each function. These functions also represent a starting point for a design (i.e. composing of technical systems). A materialization of working principles is made through models of shape, which fulfil subfunctions of the future technical system (Fig 2). Choosing of working principles and appropriate models of shape depends also on auxilary functions, which are used to express basic and binding functions precisely. Binding functions serve as some sort of links between models of shape (Fig. 3).

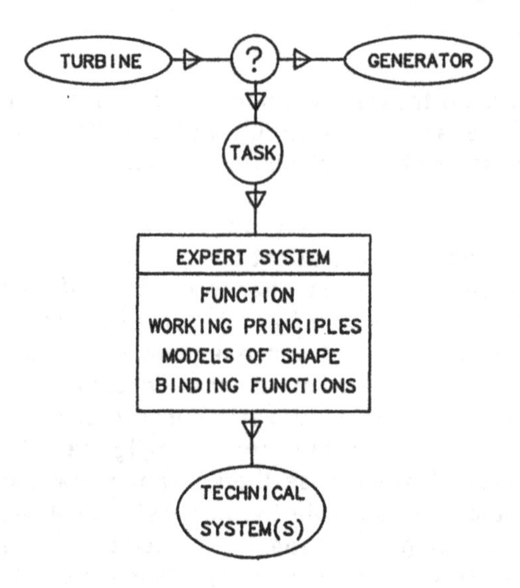

Figure 1. The purpose of model-based expert system

Basic and binding functions are described, e.g. Duhovnik [2], Duhovnik and Žavbi [8], in terms of statics, dynamics, thermodynamics, electrics, space conditions etc. (for example: transformation of mechanical energy in 3D, transmission of mechanical energy in 1D, withstanding of reaction forces, heat transfer, electrical isolation etc.).

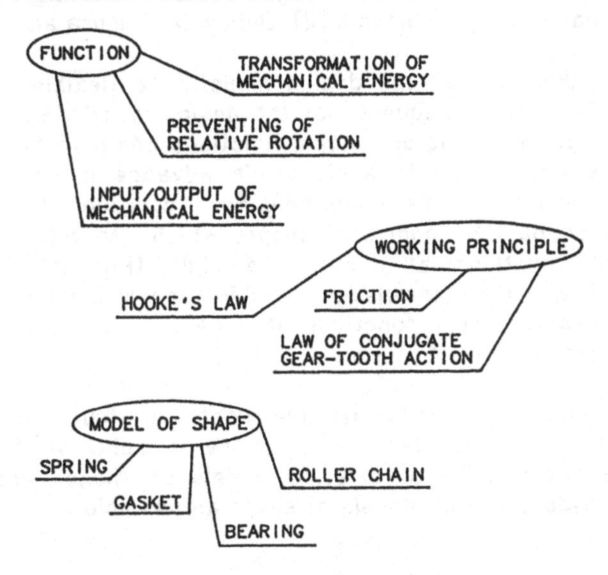

Figure 2. Examples of functions, working principles and models of shape

Auxilary functions are described in terms, which determine the basic functions more precisely (for example, small volume/power ratio, smooth running, small center distance, preservation of the relative direction of rotation etc.). Working principles are described in terms such as law of conjugate gear-tooth action, friction couple, Hooke's law etc. and models of shape are presented with their names (for example: a flat belt, an involut gear , an asyncron AC motor, a tangential key etc.)

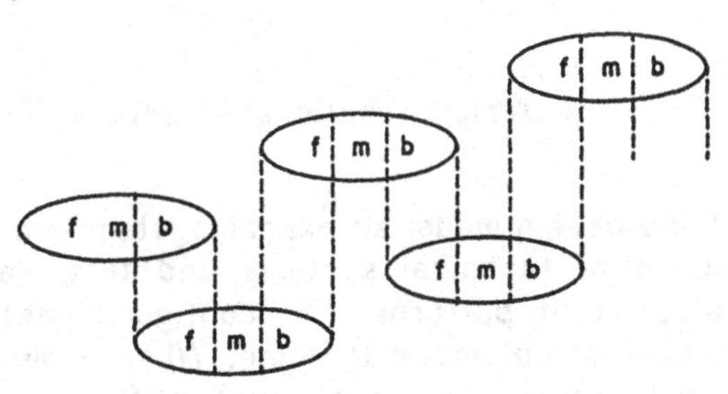

Figure 3. Binding functions

USAGE OF THE DESIGN MODEL IN EXPERT SYSTEM

The above described model (the model is qualitative) is characteristic for an innovative (i.e. variation of working principles) design; other types of design (new design (most creative design), variation design, adaptive design) require different models, e.g. Duhovnik [2], Duhovnik, Kimura and Sata [3].

The main advantage of the design model is a flexible functional structure (it is a consequence of the design model), which grows successively, depending on binding functions and models of shape. So we are not limited to a rigid , in advance given functional structure, which can be used only for variation of equivalent models of shape (i.e. models of shape, which can fulfil the same number of functions at once). The rigid (Fig. 4) functional structure (i.e shallow knowledge) fails in case of models of shape, which can satisfy more functions at once (Fig. 5), because they cannot be managed by it.

But to take advantage of the flexible functional structure, we need appropriate functional description (more in subtitle: A form of functional description) of existing models of shape, which must enable as wide usage of models of shape as possible.

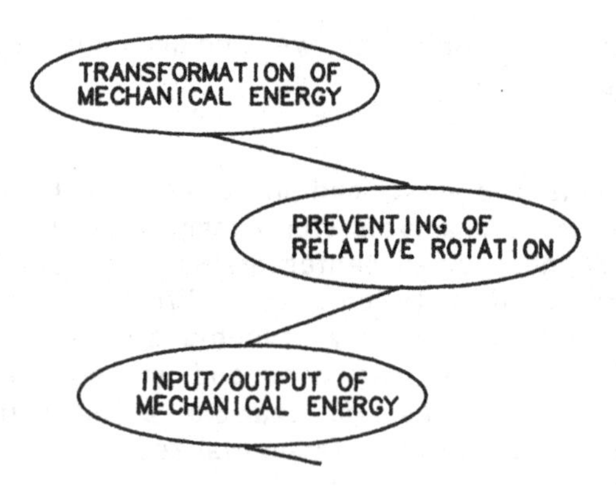

Figure 4. A rigid functional structure

On the base of the design model an expert system can synthesize new configurations of technical systems and they can be then stored into database of patterns (i.e. configurations) ; we can speak of some sort of automatic learning. When a designer/user defines a basic function, an expert system first tries to find already synthesized appropriate configurations (which correspond

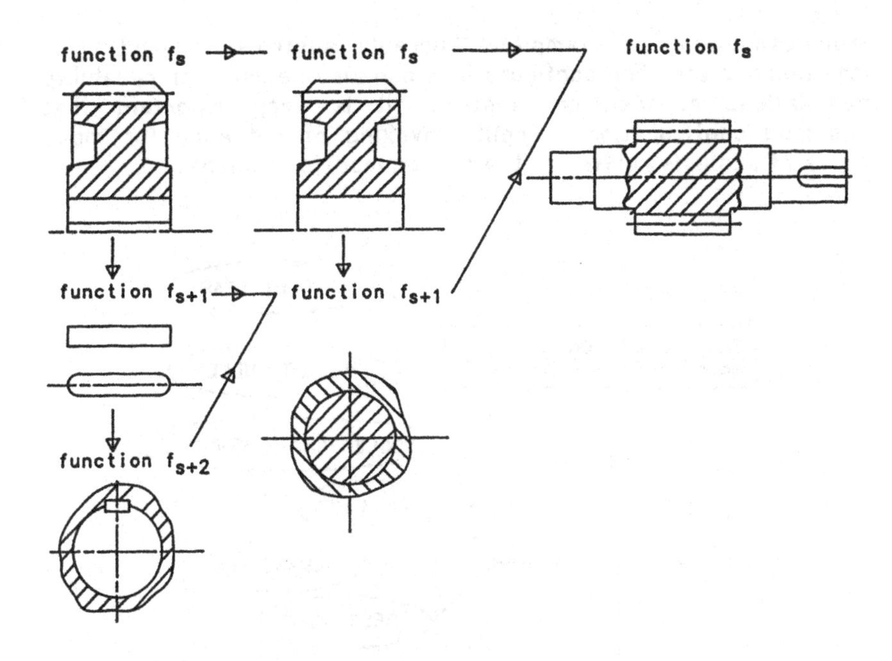

Figure 5. Flexible functional structure in case of different models of shape

to the entered basic function) in a database of patterns and functional structures. If neither satisfies the basic function an expert system then composes (through invoking of the design model/deep knowledge) configuration(s). The design model enables an expert system to, indirectly, make explicit functional structure(s), which can serve for variation of working principles and models of shape (Fig. 6). It can be seen (Fig. 7) how many alternatives are possible for relatively simple machine part. The alternatives (i.e. models of shape) perform the same function with very different degree of satisfaction, allowing the designer to choose appropriate model(s) of shape, using auxilary functions.

The design model can also be used for the design of hydraulic, pneumatic, cooling, air-conditioning and similar systems (with appropriate components and their functional descriptions, of course) and such configurations (i.e. functional structures with models of shape) composed by an expert system can also be used as device-models for model-based diagnostic expert systems, which must possess relevant description of components needed for diagnostic purposes. Diagnostic in technical systems is used for troubleshooting (for example in hydraulic systems), predicting of maintenance etc. The configurations can also be used for reliability calculations of technical systems. It is very important that designed configurations simplify invoking of software for many types of analysis related to the chosen models of shape.

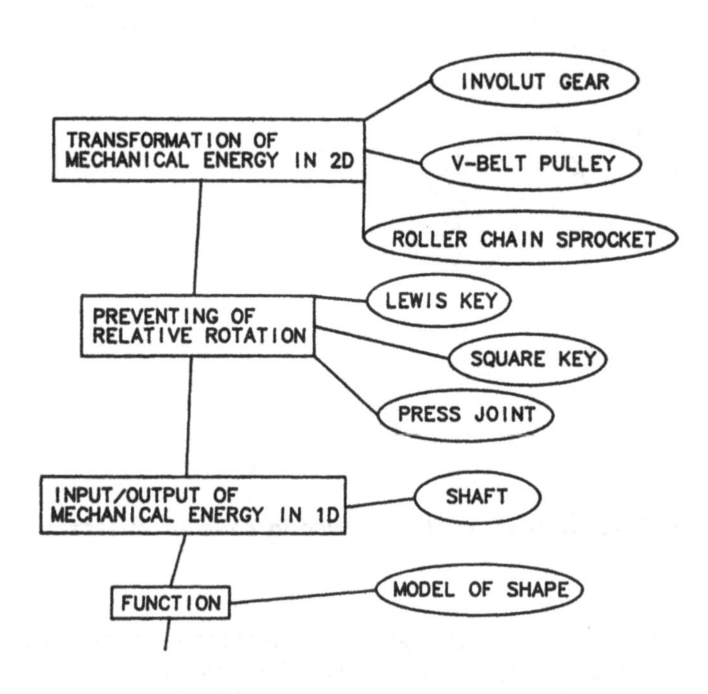

Figure 6. A rigid functional structure and variation of equivalent models of shape

Example

When the expert system is established, the user must enter the initial state - basic function which will be satisfied through futur assembly. Then the matcher, e.g. Patterson [13], Rich [16], searches through the knowledge base to find the models of shape whose function matches the entered basic function. After that, all appropriate (i.e. working principles which belong to the entered basic function) working principles are shown on the screen. Then the expert system requires additional data - the user has to enter auxilary functions to express further design needs (the user is also allowed not to enter any of the auxilary functions). The expert system checks then only the auxilary functions of the chosen working principles and shows the percentage of the auxilary functions which are satisfied. The user uses that information to decide which working principles and models of shape will be chosen for further composing of the assembly. We want that a designer has an active role in decision making, so we do not use any kind of automatic conflict resolution. But here we must emphasize that all proposed models of shape can satisfy the basic function. The user's decision also means automatic selection of binding functions which belong to chosen models of shape. If none of the working principles can satisfy the design needs 100 %, it means that they are to strong and could not be satisfied.

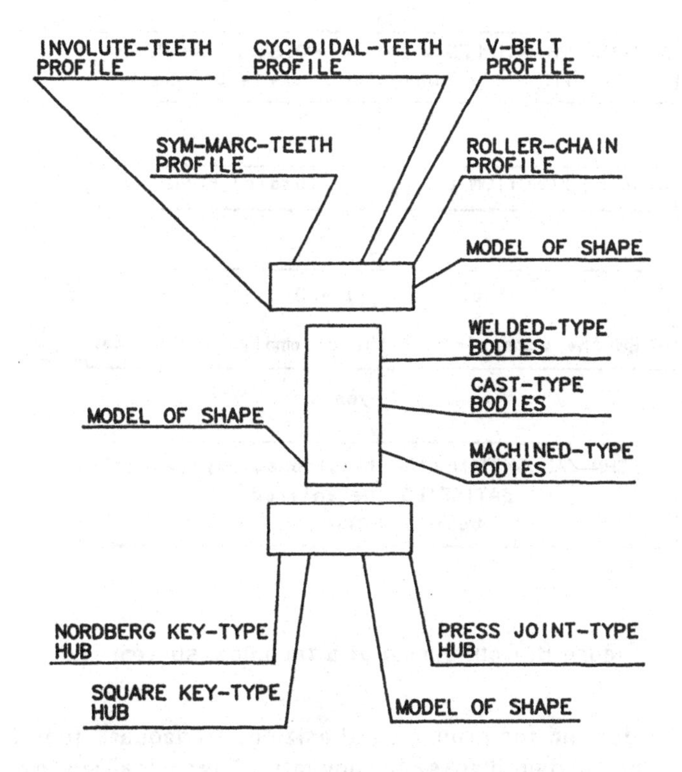

Figure 7. A wheel – the richness of alternative models of shape

Binding functions , from the standing point of the expert system, become basic functions (i.e. initial state for further reasoning – searching for appropriate working principles, models of shape and binding functions). Reasoning is then continued in such manner until the assembly, which satisfies the entered basic function, is established (Fig. 8).

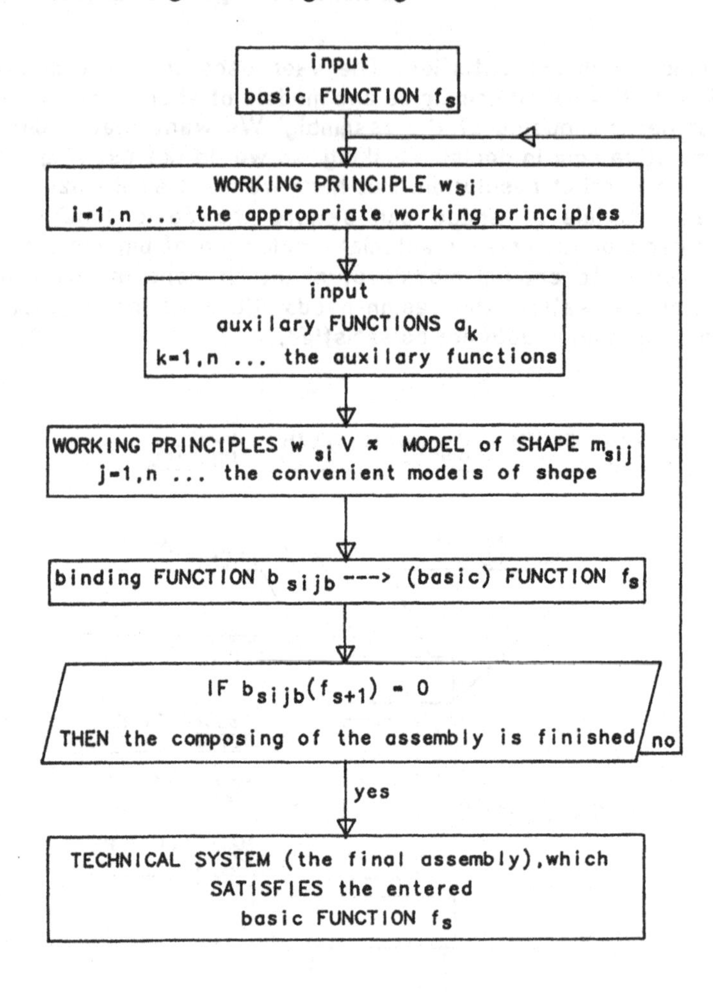

Figure 8. Configuring of a technical system

The crucial demand for proper synthesizing is adequate functional description of building blocks/components of technical systems. It depends on the complexity level of components which can be leveled as elements (a bearing, a gear, a spring, a gasket, etc.), parts or higher (an AC motor, a gearbox, a belt conveyor, a crusher, etc.); building blocks of higher complexity level are composed of building blocks of lower complexity level (for example: a bearing is a part of a gearbox, the latter is a part of a crusher, etc.). So, the complexity level of described components is adapted to the design level of technical systems - a technical system can be a clutch, or a gearbox, a crusher, a crane, etc.

A FORM OF FUNCTIONAL DESCRIPTION

We must be aware of the fact, that more general use a component has, more demanding is its description. An unadequate description would limit the usage of a component (Fig. 9). As an example we can take a gear as a rather special machine element and a washer, which is a quite general machine element and it has much wider usage spectrum. Another problem might be the fact, that functional description of components is not familiar yet, although it is used by designers subconsciously everyday .

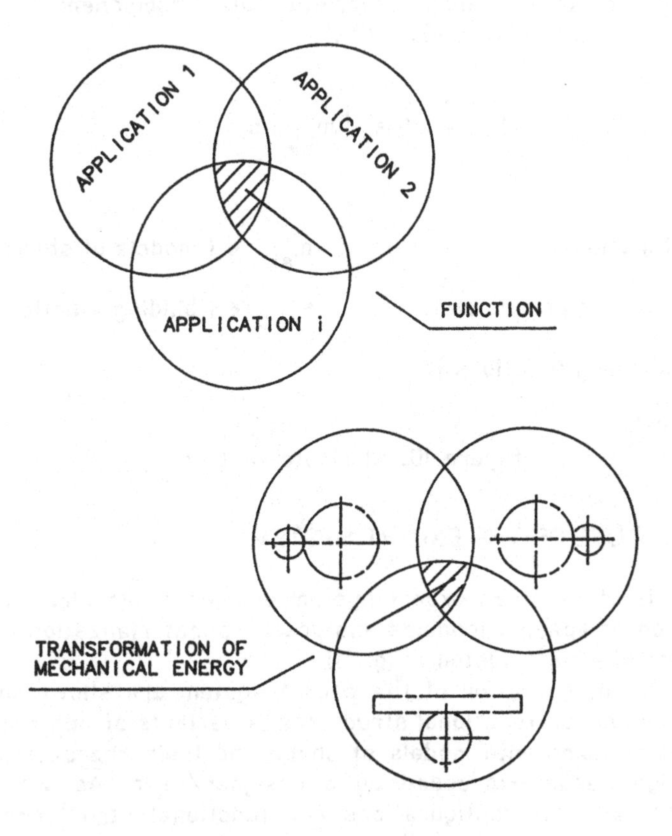

Figure 9. Importance of a general functional description of building blocks

The rigid and space consuming functional description would be something like this :

1st description: increasing of torque
2nd description: decreasing of torque
ith description: transformation from rotation to translation

and general like this :

transformation of mechanical energy

The rigid functional description will also prevent new applications of described models of shape.
By transformation we mean variation of parameters of mechanical energy and not transformation of mechanical energy to other types of energy.

The structure of functional description of a component is made through attribute vectors (Fig. 10).

$$f_s \mid w_{si} \mid a_k \mid m_{sij} \mid b_{sijb}$$

f_s ∈ { functions }

w_{si} ∈ { working principles }

a_k ∈ { auxilary functions }

m_{sij} ∈ { models of shape }

b_{sijb} ∈ { binding functions }

Figure 10. Attribute vectors

ARCHITECTURE OF THE EXPERT SYSTEM

The architecture of an expert system enables communication and integration of specific modules and presents materialization of the model-based expert system (Fig. 11).
The results of operation of the expert system are shown in an explanation modul; functional structures of variants of appropriate technical systems with models of shape and their characteristics and design needs expressed by a designer/user. As we have mentioned, all new configurations and functional structures are stored in separate database/knowledgebase.

CONCLUSION

We must emphasize, that an expert system, which is based on the design model is not meant to replace a designer, but to assist him/her to design technical systems. An average designer, unfortunately, does not know the richness of building blocks and this is the reason, that the number of proposed variants of technical systems is rather small. A model - based expert system

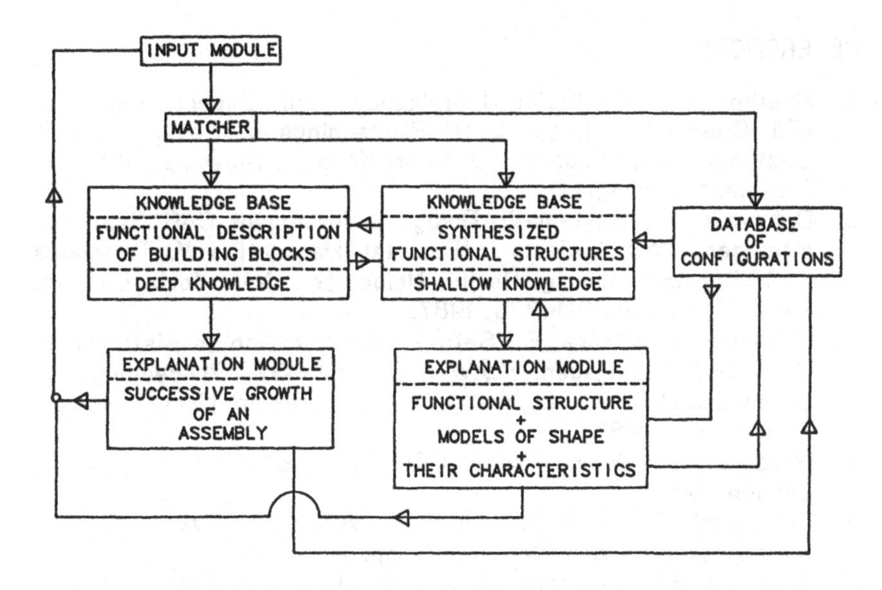

Fig. 11. Architecture and information flow in the model-based expert system

can also assist novices in design.

In the article we have mentioned many times, that the cruical demand for efficient operating of the model-based expert system is corresponding functional description of components of technical systems, which we want to design. Functional descriptions should be made by design experts for technical systems of interest.

In future, a lot of effort must be put just into the functional description of building blocks. A confirmation of suitableness of the functional description and the design model will be possible through a great number of successfully synthesized configurations of technical systems.

At the end we would like to emphasize, that the presented design model is just one possible view of design.

REFERENCES

1. Bratko, I ., Artificial Intelligence and Expert Systems, (Ed. Duhovnik, J.), pp. 2-11, Proceedings of Design: Expert Systems in CAD Process, Brdo pri Kranju , Slovenia, 1990, (in Slovenian language).
2. Duhovnik, J ., Systematic Design in Intelligent CAD Systems, Intelligent CAD Systems I, (Eds. ten Hagen, P.J.W., Tomiyama T.), Springer-Verlag, Berlin, Heidelberg, New York, London, Paris, Tokyo, pp. 211-226, 1987.
3. Duhovnik, J., Kimura, F., Sata T., Contribution to Methodic in CAD, (Ed. ten Hagen, P.J.W.), pp. 113-132, Proceedings of Eurographics '83, Zagreb, YU, 1983, Elsevier, North-Holland, Amsterdam , 1983.
4. Begg, V., Developing Expert CAD Systems, Kogan Page Ltd., London, 1984.
5. Cremonini, R., Lamma, E., Mello, P., ADES : An Expert System for ATP Design, AI EDAM, Vol. 3, pp. 1-21, 1989 .
6. Keravnou, E.T., Washbrook, J., What Is a Deep Expert System ? - An analysis of the architectural requirements of second-generation expert systems, The Knowledge Engineering Review, No. 4, pp. 205-233, 1989.
7. Kunz, J.C., Stelzner, M.J., Williams, M.D., From Classic Expert Systems to Models: Introduction to a Methodology for Building Model-Based Systems, (Eds. Guida, G., Tasso, C.), Topics in Expert Systems Design, Elsevier, North-Holland, pp. 87-110, 1989.
8. Duhovnik, J., Žavbi, R., Expert Systems in Machine Design, (Eds. Hubka, V., Kostelić, A.), Vol. 2, pp. 1038-1049, Proceedings of ICED '90, Dubrovnik, YU, 1990, Heurista, Zürich, 1990.
9. Brown, D.C., Chandrasekaran, B., Knowledge and Control for a Mechanical Design Expert System, Computer, pp. 92-100, July 1986.
10. Mittal, S., Dym, C.L., Morjaria, M., PRIDE: An Expert System for the Design of Paper Handling Systems, Computer, pp. 102-114, July 1986.
11. Adeli, H . (Ed.), Knowledge Engineering Vol.1 - Fundamentals , McGraw-Hill, New York, 1990.
12. Bratko, I., Mozetič, I., Lavrač, N., KARDIO: A Study in Deep and Qualitative Knowledge for Expert Systems, MIT Press, 1989.
13. Patterson, D. W., Introduction to Artificial Intelligence and Expert Systems, Prentice-Hall, New Jersey, 1990.
14. Tessa, D., Trucco, E., Functional reasoning for flexible robots, (Ed. Gero, J.S.), Artificial Intelligence in Engineering: Robotics and Processes, Elsevier, Amsterdam, Oxford, New York, Tokyo, pp. 3-19, 1988.
15. Suh., N.P., The Principles of Design, Oxford University Press,

Inc., New York, 1990.
16. Rich, E., Artificial Intelligence, McGraw–Hill, New York, 1983.

Numerical Methods in AI-Based Design Systems

K. Furuta (*), T. Smithers (**)

(*) Department of Nuclear Engineering, University of Tokyo, 7-3-1 Hongo, Bunkyo-ku, Tokyo 113, Japan

(**) Department of Artificial Intelligence, University of Edinburgh, 5 Forrest Hill, Edinburgh EH1 2QL, UK

ABSTRACT

Numerical methods and artificial intelligence are primary ways to utilise computer powers in modern design practice, but the interface between them has not been settled in a comprehensive manner. The aim of this paper is to present a methodology to combine both technologies so that their powers can be made more available to support designers. The authors first discuss the role of knowledge underlying numerical methods to suggest how computers can help designers in using these methods. This is then followed by the considerations in controlling design activities of using numerical methods, which are highly domain-specific and goal-directed. Finally, special support sub-systems for numerical methods are incorporated into the Edinburgh Designer System, the AI-based design support architecture based on a exploration model of design.

INTRODUCTION

When no direct means exist to obtain a design solution from design requirements, generate and test is a relevant strategy of design. Numerical methods are used in the course of this process to predict the behaviour of proposed mechanisms placed in a particular environment and to check the derived attributes against the design requirements. This approach is based on the assumptions that design process can be decomposed into discrete parts of subtasks, and that design products can be constructed from the results of these subtasks. Then, computer programs are developed and used for the tasks which are categorized as amenable to computer processes. Though this view of design has some crucial limitations as a general formalism [2], it still provides the most useful strategy of the design for which several programs already exist and are used extensively. A lot of numerical methods such as the finite difference method, the finite element method, Monte Carlo simulation, etc. are actually adopted in modern design practice. Moreover

numerical methods have come to be essential instruments in designing large, complex, and expensive artefacts, because mock-up test or experiment is too expensive and impractical.

Another trend of modern design practice is in the application of artificial intelligence (AI) techniques, which has been demonstrated by many works as a promising technology to provide intelligent aids to designers. In spite of their increasing role, the methodology to integrate numerical methods and AI-based design support is still immature. The aim of this paper is to present a sophisticated integration methodology of numerical methods and AI, which is desired for further utilisation of the both technologies. In the following section we focus on the knowledge underlying numerical methods, which design support systems should be aware of to help designers in all phases of the task. We then discuss the issues of control to reveal what tasks should be done for building a plausible plan of design activities. Finally, the implementation of these ideas in the Edinburgh Designer System is presented.

MATHEMATICAL MODEL

In domains of engineering design where theoretical backgrounds are formalized comparatively well, numerical methods are widely used in the following manner:

1. selection of a mathematical model,
2. specification of the selected model,
3. execution of the computational procedure,
4. interpretation of the results.

Computer programs are the sequence of instructions to carry out the above steps. In these domains, however, most efforts have been devoted to develop computer programs to perform the third step, and designers have been responsible for the other tasks.

Every numerical method is based on certain theoretical principle of the reality quantitatively represented, and we call such a principle a mathematical model. The knowledge on mathematical models, which is necessary for designers to accomplish their tasks, primarily consists of the following assumptions and approximations:

- assumptions on the structure of the world where the process of interest occurs,
- assumptions on the relations among different parts of the structure,
- approximations necessary to simulate the process as a sequence of numerical operations.

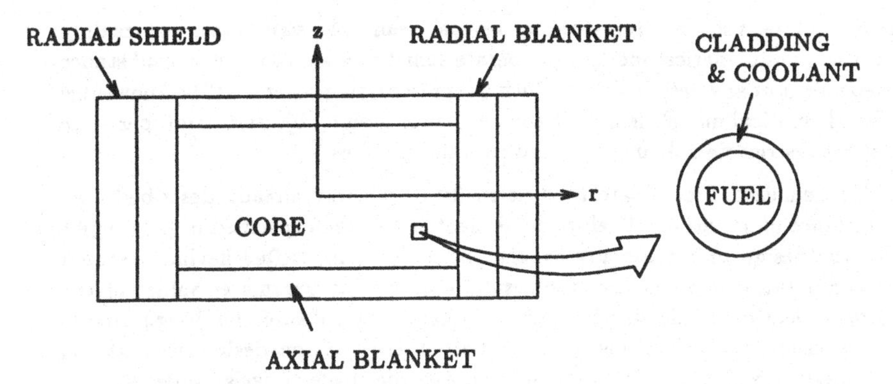

MULTI-GROUP DIFFUSION EQUATION

$$-D_g\nabla^2\phi_g + \Sigma_g\phi_g = \frac{\chi_g}{k}\sum(\nu\Sigma_f)_h\phi_h + \sum\Sigma_{h\to g}\phi_h$$

Figure 1 An example of mathematical model used in nuclear reactor design.

Figure 1 shows an example of mathematical model, which is typically used in nuclear reactor design. In this example, multi-group diffusion equation is solved numerically in the r-z coordinates over homogeneous media, which are modeled further as infinite cylindrical cells of a fuel pin and a surrounding coolant channel. Such knowledge is not stated explicitly in programs and is useless for computers, though some information is available for designers as documents or comment lines. Only experienced designers, who have acquired the knowledge through education and experience, can carry out the tasks successfully. In order that design support systems should be more helpful for designers in using numerical methods, a methodology to represent and utilise the knowledge on mathematical models has to be established. Then, program developers can provide the knowledge encapsulated in their programs based on the established framework.

Parameters and constraints on design objects must be declared as a design knowledge base in design support systems. This knowledge base is usually organized in hierarchical class definitions, and instances of design objects are generated from the general object classes and operated on in the course of the design process. It is possible to include the knowledge on mathematical models in the same framework, but mathematical models should be defined separately from any particular design object, though each model is aimed to represent a designer's view of some design object. It is partly because a single mathematical model can be used for representing several realizable structures, but mainly because a mathematical model can operate sometimes as a conceptual substance independent of any design object. Such a representation scheme as the knowledge on mathematical models is buried among that of design objects is improper since it might restrict flexible usage of mathematical models.

Mathematical models are used normally in the way already described, i.e., an instance of a mathematical model is created and then specified in accordance with the design object already existent in the system. After having executed the computational procedure, the results obtained as parameter values of the mathematical model are used to evaluate some attributes of the design object. In this usage, numerical methods function as critiques on design alternatives, while design exploration process takes place in the space of design objects.

Each mathematical model represents restricted designer's perception of a design object relating to a particular goal in design process. Which model is to be used at a particular point in the process is dependent on situations. For example, pruning the space of possible designs rapidly by an inexpensive method is preferred in the early stage, but validating the obtained design solution by an accurate but expensive method is required in the final stage. Many models can exist for a single design object, and one must chose the model satisfying the following criteria in order that the task is useful.

- The model covers all the aspects of the design object which are necessary for the current purpose.
- The behaviour of the design object can be predicted by the model accurately enough for the current purpose.
- Sufficient and proper data is available to specify the model.
- The cost of carrying out the computation is acceptable.

It can be said that mathematical models themselves are objects to be designed, and the above criteria provide design requirements. Redesign may be necessary if the selected model does not give satisfactory results according to the above criteria.

The usage of mathematical models described so far is not always the case. Design starts sometimes by creating instances of mathematical models with ambiguous associations with real objects, and design exploration is carried out first in the space of mathematical models. Then, after some proper model satisfying design requirements has been found, generation of a realizable structure is pursued referring to the conceptual substance. Mathematical models are used here as schemata which accommodate between design requirements and a design solution. Such usage is observed especially in conceptual design of new artefacts. Defining mathematical models in a knowledge base independently of design objects enables designers to adopt such a strategy of design exploration.

For example nuclear reactor designers sometimes start calculation with homogeneous reactor model already shown in Figure 1. If the conventional structure of a hexagonal fuel assembly with a channel box is the only choice, the fuel volume fraction in this model will never exceed nearly 45 percents. This limitation, however, is essential to the structure but not to the model. A designer who found that a higher fuel volume fraction results in attractive features of a metalic fuel fast reactor abandoned the conventional structure and proposed a new concept of tube-in-shell type fuel assembly, which seems not feasible with ceramic fuel used in the conventional design. In this example use of the mathematical model independent of the conventional structure led to the innovation in fuel assembly design.

CONTROL OF DESIGN ACTIVITIES

Since each numerical method deals with restricted aspects of a design object, cooperative works of several methods may be necessary to cover the whole aspects of interest in design. As a consequence, control of different design activities comes to be a relevant matter. The problem of control is sometimes solved by standardising the sequence of design activities based on the experience. Design code systems, which are collections of programs organized in particular schemes, have been developed, but this approach has two major limitations. Firstly, it is valid only for routine designs, where the sequence of design activities has been well established and will not change greatly in the future. Secondly, the interface between one program with another may leave considerable works to designers, which become more serious with the increase of system complexity. In AI-based design support systems, the problem of control should be solved explicitly, while the freedom is greater than conventional design code systems. What factors should be considered then in dealing with this problem?

The motivation to use numerical methods in design is based on the expectation that they can do some special works useful for finding a design solution. What is useful, therefore, is the first question to be answered in controlling these activities. The usefulness depends highly on the context of design process as well as design requirements. Though the system can answer this question to some extent, intuitive or heuristic judgement by an experienced designer is sometimes inevitable. Consequently, a control plan of design activities is first subject to designer's intention, which sets the preliminary goal for the activities. Secondly, the plan is also determined by goal-subgoal relations among different activities. Each numerical method requires some input data to specify the mathematical model to be used in the analysis and yields other attributes of the model as output.

In nuclear reactor design, for example, control rod worth, on which several design criteria are imposed for nuclear safety, will be evaluated by the computation scheme illustrated in Figure 2. The primary data shown in the ellipses are obtained as a result of some numerical method and passed to the next step. Such an aspect of numerical methods defines the goal-subgoal relations and constrain control plans of design activities. The system that is aware of these relations will enjoy more independence from designers while building a control plan by

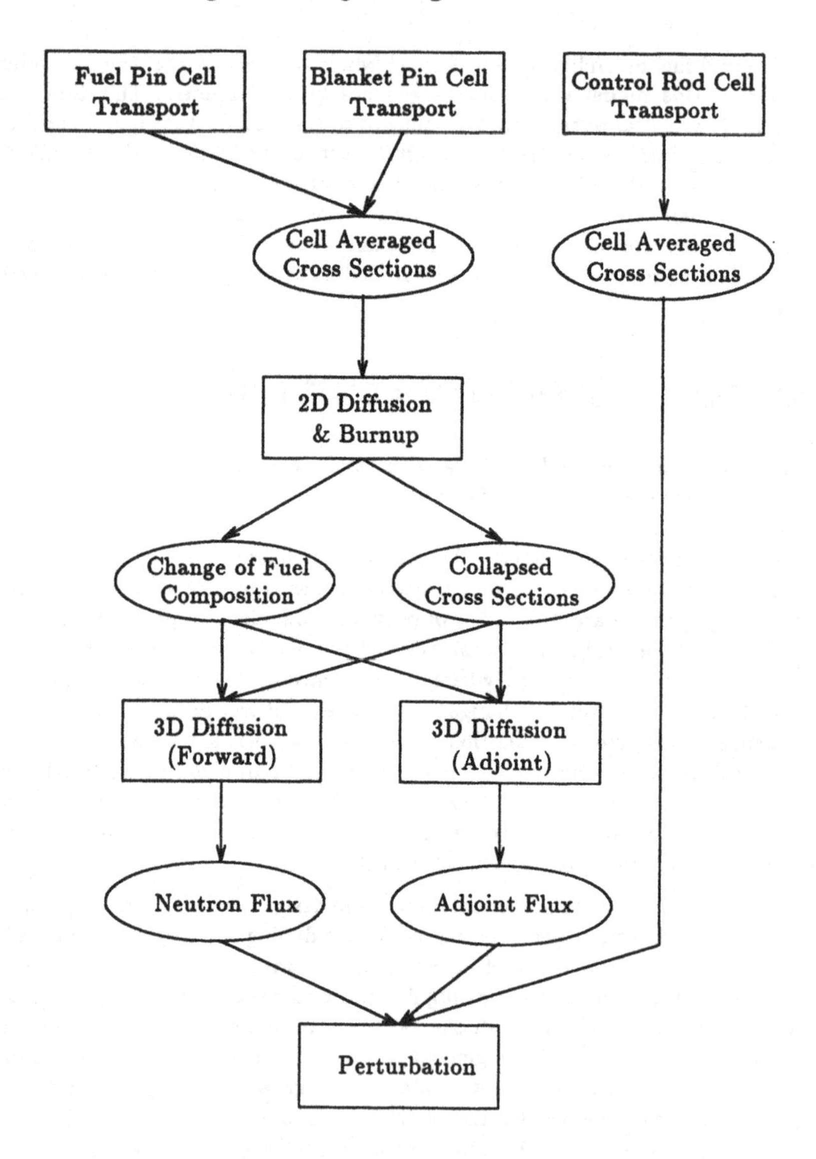

Figure 2 A computation scheme to evaluate control rod worth.

relying on them. What one should bear in mind, however, is that the validity of any numerical method depends on several assumptions on the design object and the design process. For the example of Figure 2, another scheme may be more suitable, if the configuration or the composition of reactor core has been revised.

If both the designer's intention and the goal-subgoal relations are represented overtly, the system can build a plausible control plan of design activities by doing the tasks as follow:

- **Selection of a method** by means-ends analysis, where a purpose of the method is matched with the design goal claimed by the designer or by the system. The trade-off between accuracy and cost should be considered here within the current context of design process. This action generates a task to execute the method in the design plan.
- **Creation of a mathematical model** which is necessary to carry out the invoked task. The model is created as an instance of some object class, and specified according to the design object under consideration if any.
- **Declaration of new goals** to achieve the prerequisites of the invoked task. This action generates a hierarchy of the design plan.
- **Arrangement of the tasks** in the same level of the hierarchy. The least commitment strategy will be suitable for design planning, i.e., the system is allowed to arrange the tasks only when sufficient information for the arrangement has been obtained.
- **Elimination of redundancies** in the design plan. Any detectable redundancy is eliminated automatically by the system.
- **Execution of the tasks** whose prerequisites have all been satisfied. The system executes the computational procedure as either an internal or an external process. Further, execution on a remote computer, which is more suitable for the job, will be possible.

Since the above tasks have little difference from those seen in general planning, the most architectures developed for AI-based planning systems [13] will be effective for design planning as well. Considerable parts of control decisions can be solved by the system that knows sufficiently about its own works. Since it is impossible, however, to represent every controlling knowledge in a symbolic form, the system cannot always make proper decisions on control problems. To supplement this inability, the system should follow the least commitment strategy and should be ready for designers' intervention.

EDS AND NUMERICAL METHODS

The Edinburgh Designer System (EDS) is the architecture of AI-based design support systems based on an exploration model of design [10,11,12]. The EDS primarily consists of the following sub-systems:

- **Domain Knowledge Base (DKB)** represents domain knowledge hierarchically organized by two relationships: kind-of and part-of. Each module of the DKB declares a class of parameters and constraints on design objects in accordance with the way that designers usually think about them.
- **Design Description Document (DDD)** is a knowledge base representing the knowledge generated in the course of the design process. The consistency of this knowledge base is maintained by the Assumption Based Truth Maintenance System (ATMS) [4].
- **Inference engines** carry out useful works to support the designer in design exploration process. General support sub-systems carry out tasks such as algebraic equation solving, geometric reasoning, table manipulation, etc., which are common to several tasks of design. Special support sub-systems carry out goal-directed tasks specific in a particular domain under close supervision of the designer.
- **A control system** based on the blackboard model [5] is adopted in the EDS. It makes an agenda of the tasks that any inference engine can do, selects one item with the highest priority from the agenda and executes the selected task.
- **A user interface** is provided for the designer to communicate with the system.

Programs of numerical methods can be thought as special support sub-systems in the above framework, since they perform highly domain-specific and goal-directed works. It seems insufficient, however, just incorporating new sub-systems and commands to invoke them using data in the DDD. In the following we discuss the integration of numerical methods and the EDS based on the ideas presented so far.

MATHEMATICAL MODELS IN DKB

Parameters and constraints on mathematical models are also declared in the DKB, as if they are realizable objects. These declarations are hierarchically organized just in the same manner as ordinary class definitions. A specific model is placed in a lower level of the kind-of hierarchy than more general ones. For example, the two-dimensional diffusion perturbation analysis model is to be defined as a sub-class of the two-dimensional diffusion analysis model. The two kinds of class definitions, design objects and mathematical models, have no difference in appearance, but mathematical models are necessarily associated with particular sub-systems for domain-specific works.

In the ordinary use of mathematical models, some attributes of a design object are mapped onto some attributes of its model, and the results of computation are mapped vice versa. Constrains representing such bi-directional mappings between a design object and its model are declared separately as an interface class definition in the DKB. The interface classes of the EDS are originally used to declare the constraints between objects with some relation other than kind-of and part-of, for example connected-with. Separating definitions of design objects, mathematical models and the relations between them increases flexibility in the usage of mathematical models as already described.

TASK DEFINITIONS IN DKB

In order that the system can control several support sub-systems, it must know about the functions which these sub-systems can provide. As for general support sub-systems, each task can be executed opportunisticly and independently of other tasks. For this reason, most general support sub-systems can be built tightly into the system and invoked automatically. On the contrary, special support sub-systems should be used under close supervision by the designer due to their dependence on designer's intention and the context of design process. Consequently, it is desirable that the tasks that special support sub-systems can do are defined explicitly, and that the definitions are accessible from the designer.

Domain-specific tasks of using numerical methods are defined in the DKB so that the system can perform the design planning tasks described in the previous section. Figure 3 shows an example of task definition declared in DKB. A planning engine, a kind of general support sub-system, creates a control plan of design activities referring to the above task definitions.

PLANNING IN CONTROL DDD

The propositions concerning control of the system as well as evolving design specifications are asserted in the DDD, which is divided in two parts, control DDD and object DDD, to keep and process each type of knowledge. The problem of system control are solved in the control DDD. The control DDD as well as the object DDD is truth maintained by the ATMS, so that switching of contexts in design exploration process is reflected properly in controlling the system.

The problem of control is solved as follow. A Knowledge Source Activation Record (KSAR) is created in the control DDD when any sub-system, a knowledge source, is invoked to do some useful work for design. In the blackboard model of control, strategic control of the system is usually performed by ordering an agenda of KSARs with some measure of task priority. In addition to this mechanism, a planning engine, which is a nonlinear planner like NOAH [8], works on DDD records to create a control plan of design activities, and the results are asserted in the control DDD. The KSARs in the bottom level of plan hierarchy are ordered to form the agenda, while KSARs for general support sub-systems, which operate opportunisticly, always appear in the agenda making level. Figure 4 schematically shows a snapshot of control DDD in making a design plan to evaluate the control rod worth of a nuclear reactor as shown in Figure 2.

The KSAR at the top of agenda is executed, when all of its preconditions are satisfied. It is normally the system that makes and executes subplans to satisfy these preconditions. If the system does not know what to do with some

```
/*************************************
 * One-dimensional burnup calculation *
 *************************************/

/* identifier of the task */
method burnup_1D(X, C) has

/* proposition in the DDD which activates the task */
  trigger
     X isa Y <- X isa burnup_1D_model;

/* attributes to be obtained by the task */
  purpose
     C:excess_reactivity <- C isa fbr_core;
     C:burnup_reactivity <- C isa fbr_core;
     C:max_power_den     <- C isa fbr_core;
     C:ave_power_den     <- C isa fbr_core;

/* procedure called to execute the task */
  procedure
     burnup_1D(X);

/* class name of the mathematical model */
  component
     [C, X] isa core_1D_mapping;

/* attributes required to execute the task */
  precondition
     X:geometry, X:height, X:extrapolation,
     X:power,    X:operation_period,
     X&zone(I):radius  for (I=1, X:zones),
     X&zone(I):regions for (I=1, X:zones),
     X&zone(I):meshes  for (I=1, X:zones),
     X&zone(I):atomic  for (I=1, X:zones);

/* attributes obtained by the task */
  effect
     C:excess_reactivity, C:burnup_reactivity,
     C:max_power_den,     C:ave_power_den,
     X:max_keff,          X:min_keff,
     X:max_power_den,     X:ave_power_den;

/* data to determine the priority of the task */
  control
     priority = 120,
     accuracy = 50,
     cost     = 200;

end.
```

Figure 3 An example of task definition in DKB.

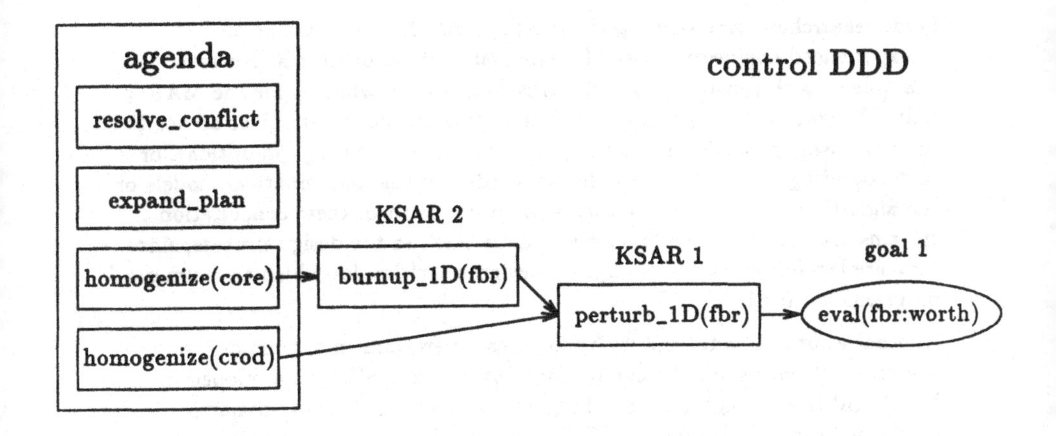

Figure 4 A schematic diagram of the control DDD.

preconditions, it informs it to the designer and suspends the KSAR until the designer takes sufficient actions to satisfy these preconditions.

There are three possible ways to invoke design support sub-systems. Firstly, a sub-system will be invoked automatically by the system, when some triggering event has occurred in the DDD. Secondly, the user can start a preferable sub-system by issuing a command. Thirdly, the user can claim a design goal as an assumption in the control DDD, and the system invokes a proper sub-system to achieve this goal. In any case, a plan to execute the invoked sub-system is created, and then KSARs in the plan are executed according to the scheduled order. The designer can insist controlling decisions such as task orders or method preferences by creating assumptions either in the control or in the object DDD, when the system cannot decide these issues properly.

RELATED WORKS

It is recognized that the coupling of numerical and symbolic computing will provide powerful solution techniques for some of the problems currently considered intractable [6], and many systems have been developed in search of a good methodology of this coupling.

Some researchers have developed expert systems for consultation to unexperienced engineers in use of numerical analysis programs. An early example of such consultation systems is SACON [1], which is for the MARC finite element analysis program. In this approach, the system gives advice to program users in setting up input data, chosing proper program options, or understanding output by using the knowledge on the mathematical models or the algorithms adopted in the target program. Most of these consultation systems are, however, helpless for synthetic tasks in the design process, since they are lacking in the knowledge on design activities, for which the numerical programs are used.

Another approach is to deal with numerical operations in a common environment with symbolic computing. For example SEE [7] developed in MACLISP is an integrated modeling environment for studying dynamic problems in interactive fashion. This approach is based on the arguments that the mathematical models underlying numerical methods should be operated on interactively by symbolic reasoning processes and that the conventional batch style of running programs is too restrictive. Many AI systems have been developed along a similar line, but numerical methods incorporated in most of them are too simple compared with those of design programs in practical use. The above arguments hold in such fields as cognitive science, social science, economics, etc., where considerable ambiguities still exist in mathematical modeling. In the principal area of engineering design, however, many numerical methods have already been established and sophisticated to such an extent that interactive modification of mathematical models is less necessary and that it is hardly beneficial to make all aspects of numerical methods amenable to symbolic reasoning processes. Based on such considerations the knowledge on mathematical models is represented apart from the algorithmic procedures in the present system.

As for the control of design activities, Chalfan developed an expert system for using a set of numerical programs in designing aerospace vehicles [3]. Her system showed that a highly generic and powerful tool is obtainable if the knowledge of how to execute a set of numerical programs and the facts about these programs are represented separately in symbolic form. The present system is similar to her system in organizing the functions of several sub-systems by input-output requirements, but the latter system works only on a single combination of mathematical models and has no representations for design objects. Simmons and Dixon argued that three important knowledge on a quantitative method are the effect, the preconditions, and the postconditions of the method to formalise quantitative methods in redesign [9]. In addition to these aspects, the present system takes some more context dependent factors into consideration in planning design activities.

CONCLUDING REMARKS

We proposed a methodology to integrate numerical methods and AI-based design support, two major usages of computers in modern design practice. It was first pointed out that the knowledge on mathematical models underlying numerical methods plays an important role to perform the tasks for which designers have been responsible, and then we discussed how such knowledge should be represented and organized in design support systems. Control of design activities concerning numerical methods requires the system to perform several tasks common to general AI-based planning due to its domain-specific and goal-directed natures, and design support systems should be aware of designer's intention and task definitions to build a proper control plan. Finally, the symbol level issues to incorporate numerical methods as special support sub-systems into the Edinburgh Designer System (EDS) were discussed. Simulation models and domain-specific tasks are defined in the domain knowledge base, and a planning engine works in the design description document for controlling to plan design activities. The EDS architecture is so flexible that the ideas presented in this paper could easily be adopted.

The approach outlined in this paper is still incomplete for the fusion of the two technologies. The approach relies on the assumption that designers can always externalise their intention to set a preliminary design goal. It is not the case, however. Inferring their intention from some evidence expressed implicitly is left for the future research. A method to interpret numerical results in qualitative terms according to the mathematical model is another research topic. One can expect some progress in this area, if the ideas proposed in qualitative reasoning are applied.

ACKNOWLEDGEMENTS

This study was performed during the first author's work as a visiting fellow at the Department of Artificial Intelligence, University of Edinburgh. The authors wish to express their thanks to the British Council, who funded the fellowship. The authors also appreciate the past and the current members of the Design Research Group for valuable discussions: Karl Millington, Brian Logan, Bing Liu, Ming Xi Tang, Nils Tomes, Hideo Ohata and Amaia Bernaras.

References

[1] Bennett, J., Creary, L., Englemore, R. and Melosh, R. *SACON: A Knowledge-Based Consultant for Structural Analysis*, Technical Report STAN-CS-78-699, Department of Computer Science, Stanford University, 1978.

[2] Bijl, A. *Function-Oriented Systems*, Computer Discipline and Design Practice, pp. 116-135, Edinburgh University Press, 1989.

[3] Chalfan, K.M. *An Expert System for Design Analysis*, Coupling Symbolic and Numerical Computing in Expert Systems, Elsevier Science Publishers, pp. 179-190, 1986.

[4] de Kleer, J. *An Assumption-based TMS*, Artificial Intelligence, Vol.28, pp. 127-162, 1986.

[5] Hayes-Roth, B. *A Blackboard Architecture for Control*, Artificial Intelligence, Vol.26, pp. 251-321, 1985.

[6] Kitzmiller, C.T. and Kowalik, J.S. *Symbolic and Numerical Computing in Knowledge-Based Systems*, Coupling Symbolic and Numerical Computing in Expert Systems, Elsevier Science Publishers, pp. 3-17, 1986.

[7] Lounamaa, P. and Tse, E. *The Simulation and Expert Environment*, Coupling Symbolic and Numerical Computing in Expert Systems, Elsevier Science Publishers, pp. 83-99, 1986.

[8] Sacerdoti, E.D. *The Non-linear Nature of Plans*, Proceedings of the Fourth International Joint Conference on Artificial Intelligence, Tobilisi, USSR, 1975.

[9] Simmons, M.K., Dixon, J.R. *Reasoning about Quantitative Methods in Engineering Design*, Coupling Symbolic and Numerical Computing in Expert Systems, Elsevier Science Publishers, pp. 47-57, 1986.

[10] Smithers, T. *Intelligent Control in AI-Based Design Support Systems*, DAI Research Paper No. 423, 1989.

[11] Smithers, T., Conkie, A., Doheny, J., Logan, B. and Millington, K. *Design as Intelligent Behaviour: An AI in Design Research Programme*, Artificial Intelligence in Design, pp. 293-334, Proceedings of the Fourth International Conference on Application of Artificial Intelligence in Engineering, Cambridge, UK, 1989.

[12] Smithers, T. and Troxell, W. *Design Is Intelligent Behaviour, But What's the Formalism?*, Proceedings of the First International Workshop on Formal Methods in Engineering Design, Manufacturing, and Assembly, Colorado Springs, USA, 1990.

[13] Tate, A. *A Review of Knowledge-Based Planning Techniques*, Expert Systems 85, pp.89-111, Proceedings of the Fifth Technical Conference on Expert System, Cambridge, UK, 1985.

A Knowledge Based System to Automate Manufacturing and Design Engineering

A. García-Crespo, M. Cortina

Intelligent Decision Systems, S.A., Bertendona, 4-5ªpl, 48008 Bilbao, Spain

ABSTRACT

The paper presents our work concerning the INDO (INtelligent Design Optimizer) system. This system is currently being developed trying to automate to the maximum level the design labours in engineering.

A first prototype is already being developed with Smalltalk V in a personal computer showing almost all capacities of the final system. Details and examples about the prototype are shown in the final paragraphs.

INTRODUCTION

The INDO (INtelligent Design Optimizer) system is defined as the system to automate the manufacturing and design engineering with knowledge based techniques.

It represents the last technology in engineering automation. The INDO system enlarges the desing process over the graphical model, allowing the user to build an intelligent model describing the design and manufacturing knowledge about the product.

INDO modernizes the automation of engineering supplying a new and powerful approach to the modeling of new products.

This new technology will change the way that engineers follow to carry out their work, expending more time in the rationalization of design engineering, remaining behind the geometrical design.

INDO system consolidates and applies the engineering experience accumulated in a company through the years, preserving the knowledge when employees retire or change their job.

It allows a company to automate repetitive engineering tasks with the consequent saving of time. It captures the manufacturing and engineering knowledge, avoids the repetition of mistakes and does all changes in the products.

DESIGN EVOLUTION

The automatized design evolution has followed the following sequence in time:

- 2D desing in the 60s.
- 3D wiring designs in the 70s.
- Surfaces modeling appeared in the 70 to guide the necessity of complex representations, boundaries and surfaces in a complex way, particulary in the aerospatial and automobile industries.
- Solids increased design productivity on a basis of assemblies.
- Parametric modeling appeared in the first 80 as soon as dimensions of solid objects became easy to modificate.
- Each of these technologies previously mentioned has evolved by improving user interaction with the design system. Formerly data were stored in geometrical data bases.
- Knowledge based engineering began a few years ago to guide the necessity of automating both engineering and design.

This gave the engineers the first design technology where product models don't have the geometry, but do have the engineering rules to create that geometry.

As in the previous evolution, existing technology is oriented to solve simple design problems. In the 90 an excision appears between knowledge based engineering for complex design problems and geometry based systems to simple assemblies and little parts.

PARAMETRICAL MODELING VERSUS KNOWLEDGE BASED ENGINEERING

With a parametrical modeling system, the user creates a geometrical model by using an interactive base of features. That is, to adapt capturing the dimensional relation in parts and little assemblies. Parts that can be easily changed in dimensions, both in scale and size.

With a knowledge based system, the user builds a model of the engineering of the product, using a descriptive language, expressing the engineering knowledge behind the geometrical design. Products are automatically redesigned by changing inputs to the model.

As an example, when a dimension is changed, the engineer wants to evaluate the effects of changing a part (as increase height, pressure, cost, etc). These rules can be built inside a model to create the features in an intelligent way.

With a knowledge based system, the structure of the product can be changed automatically from the inputs to the model. This is not possible with a parametrical modeling system.

Parametrical modeling systems are used to design parts of components and little assemblies. In exchange, knowledge based system INDO can be used for large assemblies and complex products.

FEATURES OF THE INDO SYSTEM

The INDO system complements traditional CAD technology by extending the design process beyond the graphical modeling.

With a CAD system, the user completes a detailed geometrical design using an interactive graphical modeling. These geometrical data can be used as a geometrical restriction when is sent as an input to the INDO system. With INDO the user has a model of functional engineering of the product by the use of rules of representation that describe the design and engineering of the product.

When you have finished the design with INDO you can send the product to a CAD system to:

- Complete details in the design (if necessary).
- Obtain a great number of properties of volumetric data.

- Do a finite element analysis.
- Generate NC programs.
- Generate detailed drawings.

The outputs from the INDO system include reports as material lists, costs analysis, engineering reports and manufacturing process plan.

When we want to automate the design process, engineering information must be captured, with the product geometry. If this is not achieved, engineering calculations must be altered when changes are done, in spite of the long design cycle.

A design system for large scale and complex products is necessary, not only for few components and little assemblies, but also for large series of components with important dependencies between them.

Often, complex products need multiple and different engineerings with extension of the design cycle when changes are done; for example, in a style change in the exterior design of a car hood the structure engineers must evaluate the design of the support, the design engineers must redesign the structure and the aerodinamical engineer must evaluate the performance of the design. This is a typical sequence that causes through the cycle different changes in the final design.

INDO allows multiple different engineerings focused in an only model. In this case, sufficient flexibility is provided to achieve easy adaptations to the changes in the products that can be claimed by the specifications or requirements of clients.

With knowledge based engineering, the engineering and design process has changed from manual generation and manipulation of geometrical design data to automatically generated models from design data.

With the knowledge based engineering system:

- Design engineer knowledge is captured in an intelligent model of the product, with gneration rules of the geometrical design, automatically from a set of inputs.
- Knowledge engineering comprises topics from different engineerings, that are included in the model of the product and are evaluated

simultaneously; in this way, concurrent engineering is promoted. Both manufacturing and design rules are built inside the model generating geometrical design data.

- Changes in design and optimization process are automatic, as the optimization rules are built inside the product model.

As an example, the kind of knowledge that is included in the model by the designer can be:

- The structure of the product and rules to alter given new inputs.
- Dependencies between parts of the products.
- Rules created of geometrical parts from input specifications.
- Rules of engineering design to optimize the cost of the product, the making process and the quality.
- Decision criteria to extract reports from external data bases, tables of materials and existing designs.
- Rules to analyze results from engineering analysis programs and automatically evaluate the model with new inputs.

REASONS FOR A KNOWLEDGE BASED SYSTEM AND NOT AN EXPERT SYSTEM

Expert systems don't have incorporated primitive geometries to describe mechanical products easily. The knowledge based system INDO incorporates geometrical features to describe complex products, including a complete set of rules to represent the most important primitives.

Expert systems work oriented to a collection of rules; all the engineering knowledge must be put in this format. This requires a complex control of the specifications flow and an experienced programmer.

The knowledge based system is designed to be used by engineers.

Expert systems are limited in the level of complexity that they can handle, due to the necessity of using all the rules in the determination of results.

The INDO system is built taking into account that only the rules answering to a particular requirement are going to be executed.

SOME DETAILED FEATURES OF INDO

When using INDO the design is done by means of a descriptive language, object oriented. The components are treated as objects plus some descriptive information, called properties -that can be applied to each object.

INDO, indeed, is more than a declarative language: it is a processal one, and it is designed to be used by engineers, not by programmers. Product design task is essentially a specifical one, analog to financial worksheets, where the user specifies the relation among the different elements, but not the running order.

Users may define the product or the structure of th process in a natural way, hierarchical, through the use of the structgure of the product tree, changing it automatically, by altering model inputs.

The features of the object oriented language allows the users to establish relations and dependencies between parts, to such an extent that when a part changes, all parts depending on it change automatically.

For example, users may define classes of parts through which common information to many parts can be extracted and shared, using the concept "class of inheritance".

Dependencies between parts can be done through another concept: "part of inheritance", by which the parts of the structure of the product tree inherit features from upper parts in the assembly tree. Another concept useful in this way is the "symbolic reference", by means of which the parts can reference explicitly features of another parts not directly related by inheritance.

Tree structure of the product is the basis for a product design organization in a logical hierarchical way. Each object in the tree represents a part ob subpart in all the assembly or process.

Using INDO users can include rules in the model of the product that affect to the structure of the product. For example, an input for a model could be whether we need a table for a hotel or for a little house. The parts of the system would be modified and the number of legs could change by altering the inputs to the model.

The Figure 1 shows the tree structure in this simple case.

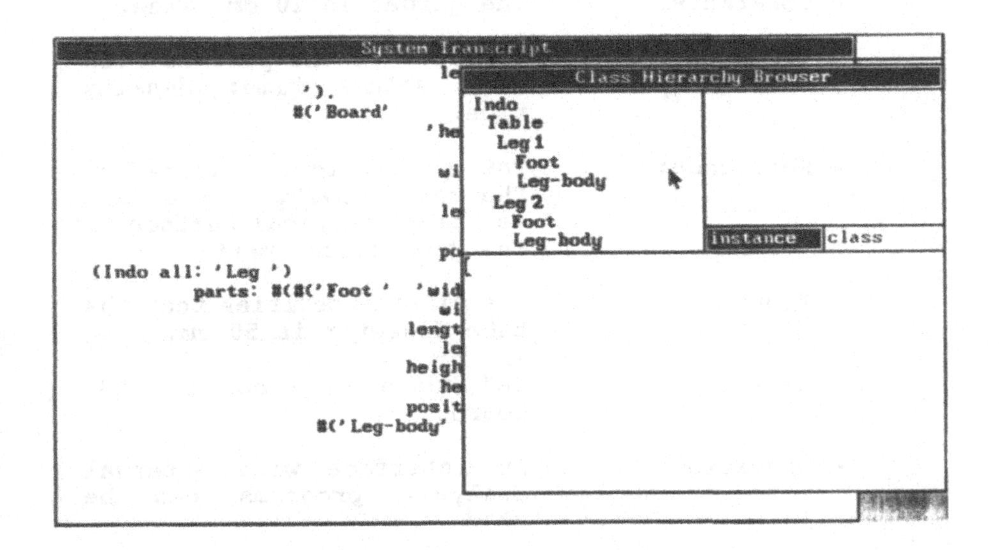

Figure 1. Design tree

Properties can be attached to objects, for example the leg width.

Parts can inherit features of upper parts in the product tree through the part of inheritance, as in the case of the height of the leg: it is the same as the height of the table.

The nature of the object oriented language of INDO allows users to describe the sets by generic definitions of classes of parts.

With this classification by parts, the information that is common to a set of parts is extracted and can be shared. If one of the classes changes, these changes are automatically inherited by the sets.

In our example of Figure 1, various kinds of legs are created from a base leg which has the common characteristics to each leg. If the base leg changes, all of the new legs inherit its changes.

With INDO the users can write rules with expressions about the engineering knowledge related to how the properties are determined or evaluated. These rules can be of various kinds:

- Constants: The girder is 10 cm. wide.
- Calculations: The cost of the part is 1325 Pts; volume time; density time.
- Reference: The geometrical surface of the mould cavity is the same as the geometrical surface of the geometrical part.
- Input: The client specifies that the tube diameter is 50 cm.
- Geometry: "A" board is a box for "B" board.
- Operation: An interface with external analysis programs can be used.

PROTOTYPE AND EXAMPLES

The prototype of the INDO system has been developed with Smalltalk V in a personal computer. This allows the user to have a language able to perform all kind of arithmetical calculations. A table would be modeled in the following way.

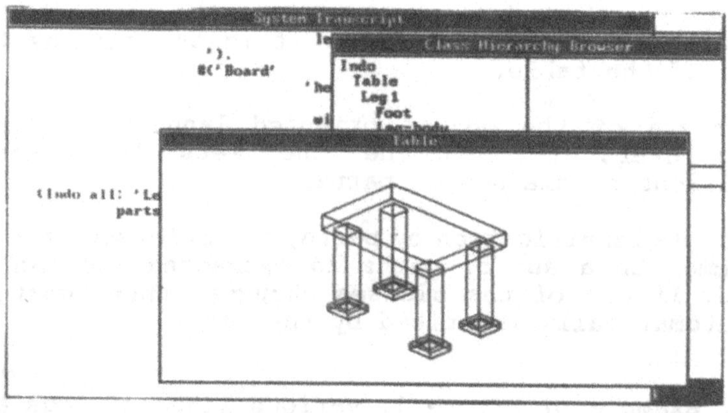

Figure 2. Table

A table would have some general features:

- width
- height
- lenght
- center.

Such a table would also have essentially four legs with some values of width and length determined by the table dimensions and a board with some values of height, lenght and width in function of those of the table and will be located up in the table. In the figure 3 all these basical properties are shown:

```
System Transcript

    Indo parts: #(#('Table'
                    'width: [10];
                     height: [10];
                     length: [10];
                     center: #(0 0 0)')).
    Table parts: #(#('Leg '
                     'number: 4
                      width:  [(Table instancia
                               width value)/10] ;
                      length: [(Table instancia
                               length value)/10]
                    ').
                   #('Board'
                               'height:  [(Table instancia
                                           height value)/10] ;
                                width:  [(Table instancia
                                          width value)+2] ;
                                length: [(Table instancia
                                          length value)+2];
                                position: #('on-top' 0)')).
  (Indo all: 'Leg ')
        parts: #(#('Foot '  'width:  [(Leg instancia
```

Figure 3. Table parts and properties

All legs will have a foot located at the bottom and with some values of width, lenght and height depending on the values of the leg or of the table. They also have a leg body with a lenght being the heigth of the table minus the heigth of the foot. In figure 4 we can appreciate the code for this:

```
System Transcript
                                length value)/10]
                  ').
                 #('Board'
                               'height:  [(Table instancia
                                           height value)/10] ;
                                width:   [(Table instancia
                                           width value)+2] ;
                                length:  [(Table instancia
                                           length value)+2];
                                position: #('on-top' 0)')).
(Indo all: 'Leg ')
        parts: #(#('Foot '   'width:  [(Leg instancia
                                 width value)+1] ;
                              length: [(Leg instancia
                                 length value)+1];
                              height: [(Table instancia
                                 height value)/10];
                              position: #('below' 0)').
                 #('Leg-body' 'length: [(Table instancia
                                     length value) -
                                        (Foot instancia
                                     length value)]').
```

Figure 4. Leg and foot definition

As we can see, the design of a table has been modeled, not in the way of a geometrical design, nor in the way of a parametrical design, but in the way of a specification like the one an engineer would do if a description about how is a table should be done.

This modeling, though basical, has all the necessary knowledge for the design and is perpetuated in a text file that the INDO system will process giving a graphical image as a result.

CONCLUSIONS

The INtelligent Design Optimizer was presented through the previous paragraphs. This system has been conceived to facilitate the work of design engineers in complex projects.

Actual classical CAD systems are completely overpassed with this knowledge based design technique, combining important pattern matching and frame based, object oriented artificial intelligence methodologies.

In the future we think in the final system running in workstations, with an estimated productivity of near 1000 % of the prototype shown here.

REFERENCES

1. Archer, L.B. Systematic method for designers, The Design Council, London, 1965.

2. Axe, F. CAD in British industry, J. Roy. Soc. Arts, pp. 249-261, Mar. 1988.

3. Begg, V. Developing expert CAD systems, Kogan Page, London, 1984.

4. Cross, N. Developments in design methodology, Wiley, Chichester, UK, 1984.

5. Frost, R. Introduction to knowledge based systems, Mac Millan, New York, 1986.

6. Gunn, T.G. The mechanization of design and manufacturing. The mechanization of work, (Ed. Anon), Freeman, San Francisco, 1982.

7. Krouse, J.K. Automated factories: the ultimate union of CAD and CAM, Machine Design, pp. 54-60. Nov. 1981.

8. Lawson, B.R. How designers think, Architectural Press, London, 1988.

9. Lenat D.B. et al. Building large knowledge based systems, Addison Wesley, Reading, 1990.

10. Majchrzak, A. et al. Human aspects of computer-aided design, Taylor & Francis, London, 1987.

11. Rolston, D. Principles of artificial intelligence and expert systems development, McGraw Hill, New York, 1988.

12. Smalltalk V, Tutorial and programming handbook, Digitalk, Inc., Palo Alto, 1988.

13. Smithers, T. Product creation: an appropriate coupling of human and artificial intelligence, AI and society, Vol. 2, pp. 341-353, 1988.

14. Walter, J.R. et al. Crafting knowledge based systems: expert systems made realistic, Wiley-Interscience, New York, 1988.

15. White, M. Towards conceptual integration of knowledge engineering, Heuristics, the journal of knowledge engineering, p. 25, Sep. 1988.

HYPERARCHITECTURES - A New Multidimensional and Humanistic Scheme for CAAD

T. Oksala

Department of Architecture, Helsinki University of Technology, SF-02150 Espoo, Finland

ABSTRACT

A new concept called hyperarchitecture for the representation of complex and hierarchical structures in architectural design is presented especially in view of CAAD-applications. Analogies between linguistic, pictoral and architectural generation and organization are pointed out under conceptions of hypertext, hypermedia and hyperreality. The need for hierarchical structures in architecture is discussed under the notions of hypergeneration and categorial hyperarchitectures. Use of complex building models in CAAD applications is discussed and the concept of hypermodel is introduced. The exact manipulation of architectural high-order structures allows rationalization of complex knowledge used in the design and planning process by the participating individuals. A hypertext approach should offer new possibilities to teamwork and knowledge maintenance and support decision making in CAD. Synergetic use of multiexpert systems are discussed under the concept of master systems guiding architectural synthesis. In the study of complex man-machine systems the level of hyperarchitectures extends the scope of discussion from architectural machines radically to a new and more humanistic direction.

INTRODUCTION

Recent research in artificial intelligence makes it possible to apply [1] ideas owing traditionally philosophical status to technology and design. In this process knowledge engineering, logic programming and qualitative modelling are offering conceptual bases for CAD-CAM activities in production work. Although the use of scientific techniques based on mathematical optimization and first order logic have given revolutionary advantage to the possibilities of mechanization in engineering design, however, a lot of critique can be directed to this development from the point of view of the quality of design results. This critique includes, without doubt, discussions concerning the richness of human thinking and its expression. Human skill, expertise and craftsmanship are based on tacit knowledge and background theories [2], [3] with varying levels and kinds, and is thus not always easily extendable. On the other hand the possibilities of logic and mathematics in the description and modelling of phenomena are much more richer, than the so called first order techniques that are mainly used. A lot of experimental work can be done based on multivalued logic and type theory in order to develop discourse concerning design automation [4, p 59]. In fact this field of study is too large to be mastered systematically and the field of study might be discussed in fragments. In the following we characterize certain higher order relational structures in architectural design and planning, which might be called hyperarchitectural phenomena in analogy with the so called hypertext conception familiar in the micro computer and AI field of today [5].

HYPERPHENOMENA IN LINGUISTIC, PICTORIAL AND ARCHITECTURAL REALMS

The development of design technology since the 60's has proceeded through data processing, picture processing, and design programming. This development is largerly based on the conception of formal language and corresponding generative devices as they are realized in formal grammars, picture grammars and design grammars [6, pp 108-109]. Certain progress in the use of this oversimplified model of language has meant the introduction of logic articulation and semantics as well as logic programming techniques. All this has lead to the use of metalanguages and metatechniques in design (discussing design of design) in order to combine higher order human capabilities to the formal systems development. In practical formation of organized knowledge storages the concept of formal language or language as a sequential medium of expression in general has obvious limitations. Analogously techniques like relational data bases and parallel processing have been largerly discussed in order to achieve more powerful tools in handling complex relations. These developments have their relevance especially in graphic and design applications due to the complexity of those areas but remain of course formally restricted.

A logically simple and promising field of techniques to organize information in text like form is the concept of multigraph as applied to meaningful parts of the original body of the text. The content of relations might vary from associations to references etc. Such structures have been extensively studied under the conception of hypertext [5]. In fact every well-articulated dictionary as understood intentionally and deeply contains such articulation, when manipulated by humans. Thus one advantage of the hypertext conception in knowledge-engineering is without doubt its natural connection to the human way of mastering large chunks of complex knowledge. This idea of articulating already articulated text anew might be applied to information management of other tools having natural basic articulation like pictures or other media. In this case we naturally speak about hypermedia as convenient. Moreover, continuing this line of thinking we might speak about hyperrealities as articulated fields of texts, visual media, and artifacts or parts of environments. Especially the concept of hyperarchitecture can be understood as a hyperreality, in which the artifacts used are meaningful parts of some architectural reality. In the following we attempt to render content to this interpretation by showing that in fact in the field of architecture and especially in the field of CAAD a lot of work has been done in order to develop highly articulated fields of architectural environments. In fact architecture might be methaphorically understood as a means of expression, a language in itself [7]. This naturally leads to analogous hyperconceptions. In the following we make, however, a sharp distinction between language-like and reality-like entities. The concept of hyperreality extends reality, because illusions and fictive meanings can equally well be connected to real objects by interpreting the pictorial and linguistic parts of it in a suitable way. The connection to fictive cognitive images extends our discussion towards architecture as art and poetry of its own kind.

MULTILEVEL MANIPULATIONS AS ARCHITECTURAL HYPERPHENOMENA

State descriptions of state descriptions

Elementary architectural generation has been largerly studied under notions of formal grammars and corresponding pictorial and architectural devices (Figure 1). In logic terms such generations might be discussed in terms of state descriptions [4]. In such a partial but complete description is made with respect of the expressional resources available for the part of possible environment. The solution base of state descriptions can serve as a basis for simple design models, in which solutions are filtered out of them. Theoretically interesting abstractions of state descriptions include structural descriptions and constituents as they correspondingly tell only the distribution of features or the spectre of features used in the particular design. In practice the set of state descriptions is too large for serious production of artifacts. One way of correcting this inadequacy is to use hierarchical or meta-level generation. Then we think that for every individual object we represent a state description and then tell by a new state description how the objects are or should be located in space. This in fact corresponds well to the common sense experience and practical know-how about our everyday environment.

In architecture we are not working in an uniform universe of science, but rather in a configuration of environments of different nature and scale. When applied once, the idea of hierarchical state descriptions can be extended to n-level state description models of the environment. Then in the production lattice the nodes are generation devices and the arcs meta-generation rules telling how to replace generation devices with each other. In fact, fractal generations are geometric examples of this principle in their recursion. The way of reading architecture is hierarchical. We may look how the rooms are distributed and in the next moment how the window is produced etc. Typically the points of interest are associated to a suitable generation system and the total field of interest corresponds to a certain multigraph of generation systems as hypergeneration. In a certain sense laws at hyperlevel correspond to the predicates on the first level. In that case hyperconstituents tell what laws are to be considered and hyperstructure descriptions tell how strictly the laws hold.

Typical configuration of types

Possible abstractions in the set of state descriptions include architectural types, rules, principles etc. They play important roles in composition of

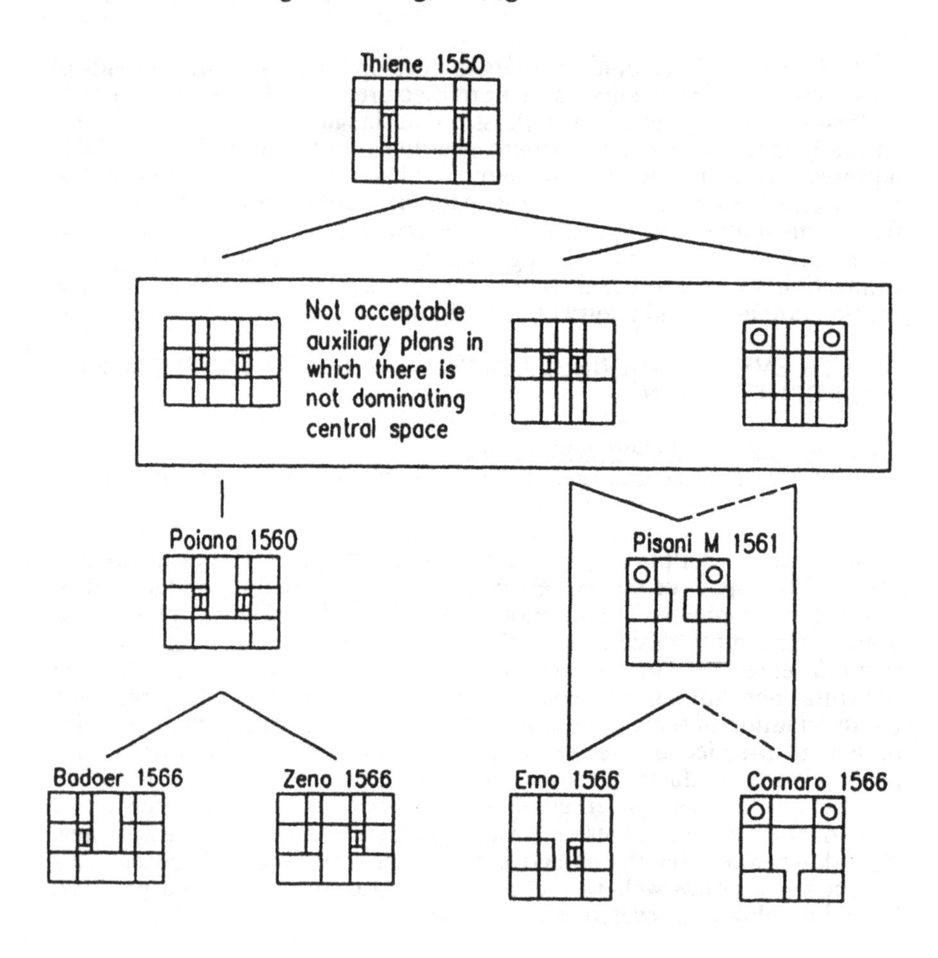

Figure 1

A category of Palladian villas generated by a design grammar and organized as an evolution semilattice of Palladian style. The figures are interpreted as an illustration of metageneration. Nodes are examples of outputs of subgrammars and arcs are transitions between them. The figure is easily extended with association relations to other styles and buildings eg. in neoclassic period.

items suitable for human experience and use of environmental items. By a typology [8] one can master partial or total solutions in design problem solving. The innovation of new instances in an architectural typology is a characteristic of creative design activity. Architectural environment is, however, experienced totally and we are perhaps looking at objects by classifying their features and types, but at the same time we see the total environment more or less as a typical configuration of its elements of varying level. In this process every meaningful part of architectural environment may have its internal architecture,while it is also interlinked by compositional rules to other parts in a metacomposition. These phenomena can be projected from architectural reality to future designs as they apply to the field of prototypes [9]. In architectural practice we often use simultanously a set of typologies hierarchically. This set is articulated with associations noticing e.g. common or contrary style features and other thematic factors. Such a body of knowledge forms so to speak a hypertypology.

Categorial hyperarchitectures

In the generation of architectural items and in the perception or transformation of architectural types we are facing the situation, that human conception tends to articulate the field as a system of interlinked subparts. Parts in an architectural work are perceived or formed under pattern laws or other more specific meaningful laws. The relations of articulation vary from multidimensional dependencies to quite special generation or composition rules. In order to describe such phenomena formally we again need a multigraph. The connections often have transformation like character. Such articulations of architectural phenomena can well be seen in analogy through concepts of mathematical category theory [10]. A category formally consists of objects of arbitrary complexity and arrows mapping transformations between them. Meaningful examples of such structures include time-evolution lattices (Figure 1) of architectures [4, p 37] and spatial multi-level networks of buildings.

On higher abstraction levels architectures themselves can be objects of manipulation. Thus architectures of certain authors might be located as totalities to sites in the crucial moment when one selects designers to some region. In discussing architectures we interlink the works of different architects, time-periods, cultural environments, styles etc. Developed know-how about building base or especially about the so called historically remarkable architecture is based on a variety of such categorizations. In historical discussion buildings are followers or not of each other. The quality and strength of this relation may of course vary in many ways. In experiencing architecture we can freely associate according to our education some parts of a certain building to other buildings. On the other hand we can make style loans, interpretations, adaptations, modifications and intended references. In any case when forming large knowledge-bases of existing or future architecture we are facing problems of categorization. Already the organization of products in a CAAD-work station after some years work is a suitable problem to be handled with categorization methods. Generated parts of architectural expression of some architect or design team form a complicated network as they are used or not maybe in transformed forms in a variety of projects. Know-how about such phenomena form the basis of fictive design experiments having suitable propability to become accepted as a basis for future realizations. A full-fledged form of hypermedia-aided design system is associated to a "multicategory" of products and processes. In such a system one can move freely via varying strategies from one solution space to another (Figure 2).

HYPERCONCEPTS AS APPLIED TO CAAD

Toward building hypermodels

Models can be conceived as nonlinguistic entities falling in the scope of linguistic interpretation. At the same time they can be used for descriptive purposes and thus posses a strictly linguistic character. In architectural design really applicable building models are in extreme cases very complex entities, to be handled in practice. That is why their use falls in to many categories [6, pp 110-111] and involves different abstraction levels. For example purely geometric models are often inadequate, although useful for special purposes eg. for calculation of volumes etc. In some studies it has been shown that effective modelling of natural buildings presupposes a family of models at the level of mass formation, functional layout, and structural detailing. Such partial models of different abstraction level form naturally a most effective totality if they are interlinked with predecessor relations, zoomings, associations etc. suitable referential "traffic guides". Be it as it may, hypermodels might be formally conceived as sets of submodels suitably linked together. In order to concretize hyperphenomena in architecture such constellations offer a natural field of discussion due to the concreteness of models and due to the familiarity of building models especially in CAD [11]. The basic set of models used in a building project enterprise thus could naturally be developed to a partial hypermodel. In addition, modern technology offers new possibilities to build models of large environmental areas with free transitions between real and fictive domains making it possible to change aspects, outlook or structure of architectural items, in question.

Intentionality as an intention

The use of first order logic models in architectural generation already means combinatorial explosion of both solutions and expressions. This means difficulties in mastering formal architectures without speaking about the combinatorial complexity of natural architecture. Thus we need various kinds of abstractions, predicates, types, laws, principles, themes, styles etc. in order to master the richness of possible realizations. In the rational problem solving or action theory approach when used as a model for design thinking and action, we need knowledge about our goal hierarchies [12]. In certain extent planning and design can be seen as a special case of intentional action. The same holds true for architecture as art although the status of deepest intentions remain problematic. Be it as it may, in order to rationally reconstruct architectural action, in any case, we have to try to identify our intentions in building design as clear as possible. Such an enterprise is easily seen as a basis for building relevant AI-models to help CAAD development and acceptable design automation. This program clearly has as its intention the intentionality of design.

The hierarchy of goals, criteria and solutions is many splendored

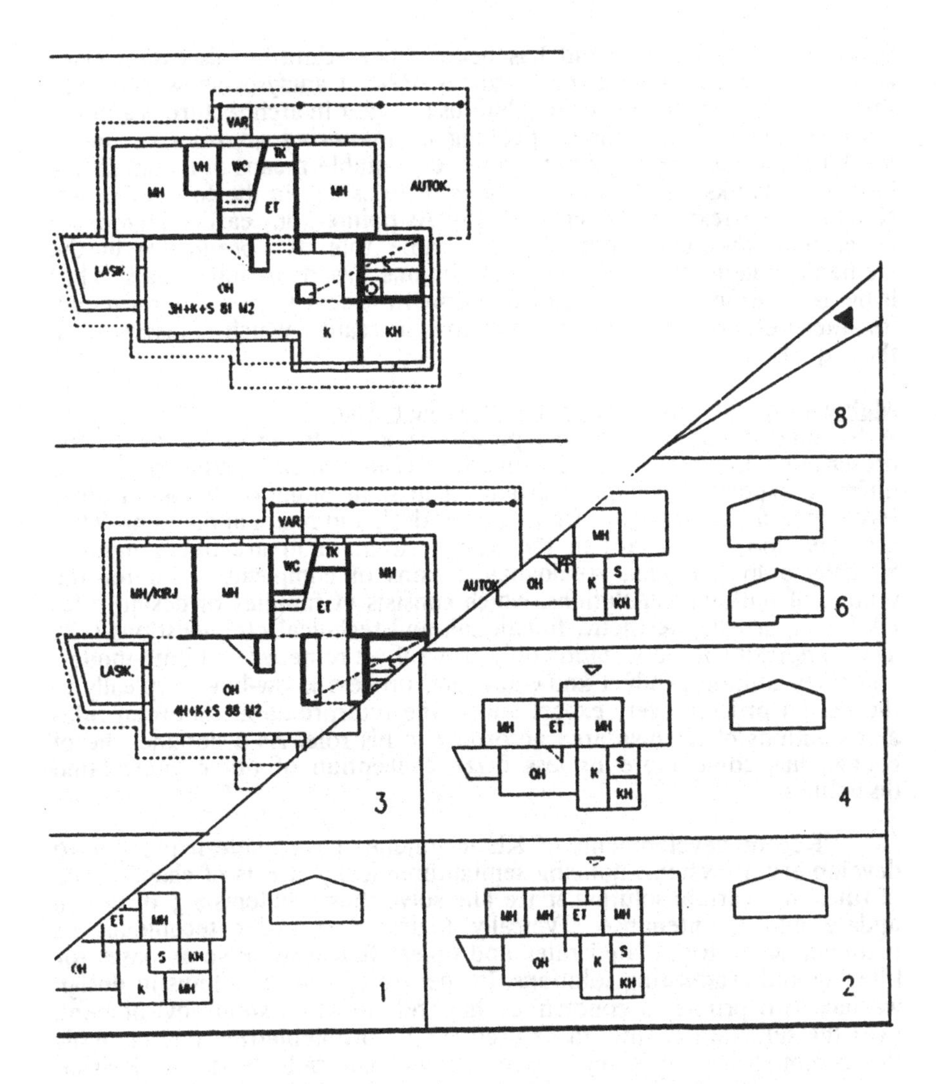

Figure 2

A Finnish small-house design system embedded in a hyperspace in which changes between size, adjustment to site, functional details, construction solutions etc. are free for user participation. The system produces 100 type-solutions, 1000 refinements, and about one million packages of drawings for realization. All packages might be kept under control of state subsidized production. Mathematically it is interesting to note that the hyperarchitecture above can be represented in a n-dimensional Hilbert space.

typically in architecture and has been under scientific discussion since antique eg. in the writings of Vitruvius. Recent analyses show that even the covering classification of regulations relevant in architecture is a major research enterprise, without speaking of mastering all rules in detail. Modern CAD technology offers, however, suitable means of manipulating large data banks and knowledge bases of rules e.g. in the form of expert systems, hypertext and hypermedia [5]. As pointed out earlier high order generation laws and rules play the same role as predicates in the elementary generation of state descriptions based on individuals. Thus hypergenerations and rules articulated by hypertext form a natural model for interpretation of building regulations in reality, which is discussed in the sequel.

Multiexpert and master systems in building design

A building design enterprise typically develops by investigating partial prescriptive models, which document decisions made under control of different experts and interest groups. The same situation holds on other levels of architectural work from interior design to regional planning. Thus the total model relevant to the design situation in architect's mind is typically a hypermodel. Analogous notions of complexity hold for the totality of building regulations, which consists of families of design rules discussing artistic, aesthetic, functional, and technical etc. questions [12]. The complexity of the situation only grows if we remeber that both models and rules contain product know-how and process know-how. Typically in the design process every expert knows the interpretations between rules and solutions of his own area according to his role. As a designer, he of course, has some more or less fuzzy conception of other interrelated disciplines.

Recent developments of KE techniques have made it possible to develop expert systems noticing semiautomatically effects of changes and of rules in a certain solution space and solve thus problems e.g. of design update and maintenance. Typically logical first order techniques or mathematically formulated rules and object functions serve as bases for filtering and readjusting solutions. In the use of expert systems in design we face two problems concerning their relation to system environment. First the selection of rules to be used can be problematized and secondly the common use of many expert systems naturally leads to decision problems. In the hypergeneration model discussed above the selection of rules becomes a natural part of generation as they are controlled by entities called hyperconstituents. If we want to make conscious design we should clarify this phenomenon by specifying how and what rules are introduced in the design game. On the other hand the study of advanced design is the only way of finding regularities in the goal combination as needed in building multi-expert systems. Systems which use expert systems as advisors in higher order decision making and which can "critically" speculate with rule selections, might well be called master systems at least in architectural discourse [13]. The concrete realization of such systems is a great challenge and to a great extent still an open problem. The concepts of hypergeneration and hyper-rule with their interplay with the designer form in any case a basis for opening the development in this direction.

CONCLUSION

The idea of automatic generation of design is age-old. In the era of enlightenment automatic generation of drawings was possible and also algorithmic design was performed by humans. In this century the development since the 60's has lead to semiautomatic generation in architectural design. This has been mainly effected under mathematical and logical first order techniques. On this basis it is possible to build conceptual extensions noticing higher order phenomena. As is well known it is not possible to master the field completely mechanically, not even in principle. In design work, as art, it is, however, possible to make experiments and to teach the results of these experiments not only to the human designer, but also to incorporate them in the system as well. In this development hyperconceptions offer a natural extension to first order-logic. The field of possible extensions is large, but problems can be handled in a natural way by usign two-level problem solving techniques, if the interpretations in the choise of levels are made freely. Especially in design the concepts of hyperarchitecture, hypermodel, hypertypology, etc. are natural formalizations of the real situation. The problem of concept formation in language is outside the scope of normal design, but analogous ideas in the manipulation of written knowledge are applicable when discussing the interpretations of planning and design rules. This makes it possible to develop high-order decision support systems for CAAD, which as totalities might be called master systems. Such program means a radical change of paradigm in CAAD. The main goal to build design machines is realized and the new goal is to interlink such machines with the human design.

ACKNOWLEDGEMENTS

I would like to thank Mr Jouko Seppänen from the Institute of Industrial Automation of the Helsinki University of Technology for discussing the CAD and KE aspects of the subject. This work is made as a part of my research work at Department of Architecture in the Helsinki University of Technology.

REFERENCES

1. Gero, J. S. (Ed.). Artificial Intelligence in Engineering Design, Computational Mechanics Publications, Springer-Verlag, Berlin, 1989.

2. Wåhlström, O. Sketching and Knowledge in the Design Process, In TIPS'86 (Ed. Oksala, T.), pp. 6-11, Proceedings of the 2nd Knowledge-based design symposium, Otaniemi, Finland, 1986, RYT, Otaniemi 1988.

3. Lundequist, J. Do Computers Increase the Architect's Distance to Architecture. Chapter 11, Knowledge-based Systems in Architecture, (Ed. Gero J. S. and Oksala T.), pp. 109-118, Acta Polytechnica Scandinavica, Ci92, Helsinki, 1989.

4. Oksala, T. Locigal Aspects of Architectural Experience and

Planning, Helsinki University of Technology, Research papers 66, 1981.

5. Nelson, T. H. Literary Machines. Edition 87.1. San Antonio, TX, Project Xanadu, 1987.

6. Oksala, T. Logical Models for Rule-based CAAD. CAAD'87 Futures (Ed. Maver, T. and Wagter, H.), pp. 107-116. Elsevier, Amsterdam, 1987.

7. Coyne, R., Radford, A. Knowledge-based Design Systems in Architecture: A linguistic perspective. Chapter 4, Knowledge-based Systems in Architecture (Ed. Gero, J. S. and Oksala, T.), pp. 27-38, Acta Polytechnica Scandinavica, Ci92, Helsinki ,1989.

8. Oksala, T. Typological Knowledge in Computer-aided Housing Design. Chapter 6, Knowledge-based Systems in Architecture (Ed. Gero, J. S. and Oksala, T.), pp. 59-68, Acta Polytechnica Sacandinavica, Ci92, Helsinki, 1989.

9. Gero, J. S. Prototypes: A Basis for Knowledge-based Design. Chapter 2, Knowledge-based Systems in Architecture (Ed. Gero, J. S. and Oksala, T.), pp. 11-18, Acta Polytechnica Scandinavica, Ci92, Helsinki, 1989.

10 MacLane, S. Categories for Working Mathematician, Springer-Verlag, Berlin, 1971.

11 Lundequist, J., Kjelldahl, L. Models in Computer-aided Architectural Design Work. Chapter 3, Knowledge-based Systems in Architecture (Ed. Gero J. S. and Oksala T.), pp. 19-26, Acta Polytechnica Scandinavica, Ci92, Helsinki, 1989.

12 Oksala, T. Toward Intelligence in CAAD by Using Quality Knowledge. Artificial intelligence in engineering design (Ed. Gero, J. S.), pp. 143-158, Computational Mechanics Publications, Springer-Verlag, Berlin, 1989.

13 Oksala, T. COMPOSITA - Towards Architectural Expert Languages. In EuropIA 88, pp. 247-260, Proceedings of the Int. Conf. on Applications of Artificial Intelligence to Building, Architecture and Civil Engineering, Paris, France, 1988. Hermes, Paris, 1988.

An Integrated Approach to the Automation of Class Three Electronic Design Problems

T. McNeill

Measurement and Instrumentation Systems Centre, University of South Australia, P.O. Box 1, Ingle Farm, SA 5098, Australia

ABSTRACT

Over the years in an effort to shorten the lead time to production in electronic systems design the various steps of the process have been progressively automated. In recent times this short lead time has become increasingly important with time to market being the significant factor in product success in the 1980s and 1990s. In the production of working prototypes and 'one off' products, a short lead time is often more important than a design that is optimised in terms of parts cost.

The paper investigates an area of electronic systems design that is yet to be automated. The area of block diagram instantiation. It is possible to capture the process by which the experienced engineer, given a specification, arrives at a block diagram for an electronic system. The automation of this step could result in significant reductions in lead time.

INTRODUCTION

The engineering of electronic systems has been extensively developed since its emergence in the early twentieth century. Much of this development in more recent times has been in the automation of the process. The first steps to be automated were those that involve repetitive solutions usually algorithmic in nature.

Research presently occurring into automating the more intelligent parts of the design process concentrated on the fields that have rapidly expanded in complexity over the last few years. Very large scale integration (V.L.S.I.) integrated circuits (I.C.s) is one such field. Advances in manufacturing processes has meant that it is possible to fabricate very complex designs on I.C.s and it is necessary for the design process to to keep pace with these advances.

There are many situations where a problem requires a low volume electronic solution and the use of a custom designed I.C. is not suitable. This paper is concerned with the automation of the intelligent steps in the design of relatively simple electronic systems for implementation in printed circuit form. Much of the time in developing these systems sees the engineer involved in regurgitating simple block designs and combining these blocks to meet the requirement. It should be possible to capture this process and implement it in a knowledge based system.

This has been the topic of ongoing research within the Measurement and Instrumentation Systems Centre (MISC) at the University of South Australia (UniSA). This paper summarises the research to date and looks at some strategies for the continuation of the research.

THE STATE OF THE ART

Engineering An Electronic System

The engineering of electronic systems starts with the user having a requirement, Fig.1. The user is often an expert in a field other than electronics. It is important that the user communicates the requirement effectively and without ambiguity to the electronic system designer as is the case with other forms of engineering. [1,2] Hence the need for a detailed specification document. This is a step in the process that is often neglected in a bid to shorten the lead time. M'Pherson [3] sees the fact that the customer's real objectives are not met as being one of the four main reasons for otherwise good system designs failing. Berlin [4] stresses the importance of this step and points out that "without an understanding of needs, one may solve the wrong problem." She later stresses the need to take time to probe and listen with an open mind.

The design step is discussed in greater detail later. Simulation and analysis then follow to check that specifications are met. The evaluation process is very dependent on the specification step. [3] If the circuit fails to meet specifications it will undergo a redesign step.

Once the specifications have been met there are a number of ways in which the circuit can be implemented ranging from custom designed integrated circuit through standard cell approaches, thick film implementations and printed circuit board (P.C.B.) to wire wrapped implementations.

The most common of these in instrumentation is the P.C.B. implementation. The P.C.B. layout and fabrication steps follow with the necessary components being loaded onto the board in the

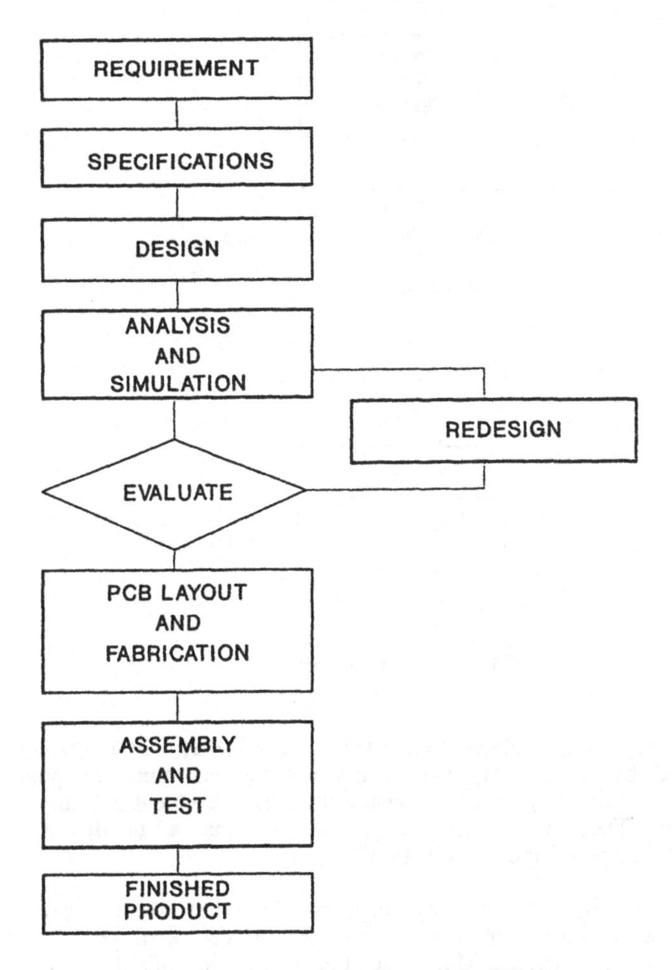

Fig. 1 The Engineering of Electronic Systems.

assembly phase. The final step involves the testing of the circuit to confirm that the analysis and simulations have been appropriate.

The Design Step

A closer look at the design step reveals that it consists of a number of distinct parts and these are shown in Fig. 2. The process begins with finding an appropriate block diagram to represent the system. Rowles and Leckie [5] indicate that this strategy enables the designer to produce a partially optimum design at a functional level without it being necessary, in the first instance, to have a rigid plan at the circuit level. At this stage it is easy to think of the solution in general terms without being concerned with the specifics. This is consistent with the

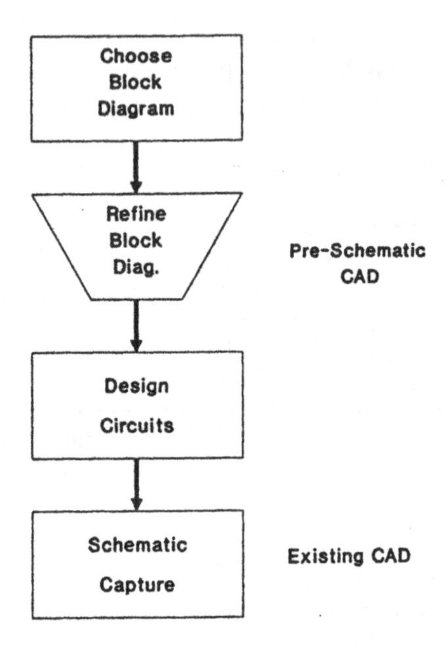

Fig. 2 The Design Step

Finkelstein and Finkelstein [6] view that best results are achieved by formulating the design problem as much as possible in abstract terms, ignoring constraints and convention in a first instance. They view the design process as a result of "lateral creative jumps of the imagination".

The block diagram refining process involving further dividing of the block diagram into simpler blocks that can then be implemented at the circuit level. Much of the work in the first two parts relies on the designer's experience, domain knowledge and design heuristics. The circuit is then ready to be captured in schematic form, that is converted into circuit symbology.

The Automated Steps

The various steps in the overall design process have progressively been automated with resultant reductions in lead times. A brief discussion of the areas that have been automated to date will follow. There are papers that provide more detail on the progress in the past twenty years and the present state of the art in CAD. [7,8,9]

Although automated specification for electronic systems has not as yet been attempted, the same has been developed for sensor systems by another member of this research team. [10,11] That work and how it relates to this study is discussed in more detail later.

The first parts of the design step are yet to be automated and form the main body of the discussion of this paper. The circuit level implementation is discussed in more detail later. The schematic capture component of the design step has been successfully automated in commercially available software packages which generally produce output files that can be used by simulation packages and P.C.B. layout packages. Once the design is captured in schematic form the circuit can automatically be simulated and analysed.

The evaluation step is often a vague process where the various constraints are traded off against each other. This will possibly be the last area to be automated although Finkelstein and Finkelstein [6] have offered insight into the formalisation of the process using utility analysis. The design can then be automatically transferred to P.C.B. The quality of this automatic P.C.B. layout is not as yet up to the standard of an experienced human operator. At present there is much industry based research into improving these packages with the work often involving the use of knowledge based systems. The automated fabrication and assembly of circuits is used in situations involving large production runs. In small runs the assembly is economically achieved manually.

There are three distinct areas that have yet to be automated. These are the specification step, the earlier parts of the design/redesign step and the evaluation step. When these areas are tackled further reduction in lead time will be achieved

AUTOMATING CIRCUIT LEVEL DESIGN

MISC has been developing a block design package called AJITA. [7] This package is being developed as an aid in the design of analogue and digital blocks in electronic systems design and is summarised below.

Once an appropriate block diagram has been produced, design on the circuit level can take place, often involving modifying existing standard designs originating either in applications notes or simply remembered from previous work by the designer. This circuit level design often involves repetitive and time consuming calculations. It is in this area that AJITA reduces lead times.

AJITA contains some 30 analogue and 15 digital blocks in an undimensioned form. Circuit topologies are predetermined and when the appropriate information is supplied by the user the package dimensions the block by calculating and selecting values for components.

The majority of analogue functions can be achieved with a combination of relatively few different blocks. Most functions fall into one of the following categories:

Input Circuit.
Amplifier.
Filter.
Wave shaping.
Modulator.
Demodulator.
Power Supply.
Output Circuit.

The AJITA menu system groups the blocks in an hierarchical fashion under these headings, with a similar menu for digital blocks, so as to facilitate block selection. With these building blocks AJITA can be used to build a large part of the systems typical of instrumentation needs.

The package targets experienced electronic engineers who would occasionally be involved in reasonably routine circuit design. It is a circuit level design package intended to reduce design lead times for the engineer. It uses dialogue familiar to the engineer when requesting information for the dimensioning of the blocks.

AUTOMATING THE DESIGN STEP

Design of electronic systems is a complicated process that involves a number of problem solving strategies and many kinds of knowledge. In general, electronic design can range from highly innovative to routine. Brown and Chandrasekaran [12] suggest that mechanical design falls into three distinct classes and it can be seen that the same applies to the electronic design activity. The three classes are summarised below.

Class 1 design is a highly innovative process that does not employ predetermined problem solving strategies or fields of knowledge. It would, for this reason, be the most difficult class to automate. Major inventions and new contributions to a field fall into this category.

Class 2 design involves substantial innovation with some routine work. In this class the problem solving strategies are difficult to predict but the knowledge sources are predetermined. This class of design can often arise from more routine design work when requirements change.

Class 3 design involves the process of selecting suitable solutions from a set of design alternatives. There can still be a level of innovation in finding the optimum solution in a very wide search space but class 3 work is often routine.

The automation of class 3 design processes is seen as the next logical step in the automation of electronic system design. It is essential to tackle this before working on class 2 and class 1 design. A system to perform class 3 design (named CAENIC, Computer Aided Engineering of electroNIC systems) is in its early stages of development [5] and it is intended that it incorporate AJITA to perform the circuit level design. Since the blocks available in AJITA make it very useful for instrumentation applications a first prototype of CAENIC would concentrate on class 3 design of instrumentation applications.

In system design lead time to production can often override the need for an otherwise optimal design [9] in the minimal cost sense (circuit reliability and maintainability should never be compromised). In instrumentation there is an area of applications where design costs outweigh production costs and a non optimised design (in the minimal cost sense) is acceptable if it can be produced with reduced lead time. CAENIC aims to provide fast reliable designs that are optimised in terms of overall cost but not necessarily minimised in terms of parts cost.

An Intelligent Front End

The first of the steps in the engineering of electronic systems requiring automation is the specification step. This would open the way for an integrated package that automatically takes the design from the user requirement to a level that can be picked up by AJITA and subsequently used by commercial schematic and P.C.B. layout packages.

An area of ongoing research in MISC is defining a need and development of a strategy for measurement. This is achieved with the MINDS software [13,14] which, by collecting information from the user about the measurement task at hand, directs the user in developing a clear understanding of the task. This results in the need to develop sensors and instruments to measure several different measurands. The user can then use an automatic sensor specification writing package, SPECRITER [10,11], to generate the specification documents for each sensor.

A system similar to SPECRITER that encompasses general electronic systems rather than specifically sensor systems, (OP-SPEC), is under development. This will be used to generate specifications for electronic systems thus reducing the lead time in this step of the design process and minimising the misunderstandings and problems in the design step discussed above. In the overall intelligent designer system OP-SPEC will be the intelligent front end that interfaces with the user.

The knowledge base of the latest form of SPECRITER [11] is divided into two distinct areas with the knowledge about how to extract a specification from an application expert being quite separate from the knowledge of sensor systems. Therefore the process of constructing a specification package for electronic systems would involve exchanging the sensor system knowledge base for a knowledge base of electronic systems.

An Intelligent Designer

A model of CAENIC now emerges, (Fig. 3.), that incorporates OP-SPEC as a front end and uses AJITA to perform the circuit level, pre schematic, design to a point where commercially available packages can take the design to completion.

The main interest of this study is the area of carrying a design from specification to a block diagram form suitable for use by AJITA. This will involve a knowledge based solution. The remainder of the paper concentrates on the implementation of such a knowledge based system. The process of knowledge acquisition presently underway is described and possible strategies for representing the knowledge are discussed. Final decisions on representation can not be made until knowledge acquisition is more advanced.

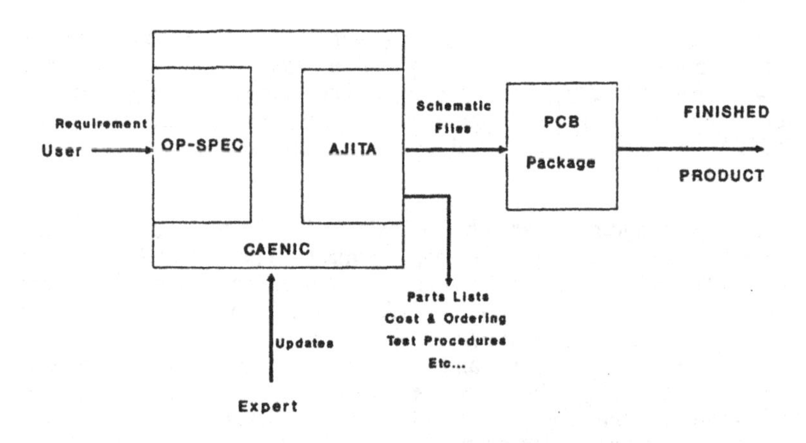

Fig. 3 The CAENIC System.

THE KNOWLEDGE BASED SYSTEM

Communication between the three parts of the system must be addressed along with the need for an 'electronic' knowledge base for OP-SPEC. An option would be that the main body of CAENIC itself forms the knowledge base of OP-SPEC, this would mean that as the dialogue is occurring between application expert and OP-SPEC, CAENIC is in the background producing possible block diagram implementations and requesting further information depending on the solution strategy under consideration. This seems to be an appropriate model of the cognitive process followed by an engineer in electronic design. An alternative approach is to keep the specification step and the design step separate and have them communicate using a specification language [15] or a hardware description language.

Knowledge Acquisition

There a several methods used to capture the way in which an expert makes use of domain knowledge. [16] These range from the simple interview to methods of machine induction.

The method most suited to the CAENIC situation is the 'talk through' interview with the various experts in the School of Electronic Engineering at UniSA. Four simple electronic design examples were chosen as follows:

10 watt audio amplifier.
Thermostatic control for a fan.
8 input mixer (audio) amplifier.
Phase locked loop circuit.

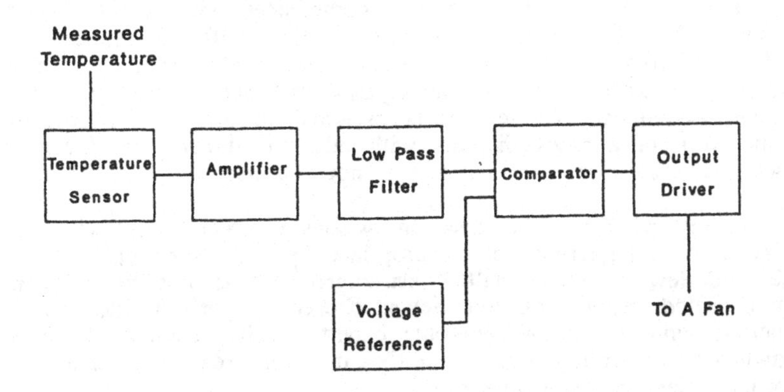

Fig. 4a A Simple Block Diagram.

FRAME: Thermostatic Control

SLOT 1: First Block
VALUE: TEMP_SENSOR

SLOT 2: Second Block
VALUE: INPUT_AMPLIFIER

SLOT 3: Third Block
VALUE: LP_FILTER

SLOT 4: Fourth Block
VALUE: VOLTAGE_REF

SLOT 5: Fifth Block
VALUE: COMPARATOR

SLOT 6: Last Block
VALUE: OUTPUT_DRIVER

Fig. 4b A Frame Representation of a Block Diagram.

These represent to an experienced engineer simple designs that would typically take a few minutes each to arrive at a first block diagram. Each design requirement has a subtle aspect that should enrich the knowledge gained in the interview. The audio amplifier represents a simple one input, one output system, the second is slightly more involved with two signal paths being combined into one output (see Fig. 4a), the third is chosen to give insight into a multiple input problem and the fourth involves a feedback system which requires reasonably deep knowledge to design. The important signal variable in this last system changes from a frequency at input to an analogue voltage in the forward path, a phase in the feedback path and a frequency as the output.

The knowledge acquisition is, in addition to acquiring the design knowledge, to be used to gain an insight into the appropriate facts that need to be captured from the user in the specification step. For this reason the interview commences with a title, given in the above list, and the expert is then left to request the further information that he or she feels is appropriate. This further information has been anticipated and set answers are kept by the interviewer. The expert is expected to arrive at a solution in the form of a block diagram with relevant 'dimensions' for each block but actual circuit level design is not expected.

The expert performs the task in whatever order suits with the interviewer interjecting at appropriate points for clarification. The interview is intentionally unstructured so as not to pre-empt any expected result. The interviewer chosen to perform the task is a person who is not an electronic expert but has enough electronic knowledge to follow the expert's line of reasoning and ask appropriate and relevant questions.

Representation

The following encompasses some initial thoughts on the appropriate knowledge representation based on the personal engineering experience of the author. AJITA through its menu driven system is inherently hierarchical which would suggest that a frame based representation is appropriate. Fig. 5 shows a typical frame for an inverting amplifier, some of the attributes are inherited from the parent frame 'amplifier'. Some of the slots are filled in by OP-SPEC where the value is a parameter to be passed to AJITA and some are filled in as values returned by AJITA. Fig. 6 shows some of the blocks in AJITA and how they are linked through inheritance.

Sullivan and Mutch [17] indicate a preference for 'templates', which is essentially an object oriented or active frame approach, to represent candidate designs. Fig. 4 shows a block diagram for a thermostatic control and the associated frame representation. Using active frames would be a useful approach to representing the AJITA blocks in the knowledge base as they can contain code to indicate when a particular block is appropriate or to provide

FRAME: INVERTING AMPLIFIER

SUPERCLASS: AMPLIFIERS
SUBCLASS: AC COUPLED, DC COUPLED

MEMBER OF: SIGNAL CONDITIONERS

SLOT 1: PURPOSE
VALUE TYPE: LABEL
VALUE: To Amplify a Voltage

SLOT 2: INPUT
VALUE TYPE: VOLTAGE
VALUE: xx

SLOT 3: INPUT CHARACTERISTIC
VALUE TYPE: IMPEDENCE
VALUE: xx

SLOT 4: INPUT CHARACTERISTIC
VALUE TYPE: LABEL
VALUE: Single Ended

SLOT 5: OUTPUT
VALUE TYPE: VOLTAGE
VALUE: xx

SLOT 6: OUTPUT CHARACTERISTIC
VALUE TYPE: IMPEDENCE
VALUE: xx

SLOT 7: OUTPUT CHARACTERISTIC
VALUE TYPE: LABEL
VALUE: Single Ended

SLOT 8: SUPPLY
VALUE TYPE: VOLTAGE
VALUE: 15v

SLOT 9: SUPPLY CHARACTERISTIC
VALUE TYPE: CURRENT
VALUE: xx

SLOT 10: SUPPLY CHARACTERISTIC
VALUE TYPE: LABEL
VALUE: Split Supply

Note: xx Denotes Values Supplied by AJITA or OP-SPEC.

Fig. 5 A Frame Representation of an Amplifier.

possible strategies should the block prove inadequate. ie: In a frame called Inverting amplifier there may be a slot called **match_impedance** which contains the following rule (written in pseudo_code):

IF input_impedance OF inverting_amp
< output_impedance OF previous_block
THEN insert_buffer(previous_block, inverting_amp)

The **insert_buffer** routine would introduce an additional block such as a voltage follower to match the impedances of the two blocks.

Another advantage of this active frame based approach is that common block diagram configurations could be retained to be invoked when the appropriate situation arises. This would be consistent with a model of the design process [18] that suggests that 'design synthesis is a process of accessing relevant aspects of past experience'. It is likely that the experienced engineer recalls whole block diagrams at times.

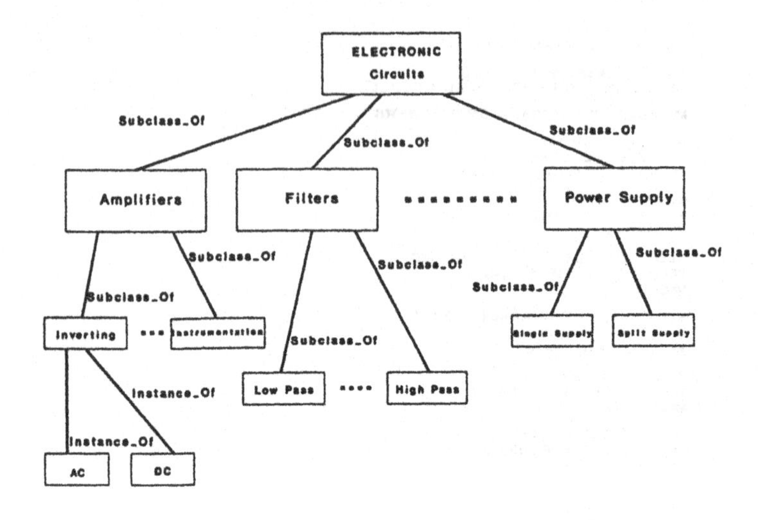

Fig. 6 Some of the AJITA Hierachy

CONCLUSION

A number of areas for on going work in the CAENIC project have been identified and three areas have emerged.

- A knowledge acquisition phase to positively identify the process used by the typical engineer to instantiate the block diagram of an electronic system. This is well underway at the time of writing and will inherently answer questions about representation.

- The development of the intelligent electronic systems design package with subsequent testing of the system. This development would need to take into account the intelligent front end and define a format for the interface with OP-SPEC.

- The development of OP-SPEC. Attention would be given to interaction with the application expert and the interface with the bulk of the CAENIC program.

The development of such a system represents the first step in investigating the cognitive processes involved in the design of electronic systems. The understanding of the processes involved in the more advanced areas of design, class 2 and class 1, depends on an understanding of the more routine processes of class 3 design. The development of an automatic class 3 design system will form the necessary basis for a class 2 system.

The targeted gain is shortened lead times in design of electronic systems through an integrated design automation approach. In the present economic climate, with new technology appearing at an ever increasing rate the shortening of lead times is taking on more importance. CAENIC, with its use of knowledge bases to replace application manuals, should be a valuable tool in achieving reduced lead times.

As is the case with much of artificial intelligence research, by investigating the processes at work, gain may be found in the education of engineers. In the conventional teaching approach, knowledge of electronics is separated into functional areas and these are taught while the overall 'art' of engineering is left to a master-apprentice approach to learning. An alternative to this approach, as suggested by Sydenham et.al. [19], may be found in the CAENIC work.

ACKNOWLEDGEMENTS

The author would like to thank the University of South Australia for its support in the way of a scholarship and Prof. Peter Sydenham, Academic in Charge, MISC, for guiding comments in the preparation of this paper. The project work of several undergraduate students proved useful in investigating some of the concepts presented.

REFERENCES

1. HOWE A., COHEN P., DIXON J. and SIMMONS M.: 'DOMINIC; A Domain-Independent Program for Mechanical Engineering Design.' Proc. 1st Conf. on Applications of Artificial Intelligence in Engineering Problems., Southampton, England., April 1986.

2. SYDENHAM, P.H. and JAIN, L.C.: 'CAENIC - User-characterised Electronic Systems Development Software.' Computer-Aided Engineering Journal, 1988,5,(5),pp. 200-205.

3. M'PHERSON, P.K.: 'Systems Engineering: an Approach to Whole-System Design.' The Radio and Electronic Engineer, 50, (11/12), pp 545-558, Nov/Dec 1980.

4. BERLIN, L.M.: 'User-Centred Application Definition: A Methodology and Case Study.' Hewlett-Packard Journal, October 1989.

5. ROWLES C.D. and LECKIE C.: 'A Design Automation System Using Explicit Models of Design.' (submitted to) Int. Journal for Engineering Applications of A.I., July 1988.

6. FINKELSTEIN, L. and FINKELSTEIN, A.C.W.: 'Review of Design Methodology.' IEE Proceedings, 130, Pt A, (4), June 1983.

7. JAIN L.C. and HARPAS P.: 'PC Software For Electronic Circuit Design in Printed Circuit Form.' Computer-Aided Engineering Journal, 1988,5,(4),pp 148-152.

8. SHELDON, D.F.: 'The Present State of the Art on Computer Aided Draughting and Design. ' IEE Proceedings, 130, Pt A, (4), June 1983.

9. ROHRER, R.A.: 'Evolution of the Electronic Design Automation Industry.' IEEE Design and Test of Computers, Dec 1988.

10. COOK S.: 'Automatic Generation of Measuring Instrument Specifications.' Measurement, 1988,6,(4).

11. COOK, S.: 'Automatic Measuring Instrument Specification Generator.' PhD Thesis, School of Engineering, City University London, (Submitted) Jan 1991.

12. BROWN, D.C. and CHANDRASEKARAN, B.: 'Expert Systems for a Class of Mechanical Design Activity.' I.F.I.P. Working Conf. on Knowledge Engineering in Computer-Aided Design, Budapest, Hungary., September 1984.

13. SYDENHAM, P.H.: 'Structured Understanding of the Measurement Process. Parts 1&2.' Measurement & Control, 1985,3,(3),pp. 115-120. & 1985,3,(4),pp. 161-168.

14. SYDENHAM, P.H., HARRIS, D.D. and HANCOCK, N.H.: 'MINDS A Software Tool to Establish a Measurement System Requirement.' (accepted for publication) Measurement & Control, 1990.

15. BLACKLEDGE, P.: 'Specification Languages.' IEE Proc., vol 130, Pt. A, no 4, June 1983.

16. HART, A.: 'Knowledge Acquisition for Expert Systems.' Mc Graw-Hill, New York., 1986.

17. SULLIVAN, G.A. and MUTCH, K.M.: 'Issues for the Construction of an Automatic Analog Circuit Design System.' Fifth Annual Pheonix Conference on Computers and Communications:PCCC'86, March 1986.

18. Mc LAUGHLIN, S. and GERO, J.S.: 'Requirements of a Reasoning System that Supports Creative and Innovative Design Activity.' Knowledge Based Systems, 2, (1), Mar 1989.

19. SYDENHAM, P.H., HANCOCK, N.H. and THORN, R.: 'Introduction to Measurement Science and Engineering.' John Wiley & Sons, Chichester., 1989.

Design Brief Expansion Tool

J. Forster (*), P. Van Nest (*), M. Cartmell (**), P. Fothergill (*)

(*) Department of Computing Science, King's College, University of Aberdeen

(**) Department of Mechanical Engineering, University College, Swansea, UK

ABSTRACT

This paper describes our work to investigate and support the process of Design Brief Expansion. It describes how the introduction of an advanced AI technique, such as HyperText, can assist with methodological design. The existing prototype system consists of HyperNeWS, a C-program and a parsing engine which uses a semantically annotated grammar. The system generates an expanded specification which can be used by a design system support tool.

INTRODUCTION

Our Research group is concerned with investigating the introduction of AI tools and techniques to foster the application of methodological design approaches. In the course of this work (SERC/ACME Project: "Introduction of formal Design Methodologies Supported by Artificial Intelligence Techniques" GR/F/56437), we have decided to focus first on the careful expansion of design briefs since we feel a clear elaboration of a design specification is basic to the application of systematic design methodologies and the production of good design solutions. As part of our experimental approach we have built a Design Brief Expansion Tool which uses Hypertext.

Traditionally, HyperText systems have been used mainly for:

- Intelligent Tutoring Systems (ITS)
- Intelligent Interfaces

Both of these domains utilize the most important features of HyperText systems: to have knowledge available in a pre-structured, but easily explored fashion and to facilitate the rapid building of user models and interfaces. However, recent work has emerged which argues for the use of HyperText in expert systems [12] and the design domain (Mechanical Advantage Interface [2], 'WOMBAT' [1]). As with these, our Design Brief Expansion Tool can been seen as an example of how to map a complicated, real-world operation onto the HyperText idea.

METHODOLOGICAL DESIGN

The key idea behind methodological design is to improve the **quality** of the design. In this context the term **quality**, refers to both the quality of the designed artifact and the quality of the design description which is generated. Since the quality of the artifact is judged by how well it conforms to the design specification, it is important to make all requirements in the specification accessible, whether they are explicit or implicit. In this manner, the quality of the artifact and that of its description are closely related.

The main aims of methodological design are to:

- optimize the overall performance
- reduce the likelihood of omitting particular points
- require the designer to make and *record* design decisions
- provide guidance for the inexperienced designer
- facilitate communication within the design team

The application of a methodology should in turn lead to a better designed artifact. In order to achieve this, one could attempt to:

1. define a procedure which automatically leads to the optimal design;
2. generate a solution field and then assess each of the derived solutions.

Systematic design approaches employ the second approach to arrive at the optimal design. In [3] several European methodologies have been reviewed (from Rodenacker, Hubka, Leyer, Pahl & Beitz, etc.). They all identify certain stages in the design process and define information and methods allocated to these stages. The design process is described by a 'process model' with different levels of complexity. (See the comprehensive approach of Pahl and Beitz [10].)

The design brief as well as the evolving specification are crucial elements in most methodologies. Constraints and requirements are extracted or generated from the brief, and these in turn guide and restrict the space of possible design solutions. This is important for both the conceptual design a large car repair workshop needs a simple rivet setting tool. the quantity is likely to be small between 5 and 10. no firm decision made at this stage. it is possible that sales to other users might be forthcoming at a later date if the original customer is satisfied.

the rivet setting tool is to be used for attaching brake linings to vehicle brake shoes. the dimensional specifications are as follows: the inside diameter of the range of vehicle brake drums likely to overhauled by the workshop is in the range of 0.24 - 0.4 m.

the rivets are to DIN 7338 with shank diameters of 3 - 5 mm and are made of copper. the maximum width of the shoe is 0.1 m and the maximum likely thickness of the parts is 3 mm. the force required is likely to be in the region of 500N. the manufacturing cost must be less than 100 pounds.

Figure 1: Original Design Brief: Rivet Setting Tool

phase as well as for the embodiment stage. Furthermore, since methodologies encourage the generation of numerous sub-solutions, it is important that they be evaluated. An 'Objective Tree' which structures the design requirements and carries weight according to the importance of the requirements is used to represent the selection criteria in a mathematical form. —

DESIGN BRIEFS

General

The first step in design is to acquire an initial design brief. This should describe the basic need for the artifact in terms of its use, manufacture and design. (See fig. 1 for an example of the specification of a rivet setting tool). This initial design brief is in turn elaborated to yield a more complete requirement specification which makes quantitative and qualitative assertions about the design.

Since approximately 90% of the later costs of an artifact are determined by the early design stages, it is important that this stage be as efficient and effective as possible. The European design schools [4] stress the importance of the careful treatment of the design brief and design specification. Examples of techniques used are:

Checklist approach of Pahl and Beitz [10]. In this there are pre-defined headings under which the expansion is carried out. The authors stress that not all of the headings are always applicable, but the use of this method reduces the likelihood of certain aspects being inadvertently omitted.

Implicit Requirements	Assessment of specified Constraints	Development Potential
Objectives to Satisfy	Solution Properties	Geometrical Aspects
Forces	Materials	Safety
Production Engineering	Kinematics	Energy
Signals	Ergonomics	Quality Control
Transportability	Maintenance	Costing

Table 1: *Assessment categories* ('Headings') used in current research

Profile approach, e.g. by Matousek [8]. This method evaluates the relative importance of pre-defined categories depending upon the intended end-use of the artifact. For example, the issue of safety would vary depending upon whether the item is to be used in an industrial or a domestic environment.

Design Brief Expansion Experiments

The checklist approach of Franke [5], as discussed in Pahl & Beitz [10], proposes a number of "main headings" and recommends that these are used as a checklist when drawing up a specification from a brief description of a problem. The original concept was that the designer, or design team, would take each main heading in turn and expand the brief in the area pertaining to that heading. Thus, when the heading "Kinematics" was reached, the designer's attention would be focussed upon the type of motion, direction of motion, velocities and accelerations if any which could or might be required in the system to be designed. Attention to this heading would result in additions to the designer's knowledge about the kinematics involved in the problem. It might also lead to decisions which could have repercussions for deliberations under other headings. By working systematically through the entire checklist of headings the designer would further his/her detailed understanding of the problem to be solved and would be able to write a full "specification" as a result.

In our work on the use of methodologies we decided to use an approach similar to that of Franke and to explore its ramifications. We have used the majority of the headings proposed by Franke, but have added some more derived from other sources. We refer to the headings as "assessment categories" since this name captures their purpose. Table 1 lists the ones currently in use.

In order to optimize and generalize such a checklist based expansion procedure, so as to produce a good specification from a relatively meagre brief, some further considerations need to be taken into account. These

Type	No. of Briefs
Testing, jig and fixture Design	5
Human activity aids	5
Domestic Aids and Systems	5
Industrial Equipment	11
Transport Related Design	5
Laboratory Equipment and Systems	3
Industrial Test and Monitoring Equipment	3
Artifact Design	11
Systems Design	9

Table 2: Design Problem Categories

include the order in which the assessment categories appear in the checklist and the interconnections between the categories. The relationship between the engineering information that has already been extracted as one proceeds in an orderly fashion down the checklist, and that which has not yet emerged requires close attention. In some cases one might wish to "jump ahead" to a connected category; in other cases one may need to "back track" to an earlier category because an amendment is needed to take account of some assertion that has just been made. Thus an assertion elicited by consideration of 'Kinematics' might alter the designer's view of what is needed under the 'Geometry' heading.

In order to investigate the expansion of design briefs in as objective a manner as possible, we have collected a large number of briefs from a variety of sources (industrial and academic) and a variety of areas. Different, independent individuals then assess each brief using the checklist, expand them under each category and record the interconnections noticed during the process. This work is being carried out by hand.

The results from the study will be collated. One of the major features affecting the optimal order of items in the checklist may be the type of design problem that is being considered. Table 2 lists the different "problem categories" that we have assigned to our example briefs. These categories are broad and in some cases a brief may justifiably be seen to fall into two or more of them.

As we have already pointed out, the different assessment categories are not entirely independent. We can therefore envisage hierarchies of categories, and different hierarchies for different problem categories. One outcome of our research should be a mapping of problems to hierarchies which can be used in a design brief expansion tool. Another outcome should be a better understanding of the dynamic nature of the process and the pattern of revisions and amendments.

The need for a tool

When a designer performs this expansion, he has an initial design brief and certain entities which suggest elaboration in terms of kinematics, forces, constraints, etc: he links a semantic entity in the brief to an assertion in the requirement specification. However, the new assertion is important not only in itself, but also in the context in which it takes place. The dependencies which eventually arise are also important. There is also the possibility that certain new assertions may require that earlier ones be reconsidered and/or refined. While examining this elaboration process, it became apparent that this method of expansion is not only tedious but difficult to monitor since not only the endpoints, but all steps and operations between are of interest. We therefore felt that it is necessary to provide a tool which would serve as:

- a useful support tool for the designer
- a method to test our ideas for design brief expansion
- a vehicle for acquiring knowledge as to how designers actually perform this design expansion

IMPLEMENTATION

General

Our experimental prototype has been built while our research into the theoretical aspects of design brief expansion (DBE) is in progress. We shall use the system not only to embody our ideas about a suitable tool for DBE support, but also as a 'knowledge acquisition' tool to generate more data to use in our paper and pencil study.

In order to facilitate this, we have built the system in such a way that we can easily modify and extend it. Our main aim is to understand how to provide methodological support for DBE; in other words how to enhance the manual framework. While the kernel must be capable of representing the basic paper and pencil approach, other tools might be used by the designer to illustrate and document the expansion; on the top level methodology knowledge might be employed to *direct* the Design Brief Expansion.

Figure 2 illustrates the structure we have used to implement these layers.

The Kernel

The basic function of the Kernel is to allow the designer to examine the design brief and to relate fragments of the text of the brief (elements) to more elaborate or clarified statements. This is analogous to the designer

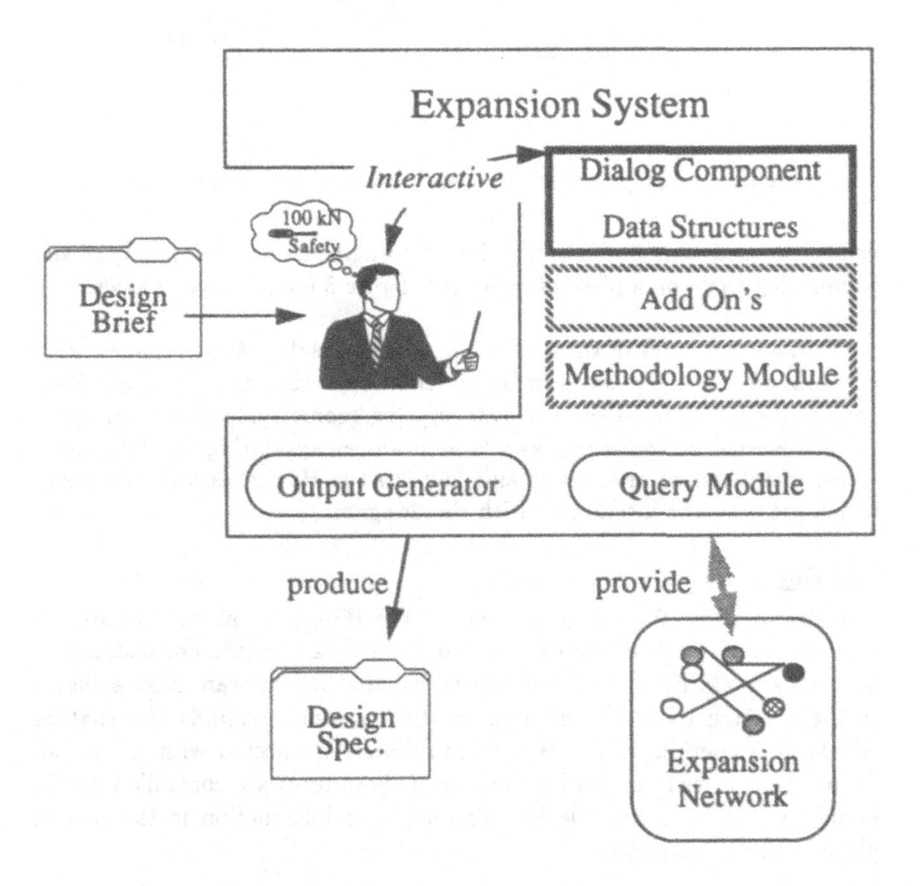

Figure 2: Structure of Expansion System

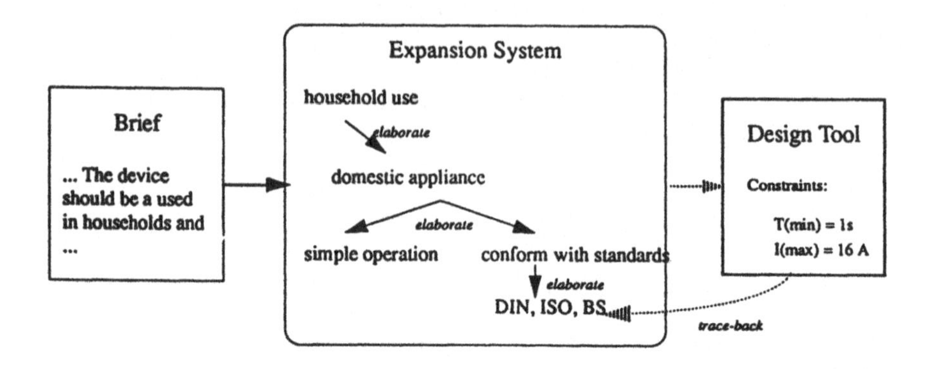

Figure 3: Applying the Expansion System to a Design Brief

using 'coloured pens' and marking different fragments in the design brief and elaborating them on a piece of paper (see figure 3 for example of assertion).

It also keeps track of the context in which an elaboration is made and of eventual dependencies between the different elaborations. The elaboration can also be given an assessment category (eg geometry, manufacturing).

The Kernel is represented as a C-program encapsulating the data structures, maintenance procedures and links to the HyperNews environment, which provides the interaction with the designer.

Add Ons

In order to allow the designer to do all the things he might normally do when expanding a brief we have included some "add-ons". For instance, if he wishes to clarify a geometric constraint in a brief he can make a sketch (using a sketch tool); if some item in the expansion reminds him that he should do something which is not immediately connected with the elaboration, he can write a post-it. The use of these tools are controlled by the kernel, and it is responsible for 'bundling' the information in the correct place in the elaboration.

The Chunker - a special Add On

Preliminary work with the DBE tool showed that it was a burden on the designer to have to be responsible for choosing and 'highlighting' the fragments of text from the brief on which he wished to work. It was therefore decided to pre-process the original brief so that it is broken into smaller, more manageable *chunks*. While these *chunks* are based on syntactic structures, such as prepositional phrases, noun phrases, etc., they carry additional semantic information such as Physics, Marketing, etc., reflecting the 'assessment categories' mentioned earlier. The designer is presented with these much smaller chunks, and can then accept, reject, or relabel them and use them in his elaborations.

Although the categories used by the *chunker* are similar to the 'assessment categories' described in table 1 they are not the same, but cover broader divisions. It was found that as the categories became more and more specific, the phrases extracted by the *chunker* become more numerous, smaller, and therefore less meaningful in a syntactic sense. Since the purpose of the *chunker* is to assist the designer with categorization rather than to perform the task automatically and independently, it was felt that it was more useful to have sensible sized fragments of text which the designer could allocate to actual assessment categories, than to have too many small fragments.

An additional consideration is performance. To be able to handle categories as semantically diverse and ambiguous as those represented by the 'assessment categories' would require a semantically sophisticated parsing system. The system used at present is limited in the amount of semantic information it can represent, but, unlike more complex systems, it is able to process text at speeds which are compatible with interactive operation.

Methodology Module

Of crucial importance for us is the support of a 'systematic' working approach. We do this by two means:

Checklist : The system has a list of 'assessment categories' the designer can use to label the expansions. Furthermore, we are looking into dependencies between expansions and have a provision for a 'link-map', representing certain expansions in the light of others (e.g. once a implicit requirement is brought up, it might trigger safety and material considerations). This link-map can be easily visualized by giving the kernel control over the HyperText links.

Profile : A 'Profile' of an expanded (or partially expanded) design brief can be generated in order to provide visual feedback. This profile enables us to see how many instances (atomic assertions) have been generated for each of the different assessment categories. In turn, this then can be overlaid with the max/min number of instances for each category for all designs in the same domain. It is therefore possible to highlight where the designer is under- or over-specifying the design and enable him to take account of this (see figure 4 for a sample profile).

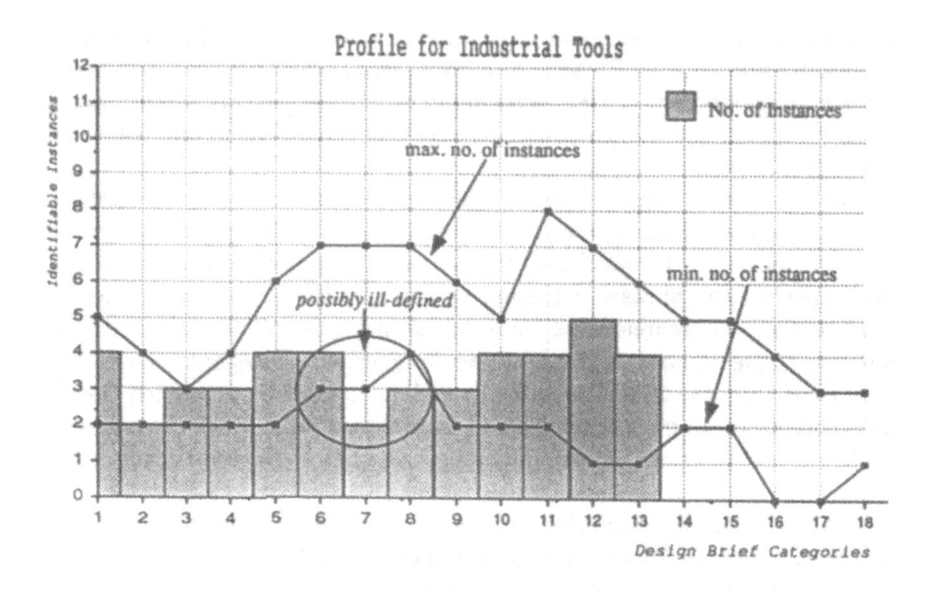

Figure 4: Sample Profile

Output system

The expanded design brief sets out the key factors of the design problem. It represents important information, which might be part of a legal document or tendering. Therefore it is necessary to provide an output system generating 'paper'. In the DBEsys, this is done by turning the tree-like data structures into a sectioned LaTeXdocument. This LaTeXdocument bundles the assertions, the references to the original brief as well as sketches and text-notes together. See table 3 for an extract of a document so generated.

The DBE system is expected to feed in to tools that support the later stages in design. The information it generates is available via a 'link module' which gives other design support systems access. They can ascertain the origin of design variables and trace elaborations back to their original design brief sources.

Physical structure of the Implementation

In order to implement the system, the development has been divided in three physical components:

HyperText is used to construct the user-dialogue and to represent the links of the different specification items. The HyperText system we are using is HyperNeWS, a software system developed by the Turing

2 Expanded Brief

2.1 Implicit Requirements

2.1.1

Unskilled labour might handle the tool.
(*a simple rivet setting tool*)

2.1.2

Has to be a design fit for in-house manufacturing.
(*a simple rivet setting tool*)

2.1.3

rule out material interferences with other parts.
(*are made of copper*)

2.2 Assesment of Spec. Constr.

2.2.1

Design should be general for brake drums.
(*vehicle brake drums*)

2.10 Prod. Eng. Aspects

2.10.1

Do not use casting designs.
(from 2.1.2: *Has to be a design fit for in-house manufacturing.*)

2.11 Kinematic Considerations

no instances

2.12 Energy Requirements

no instances

2.13 Signals/Indicators

no instances

2.14 Ergonomic factors

2.14.1

Must have built in safety aspects.
(from 2.1.1: *Unskilled labour might handle the tool.*)

Table 3: A sample LATEXoutput.

Institute [7] which runs under OpenWindows on Sun Sparcs and Sun-3's. It is functionally similar to MacIntosh's HyperCard.

HyperNeWS is written in PostScript and has a powerful object-oriented environment. Furthermore, it has 'C' binding so that actions can be associated with 'C' programs. It is useful for AI applications in that it comes with bindings for Sun Common Lisp and Quintus Prolog. The cards are interactively designed and then scripts are associated with certain objects contained in them. These scripts either explore certain links to other cards which have been created by the C-system or transmit/receive information from the *chunker.*

C-System is used to drive the HyperText system and store information regarding the elaboration of the original brief. Furthermore, it provides the link to Nexpert, which contains the rules for the domain and the hierarchy of the brief categories. The domain specific knowledge about expansion and interacting parameters has been studied as an MSc. project [14] which has provided a prototype system, written in 'Nexpert'. The C-System is linked via *message-passing* to the objects in the HyperNeWS system.

Chunker is used as a preliminary method of classifying phrases in the original design brief under certain headings. It consists of a chart parser,

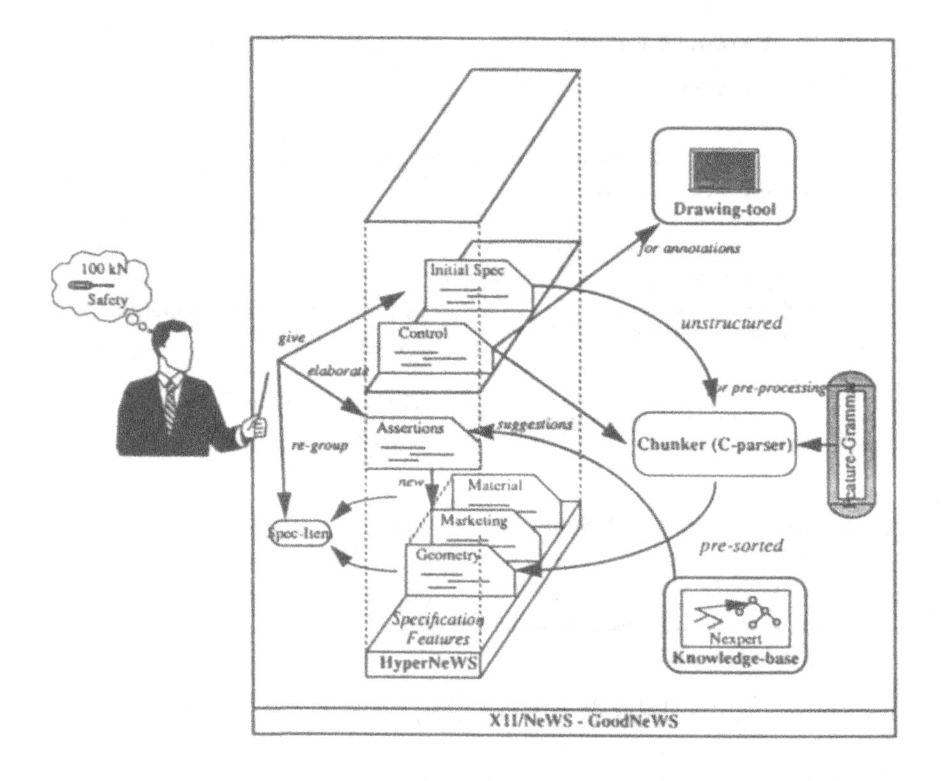

Figure 5: The Design Brief Expansion Tool

written in C, and a context-free, phrase structure grammar. The parsing system was originally developed as a syntax checker [9], but has been modified so as to enable it to cope with the additional semantic information required. It differs from traditional parsing systems in that instead of returning a completed parse tree, it only returns those phrases which contain relevant semantic information.

See figure 5 for how the different elements interact.

DISCUSSION

Our work on design brief expansions, which has been based on the ideas of Franke and of Pahl and Beitz, has followed two paths. On the one hand we have been carrying out extensive "paper and pencil" studies of actual expansions. On the other hand, we have developed an experimental software system to support the process. Our paper and pencil studies, which are still being continued, have revealed that a straightforward checklist approach, starting at the beginning and going systematically through to the end, is unsatisfactory because of the as yet unknown significance of the order in the list, and because of the dependencies between assessment categories. We need to do more work to investigate these phenomena.

The software system we have developed should help us greatly in our investigations. The way in which it has been built using Hypertext means that we can easily modify it and extend it. We still need to decide exactly how the observed dependencies between assessment categories should be represented, and how new ones should be captured. We are, however, confident that structure we have built will allow us to carry out suitable experiments.

In the present software system the checklist is not used to force the user into following any particular order. It is merely there as a list of possible labels for the expansion process, and also for gathering statistics. If we are to develop a system which properly supports a methodological approach, then we shall need to define clearly the scope of the checklists and whether they need to be enforced.

The system as it stands today does not take account of any specialised knowledge, for instance about particular company routine design tasks. A slot for a knowledge base to make suggestions about the course of the expansions does exist (see figure 5). We are investigating how we can use this slot to link our system to a specialised expert system such as ELABSPEC [14] which has a deep and rich model of how a brief should be expanded in one particular domain (pumps).

The statistics gathering mode of the system is already used in the "profiling component" to guide the designer in the course of an expansion. Similar methods can be used to assess the quality of the original brief, and of the final expansion.

We regard the 'traceability' of all assertions and decisions as an important feature of the system we have developed. This traceability extends through to the documents produced and the links to other systems. It will continue to be important as the system is extended to deal with iterations over the design brief and the possibilities of re-use of designs.

Comparing our work with other work in the field (e.g. 'Kaleit' [6]), we feel that we do have adequate book-keeping functions in our tool, easing the tedious basics of the systematic approach. However, the Design Brief Expansion tool gains over similar work in this domain, in that advanced features such as the 'chunker' and the 'profiling' component help the designer carry the task to completion where other tools point to 'bookware' (the written-up methodologies).

CONCLUSIONS

Once fully implemented, our method of design brief expansion should provide a comprehensive approach which will allow the designer to obtain a complete specification without stifling his creativity. Ensuring the accountability and completeness of this preliminary step will improve the following stages of a methodological design process.

Although the technology of HyperText has traditionally been used to model tasks such as exploratory learning or rapid prototyping of interfaces, it appears to be a universal tool with which to perform operations on textual or graphical information. This holds whether the semantic relationships being manipulated are static or dynamic. When a reasoning system is given control over the links between the entities and can therefore direct the process, the significance and utility of HyperText is greatly enhanced.

ACKNOWLEDGEMENTS

We should like to thank:
SERC (ACME) for financial support (grant GR/F/5643.7, GR/G/24019); GEC Electrical Projects Ltd. and Ewbank Preece Ltd for their collaboration in the project; Andy Robertson of Lucas CAV for his help with the EDS.

REFERENCES

[1] A. Blanford, (1990), *Wombat*, Proceedings Applications of Artificial Intelligence in Engineering (V), Vol. 1 Design, p. 215, Computational Mechanics Publications, Boston.

[2] D. Deitz, (1988) *Tools for total quality*, Computers in Mechanical Engineering, July/August 1988.

[3] J. Forster (1989), *Artificial Intelligence and Design Methodologies in Mechanical Engineering*, MSc. Thesis, Dep. of Computing Science, University of Aberdeen.

[4] J. Forster, M. Cartmell and P. Fothergill (1990), *Design Methodologies in Mechanical Engineering supported by AI techniques*, Proceedings of '3rd International Conference for Industrial & Engineering Design', ACM.

[5] H.-J. Franke (1976), *Methodische Schritte beim klären konstruktiver Aufgabenstellungen*, Journal: Konstruktion 27, p. 395-402.

[6] B. Groeger (1990), *Ein System zur rechnerunterstützten und wissensbasierten Bearbeitung des Konstruktionsprozesses*, Journal: Konstruktion 3/90, p. 91-96.

[7] Arthur van Hoff (1990), *HyperNeWS Reference Manual*, Turing Institute, Glasgow.

[8] R. Matousek (1967), *Engineering Design*, Blackie & Son, London.

[9] P.L. Van Nest (1990), *Use of Syntactic Rules and Malrules for Improvement of Written English*, MSc. Thesis, Dep. of Computing Science, University of Aberdeen.

[10] G. Pahl and W. Beitz (1977/1988), *Engineering Design (Konstruktionslehre)*, Springer Verlag.

[11] R. Popplestone, T. Smithers and J. Corney and K. Millington (1986), *Engineering Design Support Systems*, Technical Report, University of Edinburgh.

[12] R. Rada and J. Burlow (1988), *Expert Systems and Hypertext*, The knowledge engineering review, Vol 3, Part 4.

[13] W.G. Rodenacker (1976), *Methodisches Konstruieren*, Springer Verlag.

[14] J.I. Rodriguez (1990), *Guiding Rules for systematic Design in Mechanical Engineering*, MSc. Thesis, Dep. of Computing Science, University of Aberdeen.

An Integrated System for Constructibility Assessment of a Design Detail

S. Alkass, A. Abdou

Centre for Building Studies, Concordia University, Montreal, Canada

ABSTRACT

In the traditional system of construction, the two main disciplines of design and construction are separated. As a result of this, construction projects are suffering from many problems such as design complexity, increasing costs and delays. Lack of experience in the part of the design engineer and the unavailability of a systematic approach to aid in the assessment of a design detail is one among other reasons for what seems to be the inherited construction problems. Knowledge based system are found to be suitable for formulating and presenting experience related knowledge and can be utilized to counter this disadvantage.This paper describes the development of a computerized system that integrates an expert system with conventional computer programs to aid the user in assessing the constructibility of a detailed design. Knowledge for this particular domain was gained from design engineers, architects, and field practitioners such as site and planning engineers. The development of the appropriate knowledge acquisition and integration techniques are covered in details. The system is tested with practitioners and novices and the concept of consultation is validated.

INTRODUCTION

The traditional approach of construction separates design from construction. Project's design is been carried out by a consultant and is constructed by a contractor. Normally these two parties do not communicate prior to the construction stage of the project, also in the main the design engineers do not have adequate construction experience (Gray[1]). As a result of these facts important construction knowledge are missing during the design stage causing problems of design complexity and increasing costs.

Constructibility was cited as being capable of improving project performance and a mechanism to overcome these problems (Vanegas[2]). Yet there is no clear understanding of how to formally incorporate the construction expertise as part of the design process (Jergeas[3]). A systematic approach for acquiring construction knowledge, storing it in a computer system and make it available when needed by the design engineers is one a way to overcome these problems.

Contractibility was first defined by the Construction Industry Research and Information Association in England (CIRIA) in 1979 (CIRIA[4]) as " the extend to which the design of building facilitates ease of construction, subject to the overall requirements for the complete building".

Also it was defined by the Construction Industry Institute, Texas (CII) in 1986, (CII[5]) as "contractibility is the optimum use of construction knowledge and experience in planning, engineering and field operations to achieve overall operations". Indeed, one of many objectives that are addressed when preparing project's master plan is the constructibility issue.

Constructibility planning pursues the optimum integration of construction knowledge and experience with the engineering design to achieve the overall project objectives (O'Conner et al.[6]). In practice, this knowledge is in the main held personally by experienced practitioners and therefore accessible only piecemeal.

An expert system is an attempt to counter this disadvantage by bringing together as many strands of expertise as possible structured in a manner that facilitates a user to steer a step by step course in learning and so solve problems which are largely judgment dependent.

During the last few years, Expert Systems have been used in the construction industry. They have been developed either as stand alone modules or integrated with commercially available algorithmic software systems, performing various functions including: scheduling, estimating, cost control, database management, electronic accounting, drafting, and word processing. This paper describes the development of an integrated prototype system that combines design procedures with the selection of design details. It is called *Constructibility Assessment for Design Details System* (CADDS).The system integrates both database information and knowledge gained from experience in designing and constructing different projects. Knowledge is first gathered from number of experienced design and planning engineers and then structured and coded in a form suitable for manipulation and processing by the various functions within CADDS . The approach relies on well founded principles and experience based on knowledge collected from literature and field surveys. An example application of a wall design selection is presented to illustrate the features of the system. The system will assist the architects and structural design engineers in selecting the most appropriate and easy to construct design details. This will contribute in saving both the owner and the contractor some time, money and anticipated legal disputes.

DESIGN DETAILS AND CONSTRUCTIBILITY

Easy to construct design details is an important factor among many others such as functionality. These must be carefully considered during the design details stage of a project. Both functional and constructional aspects of detailing influence its overall performance. For example, different forms of exterior wall construction must fulfil basic requirements pertaining to aesthetics, durability, air barrier, vapor retarder, fire resistance, acoustic barrier, thermal resistance, structural strength and stiffness. In addition, the constructibility of particular exterior wall such as fitting the surrounding components, installation, costs of production (i.e. material, equipment and labour costs) should be considered. It is not efficient enough to have details which are aesthetically pleasing, functionally satisfactory but, they should be clearly thought out in terms of all the construction-related aspects.

In spite of the importance of the overall constructibility of the design detail, it is often ignored. As a result, problems arise during the project's construction phase causing disruption to its schedules, duration and normally leading to legal disputes. This means that the extent to which the design of details facilitates ease of construction should be carefully considered. Furthermore the implications of using a specific design detail have to be identified for future improvement to obtain full performance satisfaction.

Constructibility as a concept is not new in the construction field (Eldin[7]) and the importance of the constructibility improvement has been emphasized and investigated in project plan, site plan and major construction methods (Tatum[8]). In another study constructibility improvement data collection techniques were discussed and analyzed (O'Connor et al.[9]). Although constructibility concepts have been identified to be a major factor in evaluating design alternatives, there havebeen no significant efforts utilized to devise a systematic approach to assess and catalogue construction details with respect to constructibility. The difficulty of capturing and coding the field experience and presenting it to the inexperienced design engineers in an easy to use computer programs such as expert systems is mainly the reason for that.

THE RULE OF THE KNOWLEDGE BASED SYSTEMS IN THE CONSTRUCTION INDUSTRY

Expert systems are computer programs that incorporate human expertise to provide advice on a wide range of topics developed from knowledge bases, which contain knowledge collected from all possible sources, mainly with the help of an expert practitioner. These systems function as consultants in the given domain to provide explanations of seasonings, simulating a consultation as though the computer was the tutor and the user a pupil (Alkass et al.[10]).

Because of their attributes of combining factual knowledge with judgment, including the ability to handle incomplete and uncertain data, and communicate with their users in a natural language like English, such systems could have a special appeal to the construction profession. Expert systems however, are not a total substitute for experts, but they do help to conserve expertise and are used to make expertise more widely, easily and quickly available for assistance in the decision-making process.

Expert systems and their application in construction in general have been extensively described in the literature, (Mohan[11] ; Moselhi et al.[12]; Alkass[13] ; Alkass et al.[10]). Their application to evaluate design details for building projects and evaluation of design alternatives have been also discussed (Cornick et al.[14], Maver[15]). This paper describes an integrated system designed to assist the design team selecting the most appropriate design details as related to ease of construction and optimum cost.

KNOWLEDGE BASED SYSTEMS AND CONSTRUCTIBILITY ASSESSMENT

From the definitions of Constructibility as suggested by CIRIA and CII, it has been concluded that while ease of construction may be influenced by many organizational, technical, managerial and environmental considerations, the major contribution is thought to lie in those factors which fall within the influence or control of the design team. Therefore, the emphasis is better implemented on reflecting the construction experience during the design phase.

Constructibility assessment when considered as a part of the design process is partially subjective and seems to rely mainly on personal knowledge and experience. The rules of constructibility are based upon the construction ease and performance implications of the detail components and how these satisfy various functional criteria. The assessment input could be a partially or completely specified detail design and the result is usually a recommendation for improvement. Two main problems promptly arise in developing knowledge-based systems to assess constructibility. The *first* lies in the formalization of the assessment criteria and the *second* is the interface between a syntactic approach of the graphical representation of the given design detail and the semantic representation of a knowledge-based system (Gero et al.[16]). This means that an approach to translate design description in the form of graphical representation into appropriate form of knowledge base description (i.e. from syntactic to semantic representation) is greatly required to facilitates the development of an integrated system.

A NEED FOR AN INTEGRATED COMPUTER-BASED SYSTEM

In the construction industry, much appropriate software already exist to aid the personnel involved in decision making process for example design, drafting, scheduling, estimating, cost control, database management, electronic accounting, and word processing which could usefully be incorporated in any future new system and so possibly avoid some of the additional retraining of staff and new equipment when introducing a new system and also help counter the negative momentum generally engendered with the introduction of new technology. An integrated system might also include an Expert System containing experimental knowledge and engineering judgment to facilitate the decision making process. Clearly a successful outcome would depend on the economic integration with existing systems.

A PROPOSED INTEGRATED SYSTEM

The proposed system is designed to demonstrate the concept of how construction knowledge and experience may be most effectively presented and used during the detailed design phase of the project. CADDS is designed in an integration environment. It combines design procedures with construction methods to arrive at the selection of the appropriate design details for a specific case based on ease of construction at optimum cost. The system comprises the integration of expert system with existing management software tools.

The knowledge-base for CADDS is build by combining experienced judgment extracted from a lengthy consultation with architects, structural engineers and construction practitioners with known facts on construction methods, weather conditions and cost data. The essence of the system as in other expert systems, is in encoding of expert knowledge in a form usable by non-expert, the inference mechanism being forward-chaining and knowledge being represented as rules. The mode of operation consists of series of questions linked by IF THEN logic, the logic tree being set of rules arranged to reflect the reasoning of the expert practitioner. The approach relies on well founded principles requiring the contractibility of the design be carefully evaluated after which the appropriate design details are selected using rules of thumb ie. type of knowledge acquired from the domain experts and stored in an expert system. Data base information, graphical knowledge representation and algorithms for making routine calculations are also linked to the system.

CADDS is divided into two main modules, the *first* deals with the selection of a detail (ie. a wall system) to conform with particular requirements input by the user during the query session whilst the *second* assess the constructibility of a selected detail. This is done based upon the detail characteristics and its construction requirements, the system rules and the user preference of a set of goals. The system also identifies the implication of using the selected detail upon a request from the user. Fig.1, illustrates the system's architecture while Fig.2, shows the integration of the expert system with other programs.

During the selection criteria of a detail (awall system in this case), attention is paid to set of characteristics pertaining to performance, construction aspects and constructibility which may include :

* The wall system self weight (e.g. low .. large)
* Stiffness to lateral loads.
* Support of insulation , air/vapor barriers and interior finishes.
* Deterioration of the wall components.
* Installation of electrical services within the wall
* Dependence of the wall construction on weather conditions.
* Degree of inspection required.

Designers first specify the characteristics of the desired wall system and according to their preferences with respect to previously mentioned selection criteria, CADDS then will advise on the selection of the best wall system. Once the user is satisfied with the decision, he may acquire more information on the detail by selecting an icon from a menu. CADDS will provide him with the following informations:

* The outline geometry of the detail
* The position, shape, size of each component in the assembly.
* The attachment type of each component.

This information is held in a file within the data base accessible by CADDS. In this file, the syntactic representation of the drawing is converted to semantic representation so as to be accommodated in the expert system. This approach minimize the number of question posed by the system. Information attached to each detail which is stored in the file may include the followings:

* Number of components.
* The attachment description of each component.

* Number of trades involved in construction of the detail assembly
* Type of required site supervision and inspection effort.
* Number of key features in the detail which become quickly hidden behind other layers of construction .
* Advantageous and disadvantageous characteristics of the detail
* The detail construction implication .
* Total costs of construction (i.e. material and labour costs).

The user is requested to input his relative importance of the goals by providing numerical values or selecting from multiple choice criteria presented by the system in the form of *Great Importance, Fair, Moderate, Minor and, Not at all* . Fig.3 illustrates the constructibility assessment for a wall system.

At the end of the consultation, the advice on the selected detail is presented to the user followed by three explanation reports. The first deals with the detail selection, in this case a wall system, the second report deals with constructibility assessment of a given detail which is presented as a ratio of the total relative importance input by the user for all desired goals. The third report shows the implications caused by using the specific detail design. Fig. 4 shows a sample of thes reports.

IMPACT OF CONSTRUCTION DESIGN DETAILS UPON CONSTRUCTIBILITY AND PERFORMANCE (A Case Study)

To illustrate the manner in which construction details can influence constructibility and performance, the following case study [(adapted form (Drysdale et al.[17])] of designing a simple detail is presented.

A major decision in the design of the shelf angle carrying the brick veneer in a cavity wall system is to what extent its position should be made adjustable. If the shelf angle is fixed in position at the time of concrete placement for the structural frame, its location both vertically and horizontally is tied to the construction tolerances of the frame. Such a method of shelf angle attachment presents some advantages and disadvantages with respect to both constructibility and required performance. On one hand, attachment to the frame usually by means of a cast in strap anchor is secure (i.e.performance). Also the construction is simple and straight forward and hence can proceed with a minimum of potential slipups and a minimum of site supervision and inspection effort (i.e. constructibility).

On the other hand variation in the vertical position of the shelf angle can result in either too large or too small a gap for the movement joint beneath the angle and thus, violating the structural requirements for movement joints. Such variation in the horizontal position of the shelf angle can be accommodated by varying the bearing area of the brick on the shelf angle from one floor to another which in turn, is constrained by a limited projection which is 1/3 brick width (i.e.Building Codes). In addition it is important that existing construction be measured up (e.g. Concrete slabs edges, Alignment and Spacing) and problems resulted be resolved before actual construction of the wall system begins (i.e. constructibility).

Alternatively, an adjustable shelf angle attachment is possible by means of cast in-anchors or drilled-in anchors. By using this method vertical adjustment is possible by means of slots in the shelf angle and horizontal adjustment is achieved by means of shims. However, cast-in anchor bolt location may not match shelf angle slots (i.e.constructibility). Consequently, greater construction care and inspection effort are required to assure proper alignment of the shelf angle, proper use of shim plates and proper torque of friction connection created by tightening the nuts on the anchor bolts (i.e. constructibility).

Out of the case study presentation of a simple detail alternatives of the shelf angle attachment, it could be seen that to select which type of attachment should be used with respect to constructibility and performance, several aspects or goals have to be considered by the designer with construction knowledge. Therefore it is quite beneficial to base the selection on a thoughtful assessment. A consultation with will advise on the best alternatives and the results are presented within the three reports shown in Fig.4. Moreover, the system will present the detail graphically as illustrated in Fig. 6.

CONCLUSIONS

Design details for construction projects depends greatly on skilled judgment taking account of all the likely variables. Much of this knowledge is held by experienced practitioners and unlikely available for inexperienced personnel.

Constructibility improvement of a design details in a project can be achieved by effective design and construction integration. This requires either that construction experts become involved from the outset, ie. from the conceptual development to scheduling and cost estimating or their knowledge made available for the inexperienced design engineers. CADDS is an attempt to achieve the lateral by capturing the knowledge and presenting it to the user in a way of dialogue and advice presenting the final solution in a graphical representation.

By way of concluding remarks, the potentials of the prototype system for design detail assessment and selection are clearly manifold, but most importantly, results of trails with users indicate that the concept provides a disciplined method of transferring knowledge and expertise to young and untrained design engineers.

REFERENCES

1. Gray, C. "Buildability - the construction contribution. Occasional paper No. 29, The Chartered Institute of Building 1983, England.

2- Vanegas J. "A Model for Design /Construction Integration During the Intial Phases of Design for Building Construction Projects", a Ph.D thesis, Civil Engineering Department, Stanford University 1987, USA.

3. Jergeas, G. " Detailed Design and Constructibility" a Ph.D thesis, Civil Engineering Department, Loughborough University, 1989, England.

4. CIRIA (Construction Industry Research and Information Association), Buildability: An Assessment, Special Publication No. 26. England 1986.

5. CII (Construction Industry Institute) Constructibility: A Primer Publication Bureau of Engineering Research, The university of Texas at Austin, 1986.

6. O'Connor J. and Tucker R., " Industrial Project Constructibility Improvement" , Journal of Construction Engineering and Management, vol. 112 1986.

7. Eldin, N., " Constructibility Improvement of Project Designs " , Journal of Construction Engineering and Management , vol 114 No. 4. Dec. 1988.

8. Clyde B. Tatum ," Improving Constructibility During Conceptual Planning", Journal of Construction Engineering and Management , vol 113 No. 2. Jun. 1987.

12. O'Connor J. Larimore M., and Tucker R., " Collecting Constructibility Improvement Ideas", Journal of Construction Engineering and Management, vol 112 No. 4. Dec. 1986.

10. Alkass S and Harris F " An Expert System for Earthmoving Equipment Selection in Road Construction"Journal of Construction Engineering and Management , vol 114 No. 3 1988.

11. Mohan S " Expert Systems Applications in Construction Management and Engineering ", Journal of Construction Engineering and Management , vol 116 No. 1990.

12. Moselhi M and Nicholas M "Hybrid Expert System for Construction Planning and Scheduling"Journal of Construction Engineering and Management, vol 116 No. 2 ,1990.

13. Alkass S " An Expert System for Earthmoving Equipment Selection in Road Construction" a Ph.D thesis, Civil Engineering Department, Loughborough University, 1988, England.

14. Cornick , T. and Bull, S.," Expert Systems for Detail Design in Building ", CAAD Futures '87, Eindhoven, The Netherlands,1987.

15. Maver, T.," Software Tools for the Technical Evaluation of Design alternatives", CAAD Futures '87, Eindhoven, The Netherlands,1987.

16. Gero, J., and Maher, M., " Future Roles of Knowledge-based Systems in the design process " , CAAD Futures '87, Eindhoven, The Netherlands,1987.

17. Drysdale, R.G., Suter, G.T., Advisory Document , part 2 " Seminar on Brick Veneer Wall systems ", sponsored by CMHC CCRB, ABGC, MBEC, OBEC, June 12-14, 1989.

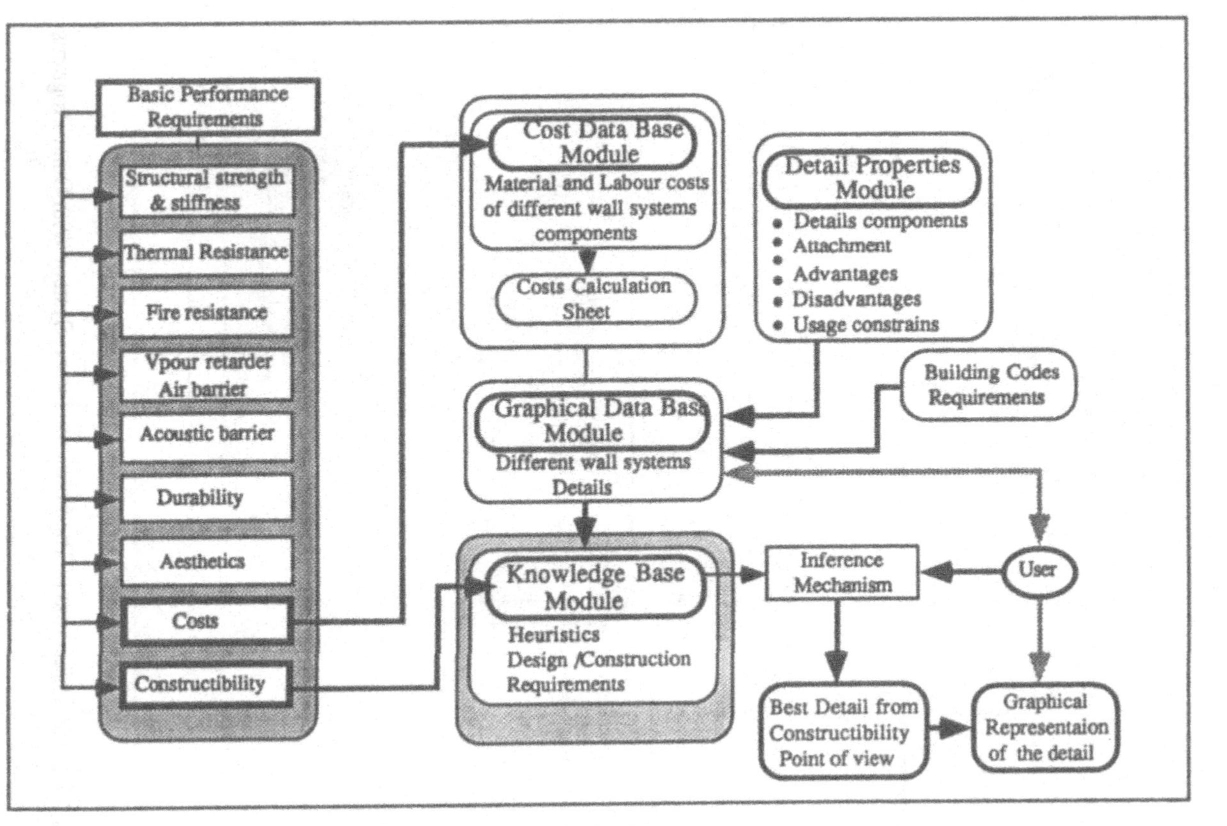

Fig.(1) Schematic Diagram of " CADD " Prototype System Architecture

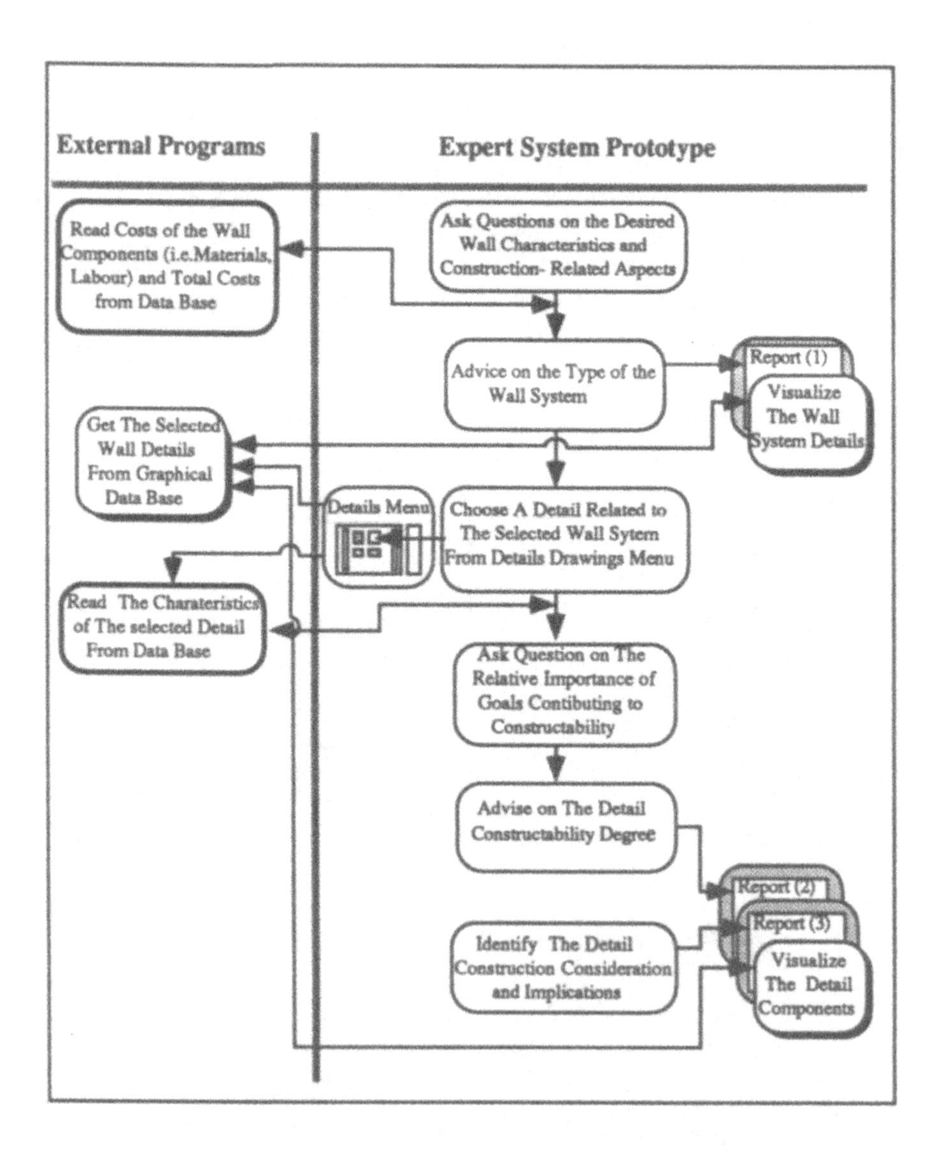

Fig. (2) The Interaction of The Prototype Expert System With Other Programs

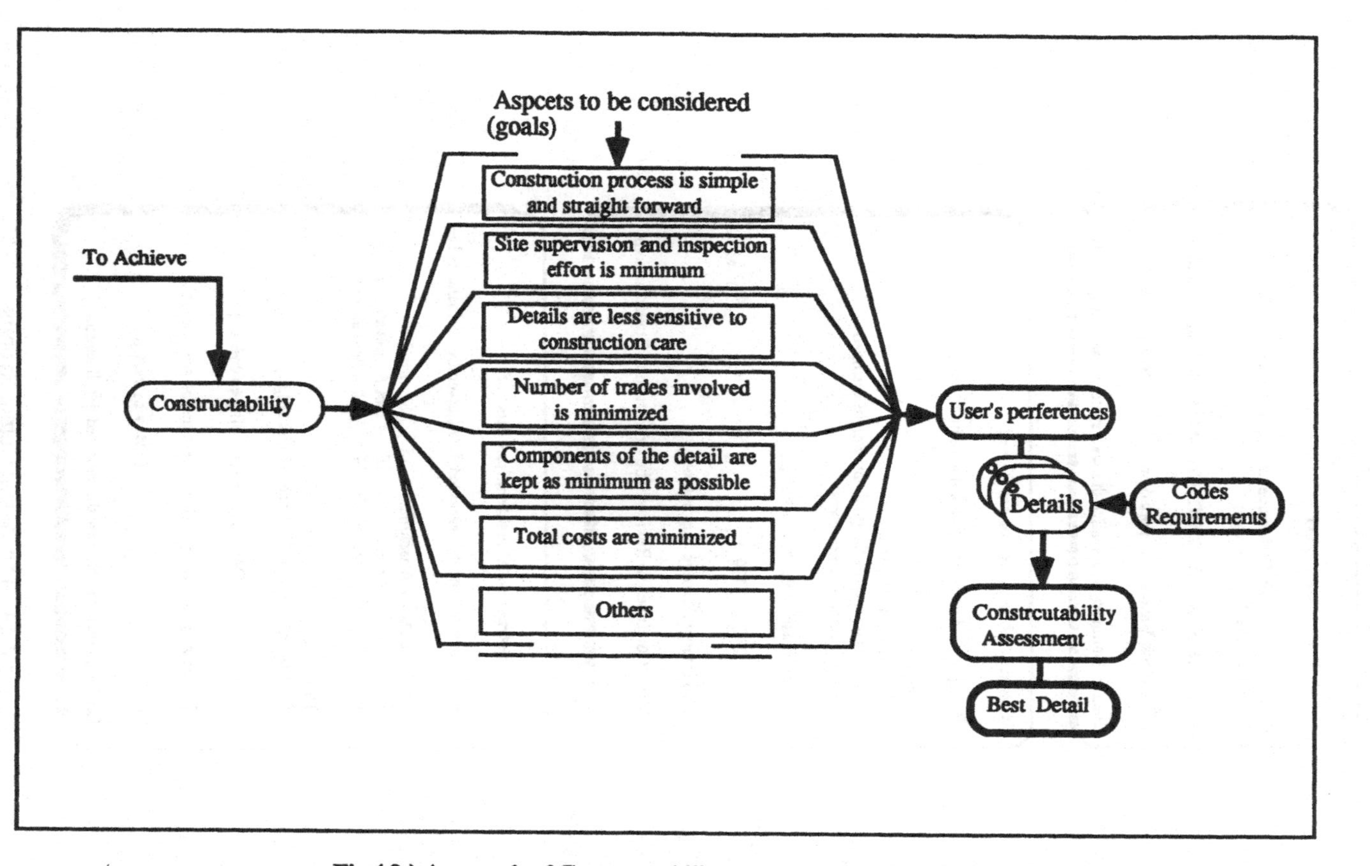

Fig.(3) Approach of Constructability Assessment of wall systems Details

The Wall Selection Report

Based on Your Own Preference of The Selection Criteria
We Recommend you to use :

* [BRICK VENEER_CONCRETE MASONRY backup wall]

* The Brick Veneer : Supported on [Shelf Angle]

Click <Continue> to visualize the wall system main components

Constructability Assessment Report

Based on : The selected detail assembly and its

- Characteristics.
- Construction requirements .
- Your perference of the goals by weight assinged by you prior to assessment.

The detail assembly has achieved

[130] / [130]

of your Goal Relative Importance

Click <Continue> to Review The Construction Considerations

The Construction Considerations Report

The performance of the assessed detail assembly is found to be :

- The Shelf Angle Attachment IS : [**Secure**]
- Horizontal Adjustment IS : [**Not Achievable**]
- Vertical Adjustment IS : [**Not Achievable**]

Therefore :
- Resulted Movement joint beneath the anlge :

[**May not be according to codes**]

- The bearing area of the brick veneer on the shelf angle :

[**May be varied**]

Click <Continue> to view the selected detail assembly

Fig. (4). The system Output Reports

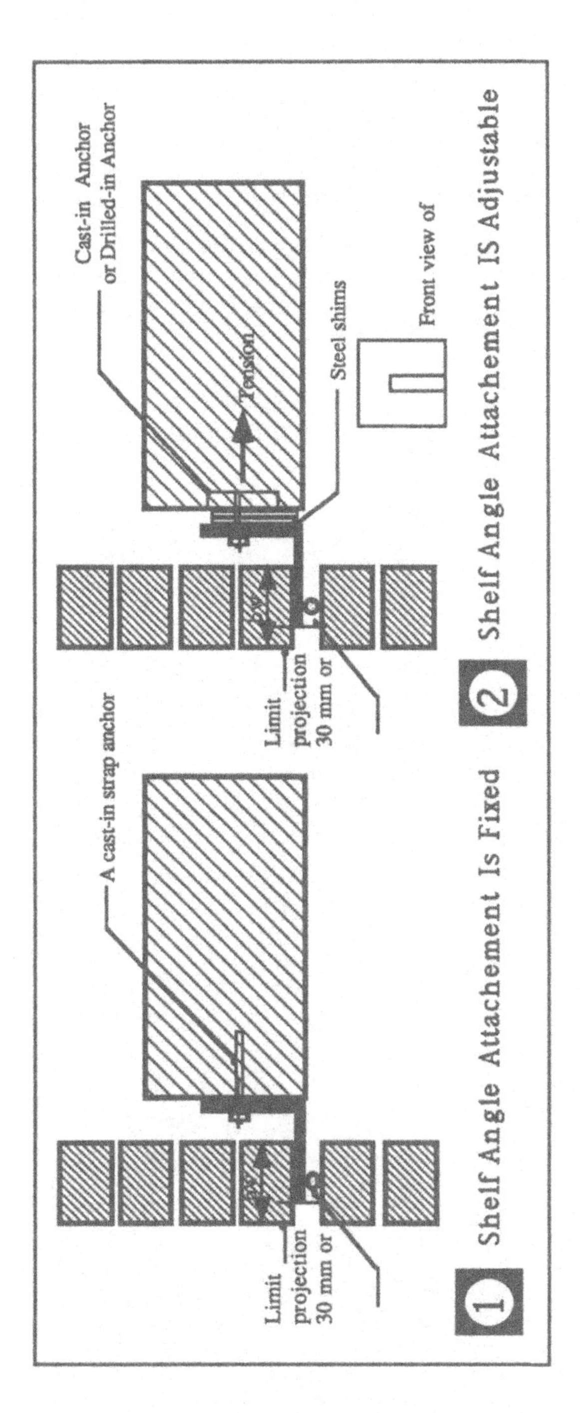

Fig. (5) The Detail Alternatives for The Shelf Angle Attachment

Knowledge Based System for Geometry Features Reasoning from Three-Dimensional CAD Data

B.V. Jerbic, B.R. Vranjes, Z.A. Kunica
Faculty of Mechanical Engineering and Naval Architecture, University of Zagreb, Salajeva 5, 41000 Zagreb, Yugoslavia

ABSTRACT

Integrated planning of flexible assembly combines the designing of product, process and system into a single computer aided system. Such an approach can significantly enhance planning results. But, computer reasoning on product geometry appears as the primary problem for computer implementation in assembly planning. In the present paper the knowledge based system for geometry features extraction from 3D CAD data provides formalization of a product geometry in order to enable the processing of its geometry properties. Implementations in assembly sequence planning, mating features deducing and automatic robot programming are presented.

INTRODUCTION

Competition and changing needs in the international market constitute a background that leads to flexible and fully automated assembly systems even in the case of small or medium batch production. The development of flexible assembly, until recently often overshadowed by the development of other manufacturing domains, is acquiring a major production and economic importance today.

A assembly automation has aim to imitate and substitute human skills which is an extremely complex task. Thus, emphasis is placed on the efforts to automate and to enhance the process of flexible/robotic assembly system planning by facilitating a computer aided approach. Vranjes, Jerbic and Kunica [1] have established the concept of an integrated system for planning of automatic assembly (Figure 1.). This concept combines product design, analysis and production engineering into a single system which can thereby increase the efficiency of the entire system.

The planning of a robotic assembly cannot be effected by the conventional algorithmic computing approach because it is primarily based on complex analysis of product geometry and physical features. A computer aided planning tool has to possess and process a three-dimensional geometric knowledge about products, a task requiring the most sophisticated intellectual and visual capabilities. These capabilities are common to humans rather than to computers, indicating difficulties for computer implementation in this domain.

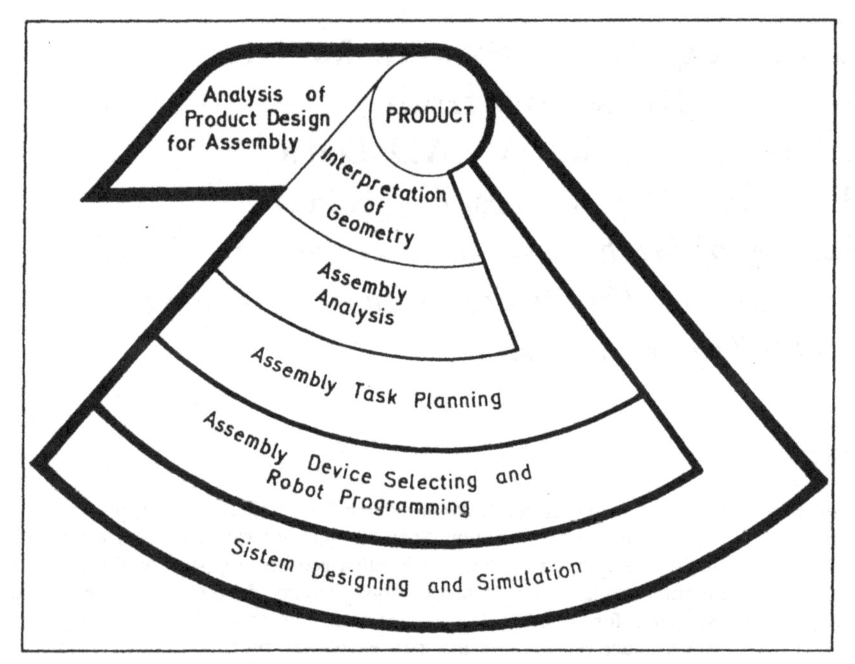

Figure 1. Concept of integrated planning of automatic assembly

However, artificial intelligence techniques and the knowledge based approach enable the computer support to be involved even in this 'human intellectual' sphere.

Computer implementation of representation and reasoning of product geometry is seen as a fundamental problem in assembly planning.

REPRESENTATION OF PRODUCT GEOMETRY

Concerning the product geometry representation various methods could be applied :

- coding system [2],
- description language [3, 4, 5, 6],
- vision system [7],
- CAD-model [8].

The coding system and description language require the tedious translation of a product design into descriptive form which is usually the source of many mistakes.

The vision system very closely simulates human visual perception abilities, but for the time being does not enable acquisition of sufficiently exact three-dimensional geometry data of complex parts end assemblies.

The 3D CAD-model describes all the detailed geometry, topology and physical data of a product design and avoids the cumbersome additional description by either coding or dedicated computer language. Therefore, the CAD-model imposes itself as the most powerful solution like a bridge between human visuality and the computer's numerical abilities.

CAD-MODEL REPRESENTATION The CAD representation of a model is stored in the computer memory by unrelated geometric entities expressed by numeric codes. Obviously, these CAD data cannot be directly applied to any reasoning on a model geometry. They should be interpreted through geometry and logic synthesis to provide the comprehensive understanding of a model design and of its relevant geometry features.

This work attempts to solve the problems of CAD integration in assembly planning by the means of a knowledge based expert system for CAD-model geometry interpretation . Since the incompatibility of various CAD systems and modelling methods appears as a serious barrier to application, the proposed interpreter is founded on the IGES (International Graphics Exchange Specification) processor [9] for CAD data acquisition.

IGES The IGES standard, established in the U.S. Department of Commerce, National Bureau of Standards, permits the compatible exchange of product (model) definition data used by various CAD/CAM systems. The IGES specification defines a standard ASCII file format to represent structure, language, topological, geometric and non-geometric data. It has been written so as to cover a large number of data structures by numerous software vendors for wireframe and surface CAD/CAM systems. There is currently no IGES support for the definition of solid geometry. Low level geometric entities with no interconnecting relationships are used to represent 3D geometric information. The IGES definition space is three-dimensional Euclidian space with right-handed Cartesian X, Y, Z coordinate system.

The file format defined by this Specification treats the CAD-model as a file of entities categorized as geometric and non-geometric.

A file consists of five subsections :

- start,
- global conditions,
- directory data,
- parameter data and
- terminator.

The start section is a human-readable prologue to the file.

The global section contains information describing the pre-processor and information needed by the post-processor to handle the file.

The directory entry section has one directory entry for each entity in the file.

The parameter data section of the file contains the geometric parameter data associated with each entity.

The terminate section indicates the end of the file.

The IGES includes more than 20 various types of geometric entities. In this work the analytical representation of both curves and surfaces is used, covered by just nine entity types (Table 1.).

Entity Code	Entity Type
100	Circular Arc
102	Composite Curve
108	Plane
110	Line
112	Parametric Spline
118	Ruled Surface
120	Surface of Revolution
122	Tabulated Cylinder
124	Transformation Matrix

Table 1. IGES entities for analytical representation of curves and surfaces.

EXPERT SYSTEM FOR CAD-MODEL GEOMETRY INTERPRETATION

Since the IGES representation of CAD-model consists of low level unrelated geometric entity definitions, the interpretation of CAD-model geometry should be based on the recognition of the resultant primitive objects (cylinder, block, wedge, etc.) that construct the model geometry. The appropriate expert system written in Prolog is presented in this section.

The fundamental element of any expert system is a knowledge which must be structured and objectified in order to enable deduction in the problem domain.

KNOWLEDGE

There are two major aspects of knowledge that appear to be relevant to CAD-model geometry interpretation :

- input knowledge : about the subject (CAD-model geometry) and
- procedural knowledge : required for recognition and interpretation.

INPUT KNOWLEDGE contains a CAD definition of a model derived from IGES representation which must be translated into the corresponding Prolog data structure.

Unit clauses that build up the model knowledge base are specified as follows :

assembly(a_i, PART_list([p_{i1}, ...])).

part(p_{in}, ENT_list([e_{n1} ...]), _).

entity(p_{in}, e_{nm}, ENT_type([parameters])).

where

a_i	is the assembly name.
PART_list([p_{i1} ...])	is the list of the integral parts.
p_{in}	is the part number.
ENT_list([e_{n1} ...])	is the list of the entities which form the part geometry.
e_{nm}	is the entity number.
ENT_type([parameters])	is the definition of the appropriate entity depending of the entity type (*line, arc, tabulated cylinder*-projection vector,... etc). The *parameters* corresponds to the IGES geometric parameter data.

Some *entity* clauses are explained below :

entity(p_{in}, e_{nm}, *line*(X_1, Y_1, Z_1, X_2, Y_2, Z_2)).

where

X_1, Y_1, Z_1 are the coordinates of the starting point of the line.
X_2, Y_2, Z_2 are the coordinates of the ending point of the line.

entity(p_{in}, e_{nm}, *arc*(*Xc, Yc, Zc*, X_1, Y_1, Z_1, X_2, Y_2, Z_2)).

where

X_c, Y_c, Z_c are the coordinates of the center point of the arc.
X_1, Y_1, Z_1 are the coordinates of the starting point of the arc.
X_2, Y_2, Z_2 are the coordinates of the ending point of the arc.

entity(p_{in}, e_{nm}, *tab_cyl*($e_{n'm'}$, *X, Y, Z*)).

where

X, Y, Z are the projection vector of the entity $e_{n'm'}$.

PROCEDURAL KNOWLEDGE FOR RECOGNITION OF PRIMITIVES To recognize the primitive objects that construct the CAD model, the expert system must possess the knowledge about geometry of primitives. In this work, 12 types of primitives shown in Figure 2. are comprised and their geometric definitions are embedded in the knowledge base of primitives.

Each primitive object is defined by specific geometric entities and their interconnecting relationships. The definitions of primitives should correspond to different analytic representations which vary depending on the modelling method applied. As shown by example in Figure 3., two cylinders might look the same when displayed on the screen, but if they have been designed by different CAD tools their representations can be diverse in some aspects. The cylinder designed as a wire-frame (Figure 3a.) is represented by two circles and a connecting line, but one designed as a solid or by 'sweeping' (figure 3b.) is given by a circle and the corresponding projection vector (*tab_cyl*).

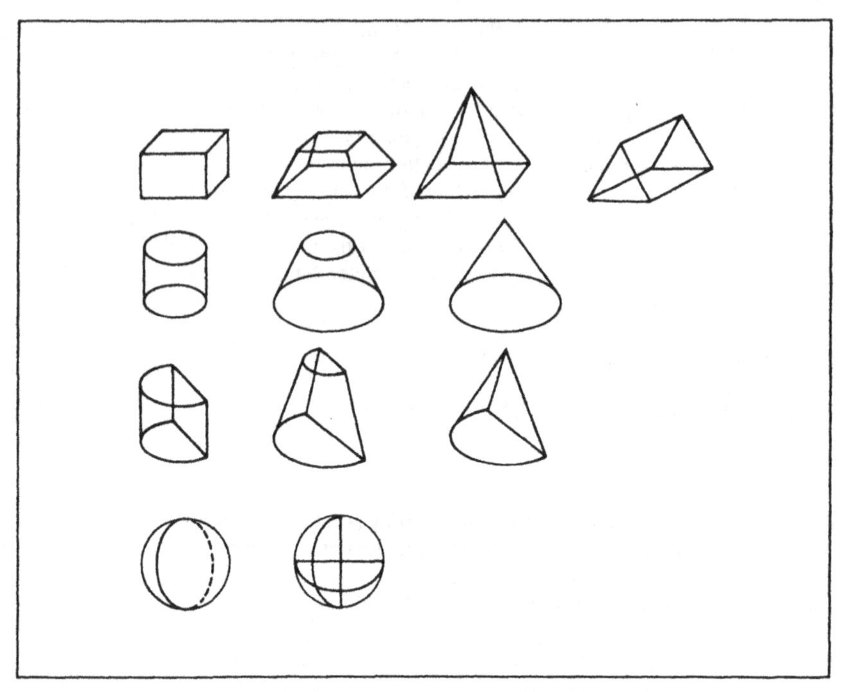

Figure 2. Primitive objects

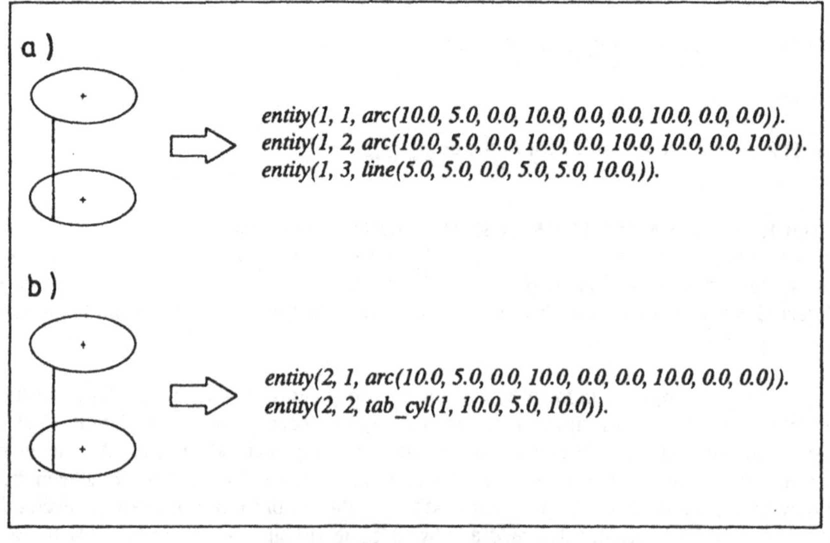

Figure 3. Variations of cylinder analytic representation

Obviously, all possible variations of definitions must be anticipated in the knowledge base of primitives to provide the recognition of primitives independently of the used modelling methods.

The situation in Figure 3b. is defined in the knowledge base of primitives by Prolog as follows :

```
%       Tabulated cylinder

cylinder_s(Layer, C) :-
        entity(Layer, Ptc, _, tab_cyl(P, X, Y, Z)),
        entity(Layer, P, _, sl_kriva(L)),
        circle(Layer, L, Z, _, _, _, _, _, C, L_del),
        retract(entity(Layer, Ptc, _, tab_cyl(P, X, Y, Z))),
        retract(entity(Layer, P, _, sl_kriva(L))).

cylinder_s(Layer, C) :-
        entity(Layer, Ptc, _, tab_cyl(P, X, Y, Z)),
        circle(Layer, [P], Z, _, _, _, _, _, C, L_del),
        retract(entity(Layer, Ptc, _, tab_cyl(P, X, Y, Z))).

%       Circle

circle(Layer, [ ], Z, Z1, Xc, Yc, X1, Y1, _, L_del) :-
        [!recorded(prim_t, prim_t(X, Y), _)!],
        X11 is round(X, 2), Y11 is round(Y, 2),
        X22 is round(X1, 2), Y22 is round(Y1, 2),
        X11 == X22,
        Y11 == Y22,
         Rx1 is abs(Xc - X1),
         Ry1 is abs(Yc - Y1),
         R  is round(sqrt(Rx1 ^2 + Ry1 ^2), 2),
         V  is abs(Z1 - Z),
         Fi  is R*2,
        ident_cylinder(Fi, V, Xc, Yc, Zc),
        erase_p(Layer, L_del), erase_p_t.

circle(Layer, [ ], Z, Z1, Xc, Yc, X1, Y1, C, L_del) :-
        entity(Layer, Ptc, _, tab_cyl(P, X, Y, Z)),
        entity(Layer, P, _, sl_kriva(L)),
        notmem(L, L_del),
        circle(Layer, L, Z, Z1, Xc, Yc, X1, Y1, C, L_del),
        retract(entity(Layer, Ptc, _, tab_cyl(P, X, Y, Z))),
        retract(entity(Layer, P, _, sl_kriva(L))).

circle(Layer, [ ], Z, Z1, Xc, Yc, X1, Y1, C, L_del) :-
        entity(Layer, Ptc, _, tab_cyl(P, X, Y, Z)),
        notmem( [P], L_del),
        circle(Layer, [P], Z, Z1, Xc, Yc, X1, Y1, C, L_del),
        retract(entity(Layer, Ptc, _, tab_cyl(P, X, Y, Z))).

circle(Layer, L, Z, Z1, Xc, Yc, X1, Y1, C, L_for_del) :-
        (entity(Layer, Px, _, arc(Z1, Xc, Yc, X1, Y1, X2, Y2));
```

```
entity(Layer, Px, _, arc(Z1, Xc, Yc, X2, Y2, X1, Y1))),
member(Px, L),
del(Px, L, Ldel),
append(L_for_del, [Px], L_del),
inc(C, C1),
ifthen((C1 == 1), recorda(prim_t, prim_t(X1, Y1), _)),
circle(Layer, Ldel, Z, Z1, Xc, Yc, X2, Y2, C1, L_del).
```

However, these original definitions are not sufficient to enable the thorough recognition of a model which is commonly designed by Boolean combinations of primitives (cross section, union or subtraction). The merging of primitives inevitably causes the disappearance of some geometric entities (Figure 4.). For this reason, consistent recognition is not generally possible without hypothetical definitions embedded in the knowledge base of primitives. Such approach in geometry interpretation is closely related to the theory of human vision relating 'subjective contour' illusion. It implies human illusion capability that is manifested as the capacity to perceive vanished contours so as to infer the geometric properties of a scene. Thus, hypothetical definitions must be designed to describe the existence primitive objects on the basis of least possible geometric determination. Some situations where a block is cross-sectioned by other primitives are illustrated in Figure 4. Even border-lines are lacking it is possible to identify a block using corresponding hypothetical definitions that are based on either eight connected lines (Figure 4a.), or on three extruded lines (Figure 4b.), or on two rectangles connected by a line (Figure 4c.).

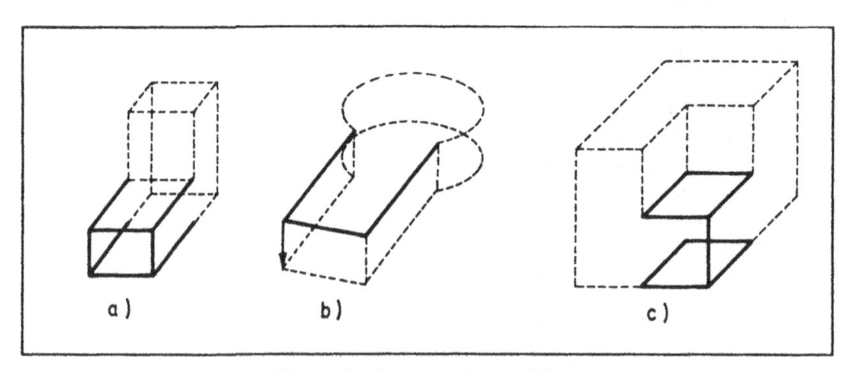

Figure 4. Cross-sections of block

RECOGNITION OF PRIMITIVES

The method for recognition of primitives establishes the inference engine founded on the structure of the previously described input and procedural knowledge. Figure 5. illustrates the method of recognition.

Besides the knowledge base of primitives, the inference engine requires auxiliary procedural knowledge for checking and proving some conditions and constraints such as entity parallelism, direction, perpendicularity and other utilities stored in it.

The method works in recursive procedure attempting in each step to explore the existence of certain primitives. As can be seen in Figure 6., the identified entities, proved to form a primitive,

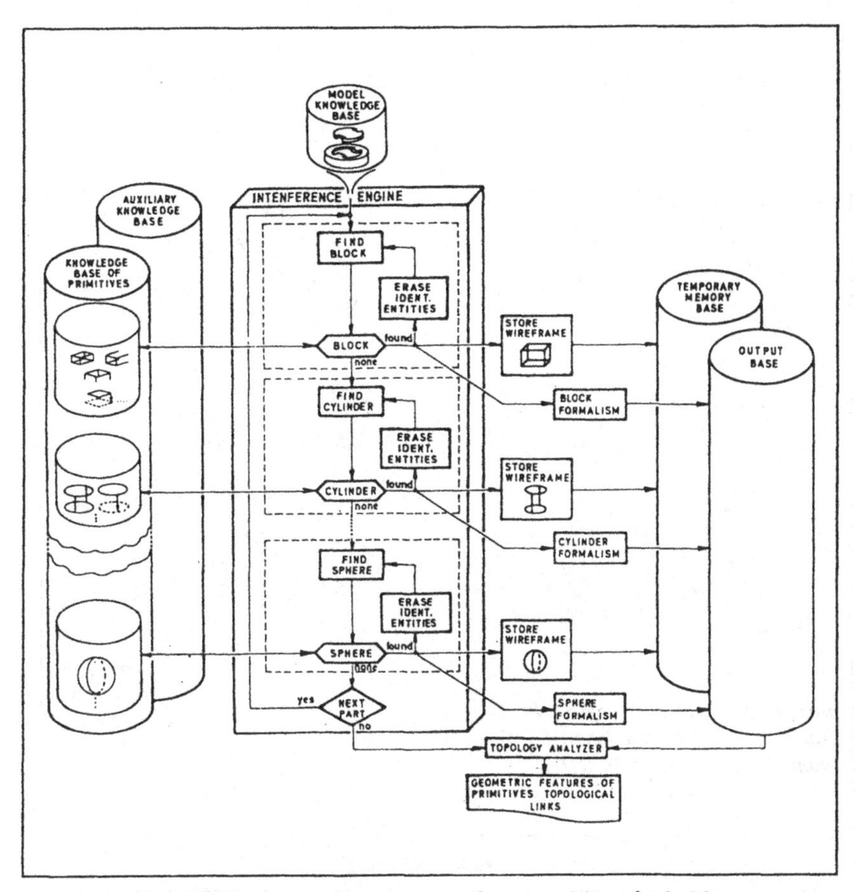

Figure 5. Structure of expert system for recognition of primitives.

should be removed from the model knowledge base to avoid ambiguous interpretation in the steps that follow. However, the removal of identified primitives 'strips away' the model geometry, causing the loss of the relevant entities and thus preventing consistent recognition. Therefore, if missing entities are discovered they have to be inserted as potentially linking entities of contiguous primitives (Figure 7.).

The example in Figure 8. shows the situation when, in spite of the insertion of missing line entities, two unrelated rectangles have remained. There are no conditions to deduce whether this is actually a block. Therefore, each identified primitive has to be stored as a wire-frame model in the temporary memory base. Its structure is similar to the model knowledge base but located in a separate expert system world. In this way, all possible linking entities are saved. In situations as the one shown in Figure 8., additional searching of temporary memory provides consistent recognition.

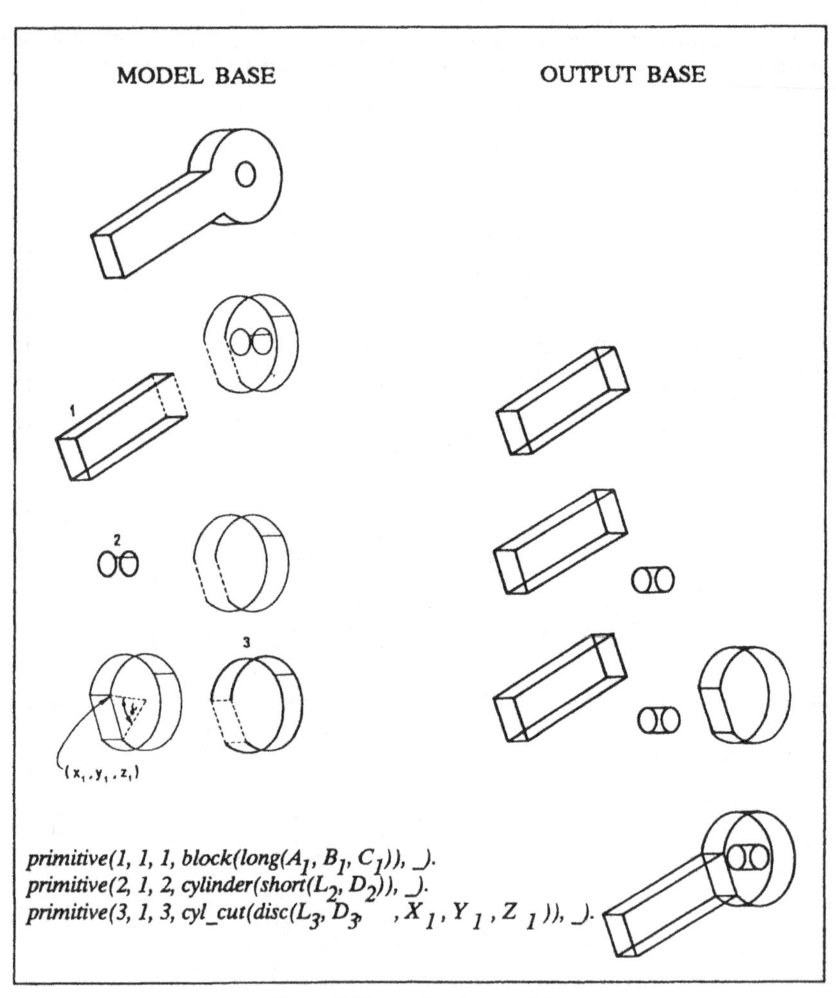

Figure 6. Example of recognition.

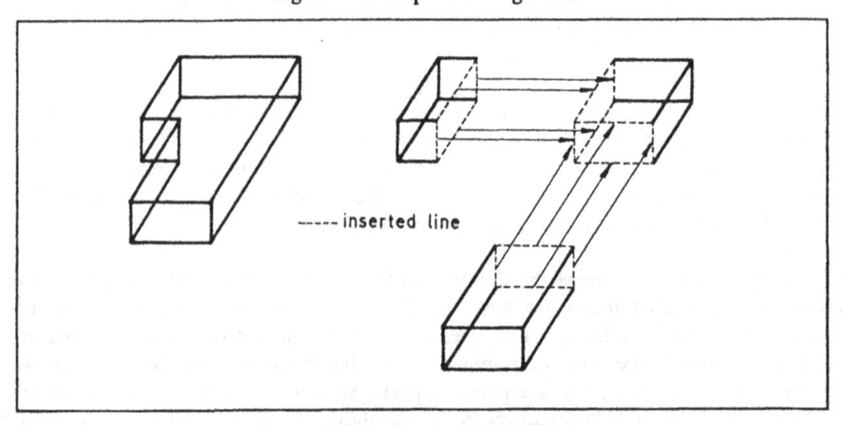

Figure 7. Insertion of linking entities.

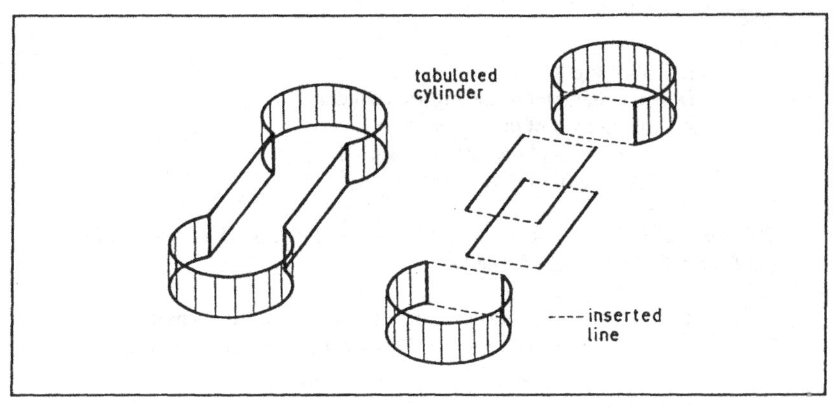

Figure 8. Inconsistent recognition situation.

The identified primitives are recorded in the output object world. Information about primitives is stored as Prolog clauses in the following manner :

part(p_{in}, _, s_n).

primitive(b_{nk}, p_{in}, s_k, *PRIM_type(SUB_type, [parameters]),_*).

transformation(s_k, *mat*($n_x, o_x, a_x, p_x,$
$n_y, o_y, a_y, p_y,$
n_z, o_z, a_z, p_z)).

where

s_n is the pointer to transformation matrix of the part.
b_{nk} is the primitive number.
s_k is the pointer to transformation matrix of the primitive.
PRIM_type(SUB_type, [parameters]) is the primitive type definition where *SUB_type* denotes the 'subtype' of the primitive. There are three different subtypes for both rotational and non-rotational primitives as given by Boothroyd [10] (Table 2.).

Rotational	
1. Disc 2. Short cylinder 3. Long cylinder	L/D < 0.8 0.8 < L/D < 1.5 L/D > 1.5
Non-rotational	
1. Flat 2. Long 3. Cubic	A/B < 3 and A/C > 4 A/B > 3 A/B < 3 and A/C < 4

Table 2. Subtypes of rotational and non-rotational primitives

where

L is the length.
D is the diameter of the biggest cylinder.
A is the length of the longest side.
B is the length of the intermediate side.
C is the shortest side.

The *parameters* depend on the type of primitive. For a block they are A, B and C, but for a cylinder they are L and D.

$\vec{n}$, $\vec{o}$ and $\vec{a}$	are the rotation vectors of the transformation matrix.
$\vec{p}$	is the translation vector of the transformation matrix.

Homogeneous transformation [11] should be assigned to a primitive in order to define the location and orientation of the primitive with respect to the reference coordinate system of the part considered. The coordinate system of the primitive is attached to the characteristic position depending on the type of primitive (Figure 9a.).

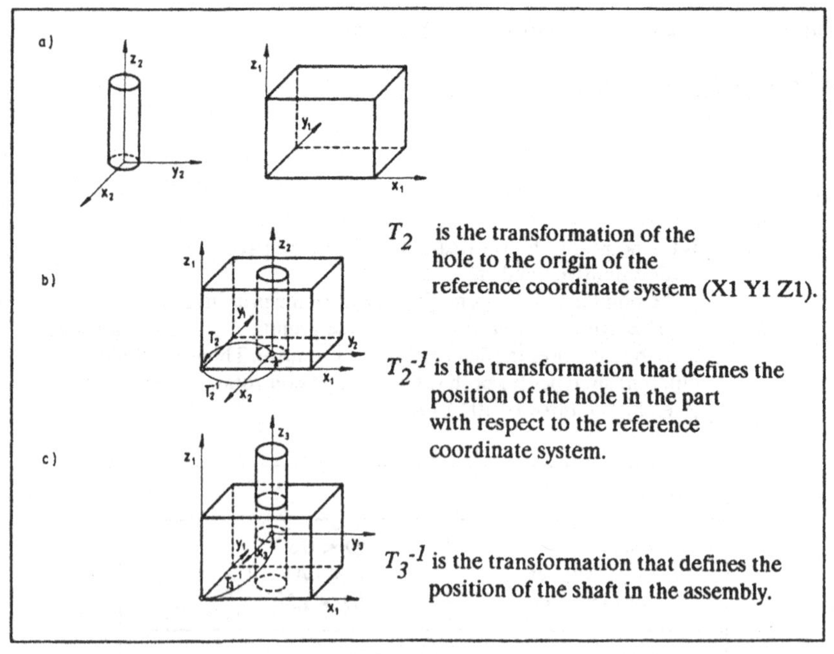

a)
b)
c)
T_2 is the transformation of the hole to the origin of the reference coordinate system (X1 Y1 Z1).
T_2^{-1} is the transformation that defines the position of the hole in the part with respect to the reference coordinate system.
T_3^{-1} is the transformation that defines the position of the shaft in the assembly.

Figure 9. Coordinate systems.

The coordinate system that belongs to the integral primitive with the largest volume is the reference coordinate system of a part (Figure 9b.). Each part has its own transformation to define its position in the assembly (Figure 9c.).

The transformations are derived from coordinates using the Euler general rotation transformation [12]:

$$\mathrm{Euler}(\phi,\theta,\psi) = \mathrm{Rot}(z,\phi)\,\mathrm{Rot}(y,\theta)\,\mathrm{Rot}(z,\psi) \qquad (1)$$

$$\mathrm{Euler}(\phi,\theta,\psi) = \begin{bmatrix} \cos\phi\cos\theta\cos\psi - \sin\phi\sin\psi & -\cos\phi\cos\theta\sin\psi - \sin\phi\cos\psi & \cos\phi\sin\theta & 0 \\ \sin\phi\cos\theta\cos\psi + \cos\phi\sin\psi & -\sin\phi\cos\theta\sin\psi + \cos\phi\cos\psi & \sin\phi\sin\theta & 0 \\ -\sin\theta\cos\psi & \sin\theta\sin\psi & \cos\theta & 0 \\ 0 & 0 & 0 & 1 \end{bmatrix} \qquad (2)$$

$$T = \mathrm{Euler}(\phi,\theta,\psi)\cdot\begin{bmatrix} 1 & 0 & 0 & P_x \\ 0 & 1 & 0 & P_y \\ 0 & 0 & 1 & P_z \\ 0 & 0 & 0 & 1 \end{bmatrix} \qquad (3)$$

INTERPRETATION

The identification of constitutive primitives enables to conceive the shapes that participate in forming the model design. But this is not sufficient in order to understand the geometric features of a model. As shown in Figure 10. *primitive_1* formed by a block and a cylinder is actually a solid block with a cylindrical hole. Thus, the cylinder is obviously a blank primitive having a particular meaning. Therefore, geometric features must be deduced through topology analysis. The aim is to find out what primitives are mutually related and what is the character of their spatial relationship - cross section or link.

To accomplish this task the topology analyzer generates the reference points of primitives. The reference points can be vertices of faces and/or points of axis symmetry. The appropriate relationships of primitives are deduced by comparing the spatial positions of the reference points.

If the cross section of two primitives is indicated, the one with the smaller volume would be treated as a blank. It could describe either a hole or a slot or a step or some other similar geometric feature.

The link means that two primitives lean against each other describing the complex topology characteristic of the part.

Thus, relationships are divided into the ones which represent 'local' geometric features of primitives and those which define overall topology as links of relevant primitives.

The example in Figure 10. shows the formalization of geometry obtained after topology analysis of the rotational part composed of four primitives. The subtraction of *primitive_1* (*block_cuboid*) and *primitive_2* (*cylinder_long*) is represented as a block with a hole as its local feature. Other primitives (*3* and *4*), together with *primitive_1*, build the topology of the part which is described by means of their face links.

IMPLEMENTATION IN ASSEMBLY PLANNING

Within the scope of the proposed concept of integrated computer aided approach to the planning of flexible assembly systems, this work has aim to provide all required geometric information about a product. The CAD-model geometry formalized above is actually the groundwork for the implementation of assembly sequence planning, mating features deducing and automatic robot programming. The mentioned integrated system for assembly planning is still in the development stage [13, 14] and this section of the paper presents the progress made to date.

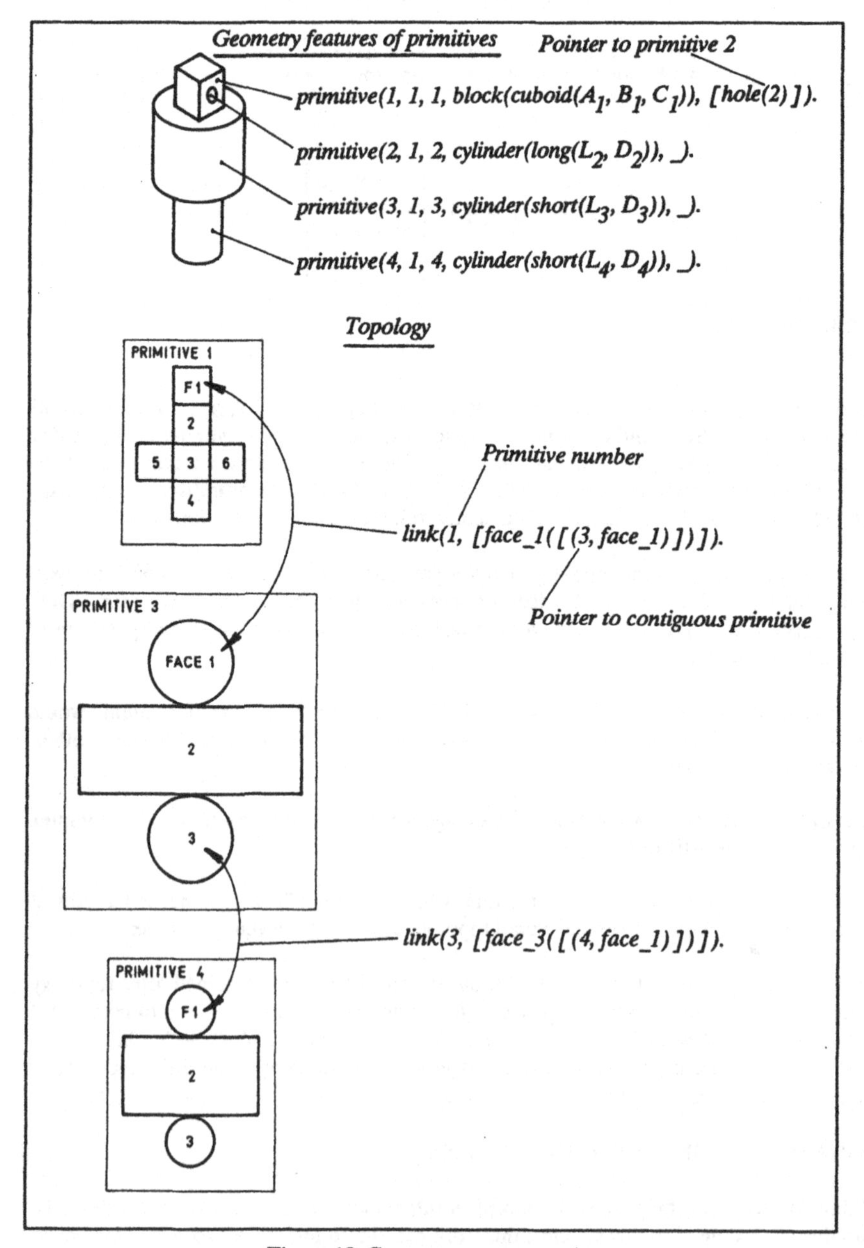

Figure 10. Geometry representation

ASSEMBLY SEQUENCE AND MATING FEATURES

Assembly sequence planning provides a basis for scheduling of operations and system resources, and permits evaluation of device and robot implementation alternatives. For this purpose the applied approach is based on the assembling/disassembling simulation algorithm [15, 16]. Homogeneous transformations are used to simulate the joining of the parts, their integral primitives in fact. In order to deduce corresponding sequence and mating features (fit, plane contact, clearance fit, ...), the system inquires for possible spatial interferences. The obtained results are stored in relationship matrix.

ROBOT PROGRAMMING

The system for automatic robot programming should enable generating of a robot programme for any given assembly task. The intention is not to develop the support for a specific robot, but a general assembly post-processor which can provide an integral flexible approach to assembly planning what is especially important in cases of smaller product quantities and from the standpoint of the assembly system flexibility.

Obviously, the assembly task has to be identified first. A suitable system, based on the generative approach, has been developed by Kunica [17].

The assumption is that the automated assembly process for any part mostly includes three basic operations :

- preparing (feeding and orienting),
- transferring and
- joining.

Thus, possible process structures such as linear, cyclical, parallel etc. are generated. For example, concentrated structure means that all parts are sequentially mounted and the basic part is not moved during mounting. The structure determines the assembly workspace. But the assembly workspace cannot be exactly determined without consideration of devices required for the assembly operations performing.

Having the geometry formalism, assembly sequence and mating features, appropriate assembly devices (robots, manipulators, feeders ...) may be associated for generated structures using the data base of assembly devices.

When the process structures and the main assembly equipment (their type and number) are known the most suitable structure and manufacturer can be chosen by simulation.

Every associated device has its own operation sequence and it is thus possible to form a list of necessary operations (allocating, magazining, orienting, inserting ...), which presents, together with the defined assembly workspace, the template for the automatic derivation of motion control programs and robot programming as the next line of research.

CONCLUSIONS

An integrated approach to planning of flexible assembly systems requires knowledge about product geometry. The expert system for geometry properties interpretation from 3D CAD data presented in this work provides the formalization of CAD model geometry in a way that enables the processing of product geometry in assembly planning. The interpretation is based on IGES representation of CAD-models, making its application independent of the CAD system used. Some elements of human illusion capabilities for understanding discrete geometric characteristics of things are implemented so as to recognize the resultant primitive objects that form the model geometry.

To conceive the global shape characteristic of a model, the topology analysis enables the identification of spatial relationships of primitives. The obtained geometry formalism includes information about local geometric properties of each primitive and the description of topology of a part. Such geometry representation of parts is used as input into a computer aided assembly planning system. Implementation is especially aimed toward realization of automatic assembly robot programming system which is the main subject of the work to follow.

REFERENCES

1. Vranjes B., Jerbic B., Kunica Z. Knowledge Based Approach to Automatic Assembly System Planning, Proceedings of International Symposium "Automatization and Measurement Technique", Vienne, Austria, pp. 134-144, 1990.
2. Peklenik J., Grum J. and Logar B. An Integrated Approach to CAD/CAPP/CAM and Group Technology by Pattern Recognition, Proc 16th CIRP Seminar on Manufacturing Systems, Tokyo, Japan, 1984.
3. Lieberman L.I. and Wesley M.A. AUTOPASS : An Automatic Programming System for Computer Controlled Mechanical Assembly, IBM J. Res. Development, Vol. 21, pp. 321-333, 1977.
4. Lozano-Perez T. and Winston P.H. LAMA : A Language for Automatic Mechanical Assembly, Proc. Fifth Int'l Joint Conf. Artificial Intelligence, Morgan Kaufmann, Los Altos, USA, pp. 710-716, 1977.
5. Popplestone R. J., Ambler A. P. and Bellos I., RAPT : A Language for Describing Assemblies, The Industrial Robots, pp. 131-137, 1978.
6. Hartquist E. E. and Marisa E. A., PADL-2 Users Manual, Production Automation Project, College of Engineering and Applied Science, The University of Rochester, Rochester
7. Winston P. H. Artificial Intelligence, Addison-Wesley Publishing Comp., Massachusetts, 1984.
8. Bartholomew O. N. and Hsu-Chang L. Feature Reasoning for Automatic Robotic Assembly and Machining in Polyhedral Representation, Int. J. Prod. Res., Vol.28, pp. 517-540, 1990.
9. Initial Graphics Exchange Specification (IGES), Version 3.0, U.S. Department of Commerce, National Bureau of Standards, Washington D.C., 1986.
10. Boothroyd G. and Dewhurst P. Design for Assembly, Department of Mechanical Engineering, University of Massachusetts, Amherst, 1983.
11. Plastock R.A. and Kalley G. Theory and Problems of Computer Graphics, McGraw-Hill, New York, 1986.
12. Paul R. P. Robot Manipulators, Mathematics, Programming and Control, MIT Press, Massachusetts, 1981.
13. Vranjes B., Jerbic B., Kunica Z. The Contribution to Automated Assembly System Planning, Strojarstvo, Journal for the Theory and Application in Mechanical Engineering, Zagreb, 1991. (in print)
14. Jerbic B. The Development of Knowledge-Based Expert System for Mechanical Assembly Operations Planning, Technical Report 01/89, FAMU/FSU College of Engineering, Tallahassee, 1989.

15. Jerbic B. Knowledge Based Approach to the Interpretation of CAD-Model Geometry, Proceedings of International Symposium "Automatization and Measurement Technique", Vienna, Austria, pp. 61-62, 1990.
16. Sekiguchi H., Kojima T., Inoue K. and Honda T. Study on Automatic Determination of Assembly Sequence, Annals of the CIRP, 32, pp. 371-374, 1983.
17. Kunica Z. Expert System for Defining Functional Structure of Automated Assembly Systems, Proceedings of 2nd International Conference on Advanced Manufacturing Systems and Technology AMST'90, Vol. 2, Trento Italy, pp. 393-400, 1990.

Insights into Cooperative Group Design: Experience with the LAN Designer System

M. Klein (*), S.C.-Y. Lu (**)

(*) c/o Kish Sharma, Boeing Advanced Technology Center, The Boeing Company, P.O. Box 3707, Seattle, WA 98124-2207, USA

(**) Knowledge-Based Engineering Systems Research Laboratory, Department of Mechanical and Industrial Engineering, University of Illinois, Urbana, IL 61820, USA

ABSTRACT

This paper discusses lessons learned about cooperative group design and how to support it effectively, based on experience in implementing a cooperative group design system in the Local Area Network (LAN) design domain. These lessons can be summarized as follows. Conflict resolution plays a central role in cooperative group design. Conflict resolution expertise can be effectively operationalized if it is given an identity distinct from other domain expertise. Cooperative group design systems should use a design model designed from the ground up to support conflict avoidance detection and resolution. Design agents must be able to explain and modify their actions, as well as critique abstractly described designs, in order to support such conflict resolution activity. How the LAN Designer system incorporates these lessons, as well as the strengths and limitations of relevant current research, is discussed.

INTRODUCTION

The design of complex artifacts has become, increasingly, a *cooperative* endeavor carried out by multiple agents with diverse kinds of expertise. For example, the design of a car may require experts on potential markets, function, manufacturability, safety regulations and so on. The development of tools and underlying theories for supporting cooperative group design has lagged, however, behind the growing needs implied by this evolution; see Klein [19]. In particular, while conflict-free cooperation has been well-studied (e.g. Smith [34], Malone [24], Lenat [22], Balzer [2]), how design agents can effectively interact when conflict occurs has received relatively little attention.

The goal of our research in this area has been to develop a system for supporting cooperative group design based, from the ground up, on a model of how human design agents actually interact, and in particular on how they cooperatively detect and resolve conflicts. This development has consisted of two major phases. The first phase involved studies of the cooperative group design process in human groups in two different domains (Architectural and Local Area Network design). These studies led us to develop a model of the cooperative group design process that, in the second phase, was realized as an implemented cooperative group design system (the LAN Designer) that designs Local Area Networks (LANs) using machine-based design agents.

The purpose of this paper is to describe what our experience with designing and implementing the LAN Designer system has revealed about conflict resolution (CR) in cooperative group design and how computers can support it. These insights can be summarized as follows. Conflict resolution plays a central role in cooperative group design. Conflict resolution can be effectively operationalized by instantiating general CR expertise maintained distinctly via interaction with domain expertise. Cooperative group design systems as a whole should use a design model designed from the ground up to support conflict avoidance, early conflict detection, and the effective use of CR expertise. This should include the ability of design agents to explain and modify their design actions.

In the remainder of this paper we describe these insights, including the evidence supporting them, the implications for cooperative group design system development, the strengths and deficiencies of relevant research to date, and how we incorporated each insight into the LAN Designer system. We conclude by discussing directions for future work.

THE CENTRAL ROLE OF CONFLICT RESOLUTION

Studies we performed of human cooperative group design in the domain of Architectural design (described in full in Klein [20]) suggest that conflicts, rather than being avoided at all costs, actually play a central role in the cooperative group design process. In these studies, statements made by architects cooperating to design a house (i.e. components of the design *protocol*) were collected and then categorized. These statements fell into four major categories:

- commit: A tentative design commitment made by a design agent.
- pro: An identification of a positive aspect of a design by a design agent (i.e. a positive critique of the design state).
- con: An identification of a conflict with a design by a design agent (i.e. a negative critique of the design state).
- res: A list of candidate design commitments for resolving a conflict.

A tabulation of the number of design statements in each category revealed the following pattern (Figure 1).

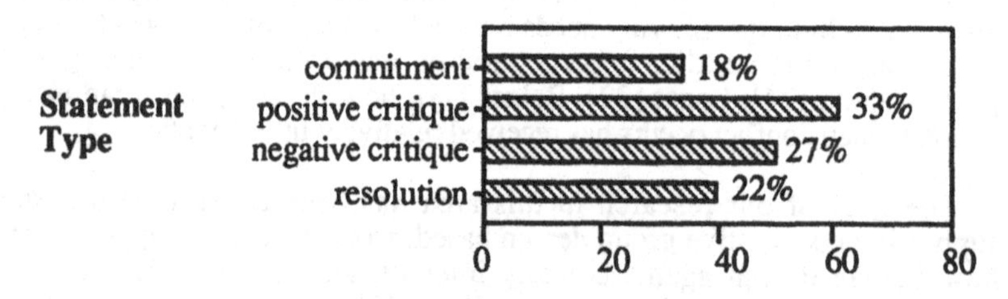

Figure 1. Counts of statement types.

There were actually more critiquing statements (positive *and* negative) as well as more conflict resolution statements than there were design commitments. Conflict identification (negative critiques) and conflict resolution together accounted for *half* of the statements made by the architects during the design sessions.

Close analysis of the design protocols suggests that the architects adhere to the following model of interaction in group design:

1. Design agents generate potential solutions for a given design subtask, usually based on "default" or "standard" solutions for that kind of problem.
2. They then evaluate the design, identifying its pros and cons. Cons include conflicting design commitments from different agents as well as negative critiques by one agent of commitments made by other agents. Pros are added to the rationale supporting the positively evaluated design decisions.
3. The design is then modified to resolve the conflicts identified.

The design process thus can be viewed as an iterative generate-and-test process, wherein candidate designs are generated using default knowledge, evaluated, and then "tweaked" as needed, in response to conflicts, to make them consonant with each agent's view of the specific demands of the given task (Figure 2).

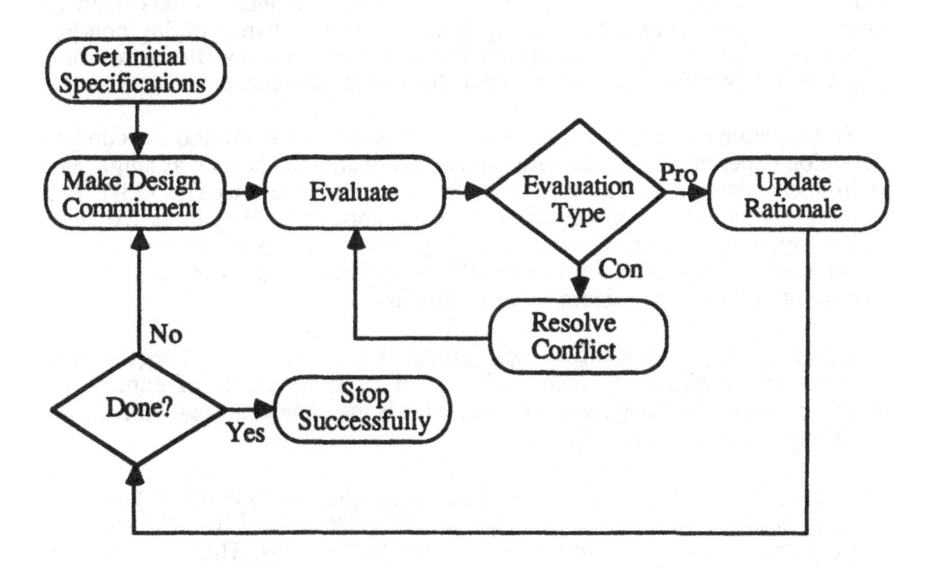

Figure 2. Model of the cooperative design process.

A related study (Klein [19]) suggests that cooperative group design of Local Area Networks is also characterized by a central role for conflict resolution. We believe that this is true, in fact, for a wide range of cooperative group design domains. Giving conflict resolution a central role can significantly improve the productivity of the design process. If we do so, design expertise can be divided up into relatively small, internally self-consistent bodies that each form an autonomous design agent. Individual design agents do not have anticipate and avoid conflicts with all other potentially relevant design agents. They can focus solely on representing their own concerns as ably as possible, relying on the conflict resolution mechanism to handle disagreements with other design agents as they occur (i.e. at run-time).

Computer systems that aim to support cooperative group design, therefore, must support the process of conflict resolution. In particular, they must support the kind of design process diagrammed above, including providing design agents with the ability to make and critique design commitments, detect and resolve inter-agent conflicts. Existing work on supporting conflict resolution in cooperative problem solving comes from AI and the social sciences. A comprehensive review is included in Klein [19]; this literature is summarized briefly below.

A large body of work is devoted to *analyzing* human conflict resolution behavior (Coombs [9]). This work highlights the importance of conflict in group interactions, but provides few prescriptions for how conflict resolution can be facilitated. In addition, much of this work focuses on issues specific to the psychology of human participants, rather than on the general nature of conflict resolution. There is in addition some work on *supporting* human conflict resolution, which includes research on group consensus building and group decision support systems, or GDSSs (e.g. Johansen [16], Nunamaker [30]) . This work focuses, however, on competitive conflicts and/or limits itself to structuring interactions among group members, rather than applying conflict resolution expertise to help resolve the conflicts. The conflict resolution expertise is thus still expected to reside in the human participants.

To find work on *computational models* that actually encode and use conflict resolution expertise, we need to turn to AI and related fields such as single and multi-agent planning/design as well as concurrent engineering (Lu [23]). The relevant literature can be grouped into three categories according to the extent to which conflict resolution expertise is given "first-class" status, i.e., is represented and reasoned with explicitly using formalisms as robust as those used for other kinds of problem-solving expertise:

- <u>Development-Time Conflict Resolution:</u> Systems of this type require that potential conflicts be "compiled" out of them by virtue of exhaustive discussions when they are developed. This is the approach used with almost all expert systems currently.

- <u>Knowledge-Poor Run-Time Conflict Resolution:</u> In systems of this kind (e.g. Marcus [25], Stefik [36], Fox [11], Descotte [10], Brown [5]) conflicts are asserted and resolved as the system runs. These approaches incorporate little CR expertise and use restrictive formalisms to represent it.

- <u>General Conflict Resolution:</u> Work in this class come closest to providing conflict resolution expertise with first-class status, and includes implemented systems (e.g. Sussman [37], Goldstein [13]) as well as unimplemented proposals (e.g. Hewitt [14], Wilensky [40]). None of this work, however, constitutes a comprehensive computational model of conflict resolution for cooperative group design.

In general, work on conflict resolution has evolved towards making CR expertise more explicit and using it to support cooperative problem solving. The LAN Designer system is the first system, however, to provide first-class knowledge-intensive support for conflict resolution in cooperative group design contexts. How this system does so is described in the sections below.

THE EXISTENCE OF GENERAL HEURISTIC CR EXPERTISE

The instances of conflict resolution identified in the Architecture and LAN domains were analyzed, as part of the above-described studies, to see if we could extract any general principles of how the CR process works. It was found that CR suggestions can be viewed as instantiations of domain-independent general strategies. Consider the following examples:

1. Jan and Mark are designing a bicycle. Jan specifies that a brace should connect the two sides of the handlebar to improve stiffness, but Mark objects because the brace will make it impossible to attach standard handlebar bags to the bike. They decide to use thicker tubing in the handlebars as an alternate scheme for improving stiffness.

2. Bob and Mike are designing a home. Mike would like large windows on the south-facing front facade of the house because of their aesthetic effect. Bob is concerned that the summer-time insolation through these windows will lead to excessive cooling costs. They decide to add an overhang over the windows to provide shade from the high summer sun.

3. Jeff and Arthur are designing a Local Area Network (LAN) for some clients. Arthur suggests a simple design that involves a single LAN trunk interconnecting all the workstations and servers at the site. Jeff is concerned that a single failure at any point along the trunk will cause the entire site to cease functioning. They decide to add several repeaters to the trunk to prevent the propagation of failure from one LAN trunk segment to another.

The conflict resolution instances given above can be thought of as instances of the following general conflict resolution strategies:

1. **If** two plans for achieving two different agents' goals conflict
   Then find an alternate way of achieving one goal that does not conflict with the other agent's plan for achieving its goal

2. **If** excessive summer insolation through south-facing windows is a concern
   Then provide overhangs to block the sun

3. **If** propagation of some unwanted entity over a shared conduit is a concern
   Then add a filter to the conduit to reduce or eliminate the propagation of that entity

As we can see, such general conflict resolution expertise can be expressed in a way specific at most to *classes* of conflicts. Strategy 1, for example, applies to any conflict where the design agents involved have alternate plans for achieving their design goals. Strategy 3 applies to any conflict where a design component that can be abstractly viewed as a conduit currently allows propagation of some undesired but filterable entity.

Another outcome of these studies was the realization that CR expertise is largely *heuristic*. This is inescapable since CR expertise deals with the interaction of internally consistent but mutually inconsistent bodies of domain expertise encoded in different design agents (Hewitt [14]). Heuristic expertise can be incorrect, so we must be able to respond appropriately if heuristic CR advice fails.

The implication of these insights is simple: rather than building design agents that inextricably mix CR and design domain expertise in their knowledge bases, it makes more sense to separate these into design and CR components with distinct roles. The division of labour allows a relatively compact corpus of general CR expertise to be applied to a wide variety of design conflicts. The CR expertise can be augmented or changed easily at any time without requiring coordinated changes in the domain expertise of all potentially affected design agents. CR expertise acquired for one class of problems is applicable to other problems with similar abstract characteristics. Existing design systems, for the most part, conflate design and CR expertise (e.g. Fox [12], Marcus [25], Brown [5]). Approaches that give CR expertise a distinct identity also exist (e.g. Sussman [37], Goldstein [13]), but suffer from a number of important limitations including trying to encode fundamentally heuristic CR expertise in deductive form, using only a very small collection of CR expertise, and representing CR expertise using formalisms idiosyncratic to a particular problem domain.

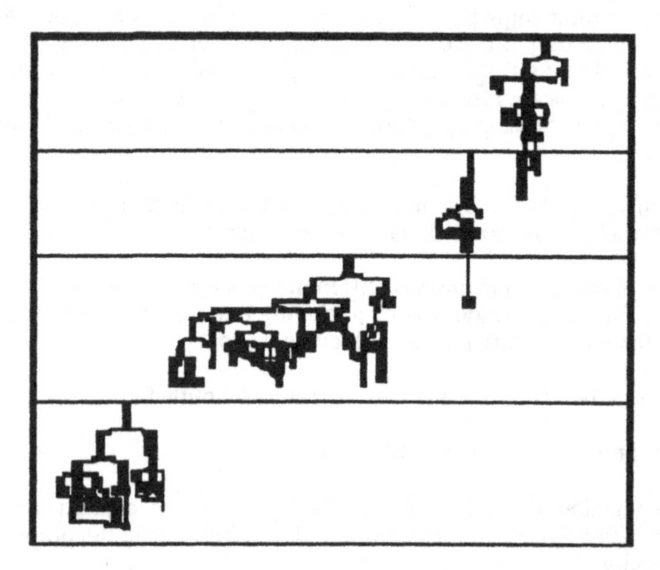

Figure 3. The CR component's conflict class taxonomy.

Our insights into cooperative group design were elaborated into a computational model implemented in the LAN Designer system (this model is described in detail in Klein [21]). In this model, general CR expertise is instantiated, via interaction with domain-level design expertise, to produce specific suggestions for resolving a given conflict. The LAN Designer's conflict resolution component thus effectively resolves conflicts in the LAN domain while including *no* LAN-domain-specific expertise. The LAN Designer's CR knowledge base is represented as a taxonomy of conflict classes with associated advice for resolving conflicts in each class. A reduced view of the current conflict taxonomy, with 115 conflict classes, is given in Figure 3.

The CR expertise currently included in the LAN Designer covers a wide range of conflict types, oriented mainly towards supporting conflict resolution in systems that design artifacts for managing and transporting resources (such as data on a Local Area Network). Conflict classes for dealing with failed conflict resolution advice are also included.

The LAN Designer system consists of a collection of design agents, each of which includes a design component that critiques and updates designs as well as a conflict resolution component that resolves conflicts (Figure 4).

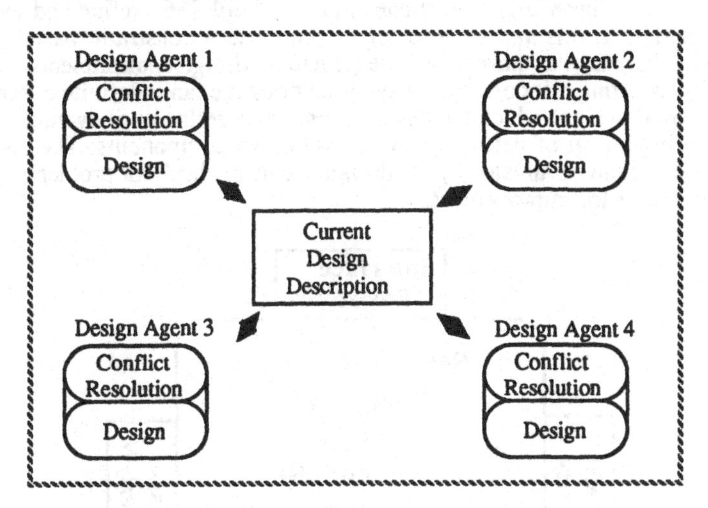

Figure 4. Architecture of the LAN Designer system.

The design and CR components are implemented distinctly, using different problem solving models (heuristic construction and heuristic classification, respectively; see Clancey [8]) and separate knowledge bases.

A DESIGN MODEL THAT SUPPORTS CONFLICT RESOLUTION

A cooperative group design system that supports conflict resolution should, ideally, provide support for all aspects of conflict management, including avoiding unnecessary conflicts and detecting them when they do occur. We have found, through experience in implementing the LAN Designer system, that it is possible to provide knowledge engineers developing a cooperative group design system with a generally-applicable conflict avoidance and detection techniques. These techniques assume the particular design model used in the LAN Designer; we briefly describe this model and then consider the conflict detection techniques based on it; see Klein [19] for a complete description.

The LAN Designer uses a *goal-driven least-commitment routine* design model that synthesizes state-of-the-art models of the design process with some novel features involving conflict management. The model is *goal-driven* in that all design actions are made in response to explicit goals asserted by the design agents. A uniform meta-planning model (Wilensky [40]) is used for both domain and meta-level reasoning. This is described in more detail in the section below on "Improved Design Agent Models".

The LAN Designer supports *routine* design. Design tasks can be divided into several categories according to the innate difficulty of the task (Brown [6]), ranging from "routine" design (where the components and plans for combining them are known) to "innovative" design (where neither components nor plans are available). To date, only routine design models have been implemented with any success for real-world design problems (e.g. R1 [26], VT [25], AIR-CYL [5] and PRIDE [28]). Fortunately, many important real-world design tasks fall into this category (Brown [6]).

The LAN Designer uses a *least-commitment* (Stefik [36]) refine-and-evaluate design model. In this approach, design agents use a constraint-based design description language to make indefinite (tentative) design commitments. Design begins by asserting a set of abstract specifications (i.e. an indefinite description of the desired design) which the design agents then refine until eventually they have specified a set of definitely-specified known components. The evolving design is constantly evaluated by the design agents to check for problems as well as opportunities for improvement.

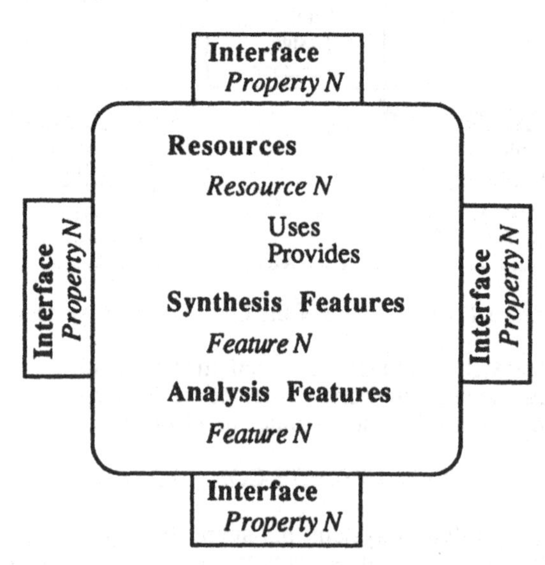

Figure 5. A component description.

The LAN Designer implements indefinite design descriptions as follows. All designed artifacts are represented as collections of known components each with characteristic features, connected to each other via a defined set of interfaces with known properties. There are two varieties of design features; *synthesis* features describe how a particular component is configured (e.g. the output frequency of an adjustable oscillator), while*analysis* features describe the outcome of analysing some aspect of a component (e.g. the expected-mean-time-to-failure for a given part). Components may provide as well as use up different kinds of resources (Figure 5).

Components are represented by indefinite descriptions of the component's *class* and *features*. The former is supported in the LAN Designer by building an abstraction hierarchy of components that ranges from abstract components at the top to existing "primitive" components at the bottom. An indefinite component class description thus involves selecting a non-primitive (i.e. non-leaf) component class. Component features are described indefinitely using the constraint-based language provided by the LAN Designer system. The constraint types currently supported include ranges, inequalities and sets as well as algebraic and boolean equations.

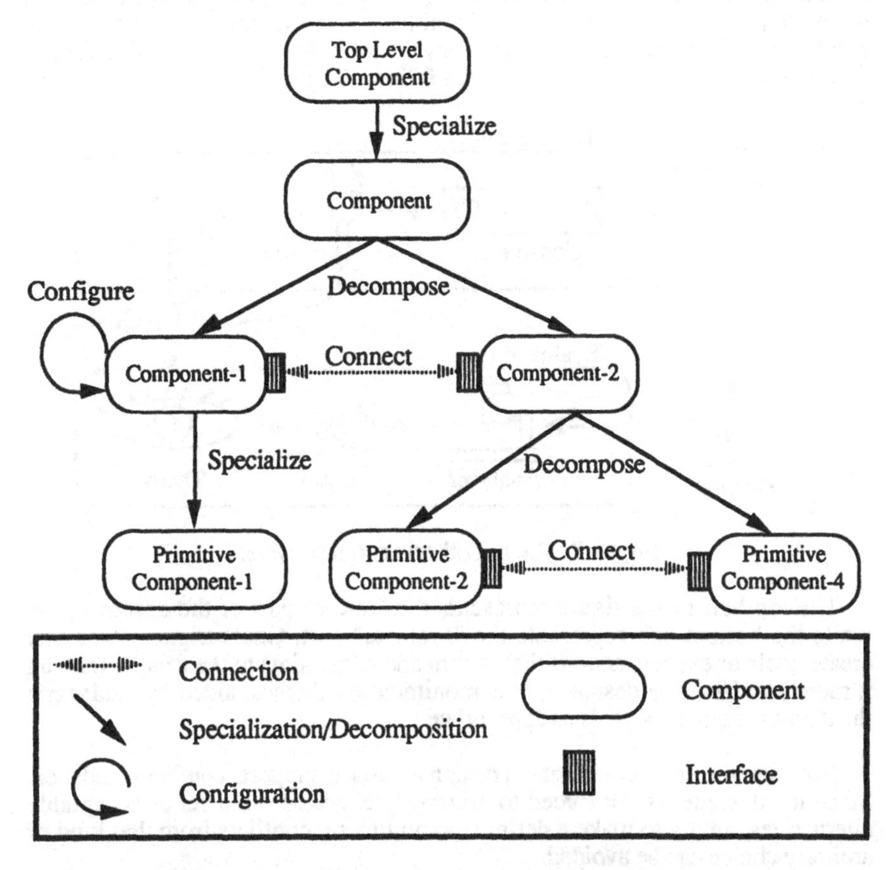

Figure 6. Refining a design description recursively.

The "refine" part of the design process begins by creating an instance of a top-level component that represents the entire design. For the LAN domain, for example, this top level component is called "the-site", and represents the entire site for which the Local Area Network is being designed. Whenever an abstract or composite component is created, the goal to instantiate this component is asserted. This goal is mapped by design plans to design actions that transform the component into more completely described components (Figure 6). This can involve either configuring the component by constraining the value of its features (perhaps relative to other features), connecting the interfaces of two components, decomposing a component into sub-components, or specializing it to a more specific component. When new components are produced, either by specialization or decomposition, goals to instantiate these components are asserted in turn. As a result, once the top-level component is created, it is recursively refined all the way down to a collection of connected primitive components. The different design actions can be interleaved arbitrarily during the design process.

The "evaluate" part of the design process allows design agents to continually check the viability of the current design. Entities known as "themes" look (with the help of "analyzers" that produce design state analyses) for design states that represent problems or fortuitous opportunities and create goals accordingly. The combined refinement-and-evaluation cycle underlying the LAN Designer's design model can thus be summarized as follows (Figure 7):

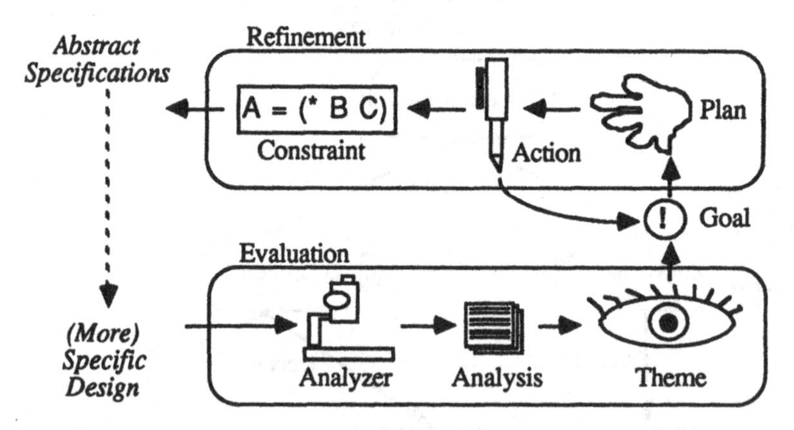

Figure 7. The hypothesis and test cycle.

The top half of the figure represents the "refine" part of the design cycle, while the bottom half represents the "evaluate" part. Goals trigger plans that create goals or execute actions that in turn add constraints to the design, making it more specific. The design state is monitored by themes, aided by analyzers; the themes create new goals as appropriate.

Note that this least-commitment design approach supports conflict avoidance. Since the designer is not forced to arbitrarily choose from a set of acceptable alternatives simply to make a definite commitment, conflicts from that kind of arbitrary choice can be avoided.

The LAN Designer also supports conflict detection (CD). In general, when different agents give incompatible specifications for a given design component, or one agent has a negative critique of specifications asserted by another agent, we can say that a conflict has occurred. Some design conflicts can be detected in a domain-dependent fashion. Design experts have in addition a lot of domain-specific expertise concerning the kinds of design situations that represent problems from their particular perspectives. Support for conflict detection thus requires both providing a complete domain-independent conflict detection mechanism as well as making it easy for domain experts to express their domain-dependent conflict detection expertise.

The LAN Designer provides tools for both these purposes, i.e. domain-independent CD methods as well as CD "idioms" built on top of these methods that can be instantiated by design experts into domain-specific CD methods (Figure 8).

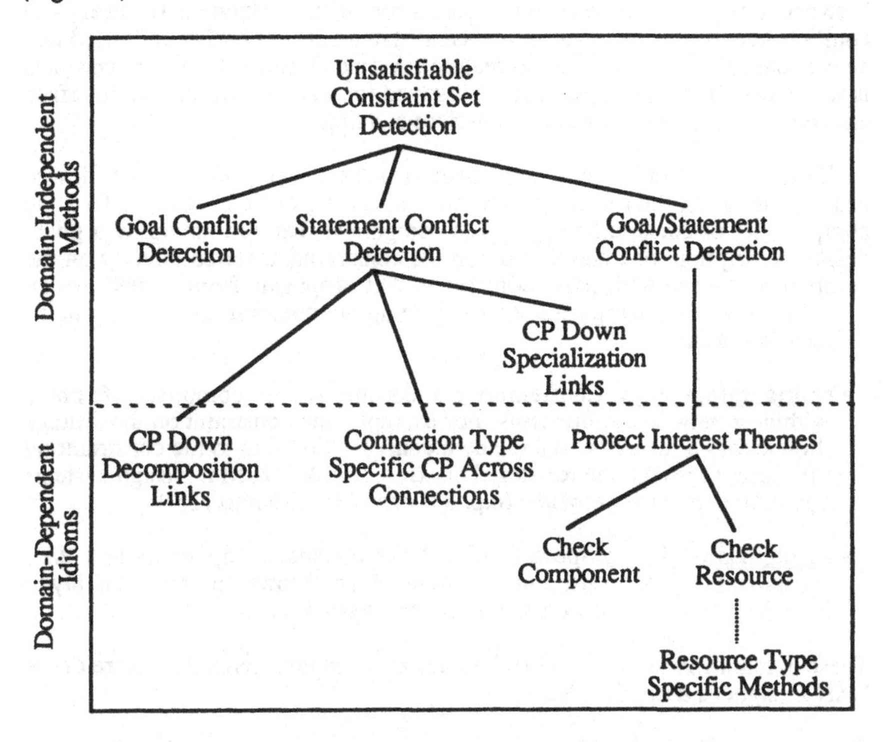

Figure 8. The LAN Designer's conflict detection tools.

<u>Domain-Independent Methods</u>
In a least-commitment model such as that used in the LAN Designer, all conflicts eventually manifest as unsatisfiable constraint sets on a given design parameter. Underlying all CD methods, therefore, is an unsatisfiable constraint set detection mechanism. This is implemented as constraint combination (i.e. simplifying constraint sets by finding their consequences) coupled with detection of individual unsatisfiable constraints. For example, the two range constraints (from 1 to 10) and (from 5 to 15) can be intersected to produce the new constraint (from 5 to 10). The range constraints (from 10 to 12) and (from 14 to 18) intersect to produce the unsatisfiable constraint (from 14 to 12). Similar constraint intersection and unsatisfiable constraint detection techniques are available for other constraint types. Note that, using a least-commitment model,

conflicts can be detected while the design is still indefinitely described, thus avoiding investing design effort uselessly on inconsistent designs.

Built on top of this is support for detecting conflicts between the two kinds of assertions made by design agents: constraints on component features ("statements") and goals. The LAN Designer's design model uses three types of goals: refine-component (the goal to refine an indefinitely described component), constrain-value (the goal to constrain a given component feature), and achieve-value (the goal to achieve a given value for a given component feature). There are thus three kinds of conflicts at this level: conflicts between achieve-value goals (inconsistent desired constraints on a component feature), conflicts between statements (inconsistent constraints on a component feature), and conflicts between achieve-value goals and statements (inconsistent desired and actual constraints on a component feature). A special case of statement conflicts is where two or more connections are asserted for a given interface; an interface can be connected to only one other interface at a time.

Constraints asserted by design plans can be propagated through design description along the different links created during design. Thus, to find the complete set of constraints on a given component feature implied by a given design description, one needs to union the constraints asserted directly on the component feature with those constraints that propagate from related design features. The following link types propagate constraints in a domain-independent fashion:

- configuration links: Configuration links inter-relate component features within or between components. For example, the constraint on the voltage drop across a resistor is related by a configuration link to the constraint on the current through the resistor. This kind of link is used in design systems based on constraint networks (e.g. Sussman [38], Marcus [25]).

- specialization links: Specialization links connect components and their specializations. All constraints that apply to an abstract component apply to its specialization, so all constraints are propagated down.

These domain-independent CD techniques are complete given the nature of the LAN Designer's design model.

Domain-Dependent Idioms

Domain-dependent CD idioms represent stereotyped techniques domain experts use to check designs. CD idioms make their job easier by requiring merely that they fill in some domain-dependent details, rather than create appropriate CD methods from scratch. While the idioms used probably vary somewhat from domain to domain, we suspect that at least some idioms are likely to have wide applicability. As a result, a library of such idioms may prove useful. The CD idioms found useful in the LAN design domain are described below:

• propagation across connection links: Connection links connect the interfaces of components to each other. The constraints that propagate through a connection depend on the kind of connection. An electrical connection, for example, propagates constraints on voltage and current. A physical connection between rotating axles propagates constraints on rotation speed and torque. For this purpose, the LAN Designer includes a hierarchy of connection types along with a description of what constraints propagate through each type. The knowledge engineer can use an existing connection type if appropriate, or else define a new one as a refinement of an existing connection type. The connection types currently understood by the LAN Designer include the following (Figure 9):

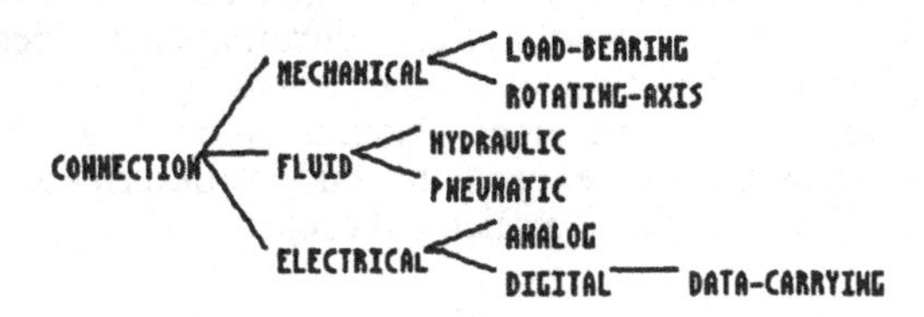

Figure 9. The connection types hierarchy.

• propagation down decomposition links: While the propagation of constraints down decomposition links is domain-dependent, one can often use the mapping between abstract and more specific connections to simplify describing such propagation. For example, an abstract LAN trunk component is usually connected to abstract LAN trunk connector components. When the abstract LAN Trunk is refined to a particular trunk technology, the constraints that propagate over connections to the abstract component's interfaces also apply to the corresponding connections in the specific Trunk instance. Domain-specific knowledge of how connections are mapped over a decomposition step is called "Structural" knowledge in MICON (Birmingham [4]).

• protect interests themes: Protect interests themes monitor the current design state for a state threatening to the interests of the design agent, and make assertions that lead to a conflict being asserted if a threatening design state should ever come to pass. A protect-interest theme works by creating two new goals: (1) an achieve-value goal that must be satisfied for interest to be safe, and (2) a constrain-value goal to find whether the design state satisfies the achieve-value goal.

We have identified two kinds of useful protect-interest CD idioms to date: *check-component* and *check-resource*. The former kind of idiom checks that interests relating to individual components are satisfied. Consider the following simple example of a check-component CD idiom:

```
(theme check-for-adequate-performance
       actions ((create-goal (achieve the bandwidth
                                      of ?lan-trunk
                                      is high))
                (create-goal (constrain the bandwidth of ?lan-trunk))))
```

This theme protects the agent interest of maintaining adequate performance for the LAN Trunk component. To check for violation of this interest, it creates the subgoal of achieving high bandwidth for the LAN trunk, and also creates a goal to see what the actual LAN trunk bandwidth for the design is. The latter goal triggers analyzers designed for this task, eventually leading to an analysis of the LAN trunk bandwidth that may conflict with the original achieve-value subgoal.

Check-resource idioms look for situations where some kind of resource budget is exceeded. Resources in most design situations are limited in some way, and it is useful to create budgets representing limits on how much of a given kind of resource should be utilized for a given portion of the design. If any of these budgets are exceeded, it is an indication that the design may be on the wrong track. This kind of theme works by creating a goal to summarize the total usage of a given resource, as well as a goal that the actual utilization should not exceed the budget described. If resource utilization exceeds the budget, a conflict will occur.

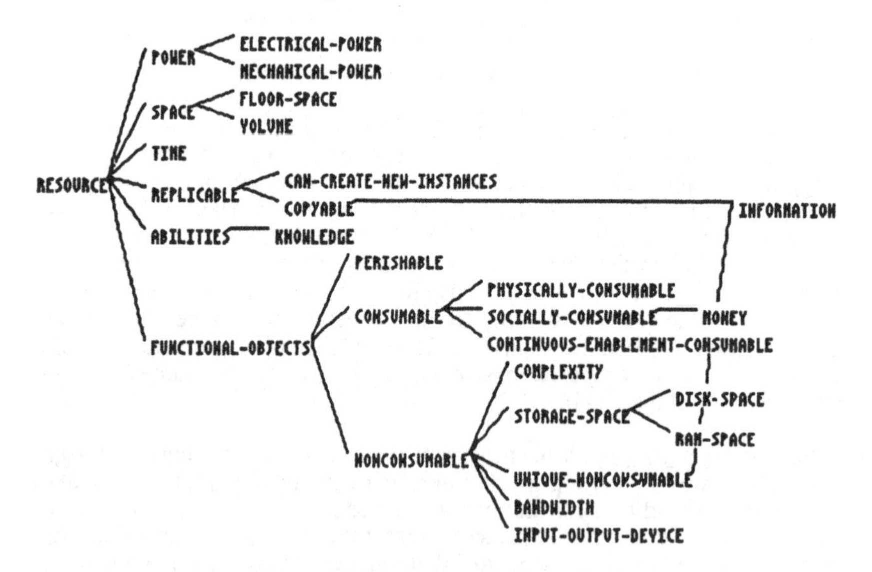

Figure 10. The resource hierarchy.

Resource over-utilization, however, is detected differently depending on the kind of resource involved. For example, a monetary budget is a fixed amount that does not increase with time; excessive money utilization can thus be detected simply by comparing the sum of the individual expenditures with the monetary budget. Checking for space over-utilization is somewhat more complicated; even in the two-dimensional case (i.e. checking for adequate floor space); equipment whose total area does not exceed the available floor space may not fit in a given area due to the particular shapes of the equipment involved. A functional resource that is not used up over time but can only support a finite number of users (e.g. a computer terminal) has resource over-utilization detected in yet a third way. The LAN Designer supports this by maintaining a taxonomy of the different abstract types of resources (Figure 10) and associating resource over-utilization detection templates with each type.

When a knowledge engineer wishes to add the ability to detect over-utilization of a given kind of resource, then, he need merely specify the resource type and limit.

Note that conflict detection idioms can be heuristic in nature. For example, the analyzer rules used by themes that check for threatening design states may perform analysis based on a shallow model. This can be useful if more exhaustive analysis is computationally expensive, but raises the possibility that the conflict may be spurious, resulting from the limitations of the analysis procedure. CR expertise that checks for spurious conflicts caused by shallow analyzer plan models is thus included in the LAN Designer's CR knowledge base.

The methods described above represent a superset of the conflict detection methods utilized in the other approaches to conflict management (e.g. those incorporated in TROPIC, VT, AIR-CYL, MOLGEN or BARGAINER), and provide a complete set of basic mechanisms for conflict detection given the nature of the cooperative group design model used in the LAN Designer. The notion of an abstract resource-type hierarchy has been previously discussed (Wilensky [40]) as has the conflict-detection advantages of using a least-commitment design model (Stefik [36]). Previous work has not discussed, however, the use of domain-independent connection-type and resource hierarchies to support conflict detection, nor the notion of instantiable CD idioms.

Where does the LAN Designer fit into the existing body of design system implementations? Design systems can be divided up into three categories according to their underlying design model, i.e. according to whether they use a Match, Hypothesis and Test, or Least Commitment model.

The *Match* design model is used by XCON and its predecessor R1 [26]. This design model assumes that the appropriate plan for a given design situation can always be found without the need for any search. While this approach was successful in its domain (verifying and elaborating computer equipment orders) it cannot be applied to design domains where the correct decision is not always clear at every decision point.

The *Hypothesize and Test* design model is used in a wide variety of systems, including VT [25], PRIDE [28], AIR-CYL [5], and MAPCon [31]. These systems proceed by making tentative design commitments and exploring the consequences of these commitments, backtracking as necessary when a given design branch is less than satisfactory. These kinds of systems can engage in needless backtracking due to the need to express arbitrary specific design commitments when only a constraint on the form of the solution is known.

The *Least Commitment* approach (also known as the Refinement plus Constraint Propagation approach) is a variant of the hypothesis-and-test approach that has been used widely in such areas as the design of VLSI circuits (Mitchell [27], Tong [39]) genetics experiments (Stefik [36]), alloys (Rychener [32]), buildings (Sriram [35], Chan [7]) single-board computers (Birmingham

[4]) and V-belts (Howard [15]) among others. It avoids the problems faced by the first two approaches and seems to represent the state-of-the-art in terms of AI models of the design process. The LAN Designer's design model fits in this category. This model, in general, is suitable for a wide variety of design tasks but is probably less well-suited for artifacts where components are ill-defined and the interactions among the components are difficult to model as constraints propagating between component interfaces.

IMPROVED DESIGN AGENT MODELS

The demands of cooperative group design imply the need for improved computational models for how individual design agents work, so we can build machine-based design agents that effectively cooperate with each other as well as interfaces that help human design agents do the same. We have identified four major challenges that these improved design agent models should meet:

- Representing Design Rationale: In order to find a suitable CR strategy for a given conflict instance, we have found that design agents must be able to answer questions from the conflict resolution component concerning the reasoning underlying the conflict. This requires that agents have a rich representation of their own design rationale, which in turn requires explicit inspectable representations of domain expertise, e.g. as mathematical equations, qualitative process models, goal/plan/theme graphs, and so on. The LAN Designer represents design rationale using meta-plan structures and design dependencies. *Meta-plan structures* describe the links from top-level goals through subgoals and plans to design actions, as well as theme-based links from design states to the goals they threaten or support (Wilensky [40]). Since meta-level (i.e. control and conflict resolution) decisions can lead to conflicts, meta-versions of the plan structure entities are used to store the rationale behind such meta-level decisions. *Design dependencies* store the design feature values supporting other such values, and are represented using a combination of constraints and truth-maintenance techniques (see de Kleer [18] and Klein [19]).

  Another way in which design rationale is maintained in the LAN Designer system is via typing of resources, interfaces, components and synthesis/analysis features. Each of these parts of a design description have their own taxonomy so that some idea of the nature of these descriptions can be derived by looking at where they fit in the commonly-understood taxonomy for that description type.

- Flexible Redesign: Once the CR component of a design agent has produced a suggestion for resolving a conflict, the domain component has to be able to respond to this suggestion and produce a new, internally consistent design commitment that implements the suggestion. There are four primitive suggestion types used by the CR component to express CR suggestions: try-deeper-model, try-new-plan-for-goal, modify-value and add-component. Try-new-plan-for-goal is implemented simply by running a different plan for a given triggering goal. Add-component suggestions are implemented by creating an add-component goal, which triggers a plan that changes the appropriate parameters to the refinement plan for the component to which another component is to be added. The other two CR primitives are currently handled in a domain-idiosyncratic way.

- Bidirectionality: Due to the asynchronous and unpredictable nature of design assertions made by different agents in a cooperative group design context, individual design agents need to be able to both add commitments to as well as critique existing design commitments, which implies being able to use the same domain expertise in both directions. i.e. for both analysis *and* synthesis. The LAN Designer implements this by essentially duplicating versions of the same expertise as design plans (for commitment-making) and themes (for design-state critiquing).

- Analyzing Indefinite Descriptions: Design agents should be able to operate on indefinite descriptions. While this is often fairly straightforward for the "hypothesize" part of the design cycle, the need to be able to *analyze* abstractly defined components during the "test" part of the design cycle represents a significant challenge. How can commonly-used techniques such as finite-element analysis, for example, be applied to indefinitely described designs, if at all? The LAN Designer currently uses domain-idiosyncratic techniques for this problem.

These challenges seem to imply a central role for design plans in the design process. In the LAN Designer, plans represents the smallest unit that the CR component asks questions about. Plans are also the smallest unit that can make consistent additions to the design, and should be able to change in response to suggestions to produce different self-consistent design additions. This suggests, in general, that plans take on a rich internal structure, acting more as small design "agents" rather than simple static structures.

Our work on the LAN Designer has focused more on the demands of cooperation rather than on perfecting single-agent design models, and accordingly raises more questions than answers regarding how to meet the challenges described above. There is a sizable body of relevant existing research in this area, however. Promising work on representing design rationale is described in Kellog [17] and Batali [3]. Redesign is an active research area; see for example Acosta [1], Brown [5], Simoudis [33], Mostow [29] and Howard [15]. The view of design as the interaction of small agent-like plans has been explored in Brown [6].

CONCLUSIONS

Our experience with implementing the LAN Designer system for cooperative group design of local area networks has led to a number of insights into how cooperative group design systems can be structured to effectively support this process. In particular, we have found that:

- support for cooperative conflict detection and resolution appears to be critical for real-world cooperative group design tasks
- using general heuristic conflict resolution expertise represented distinctly is an effective approach to conflict resolution
- conflict management (in particular, conflict avoidance and detection) can be supported by domain-independent methods and idioms built on a least-commitment design model
- future progress in supporting cooperation between machine-based and human design agents requires computational design models that can

adequately meet some new challenges such as self-understanding (i.e. rich design rationale) and self-modifiability (i.e. a powerful replanning facility) as well as the ability to analyze and critique indefinitely described designs

The LAN Designer system is an instantiation of a generic cooperative group design collaboration support shell we have developed. We are currently extending this shell to support the interaction of both human and machine-based design agents, and to learn from the process of doing so. We plan to instantiate the shell in a number of different cooperative design and planning domains in order to evaluate and enhance its effectiveness and breadth of applicability.

ACKNOWLEDGEMENTS

The assistance of Profs. A.B. Baskin and R.E. Stepp at the University of Illinois as well as Drs. H. Motoda and A. Sakurai at the Hitachi Advanced Research Lab is gratefully acknowledged.

REFERENCES

1. Acosta, R.D., Huhns, M.N., and Liuh, S.L. Analogical Reasoning for Digital System Synthesis, Proceedings ICED, pp. 173-176, November 1986.

2. Balzer, R., Erman, L., London, P., and Williams, C. Hearsay-II: A Domain-Independent Framework For Expert Systems, Proceedings of the First Annual National Conference on AI, pp. 108-110, 1980.

3. Batali, J. Dependency Maintenance In The Design Process, IEEE Int Conf Computer Design: VLSI in Computers, pp. 459-462, 1983.

4. Birmingham, W.P. and Siewiorek, D.P. Automated Knowledge Acquisition for a Computer Hardware Synthesis System, Knowledge Acquisition, Vol. 1, pp. 321-340, 1989.

5. Brown, D.C. Failure Handling In A Design Expert System, (Ed. Gero, J.S.), Butterworth and Co., November 1985.

6. Brown, D.C. and Chandrasekaran, B. Expert Systems For A Class Of Mechanical Design Activity, Proc of IFIP WG5.2 Working Conference on Knowledge Representation in Computer Aided Design, 1984.

7. Chan, W.T. and Paulson, B.C. Exploratory Design Using Constraints, AI EDAM, Vol. 1, No. 1, pp. 59-71, 1987.

8. Clancey, W.J. Classification Problem Solving, AAAI, pp. 49-55, 1984.

9. Coombs, C.H. and Avrunin, G.S. The Structure of Conflict, Lawrence Erlbaum Associates, 1988.

10. Descotte, Y. and Latombe, J.C. Making Compromises Among Antagonist Constraints In A Planner, Artificial Intelligence, Vol. 27, pp. 183-217, 1985.

11. Fox, M.S., Allen, B., and Strohm, G. Job-Shop Scheduling: An Investigation In Constraint-Directed Reasoning, AAAI-82, pp. 155-158, 1982.

12. Fox, M.S. and Smith, S.F. Isis - A Knowledge-Based System For Factory Scheduling, Expert Systems, 1984.

13. Goldstein, I.P. Bargaining Between Goals. Tech. Report Massachusetts Institute of Technology Artificial Intelligence Laboratory, 1975.

14. Hewitt, C. Offices Are Open Systems, ACM Transactions on Office Information Systems, Vol. 4, No. 3, pp. 271-287, 1986.

15. Howard, A.E., Cohen, P.R., Dixon, J.R., and Simmons, M.K. Dominic: A Domain-Independent Program For Mechanical Engineering Design, Artificial Intelligence, Vol. 1, No. 1, pp. 23-28, 1986.

16. Johansen, J., Vallee, V., and Springer, S. Electronic Meetings: Technical Alternatives and Social Choices, Addison-Wesley, 1979.

17. Kellog, C., Jr., R.A.G., Mark, W., McGuire, J.G., Pontecorvo, M., Sclossberg, J.L., and , J.W.S. The Acquisition, Verification and Explanation of Design Knowledge, SIGART Newsletter, No. 108, pp. 163-165, 1989.

18. de Kleer, J.D. An Assumption-Based TMS, Artificial intelligence, Vol. 28, pp. 127-162, 1986.

19. Klein, M. Conflict Resolution in Cooperative Design, Ph.D. dissertation, University of Illinois at Urbana-Champaign, January 1990.

20. Klein, M. and Lu, S.C.Y. Conflict Resolution in Cooperative Design, International Journal for Artificial Intelligence in Engineering, 1990.

21. Klein, M. A Computational Model of Conflict Resolution in Integrated Design, Proceedings of the ASME Symposium on Integrated Product Design and Manufacturing , November 1990.

22. Lenat, D.B. Beings: Knowledge As Interacting Experts, IJCAI-75, pp. 126-133, 1975.

23. Lu, S.C.Y., Subramanyam, S., Thompson, J.B., and Klein, M. A Cooperative Product Development Environment To Realize The Simultaneous Engineering Concept, Proceedings of the 1989 ASME Computers in Engineering Conference , Anaheim, CA, July 1989.

24. Malone, T.W., Fikes, R.E., and Howard, M.T. Enterprise: A Market-Like Task Scheduler For Distributed Computing Environments. Tech. Report Cognitive and Instructional Sciences Group, Xerox Palo Alto Research Center, October 1983.

25. Marcus, S., Stout, J., and McDermott, J. VT: An Expert Elevator Designer, Artificial Intelligence Magazine, Vol. 8, No. 4, pp. 39-58, 1987.

26. McDermott, J. R1: A Rule-Based Configurer Of Computer Systems, Artificial Intelligence, Vol. 19, pp. 39-88, 1982.

27. Mitchell, T.M., Mahadevan, S., and Steinberg, L.I. LEAP: A Learning Apprentice For VLSI Design, Proceedings of .IJCAI , pp. 573-580, 1985.

28. Mittal, S. and Araya, A. A Knowledge-Based Framework For Design, AAAI, Vol. 2, pp. 856-865, 1986.

29. Mostow, J. and Barley, M. Automated Reuse of Design Plans, Proceedings ICED , pp. 632-647, August 1987.

30. Nunamaker, J.F., Applegate, A., and Konsynski, K. Facilitating Group Creativity: Experience with a Group Decision Support System, Proceedings of the Twentieth Hawaii International Conference on System Sciences , pp. 422-430, 1987.

31. Parunak, H.V.D., Kindrick, J.D., and Muralidhar, K.H. Mapcon: A Case Study In A Configuration Expert System, Artificial Intelligence EDAM, Vol. 2, No. 2, pp. 71-88, 1988.

32. Rychener, M.D., Farinacci, M.L., Hulthage, I., and Fox, M.S. Integration Of Multiple Knowledge Sources In Aladin, An Alloy Design System, AAAI, Vol. 2, pp. 878-882, 1986.

33. Simoudis, E. Learning Redesign Knowledge, Banff Knowledge Acquisition for Knowledge-Based Systems Workshop , November 1988.

34. Smith, R.G. and Davis, R. Cooperation In Distributed Problem Solving, IEEE Proceedings of the International Conference on Cybernetics and Society, pp. 366-371, 1979.

35. Sriram, D. All-Rise: A Case Study Of Constraint-Based Design, Artificial Intelligence in Engineering, Vol. 2, No. 4, pp. 186-203, 1987.

36. Stefik, M.J. Planning With Constraints (Molgen: Part 1 & 2), Artificial Intelligence, Vol. 16, No. 2, pp. 111-170, 1981.

37. Sussman, G.J. A Computational Model Of Skill Acquistion. Tech. Report PhD Thesis. AI Lab, MIT, 1973., 1973.

38. Sussman, G.J. and Steele, G.L. Constraints - A Language For Expressing Almost-Hierachical Descriptions, Artificial Intelligence, Vol. 14, pp. 1-40, 1980.

39. Tong, C. AI In Engineering Design, Artificial Intelligence in Engineering, Vol. 2, No. 3, pp. 130-166, 1987.

40. Wilensky, R. Planning And Understanding, Addison-Wesley, 1983.

Decomposition of Design Activities

A. Kusiak, J. Wang

Intelligent Systems Laboratory, Department of Industrial Engineering, College of Engineering, The University of Iowa, Iowa City, IA 52242, USA

ABSTRACT

Concurrent design may result in reduction of the duration of a design project, cost savings, and better quality of the final design, however, it may increase the complexity of the design process and make it more difficult to manage. One way to reduce the complexity of a large scale design project is to decompose it into subsystems. In this paper, three types of decomposition are discussed:

- decomposition of module (component) - activity matrix
- decomposition of procedure (formula) - parameter (variable) matrix
- decomposition of activity (variable) - activity (variable) matrix.

The concepts presented in the paper are illustrated with examples.

INTRODUCTION

Design of products has been an important task in business for many years. Due to decreasing life cycle of products, it is important to reduce the time and cost of product development. Thus, product design has become a focus of competition in many industries. A concept of concurrent design has emerged in recent years. It attempts to incorporate various constraints related to the product life cycle, i.e., manufacturability, quality, reliability, and so on, in the early design stages. Concurrent design aims at the improvement of the product quality, reducing the development time and cost. However, due to interactions between the various facets of design, the complexity of the design process increases and makes the process more difficult to manage. To simplify the design problem, one needs to decompose an overall design task into groups of activities or modules.

In this paper, the following three types of incidence matrices are considered:

- module (component) - activity matrix
- procedure (formula) - parameter (variable) matrix
- activity (variable) - activity (variable) matrix

A non-empty element in an incidence matrix represents a relationship between the corresponding activities (row and column). The objective is to organize the incidence matrix so that the design task can be simplified. The organized matrices can be categorized as follows (see Figure 1): uncoupled matrix, decoupled matrix, and coupled matrix. An incidence matrix is uncoupled if its rows and columns can be ordered in such a way that the matrix separates into mutually exclusive submatrices (see the matrix in Figure 1(a)). Analogously, an incidence matrix is decoupled if it can be rearranged in a triangular form (see the matrix in Figure 1(b)). An incidence matrix in Figure 1(c) is coupled if it is not decomposable.

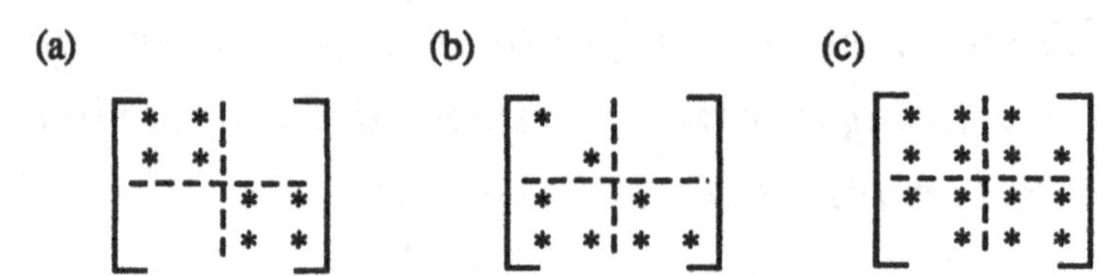

Figure 1. Three types of matrices
(a) uncoupled matrix
(b) decoupled matrix
(c) coupled matrix

Since the design activities associated with the upper left corner submatrix in Figure 1(a) are independent of the activities corresponding to the lower right corner submatrix, they can be performed simultaneously. The activities in the decoupled matrix (eg., Figure 1(b)) are dependent so that they can be performed partially in parallel and in part sequentially. The degree to which some tasks can be performed in parallel depends on the sparsity of the matrix. In the coupled matrix, the activities are strongly interdependent which may occur in concurrent engineering. They may reflect an iterative nature of design or a negotiation process.

Decomposition of design activities simplifies the design task and may reduce the design cycle. The challenge is to decompose the system of design activities into subsystems that are of acceptable size, can be easily solved, and the overall solution is consistent with the partial solutions.

Rather than building an incidence matrix for design activities one could consider modules (eg., systems, components), procedures, or formulas. In this paper, three types of decomposition in the product design environment are discussed. Module (component) - activity decomposition is presented in Section 2. Procedure (formula) - parameter (variable) decomposition is discussed in Section 3. Activity (variable) - activity (variable) decomposition is presented in Section 4. Conclusions are presented in Section 5.

DECOMPOSITION OF MODULE-ACTIVITY MATRIX

A product or system (mechanical, electrical, etc.) can be decomposed into subsystems and these in turn into modules or components. Design of each module involves a set of design activities. Similar or identical activities may be performed in the design of different modules.

To define modules, consider design of a vehicle (Kusiak and Park [4]) that can be decomposed into subsystems as shown in Figure 2. Each subsystem is further decomposed into new subsystems and modules. For example, the carriage unit can be decomposed into modules as presented in Figure 3. A large number of design activities and procedures are involved in design of each module. Therefore, many activities need to be scheduled efficiently in a product design project. Reducing the complexity and development time of a design task is an important issue.

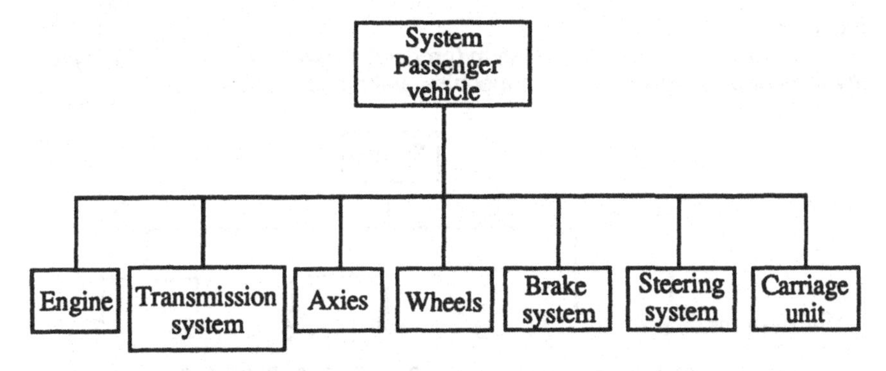

Figure 2. The passenger vehicle and its subsystems

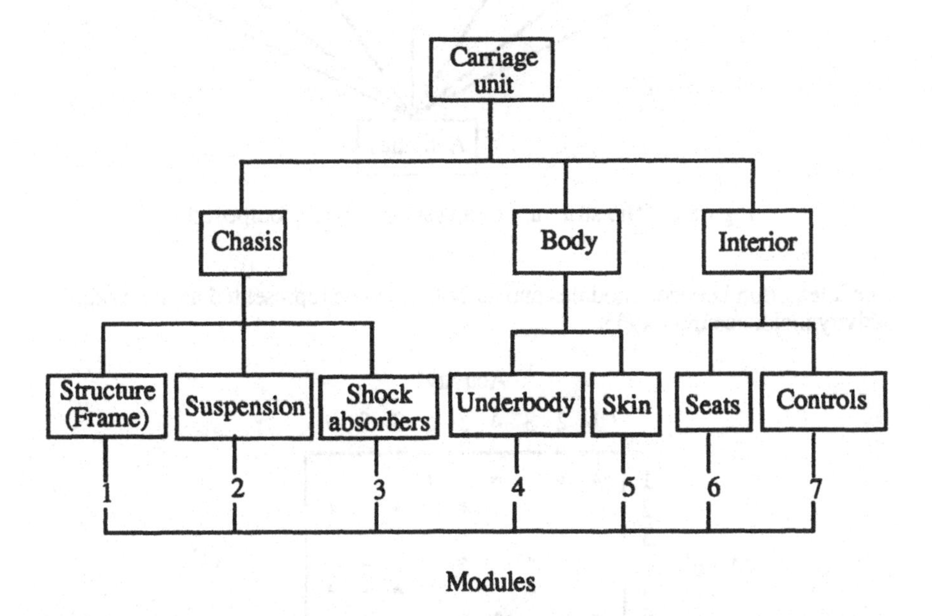

Figure 3. The carriage unit and its decomposition into subsystems and modules

Clustering of activities involved in the design process allows one to determine a potential group of activities that might be scheduled simultaneously. The clustering problem to be solved is to maximize the number of mutually separable clusters in the module-activity incidence matrix, subject to constraints, for example, limited size of some or all clusters and inclusion of specific activities in some clusters.

There are several clustering algorithms existing in the literature. In this section, the clustering algorithm presented by Kusiak and Cheng [3] is employed to decompose the module-activity incidence matrix. An example of system decomposition is shown next.

Example 1

Consider a hypothetical system (product) shown in Figure 4. This system involves three subsystems, seven modules, and nine activities.

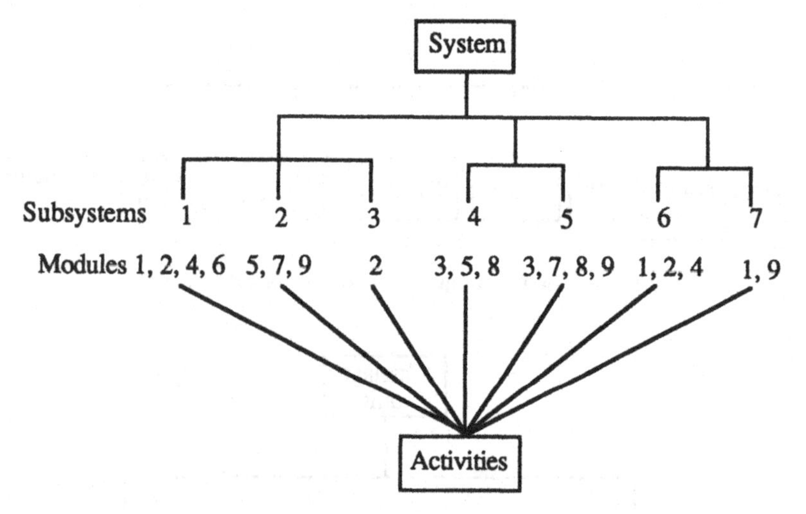

Figure 4. The structure of a system to be decomposed

The interaction between modules and activities can be represented as the module-activity incidence matrix (1):

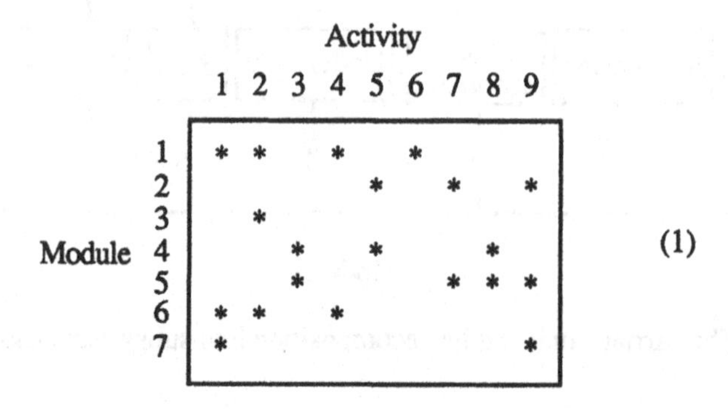

Module \ Activity	1	2	3	4	5	6	7	8	9
1	*	*		*		*			
2					*		*		*
3		*							
4			*		*			*	
5			*				*	*	*
6	*	*		*					
7	*								*

(1)

Solving the problem represented in matrix (1) with the clustering algorithm (Kusiak and Cheng [3]) results in matrix (2).

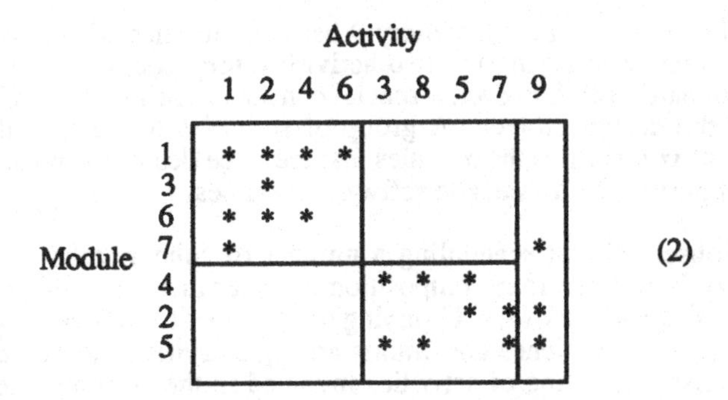

Two groups of modules GM-1 = {1, 3, 6, 7}, GM-2 = {4, 2, 5}, and two groups of activities GA-1 = {1, 2, 4, 6}, GA-2 = {3, 8, 5, 7} are visible in matrix (2). Activity 9 overlaps with the two modules obtained. Another solution generated by the clustering algorithm is presented in matrix (3), if the size of each cluster are limited to three.

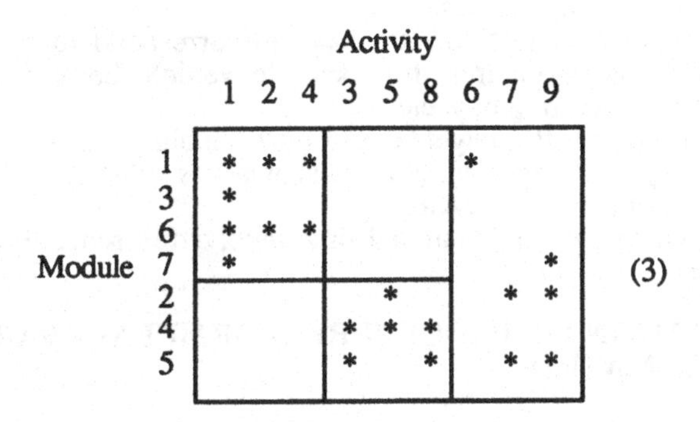

Two groups of modules GM-1 = {1, 3, 6, 7}, GM-2 = {2, 4, 5} and two groups of activities GA-1 = {1, 2, 4}, GA-2 = {3, 5, 8} are visible in matrix (3). Activities 6, 7, and 9 overlap with the two groups of modules. Introducing additional resources for performing the activity 9 allows decomposition of matrix (2) into two mutually separable submatrices shown in (4).

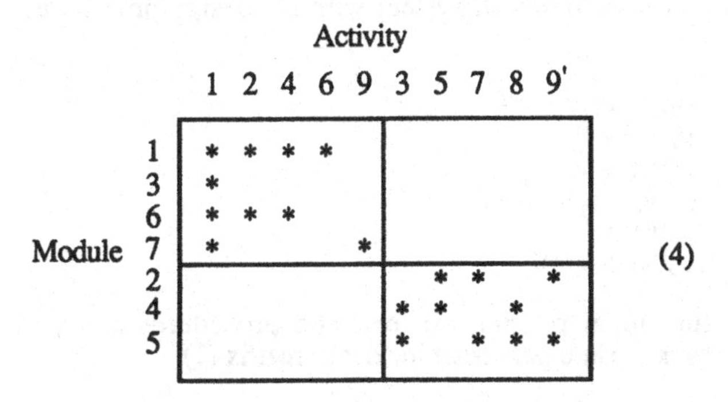

As shown in matrix (9), activity 9 has been duplicated (activities 9 and 9'). Rather than multiplying the two activities, they could have been left as presented in matrix (2); however, a special consideration should be given while scheduling the design project. A group of activities in the module-activity incidence matrix identifies the modules that are to be designed and analyzed by the required human, hardware, and software resources.

It is observed that scheduling a number of subnetworks with a small number of activities due to decomposition is easier than scheduling the entire network (Kusiak and Park [4]). Grouping of activities simplifies the process of generation of the precedence constraints among activities and duration of the activities. Also, clustering of activities involved in the design process allows one to determine a potential group of activities that might be scheduled in parallel.

Application of the decomposition concept to the design problem has the following advantages:

1. Separation of the overall design task into groups of modules (components) and activities.
2. The group of modules do not have to correspond to the traditional organizational structures, for example vehicle body design group, transmission design group, etc.
3. Potential activities that might be performed simultaneously are detected.
4. Complexity of management of the design task is reduced.
5. Reduction of the design cycle.
6. Reduction of the computational time involved in scheduling of design activities.

DECOMPOSITION OF PROCEDURE (FORMULA) - PARAMETER (VARIABLE) MATRIX

Complex designs may involve a number of design procedures (formulas) and parameters (variables). In most situations, design procedures exhibit some degree of coupling which tends to complicate the design process and tools used. To simplify the complex design task, the coupling procedures have to be recognized and grouped together into clusters.

Consider a hypothetical system with six design procedures and seven parameters:

$$p_1(u_2, u_5, u_7) = 0$$
$$p_2(u_1) = 0$$
$$p_3(u_3) = 0$$
$$p_4(u_2, u_5, u_7) = 0$$
$$p_5(u_4) = 0$$
$$p_6(u_4, u_6) = 0$$

The functional relationship between procedures and parameters is represented as procedure-parameter incidence matrix (1).

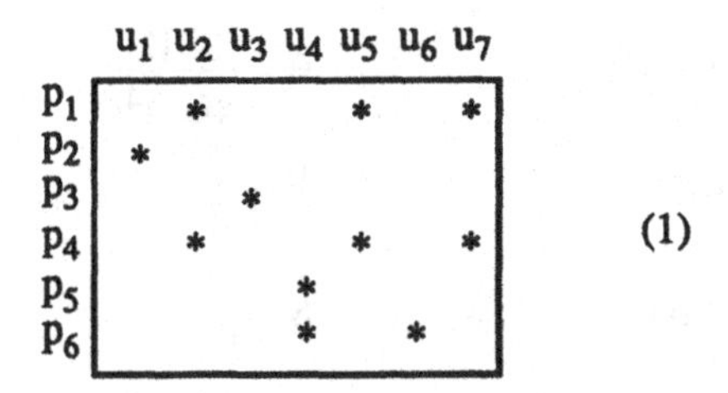

(1)

Matrix (1) is nonstructured and it is difficult to recognize any groupings among procedures and parameters. The problem to be solved is to cluster the incidence matrix into groups of procedures and parameters.

Solving the problem represented in matrix (1) with the clustering algorithm presented in Kusiak and Cheng [3] results in matrix (2):

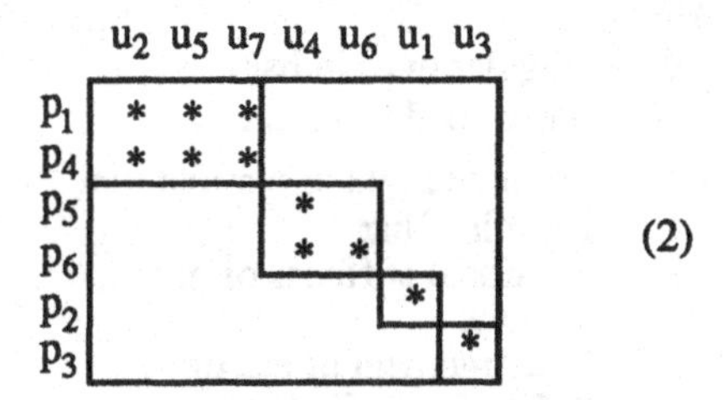

(2)

Four groups of procedures GP-1 = {1, 4}, GP-2 = {5, 6}, GP-3 = {2}, and GP-4 = {3}, which are completely independent of each other are visible in matrix (2). The procedures grouped can be applied in parallel.

Example 2

Consider design of the torsion bar spring in Figure 5.

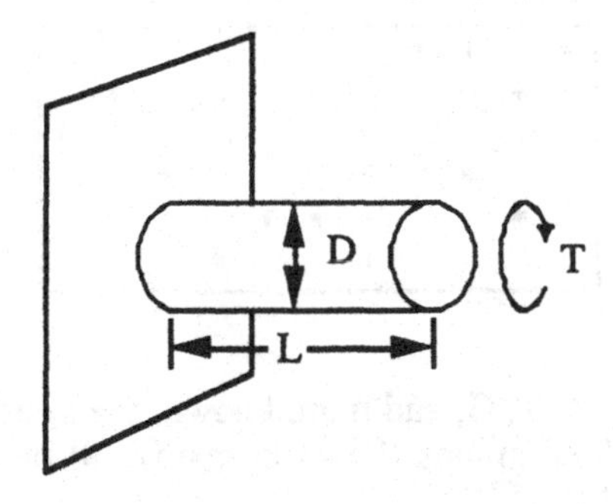

Figure 5. Example of a torsion bar spring

The equations pertaining to design of the spring are as follows (Srinivasan et al. [9]):

f_1: Twist angle $\theta = \dfrac{32TL}{\pi G D^4}$

f_2: Stiffness $K = \dfrac{T}{\theta}$

f_3: Stress rate $S = \dfrac{\tau}{\theta}$

f_4: Volume $V = \dfrac{\pi D^2 L}{4}$

f_5: Stress $\tau = \dfrac{16T}{\pi D^3} n$

f_6: Polar moment of inertia $J = \dfrac{\pi D^4}{32}$

where:

θ : Angular twist in bar
G : Shear module of bar
τ : Shear strength of bar material
L : Length of bar
K : Torsional stiffness of bar
V : Volume
J : Polar moment of inertia
n : Safety factor for bar
D : Diameter of bar
S : Stress rate in bar
T : Torque on bar

The system can be represented with formula-variable incidence matrix (7):

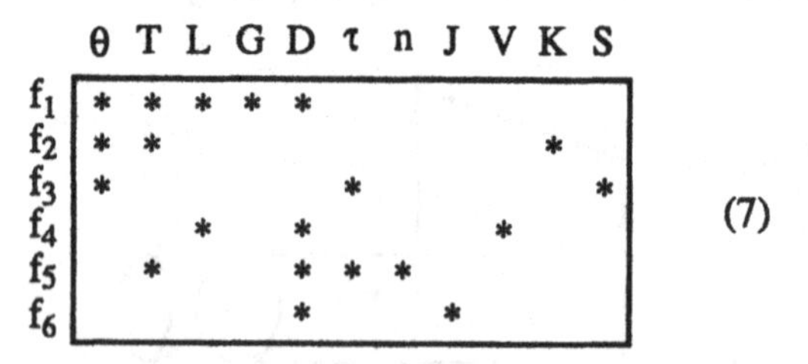

	θ	T	L	G	D	τ	n	J	V	K	S
f_1	*	*	*	*	*						
f_2	*	*								*	
f_3	*					*					*
f_4			*		*				*		
f_5		*			*	*	*				
f_6					*			*			

(7)

If variable θ, τ, T, G, and n are known, the system can be transformed into independent three groups shown in matrix (8) by a clustering algorithm (eg., Kusiak and Cheng [3]).

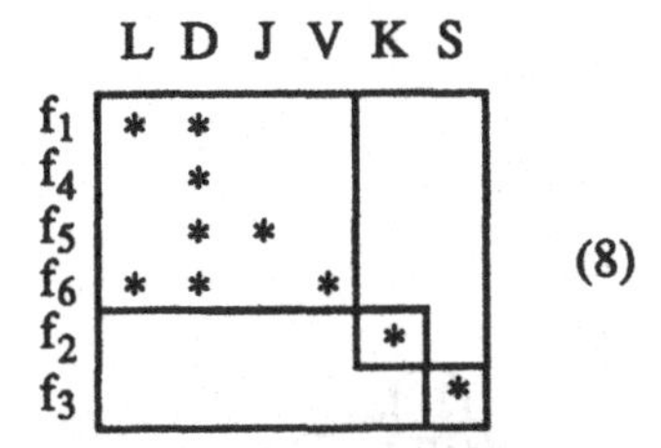

As shown in matrix (8), equations f_1, f_4, f_5, and f_6 are coupled.

The advantages from the procedure (formula) - parameter (variable) decomposition are similar to the module-activity decomposition. It also simplifies the complex design project and reduces the product development time.

DECOMPOSITION OF ACTIVITY (VARIABLE) - ACTIVITY (VARIABLE) MATRIX

In this section, an activity-activity incidence matrix and the corresponding directed graph (digraph) are used to represent design activities and the relationship among them (Warfield [11], Steward [10]). The triangularization algorithm presented in Kusiak and Wang [5] is employed to decompose the activity-activity incidence matrix. The triangularization algorithm decomposes an overall design task into groups of activities, simplifies the entire design process, and sequences all design activities to speed up the product development time. In order to reduce the number of cycles in the design process, the algorithm minimizes the number of non-empty elements in the upper triangular incidence matrix. Two examples illustrating the decomposition of activities are are shown next.

Example 3

Consider a digraph of design activities and the corresponding activity-activity incidence matrix presented in Figure 6.

Solving the problem represented in incidence matrix in Figure 6 with the triangularization algorithm presented in Kusiak and Wang [5] results in the matrix in Figure 7.

The triangularization algorithm has produces the following sequence of activities (2, 3, 1, 11, G_1, G_2, 7). The matrix in Figure 7 shows that activity 2 has to be performed prior to activity 3. Activities 1 and 3 are independent so that they can be accomplished in parallel. Also, there are two groups of coupled activities G_1 and G_2. In order to minimize the number of cycles, the activities in group G_1 should be performed in the sequence (12, 6, 10, 9).

There are a number of ways of decoupling design activities. For example, in order to decouple the activities included in group G_2, one could introduce a new activity 13. The new activity might remove the precedence relationship (8, 5), i.e., element (8,5) in the incidence matrix. The resulting matrix is shown in Figure 8. The decoupling process may decrease the duration of the design project, however, some additional information about the activities is required.

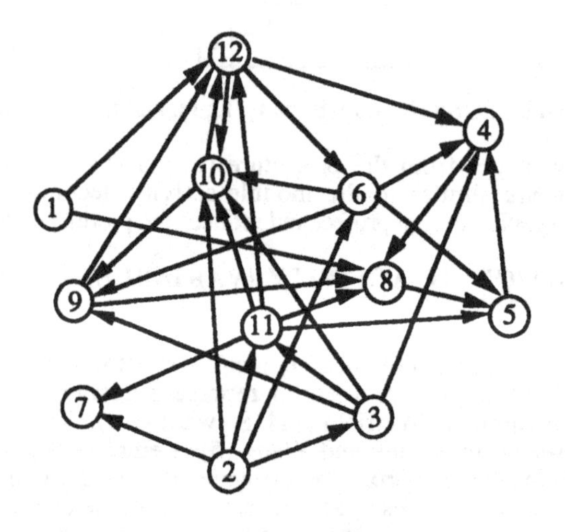

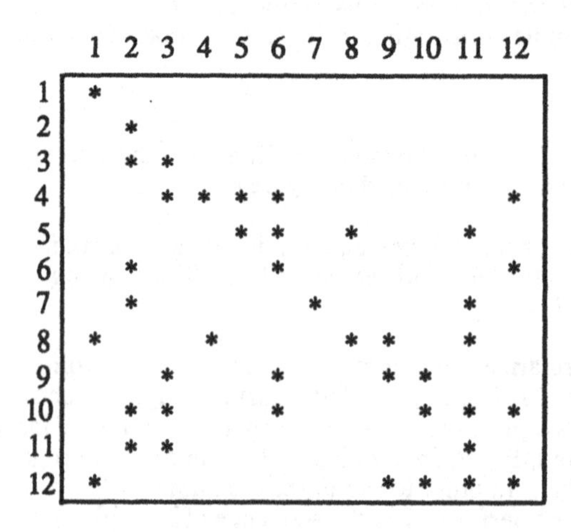

	1	2	3	4	5	6	7	8	9	10	11	12
1	*											
2		*										
3		*	*									
4			*	*	*	*						*
5					*	*		*			*	
6		*				*						*
7		*					*				*	
8	*			*				*	*		*	
9			*			*			*	*		
10		*	*			*				*	*	*
11		*	*								*	
12	*								*	*	*	*

Figure 6. A digraph and the corresponding incidence matrix with 12 activities

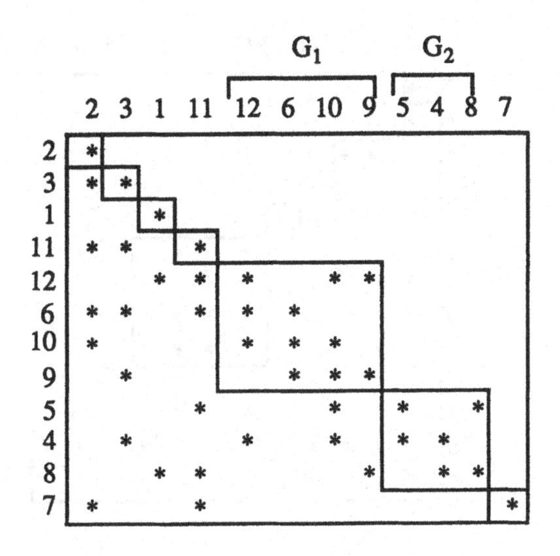

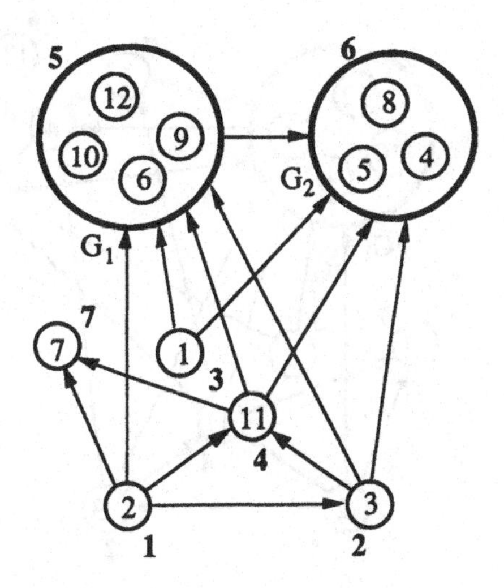

Figure 7. The condensation digraph and the corresponding reordered incidence matrix

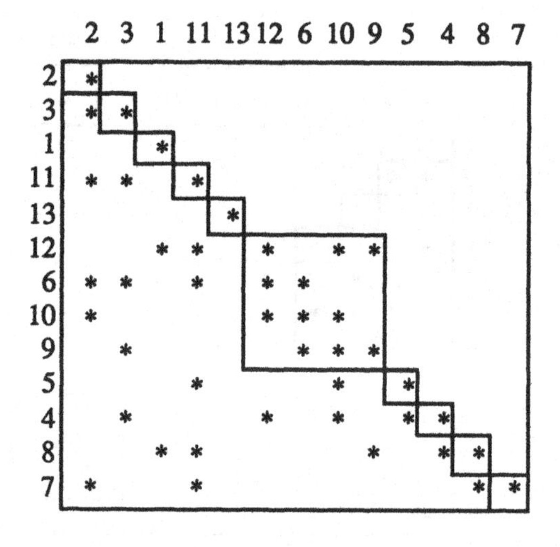

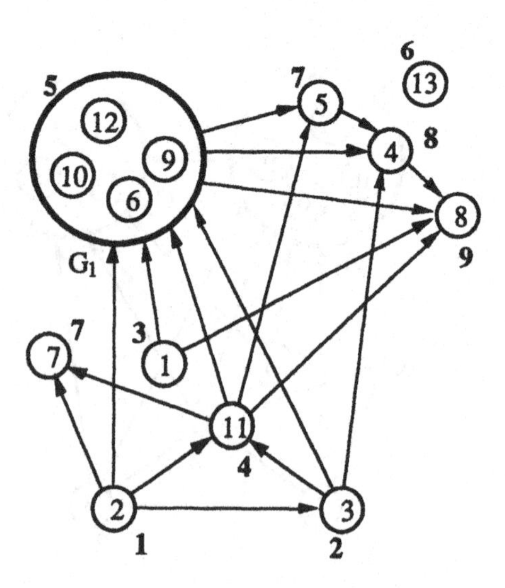

Figure 8. Digraph and the corresponding matrix with an additional activity

Example 4

Consider design of the cantilever beam in Figure 9.

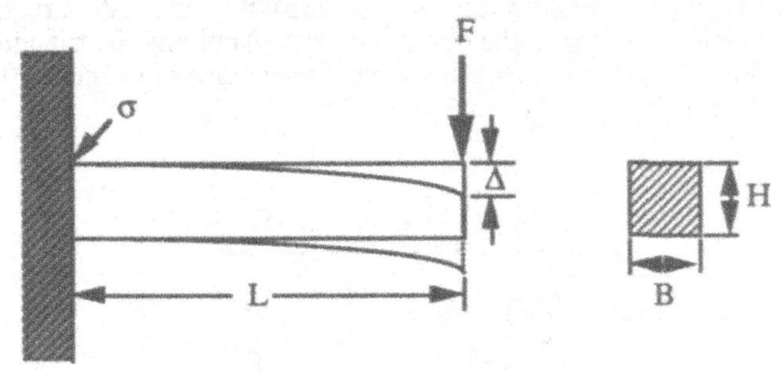

Figure 9. Cantilever beam

The equations pertaining to design of the cantilever beam are as follows (Shigley and Mischke [8]):

f_1	$\sigma = \frac{MY}{I}$	=>	$\sigma = \sigma(M, Y, I)$
f_2	$M = FL$	=>	$M = M(F, L)$
f_3	$I = \frac{BH^3}{12}$	=>	$I = I(B, H)$
f_4	$Y = \frac{H}{2}$	=>	$Y = Y(H)$
f_5	$\Delta = \frac{FL^3}{3EI}$	=>	$\Delta = \Delta(F, E, I, L)$

where:

- σ : the maximum bending stress occurring at the section closest to the support
- M : bending moment
- L : length of beam
- H, Y : geometry of beam
- F : applied force
- Δ : deflection parameter
- E : modulus of elasticity
- I : second moment of area about the neutral axis

Using the design equations above, a designer can define variables of interest to obtain the design required. In this example, F, E, H, B, and σ are designated as the set of known variables. The set of unknown variables is M, I, Y, L, and Δ. In order to evaluate the set of design equations, the above equations should be transformed so that the known variables appear on the right-hand side. The transformed equations are as follows:

$$M = M(\sigma, Y, I) \qquad Y = Y(H) \qquad I = I(B, H)$$
$$L = L(M, F) \qquad \Delta = \Delta(F, L, E, I)$$

A set of equations can be represented with an incidence matrix in which the non-empty elements of the incidence matrix represent the relationship between variables (Figure 10). For example, in order to evaluate the bending moment M, variables σ, Y, and I have to be obtained. If the design variables could be sequenced so that each would be able to receive all the information required, then there would be no coupling in the design process. Applying the triangularization algorithm (Kusiak and Wang [5]) to the incidence matrix in Figure 10 results in the matrix shown in Figure 11.

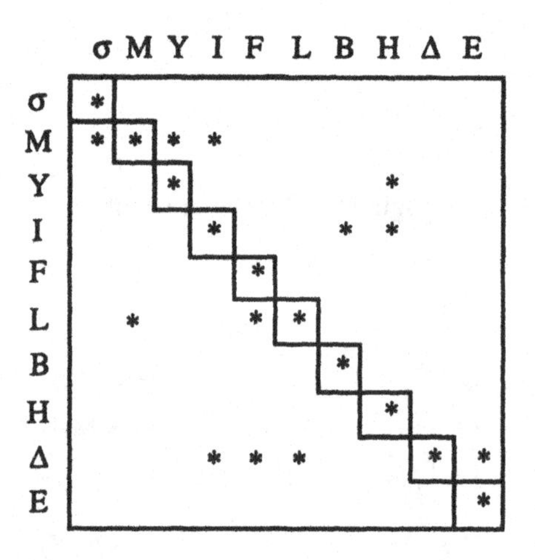

Figure 10. The incidence matrix corresponding to the set of transformed design equations

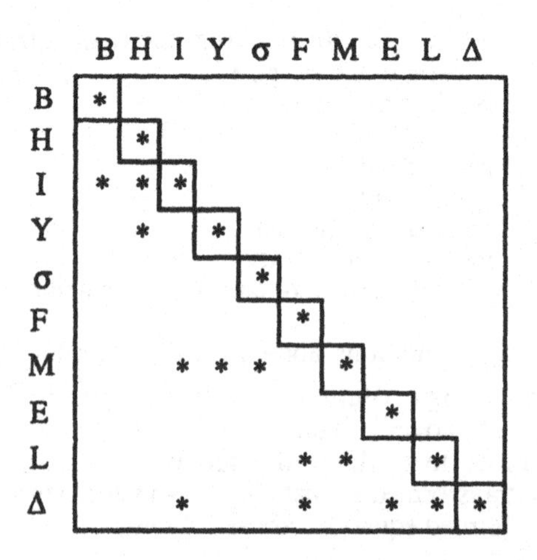

Figure 11. The ordered matrix from Figure 10

An alternative method that evaluates a set of design formulas, is to assign (match) unknown variables to formulas (Serrano and Gossard [7]). The assignment problem has been studied extensively in the literature (Even [1],

Sedgewick [6]). For a given set of design equations more than one assignment might be possible. Figure 12 shows one of the possible assignments which can be converted to a digraph where the square nodes denote the known variables.

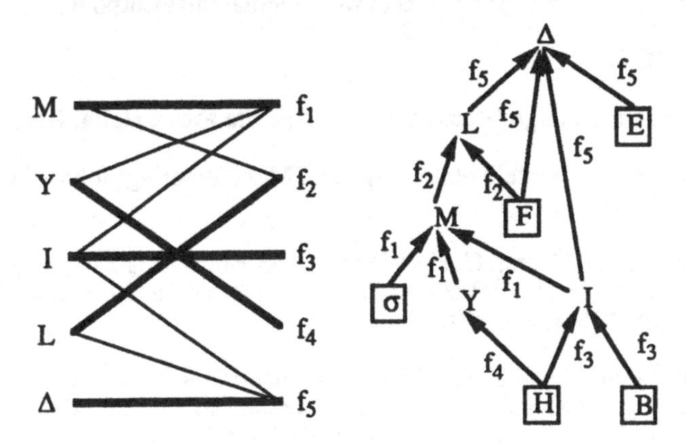

Figure 12. Bipartite graph and the corresponding digraph

Concurrent engineering may increase the number of coupled design activities. Coordinating the uncoupled or decoupled activities is quite straightforward. However, it is much more challenging to organize the coupled activities.

The triangularization algorithm (Kusiak and Wang [5]) employed in this section identifies uncoupled and decoupled activities as well as those to be performed in groups with the minimum number of cycles. The algorithm can significantly reduce the overall project complexity and product development time. For details of other approaches to concurrent engineering, see Kusiak [2].

CONCLUSIONS

In concurrent engineering, an attempt is made to perform the design activities simultaneously rather than in the series as in the case of traditional design. This results in reduction of the duration of the design project, cost savings, and better quality of the final design. However, the concurrent strategy might increase the complexity of the design process and make it more difficult to manage. One way to reduce the complexity of a large scale design project is to decompose it into subsystems.

In this paper, three types of decomposition were discussed:

- decomposition of module (component) - activity matrix
- decomposition of procedure (formula) - parameter (variable) matrix
- decomposition of activity (variable) - activity (variable) matrix.

Two algorithms (Kusiak and Cheng [3], Kusiak and Wang [5]) were applied for decomposition and simplification of the design process. It is observed that the decomposition not only reduces the complexity of the overall design project, but also speeds up the product development.

ACKNOWLEDGEMENT

This research has been partially supported by grants for the National Science Foundation (DDM-9007158) and Rockwell International Corporation.

REFERENCES

1. Even, S. Graph Algorithms, Technion Institite Press, Haifa, Israel, 1979.

2. Kusiak, A. (Ed.). Intelligent Design and Manufacturing, John Wiley, New York, 1992.

3. Kusiak, A. and Cheng, C. H. A Branch-and-Bound Algorithm for Solving the Group Technology Problem, Annals of Operations Research, Vol. 26, No. 4, pp. 415-431, 1990.

4. Kusiak, A. and Park, K. Concurrent Engineering: Decomposition and Scheduling of Design Activities, International Journal of Production Research, Vol. 28, No. 10, pp. 1883-1900, 1990.

5. Kusiak, A. and Wang, J. An Efficient Algorithm for Organizing of Design Activities, Working Paper #91-31, Department of Industrial Engineering, The University of Iowa, Iowa City, IA, 1991.

6. Sedgewick, R. Algorithms, Addison-Wesley, Reading, Massachusetts, 1984.

7. Serrano, D. and Gossard, D. Constraint Management in Conceptual Design, Knowledge Based Expert Systems in Engineering: Planning and Design, (Ed. Sriram, D. and Adey, R. A.), pp. 211-224, Elsevier, Amsterdam, 1990.

8. Shigley, J. E. and Mischke, C. R. Standard Handbook of Machine Design, McGraw-Hill, New York, 1986.

9. Srinivasan, R. V., Agrawal, R., and Kinzel, G. L. Design Shell: A Framework for Interactive Parametric Design, (Ed. Ravani , B.), pp. 289-295, Proceedings of the 1990 ASME Design Technical Conferences - 16th Advances in Design Automation, Chicago, IL, 1990.

10. Steward, D. V. Systems Analysis and Management: Structure, Strategy, and Design, Petrocelli Books, New York, 1981.

11. Warfield, J. N. Binary Matrices in System Modeling, IEEE Transactions on Systems, Man, and Cybernetics, Vol. SMC-3, No. 5, pp. 441-449, 1973.

Mixpert: A Maintainable Configurer's Assistant

C.E. Johnson, M.D. Cope

Pervasive Technology Department, Shell Research Ltd., Thornton Research Centre, P.O. Box 1, Chester, CH1 3SH, UK

ABSTRACT

Configuration of systems requires their components to be selected and arranged according to a schema of some kind. This task demands a lot of skill and experience since typically many components require setting up and dependencies exist between them. To assist the configurer's task, computer systems have been developed. These advise on the setting of new parameter values and check that a configuration is complete and consistent, possibly flagging non-optimal parameter settings and suggesting more suitable values.

Complex systems may have many possible configurations, a combinatorial explosion being likely to result if one tried to describe them all. It is for this reason that production rules rather than procedural languages have usually been used to represent configuration systems [1]. A problem with the rule-based approach is that the systems are difficult to maintain, a team of computer specialists being required to do this. Since systems requiring configuration frequently change, maintenance of configuration systems is a big issue and one which is addressed by the prototype system described in this paper.

Within the Shell Group of companies, computer systems have been developed to support its worldwide operations. A prototype computer-based configurer's assistant, which has generic application, is described; it has been developed to assist in the configuration of one such system. The prototype, called Mixpert, can be updated by configurers, who are not programmers, to coincide with upgrades to the system to be configured, knowledge about early versions of the system being retained if required. In this way, Mixpert can process configuration settings for operating companies running different versions of the system. Within the prototype, an integrated set of tools is provided, bringing together disparate sources of information for use in defining and maintaining a configuration's parameters. The system has been developed in the object-oriented programming language Smalltalk/V286 and runs on a PC.

1. INTRODUCTION

Computer systems have been developed in the Shell Group to address problems that affect its worldwide operating companies (OPCOs). To allow for variations in OPCO's needs, arising from their local operating environments, these systems must be customised. This often results in them having a complex and interrelated set of configuration parameters which have both technical and business implications.

Configuring such systems demands a lot of skill and experience. Owing to the transfer of personnel, it is difficult to retain this expertise although documentation can partially capture this knowledge. However, in practice, the documentation is not always up-to-date, accessing information distributed in various manuals is time-consuming and novices are likely to have difficulty understanding the detailed information.

Object-oriented techniques have been used to develop Mixpert, a configurer's assistant. Mixpert provides a user-friendly support system for one business system which shall be referred to as the 'target' system. It comprises various tools including a database for archiving a particular OPCO's parameter settings; a configurations browser which may be used to view or modify existing configurations imported from the 'target' system or to setup new configurations with generic settings based on OPCO archetypes, advice being given on the selection of particular values and inconsistencies being revealed in real time; an invalidation browser which shows and explains inconsistent parameter setting; an auditing facility to log all changes made to a configuration; a report generator to print a configuration with or without supplementary information describing why settings have been chosen and their implications; a reference browser which provides help and description information for each parameter; and an on-line manual, the 'target' system's User Guide. All the knowledge contained within Mixpert can easily be modified using the Mixpert Builder. This facility enables Mixpert to be kept up-to-date with amendments and updates made to the 'target' system.

Mixpert has been implemented in the Smalltalk/V286 programming language [2]. This language was chosen because of its advanced user interface, its object-oriented paradigm for logically structuring Mixpert's configuration knowledge, powerful development environment and PC compatibility. Smalltalk's dependency mechanism has been modified to provide propagation of dependency information about Mixpert's configuration objects and unloading of this information and the objects belonging to a configuration set into a file [3]. The strength of untyped languages with dynamic binding such as Smalltalk, where the type of a variable is determined at run time rather than compile time, is clearly demonstrated. Mixpert is a highly dynamic application for which it is impossible to predict what classes and methods will be used or needed, making it difficult to implement in a strictly statically typed language.

Mixpert is a prototype system. It has been developed in the research laboratory in conjunction with the business but has not yet undergone extensive business evaluation. Although the motivation behind the system was to assist in configuration of the 'target' system, its design would map to other configuration problems, the builder facilitating input of their knowledge into the system.

2. OVERVIEW OF THE CONFIGURATION PROBLEM

The 'target' system is an on-line, multi-user Depot Administration System (DAS), which can perform various business processes. It can run on a range of hardware, using identical software and can link to other business systems. It may be run as a stand-alone installation, or over a network, linking an OPCO Head Office and satellite depots. Provided it is configured appropriately, the 'target' system may be used in a variety of operating environments. The operating companies' functional requirements, hardware and networking needs drive this configuration process.

Configuration of a DAS requires values to be assigned to approximately 1000 parameters. These parameters are grouped in about 127 entities, which are termed *tables* or *switches* by convention. Tables are used to describe entities that can have multiple occurrences, for example, the parameters despatch code and description reside in the table despatch codes, which can have associated many instances each describing a different code. Switches describe entities that can have only one occurrence, for example, standard invoice layout having associated parameters describing a page layout. The aggregation of parameters into tables and switches is key to the configuration procedure. It is a "natural" process and could be used to represent other configuration problems.

Dependency relationships exist between parameters. Two kinds of dependencies have been identified, hard and soft. Hard dependencies describe relationships between parameters that must be satisfied for a configuration to be consistent, soft dependencies describe relationships between parameters that usually would be satisfied but would not make the system inconsistent if they were not.

Some parameters have values that rarely vary between different operating companies. These parameters are assigned default values, which may be changed by the configurer. Parameter values may be constrained to be of a fixed type (for example, a string or an integer), satisfy a specified range (for example, an integer between 1 and 10), be unique and not exceed a size which is defined as an integer representing the maximum number of characters the value can comprise. Parameter values may also be mandatory.

The DAS is still subject to ongoing development and maintenance. Subsequently, the documentation can get out of date. A paper-based manual describing how the tables and switches relate to the function of the system is available but it does not have an index, making it difficult to locate relevant information.

Mixpert provides an intelligent front end to the 'target' system's tables and switches. The system can be used to set up new configurations in which case the configurer is assisted in assigning parameter values in a logical order or it can be used to amend existing configurations. When amending configurations, changes made to a part of a configuration may have repercussions elsewhere and invalidate other values. To overcome this, the system directs the user interactively to any inconsistent parameter values and logs any changes made to parameters. Up-to-date help and description information is provided for all the tables and switches and their parameters. The user manual has been incorporated in electronic form, the system providing an indexing facility to enable rapid access to required information.

Mixpert integrates with the DAS by exchanging information, which includes sets of configuration values using a file transfer mechanism. General table and switch information can be uploaded from the DAS to Mixpert, where it must be supplemented. In particular, further dependency, help, description and form layout information must be added. Configuration values are saved by Mixpert in a database as unloaded objects and saved in a text file for downloading to the DAS. By storing all configuration sets excluding the current one in a database, free memory is kept to a premium, ensuring Mixpert's runtime performance does not degrade as more configurations are defined.

3. REPRESENTATION OF CONFIGURATION PARAMETERS AND THEIR DEPENDENCIES

The distribution of the DAS's parameters into switches and tables provides a basis for conceptualising the domain. Further structure has been added by organising the switches and tables into *groups*, each group describing a different functional area, such as order capture. The groups are organised into a group set. Figure 1 shows the configuration structure, which is essentially a part-of hierarchy. For each group, the mix switches and tables were ranked according to the domain expert's recommended order of completion to determine the order in which they should be presented to the user. This takes into account the expert's perception of the importance that each switch and table has on the group's overall function.

Each table and switch has parameters that have associated values as shown in Figure 1. For switches, each parameter has a maximum of one value unlike tables which may have multiple values, each corresponding to a different instance (or record) of the entity it represents. For example, if table 1 (T1) describes vehicles and has three parameters P1, P2 and P3 representing registration, make and model, the values of two instances of vehicles could be described by the values V1 and V2 respectively in P1, P2 and P3. This is illustrated in Figure 2 where record 1 represents a Ford Escort with registration number B195 ABP

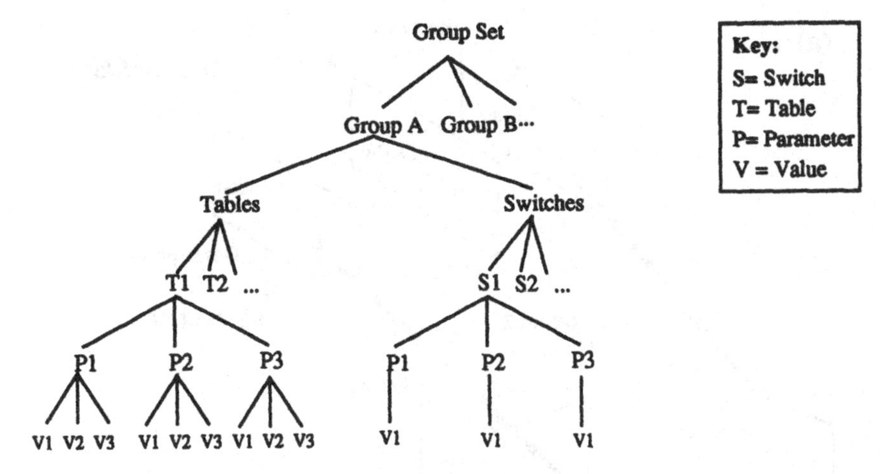

Figure 1 Configuration structure

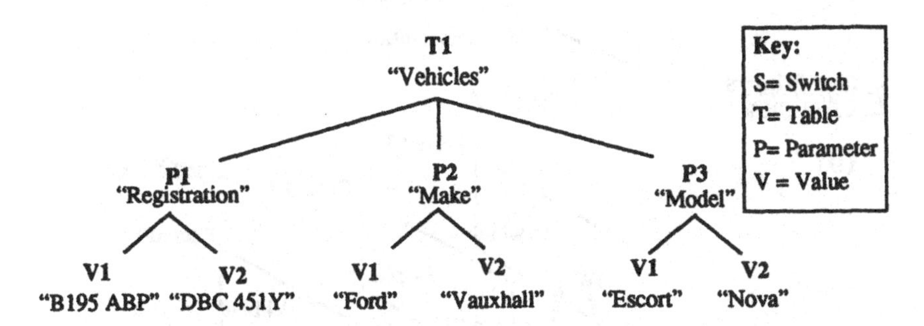

Figure 2 Representation of knowledge in a table describing vehicles

Certain parameters' values are dependent on values that may be defined by other parameters or by lists. These are 'hard' dependency relationships. To describe the 'hard' dependency relationships existing between parameters, fixed and linked dependency types have been defined. Where these dependencies occur, the system constrains the user to define the determinant parameter's values before the dependent thereby introducing a temporal ordering.

i) Fixed dependencies. These represent one-to-one or many-to-one relationships between a determinant and a dependent parameter, depending on whether the dependent parameter is, or is not, unique-valued. It is assumed that the determinant is a unique-valued parameter. The dependent can take only values that have been assigned to the determinant and, if unique, values that have not been defined in another occurrence of the dependent. Figure 3 illustrates two fixed dependencies.

ii) Linked dependencies. These are used to represent a many-to-one associative relationship between a determinant and a dependent parameter. Both parameters reside in the same table or switch, the determinant having associated a fixed

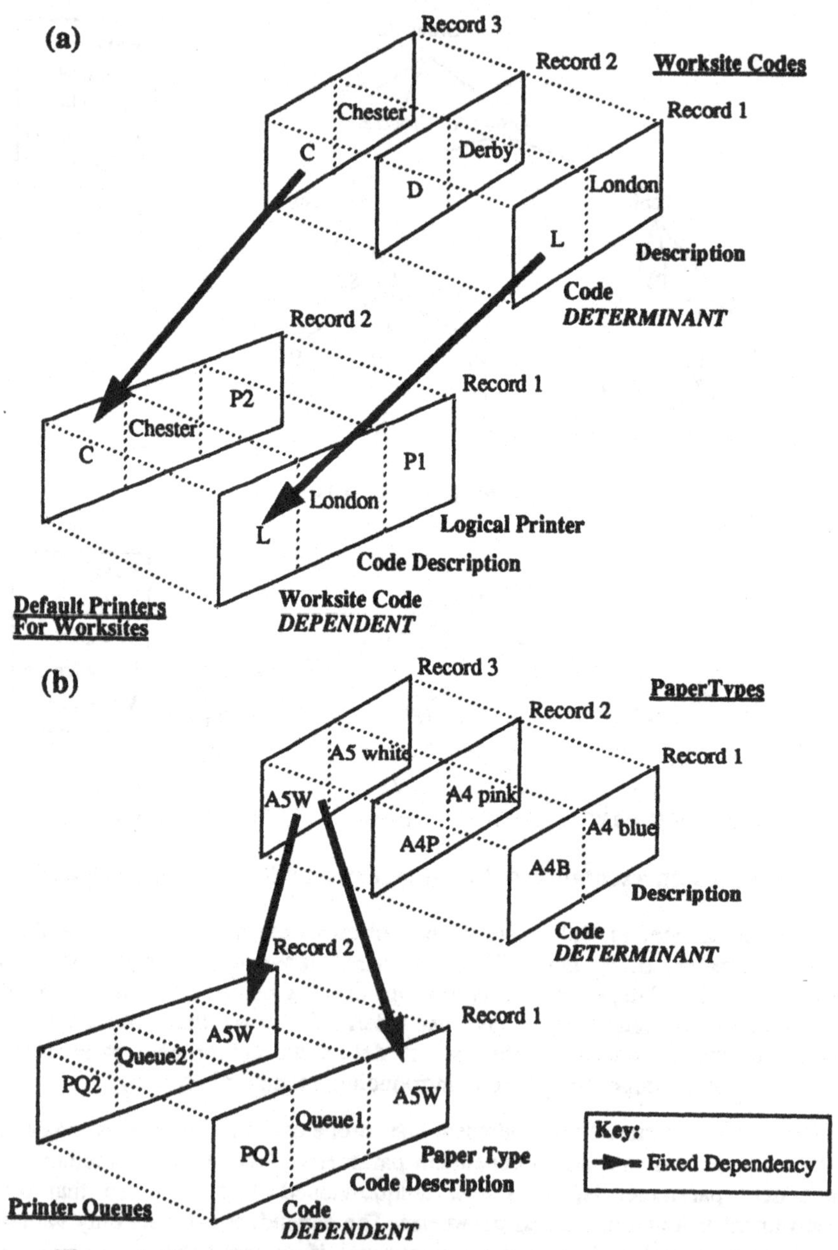

Figure 3 Examples of Fixed Dependencies. In (a) the dependent parameter is unique-valued, in (b) it is not.

dependency relating it to another table. The table that contains the determinant parameter, associated with the fixed dependency, contains additional parameters, one of which (call it 'PL') is associated with the linked dependency's dependent parameter, this association being defined by a 'linkedReference' dependency. The record in this table containing the value of the determinant parameter, which it is assumed is assigned a value before the dependent parameter, provides the value of the linked dependencies dependent parameter, this value being described by PL. A linked dependency is illustrated in Figure 4. The dependent and determinant parameters, 'description' and 'paper code', respectively, are held in a table called 'printers'. The dependent parameter is also associated with a parameter that is defined in the 'paper types' table, called 'description'. When the determinant parameter is assigned a value, the dependent parameter is given the value of the 'description' parameter defined in the same record of 'paper types'. For example, if the determinant parameter takes the value 'A4W' the dependent parameters value will be 'A4-White'.

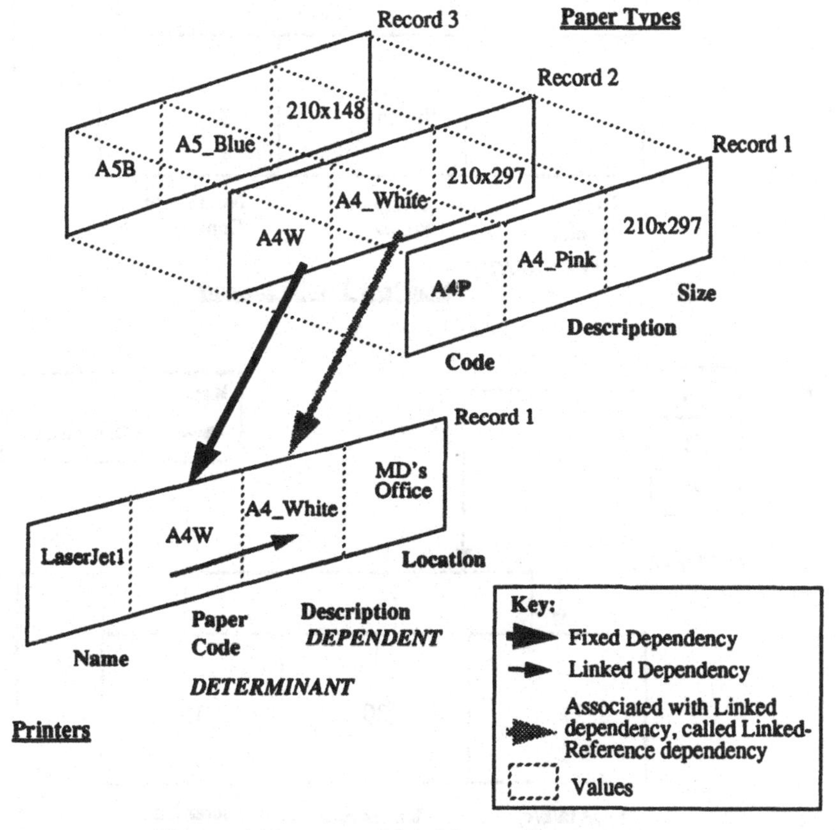

Figure 4 Example Of A Linked Dependency

'Hard' dependencies existing between lists of values are of two types: list and lookup.

iii) List dependencies. These dependencies relate a parameter value to any value contained in a list. One of the simplest examples of this is a 'yes' or 'no' dependency where a parameter value can take either of the values held in the list '(Yes No)'. See Figure 5 for other examples.

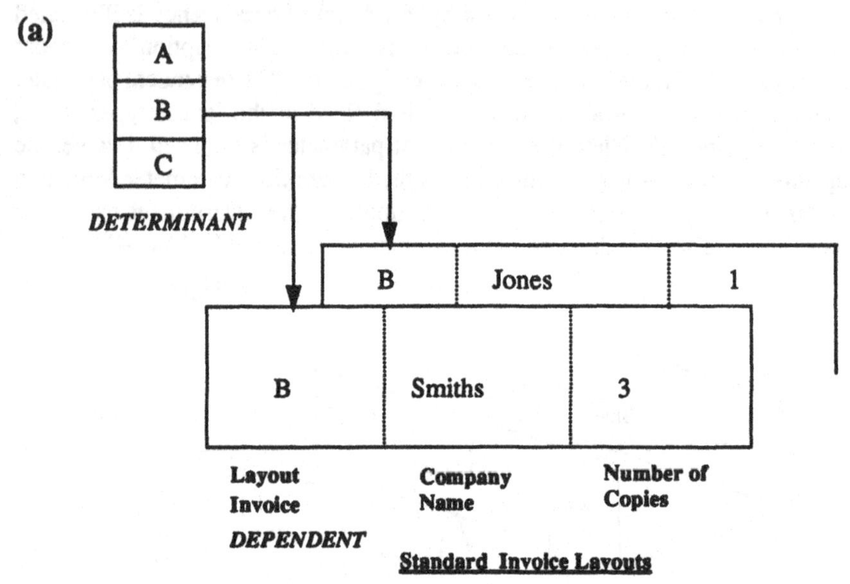

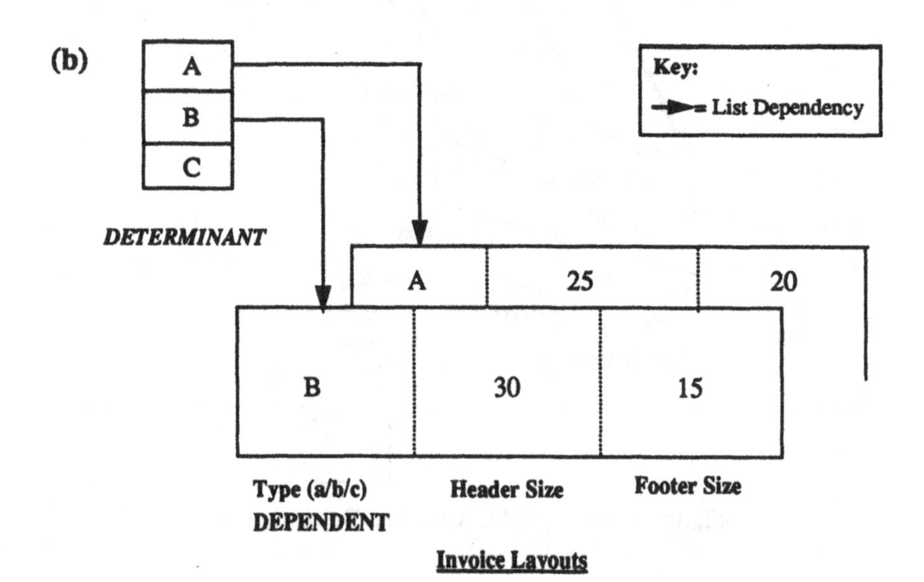

Figure 5 Examples of List Dependencies. In (a) the dependent parameter is not unique-valued, in (b) it is.

iv) Lookup dependencies.These are similar to linked dependencies but the determinant parameter associates a list of potential values to the dependent parameter rather than a single value. Figure 6 shows an example of a lookup dependency. The table 'delivery times' has a parameter 'zone' which is a character representing the distance to a distribution depot. The 'delivery time' is dependent on the 'zone' value and is described by the dependency. If the zone is 'A', the time is either '24' or '48', if 'B' it can be anything.

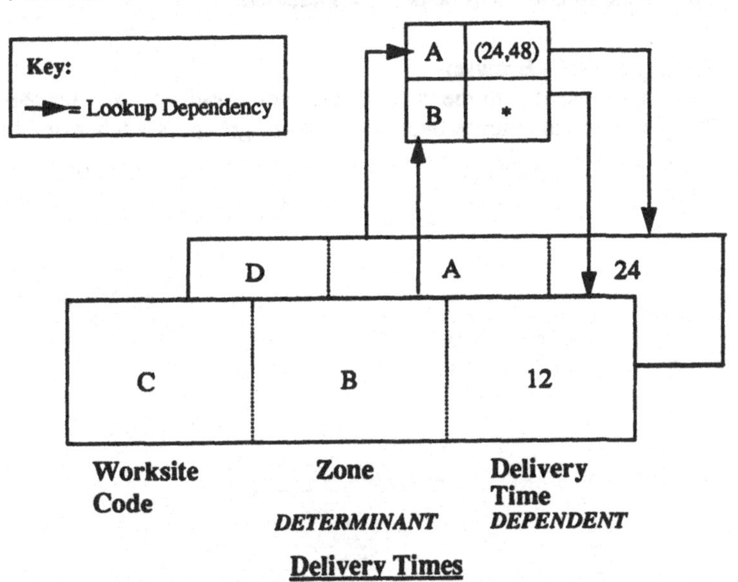

Figure 6 Example of a Lookup Dependency

For fixed, list and lookup dependencies, there may be several possible values for a dependent parameter. For linked dependencies, there will only be one value. Mixpert automatically assigns linked-dependencies-dependent parameter values. For the other dependency types, a menu of values is generated, the user being required to select an appropriate one.

Besides the types of 'hard' dependencies that have been described above, the Mixpert system has been designed to include a powerful 'soft' dependency facility. 'Soft' dependencies are defined as those types of fuzzy inter-relationships that are typical of business constraints. For example, the optimum allocations of processing and storage capacities for different products may depend upon many factors, some parameters will be defined within Mixpert, others, relating to the general business environment, will have to be obtained from the user. In such cases, the data would be gathered from the various sources and processed by a knowledge-base to provide a recommendation. This is clearly beyond the capabilities of a simple menu or dependency facility and requires a small self-contained knowledge-base embedded within the table or switch.

4. THE MIXPERT SYSTEM

A description of the tools provided by Mixpert follows. A discussion of the builder is deferred until after the implementation of Mixpert has been described in order that its functionality will be better understood.

4.1 The Configuration's Browser

Configurations are presented to the user via the configuration's browser shown in Figure 7. By clicking the mouse over the names of groups, switches or tables,

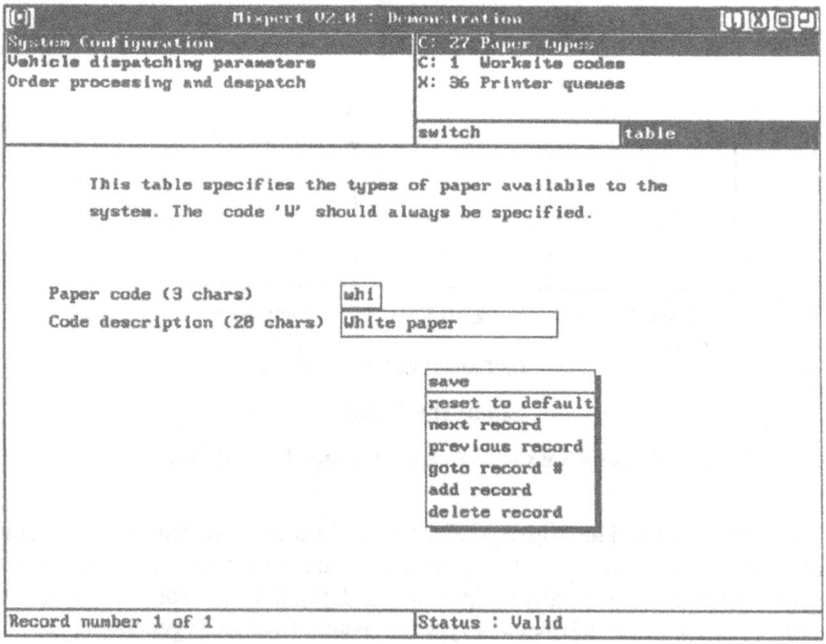

Figure 7 Configuration Browser

the user may select a switch or table to complete. Switch and table names are displayed according to the format:

<status>: number <item title>

where <status> is I, C, or X,

number is a switch or table number, and

<item title> is a switch or table title.

If the status of a switch or table is I, it is initialised and its parameters take default values. Switches and tables that have been modified and are consistent have the status C; others that are inconsistent have status X. Switches and tables that are dependent on other switches and tables that have not been completed are greyed out. The user is prevented from accessing such items until their determinants have been completed.

Switches and tables are displayed using dynamically generated forms. A general description of the switch or table is displayed as a heading, followed by parameter prompt strings and fields containing default values. Values may be assigned to fields by clicking the mouse over them. If the field has associated a non-linked dependency, a menu is activated, otherwise a text selector character appears. For menus, the configurer can select an item using the mouse, control then passes to the next field. Fields displaying a text selector require input from the keyboard. As the user enters characters into the field they are validated, as are the entire contents when the user presses 'return' to move to the next field.

4.2 Invalid Switch and Tables Browser

This is accessible from the configuration's browser. Invalid switches and tables are displayed in the top part of the browser. Clicking the mouse over a switch or table results in graphical explanation of the inconsistency being displayed in the lower part of the browser (Figure 8).

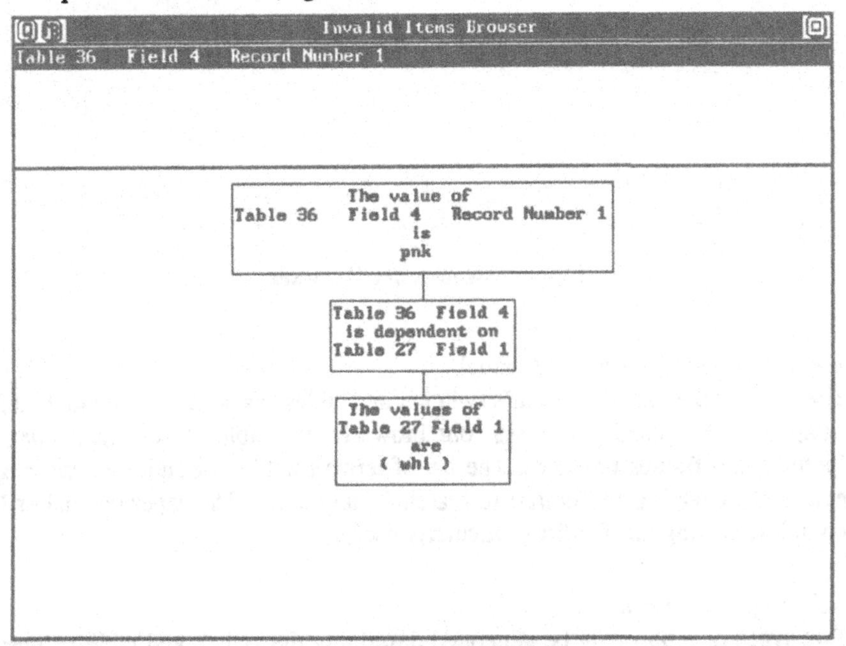

Figure 8 Invalid Items Browser

4.3 Reference Browser

This provides description and help information for each field in a table or switch. The description information provides an overview of the significance of the field and the help information advises on the choice of field values. Using the browser, the configurer can toggle between help and description information, move forwards and backward between fields, and move between groups, switches and tables (Figure 9).

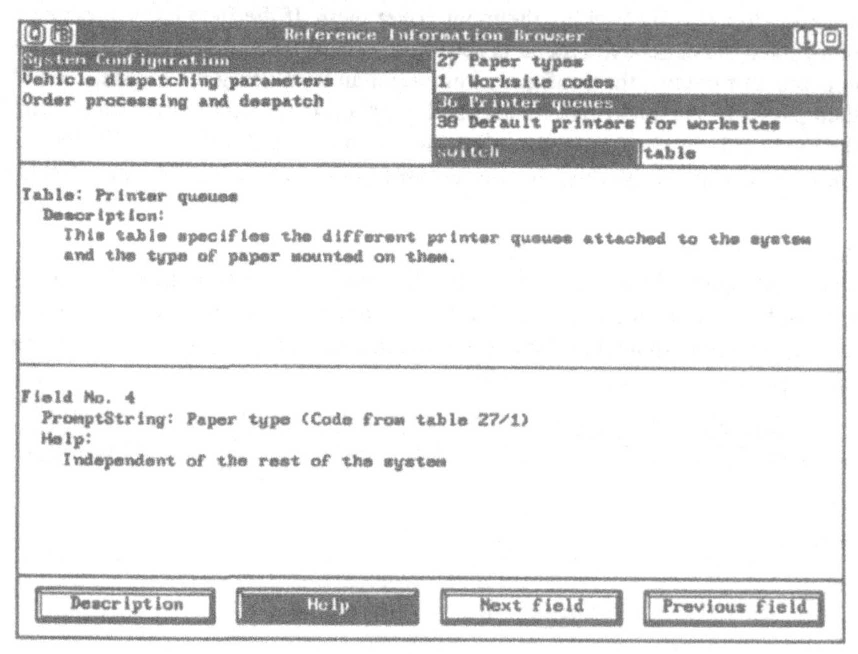

Figure 9 Reference Browser

4.4 Embedded Automated Manual

The guide to the 'target' system's switches and tables has been incorporated into Mixpert. The manual comprises four browsers, the table of contents, index, footnote and figures browsers. The list of terms used by the index browser is extensible, enabling configurers to search for any terms. The paper manual had no index, making this facility particularly useful.

4.5 Report Generator

Two types of report may be generated describing the switch and table values; concise or detailed. The difference between these formats is that the detailed report contains help and description information for each parameter value, whereas the concise form does not.

4.6 Audit Trail
An audit trail logs any of the changes to the configuration. The reason for the change, old and new values, the date and the configurer who made the change are recorded. This should be useful when faults occur during the operation of the 'target' system. By knowing the date when a fault first manifests itself and the parameter dependency information, the audit trail will allow more effective diagnosis of the cause of the fault by highlighting those parameters that were changed when the fault first manifested itself.

5. SYSTEM IMPLEMENTATION

5.1 Representation of knowledge in objects
Configuration information is represented by the class hierarchy shown in Figure 10. At the top level, a configuration is described by the object GroupSet. This has associated Group objects, each of which has associated Table and Switch objects, both types of Items. Items have parameters that are represented by FieldData objects. The value of each parameter is described by a ValueDescriptor object. Attributes common to all objects, such as help and description information, are passed down the hierarchy, via inheritance. Table and switch objects inherit common behaviour, such as the notion of records, from the class Items. Records are used to index entity instances.

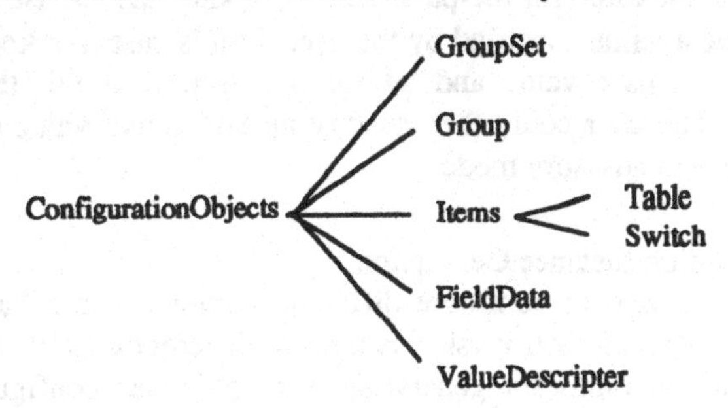

Figure 10 Class Hierarchy

Dependencies are described by instances of the classes, Fixed, Linked, List and LookUp. These objects contain knowledge of how parameters are related in general. For fixed dependencies, the table, or switch, and field defining the determinant parameter is defined. For linked dependencies, this information is supplemented by the parameter associated with the linked field and, for lookup dependencies, a dictionary containing the associations where the key describes the determinant's value and the value the dependent's value. List dependencies require a collection of selections to be defined. Each dependency object is held in the FieldData objects representing its dependent parameter. When the parameter is being assigned a value, the dependency is invoked to infer the values it can take.

Smalltalk V/286 contains an embedded Prolog system, which runs on top of the Smalltalk language. This implementation of Prolog is fairly basic with a somewhat idiosyncratic syntax. The basic structure of the Prolog facility is similar, however, to the Edinburgh syntax [4]. A major advantage of Smalltalk's Prolog facility over more traditional Prolog implementations is the ability to structure and segment a particular knowledge-base into different objects. This facilitates the independent development of small self-contained knowledge-bases for any table or switch which needs them in order to define 'soft' dependencies. This approach considerably simplifies the maintenance of the system and the addition of new knowledge-bases for other tables and switches. The availability of the Smalltalk Prolog facility was a major factor in our selection of Smalltalk for the implementation of the Mixpert system.

Embedded knowledge-bases may be attached to individual parameters within tables and switches, in a similar way to the hard dependencies described earlier. Each embedded knowledge-base is a self-contained encapsulated object with a simple communication protocol to other objects within the system. The embedded knowledge-base is triggered when a user activates a particular table or switch parameter. If the user is able to enter a legitimate value for the parameter, then this would be accepted, otherwise the knowledge-base would begin to gather data from both the configuration database and the user to enable it to recommend a value for the parameter. The knowledge-base could also be used to critique a value specified by the user. In this case, the knowledge-base would analyse a user value and advise the user, if it felt the value was inappropriate. The user could then specify an alternative value or invoke the knowledge-base in advisory mode.

5.2 Configuration Instance Generation

Mixpert may be used to configure different versions of the 'target' system. General knowledge of each version is held in different classes. Each of these classes contains methods for generating new initialised configurations, each comprising a different instance of the groupSet object. It is to an initialised configuration set (a groupSet object) that a configuration's details are added. Initialisation methods are of four types, to:

1. Create a groupSet object referencing group objects.
2. Create group objects referencing switch and table objects.
3. Create switch and table objects containing fieldData objects.
4. Create a collection of fieldData objects.

To define a fieldData object, the following information is required:

- default value,
- whether or not it is unique,

- whether or not it is mandatory,
- dependency relationships,
- a textual description,
- help information describing the significance of parameter values,
- validation information, and
- data including a prompt string, field-width and position for displaying it on a form.

Subdivision of the initialisation procedure into different method types allows changes to the "general" configuration knowledge to be made without rewriting vast amounts of code. Each method type defines how to create a specific instance of a particular class, for example, an initialised instance of the table class corresponding to the table describing despatch codes. This feature of the design has been used by the Builder and is discussed in Section 6.

All configuration information conforms to an explicit predefined format. Using the configuration classes, multiple instances of the same configuration can be generated, each pertaining to a different installation of the DAS.

5.3 Consistency Maintenance

Smalltalk has a dependency mechanism which allows objects to activate behaviours in other dependent objects. The mechanism essentially maintains a dependency list of dependent objects for each object in the system. Messages can be passed to an object's dependents together with arguments. Smalltalk's dependency mechanism has been adapted to allow selective activation of an object's dependents. For each object used to define a configuration, an attribute or instance variable called 'dependents' has been defined. This is a dictionary object whose keys include 'default', 'fixed', 'linked' and 'linkedReference' and whose values are lists of dependent objects. Messages can now be sent to all objects associated with a key, for example, all 'fixed' dependents. Since the dependents are held in an instance variable, rather than a class variable in the class Object, they can be filed out using the Smalltalk loader [1].

Instances of dependency objects are created when a configuration is initialised. These are referenced by the 'dependent' parameter's fieldData object. In addition, they are made dependents of the 'determinant' parameters' fieldData objects using the modified Smalltalk dependency mechanism. Thus the determinant objects can then identify the objects that are dependent on them.

When a switch or table has been completed, all its mandatory values having been defined, it may be saved. Each of its fieldData objects is given a value which is stored in a ValueDescriptor object. ValueDescriptors have various properties including status, value, record and dependency information. If a valueDescriptor has not been created, corresponding to the current record, a new one is created. The valueDescriptor's status is made valid. If the fieldData object has a dependency object defined of type, linked or fixed, the record in the determinant fieldData object associated with the value is found and stored in the valueDescriptor together with the dependency object. A dependency is set up between the valueDescriptors by calling the dependencies 'makeMeA-Dependent' method. This adds the dependent valueDescriptor object to the appropriate dependency list of the determinant object, the lists being fixed, linked or linkedReference.

If values are changed or a record is removed from the system and other parameter values are dependent on it, it will become inconsistent. To illustrate this, consider the following situation. Two tables, 'paper types' and 'printers' have associated the fixed and linked dependencies shown in Figure 4. If the code in record 2 of paper types is changed to 'A4B', the following actions will occur. All records in 'paper types' will remain valid. All valueDescriptors having a fixed dependency relationship with the code defined in record 2 (those on the fixed dependency list) will be sent the message 'valueFromFixedDependency-Changed' with the new value 'A4B' as an argument. ValueDescriptors that are related via linked or linkedReference dependencies would be sent the messages 'valueFromLinkedDependencyInvalidated' and 'valueFromLinkedReference-DependencyChanged' with argument 'A4B', respectively.

The valueDescriptor associated with the paper code in record 1 of the table 'printers' has a fixed dependency relationship with the code that was modified in table 'paper types'. This value descriptor will become invalid and be added to the invalidation queue, which is a list of invalid valueDescriptors held by the invalid items browser. The messages 'valueFromFixedDependencyInvalidated', 'valueFromLinkedDependencyInvalidated' and 'valueFromlinkedReference-Invalidated' will be sent to the paper code's 'fixed', 'linked' and 'linked-Reference' dependents, respectively. The message sent to the 'linked' dependents would activate the valueDescriptor having the paper description 'A4_White'. The valueDescriptor would become invalid, it would be added to the invalidation queue and the messages 'valueFromFixedDependency-Invalidated', 'valueFromLinkedDependencyInvalidated' and 'valueFrom-LinkedReferenceDependencyInvalidated' would be sent to its fixed, linked and linkedReference dependents, respectively.

If the code in record 2 of 'paper types' was restored to 'A4W', then its valueDescriptor would send the message 'valueFromFixedDependentChanged' with argument 'A4W' to its fixed dependents. The valueDescriptor in 'printers' having associated the fixed dependency would then become valid, it would remove itself from the invalidation queue and it would send the message 'valueFromFixedDependencyValidated', 'valueFromLinkedDependencyVal-idated' and 'valueFromLinkedReferenceDependencyValidated' to its fixed, linked and linkedReference dependents, respectively. The 'valueFromLinked-DependencyValidated' method in the valueDescriptor containing the value 'A4_White' would then be activated. It would remove itself from the invalidation queue, make its status valid and send the messages 'valueFromFixedDependencyValidated', 'valueFromLinkedDependencyValidated' and 'valueFromLinkedReferenceDependencyValidated' to its fixed, linked and linkedReference dependents, respectively.

6. THE BUILDER

The builder provides a forms interface for adapting Mixpert to handle changes made to the structure of the 'target' system, Figure 11. It has been designed for

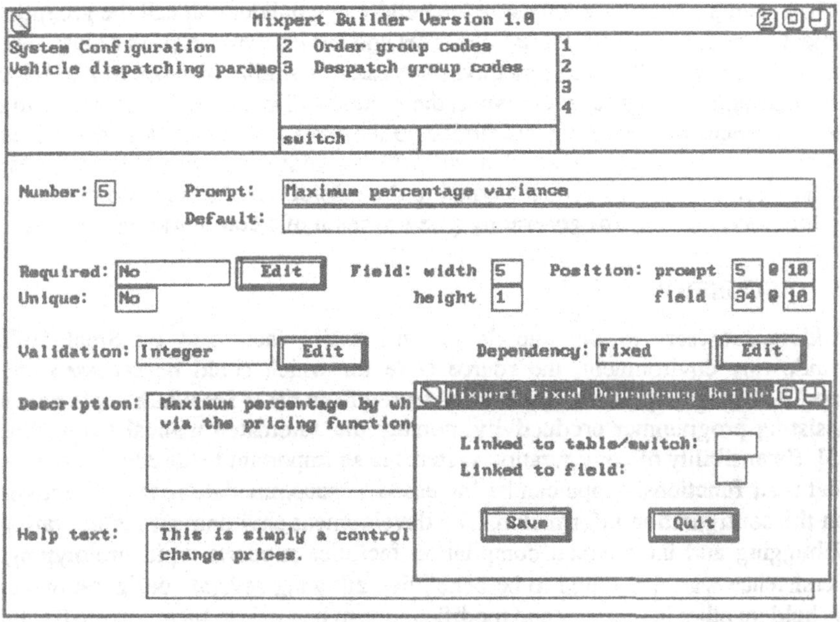

Figure 11 Builder

use by a configurer who has no knowledge of programming but is familiar with group, switch, table and field concepts. The builder can create new configuration classes corresponding to new versions of the 'target' system and write Smalltalk code to them, providing code for the generation of new groups, tables and switches, and fieldData items (see section 5.2). Knowledge of configuration structure can be extracted from the 'target' system automatically. The builder allows the configurer to add additional information, such as help and form layout information, to it.

The builder requires an initialised configuration to be defined in order to operate. This configuration, being represented by objects of type GroupSet, Group, Switch, Table and FieldData, provides the builder with the knowledge of the system being built. The user must adapt the initialised configuration.

These changes are made to the configuration objects in memory. For example, if the user adds parameters to a switch, the switch's Switch object will become associated with more FieldData objects or, if the user changes a parameter's prompt string, its FieldData's promptString attribute will be assigned the new value. The forms display of modified mix items can be inspected by opening a configuration's browser on the configuration. Changes made to a parameter's prompt string, prompt string position, field width or position, can be viewed before they are saved.

Configuration objects know how to save themselves as source code methods. When saving a configuration using the builder, the builder will call the groupSet object associated with its configuration set and tell it to save itself in a class. The groupSet object will create a method for creating a similar instance of itself and will call each of its groups. Likewise, the groups will write methods for creating new instances of Group objects similar to themselves and call their associated switch and table objects. These too will write methods to redefine objects similar to themselves and propagate a method to their fieldData objects to do the same. In this way, the code for generating a new version of a configuration is saved.

7. DISCUSSION

Adding browsers to the knowledge was easily facilitated by Smalltalk's windowing environment, the source code for which could be accessed and modified as appropriate. Objects were reused from other Smalltalk applications assisting programmer productivity, notably the automated manual from Elias [5]. Extensibility of configuration systems is an important feature [6], to ensure that their functional scope can be increased to accommodate new perspectives on the configuration information. The development environment incorporating debugging and incremental compilation facilities supports rapid prototyping. Persistence was also found to be beneficial, allowing several configurations to be held together in memory and modifications to be made to them 'on the fly' by the builder.

Although Mixpert has not been field tested, the functionality it offers has been demonstrated to configurers. The configurations browser was considered to be more user friendly than that provided by the DAS.The presentation of possible values for parameters having associated dependencies, the on-line detection and conveyance of inconsistencies and reference information were thought likely to benefit productivity. The features that were considered to be of added value to the DAS were the automated manual and the report generator. The builder has not been tested, nor have the 'soft' dependencies.

Several important points have come out of the project. Using an existing system to develop the prototype had the benefits that domain concepts for communicating knowledge during the acquisition stages of the project were established and the system's interface provided a benchmark for comparison. However, the weaknesses of this approach are that, when developing the representation, the domain expert has to discuss a lot of the knowledge that has already been encoded in the existing system. This stage is very important to the knowledge engineer for deciding what the knowledge requirement is and how it can be structured to allow a general representation to be developed for use in acquiring further knowledge. However, it is very time consuming and must be followed by much programming of the kernel system before any pay-offs are realised by the domain expert. During this time, it is very important that the domain expert is kept involved with the project to avoid his/her motivation decreasing. The object-oriented programming style can help this by allowing interfaces to be developed to demonstrate the 'look and feel' of the system as its underlying knowledge evolves.

8. CONCLUSION

Configuration systems that are to be used routinely by configurers of business systems, which are subject to development and maintenance, need to be maintainable. Those that are rule-based are maintainable only by programmers who are familiar with the representation and content of the knowledge in the system. The network dependency-mechanism paradigm that underlies Mixpert is clearer than the rule-based paradigm, enabling the system to be maintained by the configurers themselves. Within Mixpert, rule-based techniques may be used to implement 'soft' dependencies but their use is tactical rather than strategic.

The Mixpert system addresses a specific configuration problem. Concepts have emerged from its development that could be applied to other heavily parameterised configuration problems. These are the aggregation of parameters into tables and switches that have associated dependencies describing how they are related. Four types of 'hard' dependencies are used in Mixpert, fixed, linked, list and lookup. These could be extended to cover other types of dependencies, for example, mathematical relationships between parameters, if required. Object-oriented programming allows general and specialist behaviours to be described by 'abstract' and 'specialist' classes, respectively. Abstraction and specialisation have been used in the design of Mixpert's dependency classes. A benefit of using this approach is that new dependency types can easily be implemented.

9. REFERENCES

1. Barker, V.E. and O'Connor, D.E., Expert Systems For Configuration At Digital: XCON and Beyond, Communications Of The ACM,1989, 32(3) pp. 298-317.

2. Smalltalk/V286, Digitalk Inc., 9841 Airport Blvd., Los Angeles, CA 90045.

3. Goodies#1 Application Pack, Digitalk Inc., 9841 Airport Blvd., Los Angeles, CA 90045.

4. W.F. Clocksin and C.S. Mellish, Programming In Prolog, Springer Verlag, Heidelberg, 1984.

5. A. Bamigboye, H. Dorans, K. Lunn and H.L. Walmsley, An Object-Oriented Approach To The Analysis Of Incidents Involving Electrostatic Discharges, Proceedings of AIENG 90.

6. J. Liebowitz, Expert Configuration Systems: A Survey and Lessons Learned, Expert Systems With Applications, 1990, Volume 1 pp. 183-187.

Design of Breakwaters using Knowledge-Based Techniques

J. Murphy, B. O'Flaherty, A.W. Lewis
Hydraulics and Maritime Research Laboratory, Department of Civil Engineering, University College Cork, Ireland

ABSTRACT

The design process for a breakwater is unlike that of almost all other civil engineering structures. Whereas such major structures as bridges and skyscrapers lend themselves easily to precise mathematical analysis the same cannot be applied to breakwaters. What does exist is a multitude of empirical formulae and heuristic methods on which the engineer must base his design. Very often the engineer is not aware of methods and techniques that could be of help in the design. This lack of knowledge can be addressed by developing an expert system that can store much of the information presently scattered in numerous design manuals and research papers. At the Hydraulics and Maritime Research Laboratory (HMRL) in Ireland such a system is being developed. It is envisaged that the completed system will act as a prototype for use in the design office of the future.

INTRODUCTION

The sea since time immemorial has been ravaging the coastlines, claiming countless lives and doing untold damage. Man over the centuries has sought to understand and control it's power but has never quite succeeded. With every passing year the sea reaks more and more havoc aided now by such influences as climatic changes and sea level rises induced by the greenhouse effect. Many coastal structures and sea defences designed to last for hundreds of years have been crumbling under the constant barrage of the indefatiguible sea. On the front line of attack is the breakwater structure. This man made structure in its most typical form projects seawards from the coastline and acts as a barrier designed to bring onrushing waves to a halt. Figure 1, for example shows the layout of the breakwaters at port of Zeebrugge. On the leeside of the breakwater there is an area of calm water suitable for ships to navigate and moor. The importance of breakwaters can easily be understood, for

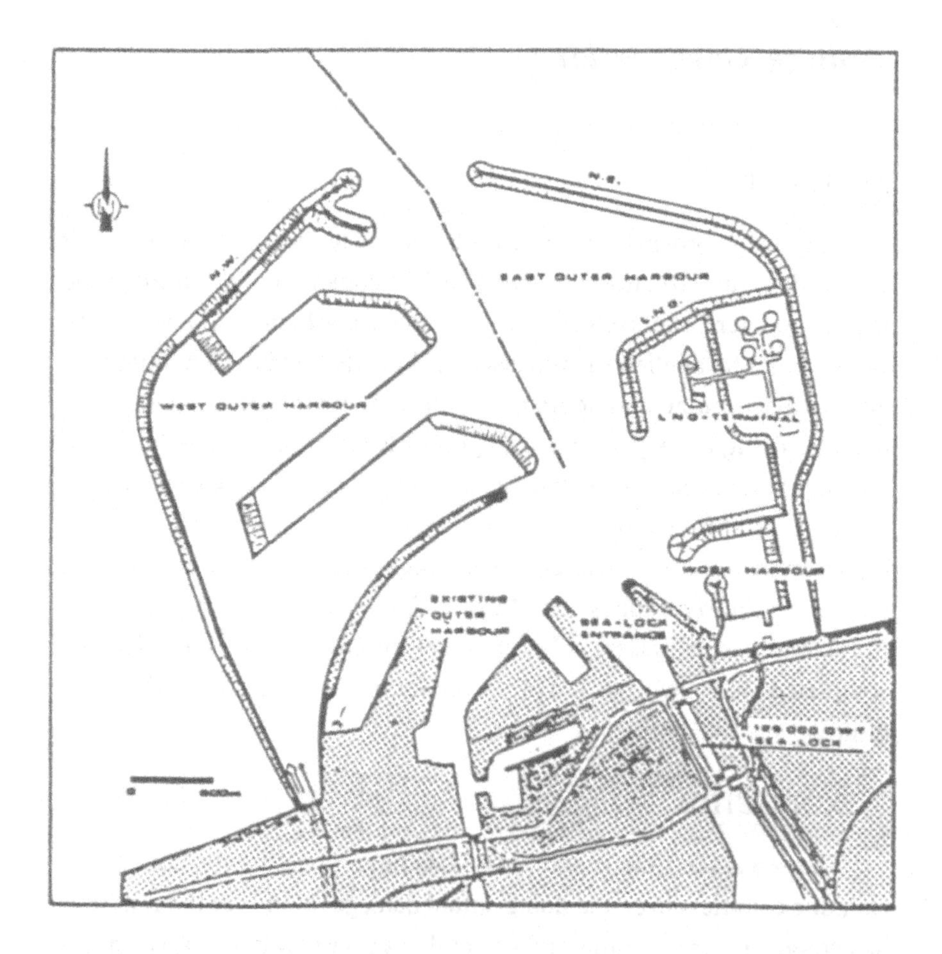

Fig. 1 Breakwater layout at port of Zeebrugge.

without them harbour areas would be subjected to excessive wave action that would inevitably lead to delays in the mooring of vessels and the offloading of cargo. As well as providing this safe haven for ships breakwaters can also have a number of secondary purposes, such as serving as quays, reducing dredging requirements and guiding local currents in a channel. In view of the importance of breakwater, this paper sets out to describe briefly a major type of breakwater, the expertise and main sources of knowledge used in the design process, as well as a Computer-Aided Design System (called BW-CAD) developed here at U.C.C., which attempts to encapsulate this design knowledge.

Before embarking on a description and discussion on the type of expert system presently being developed at the HMRL, it will be useful if a brief overview of breakwaters and their design is outlined. This will serve to highlight the empirical and heuristic nature of the design process

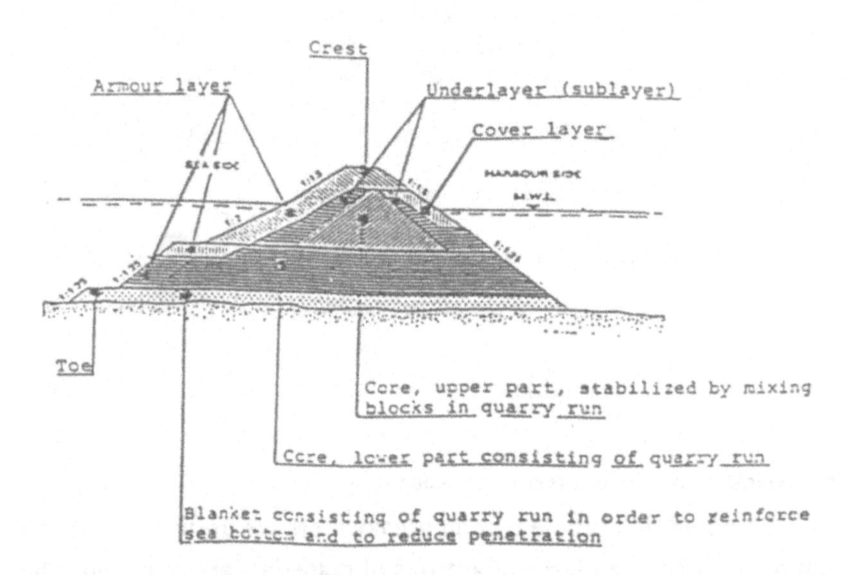

Fig.2 Typical rubble mound breakwater section.

as well as showing the degree of uncertainty that exists about the stability of the finally constructed structure. It is also hoped that such a description will adequetely show how computers can aid and improve the design process.

BREAKWATER TYPES

Breakwaters come in numerous forms, often being very site specific, but can be broadly classified into two main types, the rubble mound and the vertical faced. A typical rubble mound breakwater section is shown in Figure 2. It is a substantial structure consisting of inner layers of stone and gravel of different grading which are protected on the exposed face by one or more layers of heavy armour stones or concrete armour units. Some of the benefits of these structures include durability, flexibility accomodating both settlement and irregular bathymetry, uses local material and functions well even when severely damaged. Although it

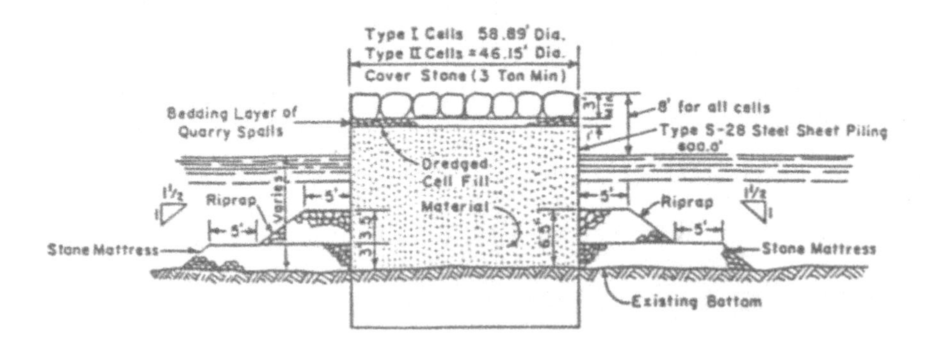

Fig. 3 Typical Vertical Faced Breakwater section.

should be noted that rubble mound breakwaters are unsuitable for soft ground and very often large quantities of materials are required. The monolithic vertical faced breakwater (Figure 3) is a much more compact structure which is most often made up of sunken caissons or large elements stacked upon each other in a regular fashion. Although appearing to be a much neater structure than the rubble mound type, it does have a number of disadvantages. These include inflexibility to settlement and uneven bathymetry, high wave impact forces, very little energy dissipation, large erosion at structure base and often catastropic failure when design conditions are exceeded.

BREAKWATER DESIGN HISTORY

Contary to popular perception breakwaters are not modern structures but can be dated back thousands of years. The oldest recorded is the remains of the Minoan breakwater, constructed over 4500 years ago. Although totally submerged now this breakwater is almost completely intact. A curious point about this and other ancient breakwaters is that their profile shape is remarkably similiar to many modern computer aided designed breakwaters. This immediately poses the question that following such a long history of design and construction of breakwaters there is still so much uncertainty today about the stability of each new structure built. In fact there has been alarming rise in the number of breakwater failures in recent years. These failures as well as being expensive to remedy have a disabling effect on the now exposed harbour as cargo handling becomes extremely difficult. Modern breakwater failures are testament to the great uncertainties that presently exist in quantifing loads and forces exerted on rubble mounds by waves.

DESIGN PROCESS

Since a breakwater in its most fundamental form is an artificial pertrusion from the shoreline it is going to affect in some way the natural processes that are inherent to all coastlines. Siting a structure on a sensitive area of coastline can have disasterous ecological and environmental effects. One common problem with breakwaters is that they prevent the natural flow of sediment along the coastline which results in erosion and accretion. As there is no universal philosophy for the design of breakwaters and the final design is as much influenced by politics as it is by experience and tradition available locally, the engineer has a very difficut task when embarking on a new project. He/She must be aware of local coastal processes as well as skills that are indigenous to that particular area and thus needs a large store of information readily

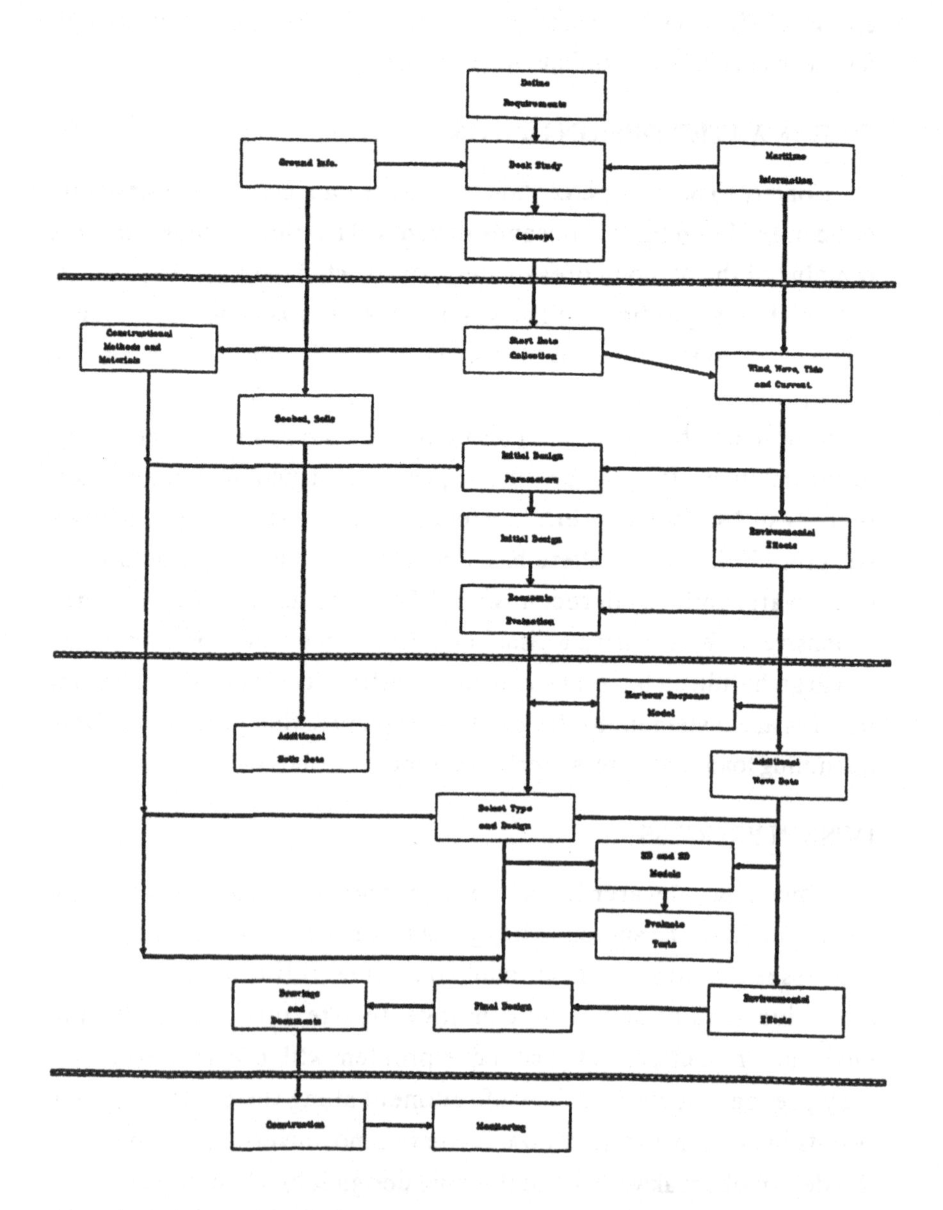

Fig. 4 Design Process of A Rubble Mound Breakwater.

available to carry out a proper design. This is where the computer can be especially useful in carrying out tedious calculations and providing information regarding previous breakwater failures and areas that may be of special concern. The complexity of the design process can be appreciated by observing Figure 4. This flow chart aptly demonstrates how each element fits into the overall design process. Two areas of special concern include firstly the overall layout of the harbour and how the shape will affect waves propagating towards the shore and secondly the stability of the armour layer. Since all aspects of design cannot be adequetely dealt with in a single paper, what will be considered now is the methodology involved in sizing an armour unit to resist wave attack on the exposed face of a breakwater. This would correspond to the 'Initial Design Stage' given in Figure 4.

ARMOUR UNIT SIZING

Waves breaking and dissipating their energy on a rubble mound slope is a complicated process and no precise mathematical formulation of the problem has yet been derived. The lift and drag forces induced in the armour unit by wave action must be resisted by its weight and friction developed between adjacent units. The displacement of an armour unit can have catastropic consequences, therefore its correct sizing is critical. Unfortunately there are no all encompassing analyical formulae available to do this so engineers have to fall back on their own design experience, empirically derived formulae and the results of scaled model tests. The procedure that is usually followed in sizing an armour unit is firstly obtain a value for its weight using an empirical formula, judge this against experience and finally carry out a model test on a breakwater section to examine the units stability against wave attack. Though not perfect this system of design has worked satisfactorily for most breakwaters.

The main formula used to calculate the breakwater armour weight is the Hudson Formula.

$$W = p\ g\ H^3\ /\ D^3 Kd\ \cot \text{Ø}$$

W = weight of the armour unit in kg.

H = incident wave height.

p = mass density of rock.

Kd = empirical stability co-efficent

cot Ø = slope of breakwater cross-section.

D = Relative mass density.

Formulae. 1 Hudson Formula

This formula was developed in the 1950's and although it has been verified in large and small scale models as well as prototypes, it does have a number of limitations which are not always considered when using the formula. It is only valid for certain slopes and for non breaking waves acting on the breakwaters. In addition to this the entire storm conditions are characterised by a single parameter, H, the wave height. No account is taken of such things as wave period, storm duration and wave grouping. The formula is very sensitive to small variations in the value of H, a 10% increase in the wave height augments the armour weight by 33%. Therefore an underestimation of the design wave height can result in armour blocks incapable of resisting real storm conditions being placed on the slope. Needless to say this has been the cause of many breakwater failures. However one cannot critizise the Hudsons Formula too harshly as for many years it has filled the breach, supplying engineers with a starting point for breakwater design where otherwise there would have been nothing.

Breakwater research is ongoing in many Hydraulic Laboratories worldwide for example Delft (The Netherlands), DHI (Denmark), HRL (England) and in a number of centers in the United States and Asia. They are endeavouring and upgrade and improve on the Hudson Formula. This research has led to the development fo a number of empirical formulae each claiming validity under certain conditions. The most prominent of

$$H_s/\Delta D_{n50} * \sqrt{\xi_m} = 6.2\ P^{0.18}(S/\sqrt{N})^{0.2}$$

for plunging waves ($\xi_m < \xi_m$ (transition)),

$$H_s/\Delta D_{n50} = 1.0\ P^{-0.13}(S/\sqrt{N})^{0.2}\ \sqrt{\cot\alpha}\ \xi_m^P$$

for surging waves ($\xi_m > \xi_m$ (transition)), with:

$$\xi_m\ (\text{transition}) = [6.2\ P^{0.31}\ \sqrt{\tan\alpha}]^{1/(P+0.5)}$$

where:

H_s = significant wave height
Δ = relative mass density = $(\rho_a/\rho) - 1$
ρ_a = mass density of rock
ρ = mass density of water
D_{n50} = nominal diameter = $(W_{50}/\rho_a)^{1/3}$
W_{50} = 50% value of the mass distribution curve
ξ_m = surf similarity parameter = $\tan\alpha/\sqrt{s_m}$
α = slope angle
s_m = wave steepness = $2\pi H_s/gT_m^2$
T_m = mean wave period
P = permeability coefficient of structure:
- P = 0.1: impermeable core (lower limit)
- P = 0.4: most multi-layer breakwaters
- P = 0.5: permeable core
- P = 0.6: homogeneous structure (upper limit)

S = damage level = A/D_{n50}^2
S = 2-3: start of damage
S = 5-8: moderate damage
S = 8-15: filter layer visible (two layer system)
A = erosion area of cross-section

N = storm duration in number of waves

Formulae. 2 Van der Meers formula

these is the Van der Meer Formula that is the result of exhaustive tests carried out at Delft Laboratory.

This formula accounts for mound permeability, storm duration and wave period and can take two forms depending on the type of wave breaking on the structure. These formulae have not been fully verified yet and take a different form depending on the type of armour unit. The above formulae are explicit to slopes protected with rock armour. Also Van der Meers formulae are not as easy to apply as Hudson's, since a larger database of information is required, so there is often a reluctance among engineers to use them when designing breakwaters.

Another aspect of breakwater design which is instructive to consider is wave runup and overtopping. Wave runup is the vertical distance from the stillwater level that the water from an incident wave will run up the structure while overtopping occurs when the runup distance is greater than the crest level. A pletera of formulae and tables presently exist that are designed to calculate these two parameters.

These formulae have very specific applications and are loaded down with empirical coefficients. It would be of great advantage to the design engineer to have these formulae together with their coefficients and guidelines as to their application were available on computer. This would eliminate much of the confusion that presently exists and also help in the optimization of the design process. In BW-CAD the above mentioned formulae plus many of the tables containing design coefficients have been input and can be accessed easily by the design engineer. Cost optimisation and risk analysis, two important areas in breakwater design are presently being added to the system.

COMPUTER AIDED BREAKWATER DESIGN

Aspects of breakwater design that could be improved by computerisation have already been touched on above. In fact at present there exists many software packages which carry out specific segments of the design. These include sophisicated finite element models which can predict wave heights inside a harbour [Yoo89] [Rot85] and internal flows within a breakwater. [Hal90] [Hol88] However there appears to be no expressed desire to bring these packages plus all the heuristic and empirical formulae expressed above into one total all encapsulating design package. Very often, the expertise available is very spread out and the final design very localised. The development of a user friendly breakwater design system would improve the productivity and effectiveness of a design engineer.

TYPES OF EXPERTISE UTILISED IN BREAKWATER DESIGN

As stated above the topic area of breakwater design is very much empirically based. The main source of expertise is found in research papers reporting the results of physical model tests. This expertise can be very local, for example, Japan, has a wealth of research findings available on vertical and hybrid type breakwaters. We firstly wish to study the requirements of an engineering software development environment. Engineering solutions require a various types of expertise and data and

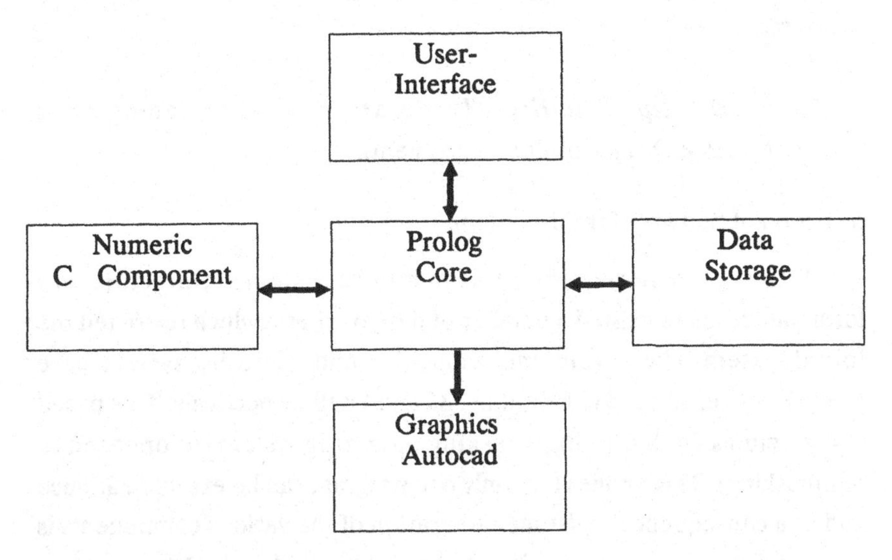

Fig. 5 Components of BWCAD.

Civil Engineering has proved to be a rich source of applications for Expert System Development, as is evident in [SRI85] that describes 64 projects worldwide which were developed using Artificial Intelligence techniques.

They could require heuristic knowledge of design practise, experience, numerical model results, economic models. Engineering solutions also require output produced in graphic form i.e. a design drawing. Therefore the software requirements of the ideal system for providing solutions to engineering design problems would contain

1. *Knowledge Representation facilities* (provided by an expert system shell).

2.*Numeric Modelling capabilities* (provided by a high-level language interface (Fortran or C) or interface to a finite-element package.)

3. *Data Storage* (interface to a database or file processing capability).

4. *User-Interface Development facility.* (Windows, Mouse, Dialog Boxes etc.,)

5. *CAD Output facility.* (Integration or Interfacing to a Computer-aided Design or Drafting system)

Selection of Software for this System.

With the above requirements in mind a system architecture was formulated. There existed a number of design criteria which restricted the initial system. The development machine and operating system were predefined i.e. an IBM compatible 386 and DOS respectively. This posed a problem as DOS is a single tasking operating system (as opposed to multitasking). This implies that only one program can be executed at once and as a consequence real-time integration of the various components is not possible. To overcome this problem Quaterdeck's Desqview was used. This package acts as a shell running on top of DOS and allows a limited degree of multitasking (2 programs at once). It also provides memory management facilities which allow the inherent DOS barrier of 640k to be broken, by allowing the utilisation of expanded and extended memory. The system is made up of the following components:

KNOWLEDGE REPRESENTATION

Because Prolog has the Procedural expressiveness of a conventional programming language, it was deemed suitable as the core control component of the system. Having reviewed the currently available software, LPA prolog was chosen. The main disadvantage of using prolog as an Expert System development environment is that it only supports one inference engine (robinson's resolution) namely backward chaining. This difficulty was overcome by using Flex the expert system tool kit which complements LPA Prolog by adding additional features to the prolog core system. This toolbox gives LPA prolog all the functionality of an Expert System shell as well as the maintaining the versatility of a programming language and includes features such as Forward chaining, Knowledge Representation mechanism (Frames, inheritance and Rules), User-Interface development functions. In fact Prolog is a popular development language and it has been used in many expert system project. [KUM89]

The following are examples of rules of thumb and contraints used in the design of rubble mound breakwaters, which are represented in Prolog.

Armour layer : Armour units are placed at a depth of between 1 to 1.5 times the design wave height below minimum design seawater level

Underlayer Thickness : Two layers of blocks at least having a block weight of about 25% of armour or cover layer block weight.

Lower part of windward side: Block weight 50% at least of armour layer block weight, slope about 1 on 1.5

Crest Width: About equal to the design wave height. It is recommended a crest width corresponding to the combined width of three armour blocks

Crest Level: 1 to 1.5 times design wave height above maximum design seawater level

NUMERIC COMPONENTS

In This BW-CAD all numeric computations are written in C and these functions can be accessed by the Prolog C interface. This interface allows the user to recompile the core Prolog system, linking in the new numerical function. The consequence of this is that these functions are now part of the Prolog built-in functions and can be accessed as conventional Prolog procedures. Various functions were implemented as

C functions i.e. Hudson's and Van Der Meers formula etc., It would be feasible to interface a finite-element model using this method.

GRAPHICS OUTPUT

A very important product of any engineering design effort is the Design drawing, which of all of the documents produced is the most useful. BW-CAD produces a 2-dimension cross-section of a breakwater using Autodesk's Autocad Version 10. The interfacing of BW-CAD written in prolog and Autocad was one of major implementation problems. It was necessary to transfer data from the expert system component to the graphics component. This was quite difficult because Autocad does not have any relevant interfacing machanism i.e. C programming language. A flat file was chosen to convey the breakwater data from BW-CAD to Autocad and routines were written in Autolisp (The built-in language of Autocad) to interpret this data as well as making geometric calculation before actually drawing the diagram. The output was then plotted on a flat-bed plotter an example of the output can be seen in figure 7. Desqview was used to give a real-time interactive design drawing, by allowing both BW-CAD and Autocad to reside in memory simultaneously. It is possible to toggle between BW-CAD and

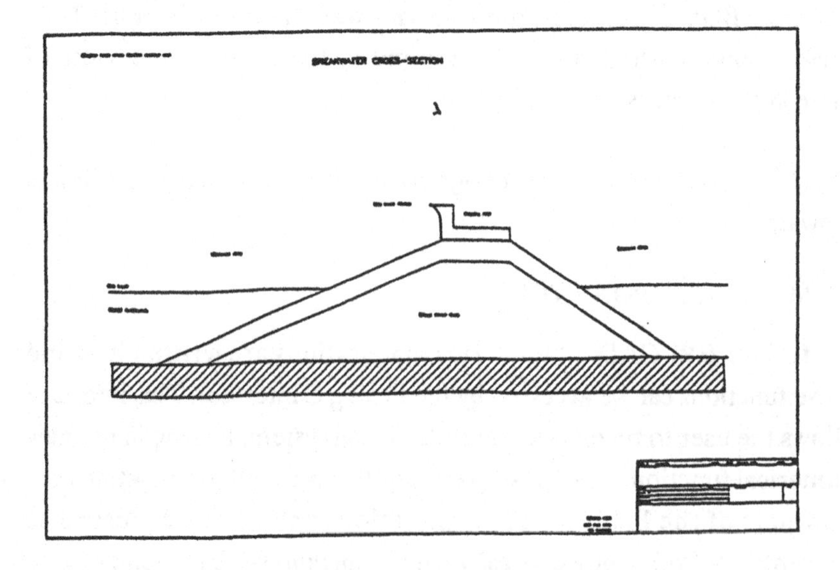

Fig. 6 Cross-section output from BW-CAD.

the Autocad with one key stroke. Using a highly developed CAD package like Autocad had advantages, apart from providing a rich collection of graphic functions (including 3-D graphics) and facilities to save and manipulate drawings, it also allowed the production of a hardcopy of the output on a variety of output devices (plotters and Printers).

USER INTERFACE

The development language provides a powerful set of user-interface development functions including Windows, pull-down menus, pop-up menus, dialog boxes etc. These functions made the process of developing a user-friendly front-end very straight-forward. One of the initial design criteria was that the interface should be very flexible, in that it should be possible to calculate a formula and then in one keystroke to backtrack, change some of the input parameters and recalculate the formula. Fig. 7 shows a typical input screen, which contains a pop-down menu bar on top as well as two dialog boxes. The first dialog box allows the user to change some input parameters like breakwater type, wave type, placement of armour units etc., while the second dialog box request the incident wave

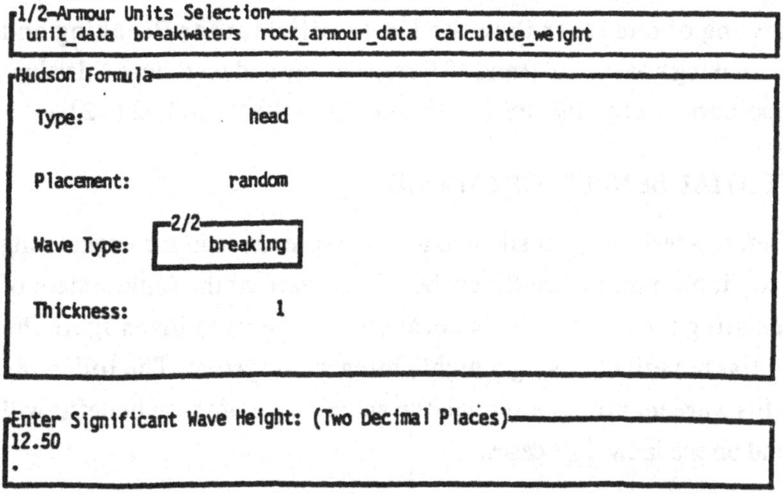

```
1/2-Armour Units Selection
 unit_data  breakwaters  rock_armour_data  calculate_weight

Hudson Formula
  Type:                 head

  Placement:            random
                    2/2
  Wave Type:          breaking

  Thickness:               1

Enter Significant Wave Height: (Two Decimal Places)
12.50
```

Fig. 7 Sample screen layout of BW-CAD.

height to be input. It is possible to backtrack and change any of these input variables and in turn observe the effect of the changes.

DATA STORAGE.

Although it is possible to interface LPA Prolog to a database, it is not necessary in the current system. It is well documented that Prolog can be used as a relation data base environment [Mai84] and that the prime disadvantages of Prolog as a relational database is the lack of secondary storage support. Prolog goal-oriented queries provide a powerful logic query mechanism. Since the data needed in BW-CAD is somewhat compact, the secondary storage facilities of a sophisticated Database Management System (DBMS) were deemed unnecessary. The following example is a Prolog fact which represents factual data on a Breakwater armour unit.

```
armour_unit_ data(dolos , trunk, random, 1, 10.0, 12.0, 1.33, 2.0).
```

This would be very similar to a database relation, if the breakwater data was stored in a DBMS. The above example indicates that the accropode breakwater unit positioned at the trunk of the breakwater and consisting of one layer then the Kd co-efficient for Breaking and Non-breaking waves is 10.0 and 12.0 respectively and that these Kd values are appropriate when the breakwater slope is in the range 1.33 to 2.0.

POTENTIAL BENEFIT OF BW-CAD

Before a technology is adopted for wide spread use in the engineering industry it must provide sufficent benefit to warrant the replacement of the existing technology. This point encourage us to investigate the potential benefits of using a highly integrated system. The following benefits were identified although the actual value of these benefits will depend on the individual cases:

1) The use of a system like BW-CAD would greatly increase the speed of a design and as a result would reduce the cost of an individual breakwater design.

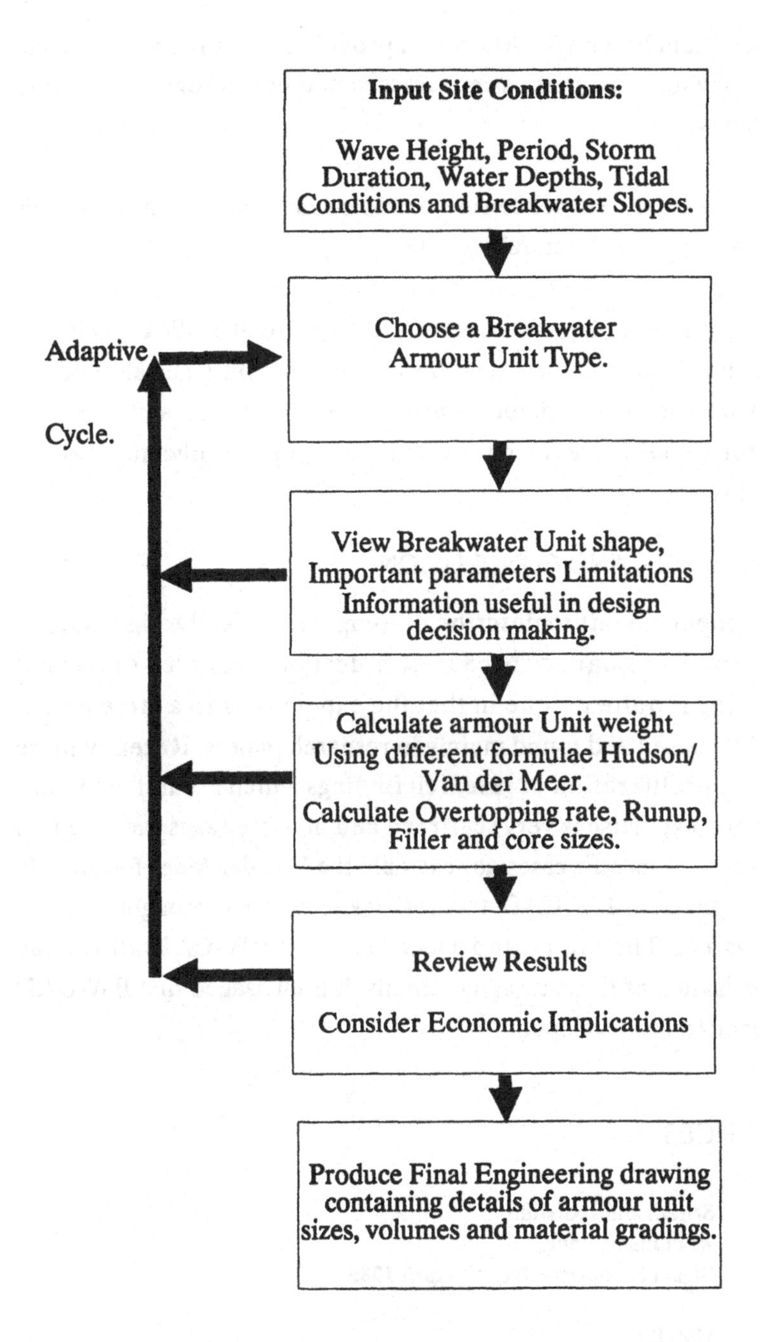

Fig. 8 BWCAD Flowchart.

2) A system like BW-CAD would provide a central pool of design expertise giving a designer easier access to uptodate design information and methods.

3) Because the designer can use more new materials and methods this should allow greater structural stability.

4) The main cost-benefit advantage of a system like BW-CAD is that it would give the design consultant the ability to optimise quickly a design using a wide range of available construction information and methods. This in turn would give the consultant a competitive advantage at the tendering stage.

FUTURE WORK AND CONCLUSION

This paper set out to describe a Computer-aided Design system, applied to the domain of breakwater design. This area of coastal engineering is quite unique in that the expertise is to a large extent empirically based and found mainly in research papers. Recently there has been a proliferation of research findings which are not yet in use because the expertise is very scattered and in some cases the research finding apply to specific cases, for example the Van der Meer formula. In the development of BW-CAD, this design knowledge is brought together in one system. The integrated architecture of BW-CAD allows the computerisation of these design methods. It is envisaged that BW-CAD will be continually developed.

REFERENCES

[SRI85] Sriram and Joobanni,
AI in Engineering.
Sigart Newsletter, No. 92. April 1985.

[Hall90] Hall,K.R,
Numerical solutions to wave interactions with rubble mound breakwaters
Canadian Journal of Civil Engineering Vol 17 no 2 April 1990

[Hol88] Holscher, P. et al,
Simulation of internal water movement in breakwaters.
Proceedings of the International Symposium on Modelling Soil-Water Structure Interactions, Delft, Sept. 1988

[Yoo89] Yoo, D. et al,
Mathematical models of wave climates for port design
Proceedings of the Institution of Civil Engineers, Paper 9407, Vol. 86 June 1989.

[Rot85] Rotteman-Sode, W. et al,
Two numerical wave models for harbours
Proceedings of the International Conference on Numerical and Hydraulic Modelling of Ports and Harbours, BHRA, Birmingham, England, April 1985.

[CIA85] CIAD Project Group,
Computer aided evaluation of the reliability of a breakwater design
CIAD Association, Zoetermeer, The Netherlands, 1985.

[SPM85] Shore Protection Manual,
U.S Army Coastal Engineering Research Centre. 1985.

[MAI84] David maier.
Databases in the fifth generation:
Is prolog a database language.
NYU Symposium New Direction for Database Systems, 1984.

[KUM89] B. Kumar and B.H.V. Topping.
A prolog-based representation of standards for structural design.
Conf. on AI techniques and applications for civil and structural engineers, 1989.

Controlling Database Integrity

R. Threadgold

RAMJET Software, 54-60 Merthyr Rd., Cardiff, U.K.

ABSTRACT

A scheme is described in which semantic nets are adapted to assist secure data acquisition. The nets are stored in a deductive database. They are normally used for system specification.

TECHNICAL OVERVIEW

A series of computer aided tools have been built. They were designed to assist the recording and presentation of a Real Time computer systems design. The record is stored in a structure of sets and relations based on set theory. The structure allows logical inferences to be drawn automatically. Several types of inference mechanism have been implemented.

The first is one which converts the record into semantic nets. These are a particularly useful way of presenting computer systems design information. The scheme is comprehensive. The user interacts with it to build up each scheme net. The dialogues of use each concern a fragment of the target net. The evolving net can be examined at each step.

The nets can be used to present several aspects of design. This versalitity has been exploited in a new application which is unrelated to the earlier subject. Rather it concerns the secure acquisition of data needed during the operation of a live computer system. The examples all relate to stock control for a printing shop. The general purposes nature of the tool means that this application is just the first of many.

INTRODUCTION

We have all laughed at the apocryphal story of the gas bill for £1,000,000. Some of us, those who have actually had to deal with computer programs, may have used the phrase: 'rubbish in, rubbish out'. The recording of credible information inside the army of computers which impact on our lives is a serious problem for database administrators everywhere.

The scheme described, was motivated by a mundane database application. This was for a printing shop. The idea was that the stock of paper, inks, etc., could be recorded and that this could be linked to control of jobs threading the printing machinery.

Management were concerned with the credibility of the record they would be using. A solution was proposed using the idea of a transaction log. All transactions affecting the contents of the database are logged. They don't alter the database itself. Instead, the log is subject to inspection. Only when management authorise the alteration, can information in the log progress forward.

Sitting within the mechanism of the log is another integrity feature. This limits access to types of information in the database. It is set up by management prior to use. It means for instance, that the order clerk cannot alter parameters such as the stock re-ordering level without authorisation.

Any alteration would normally be made through a log. Even management alterations can use one. In cases where extreme data integrity is demanded, the logs can be archived. If an inquest has to be held, pictorial logs of the contentious period can be studied.

An expert system has been developed for the specification of Real Time systems. This was described in a paper given at the Expert Systems 90 conference [1]. The system has been adapted to form a data logging mechanism. The adaptation forms the main section of the present paper.

To assist comprehension, two other sections are included. The first illustrates an application of the scheme. It includes parts of the specification drawn up on the expert system. The resultant operational databases are used by the customer during the act of stock and job control. The effects of including the integrity enhancements are discussed.

The second covers some of the foundation work. The key components have been evolved as the result of practical experience with a series of IKBS's. The section gives insights into what underlies both the expert system and its application to integrity.

THE WORKLOG

The expert system as an aid to worklog generation

An underlying aspect of the expert system is that users can construct semantic nets and have them recorded in a computer. A series of prompts and screens appears during use. The sequence has been found suitable for assisting specification of data which is to be used for updating a database.

The suitability springs from the nature of the 'messages' which form a fundamental component of any scheme produced using the system. These each comprise a message selector (or head) and parameters. When a Real Time system is specified, parameters are called 'events'. The sequence of data acquisition for an event always needs the definition of an extra piece of information. At some system levels this requires a path to be specified for the event. At others a type (as in computer data typing) is needed.

A message is formulated whenever a transaction is logged. It has a head marking which logical record is to be updated and parameters corresponding to the other data items in the record.

The recording of such a message is similar to the recording of a parameter list when a Real Time sequence diagram is specified. Particularly valuable is the part that requests definition of auxillary data. It allows the type of each message component to be controlled. This restricts the abilities of staff to alter data and helps improve database integrity.

The end result is that staff can use the expert system for logging transactions once management have set it up. The data acquisition is inherently secure.

Operations associated with message capture

The expert system blends aspects of list processing with aspects of programming in logic. It acquires information from the user by requesting him or her to input a list. It records the list as a series of facts.

As noted the data capture sequence is based on transactions in which the components are parts of a message. It starts at the head of the message and requests selection of a type. It assists selection by displaying a list of legal types. A feature of the tool is exploited here so that user definition of an illegal type is made obvious to management.

If a legal type is selected, a second list is displayed. This is used to define the content of the head of the message. If no existing data item is suitable, a new one is defined.

The same two part sequence is undertaken for each of the message parameters. As before it involves selection of a type and selection or definition of a message parameter.

When the message is complete, the facts of the transaction are saved into the database and can be displayed immediately in pictorial form. Two features of the expert system are useful if a mistake has been made. The first is a restart. This is used if the local copy of the transaction is incorrect. If the mistake isn't spotted until the pictorial record appears, the erroneous part of the picture (and the backing database record) can be amended.

Thus if stock is being ordered, the order clerk will define the item type as head of the transaction message. He will go on to define the attributes of the order such as entry date, order number, quantity etc. as the transaction parameters.

As each transaction is logged it is recorded in the staff members own worklog database. Transactions can be extracted from this and authorised when they are copied into a management database. A worked example is shown on figure 9. It was created on the net tool. Some comments have been edited on to aid clarity. The information was derived from a stock control sheet supplied by Peterborough Corporation.

When a worklog is set up prior to use, the logical data structure (LDS) of the target scheme is used. It constrains the ultimate data transfer into the target database. It defines the types of data which can be included on the worklog. It means the scheme is very general. Provided the LDS of a new target database is known, data entry transactions for populating it can be set up. The resultant worklog can be queried and data extracted in textfile form. This can be used to update a main database. The semantic net of the worklog is always available for scrutiny in cases of doubt.

ILLUSTRATIVE DATABASE

Communication model

Simplified SSADM A variant of the Structured System Analysis and Design Method (SSADM) is used for system specification. SSADM [2] is a well known method which is sponsored by CCTA, an agency of Her Majesty's Treasury. Figure 6 shows the communication model which underpins the scheme. The simplifications stem from two influences. The first is due to MASCOT [3]. This method is used by the Ministry Of Defence (MOD) for the specification of Real Time systems. It is the relationship between SSADM Data Flow Diagram (DFD) entities and both Logical Data Structures (LDS's) and Entity Life Histories (ELH's) which is simplified.

The second is due to concepts from 'Smalltalk', a computer language devised by Xerox. It includes a way of looking at programming which concentrates on the viewpoint of the 'System End User'. Programming is conceived as the act of establishing 'System Objects' (in our case the MASCOT 'processes') and of establishing the 'Messages' which flow between them (in our case the 'messages' in the SSADM Events Catalogue (EC)). The simplification is that ELH skeletons can have messages from the EC added to permit direct conversion to implementation code. A useful description of Smalltalk will be found in [4].

The simplified method has been used and a number of diagrams have been produced. The Real Time expert system was employed when this was done. Examples are referred to in the text.

Data Flow Diagram The method starts by requiring a DFD to be prepared. This identifies Real Time processes and dataflow pathways between them. Human users are treated as if they too are processes.

In the illustrative scheme there is a process for stock control aspects, one for job control aspects and one each for the two associated worklogs. Ordinary users cannot update the main databases. They only do it indirectly through the worklogs.

The scheme also includes three further databases for management use. The first controls access rights. The second records current print shop invoices (both into and out of). The other is a worklog used to archive invoices. Data flow paths are from users to worklogs, from worklogs to databases and from both worklogs and databases to users. Worklogs are called registers in the example scheme.

Logical Data Structure Processes are amplified by means of two further diagrams. The first is an LDS (see figures 4 & 5 for an example). This defines the database which is used by the process. It identifies the database structure in terms of binary relational set and relation files. The connector symbols correspond to relation files. These are named by the sets at the two ends. The LDS also identifies the type of relationships that are permitted to exist in a relation. This is useful for consistency checking.

Entity Life History The second is an ELH used to identify signalling sequences (see figure 7). A signalling sequence specifies an expected sequence of dataflows. 'Success' sequences are between a pair of independent processes or between user and process.

Event Catalogue Automatic analysis can be performed on the ELH's. This produces an ordered list of events analysed against the dataflow path used by the sequence. Event texts at this system level are the identities of the signalling sequences and each is implemented as a 'message'. The data elements carried in this, must originate from a data structure defined by an LDS. They must be stored (after transmission) in a data structure defined by another LDS.

The act of specification of data elements in messages fleshes out the events catalogue. One of the the lessons learned in earlier work concerns the need for phrase consistency in specification material - in this instance in the event texts. Phrase usage indexes are a mechanism which helps achieve this. The technique is applied during building of the EC. It complements the automatic ELH analysis.

Implementation ELH Figure 8 shows part of such an ELH. The fleshed out message specifications have been added to the original ELH framework.

Further detailing of System Requirements

The full system is quite a simple example of a Real Time system. It means that very little in the way of further, lower level, amplification has been necessary. This is mainly concentrated into the area of application specific functions (functions as in the AI language LISP [4]). Obviously, in this kind of scheme, the billing of customers and paying of suppliers is important and needs functions. There are others. These are mostly concerned with monitoring stock level changes (ie when ordered stock is received or when stock is allocated onto a print job). They also have to report when preset levels such as minimum stock are violated. Data transfers of authorised update material from a worklog use the general purpose copy function. This is described later. It is used whenever a remote database is updated.

Non success signals aren't shown explicitly. This is because existing trusted code is used to implement the signalling sequences. These possess well proved waitpoints to control unexpected signals, and include the necessary recoveries.

Reporting

An important aspect of the system is its ability to generate reports. These are wanted by ordering and by progress clerks, by print room operatives and by management alike. To be able to derive the reports means that all staff must have read only access to the databases. This involves messages from system to user and these are defined with the help of the report function.

Exploiting the Logical Data Structure The LDS's of the databases are the key to good reporting. The report writer exploits logical relationships defined by the data structures. It lets users tailor reports to their precise needs. Hypothetical Syllogism is used to extract data from a progression of linked sets. These are specified by the parameters of report requirement messages (see word REP on figure 8).

Finding the head of the wanted report Several levels of detailing are possible when report requirement parameters are specified. The first of these concern 'navigation'. This must be done when the logical record isn't at the top of the LDS. The user navigates down through the LDS by selecting screen options generated by Modus Ponens [6]. This ends when he selects the option which marks the head of the logical record he wants to extract. A logical record is the collection of set elements connected to the head element. All such elements can be reported if required. It is also possible to specify reports in which a part deals with an inner logical record. Navigation can be configured so that either a 'some' or an 'all' type of report is generated.

Data processing This is specified in the next level of detailing. The current values of report parameters can be processed. Simple arithmetic expressions are defined and the results of processing appear in the report. Data manipulation can include column width truncation (including column omission).

Message transmission between processes

Multi processing systems must be able to deal with inter process messages. Three functions are provided to assist. The first 'copies' data from one database to another when the LDS's of the logical records at each end of the transmission are the same. The second 'initiates' the data transfer implemented by the copy function when the LDS's are different. The third 'deletes' a remote logical record.

Ordinary amendment of database

The worklog scheme may not be necessary in some application schemes. An example is the supervisors ability to amend access rights. A number of functions are provided to help data acquisition in this case. They are screen based and integrity is controlled by limiting what can be amended or inserted. This is done by programming the message parameters of the 'insert' and 'amend' functions. Deletion of a local logical record can be done as well.

FOUNDATION ASPECTS

Developmental background

The computer systems which are referred to in this paper were developed over the last twenty years. Each one was designed to suit an immediate need. The key requirement was always the speed of getting the product into service. The work was always carried out as a management service or to help in the maintenance of software. It was performed for a British Defence Contractor and was usually in co-operation with an MOD department. Small teams of between two and five people were used.

A consistent approach was adopted based on set theory [5]. It was found that this discipline would give simple, resilient programs able to be developed very quickly, in 1970. The resultant systems didn't need much in the way of data preparation.

The highlights of the development progression were:

- A series of MIS schemes based on Cartesian products (mainly cost control and progressing schemes)
- A flow chart data acquisition scheme (the aim was to interface from word processor to database)
- A phrase usage indexing scheme (to help correct the poor text standards found on the flow charts)

. A flow chart to flowtree converter
(to aid users when following flow chart control threads)
. A Cartesian product mesh repository for flow charts
(the orignal DBMS broke at 40 diagrams)
. A query factory
(to aid unskilled database users)

The really difficult moment was discovering that the recommended commercial DBMS couldn't cope. There were over eight hundred flow charts and each contained about forty phrases in English. When a flow chart was destructured it generated about three hundred relationships. This gave very large numbers of consistency cross checks when a new diagram was entered and the DBMS had to be abandoned.

The original CP work was dusted off and developed to produce a practical replacement. This couldn't be done overnight. The phrase usage indexing scheme and the flowtree scheme were bespoke stop gaps to tide us over the interval.

The original query factory was written in Prolog [4]. This was a useful experience. It revealed that capturing credible data for a Prolog database was just as difficult as capturing it for any other database. In addition, it was found that the way the CP mesh was structured meant query programs from a Pascal factory could be produced just as easily.

A lot of auxilliary work had to be done which gave experience in the use of commercial PC database and spreadsheet packages. These caused questions to be asked about using them to replace the CP mesh databases. Some experiments were done with this possibility in mind. They were not successful. The commercial 4GL languages all seemed to be BASIC based and program operating times were disappointingly slow.

The last three years have been spent improving the original work and converting it into a commercial product. This was a private venture. It wasn't done for a defence contract.

The Cartesian Product as a World Model

Simple initial model A series of requirements for Management Information Systems (MIS) sparked the work. A study showed that two aspects of set theory could be exploited to produce the wanted results. The first was the use of Cartesian products (CPs) to relate two types of information into a report. The name of each set element on the CP axes is expressed in English and defines an ordinate or abcissa as appropriate. Crosses on the graph relate items in the two sets and each one allows a report entry to be deduced. Figures 1 & 2 help visualisation of this. The second was the use of co-domains (CDs) to add detail to the basic report. The detail was also in English. Each such co-domain is a set with elements on a one for one basis with one or other of the two CP sets.

The combination of CP and CDs is a 'world model' which is suitable for recording facts about certain classes of management system. Deductive reasoning can be applied to the populated model to produce reports. Several were built. For example a cost planning scheme captured and reported project budgets, a cost modelling scheme helped with the cost of materials aspects and a cost monitoring scheme helped with the assessment of performance.

Using cost modelling as an example, equipment names form one Cartesian product set and component names the second. The crosses relate equipments to the type of components within them. The co-domain sets are used to record things like specification number, cost centre number, unit cost, etc.. A pro-forma was designed for each such system and was filled out by management. Data entry needed a typist.

The end product data acquisition and reporting schemes were cheap and resilient. Both the crosses and the underlying mathematics were hidden from the end user. Each of the systems was a simple IKBS. The binary relationships (ie the crosses on the CPs) are like 'facts' in Prolog. The relationships between a CP set element and the corresponding CD set elements are like 'rules' in Prolog.

Improved model using CP meshes The success of these schemes led to demands which could only be met by improving the power of the underlying world model. This was done by giving each set the ability to have complex relationships with several others. Each such relation between a pair of sets needed a separate Cartesian product. The sets and relations form a logical data structure and the resultant database is of the form known as binary relational. Co-domains are not needed. Figure 3 helps with visualisation.

The simplification means Hypothetical Syllogism [6] can be applied to the mesh to deduce reports. In addition, the associated facts are structured by the configuration of the Cartesian products. It means the equivalent of Prolog rules is still implicit in the revised model. An example of a rule is the chain of conditions used to select a control path through a piece of program. This needs two CPs linked by the set used to record statement/statement transitions.

The attraction of the improved 'world model' is twofold. First it is extremely compact. Relationships are pairs of standard Pascal pointers. They point to set elements which are standard Pascal strings packed into coherent set files. Second it can be set up to be extremely fast. This is because all files, both set and relation can be fully sorted and binary searching can be employed.

By this time (1977) the demand was for analysis associated with software system needlines. It was required that consistent overviews should be able to be deduced from specification flow charts. To do it meant a complex mesh of Cartesian products had to be populated. The information originated as English texts in the flow diagram boxes. Hand loading of the data structure with these proved to be tedious. An automated loading aid was needed. In addition a means of presentation of the overview material had to be devised. Both problems were tackled simultaneously.

Semantic Nets as a way of improving comprehension

Directed graph overviews Graphs having annotations in English were found to be a suitable method of presentation. For consistency the flow diagrams were converted to this alternative form as well. Later all the graphs were improved by making them Petri net compatible. This had practical advantages associated with formalising system error types. It also meant the name 'net' could be used rather than graph. The structure of the underlying Cartesian products was arranged to give efficient both way conversion (ie from directed graph form to CP form and back).

The nets are sometimes called semantic nets. This is because they relate the English words in the requirements to the system control structures shown by the graphs. They are suitable for specification of software requirements at any level of system detail.

System levels The overviewing technique consolidated knowledge of relationships which exist between levels of requirement. Thus it was found that entity life histories identify the need for across the system signalling. Each need amplifies into a signalling sequence. These have to deal with both the 'success' and the 'non success' aspects. They combine into operational sequences. The steps are the system waitpoints. The expansion of the control paths from a single waitpoint gives rise to a control sequence. Each of these handles all legal signals (success and non success) at the waitpoint. Non success signals each initiate an appropriate recovery signalling sequence. Control sequences implement as code modules and include calls on the highest level of common user code module. The most important of these are the ones used to initiate an across the system signal.

Messages and Objects Control sequences respond to incoming signals at a waitpoint and cause outgoing signals to be transmitted. These signals are the messages which flow between the processes of the system. They are the key element in a philosophy which says that the act of programming is the act of identifying and sequencing the messages between system objects such as the independent processes.

Computer aided net production The construction of a semantic net is computer aided. It is accompanied directly by construction of the backing CP database. It means that relevant lists of existing phrases are presented to the designer during construction. He is able to put any which is suitable onto his net or to define a new one. By reducing the users need to rely on memory and by reducing key depressions, the aids give time savings during design.

Because the computer aid is keyed to helping the design of a real time system and is able to be used at any level, the scheme forms an expert system.

Other uses for semantic nets

Logical data structures The initial use of semantic nets was to portray the dynamic control structures needed to specify software systems. They can also be used to portray the logical data structures (LDSs) used by the implementation programs. Messages have been noted as a key element in programming. Each such message will normally carry structured information by means of parameters. These must each refer to a data element identified on the associated LDS.

Data flow diagrams The independent processes are in a structure which includes the inter-communication paths. Users and the dataflow paths needed when they communicate with the system are shown as well. Such diagrams identify dependent LDS's (one per process) and dependent sequencing diagrams (such as entity life histories). The sequencing diagrams are also on a one per process basis. The dataflow pathways on the diagram define interfaces between processes, other processes and users.

Exploitation of a requirements database

The generation of system phrase usage indexes As we have seen, a key part of requirements from the users point of view, is the use of English. The diagrams are useless without it. One of the reasons management want to capture requirements into a database is to help standardise phraseology. It is done with the aid of diagram/phrase usage indexes created by a database query program. We have noted the handling of more than eight hundred diagrams where each contained about forty phrases.

Actually having the indexes forces the adoption of good semantic standards (for instance the eight synonyms of the word timeout found in diagrams on the above project were able to be removed).

The generation of an events catalogue Indexing is a technique which is able to be used to improve the quality of a design. More important it has been adapted so it catalogues both those messages which flow between the independent processes of the system and those used to invoke lower level modules.

Automatic code generation The automatic production of directed graphs from the database showed us that another form of presentation was possible. This is source code. It can be generated direct from the database. It is derived from a family of databases which record all specification stages from the topmost requirements level. If these are related one to another in a fully consistent way, the code can be proved to be logically correct to its requirements. This is important because it gives possibilities for the safe reduction of testing.

Exploitation of code modules from the toolset

Production of the Real Time expert system and its support tools has meant that a large body of trusted code has been accumulated. Certain modules have been so designed that they are able to be put together as if they were functions in an extensible, LISP like language.

The most important of the re-usable modules are those which enable the central programming philosophy to be implemented. Each one concerns an aspect of message transmission. The specification of how the modules are put together uses sequencing diagrams.

These show the order of use. In addition, when a module call is specified, its parameterisation will be consistent with the associated LDS. Key re-usable modules include for example: Create record structure, Insert logical record, Delete remote logical record, Copy logical record to remote database, etc.

All these modules include appropriate features such as variable length parameter lists, search onto wanted record instance, associative addressing and inbuilt data processing.

The significance of the re-usable modules

Our belief is that it is the re-usable modules which are the true end product of our work. We believe that the worklog data acquisition scheme and the illustrative database between them, demonstrate just how effective it is to have a set of such modules to hand. The actual new modules that have had to be produced for the application is very small. Our view is that stock and job control is just one of a large number of similar linked applications (eg Personnel, Booked time, Cost plan, Cost monitoring, Accounts and so on). Work has been done which shows that these can all be built from existing trusted code with a minimum of new functions. Eventually, we believe that the personnel side of a firm can be modelled and integrated with the materials side (as exemplified by stock and job control).

The fact that it can be done on a distributed basis across several wordprocessing PC's means the whole scheme should be able to be purchased at a very reasonable cost.

CONCLUSION

Some applications of a new expert system are discussed. The main theme is that management can use an adaptation of the expert system to control alterations to a database. This is a necessary feature when the data integrity of the database can be called into question.

To support the theme an illustrative database is introduced. This exploits IKBS techniques which were built up during many years of production of MIS systems. The IKBS work culminated in the expert system. The illustrative database is Real Time in the sense that its orders and jobs are processed as early as can be, consistent with the data integrity issue. The section on the illustrative database mentions SSADM. Possibly the most important thing about this is the event catalogue which is exploited to eliminate coding. SSADM uses three main types of diagram (DFD, LDS and ELH). Simplified versions of all three types can be produced using semantic nets.

Finally, the development progression which led to the expert system is summarised. It is noted how a scheme evolved which has proved out to have attributes of both main line AI languages. It may be of some interest that the semantic nets generated and used by the scheme are similar to those of Micro prolog [4]. It has not been an aim to develop an AI language. The motivation from the very beginning has been that English is the only language the user must see. All the formal underpinnings are hidden from him. During operation he responds to lists of English texts which are presented to him on screen. The texts themselves will have been specified by a user.

ACKNOWLEDGEMENTS

The sections on the illustrative database and foundation aspects are to help comprehension. The copyright to these sections and to their supporting figures is with RAMJET Software. They may be freely copied providing a suitable acknowledgement is made.

REFERENCES

1 Threadgold, R., The Specification of Real Time Software, Proceedings of Expert Systems 90, the Tenth Annual Technical Conference of the British Computer Society Specialist Group on Expert Systems, London, September 1990, Cambridge University Press 1990.

2 SSADM Version 4 Reference Manual, NCC Blackwell Ltd 108 Cowley Road, Oxford, OX4 1JF. 1990.

3 The official MASCOT handbook, Computer Division, N Building, RSRE, 29 St. Andrews Rd., Gt. Malvern, Worcs.

4 Baron, N., Computer Languages, The Penguin Group, 27 Wrights Lane, London W8 5TZ. 1988.

5 Lipsshutz, S., Theory and problems of set theory and related topics. Schaum Publishing Co. New York. 1964.

6 Copi, I. M., Introduction to logic. The MacMillan Company, New York. Collier-MacMillan Ltd. London. 1972.

FIGURES SHOWING HOW CARTESIAN PRODUCTS CAN UNDERPIN AI

This figure represents the Cartesian plane. Each point P represents an ordered pair of real numbers (a,b). A vertical line through P meets the horizontal axis at a and a horizontal line through P meets the vertical axis at b.

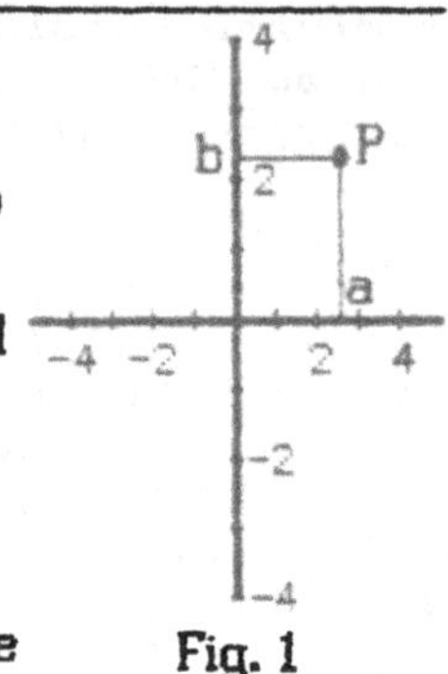

Fig. 1

The Cartesian product of two sets can be shown on a coordinate diagram in a similar way. Figure 2 shows elements of set A on the horizontal axis and elements of set B on the vertical axis. Each point P represents one relationship between elements in sets A & B.

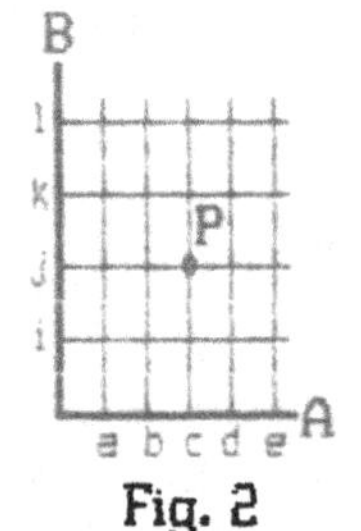

Fig. 2

Figure 2 can be exploited to give the MODUS PONENS rule of inference. This is expressed symbolically:

p implies q, p therefore q

(an element a in A implies a subset (b) in set B)

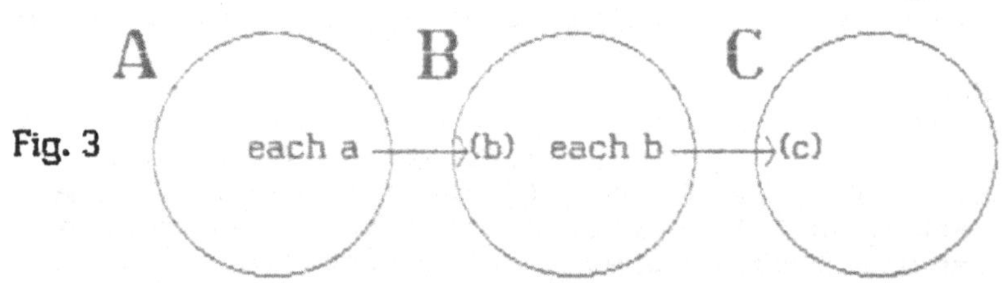

Fig. 3

Figure 3 shows one of the methods which is used to depict functions in set theory. It demonstrates the first step in building up a mesh of Cartesian products which is based on the sharing of elements in a set common to two relations. The common set is B. The relations are A x B and B x C. The symbol x reads 'cross'. The process can be continued to include further sets and relations. Repeated application of Modus Ponens can be exploited to give the HYPOTHETICAL SYLLOGISM rule of inference. This is expressed symbolically:

p implies q, q implies r, therefore p implies r

(element a in A determines a group of element subsets in C)

The 'reality' represented by a mesh of Cartesian products means other logical rules of inference can be used. Thus if Modus Ponens is used based on an element p and the q which is delivered isn't as expected, we have MODUS TOLLENS.

FIGURES SHOWING ONE EXAMPLE OF USE OF A SEMANTIC NET

The LDS in figure 4 controls the access process. It is a typical LDS. It is the core of the mechanism which controls database integrity. Passwords control commands able to be used by operators. Commands can only act on information in named sets and connected logically by named relations. They can only load and save work from or into named directories, etc.

The figure is compatible with an SSADM LDS.

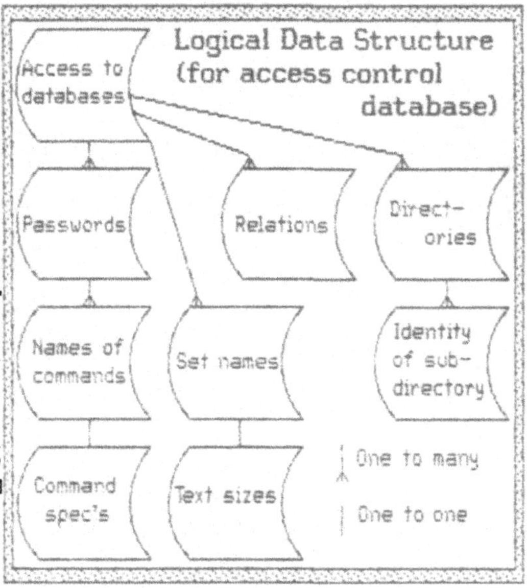

Fig. 4

There is value in recording an LDS in its own database. When this is done the user will interact with the 'net tool'. This produces the semantic net shown in figure 5. This type of net originated as a directed graph. It has since been made Petri net compatible.

```
                 0000000 Access to databases
One/many               -
                       0 Directories
One/many               -
                       0 Identity of subdirectories
One/many             -
                     0   Relations
One/many           -
                   0     Set names
One/one            -
                   0     Text sizes
One/many         -
                 0       Passwords
One/many         -
                 0       Names of commands
One/one          -
                 0       Command specs
```

Fig. 5

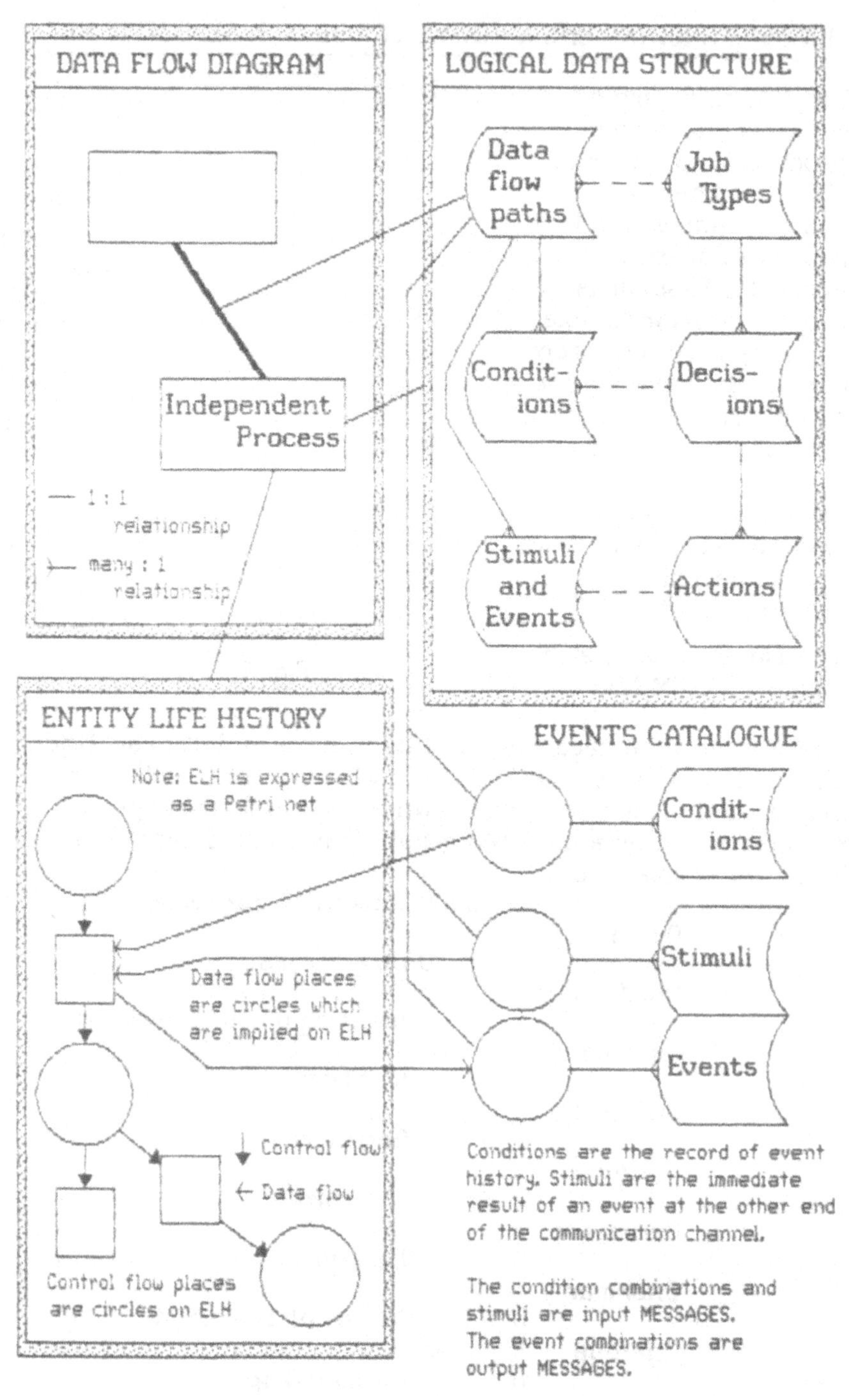

Fig. 6 Communication model for System Building

```
Order clerk prepares draft order
7 Load stock control reg   0 BUY STOCK
                           -  Report stock situation 4
7 Order requisition        0 bs2
                           -  Insert order req       7
                           -  Copy order requisition 5
  (STIMULI ie legal           (EVENTS ON RIGHT)
  commands, ON LEFT)
Supervisor confirms order
6 Load stock control db    0 bs3
                           -  Report draft orders    9
6 Load stock control reg   0 bs4
                           -  Report stock situation 9
6 Confirm order req        0 bs5
                           -  Amend to order confrmd 7
                           -  Copy order confrmation 5
Order clerk notes delivery
7 Load stock control reg   0 bs6
                           -  Report stock situation 4
7 Delivery note (INS-INV)  0 bs7
                           -  Copy supplier invoice  8
                           -  Copy supplier invoice  1
                           -  Delete order req       5
Supervisor tidies up when payment notified
6 Load invoices database   0 bs8 accounts notify pymnt
                           -  Report supplier invocs 9
6 Debit note               0 bs9
                           -  Copy 'supplier paid'   1
                           -  Delete supplier invoce 8
                           -  Print debit note       0
                           0 1 BUY STOCK
```

Fig. 7 Semantic net of ELH for buying new stock
Note: The stimulus and event texts and path reference numbers have been truncated to fit the space available. The diagram is displayed on a VDU. A help feature lets the user see the full English names of the paths. The underlined texts have been edited in.

```
7 LOA (A:\SCR1144\S2)      0 BUY STOCK
   (Remat A4, see fig 9)   -  REP L O~ B D L M       4
7 INS-ORD                  0 bs2
                           -  COP D I N X E Q B      5
                           -  INS I N X E Q B=B+Q    7
6 LOA (A:\SDB)             0 bs3
                           -  REP E: I D N X Q       9
6 LOA (A:\SCR1144\S2)      0 bs4
   (S2 means sheet 2)      -  REP L O~ B D L M       9
```

Fig. 8 Implementation ELH (part only)
The skeleton is that of fig 7.
Letters are initials from the controlling LDS

```
                         O Buy/sell stock (1144)
I   01                   -           (I = Inwards goods)
E   8:1:91               -           (E = Entry date)
N   072612               -           (N = No of order)
Q   192000               -           (Q = Quantity)
B   240000               -           (B = Balance)
                         O b/s2
O   02                   -           (O = Outwards goods)
E   21:1:91              -
W   Leisure/museums      -           (W = Where for)
Q   6000                 -
B   234000               -
                         O b/s3
O   03                   -
E   21:1:91              -
W   Architects           -
Q   9000                 -
B   225000               -
                         O 1 Buy/sell stock (1144)
```

Fig. 9 Semantic net recording stock buying/selling
Derived from manual stock sheet for Remat A4 paper.
The reference code of the item is 1144 (see fig 8).

--- by courtesy of Peterborough City Council ---

Note: Letters are initials from the same LDS as figure 8

Commentary on fig 9 The supervisor sets up the order clerks access rights so that he or she can only use relevant types of data. For example it would be arranged so that the clerk couldn't alter the maximum stock limit or change an expenditure code.

Intelligent Conceptual Design for Mechanical Systems

Q. Wang (*), J. Zhou (*), M. Rao (**)
(*) Department of Mechanical Engineering,
Huazhong University of Science and Technology,
Wuhan, Hubei, P.R. China
(**) Intelligence Engineering Laboratory,
Department of Chemical Engineering,
The University of Alberta, Edmonton, Alberta,
T6G 2G6, Canada

ABSTRACT

In mechanical system design, the Integrated Computer Aided Design (ICAD) techniques have evolved a new generation of design techniques. It also paves the way for implementing Computer Integrated Manufacturing System (CIMS). However, the key issue to accomplish the objectives in ICAD is the conceptual design automation. In this paper, we first introduce the fundamentals of conceptual design for mechanical systems design. Then, a strategy to solve the problems in conceptual design is proposed. The associated control structure and a tool for developing mechanical systems conceptual design expert systems are presented. Finally, applications cases are studied.

INTRODUCTION

Currently, the computers are widely used in the research and development of mechanical system design. As a new technological frontier, intelligent system technology has been extensively applied to design processes [2-3; 10-11]. Design is a very complex and not a completely understood process particularly since it is abstract and requires creativity. It is largely a trail and error process, in which human design experts often employs heuristics or rules of thumb to solve their problems. The Integrated Computer Aided Design (ICAD) techniques have evolved a new generation of design techniques. It also paves the way for implementing Computer Integrated Manufacturing System (CIMS) [5-6, 8, 10]. However, the key issue to accomplish the objectives in ICAD is the conceptual design automation.

Mechanical system conceptual design (or scheme design) is a very important but difficult target in CAD (Computer-Aided Design). A Mechanical system consists of many components and subsystems to facilitate numerous functions. It includes various complicated structures, such as machining centers, construction machinery, and generator units. On the other hand, a simple machine, (such as a motor, an oil pump, a washing machine, or a lawn mower) provides a single function and has simple structures. Compared with the design of a simple machine, a mechanical system design is much more challenging, complex and difficult. For example, the design and manufacturing of a mechanical system are more expensive and time consuming. Also, a mechanical system design usually contains more design procedures and experiments. A poor design may result in an incredible waste. Since a mechanical system is composed of many components and simple machines, its quality depends on not only the quality of individual parts, but also the correct connection of these parts. Thus, it often involves the cooperation of many experts from different disciplines [1, 8-9].

Mechanical system design includes three different designs: pattern design, creative design and modifying design. The pattern design combines existing components and simple machines into a large mechanical system (a pattern). This kind of design is often used in the process mechanical system design. The creative design is to produce a physical entity based on mathematics and mechanical principles, and is widely applied to the design of simple machines. The modifying design bridges the pattern design and creative design. It intends to improves the pattern design and creative design.

The conceptual design is a main process in a mechanical system design [7], in which a pattern is generated gradually, very similar to the procedure in playing chess. When playing "GO", the player, facing the situation on the chessboard, seeks an optimal strategy (function) to counterattack his opponent. To achieve the objective, the player has to formulate his own patterns, that is, the placement of the chess pieces. As there exist more patterns than one, a decision-making process to generate a final solution are required. In a mechanical system conceptual design, designers must organize patterns (structures) to achieve system functions according to the system environment and application requirements. A mechanical system conceptual design addresses the problem how to choose system structures and types of machines and then to integrate them into a unit. This is an ill-formulated problem, in which mathematical models are not amenable. Therefore, it is suitable for the application of artificial intelligence techniques [2].

The structures and subpatterns in a pattern are called concepts, such as pumps, motors, brakes, and so on. The pattern design is to put several valid concepts together. Then, the attributes of each concept must be assigned. For instance, an engine should be selected as a diesel engine, a gasoline engine or a motor; how much its power and torque should be. The determination of attribute values not only implements concepts, but also provides data for analyses and evaluations to make decision. Here, the process to determine attribute values is referred as parameter design. In summary, a mechanical system conceptual design consists of two parts: a pattern design (concepts combination) and a parameter design (attribute values determination).

DEFINITIONS ABOUT CONCEPTUAL DESIGN

In order to present conceptual design techniques clearly, it is necessary to introduce a few definitions and terminology.

Definition 1. Concept is defined as the abstract description of the natural property of an object. Each concept has its own identifier to refereed as C_i.

Definition 2. Concept Space is a set that includes all concepts in a specific domain. These concepts are not heaped up disorderly, but they are organized in a specific order and hierarchy to behave the certain relations between them. Concept Space is written as CS. Needless to say, $C_i \in CS$.

Definition 3. Function Concept is the abstract description of the features of system functions. It is expressed as FC_i.

Definition 4. Function Concept Space is a set that includes all functions in a specific domain. It is written as FCS. Similarly, $FC_i \in FCS$, and $FCS \in CS$.

Definition 5. Structure Concept is the abstract description of the essence of structures. It is represented as SC_i. There exists such a relationship that $SC_i \in CS$.

Definition 6. Structure Concept Space consists of all structure concepts. It is also a subset of CS. It is denoted as SCS. $SC_i \in SCS$, and $SCS \in CS$.

Definition 7. Effective Concept is the concept that satisfies the application environment and objectives as well as key constraints. Its notation is EC_i.

Definition 8. Effective Concept Space (ECS) consists of all effective concepts. $EC_i \in ECS$, and $ECS \in CS$.

Definition 9. Effective Function Concept is a function concept that satisfies the application specifications and constrains. It is denoted as EFC_i.

Definition 10. Effective Function Concept Space (EFCS) includes all effective function concepts. $EFC_i \in EFCS$, while $EFCS \in FCS$.

Definition 11. Effective Structure Concept (ESC) is a structure concept that satisfies the application environment, objectives, constraints and effective function concepts. $ESC_i \in SCS$.

Definition 12. Effective Structure Concept Space (ESCS) consists of all effective structure concepts. $ESC_i \in ESCS$, and $ESCS \in SCS$.

Definition 13. Design Pattern is a tree structure that consists of nodes and arcs. Each node represents effective structure concept, and each arc indicates an "AND" relation of nodes, or an "OR" relation of a single node. It is denoted as DP_i.

Definition 14. Design Pattern Set (DPS) is equivalent to ESCS. Each design pattern represents a design scheme. Therefore, a pattern set is also a scheme set that meets application environment and design specifications. $DP_i \in DPS$.

GENERAL PROBLEM SOLVING STRATEGY

A five stage problem solving strategy for general mechanical system conceptual design is described as follows (see Figure 1).

Stage 1 is a Problem Definition stage for design tasks (from application environment and purposes to functions). Functions to be used are chosen from the expertise function memory according to the application environment and purposes provided by users. For example, a wheel loader must be capable of traveling, shoveling, digging, shooting, and so on. The knowledge to define functions is shallow knowledge.

The first step in problem solving can be expressed as follows:

$$\text{STEP 1} = (FCS, \sum_{i=1}^{n} EFC_i | S_i \text{ and } T_i, EFCS) \quad (1)$$

where, S_i and T_i ($i = 1, 2, \ldots\ldots n$) represent specifications and constraints provided by users. The objective in the above expression is to find an EFC_i to satisfy S_i and T_i in FCS, then to combine these EFC_i into EFCS.

Stage 2 is an Effective Conceptual Design stage (from function to structure). The structures to execute functions are selected from structure memory. The communication between functions and structures is not an "one to one" mapping. Such a "multiple to multiple" mapping configuration (see Figure 2) indicates that a function can be realized with many different structures, and a structure may possess many functions. For example, the function for traveling can be implemented by two structures: tire and caterpillar. This "multiple to multiple" mapping makes the pattern design more complicated and diversified. Each design pattern has a design scheme. If there are more than one design patterns, we need to select an optimal (or near optimal) one. In the case of no design patterns are available, new design techniques will be used (since no existing structure can be used for the needed functions). If the existing design patterns fail to satisfy application requirements, the design has to be improved. The knowledge to formulate effective structure concepts is heuristic knowledge which can be represented by heuristic rules.

The following expression is the second step of problem solving:

$$\text{STEP 2} - (\text{SCS}, \sum_{i-1}^{n} \text{ESC}_i | S_i \text{ and } T_i, \text{ESCS}) \quad (2)$$

This expression presents how to select all ESC_i to satisfy S., T_i and EFC_i in SCS, then to combine these ESC_i into ESCS.

In general, ESCS is a set of design patterns. It is necessary to resolve them into individual patterns such that each individual pattern is one design scheme. The process is called as scheme resolving. It can be expressed as the following algorithm:

$$\text{RESOLVING} - (\text{ESCS}, \text{OP}, \sum_{i-1}^{n} \text{DP}_i) \quad (3)$$

where, OP stands for a set of resolving operations. The purpose in the above expression is to divide ESCS (or DPS) into several DP (or design schemes) with OP.

Stage 3 is a Parameter Design stage (from structure to parameter). At this stage, the detailed description of structures can be completed by using the design models stored in model memory according to the characteristics of effective structure concepts. The knowledge to determine structural attributes is deep knowledge, namely model knowledge, which is represented by object-oriented frames.

The third step can be shown below:

$$\text{STEP 3} = \left(\sum_{i=1}^{n} DP_i,\ OPF,\ \sum_{i=1}^{n} ADP_i\right) \tag{4}$$

where OPF represents an operating set to frames, and ADP_i is a design pattern with attributes and values (parameters). The expression indicates that it converts DP_i into ADP_i by OPF.

Stage 4 is an Analysis stage (from parameter to analysis). Because functions and structures share a "multiple to multiple" mapping configuration, numerous design schemes are usually produced. After parameters are given, all design schemes will be analyzed by selecting a numerical computation method (e.g. finite element analysis, optimization, and so forth) from method memory. Conventional CAD techniques can be utilized here.

The fourth step of problem solving can be expressed:

$$\text{STEP 4} = \left(\sum_{i=1}^{n} ADP_i,\ OM,\ \sum_{j=1}^{m} ADP_j\right) \tag{5}$$

where, OM represents an operating set of analysis methods, and $m \le n$. The algorithm is used to analyze every design scheme so that the feasible schemes are selected from ADP_i to satisfy the requirements of analyses. Usually, a few schemes (or patterns) is omitted, and only the practical schemes are kept.

Stage 5 is a final stage for comprehensive evaluation (from analysis to evaluation). According to analysis data, a proper evaluation index system from index knowledge memory and a comprehensive mathematical model from evaluation models will be chosen to evaluate the selected practical scheme. Techniques of fuzzy mathematics and system engineering are used in evaluation.

The fifth step of the proposed strategy can be represented as:

$$\text{STEP 5} = \left(\sum_{i=1}^{m} ADP_i,\ OE,\ ADP^*\right) \tag{6}$$

where, OE is an operating set of evaluations. $i = 1, 2, \ldots\ldots n$. This algorithm intends to find the best scheme among practical candidate patterns by using evaluation OE. ADP^* is an optimal design to be sought.

Each stage in this problem solving strategy is very important. It combines numerical calculation (such as mathematical modeling, optimization and scheme analysis) with symbolic reasoning (knowledge representation and model handling as well as scheme evaluation) to accomplish the objectives in every stage.

SYSTEM CONFIGURATION AND IMPLEMENTATION

The software system control structure for the proposed conceptual design methodology is shown in Figure 3. The structure simulates the human being reasoning behavior in design process such that it can be used as a general framework for developing expert systems to accomplish conceptual design. The system control structure uses the module technique and consists of a menu, thirteen subsystems (modules) and nine knowledge bases. The following descriptions briefly describe their functions and characteristics.

(1) Menu Management System: It can guide users to select modules and observe the performance of the modules. Since the menu system employs a tree structure (each subsystem has its own sub-menu), users can select and run modules according to the contents on screen.

(2) Task Definition Module: It is a window to input information. Users can define design tasks, application working environment, purposes and specifications through the window.

(3) Function Concept Design Module: It functions as the first step of the strategy. Its purpose is to further expand facts and information in the knowledge bases due to the existing specifications, then reduce interactive contents.

(4) Structure Concept Design Module: It can select the types of machines and potential structural configuration components to satisfy the function and key constraints. This module works in the second step and provides parameter design with a variety of features. Its knowledge base includes experts' experience and heuristic knowledge which are represented as heuristic rules. Inference employs constraint reasoning.

(5) Parameter Design module: It performs its function that is similar to the third step of design. With the structures obtained from concept design and the facts provided by users, attribute values (parameters) of the structures are determined. The knowledge of parameter design is expressed by the data structure of object-oriented frame that is operated by the problem solver.

(6) Layout Design Module: Based on the geometrical parameters, the layout design module determines positions and orientations and realizes drawing design. Its another purpose is to perform the motion simulation for testing interface.

(7) Scheme Analysis Module: It functions in the fourth step. Based on the results from parameter design and layout design, this module analyzes each preliminary design scheme and provides data for evaluation. There are two paths in the module. If there exist more satisfactory specifications, then the system selects next module. Otherwise, the local redesign is required.

(8) Scheme Evaluation Module: This module is equal to the fifth step of problem solving (comprehensive evaluation). Its purpose is to evaluate comprehensive functions. In other words, all schemes entering the evaluation module are practical ones that are different with each other only in quality. The evaluation uses the level evaluation, fuzzy mathematics and system engineering techniques. Index system and weights are selected by domain experts. There are two paths in the evaluation module. If the evaluating results satisfy the specifications provided by users, then the information is sent to the Decision Making Module. Otherwise, a global redesign will be performed.

(9) Decision Making Module: In general, it is very often to generate multiple schemes as design result. The decision making module will pick up a best scheme among these schemes. Of course, if the user is a design expert, he (or she) selects an optimal scheme in terms of his (or her) own opinion. During a mechanical system design, the index system solicits various opinions from different domain experts. Thus, the conflicting solutions are generated from the different experts opinions. It is reasonable to have experts make a final decision, rather than by a computer program.

(10) Interpretation Module: It connects with the inference engine and provides the interpretation for concepts and reasoning paths.

(11) Inference Engine: Inference engine usually performs specific tasks and formulates knowledge representation. Obviously, in order to solve the large hybrid problem in mechanical system conceptual design, a simple inference engine that provides only one reasoning technique is unsatisfactory. The integration of different inference mechanisms are very demanding in solving the real world problems.

(12) Problem Solver: It operates method knowledge base with a variety of problem solving strategies, including reasoning, calculation, table look-up, curve observation and analogy.

(13) Layout Inference Engine: Layout inference is more complicated than symbol inference because it deals with shapes, positions, orientations, constraints and interferences of bodies in the space.

(14) Static Fact Base: SFB can be used to record task definitions and specifications provided by users. The information in SFB contains the essential conditions (constraints) for function and structure concept design. The facts in SFB is expressed with vectors.

(15) Dynamic Fact Base: In a reasoning process, users must continuously provide the more detailed facts and data which are stored in DFB.

(16) Intermediate Result Base: IRB stores intermediate results in the processes of symbolic reasoning and numerical computing. There are two purposes for setting up MRB. These results are used in a continuous reasoning process. When a design fails, a backtrack (redesign) will be performed with the information stored in MRB.

(17) Final Result Base: FRB stores all acceptable schemes in the design. The decision making module refines the best one among those stored in FRB.

(18) Method Base: It records the structure descriptions for all problem solving methods. Each parameter or structure attribute has its own specific methods to be generated through reasoning, table look-up, analogy, calculation and so on. MB is operated in many different ways. It is separated from the problem solver, and is only some descriptions of problem solving procedure. When needed to handle a new parameter, users may add description to MB without changing the problem solver. All methods are described with an object-oriented frame.

(19) Knowledge Base: It consists of many files organized as production rules. Each file is generated by the knowledge acquisition module to perform a specific subtask.

(20) Model Base: Model Base stores various parametric models needed to assemble objects. A model can be placed on a suitable position if key parameters and a transformation matrix are provided.

(21) RDB: RDB is a commercial database and functions as a center to exchange information.

(22) Optimization Base: It provides seven optimization techniques, such as linear optimization, nonlinear constraint optimization, discrete optimization.

Based on the configuration discussed above, a software platform for developing mechanical system conceptual design expert systems has been implemented. This system, namely CDESTOOL, has the following features:

(1) It can be used to develop concept design (scheme design) expert system for mechanical systems or their components.

(2) It provides the problem solving strategy for design types with multiple objectives or uncertain goals.

(3) Its system configuration is an open structure. With different requirements, users can modify the system at any level of system.

(4) CDESTOOL provides a variety of inference methodology, including constraint inference, layout inference, certain and uncertain inference, etc. So far, eleven inference engines are built in this system. Each of them can be selected for a specific subtask.

(5) A variety of knowledge representation techniques are provided in CDESTOOL. Each knowledge representation technique is associated with an inference engine. Users can choose these techniques whenever needed.

(6) CDESTOOL interfaces with database.

(7) It combines numerical calculation with symbolic inference.

(8) The layout design of structures is accomplished.

(9) Decision support subsystem in CDESTOOL can evaluate all schemes.

(10) It can accomplish the redesign of optimization backtrack.

(11) The system provides a good interface for the analysis module.

(12) Because the knowledge of reasoning, methods of analyses and formula of calculation are separated from control structure, problem solver and inference engines, it is easy to modify and manage knowledge base.

APPLICATION CASE STUDY

Three applications have been successfully developed: (1) Industrial Turbine Conceptual Design Expert System (ITCDES); (2) Wheel Loader Conceptual Design Expert System (WLCDES); (3) Milling-Boring machining center Conceptual Design Expert System (MBCDES). These systems have been successfully employed in industry. For example, ITCDES has been used to test 150 real industrial problems (old products) and designed five new industrial turbines, and MBCDES is also successfully used in the real industrial design. In this paper, we briefly present WLCDES as an application illustration.

WLCDES can accomplish a variety of wheel loaders (from 1 ton to 10 tons) and is employed in the integrated WLCAD (Wheel Loader Computer-Aided Design) environment which includes hydraulic system, transfer motion system, brake system, working unit, and so on.

WLCDES is implemented on a Micro VAX-II, which is heterogeneous knowledge integration environment. Its symbolic reasoning system is developed with Common Lisp language, while geometric conceptual design and simulation systems are built in Pascal. FORTRAN is used to develop evaluation model, optimization base and analysis model. A graphic package (UIS) is run at a GPS graphics workstation. The platform (hardware and software of WLCDES is demonstrated by Table 1.

TABLE 1

Hardware	Software	Objectives
Micro VAX II	Common Lisp	symbolic reasoning
Micro VAX II	FORTRAN	optimization, analysis
Micro VAX II	PASCAL	simulation, geometry design
GPX	UIS	graphics

WLCDES environment is developed for the system conceptual design for the wheel loader equipment, it can accomplish the following tasks:

1. conceptual design
2. parameter design
3. layout design
4. pattern selection and parameter determination
5. important design parameter selection.

The implementation of WLCDES is based on the concept of Integrated Intelligent System [4], which is a large knowledge integration environment. An integrated intelligent system consists of several different intelligent systems and numerical computation packages. As demonstrated by Figure 4, these software programs may be implemented in different languages and be used independently. They are under the control of a supervisory intelligent system, namely, metaf-system. The meta-system manages the selection, operation and communication of these programs.

The key issue of constructing an integrated intelligent system is to organize a meta-system. The main functions of the meta-system is to coordinate symbolic reasoning and numeric computing processes, to distribute knowledge sourses, to find an optimal solver for conflicting solutions, to provide parallel processing capability, as well as to standardize communication information.

WLCDES is the first intelligent system developed for the conceptual design of wheel loader system. It provides the following features:

1. It is an integrated intelligent design environment [4-5]. Its systematic functions include conceptual design, layout design, dynamic simulation, performance analysis, and methodology evaluation.

2. Its modularity enables system configuration so flexible that the knowledge base is easy to be expanded and modified by the end users, rather than the original developers.

3. It can accomplish both multi-objective design tasks and uncertain multi-criteria design tasks.

4. Integrated intelligent system concept provides an open structure for organizing this intelligent system. Such a configuration allows the user to modify knowledge base at any level of the system. In WLCDES, numerical computing process is coupled with symbolic reasoning to enhance system capability.

5. Several different problem solving strategies, reasoning mechanisms and knowledge representation techniques are provided in WLCDES. A user can select the suitable knowledge processing techniques to deal with his (or her) specific design problem.

6. WLDCES interfaces with the commercial databases.

ACKNOWLEDGMENTS

The authors wish to acknowledge the financial support from the Chinese National Science Foundation and National Sciences and Engineering Research Council of Canada.

REFERENCES

1. Manoocheiri, S. and A. Seireg, "Computer-Aided Generation of an Optimum Machine Topology for Specified Tasks", CIME 6(3), pp. 10-24 (1987).

2. Rao, M., T.S. Jiang and J.P. Tsai, "IDSCA: An Intelligent Direction Selector for the Controller's Action in Multiloop Control Systems," International Journal of Intelligent Systems, 3, pp. 361-379 (1988).

3. Rao, M., J.P. Tsai and T.S. Jiang, "An Intelligent Decisionmaker for Optimal Control," Applied Artificial Intelligence, 2, pp. 285-305 (1988).

4. Rao, M., T.S. Jiang and J.P. Tsai, "Combining Symbolic and Numerical Processing for Real-time Intelligent Control", Engineering Application of Artificial Intelligence, 2, pp. 19-27 (1989).

5. Rao, M., J. Cha and J. Zhou, "New Software Platform for Intelligent Manufacturing", Proc. AAAI 90 Workshop Intelligent Manufactures, pp. 62-65, Boston, MA (1990).

6. Wang, Q., J. Zhou and J. Yu, "A Method of Product Conceptual Design Using AI Technology", Proc. 5th Intern. Conf. CAPE, Edinburgh, UK (1989).

7. Wang, Q., J. Zhou and J. Yu, "Decision Support System of Mechanical Product General Scheme Design", Proc. ICED 89, pp. 705-714, London (1989).

8. Yu, J., J. Zhou and Q. Wang, "Mechanical Product General Scheme CAD Based on Expert System Technology", Proc. Advance in Design Automation, Boston, MA (1987).

9. Zhang, Z. and S.L. Rice, "Conceptual Design: Perceiving the Pattern", CIME 11(7), pp. 58-60 (1989).

10. Zhou, J., D.Z. Duan, and Q. Wang, "A Study of Evaluation Subsystem for Scheme Design Expert System", Journal of HUST 17(2), pp. 10-24 (1989).

11. Zhou, J., Q. Wang and J. Yu, "Mechanical Product General Scheme Optimization Design and Intelligent CAD", Mechanical Engineering 89 (3), (1989).

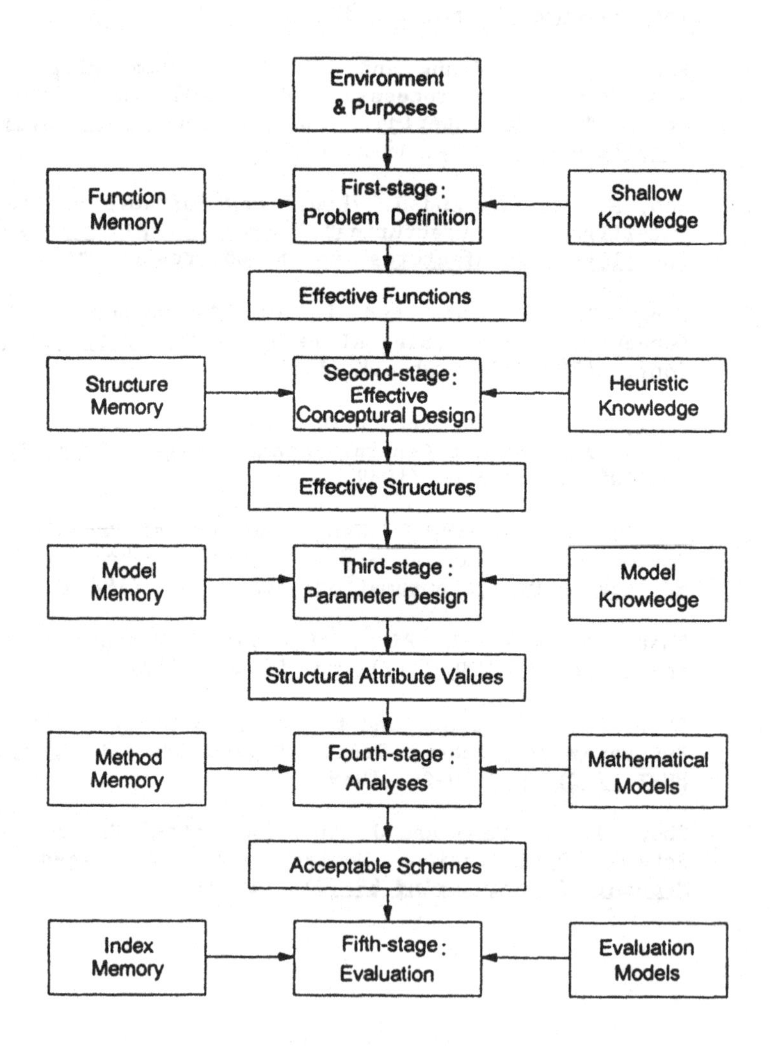

Figure 1. Problem Solving Strategy

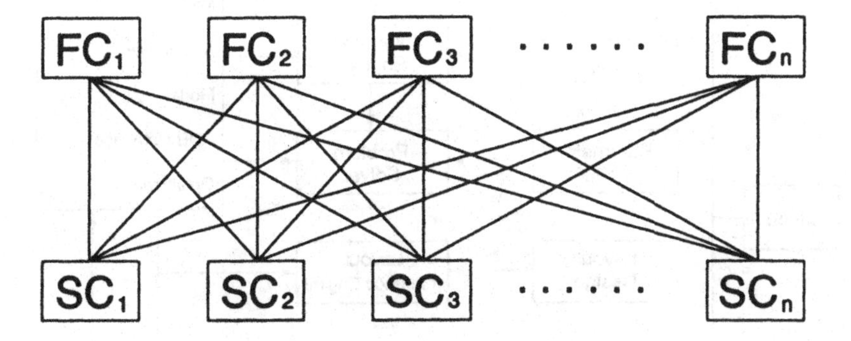

Figure 2. Configuration of Multiple Mapping

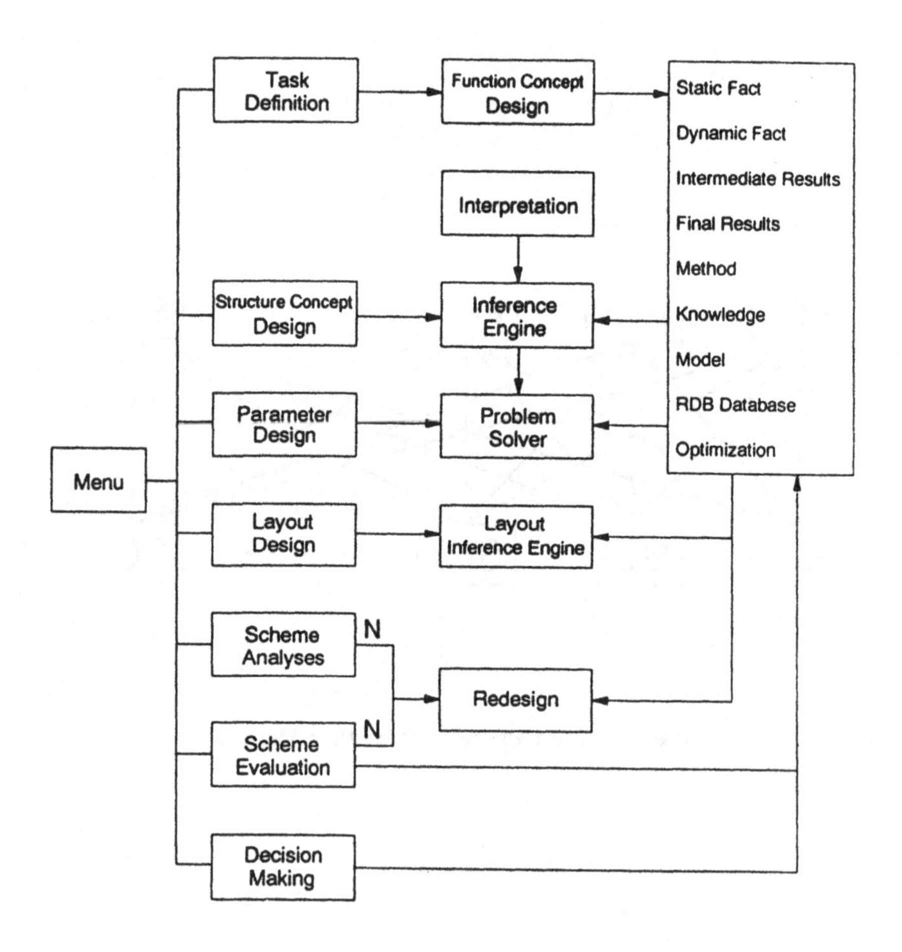

Figure 3. Control Structure of Intelligent System CDESTOOL

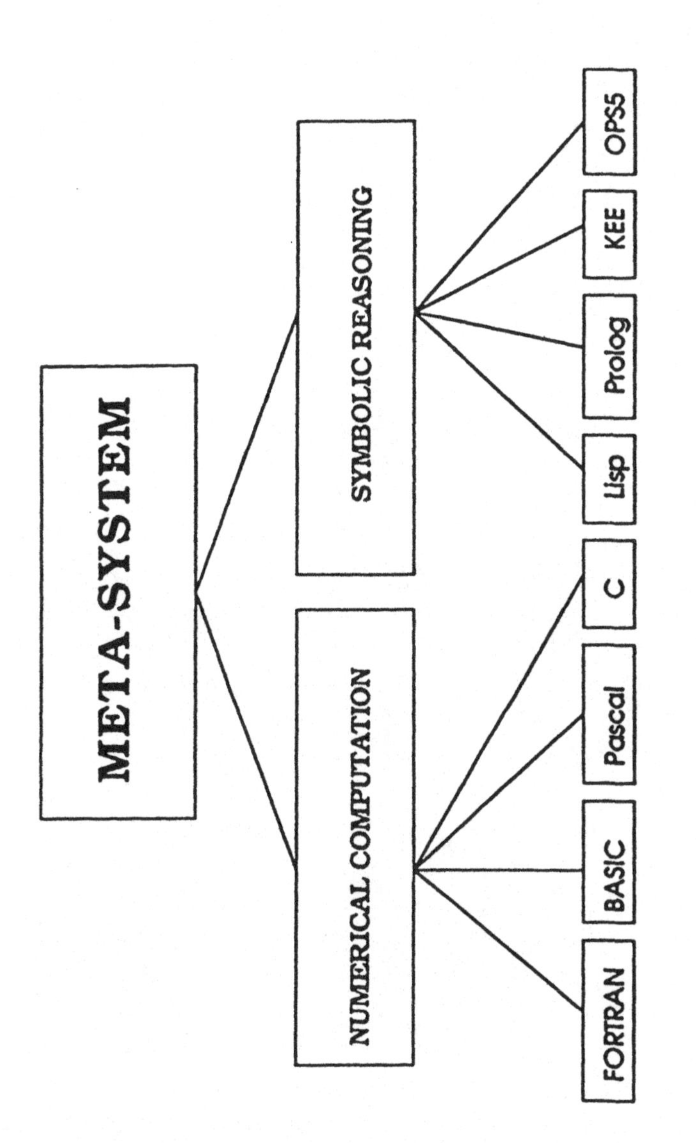

Figure 4. Architecture of the Integrated Intelligent System

Design and Implementation of a Prototype Intelligent System for Evolution Design of Dynamic Objects

V. Tsybatov

Kuibyshev Branch of Institute for Study of Machines, USSR Academy of Sciences, 1 Pervomayskaya St., 443100 Kuibyshev, USSR

ABSTRACT

Prototype for evolutional design integrating computer construction of simulation models of dynamic objects and methods of artificial intelligence is presented. System requirements and prototype system implementation using RESOURCE system are described. The system is intended for use during construction and optimization of manufacturing objects as well as objects functioning in stochastic environment.

INTRODUCTION

Design process may be considered as a sequence of transformation steps of design X performed for getting extremum of goal function Q(X) under some set of constraints to characteristics Y(X). At every of alternation k-step the decision X is made to modify the design based on the values analysis $Y(X_{k-1})$ and $Q(X_{k-1})$. The quality of the next alternation step depends on the accuracy of these values calculations. For static objects characteristics $Y(X_k)$ and $Q(Y_k)$ can be calculated analytically. For dynamic objects analytical calculations give only crude estimate. It becomes still worse when designing manufacturing systems since the latter function in stochastic environment [1].

In this paper it is suggested to make transformation at a simulation model of object. This approach makes easier the analysis of the modification quality, because the characteristics $Y(X_k)$ and $Q(X_k)$ can be observed on the simulation model. Besides it is desirable to offer the designer some assistance in making decisions allowing the computer to make purposeful suggestions [2].Then the motivation, for the work described in the present article is the need for an "intelligent" CAD system model construction system that can interact with the designer to help to construct object design with the required qualities by modifying its simulation model.

SYSTEM REQUIREMENTS

The system is to construct and to simulate models of designed objects by their description input by the user at the language of the object domain. A required software property is a useful group of alteration steps that the user can perform to change the evolving design.

Along with this group of alteration steps is reqiired a specific method for knowledge representation of "built-in designer"is required. Production rules seem the natural choice. They can be programmed to adress many different types of decisions, and they allow the knowledge base to grow incrementaly [2]. Hence the main requirement to the system comes as follows: the system should construct the "built-in designer" model by a set of rules input by the user in a usual mathematical form.

A final requirement concerns the general integration of the computer aided design software and artificial intelligence technique. The conventional approach consists in integrating commercial CAD and commercial artificial intelligent system [1,2,3]. However, such an integration requires to develop a special interface. Besides the additional expenses this interface decreases the efficiency of the "built-in designer" model and that of the designed object interaction. In this article it is suggested to integrate the model of the designed object with the "built-in designer" model in one and the same model, i.e. to create a homogeneous computer network. This would greatly improve the flexibility of models interaction and would enable to avoid major reprogamming when the information used is changed.

SYSTEM DESIGN

A system design that meets all the system requirements discussed in the preceding section is presented (Fig.1).

The design relies on resource approach to modelling [4].The design is simulated on the basis of RESOURCE system [5] which is an intelligent knowledge-based modelling system intended for models construction and investigation of complex dynamic objects. The system makes it possible to model discrete objects and those described with differential algebraic and logical equations on the common basis.

According to the resource approach an object of modeling is defined as assemblage of converters and distributors (ACD) that realizes the techology of conversion of initial resources into output product. Resource is interpreted as ''entity'' which may be : material or energy or information. This process of conversion is provided with the necessary energy and information (Fig.2). The boundary of the object of modeling is the set of generators of resources and generators of demands which are modelling the torn relations of object with environment. Generators of initial resources produce resources R_1 , being converted by the ACD into a set of output products R_2. The latter is absorbed by the user of the environment. These users also being generators of the demands produce demand S_2 for output producys. The demand S_2 is converted by the ACD into the demand S_1 for initial resources. Generators of energy convey energy E_r to the ACD according to the demand E_s arising during the process. The generators of information produce control actions X in accordance with the observable ACD parametrs Y. Parameters X are current physical parameters of ACD. The ACD is modeled by the bidirectional net of functional moduls (converters and distributors) ,further it will be called ''general resource model'' or GRM - net (Fig.2). Functional module (FM) (Fig.3) is the building block of the GRM - net. FM fulfils mutually opposite functions in processing resources (F_r) and demands for them (F_s). For this purpose functional modules are provided with the information (X) and energy (E_r).

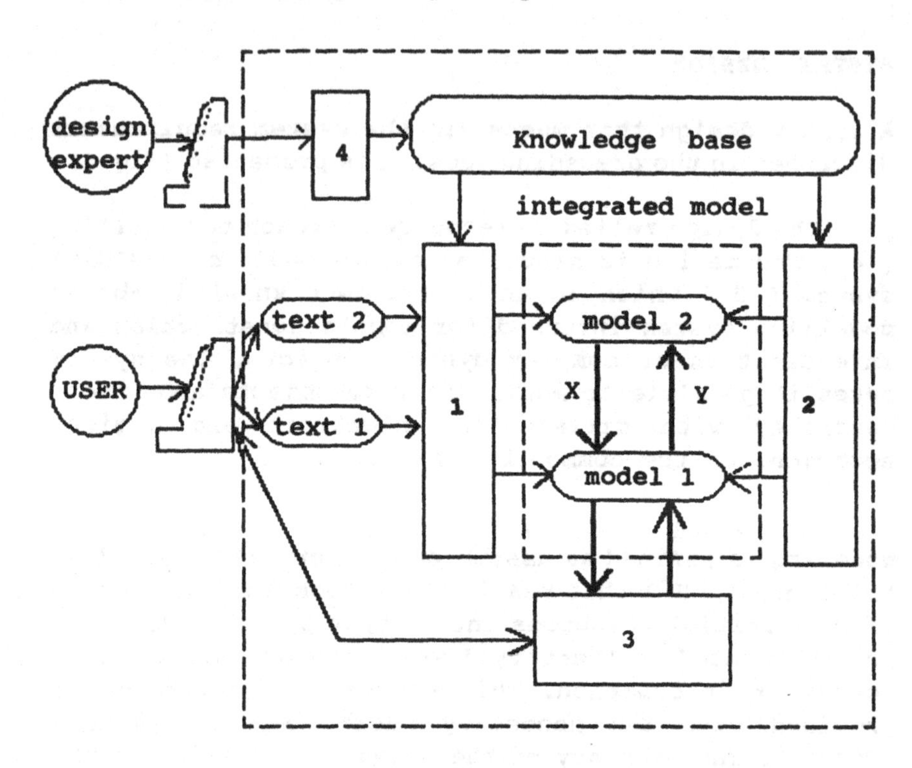

Fig.1. The Architecture of an intelligent System for Evolution design:

1 - Compiler;
2 - Simulation processor;
3 - User's Modificator of the GRM-net;
4 - Monitor;

model 1 - Designed object model;
model 2 - "Built-in designer" model;
X - controlled parameters;
Y - observable parameters.

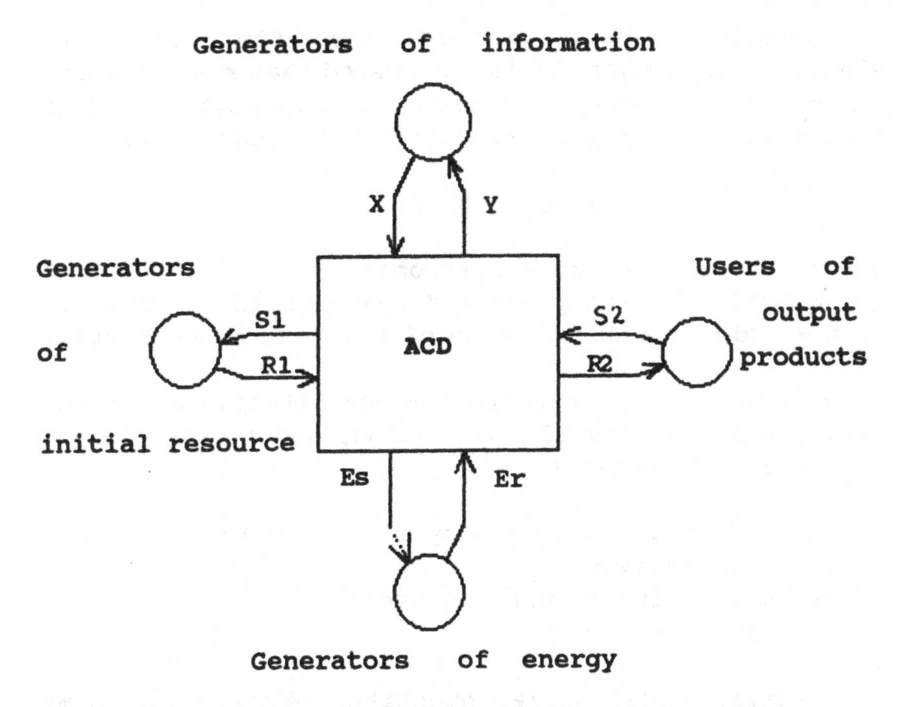

Fig.2. The resource approach to modelling

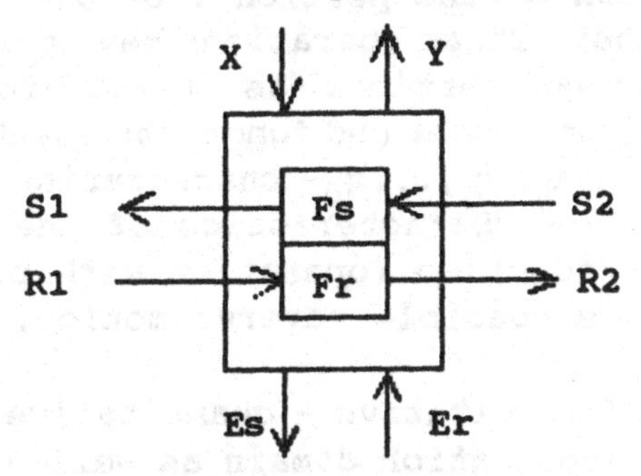

Fig,3, Functional module (FM):
Fs - function for demand transformation;
Fr - function for resource transforma
tion
R1,R2 - input and output resource;
S1,S2 - input and ounput demand;
X,Y- observable and controlable information;
Es - demand for energy;
Er - energy.

Resources and demands are the ''transient element'' of the net. It is considered that resources and demands are converted and transfered by portions, each representing a structural object with four fields:

$$r = [\ p,\ q,\ a,\ t\ ],$$

p - resources name in a portion;
q - quantity of the p - named resource in a portion;
a, t - address and birthtime of portion, respectively.

Fields p and q characterise quanlitative and quantitative properties of the portion, and fields a and t - spatial and temporal ones.

Accordingly, we divide operations with portions into two categories :

- qualitative - quantitative ;
- spatial - temporal.

Finally, qualitative - quantitative operations come to the conversion of the portion from one qualitative state into another. These operations may be described by the relations among portions as a net with two types of nodes : portions - nodes and functions - nodes (Fig.4). There are : $p_1,\ldots,p_N$, $q_1,\ldots q_N$ - characteristics of input resources, p_m, q_m - characteristics of the product of operation ; F - function (operation with portions). A dashed line shows possible reverse motion.

The range of qualitative - quantitative operations depends on the application domain as well as on nature of the handled resource. For example, if a resource is of material nature the conversion operations should not break the law of conservation of matter. Spatial - temporal operations are those which provide portions replacement along the net with the given topology.

To automate the assemblage procedure of the resource model the functions basic set is fixed . It includes the following functions: generation, absorbation, composition, decomposition, multiplexation, distribution, accumulation and expenditure [4,5]. Each of

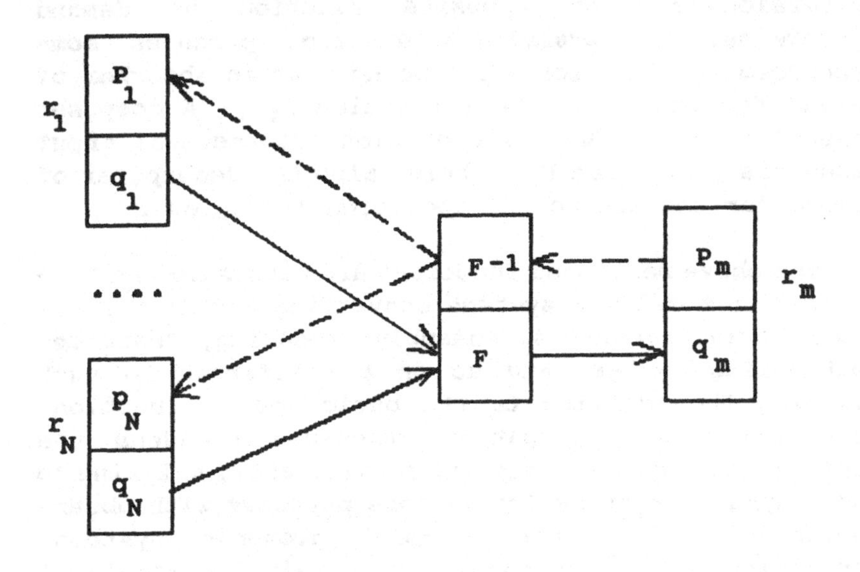

Fig.4. The model qualitative-quantitative operation with resource

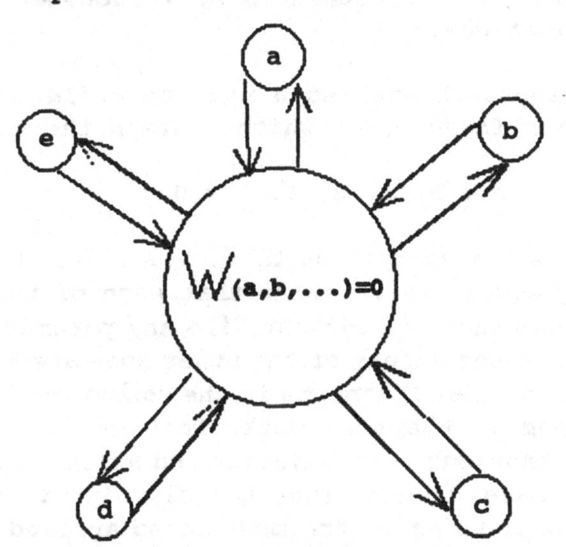

Fig.5. Fragment of computer network from the users library

these modules fulfils a certain function of resource conversion and an opposite function of demand conversion. For example, a generator produces some resources (function F_r) being also an absorber of demand for this resource (function F_s). A composer assembles the product of operation from several input resources (function F_r), being also the decomposer of demand for the product of operation (fuction F_s).

The above mentioned functions are necessary and sufficient to model the systems converting and distributing material resources. Assembly producing, resources controlling systems and so on are referred to such systems. The addition to the basic set of functions operating with information resources broadens the application domain of modeling considerably, allowing to model dynamic systems for various purposes with observation control (for example ,economic systems, automation,control systems), to sustain the models of systems expressed by differential equations, to solve algebraic equations.Additional functions are direct and reverse arithmetical operations, trigonometrical and logical functions.

This approach enables a user to build a computer network each fragment of which sustain the equation:

$$W (a, b, c,...) = 0$$

between the parameters a, b, c, ... All the parameters being equal, which means that each of them may be both a result and an argument, i.e any parameter can be calculated if the values of the other ones are known. The user can keep these fragments in the computer library and can use them in the next work. That makes easier the process of knowledge accumulation and allows the applied knowledge base to grow incrementaly. In addition that enables a user to build fragment based applied systems.

The user"s work with the system consists of three stages.

Stage 1. At the beginning user inputs initial description of the designed object that shows either the user's initial suppositions (if the object is being designed) or the present state of the object (if the object is being optimized). This description (text1) is input at the object domain language and consists in filling in tables and answering questions. The user must input the next facts:

- the list of resource names;
- qualitative - quantitative correlations between resources;
- the list and characterstics of accessible equipment;
- the list and characteristics of accessible workers.

Qualitative - quantitative correlations are input in the next form:

$$P_1(q_1) \xrightarrow{A_j(q_j)} P_2(q_2) \# P_3(q_3) \# \ldots \# P_n(q_n).$$

These are : P_1, P_2, ... P_n - names of resources;
q_1, q_2, ... q_n - quantity of resources;
A_j, q_j - name and quantity of necessary of productive resource;

$$\text{"}\#\text{"} = \begin{cases} \text{"and" - traditional composion;} \\ \text{"or" - alternation composiion.} \end{cases}$$

A compiler of the system synthesizes a computer network (model 1) [4] by the input description.All the nodes of this network are parallel processes transforming the resources and resources demand when provided with the necessary energy and infomation, all the processes having observable (Y) and controlled (X) parameters.

The observable parameters characterize the object capability to fulfil the given task, e.g., the amount of work done by the equipment, delays, storehouse contents, etc. The controlled parameters are physical characteristics of the object x_1, x_2, ... x_N which are accesible to modification. The aim of the designing is to select an optimum vector X. We refer to the controlled parameters the following ones: productivity of equipment and workers, payment, communication capacity, spaciousness of storehouses, etc.

Stage 2. After the object model (model 1) having been constructed the user inputs the ``built-in designer'' knowledge in terms of the object modification rules:

$$x_i = R (X, Y), \langle \text{antecedents} \rangle , i=1,2,... \quad (1)$$

Compiler using the text rules (text 2) constructs designer model (model 2) as an analog of computer GRM-network and connects it with the object model by common variables X and Y.

Stage 3. Then the user chooses the mode of the interaction with an integrated model (model 1 + model 2) determining his role in the modelling process and after that send the model for being operated.

A simulation processor (Fig.1) models the functioning of the integrated model, parameter Y obtained as the interference of the proceseses occuring in the object. The rules (1) operate in the well known match -resolve - execute cycle. Each rule has a set of antecedents and consequents. An antecedent matches values of a current vector $Y(t)$ against goal vectors and yields a Boolean truth value.

If all the antecedents rules yield a value of ``true'', then the rule fires and gives the computed value $x_i(t)$, which leads to the modification of the object model. The user has got the possibility to change the set of rules, removing some of them and inputting new ones.

PRACTICAL RESULTS

The system was implemented for optimization of real manufacturing objects, in particular, of assembling production. The task was set in the following way: the description of the assembling production requiring a reconstruction is given. The production is in the stochastic environment:non-rhythmic materials supply, market instability. It is required to spend a limited amount of money having a limited market of equipment and workforce in such a way that the production should bring maximum profit. As an initial variant the user inputs the description of the production under reconstruction or a project of the reconstruction. A ''built-in designer'' intelligence consists of rules changing the labour productivity of workers and equipment efficiency depending on their current engagement to provide the equilibrium.

In the simulation process some limitations are revealed that are automatically cancelled by a ''built-in designer'' in accordance with its intelligence. The quality of the final project would depend on the description accuracy of stochastic environment and on intelligence of ''built-in designer''.

There are several issues related to the design and implementation of this system that require further discussion. One of them is the idea of creation of a basic regulators set which could be added to the object model at the compilation stage according to the optimization goal. This is a function of a designing expert.

Another issue is the need for a very effective man-computer interaction. It is demanded when a further development of a project becomes impossible. This may be caused by the lack of knowledge of the '' built-in designer '', too rigid limitations to the controlled parameters X or too rigid requirements to the system quality Q.

Both of these issues concern performing effective knowledge representation studies with the system, and solutions require further work.

REFERENCES

1. Wrigt, P.K. and Bourne, D.A. Manufacturing Intelligence, Addison - Wesley Publishing Company, 1988.
2. Jakiela M.J., Papalambros P.Y. Design and

Implementation of a Prototype ``Intelligent'' CAD System. Journal of Mechanisms, Transmissions, and Automation in Design, No. 2,1989.

3. Ellsworth R., Parkinson A. The Complementary Roles of Knowledge - Based Systems and Numerical Optimization in Ingineering Design Software, No. 2,1989.
4. Tsybatov V.A., Resource approach to modelling of objects in engineering, Avtomatizatsiya nauchykh issledovaniy, Gorky, IPF AN, 1989 , pp. 47-53.
5. Vittikh V.A., Tsybatov V.A. Obobshyonne resursnye modeli sistem mashina-chelovek-sreda. -Moskva: Problemy mashinostroyeniya i nadyozhnosti mashin, No. 1,1990.

Evolutionary Inheritance and Delegation as Mechanisms in Knowledge Programming for Engineering Product Design

A. Demaid, J. Zucker

Knowledge Based Systems in Engineering Research Group, Faculty of Technology, The Open University, Milton Keynes, MK7 6AA, UK

ABSTRACT

In this paper we present some of the arguments which led to the production of an object-oriented language, Splinter, specifically intended to support experiments on the computer representation of engineering design activities. We review and criticize some other options among object-oriented languages and knowledge representation systems for supporting engineering design. We propose a software system that enables design concepts to emerge easily as the user *describes* and *modifies* them at the keyboard.

To provide a degree of coupling between objects we employ schemes of property inheritance which, in contrast to standard object-oriented systems using classes and instances, uses "exemplary" or **prototypical** objects. Our machinery yields a **class-free** inheritance system in the tradition presented in the literature of the Actor languages.

The prototype-oriented mechanism offers a means of continually replicating design concepts through object refinement: this principle of one object being specified as a refinement of another object is **specialization**. Treating a knowledge base as evolutionary encourages exploratory comparisons and supports its customization towards the needs of a particular user.

Starting from the idea of property-inheriting prototypes, extra representational capacity is achieved by way of the mechanism of object-to-object **message delegation**.

Both single taxonomic inheritance and delegation are united in Splinter to afford a higher-order prototype formalism. As a knowledge engineering tool, the two features satisfy different modelling intentions. A conceptual model achieved in terms of a higher-order prototype representation is demonstrated to be effective in combining information which originates from different perspectives. These perspectives are distinguished structurally as separate inheritance subhierarchies but are interrelated through the communications process of message delegation.

In this way our design world of discourse consists of many ad hoc groupings which can be programmed by the user, or which can interact by borrowing services from one another.

DESIGN REPRESENTATION

Engineering designers present highly challenging, information processing needs which spring from the ill-structured nature of design decision-making processes: indeed the design discipline is the study of an activity whose focus is "elaboration and exploration as opposed to any specific, well-formulated end."[1] The analysis given in this paper deals with basic issues addressed by a computing tool for creating knowledge bases which we have developed to support the activity of design reasoning. The watchwords of this tool are in accord with the sentiments expressed in the above quote, they are flexibility and experimentation.

Such a tool needs to allow the user to make a substantial redescription and remodelling of the domain during development of a knowledge base in order to investigate the effects of opportunistically refining descriptions and discriminatory judgements. Conceptual modifications to the property-structures and organizational-relationships by which sets of objects are described need to be propagated so that all objects conform to new descriptions. In effect, the logical database specification must be adapted while attempting to ensure that it remains coherent. In sympathy with this database objective of **schema evolution**[2], we expect to make far reaching changes to an evolving conceptual model of design and to execute those changes dynamically during a users interaction with the knowledge base.

We maintain that the prevailing object-oriented paradigm of representational software is ill-matched to this intrinsic evolutionary requirement. The paradigm dictates that a representation uses some form of template (class and/or type) in specifying the conceptual model. The traditional understanding of how a conceptual model is supported by any standard data model, of database theory for example, is that a logical schema produces the defining step in characterizing objects as type-occurrences in the conceptual model, because it stipulates the format to which all type-occurrences must conform. Many designs of object-oriented programming system are based on such defining, generative templates, which are theorized as stored, abstract representations of a 'class' of knowledge.

In the area of design reasoning, however, the defining-template doctrine presents a paradox if we insist that our concepts are perpetually free to change and that a concept's exemplars – which must directly change with it – are simply more detailed concepts from which further exemplars may derive.

The tension between definition of a concept and its ability to mutate poses a problem for most database and object-oriented mechanisms. For this fundamental reason, our alternative paradigm is a class-free method of representation (see e.g. Nierstrasz[3]) which avoids the use of any sort of stored software template which behaves solely as a passively instrumental, structuring device.

Prototypical objects are interrogable resources which can perform useful services if they are given useful state and behaviour properties. Behavioural properties are the prime means of giving rise to a means of **communication**, since these guide the way messages propagate from one object to another. Specifically, messaging is a technique for turning the private attribute space of objects into easily interrogated representations of real-world properties connoted by those objects.

Message delegation builds further representational capacity upon the mechanism of communicating by issuing controlled messages from one object to another. In Hewitt's original definition[4] of delegation, established as a communications paradigm within the Actor framework, the proxy object is created dynamically in the process of the delegating communication and has access to the data internal to the object which created it, the proxy is independent of the client, however, as it is not influenced by concurrent changes to the client.

We argue that object-to-object delegation is practically helpful in modelling decision-making in our domain of engineering design: strategically, it permits the user to submit a query which needs to be answered through accessing some of the properties embodied by objects *other* than the original receiver of the message. As a knowledge encoding methodology, this use of delegation differs from inheritance manifestly because the latter makes explicit a taxonomic organization of objects whereas the former does not. Despite this, both delegation and inheritance provide methodologies for programming the sharing of properties between objects. In this light the computational difference between delegation and inheritance lies in the localization of processing.

In theory, delegation is based upon implementing prototypical characteristics by allowing one object to have extended capabilities by acting as an extension of another object. This extension to a prototype arises whenever a message is forwarded to a new object, which continues to see the original properties of the object which forwarded the message. We shall illustrate the technique of delegation in the final section of this paper with reference to our own language mechanism.

We maintain that the subject of engineering design requires the plasticity of a prototype or class-free approach to create the high-level abstractions and groupings required for representing design knowledge and that the mechanism of message delegation provides a flexible method of programming by requesting services from those high-level abstractions.

PROTOTYPES

In a class/instance inheritance system, instances of classes are the *only* representations of real-world property information, whereas the sharing of property information is controlled through the relationships between classes. The usual relationship is that a class, the superclass, donates its instance information to the instances of other classes (its subclasses). Overlapping (or multiple) inheritance[5] arises when one class receives information from more than one superclass. Wegner[6] offers one discussion of the epistemological ramifications, for real-world modelling, of systems oriented to instances of classes.

In a prototype system there are *only* prototypes, i.e. sort of "exemplary individuals". Their purpose is to represent real-world information *and* to control the way this information is shared[7]. Our idea of property-sharing between prototypes corresponds to judgements that there are family resemblances between objects. This idea of property-sharing is in contradistinction to the idea that objects need to be seen as members of a definable set.

Suppose that we have created a computer representation of a complex artefact and wish to experiment with some variation in the design specification - say a larger diameter shaft to increase its whirling speed. In a class/instance system the two artefacts will be instances of an abstract concept whereas in a prototype system there are no separate class-type and instance-type representations, there are only prototypes. Consequently, using prototypes, we can directly specify that we have a conception of "stiff artefact" which shares characteristics with the prototype artefact as previously described.

Prototype-oriented knowledge representation techniques break free from the constraints of approaches based on templates. In particular, prototype objects provide a vehicle for **evolutionary inheritance**, whereby the transmission of property information is easily reconfigurable using the fundamental language mechanisms rather than a superimposed environment.

For design representation we require that both the property information associated with objects and the relationship between objects should be revisable through experiment to allow the development of a knowledge base. We maintain that the domain is sufficiently rich that the user cannot have a complete decomposition analysis of the domain at the outset, but rather such an analysis emerges as the knowledge base evolves. This policy for opportunistic knowledge base construction stands in contrast to the use of a top-down or bottom-up discipline, as we have discussed elsewhere[8].

Knowledge representation systems employing prototypes have an established ancestry following two streams of development. One, due to Borning[9], employs inheritance, the other, the long-standing Actor tradition of language design originated by Hewitt[10], employs message delegation. In much of the literature it is assumed that inheritance and delegation are alternative techniques toward the same aim: the software engineering objective of highly non-redundant, reusable, modular code. Our contrary idea, coming from a knowledge representation perspective, is that we may usefully devise and employ a kind of inheritance technique for factoring a design domain while still retaining a separate kind of delegation technique for more subtle forms of exploratory relationships between objects.

The benefits of our prototype strategy, as implemented in Splinter, include:

- The uniformity of the regime, embracing objects but no other types of structure, leads to clean visualization and simple explanation.
- Default value transmission is fairly straightforward since we assume that any property of an object which is not re-described is inherited from the object which it specializes, acting as a property-donor.

- The individualization of an object is straightforward since it arises from property modifications to that object alone.
- Object refinement, the cornerstone of an evolutionary scheme of inheritance, is supported by the language itself. The browsing amenity turns into precisely a facility to remember what the user has constructed and the various generational descriptions that the user has employed during the session.

LANGUAGE ENVIRONMENTS

Today, there are a few high-level class/instance language systems which provide some facilities to address the need for an evolutionary structuring of information. Such software products are, in reality, not only notations but also development environments. Smalltalk-80[11] is one highly influential example and the more recent Common Lisp Object System[12] (CLOS), which adopts the environment of its underlying Common Lisp, is another.

To be useful for design representation, the purpose of the development environments of class/instance languages must be to help manage the generation of large and/or complex knowledge representations. Corresponding to our demand that knowledge bases should be able to evolve, these environments must allow classes to be refined, re-related and for the changes to be observed by all existing instances of the changed class, by all descendant classes of that changed class, and by all existing instances of those descendant classes. The changed class specification has to be propagated down the inheritance pathways and must appear to upgrade all instances encountered along those pathways. In implementing such a strategy for class modification, modern class/instance language systems support evolutionary changes by virtue of the supervisory rôle of the program support environment. The programming notations are not, themselves, powerful enough to provide logical models of evolutionary change.

In the world of the class/instance Lisps the implementation problem of performing the propagation of changes to instance variables (which depict state properties, in these systems) was solved in the mid-1980s, as implemented in New Flavors[13]. The preceding Lisp-hybrid systems such as original Flavors and original LOOPS had no automatic support for evolutionary structures.

It is a convenient feature of the up-to-date object-oriented Lisps, including CLOS, that the laborious propagation of changes is performed automatically (by the setting of flags in the symbol table for classes). Smalltalk-80 provides an interesting counterpart to this environmental feature of CLOS. Here it is desirable to express, explicitly in code, the flag-consultation ingredient of the interactive search for changes to a class. An AddDependents: message (followed by an existing class as its argument) sent to a class lets that class know which classes need to experience the propagation of changed status. This is a useful delimiter of the propagation of change, when a class re-definition occurs, and a useful delimiter of the search for changes whenever an instance is accessed.

By way of contrast, our prototype-oriented approach makes an evolutionary strategy conceptually straightforward. The ability of our inheritance structure to evolve is a consequence of its representation and not a consequence of additional programming-environment features such as special editors to support the revision and re-submission of earlier program statements.

LANGUAGE PLASTICITY

Prototype mechanisms in class-free languages[14,15,16,17] improve the malleability of the way objects share property information and we have argued that this is an essential requirement for a system which has serious pretensions to modelling engineering design. The chief difference between the modern class/instance approach to effecting changes and our own prototype approach lies in the restrictions imposed by the reliance of the former upon a support environment. The advantage we claim arises from the plasticity of the prototype concept in Splinter, illustrated by the example of Table 1.

Our system of inheriting prototypes is shown to give a particularly simple and dynamic system of inheritance. In Table 1(i), we describe coffee_cup as being a specialization of container. In Table 1(iv) we redescribe coffee_cup as being a specialization of cup, which we happen to have just introduced, and rightly expect the specialization ordering container ≻ cup ≻ coffee_cup to be observed. This implies that coffee_cup inherits from cup: now coffee_cup inherits from container through the intermediary of cup.

By making the first description (of coffee_cup as being a specialization of container), we imply that coffee_cup is to be created as a software individual by some behind the scenes activity in which container is playing a generic rôle, i.e. acting as a generator for coffee_cup.

Any objects – coffee_cup, cup or container – may service queries about their properties. In Table 1(v), we may ask either coffee_cup or cup to read out its number-of-handles property information. Since any object may respond to messages which mention its properties, the objects which we have illustrated play the rôle of being concrete, responsive representations.

Our idea is that coffee_cup, cup or container may always play the rôle of being specialized or the rôle of being queried. For example we may construct a further object such as small_coffee_cup, simply by describing it as being a specialization of coffee_cup. All the while coffee_cup will continue to be responsive to messages which are submitted to it, after the : in our given syntax. Since both rôles are continually being played, in principle by any object which we choose to construct, all objects are both **generic** and **concrete** using our software policy.

Referring to Table 1(v) we see that we may supply messages to any given object, not only to ask it for information (or to create itself), but also to elaborate its properties. Further, referring to Table 1(vi) and (vii), we see that messages may be supplied which change the values (state or behaviour) expressed by those properties. During construction of new objects, any values expressed by properties therefore serve as defaults. In all cases the transmission of property information is highly dynamic: changes are propagated to the objects which inherit from the changing object.

To refine an object, only the new features of the changed object need be described. All the features which are not stated to be different should be regarded as remaining unchanged: the unaffected properties are remembered. Our policy is strictly **incremental description** of any object.

The relation of coffee_cup to cup, i.e. the specialization relation, is therefore a perpetual dependency in an evolutionary scheme of prototype-inheritance. The collection of all specialization relations, constructed during the execution of (those) programming statements (which make inheritance assertions), constitutes the organized, inheritance structure of a design knowledge base.

MESSAGING AND MESSAGE DELEGATION

A design domain factored into objects, organized by means of an evolutionary inheritance strategy, provides the high-level abstractions required to populate the design world. The mechanism for discourse is for objects to interrelate, one to another, through communicating by forwarding messages. Each time that this happens, the re-routed message is serviced by the newly contacted object as an agent, specially prepared for the task by the original object. This is the basis of message delegation[18].

We have shown that our inheritance principle gives us a tool for expressing a factoring of objects according to family resemblances between prototypes. By contrast, an auxiliary scheme of delegation is a tool for modelling more subtle connections and actions. These connections arise from the way one prototype's information affects another prototype's information without the two prototypes necessarily being at all like one another or in any other way associated.

Among prototype languages, there have emerged a number of schemes and different balances of features. For example, Agha and Hewitt[19] have advanced an interpretation of the Actor tradition of language computation which envisages the existence of "higher order" Actor languages, such as Splinter, in which prototypes engage in the activities of both message delegation and property inheritance. Our machinery for representation manifests both techniques and combines them in a way which is both practically useful and theoretically principled for the purpose of design representation.

The syntactic pattern

(⟨expression referring to object⟩ : ⟨message⟩*)

(using Backus-Naur notation extended through * to indicate the Kleene closure[20]) is our general form of messaging statement. This language syntax serves to implement an important semantic idea for the processing of objects. The object doing the receiving should be empowered to answer any general, algorithmic construction of messaging query. In doing so, the query will respect the object's memory of named properties.

The object, coffee_cup, has been described as having a property named temperature_resistance. The named property (i.e. temperature_resistance) is remembered as a local, distinguishing attribute of the object (i.e. coffee_cup). In turn, the object (i.e. coffee_cup) inherits properties from other objects up the inheritance pathway (e.g. cup, container). These more general objects serve as donors of a namespace of properties which they have, likewise, remembered. The local, distinguishing properties (of coffee_cup) are nested within the scope of the inherited properties, according to rules of static scoping, the prevalent programming language policy for controlling the visibility of variables.

Object-to-object communication is expressible in Splinter by embedding messaging statements inside messages. For example, the pattern

(⟨reference to client object⟩ : (⟨reference to proxy object⟩ : ⟨message⟩))

is the form for a single message to be forwarded from one object to another object acting as its proxy. In general, the Splinter programming language invites the user to adopt the metaphor of sequences of messaging statements nested within messaging statements, and interprets this explicit style of inter-object communication according to some particular rules of delegation. For example, from our small world of coffee cups in Table 1(vi):

```
➡(container : (coffee_cup : number-of-handles))
     1
```

The object named container has delegated the request for a number-of-handles property to a proxy. The proxy object is based upon the existing object named coffee_cup with the rider that the proxy object is especially initialized by container for the task. In particular the proxy based upon coffee_cup sees the distinguishing properties of container (whatever they may be). The properties of the client shadow (or supersede) the distinguishing and inherited properties of the object (here, coffee_cup) on which the proxy is based.

In the above example, we have assumed that container has no number-of-handles information itself. Therefore the answer has been looked up in the attribute space provided by the existing object named coffee_cup. Since container undertakes no further processing of the answer, the result returned by evaluating the messaging statement is equivalent to consulting coffee_cup without delegation:

```
➡(coffee_cup : number-of-handles)
     1
```

in terms of the answer returned. There is an underlying difference, however, in that the answer in the former case is returned from container and not from coffee_cup, under the control of the particular message-reception protocols of container.

We have said that the distinguishing properties of the client always shadow the re-used properties of the proxy. Now, number-of-handles is a distinguishing property of coffee_cup. We may, therefore, re-arrange the order in which objects communicate in the above example and obtain the same answer (this time controlled by the protocols of coffee_cup):

```
➡(coffee_cup : (container : number-of-handles))
     1
```

By contrast, if container were consulted directly – not as a proxy of coffee_cup – for its number-of-handles information, then no such information would be discovered:

```
➡(container : number-of-handles)
     #!ABSENT
```

There is generally a problem of referring to "self" in object-oriented systems employing delegation. The standard solution[21] is that during delegation the whole of the client object is bound to self and this binding appended with the forwarded message. Using the standard policy, occurrences of the variable self in the code for the behavioural properties of the message-receiving object refer to the client object during delegation or to the receiving object if it experiences no delegation.

Our knowledge-representation principle is to use delegation in order to capture shades of meaning from the domain by way of the unanticipated relationships between objects conveyed through inter-object communication. We therefore wish to distinguish: (i) the receiving object itself, (ii) the client object itself, and (iii) where the receiving object is actually a property of an object, what object it ascribes. Splinter gives three support functions — (self), (client) and (owner) — to return these three species of "self". For illustration's sake (returning the *names* of the three "self" objects in this example):

```
➔(coffee_cup :
    (container :
        (name (self))
        (name (client))
        (temperature_resistance : (name (owner))) ))
  ( CONTAINER
    COFFEE_CUP
    COFFEE_CUP )
```

The principles of inter-object communication, illustrated above, offer a rich, practical scheme of delegation, i.e. passing of responsibility for evaluation from client to proxy during messaging, which is intended to satisfy the needs of a design representation language and complement our flexible inheritance policy.

INHERITANCE PLUS MESSAGE DELEGATION

The singular advantage of our particular language strategy is that it affords a lawful combination of inheritance and delegation styles of inter-object relationship - a "higher-order prototype representation."

The conceptual modelling policy for design knowledge description offered by our formalism is:

(i) to use inheritance as an organizer of conceptual entities, to establish the anticipated connectivity between objects so as to govern the visibility of each object's memorized attributes, and

(ii) to use delegation to achieve a fluid system of coupling between conceptual entities and to establish an additional flow of property-information control at messaging-time.

When higher-order prototype relationships representing design conceptualizations exist, the use of delegation allows a process of addressing extra attribute spaces incidental to the attribute spaces of objects determined through inheritance relations. In modelling design information, our particular use of delegation has been to model **rôle combination**. Here, we imagine that the newly contacted object has not been described as related to the client object through inheritance, but nevertheless contains information relevant to the characteristics associated with a query which we may ask about this client. We use inheritance and delegation techniques, within the same software architecture, to accomplish different representational purposes.

The conventions of the Splinter language promote this division of labour. We maintain that our approach cuts across the more general arguments in the object-oriented programming literature[22,23] about the relative standing of inheritance and delegation, which address the general question of the relative powers of inheritance and delegation architectures and the extent to which they map one to the other.

MODELLING DESIGN RÔLE RELATIONSHIPS

We maintain that the problem of rôle relationships[24] is fundamentally important to getting object orientation to do useful work in the field of design representations. The observation that many of our real-world concepts are multi-faceted in their relations to other concepts is at the heart of the flexible software approach which we report in this paper. The complexity of this representational task founders on a purely hierarchical structuring of knowledge: "How anybody can get useful work done when restricted to hierarchical inheritance is beyond me; the world just doesn't work hierarchically." (Weinreb, as reported by Touretzky[25]). We shall proceed to illustrate our delegation technique — as an extension to our bare, hierarchical inheritance structure captured through incremental descriptions — in comparison to the overlapping (or multiple) inheritance strategy advocated by Weinreb and by the majority of class/instance practitioners.

Figure 1, below, depicts a portion of a knowledge base that illustrates a modelling problem which arises from the need to construct two highly distinct subhierarchies. Here one subhierarchy embodies information about domestic utensils, another embodies information about engineering machines. The pragmatics of this knowledge description activity is important to our argument.

Essentially, it is the manner of usage of the knowledge base which dictates that it entertain distinct domestic-utensil and engineering-machine information structures. It is generally the case in the conceptual design of an engineering artefact that the design information factors into different perspectives according to their particular technical specializations. In Figure 1 we present a simple distinction between subdividing information about domestic end-products and subdividing information about power-consuming apparatus. Although the distinct branching conveniently arises at the top of the complete inheritance structure (i.e. at Object), nevertheless in an evolutionary system of inheritance the exact roots of the subtrees conveying different perspectives are less important than the brand of queries that the object nodes are intended to answer.

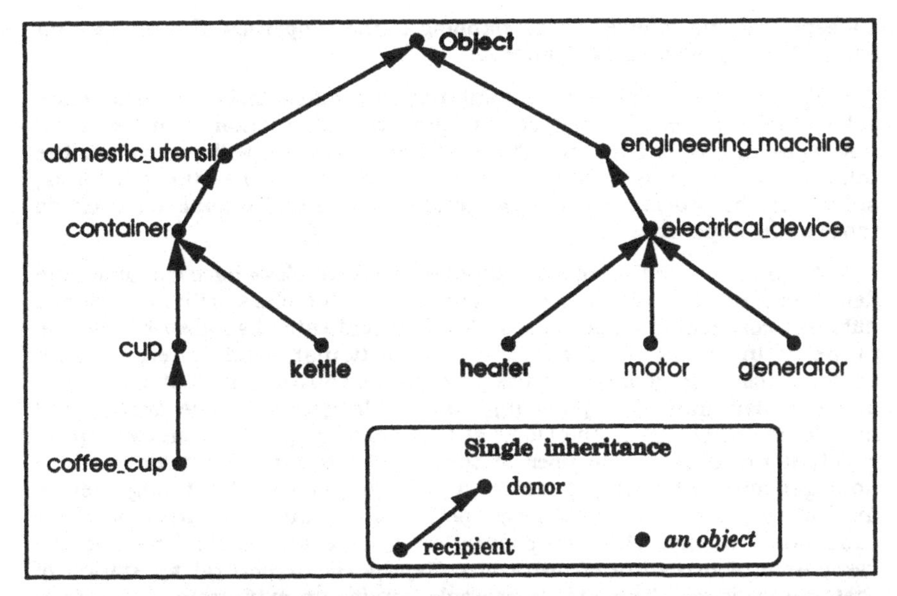

Figure 1 Skeleton of an inheritance hierarchy with distinct 'domestic utensil' and 'engineering machine' subtrees

If we imagine that our knowledge base is developing in this manner of reflecting different perspectives on the domain, then single taxonomic inheritance has a clear advantage. Each perspective may factor the knowledge independently of other perspectives. So a hierarchical ordering of objects helps enforce a sanitary segregation of different concerns. For an unsanitary arrangement see Banerjee's submarines, as reported by Mariani[26] .

There is, however, a residual problem. How do we account for a coupling that may be required between the different perspectives? In accordance with Figure 1, we propose to develop the propertied concept of kettle as a specialization of our container concept. A knowledge base designer's thinking on domestic utensils will be reflected in this way of characterizing the object named kettle. However, we suppose that a knowledge base designer has available a representation of electrical devices which leads to an emergent characterization of the concept of heater. If we construe kettle's description under the domestic utensil perspective as being the first and foremost way in which kettle is characterized, then there remains the problem that the characteristics of being electrically heated are also relevant to the queries which it should be able to answer.

Clearly, there needs to be some manner of **liaison** between kettle, under the domestic-utensil perspective, and heater, under the engineering-machine perspective. Such liaison needs to accomplish property sharing, in the sense that relevant properties which heater might possess, such as electrical- and fire-safety calculations, need to be accessible to kettle for the purposes of answering questions. One solution to the problem is therefore to say that kettle inherits from container and also inherits from heater, i.e. that our concept of kettle receives an overlapping of the properties of container and heater through inheritance, Figure 2.

Prima facie, overlapping inheritance in a class/instance mechanism solves the problem of specifying the property information from which all instances of 'kettle-ness' will be generated. Kettles will receive all the information about domestic utensils and about engineering machines, subject to the proviso that the exceptional values in the subclass override those in the superclass.

Let us suppose that our instances of the kettle class have conjoined the definitions of the container class and the heater class with no obvious difficulty concerning name spaces. We find that kettle has absorbed all the property information of container in particular, and inherited from domestic_utensil in general, all the property information of electrical_device in particular, and of engineering_machine in general. The overlapping inheritance mechanism, therefore, at best achieves precisely an overlap: in effect, it presents to the user of the representation of a given kettle a homogeneous attribute space in which properties about engineering machinery are as significant as properties about domestic utensils. As a conceptual model, this representation is a compromise between the viewpoints of two sorts of users: it collapses the structural separation of their perspectives. This simple example heralds an extraordinary potential for complex expansion as useful design perspectives are numerous. It is easy to add the perspective of a manufacturing method to those of domestic utensil and engineering machine. Further small associations which also provide useful perspectives, such as surface decoration, might well be added. In order to achieve a comparable effect using overlapping inheritance, it would be necessary continually to absorb merger upon merger of inheritance pathways. The size of each object affected by the multiplicity of superclasses would lead to great structural complexity.

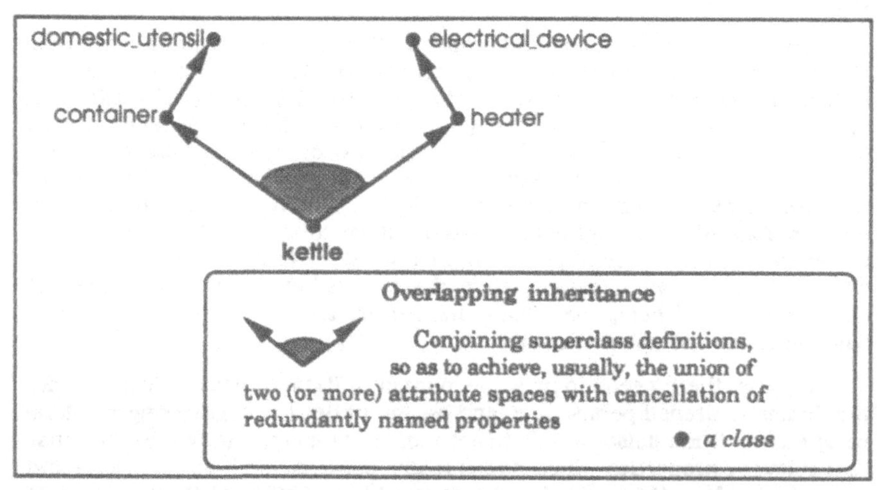

Figure 2 **An overlapping (multiple) inheritance solution to the combined rôles of the information services demanded of the kettle concept, as often implemented in class/instance systems.**

Our preferred alternative proposal is to use delegation to access services from simple inheritance hierarchies. In response to messages which relate to electrical-safety or fire-safety considerations kettle (as a client) delegates the message to a proxy based on the object named heater, Figure 3.

The delegation link of Figure 3 works by forwarding a message from kettle to heater. In message delegation, the activity of forwarding implies that responding to the entire message has become the responsibility of the agent object or **proxy** (i.e. the object delegated to), with the original object acting as **client**. During message delegation, control passes from the client to its proxy.

Notice that kettle and heater remain in the hierarchies with which they were described. This has the effect of preserving the distinct domestic-utensil and engineering-machine perspectives or rôles which we spoke about at the start of this section. The delegation link is a run-time communications link, in the sense that it is effective for the duration of a particular message.

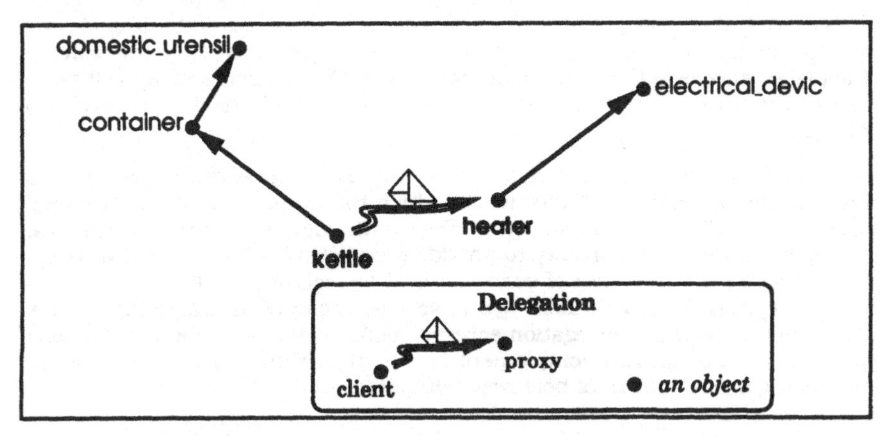

Figure 3 **A delegation solution to the combined rôles of the information services which can be required, at messaging time, of the kettle prototype**

A simple example of using delegation might be the single messaging statement:

```
➔(kettle :
    (heater :
        (< mains-voltage
            (breakdown-voltage-of ((client) : body-material)))) )
```

Here kettle is the object returned by (client) in the message serviced by heater, because heater is the object to which kettle transfers control (appearing after the colon as the receiver of the message), i.e. heater is acting as the proxy of kettle as client.

The purpose of the message received by heater, in this illustration, is to test for voltage safety, say. If it is found to be an appropriate piece of code, then the programmer's/user's next step would generally be to turn this complete messaging activity into a behavioural property stored as a property of kettle. Suppose it is described as being a behavioural property of kettle named voltage-safety? .

Once the voltage-safety? operation is described, the user is in a position to issue a simple query of kettle, or a descendant of kettle,

```
➔(kettle : (voltage-safety?))
```

with the effect that the delegatory link is used automatically as part and parcel of the behavioural characteristics of kettle.

A more sophisticated use of the delegatory process assumes that we do not or cannot specify the particular proxy object from which we require a service. We will assume, however, that a particular *hierarchy* is known to be an appropriate place to search.

A mapping/search procedure taking the object electrical_device as an argument can be used to propagate a search, recursing over all objects successively at or below this root node and, on each occasion, to delegate the voltage testing activity explicitly to the discovered object. If the voltage inequality succeeds then the usual policy of such a search activity will be to return the proxy object which gives a satisfactory result for the body-material of kettle.

Such a search is biased to reveal the **most general object** in the engineering machinery perspective for which the proposed query is suitable. Our delegation mechanism, in alliance with simple search procedures, possesses an in-built capacity to provide generalized solutions. Conversely, if we reverse the direction of search, we make use of a built-in capacity to provide **specialized** solutions. Quite varied means of interrogating objects are possible using our delegation scheme, as delegation – a communications process – is a natural complement to the algorithmic process of search, producing a mechanism of borrowed computation.

CONCLUSIONS

The evolution of objects of knowledge, whether they represent abstract or concrete concepts, is necessary if the designer is to be allowed accurately to express a developing model of the world in software and, further, to explore the effects of changes in a design environment. Property-inheriting prototypes provide a mechanism for organizing such an evolutionary structure.

Our generalized prototype-oriented representation paradigm, expressed in the language Splinter, manifests: an evolutionary strategy for interactively absorbing the user's descriptions of objects and relationships; a segregation of separate concerns; intertwining of related information services offered by different sources; and highly flexible querying through messaging expressions. A virtue of the key components which we have highlighted is that they are all supportable at the level of notation rather than achieved through an interactive support environment.

We view subhierarchies of prototypes as offering separate perspectives on the design problem. Combining perspectives using a delegation mechanism to achieve information combination is both effective and virtuous from the standpoint of the program user's point of contact with the knowledge base. In comparison, we suggest that the overlapping (or multiple) inheritance classification method sits uncomfortably with the detailed real-world judgements which a knowledgeable user may need to express.

The advantages offered by the strategy of borrowed computation in comparison to the technique of conjoining information from different sources in the inheritance graph into a single, amorphous attribute space are: enrichment of knowledge base performance; ability to express intricacy and particularity of interrelationships; independence and multiplicity of perspectives; extensibility of the conceptual model.

Inter-object message delegation, in a higher-order prototype-oriented knowledge base, has been shown to be an effective way of combining information from a variety of sources which are best regarded as belonging to separate perspectives on semantic and pragmatic grounds. Delegation has the conceptual advantage that properties are not redundantly re-stated simply because they belong to different perspectives and that the structural independence of such perspectives is preserved, whereas it is collapsed in the case of overlapping inheritance.

We have accounted for a class free approach to developing an object-oriented representation technique whose ingredients may be summarized as:

class-free prototype-orientation = *objects* + *inheritance* + *delegation.*

The embryonic mechanisms which we have illustrated are effective as a knowledge engineering vehicle for engineering design, because they facilitate an extremely dynamic style of development featuring modular development of subhierarchies, exploration of property modification through evolution in each subhierarchy and experimentation with communication between subhierarchies.

ENUMERATED REFERENCES

[1] S. Newton, The irrelevant machine, Design Studies 10(2) (1989) 118 - 123

[2] G.T. Nguyen and D. Rieu, Schema evolution in object-oriented database systems, Data and Knowledge Engineering 4 (1989) 43-67

[3 O. Nierstrasz, A survey of object-oriented concepts, in: W. Kim and F. Lochovsky (Eds.), Object-Oriented Concepts, Databases and Applications (ACM Press / Addison-Wesley, Reading, MA, 1988) 3-21

[4] C. Hewitt, P. Bishop and R. Steiger, A universal, modular Actor formalism for Artificial Intelligence, Proc. 3rd Intl. Joint Conf. Artificial Intelligence (1973) 235-245

[5] D.J. Carnese, Multiple inheritance in contemporary programming languages, MIT Technical Report LCS-TR-328, Cambridge, MA, 1984

[6] P. Wegner, The object-oriented classification paradigm, in: B. Shriver and P. Wegner (Eds.), Research Directions in Object-Oriented Programming (MIT Press, Cambridge, MA, 1987) 479-560

[7] H. Lieberman, Using prototypical objects to implement shared behaviour in object oriented systems, Proc. OOPSLA'86 ACM Conf. Object-Oriented Programming Systems Languages and Applications (SIGPLAN Notices 21(11), 1986) 214-223

[8] J. Zucker and A. Demaid, A software machine designed for selection, Knowledge-Based Systems 2(3) (1989) 178-184

[9] A.H. Borning, Classes versus prototypes in object-oriented languages, Proc. IFIP Fall Joint Computer Conf. (1986) 36-40

[10] C. Hewitt, P. Bishop and R. Steiger, A universal, modular Actor formalism for Artificial Intelligence, Proc. 3rd Intl. Joint Conf. Artificial Intelligence (1973) 235-245

[11] A. Goldberg and D. Robson, Smalltalk-80: The Language and its Implementation (Addison-Wesley, Reading, MA, 1988)

[12] S. E. Keene, Object-Oriented Programming in Common Lisp: A Programmer's Guide to CLOS (Symbolics Press / Addison-Wesley, Reading, MA, 1989)

[13 D.A. Moon, Object-oriented programming with Flavors, Proc. OOPSLA'86 ACM Conf. Object-Oriented Programming Systems Languages and Applications (SIGPLAN Notices 21(11), 1986) 1-8

[14] A.H. Borning, Classes versus prototypes in object-oriented languages, Proc. IFIP Fall Joint Computer Conf. (1986) 36-40

[15] W.R. Lalonde, D.A. Thomas and J.R. Pugh, An exemplar based Smalltalk, Proc. OOPSLA'86 ACM Conf. Object-Oriented Programming Systems Languages and Applications (SIGPLAN Notices 21(11), 1986) 322-330

[16] H. Lieberman, A preview of Act 1, MIT Memo AI-625, Cambridge, MA, 1981

[17] D. Ungar and R. Smith, Self: The power of simplicity, Proc. OOPSLA'87 ACM Conf. Object-Oriented Programming Systems Languages and Applications (SIGPLAN Notices 22(12), 1987) 227-242

[18] H. Lieberman, Delegation and inheritance: two mechanisms for sharing knowledge in object-oriented systems, Journées Languages Orientés Object (AFCET, Paris, 1985) 79-89

[19] G. Agha and C. Hewitt, Actors: A conceptual foundation for concurrent object-oriented programming, in: B. Shriver and P. Wegner (Eds.), Research Directions in Object-Oriented Programming (MIT Press, Cambridge, MA, 1987) 49-74

[20] J.C. Cleaveland and R.C. Uzgalis, Grammars for Programming Languages (Elsevier, 1977) 1-33

[21] H. Lieberman, Using prototypical objects to implement shared behaviour in object oriented systems, Proc. OOPSLA'86 ACM Conf. Object-Oriented Programming Systems Languages and Applications (SIGPLAN Notices 21(11), 1986) 214-223

[22] L.A. Stein, Delegation is inheritance, Proc. OOPSLA'87 ACM Conf. Object-Oriented Programming Systems Languages and Applications (SIGPLAN Notices 22(12), 1987) 138-146

[23] L.A. Stein, H. Lieberman and D. Ungar, The Treaty of Orlando: A shared view of sharing, in: W. Kim and F. Lochovsky (Eds.), Object-Oriented Concepts, Databases and Applications (ACM Press / Addison-Wesley, Reading, MA, 1988) 31-48

[24] C.W. Bachman and M. Daya, The role concept in data models, Proc. 3rd Intl. Conf. Very Large Data Bases (IEEE, NY, 1977) 464-476

[25] D.S. Touretzky, The Mathematics of Inheritance Systems (Pitman / Morgan Kaufmann Research Notes in Artificial Intelligence series, Los Altos, CA, 1986)

[26] J. Mariani, Object-oriented database systems, in: G. Blair, J. Gallagher, D. Hutchinson & D. Shepherd (eds.), Object-Oriented Languages, Systems and Applications (Pitman, London, 1991) 176

An Expert System for Ergonomic Workplace Design Using a Genetic Algorithm

D.T. Pham, H.H. Onder

Intelligent Systems Research Laboratory, School of Electrical, Electronic and Systems Engineering, University of Wales, College of Cardiff, P.O. Box 904, Cardiff CF1 3YH, UK

ABSTRACT

This paper describes an expert system being developed for the optimum ergonomic design of workplaces. Ergonomic knowledge is encoded in rules and frames using the Leonardo development tool, a hybrid shell with rule-based and object-oriented knowledge representation facilities and a forward /backward inference mechanism. The optimization is performed in a program external to Leonardo. The program implements a Genetic Algorithm which is an efficient directed random search procedure capable of yielding the global optimal solution in a complex search space without requiring specific knowledge about the nature of the problem to be solved.

Keywords: Expert systems; ergonomics; genetic algorithms.

INTRODUCTION

There has been increasing interest in applying expert systems to the solution of ergonomic problems. One area of ergonomics that is of special importance is anthropometric design of engineering workstations and industrial workplaces. This takes into consideration static and functional (dynamic) measurements of the operator's dimensions and physical characteristics as he occupies space, moves, and applies energy to objects in his environment [1].

The reason for considering anthropometric data in workplace design is that an industrial workplace layout should be compatible not only with a system's performance specifications but also with the user's needs. It should match the capability and limitations of the user, ensuring that he is able to see the working area clearly and that his posture is natural and comfortable. The workplace should therefore be adequate in terms of postural support, distribution of body and limb weights, and maximum reach and force requirements.

Clearly, to accomplish this ergonomic design, the designer must take into account the physiological, psychological, environmental and dimensional factors which will affect operator performance and well-being. The overall design will no doubt be dependent on the interactions among the stated factors [2]. The task of a workplace designer is thus a complex one, demanding a large amount of knowledge and experience. This paper describes an expert system being developed to assist a designer in this task. The program is to be used to give advice, during the design process, on how to achieve an optimum layout complying with basic ergonomic principles.

The system was developed with the help of Leonardo [3], an advanced expert system shell which supports the use of rules and frames to represent design knowledge and the hybrid method of inference based upon a combination of forward and backward chaining. The optimization part of the system is carried out in a program external to Leonardo. The program implements a Genetic Algorithm (GA) which is an efficient directed random search procedure.

The components of the expert system are shown in Figure 1. In addition to the Leonardo and genetic algorithm modules, a module for recording various ergonomic data (the DB module) can also be seen in the figure. Each of these modules will now be described in turn.

LEONARDO MODULE

This is the main module in the expert system. The module comprises three parts. As shown in Figure 2, these are : a knowledge base, an inference engine and a user interface.

Knowledge Base

The Leonardo knowledge base contains rules, rulesets, frames, classes, objects, and procedures, related to

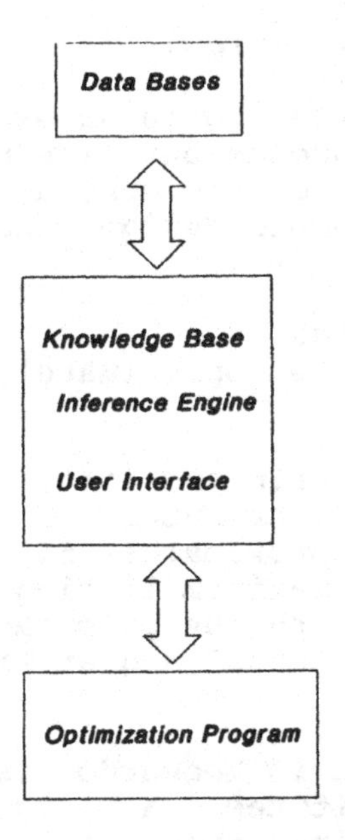

Figure 1. Components of expert system

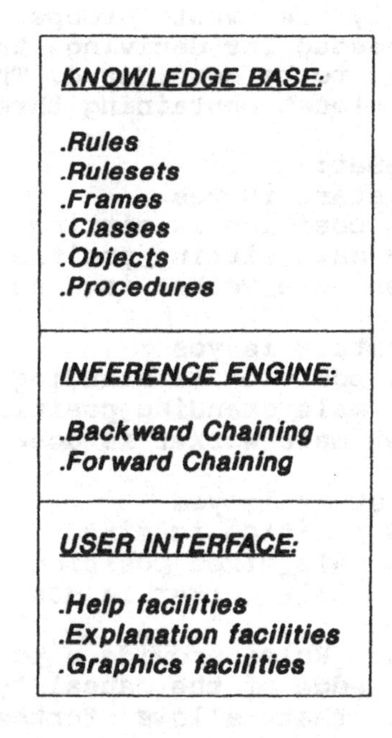

Figure 2. Details of Leonardo module

ergonomic workplace design.

Rules. A rule in Leonardo is an expression of the form "IF premise THEN conclusion". The premise contains one or more conditions for the conclusion to be true. An example of an ergonomic design rule in Leonardo is:-

```
"IF palm is supinated
THEN lifting_force is approximately_55_lbs".
```

In the above rule, palm and lifting_force are Leonardo "objects" whose attributes include the values "supinated" and "approximately_55_lbs", respectively. The rule gives the maximum lifting force that can be exerted by a hand with the palm facing upwards, when the forearm is horizontal and at right angle to the arm.

Rulesets. A ruleset in Leonardo is a collection of rules that constitutes a module of heuristic knowledge. Organizing rules into rulesets makes a knowledge base easier to maintain as related rules can be grouped together. Rulesets also increase execution efficiency as only relevant groups of rules, for instance, those needed for deriving the value of a given object, have to be considered. The following is an example of a ruleset containing three rules:-

```
RuleSet:
 if start is yes
 and position is sitting
 and male_sitting_position is done
 then male_worker is done

 if start is yes
 and position is standing
 and male_standing_position is done
 then male_worker is done

 if start is yes
 and position is mixed
 and male_mixed_position is done
 then male_worker is done
```

Frames and classes. Rules provide a powerful means for dealing with knowledge of the causal type. This is the kind of knowledge that allows further data to be inferred from the presence of some known facts. However, rules are less appropriate for representing knowledge about the properties of objects and the relationships between them. Representation of this kind of knowledge as rules typically involves adding a rule to the rule base for each property of every object, and its relationships with respect to all other objects. It is easy to see that the size of the rule base quickly becomes very large. Frames and classes provide a more compact method of handling this type of information.

An example of a frame in Leonardo is given below:-

```
       Name:worker
   LongName:
       Type:Text
AllowedValue:male,female
QueryPrompt:Please enter the gender of the worker
```

This frame encodes information about the object "*worker*". The values that this object can take are "*male*" and "*female*". These are shown in the "*AllowedValue*" slot of the frame. Whenever the value of the object is required, a query would be issued to the user as indicated in the "*QueryPrompt*" slot.

Examples of a class frame and a frame representing a member of that class are given below:-

Class frame:

```
       Name:forces
   LongName:
       Type:class
    Members:
            220,185,184,135,160,115,159,100,150,99,
            250,186,187,149,165,140,105,114,219,230
MemberSlots:
      elbow:
 right_left:
  push_pull:
```

Member frame:

```
       Name:220
   LongName:
       Type:Undefined
      Value:
DerivedFrom:
        ISA:forces
MemberSlots:
      elbow:180
```

```
 right_left:right
  push_pull:push
```

The *"forces"* class includes all the maximum push/pull forces that can be exerted by the arms of a worker in a sitting position, with the elbow at different angles. Each member of the class represents the force for a given elbow angle. The correct member is selected by specifying the elbow angle, side of the worker (left/right) and the direction of the application of the force (towards/away from the body, or push/pull). A rule is usually employed in conjunction with a class to identify the required member. An example of such a rule in the case of the *"forces"* class is :-

```
RuleSet:
       For all forces
if ques_elbow is elbow: of forces
and ques_right_left is right_left: of forces
and ques_push_pull is push_pull: of forces
then suitable_forces includes name: of forces;
```

Procedures. A procedure is a set of instructions for performing a routine task. Leonardo provides a procedural language similar to a conventional programming language for implementing procedures. The following is an example of part of a Leonardo procedure for computing the anthropometric dimensions of a worker from a knowledge of his height.

```
          Name:Anthropometric_dimensions
      LongName:
          Type:Procedure
       .......:
     LocalReal:height,st_height,st_sitting_height,
               sitting_height,eye_height,.........
       .......:
          Body:
        print('please input height in mm. >')
        finput(height,4)
          st_height=1740
          st_sitting_height=850
          st_eye_height=750
sitting_height=height*st_sitting_height/st_height
eye_height=height*st_eye_height/st_height
...........
at(10,8,'sitting height=',sitting_height)
at(11,8,'eye height=',eye_height)
...........
```

Inference Engine

The inference engine operates on the knowledge stored in the knowledge base and on the data supplied by the user to produce design solutions. Leonardo permits inferencing to be carried out in both the forward chaining and the backward chaining modes [4]. The method of inferencing adopted for this work is actually a combination of backward and forward chaining, called "backward chaining with opportunistic forward chaining". With this method, backward chaining is first performed to instantiate the objects in a rule or a user query and then forward chaining is executed to maximize the use of the data. The procedure then resumes with a return to the backward chaining mode.

User Interface

The user interface enables two-way data communication between the expert system and the user. The expert system requires from the user specific data about the operator, the task and the workplace. The user receives prompts from the expert system indicating the required information and also the results of the inferencing and design processes. The user interface comprises two main modules, a graphic display module and an explanation module.

Graphics Module. This module is implemented using Leonardo's graphics toolkit. It is provided to display to the user sketches of the workplaces designed by the expert system and other graphical information. An example of a sketch produced by this module is shown in Fig 3.

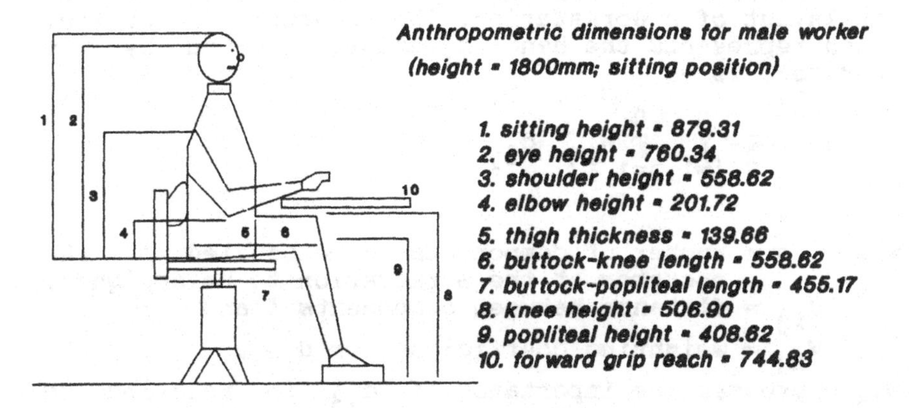

Figure 3. Sample output from Leonardo module

Explanation Module. The standard explanation facility available in Leonardo is used to give the user the reason for a particular request for information or to show him the inference chain leading to a certain result.

An example of an explanation provided by the system is shown below:-

To find if suitable_working_position was sitting
I used MainRuleSet, rule:

if workbench_height is desk_height
or both_feet_must_operate_controls is true
and required_mobility is low
then suitable_working_position is sitting

workbench_height	desk_height
both_feet_must_operate_controls	false
required_mobility	low

GA MODULE

This optimization module employs a Genetic Algorithm (GA). Details of the use of GAs to solve optimization problems can be found in [5,6]. For a summary of the GA method, see [7].

The problem to be solved is the optimization of the layout of a workstation. The objective function, which represents the handling effort exerted by the operator, is:-

$$e = \frac{1}{2} \sum_{i=1}^{n} \sum_{j=1}^{n} W_{ij} \; d_{ij}$$

where n = number of components to be handled by the operator at the workstation to be designed;
d_{ij} = distance between components i and j ;
W_{ij} = weighting coefficient for d_{ij} .

W_{ij} expresses the importance of d_{ij} in relation to other distances d_{ik} or d_{lm}. For example, W_{ij} is higher than W_{ik} if d_{ij} is regarded more important than d_{ik}, that is, if object i is handled together with object j more frequently than with object k.

For simplicity, d_{ij} is computed as (assuming a 2-D layout):-

$$d_{ij} = \left|x_i - x_j\right| + \left|y_i - y_j\right|$$

where x_i , x_j = x - coordinates of components i , j
y_i , y_j = y - coordinates of components i , j

The task of the optimization module is to select the x and y coordinates for the n components in the workstation so as to minimise e .

The module comprises eight main procedures. These are briefly described below.

Procedure Initialisation.
This procedure randomly generates the starting population of strings or genes representing different workstation layouts. Each string is of the form:-

$$x_1\ y_1 \quad x_2\ y_2 \ldots\ldots x_i\ \ y_i \ldots\ldots x_{n-1}\ y_{n-1}\ \ x_n\ y_n$$

where the coordinates x_i and y_i are themselves binary strings.

Procedure Validation.
This procedure checks that the strings generated by the initialisation procedure and other subsequent operations represent valid layouts (for example, objects do not overlap with one another).

Procedure Fitness.
This procedure computes the fitness of each solution string as the inverse of e , using the values of x and y encoded in the string and the weights W provided by the workstation designer. Fitness scaling [6] is adopted.

Procedures Random_Selection and Seeded_Selection.
These procedures implement the gene reproduction function.

Procedure Crossover.
This procedure performs the gene crossover operation.

Procedure Swap.
This procedure exchanges the coordinates of two randomly chosen objects.

Procedure Mutation.
This procedure carries out the gene mutation operation.

DB MODULE

This module is an assembly of databases storing details of the anthropometric dimensions of operators and data on the workplace and operations to be performed.

PRELIMINARY RESULTS

Figures 3-4 show the first results obtained with the Leonardo module and Genetic Optimization module.

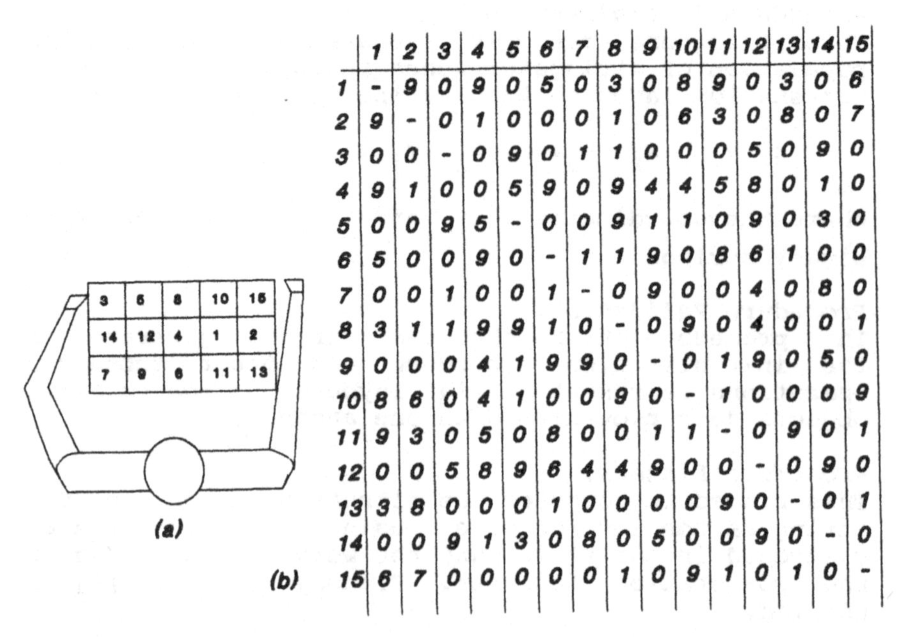

	1	2	3	4	5	6	7	8	9	10	11	12	13	14	15
1	-	9	0	9	0	5	0	3	0	8	9	0	3	0	6
2	9	-	0	1	0	0	0	1	0	6	3	0	8	0	7
3	0	0	-	0	9	0	1	1	0	0	0	5	0	9	0
4	9	1	0	0	5	9	0	9	4	4	5	8	0	1	0
5	0	0	9	5	-	0	0	9	1	1	0	9	0	3	0
6	5	0	0	9	0	-	1	1	9	0	8	6	1	0	0
7	0	0	1	0	0	1	-	0	9	0	0	4	0	8	0
8	3	1	1	9	9	1	0	-	0	9	0	4	0	0	1
9	0	0	0	4	1	9	9	0	-	0	1	9	0	5	0
10	8	6	0	4	1	0	0	9	0	-	1	0	0	0	9
11	9	3	0	5	0	8	0	0	1	1	-	0	9	0	1
12	0	0	5	8	9	6	4	4	9	0	0	-	0	9	0
13	3	8	0	0	0	1	0	0	0	0	9	0	-	0	1
14	0	0	9	1	3	0	8	0	5	0	0	9	0	-	0
15	6	7	0	0	0	0	0	1	0	9	1	0	1	0	-

Figure 4. (a) Layout of a workstation with 15 component buffers
(b) weighting coefficients

CONCLUSION

Ergonomic workplace design is a task requiring a large amount of experience. This paper has described an expert system being developed for this task. The system possesses a hybrid knowledge base, a genetically inspired optimization program and several numerical databases. This combination of knowledge-base technology, genetic optimization methods and database technology has proved to be an effective way to build powerful expert systems for solving complex problems.

ACKNOWLEDGEMENT

The authors would like to thank Mr R.A. Mansfield for his assistance.

REFERENCES

1. McCormick, E.J. and Sanders, M.S., Human factors in engineering and design, McGraw-Hill ,New York, 1982.
2. Kvalseth, T.O. (ed), Ergonomics of workstation design, Butterworth , London , 1983.
3 Creative Logic , Leonardo user manual , Creative Logic Ltd , Brunel , 1988.
4. Pham, D.T. and Pham, P.T.N., Expert systems in mechanical and manufacturing engineering. Int. J Adv Manuf Tech , 3(3) , pp 3-21 , 1988.
5. Holland, J.H., Adaptation in natural and artificial systems, The University of Michigan Press, Ann Arbor, MI, 1975.
6. Goldberg, D.E., Genetic algorithms in search optimization and machine learning, Addison-Wesley, Reading, MA, 1989.
7. Pham, D.T. and Karaboga, D., A new method to obtain the relation matrix for fuzzy logic controllers, Proc 5th Int Conf. on Artificial Intelligence in Engineering (AIENG 91), Oxford, July 1991.

SECTION 2: ENGINEERING ANALYSIS AND SIMULATION

Hierarchical Qualitative Simulation for Large Scale Dynamical Systems

K. Okuda (*), T. Ushio (**)

(*) Fundamental Research Labs., Osaka Gas Co., Ltd, 6-19-9, Torishima, Konohana, Osaka 554, Japan

(**) School of Home Economics, Kobe College, 4-1, Okadayama, Nishinomiya 662, Japan

ABSTRACT

This paper deals with a Petri net based hierarchical model for a qualitative simulation to generate concurrent behaviors. Each state in a dynamical system is represented by a dynamic variable with a pair consisting of a qualitative value and a qualitative derivative, and by a discrete variable with a qualitative value. A qualitative model is constructed by defining causal interactions among variables. A hierarchical modeling is supported to ease knowledge acquisition. A time-scale abstraction is also supported for more precise envisioning. We also apply this method to a co-generation plant.

1. INTRODUCTION

This paper deals with a framework of qualitative reasoning for large scale systems. Many modeling and/or reasoning techniques have been proposed in these few years[1,2,4,6]. But these conventional methods are mainly focused on small or medium size systems. An extension to large scale systems is proposed by Falkenhainer et al.[3] where a multiple view based hierarchy is considered using qualitative process theory. It is, however, often very difficult or even inefficient to apply such approaches to actual plants because it is quite difficult to acquire such knowledge for a large scale system. A variable-oriented approach[5] is much more suitable for knowledge acquisition of large scale systems.

From the reasoning aspect, conventional approaches often generate behaviors which never happen, because they can not explicitly express variable behaviors in such systems: concurrent, sequential, synchronized and asynchronized behaviors. In this paper, we propose a Petri net based model to resolve this problem. Hierarchical modeling and time-scale abstraction are introduced to support knowledge acquisition and more precise envisioning.

Section 2 discusses an overview of a Petri net based qualitative model. Section 3 describes a hierarchical modeling and a time-scale abstraction for large scale systems. Section 4 shows a simulation algorithm. Its application to a co-generation plant is discussed in the Section 5. Section 6 gives conclusions.

2. PETRI NET BASED QUALITATIVE MODEL

A Petri net[8,9,10] is proposed to present concurrent, sequential, synchronized and asynchronized behaviors, and conventionally applied to design and analyze sequential control, and communication systems. A Petri net model is described by

$$PN=\{P, T, I, O, \mu\},$$

where, P,T,I, and O denote a set of places, a set of transitions , a set of mapping from places to transitions, and a set of mapping from transition to places respectively. A place represents a state and a transition represents a state transition. The current state of a system is represented by the existence of a token in places, which is called a token pattern denoted by μ. In a Petri net, a place and a transition can be connected by a directed arc. If all places connected to a transition have tokens, then the transition is said to be "enabled". If the enabled transition fires, then we remove tokens from those places connected to it and put tokens to those places connected from it. A permission arc is a two way arc from a place to a transition. In a Petri net, a synchronized behavior is represented by common input places to a transition. A sequential movement of a token in a Petri net represents a sequential behavior, and a multiple movement a concurrent behavior.

2.1 Dynamic and Discrete Variables

We use a variable-oriented modeling of a system. Two types of variables are introduced: discrete variables and dynamic variables. A discrete variable is a symbol-valued variable which varies discretely over time, such as an ON/OFF switch and an OPEN/CLOSE valve. The qualitative state of a discrete variable is represented by its qualitative value (qv). A dynamic variable is an originally real-valued variable which varies continuously over time, such as the level of a tank and the flow rate of water. The qualitative state of a dynamic variable is represented by a pair consisting of a qualitative value (qv) and a qualitative derivative (qd), whose evolution is determined by the following equation:

<next qv> = <current qv> + <current qd>

Each value is specified qualitatively in terms of a set of landmark values. Landmark values may be either numerical or symbolic. In dynamic variables, landmarks of qv and qd are assumed to be numerical.

In our Petri net model, each landmark is assigned to a transition and each interval between landmarks to a place. Firing a transition corresponds to passing the corresponding landmark.

2.2 Propagation of Effect on Variables

We introduce two types of information propagation: zero order propagation and 1st order propagation. Zero order propagation consists of a qv to qv propagation, and a qd to qd propagation. 1st order propagation is a qv to qd propagation . "Exceeding normal water level causes an alarm activation" is an example of zero order propagation. The relation between acceleration and velocity is an example of a 1st order propagation because we know that the positive value of acceleration leads to the increase of velocity. The qv of the velocity will be changed due to its qd. In this sense, 1st order propagation represents a time-delay relation.

In our model, causal information like zero order and 1st order information propagation is treated as a control signal for a variable. This control signal comes from a variable which "sends" influence (this variable is called a "sender") and enters a variable which "receives" influence (this variable is called a "receiver"). In a Petri net model, the control signal comes from a place of a sender, and designated in a transition (s) of a receiver(s). Figure 1 shows a model of a dynamic variable with control signals. In this figure, qv is divided into three levels: "H" (high), "N" (normal) and "L" (low), and qd is also divided into three levels: "I" (increase), "S" (steady) and "D" (decrease). If a control input forces it to change its qd from "S" to "I", then transition t_1 will fire and a token of the qd will move from "S" to "I".

As described before, a trace of token patterns represents a behavior of a system. Transitions which do not share the same places can be fired independently. Such behaviors are concurrent behaviors. Sequential behaviors are represented by the sequential firing of transitions. Multiple places and a single place connected to a transition represent synchronized and asynchronized behaviors respectively. Thus a Petri net allows the explicit representation of concurrent/sequential and synchronized/asynchronized behaviors.

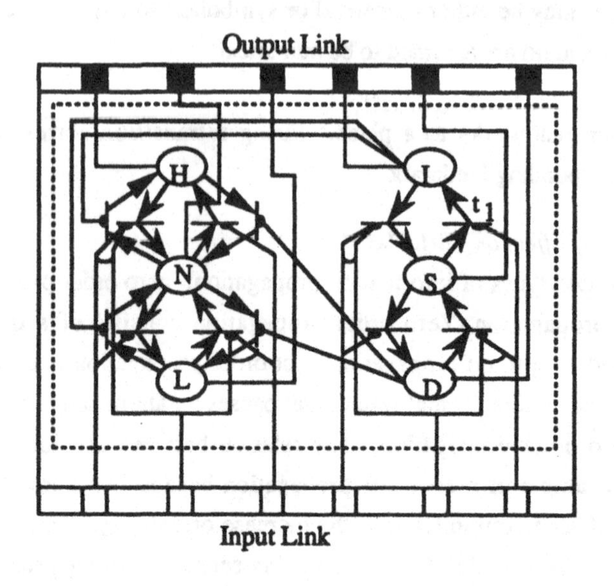

H: High, N: Normal, L: Low
I : Increase, S: Steady, D: Decrease

Permission arc

: OR link
If at least one of the places connected to this link has a token, then the transition is firable.

Fig. 1 A Petri net representation of a dynamic variable

3. HIERARCHICAL MODELING AND TIME SCALE ABSTRACTION

It may be possible, but it would be very inefficient to build a large scale qualitative model using the primitive elements introduced in the previous section. Furthermore, information on "time" is not included in our model. A concept of time is needed in order to generate sophisticated envisioning. To resolve this problem. a time-scale abstraction is introduced by Kuipers[7]. This section shows how a hierarchical modeling and a time-scale abstraction are merged in our model.

3-1. Hierarchical Modeling

An object oriented approach is used to realize hierarchical modeling, where two kinds of objects are defined: a variable object and a hierarchical object.

A variable object is a packed representation of a discrete or a dynamic variable (Fig. 1). A variable object receives control inputs through its input link, and sends control outputs through its output links. These objects can be prepared as a library for applications. A qualitative model can be constructed, using the library, by instantiating those objects and declaring their causalities.

Hierarchical modeling is considered to be a useful approach to understanding a complex system[12], and to manage a huge model without losing its consistency. We introduce a hierarchical object to easily implement this hierarchical representation. A hierarchical object consists of an internal causal model, input links and output links. An internal causal model is a network model defined by causal interactions among variable objects. Moreover a higher hierarchical object consists of an internal causal model defined as causal interactions among lower hierarchical objects, input links and output links which are defined as subsets of internal hierarchical objects' input links and output links. Figure 2 shows a hierarchical representation of objects. In it level numbers are assigned in ascending order starting with level 0 at the top. Each level contains several objects. If a level i of a hierarchy contains n objects, we call it a level i-j system ($j \leq n$).

Let's take an example of a simple gas distribution network. This is a system to send gas to appropriate places through pipes using valves and regulators. There are three variable objects: pipe, valve and regulator. A hierarchical object is constructed by defining the functional structure of those instantiated variable objects.

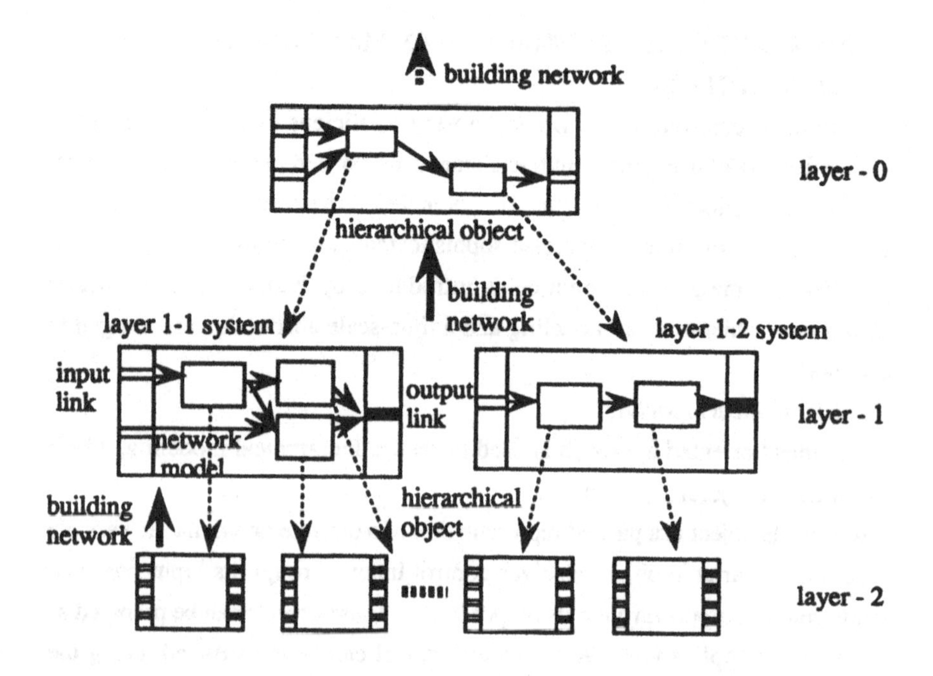

Fig. 2 Hierarchical representation

3-2. A Time-Scale Abstraction

Kuipers[7] proposed a time-scale abstraction to treat an intrinsic problem of scale by introducing a time-scale hierarchy. A time-scale hierarchy can be decomposed into "very fast", "fast" , "medium", "slow", "very slow" and so on, where a higher time-scale, i.e., faster behaviors, has higher priority to be envisioned (See Fig. 3). For example, electrical information propagates faster than fluid information.

According to the time-scale hierarchy, we prepare as many categories of transitions as the number of time-scales in the hierarchy. The faster category has the higher priority. Thus if several transitions are enabled at a certain time, those transitions which are in the faster category have higher priority to fire (We call this a "time-ordered state change") .

4. SIMULATION ALGORITHM

To simulate our model, we use an occurrence net based algorithm with some modifications. An occurrence net is suitable for explicitly tracing various behaviors like concurrence and synchronicity. An occurrence net is a directed

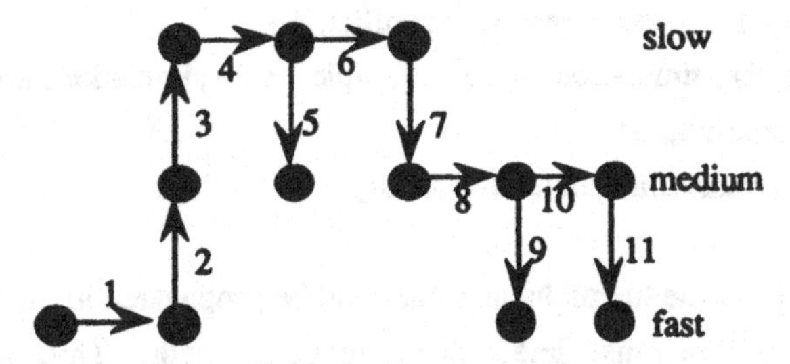

Fig. 3 Time-scale abstraction of behaviors

graph whose nodes are enabled transitions and places which caused the state change. In an occurrence net, sequential behaviors are represented by a "line", a sequential path in the net, and concurrence behaviors by a "slice", a set of places which do not form a line. Those places connected to a transition represent synchronized behaviors. A single place to a transition represents a asynchronized behavior. Some meta-rules must be added in order to generate reasonable behaviors.

Rule 1: Firing a transition which does not change a token pattern is prohibited.

In qualitative simulation, qualitative time gains one step when the state of a variable changes. On the other hand, in a Petri net, enabling a transition which does not change a marking pattern is possible. Rule 1 eliminates such behaviors in a Petri net.

Lacking quantitative information causes ambiguity which sometimes leads to forks in qualitative simulation. But some kind of ambiguity can be resolved by using the following rules.

Rule 2: Zero/1st order propagations and a variables' internal state change occur interactively.

Rule 3: For a conflict in transitions,

(1) use qualitative calculation ("add") for the resolution. (We use qualitative arithmetic expressed in Kuipers[6])

(2) apply a time-ordered state change if those transitions are classified into several time-scales.

If the above rules can not resolve the conflict, then we use the following rule:

Rule 4 : If rules 1 ~ 3 do not resolve a conflict, then

(1) fork the simulation into multiple sub- simulations according to the conflict,, or

(2) fire a transition undeterministically.

A state change in the hierarchical model will be propagated in the model. A simulation algorithm must detect the propagation path. Thus we need a searching algorithm in the simulation. A proposed search method will be explained using Fig. 4. Suppose that an input link of a hierarchical object (HO1) received a control signal to change the state (point "a" of Fig. 4). To find the state change of HO1, searching down (depth first search) to a variable object is needed (point "b" of Fig. 4). A state change of VO1 is propagated to HO2. Influence propagation inside HO2 is done in a breadth first manner (point "c" in Fig.4). Searching a complex structure of a hierarchical object to find a propagating path is often difficult, therefore a "level structuring method" is applied to resolve the problem.

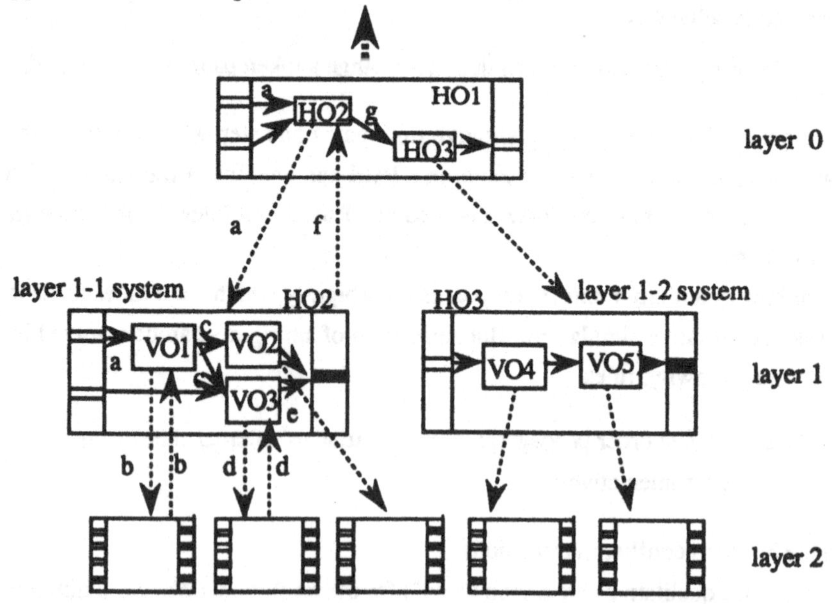

Fig. 4 Hierarchical search

We consider a directed graph G = (N, A), where N and A are the sets of nodes and directed arcs respectively.

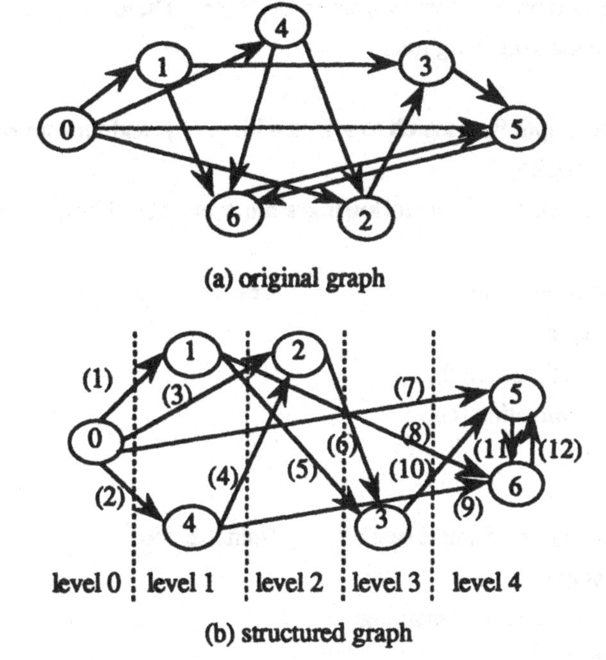

(a) original graph

(b) structured graph

Fig. 5 Structured graph

(1) Nodes from which node i can be reached are called ancestor nodes.

(2) Nodes to which node i can reach are called descendent nodes.

A level structured graph can be constructed by the following procedure.

STEP 0: j =0

STEP 1: Make a list LIST of ancestor and descendant node sets for each node. If LIST is empty, then halt.

STEP 2: The node set NS in the level j is a node set which satisfies the condition that an intersection between the ancestor set and the descendant set equals the ancestor set.

STEP 3: Remove those nodes included in NS from LIST.

STEP 4: j=j+1. Go to STEP 2.

Shown in Fig. 5 is an example of a level structured graph.

When states of VO2 and VO3 have changed (point "d" in Fig. 4), the output of HO2 is determined (point "e" in Fig. 4), whose information is also propagated to HO1 (point "f" in Fig. 4). In HO1, the output information will propagate to HO3 (point "g" in Fig. 4). The output of HO3, which is also the output of

HO1, can be determined by a similar procedure. These procedures can be summarized by the following rule.

Rule 5: Apply depth first search to search a hierarchy, and breadth first search to a hierarchical object.

Using these rules and the following simulation algorithm, we obtain an occurrence net.

```
program  main;                        {main program}
  begin
    initialize (net);
    simulation (net) ;
  end;

procedure  initialize (net);          {initialization}
  begin
    set the initial marking;
  end;

procedure  simulation (net);          {executing a simulation}
  begin
    propagate (net);
    sub_simulation (net) ;
  end;

procedure  propagate (net); {propagation in the hierarchy}
  begin
    propagate information of net in  a depth first manner;
    propagate information in each hierarchical  object of
    net in a breadth  first manner;
  end;

procedure  sub_simulation (net);  {extracting transitions}
   var  tr = empty , mark = empty;
  begin
     mark = (find current marking list) ;
```

```
    tr      = (find all transitions which can fire) ;
    action (net, mark, tr);
  end;

procedure action (net, marking, trans); {state change}
  begin
    if trans is not empty
      if  trans has conflict
        then apply rule 3, rule 4 to resolve conflict;
      if trans still has conflict
        then   fork (net, marking, trans)
      else   fire each transitions in trans,
       and expand an occurrence net;
      simulation (net);
  end;

procedure fork (net, marking, trans); {conflict resolution}
   var tr1, tr2, tr[], i=0,k;
      (k is the number of conflicting transitions)
  begin
    decompose trans as follows; trans = tr1 U tr2;
   (tr1: no conflict transitions, tr2: conflicting transition)
    find all subset of conflict-free transitions of tr2:
   tr[0] ,tr[1] ,, tr[k]
    if rule 5 (1) is selected   then
      begin
         while i < k  do:  action (net, marking, tr1 U tr[i]);
      end;
    if rule 5 (2) is selected   then
      begin
     select a transition set from { tr[0],tr[1],, tr[k]} ;
         action (net, marking, tr[i]);
      end;
  end;
```

5. APPLICATION TO CO-GENERATION PLANTS

5-1. Modeling

This section applies our approach to a co-generation plant. We consider a co-generation plant consisting of a gas turbine as its prime mover, an electric generator, and a boiler system. The turbine burns natural gas as fuel and drives the generator which produces electricity. The waste heat energy of the exhaust gases is used for generating steam in the boiler system. The co-generation plant is a very efficient way of producing energy due to its utilization of the waste heat. Because the system produces several kinds of energy (in this case heat and electricity), it is sometimes called a Total Energy System (TES)

The co-generation plant is a large scale plant and has concurrent, synchronized and various other behaviors. Figure 6 shows an overview of the plant's function. Figure 7 shows a three layered representation of the plant, and Fig. 8 shows its three layered hierarchical model. The hierarchy can be deepened if we have more information about the model. A time-scale abstraction of the model is given as follows.

"fast" mode: electrical information propagation.

"slow" mode: other information propagation.

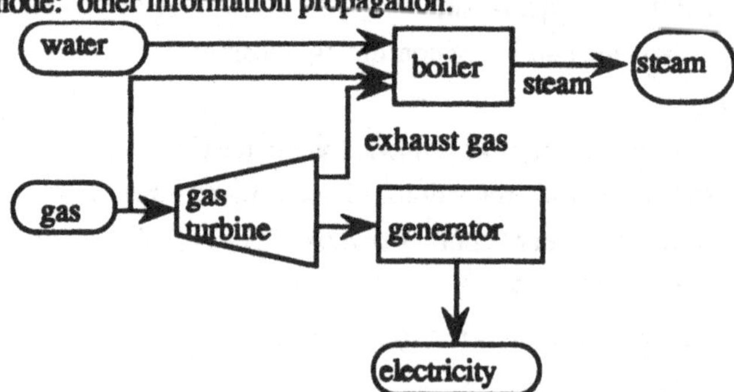

Fig. 6 Co-generation plant

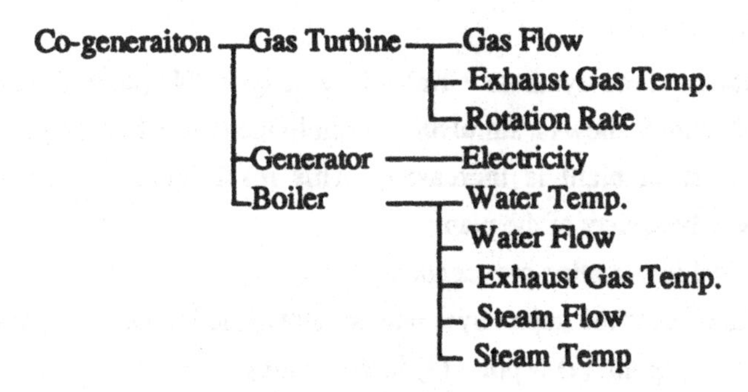

Fig. 7 Hierarchical decomposition of the plant

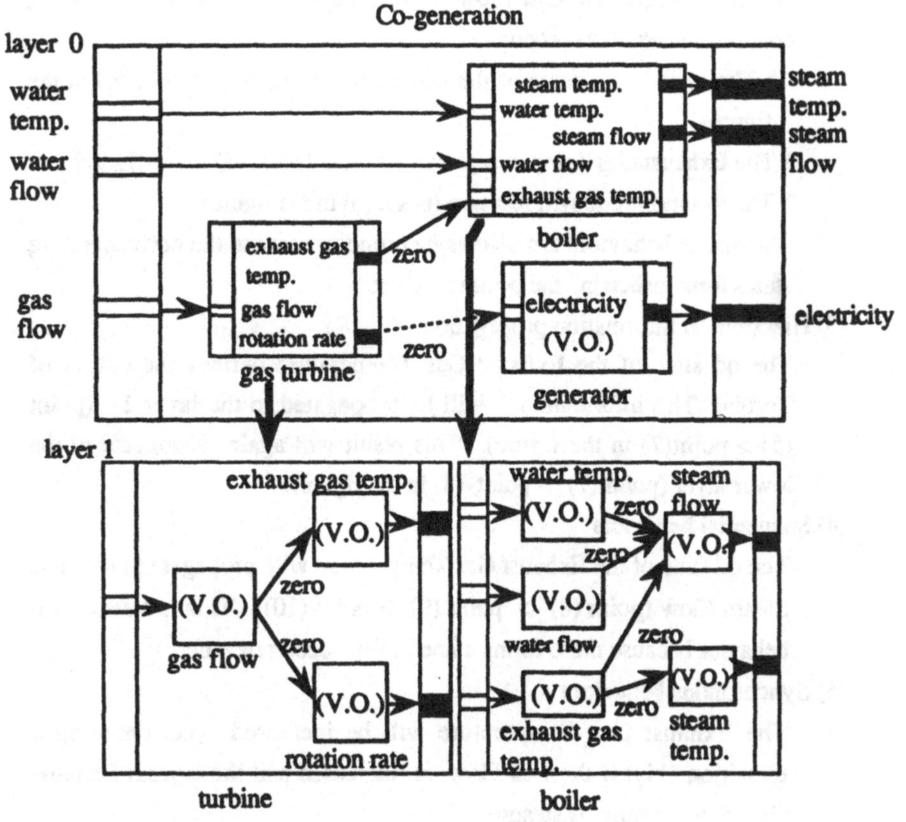

Fig. 8 Hierarchical qualitative model of the plant

5-2. Simulation

The initial state of each dynamic variable is set at qv = "N" (normal) and qd = "S" (steady). Figure 9 shows a simulation result in the case when the Gas Flow of the co-generation plant is increased. This result gives the following interpretations of behaviors of the plant:

(1) Hierarchical information propagation

An influence for the upper layer will be propagated to the lower layers. (Point (1) -> point (2)-> point (3) in the figure)

(2) Concurrent/sequential and asynchronous behaviors

In layer 2, after the Gas Flow of Turbine is increased, the following concurrent behaviors occur.

* The qualitative value of the Gas Flow will be high (point (4) in the figure)
* The Exhaust Gas Temperature will increase (point (5) in the figure)
* The Rotation Rate will increase (point (6) in the figure)

The above behaviors are also asynchronous because the corresponding states form a slice in the occurrence net.

(3) Hierarchical information propagation

The qd state of the Exhaust Gas Temperature defines the output of Turbine. This information will be propagated to the layer 1 (point (5)-> point(7) in the figure). This result will again propagate to the lower layer (point (7) -> point (8) in the figure).

(4) Sequential behaviors

The qd state of the Exhaust Gas Temperature will propagate to the qd of Steam Flow (point (8) -> point (9) -> point (10)). This is a sequential behavior because these forms a line in the occurrence net.

(5) Synchronous behaviors

The Exhaust Gas Temperature will be increased (i.e. enabling a transition (11)) if the Gas Flow is increased and the current Exhaust Gas Temperature is steady.

(6) A time -scale abstraction

A time-scale abstraction leads to the fact that the increase in Exhaust Gas Temperature occurs later than the increase of Electricity (point (12) -> point (13) -> point (14) -> point (15) in the figure).

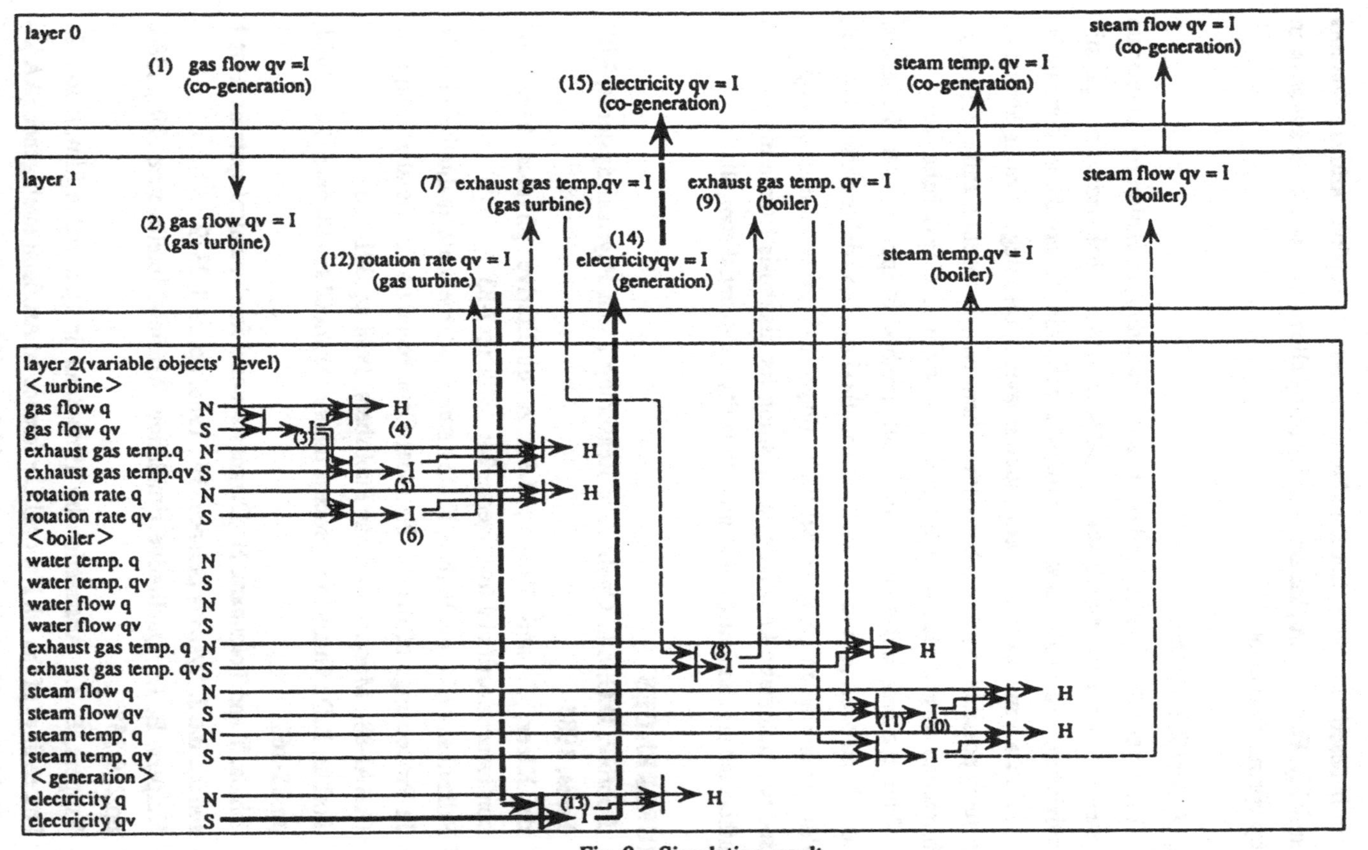

Fig. 9 Simulation result

As is described above, our simulation allows to envision various behaviors explicitly. Furthermore, hierarchical modeling allows to describe behaviors in an appropriate abstraction.

6. CONCLUSIONS

This paper introduced a Petri net based qualitative simulation for large scale systems. Concurrent, synchronized, and asynchronized behaviors are explicitly envisioned. In conventional qualitative reasoning approaches, concurrent behaviors are mis-treated as sequential behaviors, resulting in wrong behaviors or many possible sequential behaviors. Hierarchical modeling and a time-scale abstraction are also introduced for more precise envisioning of behaviors.

From the practical point of view, this method is applicable to diagnosis, control and plant operation. Since this method uses a variable centered model, which is similar to graph model based diagnosis[11], this simulation might be used for automatic knowledge acquisition for diagnosis from a model. The authors are also applying this method to verify a sequential controller.

REFERENCES

1 Bobrow, D.G. ed.: Qualitative Reasoning about Physical Systems, MIT Press, 1985

2 de Kleer, J. and Brown, J.S.: A qualitative physics based on confluences.Artificial Intelligence 24: 1986, pp.7-83

3 Falkenhainer, B. and Forbus, K.D.: Setting up large-scale qualitative models. In Proceedings of the 7th National Conference on Artificial Intelligence (AAAI-88). Morgan Kaufman Publishers, 1988, pp.301-306

4 Forbus, K.D.: Qualitative process theory. Artificial Intelligence 24: , 1984, pp.85-168

5 Ishida, Y and Tokumaru, H. Diagnosis model for dynamical systems (Part 1 and 2). Int. J. Systems Science. 19 , 1988, pp.2291-2319

6 Kuipers, B. J.: Qualitative simulation. Artificial Intelligence 29 , 1986, pp.289-338

7 Kuipers, B. J.: Abstraction by Time-Scale in Qualitative Simulation. In Proceedings of the 6th National Conference on Artificial Intelligence (AAAI-87). Morgan Kaufman Publishers,1987, pp.621-625

8 Murata, T: High-level Petri nets for logic programming and AI applications. In Tutorials of the 3rd International Workshop on Petri nets and performance Model, , 1989, pp.210-260

9 Peterson, J.L.: Petri net theory and the modeling of systems. Prentice-Hall

10 Reisig, W.: Petri nets. Springer-Verlag 1985

11 Sztipanovits, J. et. al.: Intelligent Monitoring and Diagnostics for Plant Automation. In Proceedings of IEEE International Conference on Robotics and Automation. ,1990, pp.1390-1395

12 Weld, D. S. : Explaining Complex Engineered Devices, BBN Report No. 5489, 1983

Intelligent Support of Flight Experimental Design and Analysis

L. Berestov (*), A. Kozlov (*), V. Melnik (*),
V. Vid (*), V. Denisov (**), V. Khabarov (**)
(*) Flight Research Institute, Zhukovsky 140160, USSR
(**) Novosibirsk Electrotechnical Institute, USSR

ABSTRACT

Methods of the theory of experimental design and their computer realization are used in many full-scale and simulation research programs of aircraft and propulsions. In the course of this research empirical mathematical models of flight qualities and performance, throttle and altitude/rate propulsions performance, etc. are defined and it is also estimated how varing flight conditions and various perturbing factors influence them. Depending on the complexity of linear or nonlinear models of the object and on the type of identification factors, methods of analysis of variance, regression and covariance analysis and available a priori data about parameters are used. When using such a variety of methods in research of complex multifactor dynamical objects , an experimenter faces a difficult task, which is generally unsolvable without specialist in mathematic statistics, to choose the most effective methods of test data analysis. From the list of empirical mathematical model types, analysis methods and optimum design of experiments one can get hundreds of variants of tasks setting and ways of solving them.

Due to as stated above we are forced to research and to adopt the intelligent support of flight experimental design and analysis.

INTRODUCTION

According to current practices of conducting full-scale flight experiment and simulation, it is presumed that a wide range of methods of design and analysis of experiments may be used [4,5]. We shall merely refer to some of the methods widely used in flight exploration in order to obtain the cause-and-effect empirical matematical models. Linear and nonlinear regression analysis, variance analysis, covariance analysis which, in turn, have scores of different variants depending on the strategy of conducting the experiment, character of data and type of models [1,4].

The experimenters are posed with not so simple probblem of choosing the strategy for conducting an experiment including the selection of the empirical mathematical model structure, experimental design for discriminating the competing model structures, design of refined (qualified) experiments, test of model goodness of fit, etc.

The availability of extensive literature on the problems of statistic analysis (SA) and experimental design (ED), as well as quite a wide choice of program packages [1] allow to solve local problems, but on the other hand,the variety of methods and resources causes an element of discomfort with the researcher having a pragmatic type of thinking. As a rule,the selection of statistic models is of subjective character and a statistic model involves a human error along with systematic and random ones. The systematic and random errors can be minimised by effective ED and SA methods, but the human error in selecting the models, the results of which may be quite serious, can be minimised if carried out according to scientifically-based methods with the support of intelligent program medium.

In the present paper the authors state their viewpoint regarding the intelligent program medium for SA and ED, and give the description of certain components of the intelligent system(IS) model of SA and ED.

1. INTELLIGENT SYSTEM (IS) CONCEPT

According to experimental research technology, the following stages are presumed:

* Collection, analysis and adequate representation of a priori information about the object under study.
* The experiment goal selection:
  - or detecting the extreme condition,
  - or obtaining the phenomenal model of the phenomenon,
  - or, alternatively, a local approximation dependence for further extrapolation or interpolation of the object.
* Generation of hypotheses with respect to model structure on the basis of the available a priori information.
* The choice of strategy for conducting an experiment (experimental design):
  - distribution of observations and selection of the rule to stop the experiment,
  - distribution of observation in experimental space.
* Discrimination of the competing model structures.

* Parameter estimation of the model chosen.
* Test of model goodness of fit .

Now let us state the major requirements the IS supporting the experiment technology is to satisfy to.

ADEQUATE GOAL STATEMENT. There is a mechanism allowing the experimenter correct choosing and stating the goal at each step of the experiment. This mechanism provides for generation of useful hypotheses with respect to the form of the model, points at which the experiment is to be conducted, etc.

METHOD ADAPTATION. There is a mechanism set up for analytical and numerical methods which are necessary to achieve the goal set.

PROBLEM ADAPTATION. There is a mechanism set up for the specific problem being solved (particular terminology, physical dimensional representation of factors, a priori information with respect to factor,etc).

OBJECT ADAPTATION. There is a mechanism which accumulates information on the specific object in the form of a posteriori distribution, confidence intervals, point estimation, etc. after each step of the experimental research.

TRAINING AND EXPLANATION. There is a mechanism for explaining one or another of question posed by the experimenter and for explaining a result is made. Here may be WHY-, HOW-, WHAT-, FOR- type questions.

The requirements listed allow to classify IS as an expert system. But by contrast to a classical expert system which has a component of logical inference, an explanation mechanism, knowledge base and data base [3], IS under consideration has a component of statistical inference which closely interacts with the logical inference component. Here three levels of adaptation (method, problem and object adaptation) exist which, in fact, form a part of system cognitive mechanism.

The mechanism of such system operation generally reduces to the following:
- IS is initially assumed to be an empty shell possessing a "genetic" ability to self-development;
- an expert-statistician fills the shell with specific knowledge about the necessary units of SA and ED by means of a dialogue initiated by the shell. The dialogue is of directed character. It means that knowledge with respect to the use of one or another of statisti-

cal method are matched up with the knowledge already available for completeness and non-contradiction. The syntactic form of the dialogue is dictated by the shell.The dialogue initiated also belongs to the shell;
- an expert in statistical analysis and experiment design, numerical methods (it is possible that this expert could also be a statistician) when offered a technology by IS connects the computing modules to the system;
- a general debugging of the system is carried out. At this stage, knowledge and data base are updated. Then the system is considered as method-adapted;
- further the stage of problem adaptation follows. At this stage the method-adapted system initiates the dialogue with the experimenter-expert in the particular object field. It has the goal to receiving a knowledge on this domain. At this stage, for instance, dependent and independent variables are derived, their character (qualitative, quantitative), physical dimensions, measurement range are revealed, the variables are assigned names.In other words,the system tries to obtain as much a priori information as possible and set up its work for the given object field. Then the system is regarded as problem-adapted;
- further character of system development will be determined by the aims of the experimenter and statistical data obtained during the experiment. Then, on the basis of knowledge introduced at the method adaptation stage and statistical data on the object of research, the system is trying to involve the experimenter in the process of statistical model derivation. The process of the purpose-oriented deriving the model is based on the method of optimal experimental design. Since the process is of random nature,the system generates the rule of stop. Let us call this stage an object system adaptation;
- after the object adaptation stage the system is capable of extrapolation and interpolation of the object behaviour.

2. INTELLIGENT SYSTEM ARCHITECTURE

The system is built on the principle of empty shell [3], see Figure 1 below . The frame of the shell is made up of the following components.

INITIATOR. As pointed out above, the shell has a "genetic" ability to self-development. Initiator has a "thirst for information" from outside and permanently makes the system be "curious". First, the initiator's questions are of general character, then they become more specific.It is there that the system shows its ability to be adapted to the medium.

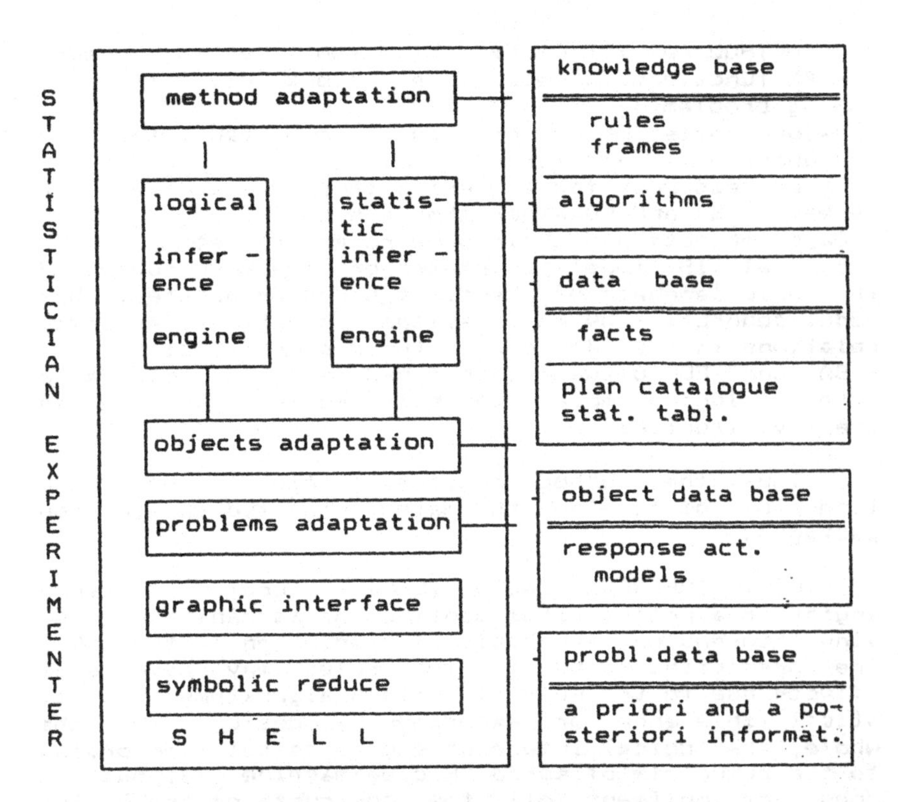

Figure 1. The architecture of the IS.

At the method adaptation stage the initiator takes over the initiative in the dialogue with the statistician. As a result, the shell is filled with a specific logical model of the statistical analysis and experimental design, as well as with SA and ED methods. The initiator "imposes" the technology of application packages for SA and ED systems in the form of prompting. Here the logical model of the system represented by the statistician in metalanguage has a priority. The initiator offers a standard for joining separate object modules to the system. Thus, at the method adaptation stage the statistician trains the system, introducing his own SA and ED concepts in the form of logical models and specific knowledge of implimentation methods.

At the problem adaptation stage the initiator activates the dialogue with the experimenter. At this stage of the system development the script of the dialogue is defined by the logical model introduced at the method adaptation stage. The same refers to the object adaptation stage. The initiator is the basis of the IS cognitive mechanism.

METHOD ADAPTATION MECHANIZM (AM). Let us consider the AM functions in detail. The block solves the following problems:
- major goals which are to be achieved within the framework of IS are set up;
- major resources for achieving these goals, i.e., the necessary SA and ED algorithms are specified;
- major objects and their properties are set up;
- general IS model is stated on the Horn subset of the logic language of the first-order predicates. This model connects the major system objects by specific relations in accordance with their properties;
- SA and ED software is refined and, in accordance with the logical model, the software is linked in an overlay structure.

Thus, the method adaptation block supports the technology of forming the method-oriented application packages.

LOGIC INFERENCE ENGINE (LIE). Logic inference engine is a program implementing the mechanizm of proving theorems on the basis of resolution method [9]. The predicates used in the rules may have side effects due to the work of certain algorithms of statistical inference or experimental design. On the whole, the logical inference engine is built according to the principle of serial Prolog-machine [7], but it does not impliment all the costructs of the Prolog language, only the necessary ones.Besides, a number of original system predicates connected with the SA and ED domain is introduced, and a dialogue mechanizm by menu is built in.

STATISTICAL INFERENCE ENGINE (SIE). SIE closely interacts with LIE and is built on the principle of Bayesian solver. The interaction is carried out as follows. The selection of statistical method and its numerical implimentation,as well as statistical models is carried out a priori by the logical inference in LIE on the basis of the statistician's expert knowledge stored in the knowledge base. Following the logical inference, a class of statistical methods or alternative models,each element of which has an expert estimate in the form of probability, is derived. This probability is considered as an "a priori" probability for the Bayesian solver. Bayesian strengthening of hypotheses or obtaining the "a posteriori" distribution for parameter estimation is carried out on the basis of the results of the experiments design suggested.

PROBLEM ADAPTATION MECHANIZM (PAM). PAM solves

a problem of structural design of empirical mathematical models of flight qualities and performances, throttle and altitude/rate propulsion performance, etc. and it also estimates how varying conditions and various perturbing factors influence them. In this case the expert system database is formed of recommendations of experienced experimenters when choosing the genetic essential factors. PAM solves a problem of integrating SA with the experimetal database. The "a priori" information about dependent and independent variables and their physical nature,etc. are allocated.

OBJECT ADAPTATION MECHANIZM (OAM). In fact, OAM controls the statistical inference engine. Besides, the block carries the statistical data base where all the experiment results, as well as the "a posteriori" information of the preceding stages of the statistical inference are stored.

GRAPHICS PROCESSOR (GP). GP implements all the graphic functions of the system: windows,icons,graphs, drawings. GP faciliates writing a script of graphic display for results of the system work on the whole. The graphic display block recieves information from the problem adaptation block data base, as well as the problem adaptation block. Besides, this block directly interacts with the statistical inference and experimental design algorithms.

ANALITICAL TRANSFORMATION PROCESSOR (ATP). This component of IS is necessary for analytical transformation of linear and nonlinear models given in the form of algebraic expressions and differential equations. Operations are carried out with both scalar objects and matrices and vectors. Operations of symbolic differentiation, substitution of expressions , reducing similar terms , deriving a common multiplier, etc. are available.

APT is used in IS for solving problems of model identifiability and discriminating the competing models given either in the form of system of differential equations or algebraic equations. ATP may be used as an independent system as well .

3. INTELLIGENT SYSTEM METALEVEL

A metalevel is allocated in IS, where the integration of the whole system is performed. In particular, the metalevel data base which defines the mechanizm of the initiator's work is generated in the IS metalanguage. The script of the dialoque and graphic interface is also generated here. At the metalevel the control is carried out by LIE.

4. METALANGUAGE and IS LANGUAGE

IS is constructed on the principle of logic programing [9]. Essentially, the metalanquage is constructed on the Horne subset of the first-order-predicates logics. The following metaconstructions are assigned: objects, goals, questions, rules, facts.

OBJECTS.Allocation of objects is initiated by the IS cognitive mechanizm. Name is generated, as well as references to the other objects, field of possible values. By way of possible values there may be references to the other objects, integer and real numbers, symbols, strings, signs. Sets of possible values may be given by properties, enumeration, interval.

QUESTIONS. Essentially, a set of questions generates the script of the dialogue. Each question is linked with a particular object. If a question is initiated, it is followed by a menu of answers. The menu is generated from a set of possible object values. A question is asked only in the case when the value for the object cannot be derived logically by means of LIE.

GOALS. Allocation of goals is necessary to define by means of metarules the mechanism of the goal statement at the metalevel. From the syntactic viewpoit goals are a clause without the left part. If the goal is not derived logically, the user is offered a menu of goals.

RULES. Rules are clauses containing predicates linked by the connectives AND,OR in the right and left parts. Instead of predicates, operators in the infix, prefix or postfix forms may be used. Rules are organized at the hierarchical system which is stated on the concept of a "possible worlds".

FACTS. Facts are either introduced in the knowledge base a priori or generated in the process of the dialogue. Generation and accumulation of facts in the knowledge base during the dialogue relieve IS of repeating the questions. Facts may have a status of temporary and permanent ones. If the mode of removing the prehistory of the dialogue is specified, temporary facts are eliminated. If desired, some temporary facts can be changed over to the status of permanent ones. The work with facts is organized by means of a "black board" tipe mechanism.

5. SA and ED ALGORITHMS

Along with declarative knowledge, the data base also

contains knowledge in the form of a procedure. These are essentially SA and ED algorithms and specific algorithms from the object field IS is set up for.

Let us enumerate the basic algorithms comprising the nuclear of IS.

< Regression analysis >:
- linear regression analysis (stepped regression, non-full rank models, models with limitations, recurrent estimation, robust method, criteria for verification of linear hypotheses, variance analysis, covariance analysis);
- nonlinear regression analysis (least-square method LSM, Bayesian estimation, LSM with parameter limitations, recurrent estimation).

<Identification of dynamic models>:
- given in the form of conventional differentional equations (analysis in the frequency and temporal fields);
- given in the form of a system of equations in partial derivatives.

<Distribution estimation>:
- for grouped observations (certain standart distributions);
- for ungrouped observations (Pearson curves, Johnson population, standard laws of distributions).

<Experimental design>:
- serial design of discriminating experiments (Hunter-Riner, Phedorov Criteria);
- serial design of refined experiments (A,D,E,L,F-criteria);
- a priori design (A,D,E,L,F-criteria for multi-response models);
- generation of combinatorial design;
- experimental design for dynamic model identification.

CONCLUSION

This works are being conducted to create new task composition technology and research methods of complex aviation objects on the basis of modern expert systems concepts. Expert system database is formed of experienced experimenters and mathematicians on the choice of object models structures, statistical methods and evaluation of identified empirical mathematical model adequacy. Broadly developed program packages for experimental design and identification serve as a functional content.

REFERENCES

1.Денисов В.И. Математическое обеспечение системы ЭВМ- экспериментатор. М.:Наука, 1977.

2.Nils J. Nilsson. Principles of Artificial Intelligence. Tioga Publishing Palo Alto, California CA 94302, 1980.

3.Hayes Roth F., Waterman D., Lenat D. Building Expert Systems. Addison Wesly Reading, M.A., 1983.

4.Летные испытания газотурбинных двигателей самолетов и вертолетов. Под ред. Г.П. Долголенко. М.:Машиностроение, 1983.

5.Берестов Л.М.,Поплавский Б.К.,Мирошниченко Л.Я. Частотные методы идентификации летательных аппаратов. М.: Машиностроение, 1985.

6.Мельник В.И., Денисов В.И., Хабаров В.И. Интегрированная система планирования экспериментов и статистического анализа. -Тезисы докл. Всесоюзной конференции "Моделирование систем информатики". Новосибирск, ВЦ СОАН СССР, 1988. с.76-78.

7.Клоксин У., Меллиш К. Программирование на языке ПРОЛОГ. М.:Мир, 1987. 336с.

8.Поспелов Д. А. Ситуационное управление. Теория и практика. М.:Наука, 1986. 248с.

9.Логическое программирование.(Пер. с английского под ред. В. И. Агафононова). М.:Мир, 1988. 366с.

A Knowledge Based System for Material Preheat in Welding

W. McEwan (*), M. Abou-Ali (**), C. Irgens (***)
(*) Paisley College Quality Centre,
(**) Department of Mechanical and Manufacturing Engineering,
(***) Department of Computing Science,
Paisley College, Paisley, PA1 2BE, UK

ABSTRACT

Knowledge based systems are suitable for applications that require a large amount of specialised knowledge to solve problems. In welding, the ultimate objective of any Welding Analyst is to produce "crack-free" weldments. One of the factors which contributes to the prevention of cracking is heating the parent metals prior to welding. In order to determine the appropriate minimum preheat temperature for a material, a certain amount of knowledge and experience is required especially for new conditions or special requirements.

This paper describes a knowledge based system for determining the minimum preheat temperature for carbon and carbon manganese steels. This system incorporates both the knowledge and reasoning processes the human expert has, together with the knowledge and information available in the Standard Codes.

INTRODUCTION

Over the last decade there has been a rapid growth in the development of knowledge based systems for engineering applications. Knowledge based systems are suitable for applications that require a large amount of specialised knowledge to solve problems. In the pressure vessel industry, the quality of weldments is critical and the ultimate objective of Welding Analysts is to produce "crack-free" weldments. One of the factors which contributes to the prevention of hydrogen cracking in weldments is heating the parent metals prior to welding. In order to determine the appropriate minimum preheat temperature, many factors must be considered. This requires a certain amount of knowledge and experience of the metallurgical behaviour resulting from welding.

In the field of welding technology conventional and intelligent computer software may be applied in many areas: from welding process control systems to prediction of weldment quality [1-3]. The avoidance of metallurgical defects in the Heat Affected Zone (HAZ) of welds such as hydrogen induced cracking is one of these areas.

PREHEAT is a commercial program developed by the Welding Institute, U.K. [4,5]. It was designed to calculate the minimum preheat temperature for carbon and carbon manganese steels with reference to BS 5135:1984 [6]. Mehrotra et al. [7] developed a micro-computer method for predicting the preheat level necessary to avoid the formation of heat affected zone cold cracks.

Although traditional preheat calculation programs have made a useful contribution to the advancement of welding application, these programs are unable to solve many problems. From the Welding Analyst's point of view, these conventional programs do not match the human heuristic thought process. The knowledge required by the Welding Analyst to support his/her decision making is buried somewhere in the program coding.

An attempt has been made by Fukuda et al. [8,9] to develop an expert system using Prolog for supporting engineers in making decisions that will avoid weld cracking. The effectiveness of this approach was verified through test sessions.

One of the obstacles which has slowed down the development of useful and usable knowledge based systems is the limitation associated with the system-building tools.
In this work, the domain of selecting the minimum preheat temperature was investigated. The necessary items of knowledge and information have been acquired from the functional experts in an industrial environment. A knowledge base was created using an object oriented building tool [14].

The work reported in the paper is a part of an ongoing research program to develop a knowledge based system that will generate optimised process plans for welding and fabrication.

AN OVERVIEW OF HAZ HYDROGEN CRACKING

Causes of Cracking

Hydrogen induced cracking can occur in steel during manufacture, during welding and in service [10]. In this paper the subject of hydrogen cracking in weldments is addressed and Figure 1 (a and b) shows a schematic diagram of the possible positions of cracks in the HAZ.

The occurrence of hydrogen cracking in the HAZ depends on a number of factors [6]:

- Composition of the steel: areas of the HAZ which undergo a rapid ferrite-austenite-ferrite transformation can have microstructures which exhibit a reduced tolerance to hydrogen embrittlement as a result of hardening caused by the associated fast cooling.
- Welding consumables: here hydrogen generated mainly from residual moisture in electrode coatings or granular fluxes, and by the decomposition of any dirt or oils in the joint or on the surface of welding wires. It is absorbed into the weld pool from the arc atmosphere and diffuses into the HAZ.
- Welding procedure: welding conditions can be selected to avoid cracking to ensure that the heat affected zone cools suficiently slowly by control of weld run dimensions in relation to material thickness.
- Stresses involved: these cannot be avoided since thermal contraction

must occur as the weld solidifies and cools. Additional stresses may also be present, arising from the interaction of the welded member with a larger structure.

Although the same four conditions listed above apply, the susceptibility of a weld HAZ to hydrogen cracking is related to the microstructural type as well as to the hardness and is therefore more complicated to predict.

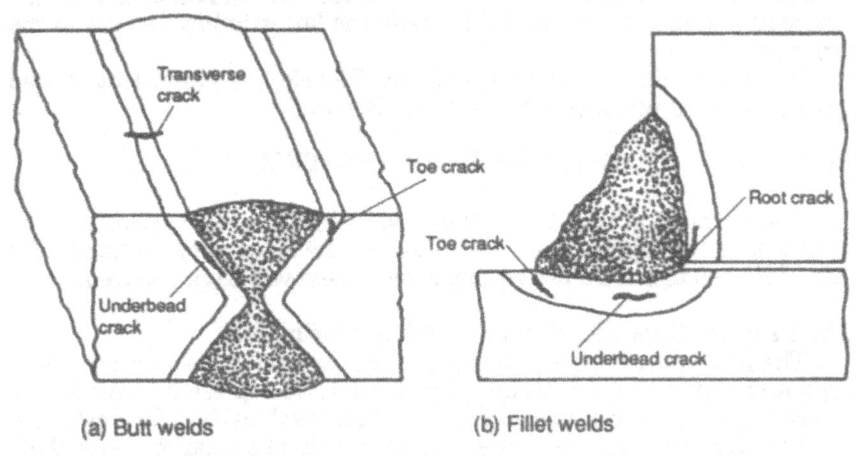

Figure 1. Position of hydrogen cracking in weldments (source: Coe, ref 10)

Prevention of Cracking

There are three main approaches to the prevention of HAZ hydrogen cracking, the most effective method being to reduce the quantity of available hydrogen. The diffusible hydrogen content of a deposited weld is classified according to one of four hydrogen scales for the purpose of predicting safe welding procedures:

Scale A - high hydrogen (> 15 ml/100g deposited metal)
Scale B - medium hydrogen (10-15 ml/100g deposited metal)
Scale C - low hydrogen (5-10 ml /100g deposited metal)
Scale D - very low hydrogen (<= 5 ml/100g deposited metal)

In practice, this means scale D could be obtained by employing a low hydrogen welding process such as Tungsten Inert Gas (TIG) welding in conjunction with thoroughly cleaned workpieces.

An alternative approach is to use a steel yielding a less hardenable HAZ microstructure, this behaviour being predicted by the carbon equivalent (CE).

The carbon equivalent can be calculated using the following formula [6]:

$$CE = C(\%) + \frac{Mn(\%)}{6} + \frac{Cr(\%) + Mo(\%) + V(\%)}{5} + \frac{Ni(\%) + Cu(\%)}{15}$$

However, in the majority of cases, the choice of parent material and welding process is fixed by metallurgical, economic and equipment considerations. Under such circumstances, the risk of obtaining hydrogen cracking may be reduced by allowing a greater time for the absorbed hydrogen to diffuse safely out of the weld region before cooling to a temperature at which cracking can occur. This can be achieved by selecting a welding procedure which employs a high heat input, or includes the application of workpiece preheating prior to welding. Either or both of these actions raises and prolongs the thermal cycle experienced by the weld metal and HAZ. The selection of the most appropriate preheat temperature requires a Welding Analyst with a certain level of experience. Therefore, developing a knowledge based system to assist Welding Analysts is of benefit to both the experienced and the less experienced. The former will get advice in a very short time and it allows the latter to have access to the experience of an expert.

In the following sections the problem of developing a knowledge based system for the preheat specifications is discussed.

BUILDING A KNOWLEDGE BASED SYSTEM

The problem of how to develop a knowledge based system may be approached from different angles [11-13]. In the following paragraphs the steps followed to build a knowledge base system will be introduced.

The Preheat Specifications - Problem Definition

The main objectives are to ensure that the system will satisfy a real need and that it will be technically feasible. The present approach to determine the minimum preheat temperature involves the following steps:

- Identify the Standard Code and/or the client's requirements. This information will be supplied by the client and it reveals specifications such as acceptance/rejection criteria.
- Investigate the joint preparation, configuration and size. At this stage, another source of knowledge is required from which the appropriate welding process could be selected.
- Investigate the parent material's chemical composition and calculate the carbon equivalent. This is the most important step which requires knowledge of welding metallurgy. In this step the Welding Analyst considers the effect of each element e.g. Carbon, Nickel etc. contents together with other factors such as the arc energy and the combined thickness and their effect on the overall quality of the welded joint.
- At the end the minimum preheat can be determined either from relevant diagrams or by utilised experience.

As described above, it is evident that the process of determining the preheat temperature involves a large number of parameters. In order to determine the preheat temperature for welding a new material or for a material not listed in any of the standards will take time and effort from an experienced Welding Analyst. If this is the case, the Welding Analyst will apply his/her knowledge and experience, using heuristic rules to determine the minimum preheat temperature, see Figure 2.

The Knowledge Acquisition

The knowledge and data necessary to build the knowledge base were acquired and collected from the functional experts in an industrial environment at the Metallurgy Department, Babcock Energy Limited, Renfrew, Scotland. A tape recorder was used to record the interviews with the functional experts.

The most common and easy method to acquire knowledge is interviewing the expert. The knowledge engineering task was performed

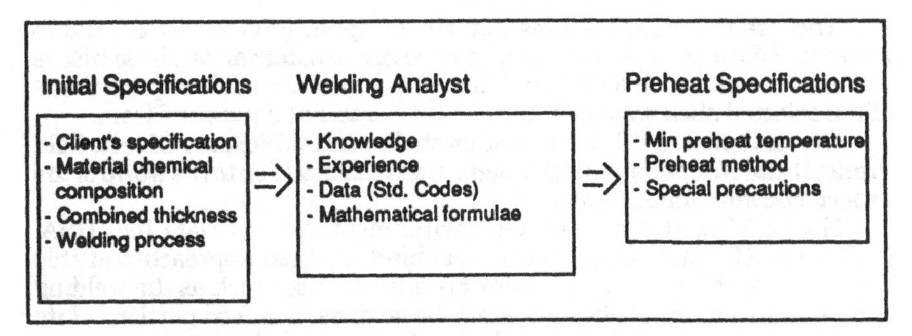

Figure 2. Generation of preheat specifications

and a set of precise questions were addressed to each expert to extract the knowledge and the reasoning criteria. The following are examples of some helpful questions: "What led you from point A to point B in your analysis? What was your mind telling you? What experience base did you address? In your judgement, what was the level of risk? What were the sources of knowledge you used?"

To disseminate knowledge from a human expert is not an easy task. In order to make the process of acquiring the knowledge efficient and fruitful, a decision tree was constructed and a procedure generation document was also created.

Selection of the Developing Environments

Based on the problem definition and the preliminary acquired knowledge the basic requirements of the expert system development environments were identified.

System hardware The general availability of PC hardware in the industrial establishment led to the selection of a PC platform for the development environment. The System was then developed on an IBM compatible PC with the following specifications: VGA coloured monitor, 286 processor, 1.2MB floppy disk and 8MB RAM (expanded and extended memory).

System software Building knowledge based systems using artificial intelligence programing languages requires a high level of computing expertise and the developed systems have many limitations regarding the memory restriction, the interfacibility with external programs and the ability to update the knowledge base. On the other hand there is significant improvement in the industry of intelligent computer software. Expert system shells have become more and more powerful and capable of building useful and usable expert systems. Therefore, there is no doubt that the use of established expert system shells will minimize the development time and enable the System Developers to utilize their time in refining and justifying their products. The choice of the most appropriate expert system shell suitable for an application requires a Knowledge Engineer (System Developer) with experience in expert systems' features and capabilities so that he/she can evaluate the nature and the problem characteristics and the user requirements.

The preheat specifications domain of application involves a large amount of data and information and so access to external database files is required to build such a system. Therefore, one of the basic requirements of the selected shell was the interface with external database files.

Object Oriented PC expert system shells such as Nexpert-Object, Gold Works II and Kappa have made a significant contribution to the world of the expert system-building tools.

Nexpert-Object expert system environment, Neuron Data Inc., USA, (written in C) places emphasis on an object-oriented approach and this makes it suitable for large, complex knowledge bases such as the welding domain. In addition, it allows System Developers to embed portions of its inference engine in their own code, combining the advantages of expert systems with algorithmic languages.

Nexpert-Object is a full-featured expert system shell that is competitive with higher-priced mainframe-based packages. It can run on a number of different machines including the IBM AT, PS/2, 386, Macintosh II , Sun and Apollo workstations, VAX and the IBM mainframe.

THE KNOWLEDGE BASE

The knowledge and information necessary for this system have been acquired from an industrial environment. Because of the complexity and interactive nature of the welding process the knowledge base (KB) was constructed in the form of modules (blocks). Each block holds a self-contained section of expertise i.e. preheat specification module, assembly information module etc. The inference engine together with the governing rules will decide which modules are required and how these knowledge bases can be linked. Figure 3 shows the basic elements of the developed system.

The knowledge base comprises four knowledge base modules, Startup KB, Assembly Information KB, Preheat Specification KB and End KB.

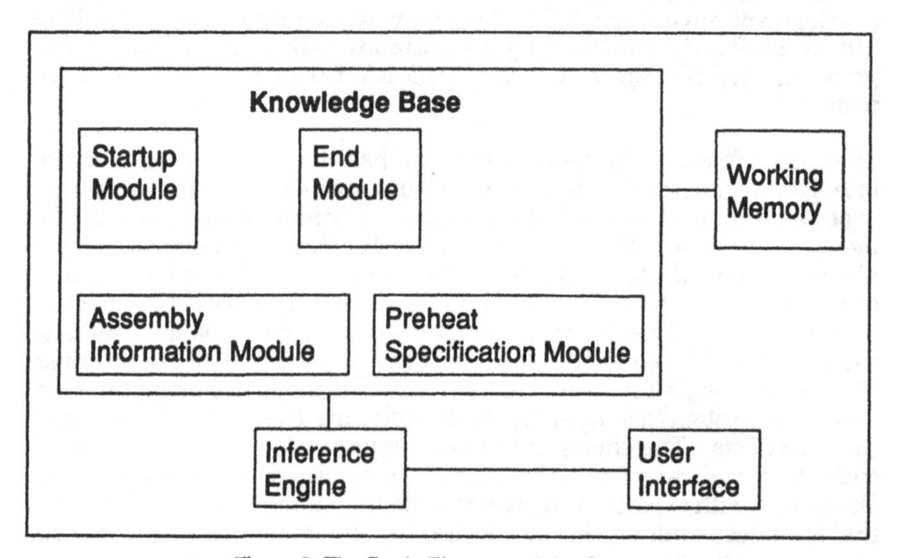

Figure 3. The Basic Elements of the System

Start up KB is responsible for interrogating the system end-user in order to identify his/her requirements. It displays some introductory screens, asks the user a number of questions if necessary, displays warning messages and at the end permits the materials information KB to be loaded.

Assembly information KB is responsible for the enquiry about the initial specifications of the assembly such as welding preparation, workpiece configuration, parent metals, etc. It also displays messages when necessary and at the end it retrieves the relevant materials information from an external database file. Figure 4 shows a portion of the Assembly information knowledge base in Nexpert program code. The program starts with defining the knowledge representation (objects, classes and properties) and then the knowledge reasoning (rules). Preheat Specification KB contains the relevant items of knowledge (rules of thumb) which are necessary to perform the reasoning process to determine the minimum preheat temperature.

End KB is responsible for ending the consultation with the end-user. It displays and/or prints the final report, and writes the consultation results in the database if that was requested by the end-user.

Knowledge Representation

Nexpert-Object is a hybrid system [14] with the ability to support both a reasoning system of 'rules' and a powerful representation of 'objects' as defined by a hybrid system.

Rules A rule is the elementary piece of knowledge. It represents a situation and its immediate consequences. In other words it links facts or observations to assertions or action. The basic structure of the rule is:

```
IF     condition    THEN    hypothesis
and  condition           and    actions
```

For example,

```
IF carbon_equiv >= 0.42                THEN  min_preheat
   And carbon_equiv <= 0.47                  Show message_4
   And presence of Sulphur                   Load end_kb
   And combined_thick <= 50.0mm
   And welding_process is TIG
   And ambient_temp >= 15 degree C
   And joint_prep is Vee
```

In the above rule, checks are made for all the conditions mentioned in the left hand side. If the conditions are fulfilled i.e. true, it sets the hypothesis to true and the set of actions in the right hand side will be executed.

Figure 5 (a and b) shows a rule printed from the development environment. Figure 5 (a) shows the rule in text mode and Figure 5 (b) shows the rule in graphic Rule Network mode. It shows how the conditions and actions are connected to the rule hypothesis.

```
(@VERSION=    011)
(@PROPERTY=   aust  @TYPE=String;)
(@PROPERTY=   c_con  @TYPE=Float;)
(@PROPERTY=   carbon_equiv  @TYPE=Float;)
(@PROPERTY=   kind  @TYPE=String;)
(@PROPERTY=   name  @TYPE=String;)
(@PROPERTY=   ni_con @TYPE=Float;)
(@PROPERTY=   strength     @TYPE=Float;)
(@PROPERTY=   type  @TYPE=String;)


(@CLASS=     materials
)

(@CLASS=     tubes_materials
)

(@OBJECT=     material_01
    (@CLASSES=
        materials
    )
    (@PROPERTIES=
        c_con
        carbon_equiv
        name
        ni_con
        strength
    )
)

(@OBJECT=     tube_1_material
    (@CLASSES=
        tubes_materials
    )
    (@PROPERTIES=
        c_con
        carbon_equiv
        name
        ni_con
        strength
        type
    )
)

(@RULE= R1
    (@LHS=
        (Is    (tube_1_material.name)  ("1Cr 1/2Mo"))
    )
    (@HYPO= get_tube_1_material)
    (@RHS=
        (Retrieve    ("c:\nexpert\nort\material.nxp")     (@TYPE=NXP;@UNKNOWN=TRUE;))
        (Do    (material_01.name)     (tube_1_material.name))
        (Do    (material_01.carbon_equiv)     (tube_1_material.carbon_equiv))
        (Do    (material_01.strength)  (tube_1_material.strength))
        (Do    (material_01.c_con)    (tube_1_material.c_con))
        (Do    (material_01.ni_con)   (tube_1_material.ni_con))
    )
)

(@RULE= R11
    (@LHS=
        (Yes   (get_tube_1_material))
        (Yes   (get_tube_2_material))
    )
    (@HYPO= get_tubes_material)
    (@RHS=
        (UnloadKB    ("c:\nexpert\nort\info\ret_data.tkb")  (@LEVEL=DELETE;))
        (LoadKB ("c:\nexpert\nort\heatment\heat.tkb")  (@LEVEL=ENABLE;))
        (Do    (get_pre_heat)  (get_pre_heat))
    )
)
```

Figure 4. An example of Nexpert program code

```
RULE: Rule 5
 If         tube_1_material . name is "1Cr 1/2Mo"
Then get_tube_material
           is confirmed.
           And Retrieve from "c: \nexpert\nort\material.nxp" @TYPE=NXP;@UNKNOWN=TRUE
           And material_01.name is assigned to tube_1_material.name
           And material_01.carbon_equiv is assigned to tube_1_material.carbon_equiv
           And material_01.strength is assigned to tube_1_material.strength
           And material_01.c_con is assigned to tube_1_material.c_con
           And material_01.ni_con is assigned to tube_1_material.ni_con
```

(a) Text mode

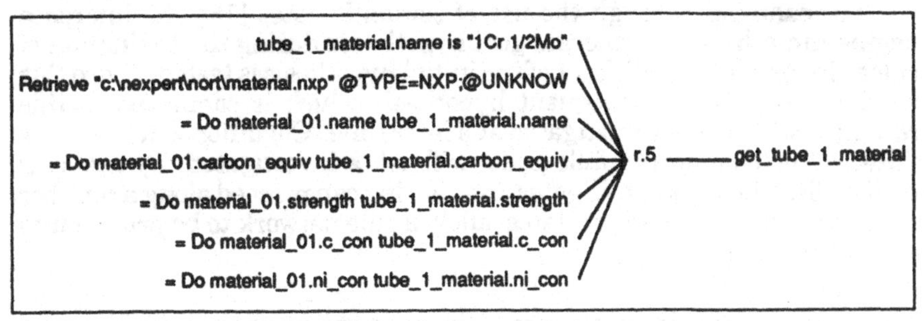

(b) Graphic mode

Figure 5. An example of a rule

Objects As the name implies, Nexpert-Object is object-oriented in nature with facilities for classes, objects, method and inheritance. Groups of objects can also be categorised into classes. An example of the structure of an object is as follows:
name : tubes
classes : hollow_cross_section, materials
subobjects: tube_1, tube_2
properties: wall_thick, outer_diameter, material

Figure 6 shows an example of a knowledge representation structure from the development environment in the graphic Object Network mode. It shows how objects, classes and properties are hierarchically represented.

Groups of objects can be networked into knowledge islands. All objects with strong links between hypotheses and conditions are grouped into the same knowledge island.

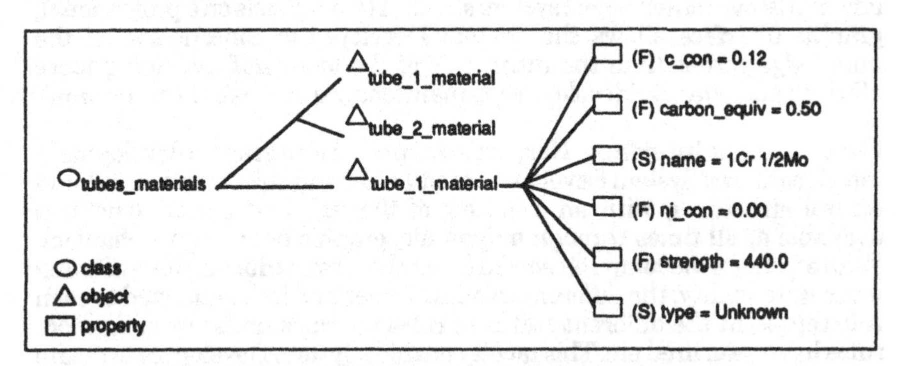

Figure 6. A knowledge representation structure - an example

Through a combination of rules and objects, a variety of methods for representing an expert's knowledge enable the System Developers to overcome the complexity and diversity of the welding domain.

Reasoning Objects may inherit properties from other objects or classes in a variety of ways. Inheritance can be changed at any time by using the built-in functions InhValueDown, InhValue Up, InhMethod, or NoInherit. Full control can be achieved on the reasoning mechanism through the four methods of inferencing on rules: backward chaining, forward chaining, semantic gates, and context links.

For example, through the use of semantic gates [15], the inference engine can selectively generate goal hypotheses during the evaluation of rules. In Figure 7, condition 1 of the initial hypothesis is tested. Since this condition is true, a subsequent hypothesis which is connected to this condition will also be investigated at a later time. Condition 2, however, is false, so the hypothesis it falls under will not be investigated. By enabling or disabling these semantic gates, inferencing can proceed along a number of possible paths. Semantic gates allow a rule network to be pruned into a smaller search space.

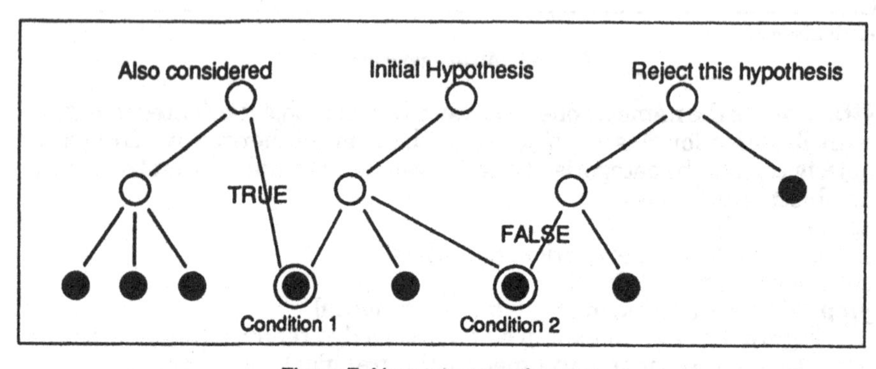

Figure 7. Nexpert semantic gates

System Interfaces

Nexpert's development and user interfaces present another competitive advantage over other expert system shells. The emphasis on a professional, graphic interface allows the System Developer to concentrate on the knowledge rather than the intricacies of the tool itself, providing more efficient and effective development, maintenance and use of the system.

Development interface Nexpert's comprehensive graphic development interface allows System Developers to edit rules and objects as well as build control structures, with an overview of the rule and object structures available at all times through a dynamic, graphic browsing mechanism. Figure 8 shows the computer screen for a rule network during the reasoning process. In reality, the different colours of rules and icons attached to each rule represent the different status of rules i.e. rules under investigation, rules have been fired etc. This facility enables System Developers to create and debug knowledge bases easily and in a short time.

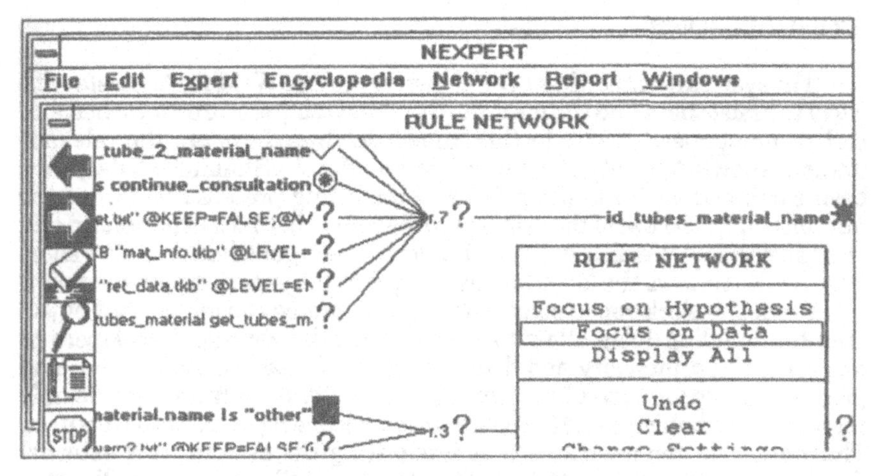

Figure 8. Development environment

User interface In the end user environment, a variety of prompts and means for user input including a default menu and custom-designed screens through a tool called NORT (Nexpert Object Runtime) are provided. By specifying Show as an action under the hypothesis of a rule, a variety of graphic formats including MacPaint, Dr Halo, Windows Paint can be displayed.

The System Integration

One of the most important feature which should be available in building tools for such application is the interfacibility with external programs and the linkage with other DBMS software. These database files could contain standard information available in the standard codes about the material's chemical composition, thicknesses etc. Figure 9 shows a Nexpert's spread sheet file (NXP) containing all necessary information about the different materials e.g. chemical compositions, mechanical properties etc. Also, by using the build-in functions available in the shell a link between the knowledge base and external computer environment such as conventional computer programs, text files and graphic files were established.

These database files could contain standard information available in the standard codes about the material's chemical composition, thicknesses etc.

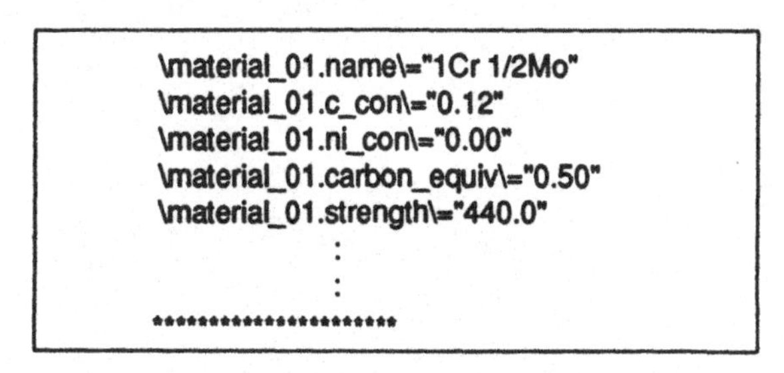

```
\material_01.name\="1Cr 1/2Mo"
\material_01.c_con\="0.12"
\material_01.ni_con\="0.00"
\material_01.carbon_equiv\="0.50"
\material_01.strength\="440.0"
:
:
**********************
```

Figure 9. NXP spread sheet file

CONCLUSIONS

The system designed has been shown to provide Welding Analysts with a tool to assist them find the minimum preheat temperature for carbon and carbon manganese steels. In this knowledge based system the relevant items of knowledge and information, available in the Standard Codes, have been combined with the expertise and reasoning process abilities of the human being. The use of the system is of benefit to both the experienced and less experienced Welding Analyst. The former will get advice in a very short time while it allows the latter to have access to the experience of an expert.

From the work reported in this paper it can be concluded that object oriented building tools allow to some extent the System Developers to overcome the complexity and diversity of the domain of application, and produce usable and useful systems. This has stemmed from their flexible reasoning mechanisms and the powerful knowledge representation features. The work also demonstrates that integrating the system with external database files enables the System Developers to utilise the space allocated for the governing rules.

The scope of the work can be broadened and linked to a CAD database through an interface so that the welding procedure could be generated directly from the CAD station.

REFERENCES

1. Street, J. Computer Technology in Welding - the First International Conference, Metal Construction, Vol.19, No.2, pp 81-83, 1987.

2. Lucas, W. Microcomputer Systems, Software and Expert Systems for Welding Engineering, Welding Journal, Vol.2, No.4, pp 19- 30, 1987.

3. Norrish, J. and Strutt, J.E. Expert Systems and Computer Software Aids for Welding Engineers, Welding & Metal Fabrication, Vol.56, No.7, pp 337-341, 1988.

4. Rodwell, M. H. Low Cost Microcomputer Assistance for Welding Engineers, 1st Int. Conf. on Computer Technology in Welding, London, U.K. 3-5 June 1986.

5. PREHEAT Program, User Guide, The Welding Institute, Abington, U.K.

6. Specification of Arc Welding of Carbon and Carbon Manganese Steels, BS 5135:1984, BSI, U.K.

7. Mehrotra, V.; Bibby, M.; Goldak, J. and Moore, J. A Micro- computer Method for Predicting Preheat Temperature, Int. Conf. on Welding for Challenging Environments, Canada, 1985.

8. Fukuda, S. and Maeda, A. Development of an Expert System for Welding Design Support: an Attempt, Transaction of JWRI, Vol.14, No.1, pp 171-176, 1985.

9. Fukuda, S.;Maeda, A. and Kimura, M. Expert System for Welding as a Better Means of Communication, The 6th Int. Workshop in Expert Systems and their Applications, France, April 1986.

10. Coe, F.R. Welding Steels without Hydrogen Cracking, The Welding Institute, 1973.

11. Wolfgram, D.D.; Dear, T.J. and Galbraith, C.S. Expert Systems for the Technical Professional, John Wiley & Sons, 1987.

12. Turban, E. Decision Support and Expert Systems, MacMillan Publishing Company, 1988.

13. Frenzel, L.E., Jr. Understanding Expert Systems, Howard W. Sams & Company, 2nd Edition, 1988.

14. Nexpert-Object User Manual, Neuron Data Inc., California, USA, 1989.

15. Aiken, M.W. and Liu-Shwng, O.R. Software Review: Nexpert- Object, Expert Systems, Vol.7, No.1, pp 54-57, 1990.

HardSys/HardDraw: A Smart Topology Based Electromagnetic Interaction Modelling Tool (NRC number 31804)

J. LoVetri (*), W.H. Henneker (**)
(*) Systems Integration Laboratory,
(**) Knowledge Systems Laboratory, Institute for Information Technology, National Research Council, Ottawa, Ontario, K1A 0R8, Canada

Abstract

An intelligent tool for the modelling and analysis of electromagnetic interactions in a complex electrical system is described. The purpose is to determine any unwanted electromagnetic effects which could jeopardize the safety and operation of the system. Modelling the interactions in a system requires the examination of the compounded and propagated effects of the electromagnetic fields. The approach taken here subdivides the modelling task into two parts: a) the definition of the related electromagnetic topology, and b) the propagation of the electromagnetic constraints. *HardSys*, a prototype object-based system implemented in Prolog, is used to propagate the electromagnetic constraints. User interaction is through *HardDraw*, a topology-drawing tool and an attribute interface.

1.0 Introduction

The effects of electromagnetic interactions in electrical systems are of concern because of the increased pollution of the environment with electromagnetic emissions and because of the increasing susceptibility of system components. The term electrical system is used herein in the general sense to include more than just networks consisting of electronic components. Systems containing biological and/or mechanical components of varying complexity are also included when reference to the term electrical systems is made. The problem is to model the electromagnetic interactions which take place between components in a complex system. The process of rendering these systems acceptably immune to the interactions is called *electromagnetic hardening* and, once achieved, the system is said to be *electromagnetically hardened.*

Theoretically, understanding the phenomena of electromagnetic interactions in electrical systems requires no more complicated theory than that explicated by Maxwell in his *Treatise on Electricity and Magnetism.* The application of this theory over the years has given insight into *mechanisms* of electromagnetic interaction. These mechanisms of interaction, associated with an electrical system, manifest a complex called the *electromagnetic system.*

From a practical point of view, it is not at all obvious how the electromagnetic integrity of systems can be *assured* even for relatively small interaction problems. *Non-algorithmic* techniques are used daily by engineers to solve electromagnetic problems in electrical systems. The purpose here is to establish an appropriate symbolic description or knowledge representation of the fundamental components in an electromagnetic interaction problem as well as the heuristics used to reason about these components.These heuristics are derived from well known engineering principles and can be viewed as *constraints* on the electromagnetic interaction problem.

The knowledge required in modelling electromagnetic systems includes the *electromagnetic topology* of the system and the *electromagnetic attributes* of each node in the topology. A prototype tool made up of an advisor and a drawing tool has been implemented in Quintus Prolog™ on a Sun™ workstation [1, 2, 3]. The advisor, *HardSys*, helps to analyze the electromagnetic attributes of an electromagnetic system. The user-interface for HardSys is a unique topology-drawing tool called *HardDraw* and is implemented with a Postscript-based user-interface toolkit called GoodNeWS/HyperNeWS (see Fig. 26 for a summary of the architecture).

2.0 Electromagnetic Topology of Systems

In order to model an electromagnetic system, it is first necessary to understand and represent the *relevant physical* attributes of the system [4, 5, 6, 7]. In this procedure, an electromagnetic system is decomposed into an *electromagnetic shielding topology* and its dual graph or *interaction sequence diagram.*

The electromagnetic topology consists of a description of the electromagnetically distinct volumes and their associated surfaces. The volumes define the

electromagnetic components involved in the interaction. The interaction sequence diagram keeps track of the ***interaction paths*** throughout the system. The two procedures are not independent of each other since the interaction sequence diagram can be derived from a given electromagnetic topology. A simple example is used to explain how the topological decomposition of systems is performed. The topology of Fig. 1 shows an extension to the theory of [7] in that circuit elements are also represented as volumes.

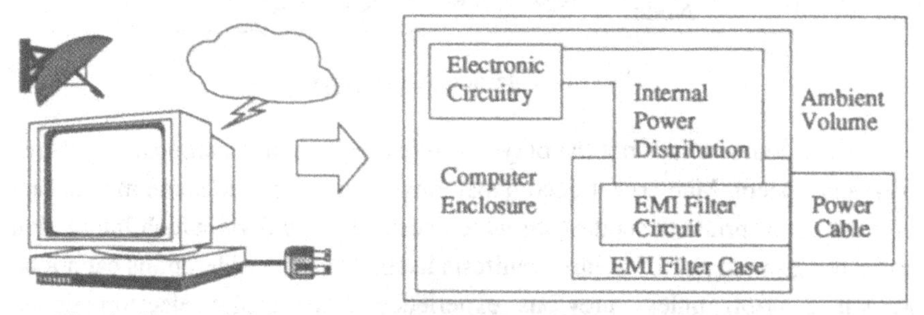

Fig. 1. Example Electromagnetic Topology Decomposition of a System

The interaction sequence diagram is obtained as a ***graph*** with ***nodes*** or ***vertices*** representing volumes, and ***edges*** representing surfaces. The graph representing the topology of Fig. 1 is shown in Fig. 2. Note the different node representation for ***field nodes, circuit nodes*** and ***interaction path nodes.***

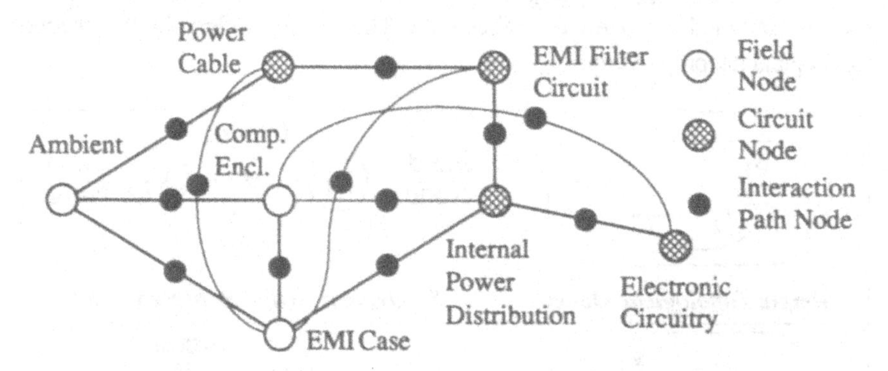

Fig. 2. Equivalent Interaction Sequence Diagram

Field nodes and circuit nodes contain attributes specific to field and circuit type quantities respectively. This leaves a possibility of four interaction path node types which are summarized in Fig. 3. Paths between field nodes, which will be denoted ***ff-paths***, take field quantities and attenuate them producing field quantities on the other side. Paths between circuit nodes, ***cc-paths***, attenuate circuit disturbances. The other two possible combinations are ***fc-paths*** and ***cf-paths*** with corresponding meanings.

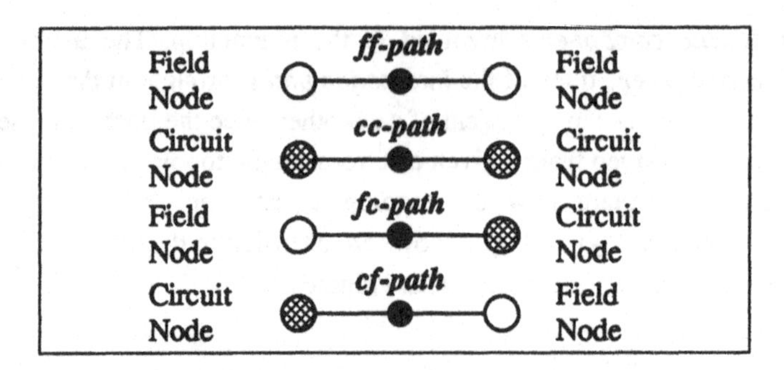

Fig. 3. Interaction path types

Of course this is not the only possible topological decomposition of this physical system. Many other decompositions are possible and some may in fact be more appropriate depending on which component and via which interaction paths the greatest risk of failure manifests itself. These considerations cannot be known a priori unless previous experience with similar electromagnetic components and topologies is available.

As an example, consider the two topological decompositions of a conducting penetration shown in Fig. 4. A conducting penetration can reduce the shielding effectiveness of an enclosure to zero in certain frequency ranges. This is because field energy can couple onto the conductor penetrating the shield and then be radiated again on the other side. This is more clear in the second topological decomposition.

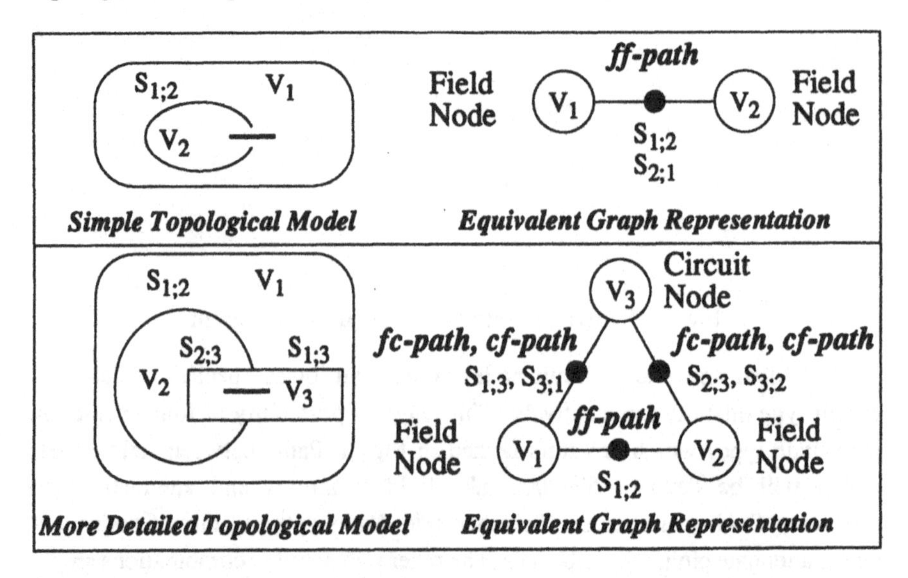

Fig. 4. Two topological representations of the same shielding imperfection

Thus a more detailed topological model may help to understand an interaction path phenomenon. Later it will be shown how a complex graph can be collapsed into a simpler representation with new composite attributes. The imposition of specific attributes on the topology components is discussed in the next section.

3.0 Electromagnetic Component and Path Attributes

The next step in modelling the electromagnetic system is to approximate the propagation of electromagnetic energy from one volume node to another as shown in Fig. 5. Electromagnetic attributes are introduced for each electromagnetic component in the topology as well as for the interaction paths between the components. These attributes *constrain* the propagation of the electromagnetic disturbances throughout the topology and represent the electromagnetic knowledge which is known about a system.

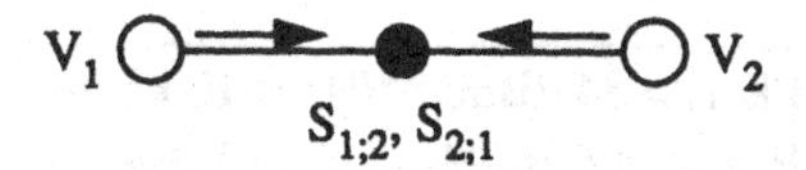

Fig. 5. Single electromagnetic interaction path

A good electromagnetic knowledge representation will have the following characteristics. It should

1) be easily derivable from available experimental/numerical data,
2) be in an easily manipulated qualitative form,
3) yield useful quantitative/qualitative results and recommendations,
4) have the capability to handle exceptions, and
5) have variable levels of approximation (coarse to fine).

Most electromagnetic interaction phenomena are calculated, measured, and reported as quantified data in the frequency domain. The reason for this is that most useful engineering information about fields, susceptibilities, and paths of interaction can be characterized *best* in the frequency domain. For example, the concept of frequency domain filtering can be used to characterize almost all *linear* paths of interaction. It is well-known that an aperture in a shield acts as a high-pass filter in the path of the electromagnetic fields [8]. Emissions from equipment are often measured using receivers with specific bandwidths of reception over large ranges of frequency [9]. Furthermore, susceptibilities of electronic components such as microelectronic circuitry are also calculated in the frequency domain [10, 11]. Thus all electromagnetic attributes are specified over quantized frequency ranges.

3.1 Electromagnetic Disturbance Representation

Each component node in an electromagnetic topology may have an electromagnetic disturbance associated with it. Field quantities and circuit quantities will describe the electromagnetic disturbance for field nodes and circuit nodes respectively. Thus an appropriate classification is to define *power density* PD with units of [W/m^2 or dBW/m^2] for field nodes, and *power* P with units of [W or dB] for circuit nodes. The magnitude of a disturbance in a *specific* frequency range can be specified as being one of several discrete values shown in Fig. 6 where the bracketed values represent the equivalent electric field for free space far fields and the voltage equivalent in a 50 Ω circuit. These ranges are chosen heuristically based on experience of electromagnetic disturbance levels and on the requirement for a useful number of ranges. Once these values are chosen, the creation of a useful database of disturbances requires that they remain constant.

extreme	if PD is > 84 dBm/m^2/Hz (>10 kV/m/Hz) if P is > 84 dBm/Hz (> 3.5 kV/Hz)
high	if PD is 44 - 84 dBm/m^2/Hz (.1 - 10 kV/m/Hz) if P is 44-84 dBm/Hz (35 V/Hz - 3.5 kV/Hz)
medium	if PD is 4 - 44 dBm/m^2/Hz (1 - 100 V/m/Hz) if P is 4-44 dBm/Hz (350 mV/Hz - 35 V/Hz)
low	if PD is -36 - 4 dBm/m^2/Hz (10 mV/m - 1 V/m/Hz) if P is -36-4 dBm/Hz (3.5 mV/Hz - 350 mV/Hz)
very low	if PD is < -36 dBm/m^2/Hz (< 10 mV/m/Hz) if P is < -36 dBm/Hz (< 3.5 mV/Hz)
nil	--> no disturbance
unknown	(propagate as unknown throughout)

Fig. 6. Field type power density and circuit type power disturbance definitions

Specification of a *unique* time domain waveform requires not only amplitude information but phase information as well. The phase information can mean the difference between a *coherent wideband* emission and an *incoherent wideband* emission [11, 12]. Thus the representation of a disturbance will also contain a *slot*, or location, for the specification of pertinent phase information.

As an example, the fields produced by a lightning strike can be simulated by the fields produced by a current pulse with pulse width of 50 μs and a 10 to 90 percent rise time of 500 ns [11]. At a distance of 100 m the field disturbance may be approximated and stored in the database as shown in Fig. 7.

Disturbance Type:	Lightning Emission (100 m distance)
Impedance:	[Plane Wave]
Magnitude:	f < 10 kHz ---> *medium* 100 < f < 400 kHz ---> *low* 400 kHz < f < 1 GHz --->*very-low* 1 GHz < f ---> *nil*
Phase:	[Coherent]

Fig. 7. Example ambient field representation stored in the database

Each node may contain many disturbance representations derived for all the sources present in that volume. The individual attributes are *frequency range normalized* to a user specified global frequency range list and *added in parallel* to determine the *total disturbance* for the node as shown in Fig. 8. For simplicity, the disturbance for the node (either P or PD) is denoted *ambient field* (AF).

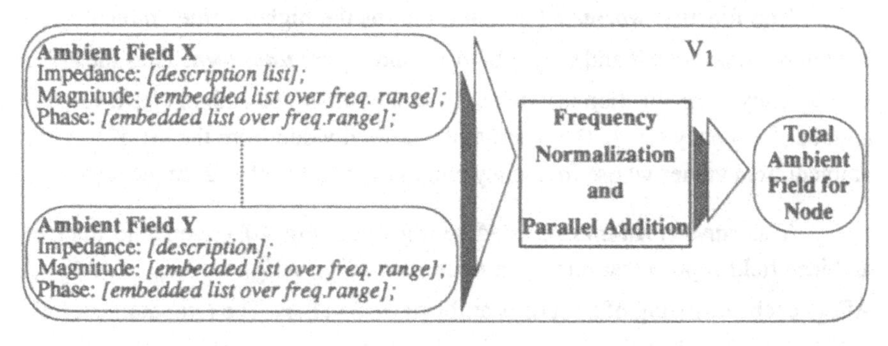

Fig. 8. Ambient field attributes for a typical volume node.

The total nodal ambient field AF_T is required over the specific *global* frequency range set F_g given by the N-element frequency range list

$$F_g = ((f_{g0}, f_{g1}), (f_{g1}, f_{g2}), \dots (f_{gj-1}, f_{gj}), \dots (f_{gN-1}, f_{gN})). \quad (1)$$

If a specific ambient field AF_x in the database is stored over a specific frequency range set F_x of say M frequency ranges, then

$$F_x = ((f_{x0}, f_{x1}), (f_{x1}, f_{x2}), \dots (f_{xj-1}, f_{xj}), \dots (f_{xM-1}, f_{xM})), \quad (2)$$

$$AF_x = ((af_{x1}), (af_{x2}), \dots (af_{xj}), \dots (af_{xM-1}), (af_{xM})), \quad (3)$$

where the af_{xj}'s represent a quantized amplitude level previously described.

Each AF_x must be normalized to the global frequency range F_g:

$$AF_{xn} = ((af_{xn1}), (af_{xn2}), \dots (af_{xnj}), \dots (af_{xnN-1}), (af_{xnN})). \quad (4)$$

This frequency normalization is performed by the algorithm given in Fig. 9.

```
                 Frequency Normalization Algorithm
Loop 1:   over the F_g ranges ( f_gj-1, f_gj), j=1, ... N;
          set af_xnj to unknown;
          Loop 2: over the F_x ranges ( f_xi-1, f_xi), i=1, ... M;
                  If f_xi ≤ f_gj-1 then end loop 2;
                  If f_xi-1 ≥ f_gj then end loop 2;
                  set af_xnj to worseAF( af_xnj, af_xi);
          continue Loop 2;
continue Loop 1;
```

Fig. 9. Frequency normalization algorithm

The function ***worseAF***(af_1, af_2) returns the higher value ambient field, but returns ***unknown*** if and only if both af_1 and af_2 are ***unknown***. Thus the effect of frequency normalization is to take, for each normalized ambient field value in a global frequency range, the worst ambient field value from the set of specific ambient field values whose frequency ranges overlap the global frequency range.

The ***Parallel Addition*** procedure is given in Fig. 10. Given k normalized ambient field representations for a node, say AF_{1n}, AF_{2n}, ... AF_{jn}, ... AF_{k-1n}, AF_{kn}, each consisting of a set of N ambient field values. The total ambient field AF_T is determined as the worst case ambient field value in each frequency range.

```
                 Parallel Addition Algorithm:
Loop 1:   over the AF_j ambient field sets, j = 1, ... k;
          set AF_jn to frequency normalize AF_j;
continue Loop 1;
Loop 2:   over the F_g ranges ( f_gi-1, f_gi), i = 1, ... N;
          set af_Ti to unknown;
          Loop 3:over the AF_jn ambient field sets, j = 1, ... k;
          if af_jni = unknown then set unknown flag;
          set af_Ti to worseAF( af_Ti, af_jni);
          continue Loop 3;
continue Loop 2.
```

Fig. 10. Parallel addition algorithm

3.2 Component Susceptibility Representation

Each volume node in an electromagnetic topology may also have a *system susceptibility* (SS) associated with it. The system susceptibility is inversely related to the level of disturbance which will cause either ***upset***; or ***permanent damage*** to the susceptible component. That is, the lower the disturbance level which will cause upset or damage to a component the higher the defined susceptibility of that component.

There are many ways to define the susceptibility of an electromagnetic component. For instance, many logic circuits will be upset by peak voltage disturbances at their input terminals [10]. In the case of analogue circuits, definition of a precise disturbance level where upset occurs is not as simple as for logic type circuits. For these types of circuit, damage level may be easier to define. Damage and upset levels of many integrated circuit technologies have been tabulated in [10].

As with the ambient field representation, system susceptibility is represented in the frequency domain as *quantized*, 40 dB levels given in Fig. 11.

very low	if SS is > 84 dBm/m^2/Hz ***or*** dBm/Hz
low	if SS is 44 - 84 dBm/m^2/Hz ***or*** dBm/Hz
medium	if SS is 4 - 44 dBm/m^2/Hz ***or*** dBm/Hz
high	if SS is -36 - 4 dBm/m^2/Hz ***or*** dBm/Hz
extreme	if SS is < -36 dBm/m^2/Hz ***or*** dBm/Hz
nil	--> not susceptible
unknown	propagate as unknown throughout

Fig. 11. System susceptibility definitions

Along with this *level representation*, the type of sensitivity and the effect of failure (i.e. upset or damage) may be given for each frequency range. An example of the SS for CMOS integrated circuits is shown in Fig. 12.

System Susceptibility: CMOS Integrated Circuit	
Level:	f < 200MHz ---> *high*
	200 MHz < f < 10 GHz ---> medium
	10 GHz < f ---> nil
Type:	[peak sensitive]
Effect:	[upset]

Fig. 12. Example system susceptibility representation

Volume nodes may contain many system susceptibility characterizations. For example, the volume node representing a circuit board may be characterized by specifying susceptibilities for CMOS, TTL, and line driver integrated circuits. These specific SS values are stored in a database and can be retrieved by the user to characterize each node in a topology. Once the specific system susceptibilities of a volume node have been defined, frequency normalization and parallel addition routines, similar to those used for ambient field with ***worseAF***(af_1, af_2) replaced by***worseSS***(ss_1, ss_2), are used.

3.3 Interaction Path Shielding Effectiveness Representation

The shielding effectiveness (SE) is a representation of the path characteristics between two component nodes and is used to determine the amount of attenuation the ambient field will encounter when crossing an interaction path. The SE is also given over discrete frequency ranges and the units for this quantity depends on the two component nodes which the path connects. The SE is defined with discrete *qualitative* levels shown in Fig. 13.

excellent	if SE > 100 dB
good	if 80 < SE < 100 dB
fair	if 60 < SE < 80 dB
not good	if 40 < SE < 60 dB
poor	if SE < 40 dB
nil	--> no shielding
unknown	propagate as unknown throughout

Fig. 13. Shielding effectiveness definitions

Each interaction path may be made up of a number of different parallel paths between volume nodes (see Fig 14). Each of these parallel paths is given an SE characterization and a total SE is obtained via the frequency normalization and parallel addition algorithms.

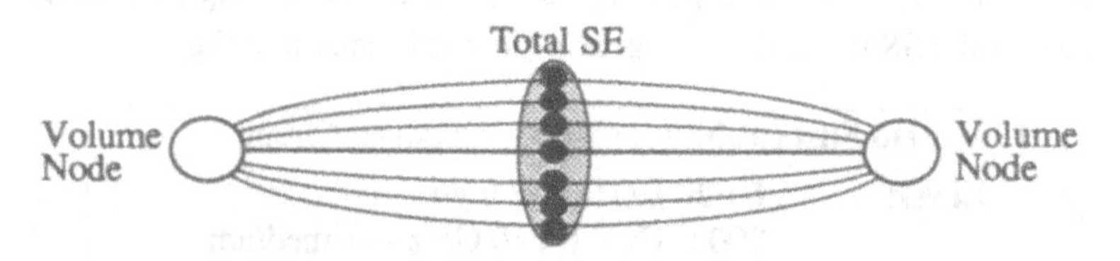

Fig. 14. Interaction Path Composed of Many Parallel Paths

As an example of multiple parallel paths between two field nodes, consider the exterior and interior volumes of the shielded enclosure shown in Fig. 15. Each shield imperfection may be characterized as one parallel path.

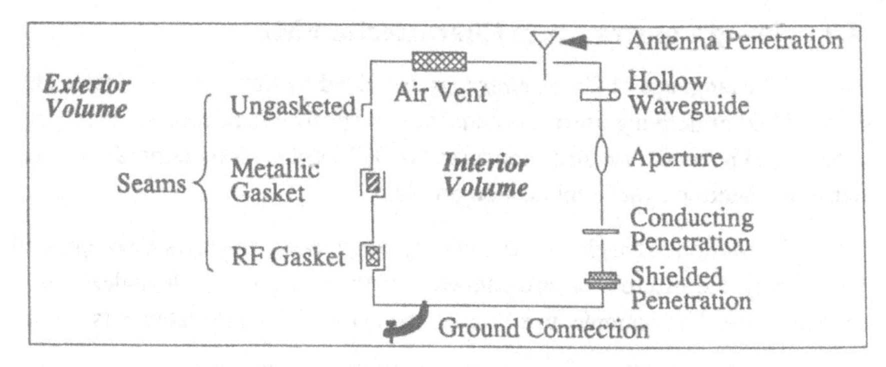

Fig. 15. Typical shield imperfections

As an example, the shielding effectiveness of an aperture [12, 13] is stored as shown in Fig. 16. This attribute would characterize one of the multiple parallel paths making up the total interaction path between the two field nodes.

Shielding Effectiveness: aperture: L = .1; W = .01;	
Level:	f < 500 kHz ---> *excellent*
	500 kHz < f < 2 MHz ---> *good*
	2 MHz < f < 6 MHz ---> *fair*
	6 MHz < f < 12 MHz ---> *not-good*
	12 MHz < f < 300 MHz ---> *poor*
	300 MHz < f ---> *nil*

Fig. 16. Example shielding effectiveness representation

Circuit-circuit path nodes are used to model any interaction between circuit nodes. For example, any electrical circuit connection can be modelled by a cc-node. The level of modelling detail required for the system will dictate the definition and introduction of cc-nodes into the topology.

Field-circuit nodes are used to define the coupling of field to circuit nodes. For example, the coupling of far fields to *printed wiring boards* (PWB's) may be approximated by the *maximum effective aperture*, A_{em}, for a half-wave dipole. This has been found to be a good approximation for frequencies ranging from 100 MHz to 10 GHz [10]. For lower frequencies an $A_{em} = 1$ is an appropriate worst case approximation.

The circuit-field nodes are used to represent the emission of fields from circuit nodes. For example, currents existing on a PWB will radiate electromagnetic energy. Some estimations for the level of these emissions are derived from approximating the sources as loop and dipole radiators [12]. More accurate models are obtainable for emissions from circuit boards [14].

4.0 Using Constraints to Characterize EMI

The language of *Constraints* was described by Sussman and Steele [15] as a method of deriving useful consequences by propagating conditions through a constraint network (see also Montanari [16]). Presently, constraints are used to define an electromagnetic interaction problem.

Definition of the electromagnetic topology for a problem is accomplished by defining the discrete electromagnetic components (i.e. each node) with a suitable name. For example, in a Prolog type syntax [17], the statements

```
node( external vol).          node( computer).        node( cpu).
node( power distribution).    node( power cord).      node( power supply).

surface( external vol, computer).            surface( external vol, power cord).
surface( external vol, power supply).        surface( computer, cpu).
surface( computer, power distribution).      surface( computer, power supply).
surface( cpu, power distribution).           surface( power supply, power cord).
surface( power distribution, power supply).
```

would declare the existence of six nodes and nine surfaces in the current topology. A surface declaration is said to *constrain* two components into sharing a surface. Similarly, the statement

```
global_frequency( frequency_range_x).
```

will constrain the global frequency variable to the frequency ranges defined in *frequency_range_x* which is stored in the database as a *list* of discrete frequency ranges. It should be noted that the number of discrete frequency ranges defined will affect the speed of computation for the 1) frequency normalization, 2) parallel addition, 3) worst case shielding path determination, and 4) risk of failure operations. Thus a coarse global frequency range set is preferred during preliminary investigations.

Specific electromagnetic attributes can be imposed over the defined topology. For example, the statement

```
disturbance( node( external vol),
                    [[standard, NEMP], [standard, LEMP], [cw, HF]]).
```

will constrain the *external vol* node to have disturbances associated with the list of disturbances contained in the list. These disturbances exist in the data base in the previously described form. Entering these constraints would trigger the *frequency normalization* and *parallel addition* algorithms producing a total disturbance for this node. Nodes which are not instantiated with specific attributes are taken to be *unknown* over all global frequency ranges. The statements

```
susceptibility( node( cpu),
        [[digital, TTL], [digital, CMOS], [analogue, line_driver]]).
susceptibility( node( power supply),
        [[analogue, volt_regulator], [digital, comparator]]).
```

constrain the total susceptibilities of the specified nodes, while the statements

```
shielding( surface( external vol, computer), [[shield, wire-mesh-gasket],
                [shield, honey-comb-cooling-vent]]).
shielding( surface( external vol, power cord), [[coupling, short-cable]]).
shielding( surface( external vol, power supply),
                [[shield, wire-mesh-gasket]]).
shielding( surface( computer, cpu), [[coupling, pcb, 30cmX30cm]]).
shielding( surface( computer, power distribution),
                [[coupling, short-ribbon-cable]]).
shielding( surface( computer, power supply), [[shield, aluminum]]).
shielding( surface( cpu, power distribution), [[filter, nil]]).
shielding( surface( power distribution, power supply),
                [[filter, feed-thru caps]]).
shielding( surface( power supply, power cord), [[filter, EMI-461]]).
```

constrain the shielding effectiveness of interaction paths in the topology.

Information regarding the interaction between any two nodes in the topology is explicitly derived by determining the worst case shielding path between them. A search for the worst case shielding path from each susceptible node to all other emitting nodes is performed using *Dijkstra's algorithm* [18, 19, 20] using the *distances* shown in Fig. 17 for each heuristic shielding level.

excellent	SE = 3.0
good	SE = 2.0
fair	SE = 1.5
not good	SE = 1.0
poor	SE = 0.5
nil	SE = 0.0 --> no shielding, and
unknown	propagate as unknown throughout

Fig. 17. Definition of SE variable distances

The worst case shielding path to the two emitting nodes is highlighted in Fig. 18 with the total shielding represented as

```
total_shield( path( v1, v4), [ ... [not-good] ... ]).
total_shielding( path( v1, v5), [ ... [good] ... ]).
```

The second argument of *total_shield* is an imbedded list of the shielding effectiveness for each frequency range. Notice that although the total SE variable for path(v1, v5) adds up to 2.5 (corresponding to slightly *above* the good level) the value of *good* is displayed for the total shielding. Internally, the value of 2.5 is maintained for the total SE variable.

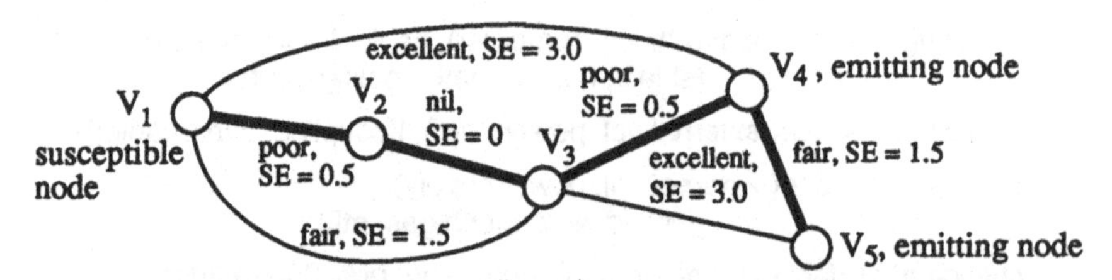

Fig. 18. Example worst case shielding path for a specific frequency

During the search, the individual surfaces on the path are sorted from poorest to best total shielding and kept as a trace. The determination of total shielding between two non-neighbor nodes is the same as imposing a surface between them and is similar to the *slices* concept in [15].

The *likelihood of failure* is determined at the susceptible nodes, by a comparison of the *propagated* electromagnetic disturbance through the worst case shielding path from an emitting node to the susceptibility of the node. The specific disturbance and susceptibility levels at a node are assigned discrete AF and SS numerical values as in Fig. 19.

extreme	AF = 5; SS = 1
high	AF = 4; SS = 2
medium	AF = 3; SS = 3
low	AF = 2; SS = 4
very low	AF = 1; SS = 5
nil	AF = -2 --> no disturbance
	SS = 8 --> not susceptible
unknown	(propagate as unknown throughout)

Fig. 19. Ambient field and system susceptibility discrete levels

The *propagated ambient field* (PAF) is determined by subtracting the total shielding effectiveness of the path traversed (i.e. the SE value) from the AF value, that is

$$PAF = TAF - TSE \tag{5}$$

where TAF is the total ambient field emitted by a node and TSE is the total shielding effectiveness of a path. Now, since non integer values of SE exist (for example a shielding effectiveness of fair ↔ SE = 1.5), it is *not* always the case that after passing through a shielded path the ambient field drops a level.

Setting an ambient field attribute to a certain level implies that there is an equal probability for the amplitude being any value in the range of amplitudes in the level. For example, if in the frequency range (1 MHz - 10 MHz) a level of *high* is specified then this approximates the disturbance with equal probability over a 40 dB ambient field range as shown in Fig. 20. If this level is propagated across a *poor* shield the resulting AF probability distribution would lie somewhere between the *high* level and the *medium* AF level as shown in Fig. 20 which, as a worst case, would be reported as *high*. Thus, a *good* shielding effectiveness would reduce the *high* AF by at least 80 dB producing a *low* propagated AF.

Once the propagated ambient fields from *all* other emitting nodes in the topology has been determined for a susceptible node they are added in parallel using the parallel addition algorithm. In doing this, a trace of the highest PAF to lowest PAF is kept for each frequency range in the global frequency range.

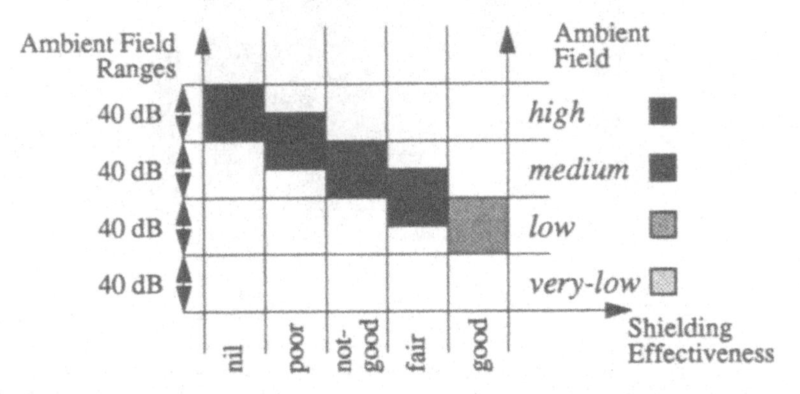

Fig. 20. High ambient field approximation propagated through good SE

The likelihood of failure is reported as being one of the discrete values shown in Fig. 21 for each global frequency range. The discrete level assigned to a frequency range is dependent on a failure index variable denoted FI. This is calculated for a susceptible node by

$$FI_{\text{for node}} = PAF_{\text{all emitting nodes}} - SS_{\text{total for node}}. \qquad (6)$$

If the likelihood of failure of any susceptor is too great, then parameters in one or all of the three constraining factors must be modified at one or more locations in the topology. Many ways will exist to reduce the likelihood of failure at a specific node, each with advantages and disadvantages. This is where the traces which were developed throughout the analysis will help.

extreme	if FI 1.5
high	if 0.5 ≤ FI < 1.5
marginal	if -0.5 ≤ FI < 0.5
low	if -1.5 ≤ FI < -0.5
very low	if -2.5 ≤ FI < -1.5
nil	FI < -2.5 or associated with a non-susceptible nodes
unknown	if either susceptibility or disturbance are *unknown*

Fig. 21. Likelihood of failure discrete levels

Visually it is convenient to refer to Fig. 22 to understand how the likelihood of failure is determined. Notice how the calculated FI is not uniquely determined by the *reported* PAF and system susceptibility (highlighted block).

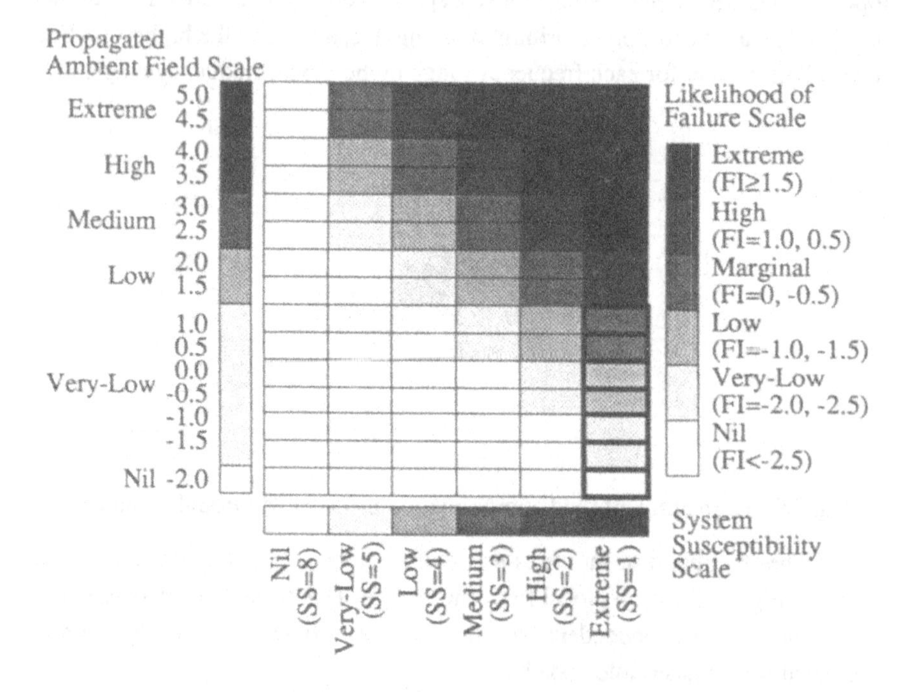

Fig. 22. Likelihood of Failure Chart

This procedure for determining the likelihood of failure is justified heuristically by an understanding of what each FI level represents. The ambient field ranges can be plotted on a graph of the susceptibility levels, as shown below in Fig. 24, where the likelihood of failure results for a system susceptibility of *medium* are shown across the top of the figure.

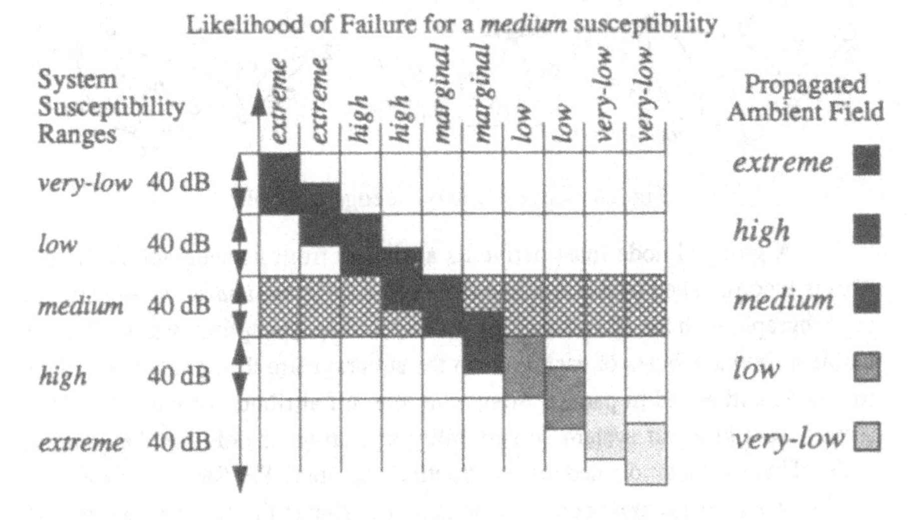

Fig. 23. Overlap of PAF ranges with SS ranges to determine FI

Note that, due to the 20 dB discretization of the shielding effectiveness, there are two ambient field ranges for each level shown in the propagated ambient field key at the right of the figure. For the case where there is an overlap between the PAF distribution and the SS distribution, the likelihood of failure is said to be *high* or *marginal* (*high* if the PAF distribution overlaps above the SS distribution). When the PAF distribution lies totally above the SS distribution then the likelihood of failure is said to be *high* or *extreme* depending on how much higher the distribution lies. Alternatively the likelihood of failure is determined as *low*, *very-low* or *nil* depending on how much lower the PAF distribution is than the SS distribution.

5.0 Grouping of Electromagnetic Component Nodes

Given a graph G for which the edge set E can be partitioned into two nonempty subsets, say E_1, and E_2, such that the subgraphs generated from these subsets, that is $G[E_1]$ and $G[E_2]$, have just a node v in common, then v is called a cut node [20]. As an example, nodes v_1, v_3, and v_5 in Fig. 24 are cut nodes. A set of nodes forms a *valid grouping* if it is equal to a node set of one of the subgraph partitions generated by a cut node. The valid grouping includes the cut node and the group is represented by a *grouped node* in place of the cut node. In the right most graph of Fig. 24, v_5, v_7 and v_6, are grouped into node v_3.

The process of grouping is used to reduce the search space in the worst case shielding path algorithm. The subgraph containing nodes which are not expected to change may be grouped as a valid grouping.

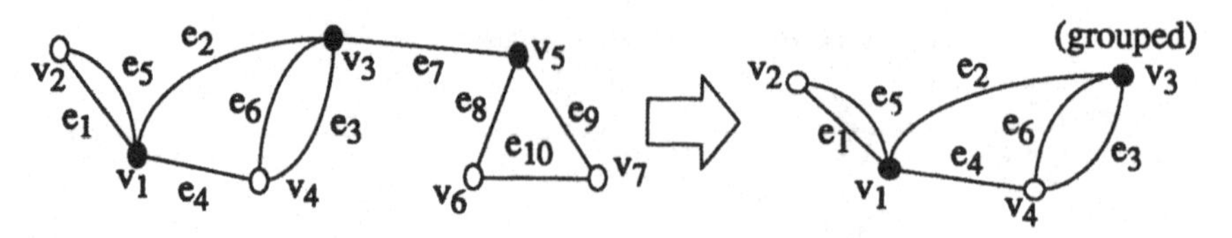

Fig. 24. Cut node and cut edge examples

A grouped node must derive its attributes from its subgraph attributes. This is accomplished by determining the *single source minimal spanning tree* for the subgraph, with the cut-node as the root node. The susceptibility as well as the ambient field attributes of each node in the subgraph are then propagated to the root node and added in parallel along with the self attributes of the cut-node to form a new grouped system susceptibility and ambient field for the grouped node. This is accomplished by subtracting the total SE for the worst case shielding path to the root node from the total SS value of the node being grouped. An example of the grouping process is shown in Fig. 25.

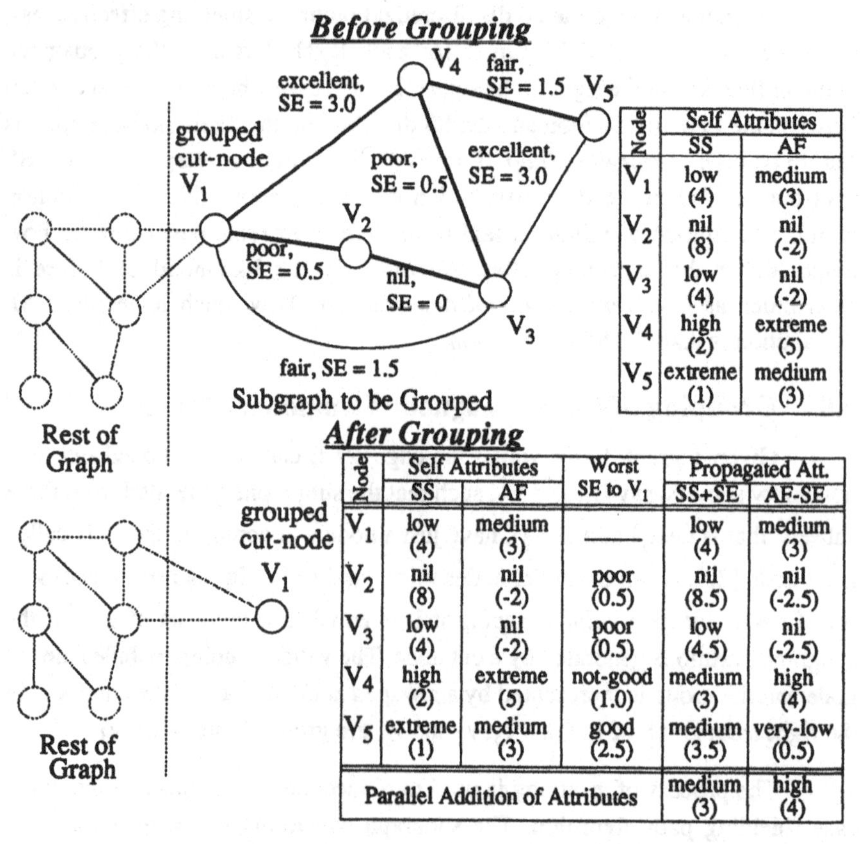

Node	Self Attributes SS	Self Attributes AF
V_1	low (4)	medium (3)
V_2	nil (8)	nil (-2)
V_3	low (4)	nil (-2)
V_4	high (2)	extreme (5)
V_5	extreme (1)	medium (3)

Node	Self Attributes SS	Self Attributes AF	Worst SE to V_1	Propagated Att. SS+SE	Propagated Att. AF-SE
V_1	low (4)	medium (3)	——	low (4)	medium (3)
V_2	nil (8)	nil (-2)	poor (0.5)	nil (8.5)	nil (-2.5)
V_3	low (4)	nil (-2)	poor (0.5)	low (4.5)	nil (-2.5)
V_4	high (2)	extreme (5)	not-good (1.0)	medium (3)	high (4)
V_5	extreme (1)	medium (3)	good (2.5)	medium (3.5)	very-low (0.5)
Parallel Addition of Attributes				medium (3)	high (4)

Fig. 25. Example of grouping a subgraph in a cut-node

In order that information is not lost about which node might be failing when the likelihood of failure of a grouped node is calculated, a *trace* is kept of the node in the group contributing the most to the grouped node susceptibility and ambient field attributes. As an example, for the grouped nodes of Fig. 25 the trace of the grouped SS attribute would be represented as the list [V_4, V_5, V_1, V_3, V_2] while the grouped AF trace would be held as [V_4, V_1, V_5, V_2, V_3]. From these traces one can immediately determine which nodes' SS to increase or which nodes' AF to decrease if an interaction problem exists within the group.

6.0 Implementation

The heuristic techniques and procedures described herein have been included in a prototype implementation on a Sun Microsystem SPARCstation-1™. The implementation consists of two main parts; a smart topology-drawing tool referred to as ***HardDraw***, and an electromagnetic interactions advisor referred to as ***HardSys***. A picture of the overall structure of the system is shown below in Fig. 26.

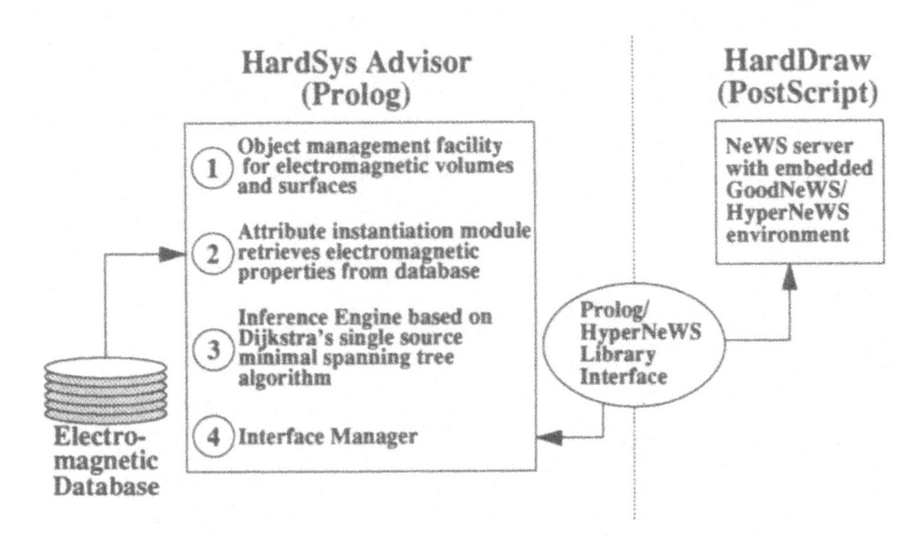

Fig. 26. Overall system architecturs

The purpose of *HardDraw* is to give the user of the software an easy way to input the electromagnetic topology of systems (see Fig. 27 below). This smart topology-drawing tool is implemented with a NeWS™ -based [21] user-interface toolkit called GoodNeWS/HyperNeWS [22, 23]. It communicates with HardSys by means of a Prolog/HyperNeWS library.

7.0 Conclusions and Future Plans for the EMI Advisor

Work on the *Electromagnetic Interactions Advisor* is ongoing in order to bring it to the level of a usable commercial software tool. The future plans in this development can be described as falling under three categories:

1) knowledge representation and search technique enhancements;
2) user interface enhancements; and
3) knowledge acquisition and validation.

The search technique used thus far is the *theoretically* most efficient [19] although implementation techniques to improve efficiency can be applied. For example, automatic grouping of nodes by the advisor is a possibility to be examined. Changes in the knowledge representation may also yield better search times.

Currently the advisor works in the *analysis mode*, that is, the physical topology along with the associated node attributes are first entered and the system then determines if any interaction problems exist. In the *design mode* the advisor would be given some global constraints to satisfy and it would be heuristically guided to design the system within these constraints. For example, a global constraint may be a weight and cost limit. The system would then choose subcomponents from a database with each having unique cost and weight attributes as well as system susceptibilities and electromagnetic emissions. Shielding components may also be chosen in a similar way.

Knowledge acquisition is the problem of how to acquire accurate EMI data from the many sources available and how to categorize it so that it can be presented to the user in a logical and efficient manner. Validation of the knowledge base, once created, requires the use of the advisor on the design of real systems.

References

[1] LoVetri, J., Abu-Hakima, S., Podgorski, A. S., Costache, G. I., "HardSys: Applying Expert System Techniques to Electromagnetic Hardening", *IEEE 1989 National Symposium on Electromagnetic Compatibility*, pp. 383 - 385, Denver, Co., May 23-25, 1989.

[2] LoVetri, J., and Graham, D. P. W., "Constraint Propagation Through Electromagnetic Interaction Topologies", *ANTEM'90, Symp. on Antenna Tech. and Appl. Electromagnetics*, Winnipeg, Manitoba, August 15-17, 1990.

[3] LoVetri, J., and Podgorski, A. S., "Evaluation of HardSys: A Simple EMI Expert System", *1990 IEEE Int. Symp. on EMC*, Washington, D.C., August 21-23, 1990.

[4] Baum, C. E., "Electromagnetic Topology, A Formal Approach to the Analysis and Design of Systems", *Interaction Notes*, Note 400, Air Force Weapons Lab, September, 1980.

[5] Baum, C. E., "On the Use of Electromagnetic Topology for the Decomposition of Scattering Matrices for Complex Physical Structures", *Interaction Notes*, Note 454, Air Force Weapons Lab, July, 1985.

[6] Messier, M. A., "EMP Hardening Topology Expert System (Hard Top)", *Electromagnetics*, vol. 6, no. 1, pp. 79 - 97, 1986.

[7] Tesche, F. M., "Topological Concepts for Internal EMP Interaction", *IEEE Trans. on Ant. and Prop.*, vol. AP-26, no. 1, pp. 60 - 64, Jan., 1978.

[8] Schulz, R. B., Plantz, V. C. and Brush, D. R., "Shielding Theory and Practice", *IEEE Trans. Electromagn. Compat.*, vol. 30, no. 3, pp. 187 - 201, August, 1988.

[9] U.S. Department of Defense, "Measurement of Electromagnetic Interference Characteristics", *Mil-Std-462*, July 31, 1967.

[10] McDonnel Douglas Astronautics Co. *Integrated Circuit Electromagnetic Susceptibility Handbook*, Report MDG-E1929, Box 516, St. Louis, Missouri, (314) 232-0232, August 1978.

[11] Duff, W. G., *Fundamentals of Electromagnetic Compatibility*, A Handbook Series on Electromagnetic Interference and Compatibility, vol. 1, Interference Control Technologies, Inc, Gainesville, Virginia, 1988.

[12] Ott, H. W., *Noise Reduction Techniques in Electronic Systems*, John Wiley & Sons, New York, 1988.

[13] Vitek, C., "Predicting the Shielding Effectiveness of Rectangular Apertures", *IEEE 1989 National Symposium on EMC*, Denver, Colorado, pp. 27 - 32, May 23 - 25, 1989.

[14] Raut, R., "On the Computation of Electromagnetic Field Components from a Practical Printed Circuit Board", *1986 IEEE Int. Symp. on EMC*, pp. 161 - 166, San Diego, CA, Sept. 16 - 18, 1986.

[15] Sussman, G. J. and Steele, G. L., "Constraints - A Language for Expressing Almost-Hierarchical Descriptions", *Artificial Intelligence*, vol. 14, pp. 1 - 39, 1980.

[16] Montanari, U., "Networks of Constraints: Fundamental Properties and Application to Picture Processing", *Information Science*, vol. 7, pp. 95 - 132, 1974.

[17] Clocksin, W. F. and Mellish, C. S., *Programming in Prolog*, Springer-Verlag, New York, 1984.

[18] Dijkstra, E. W., "A Note on Two Problems in Connexion with Graphs", *Numerische Mathematik*, vol. 1, pp. 269 - 271, 1959.

[19] Ahuja, R. K., Mehlhorn, K., Orlin, J. B., and Tarjan, R. E., "Faster Algorithms for the Shortest Path Problem", *J. of the Ass. for Comp. Mach.*, vol. 37, no. 2, pp. 213-223, April 1990.

[20] Bondy, J. A., and Murty, U. S. R., *Graph Theory with Applications*, American Elsevier Pub. Co., Inc., 1976.

[21] Arden, M.J., Gosling J. and Rosenthal, D. S. H., *The NeWS Book*, Springer-Verlag,1989.

[22] The Turing Institute, *GoodNeWS1.3 User Guide*, The Turing Institute, Glasgow, Scotland, September, 1989.

[23] The Turing Institute, *HyperNeWS1.3 User Manual*, The Turing Institute, Glasgow, Scotland, September, 1989.

[24] Quintus Computer Systems Inc., *Quintus Prolog Development System Manual (Release 2.4)*, Quintus Computer Systems Inc., Mountain View, California.

[25] Stabler Jr., E. P., "Object-Oriented Programming in Prolog", *AI Expert*, pp. 46-57, October 1986.

Drilling Tool Selection Aid System

O. Leboulleux, M. Hittinger, S. Serfaty
SNECMA - Société Nationale d'Etude et de Construction de Moteurs d'Avions

ABSTRACT:

This paper describes an expert aid system for drilling operations on metallic aircraft engine parts. To be more precise, the system defines a complete machining process (operations, tools and cutting parameters) depending on the machine selected and the geometry of the different entities to be machined. In this paper, we will describe the aims and functions of the system, together with the software techniques used.

1. MACHINE METALLIC PARTS IN THE AERONAUTIC INDUSTRY.

At SNECMA, more than 400 specialized technicians are currently involved in designing the machining processes required to manufacture the various metallic parts used in our aircraft engines. Their work can be broken down as follows:

* Choice of the Numerical Controlled Machine
* Definition of the geometry required
* Definition of the machining process
* Choice of tools
* Computation of the cutting parameters
* Encoding of NC programs
* Test of the operation on the shop floor

A special department is in charge of organizing and facilitating the work of the technicians in the whole factory. There are several differents reasons which led to the development of this drilling aid system :

1.1 Developing specialized expertise.

The manufacturing process of any aeronautical part is subject to very strict control. The accuracy of the machining operations on engine parts is higher than that required in ordinary industrial standards. This is why the expertise we use four our manufacturing processes is so specialized and is unique to our company. The use of documents, software and set manufacturing techniques from out the SNECMA is rather limited.

Furthermore, some techniques used in the development of these machining processes are constantly changing : new tools and new methods are offered by our different suppliers every day. It is difficult for this information to be circulated and used within the company as paper documentation tends to be impractical with many lengthy documents which are rarely kept together in the same place.

1.2 Expertise acquired by experience.

The technicians often use their own experience. To start a new machining process, they usually modify techniques used on similar parts. Developing a process can be a time-consuming and complex procedure which means that some new series can be started with imprecise processes which are modified by experience over the following months. The final manufacturing process is very seldom optimized (in terms of cost and time). Finally, ineffective use of some tools is sometimes transmitted from one process to another.

1.3 Team based expertise.

The technicians at SNECMA are divided into three different units (depending on the type of parts) which do not always communicate as well as they should. Each team has its own expertise and useful methods and information do not spread quickly enough between the different units. Naturally, the global performance of the company can be adversely affected. For example, managing the tool shop is very complex because some similar tools bear different reference numbers according to the unit which uses them. This type of team-based expertise increases the length and complexity of the training required by young technicians arriving in the company. It is also very important for us to safeguard the expertise acquired by our experienced technicians before they retire.

2 AIMS OF THE SYSTEM.

The situation described above leads us to define the main aims of the drilling aid system as follows:

* Capitalization,standardization and circulation of individual know-how within the company.

* Optimization of the machining process. (cost and time required)

* Improved management and use of NC machines and tools.

* Training for junior technicians. Assistance for experienced technicians.

3 TECHNICAL CONSTRAINTS.

Various technical constraints also play an important role in the definition of the project. These contraints resulted in the adoption of an expert system method and were used to define the associated development techniques. These contraints include :

3.1 Reliability of results.

The high cost of machined parts requires the results given by the expert system to be of both high and consistent quality.

3.2 User friendliness.

The definition of the geometry requires a great number and a wide range of data. The man/machine interface must be studied carefully in order to ensure a very simple and fast system for our users.

3.3 Easy maintenance.

New techniques and hardware (NC machines and tools) are constantly being developed. The system must be able to integrate these modifications rapidly and easily.

3.4 Interactive software.

The technician must not feel subordinate to the computer program just inputting data. The software must offer enough choice to let the user have some influence on the various stages of the process. This interactivity between the technician and the software is a necessary step towards the integration of a new working tool.

4. FUNCTIONS OF THE SYSTEM.

The system generates a complete drilling sequence according to the NC machine used, the shape of the part and the geometry of the the holes (see **figure 1**).

The complete generated process defines the series of basic operations to be performed as well as the tools to be used for each type of machining (reference number of the tools in the shop).

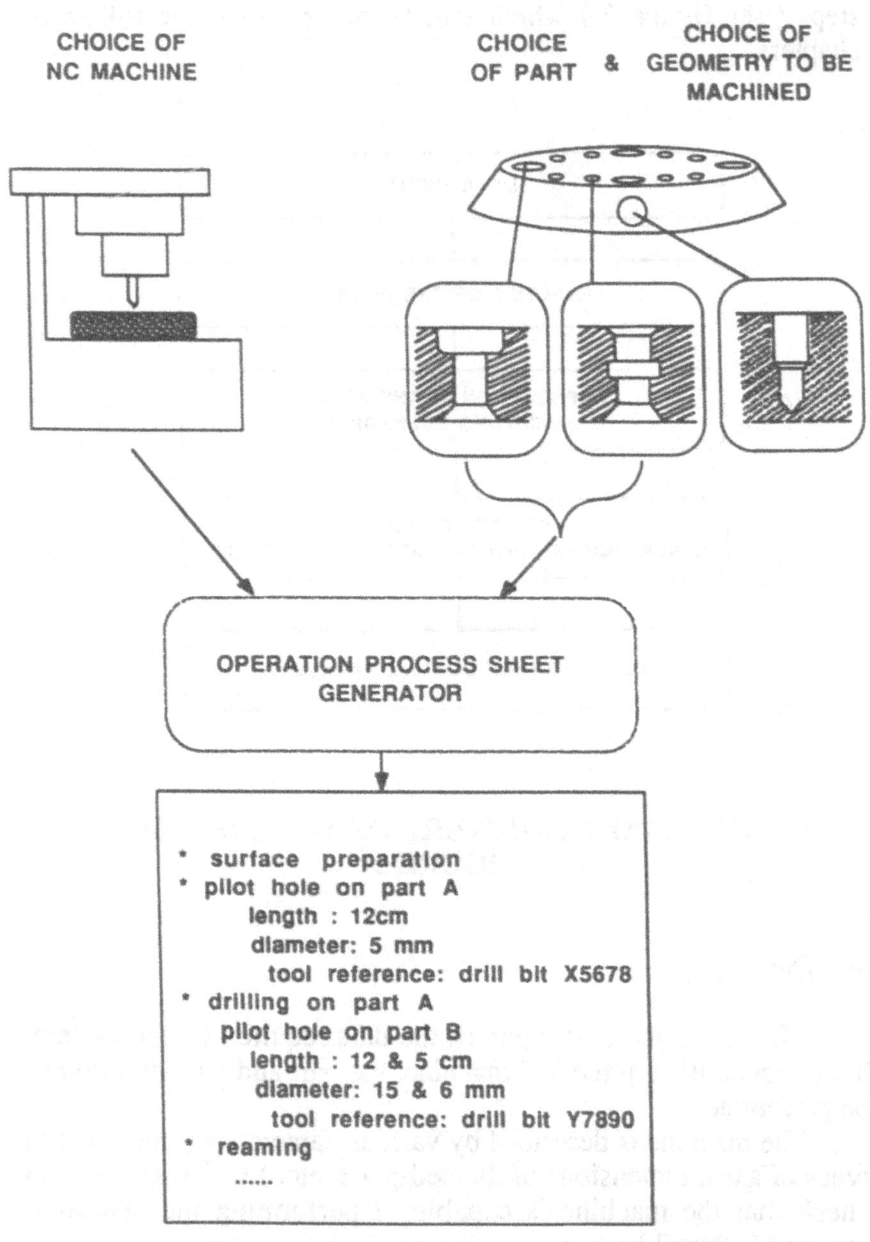

MAIN FUNCTION OF THE SYSTEM
FIGURE 1

The main function of the system can be broken down into 5 steps (see **figure 2**) which will be described in the following chapters.

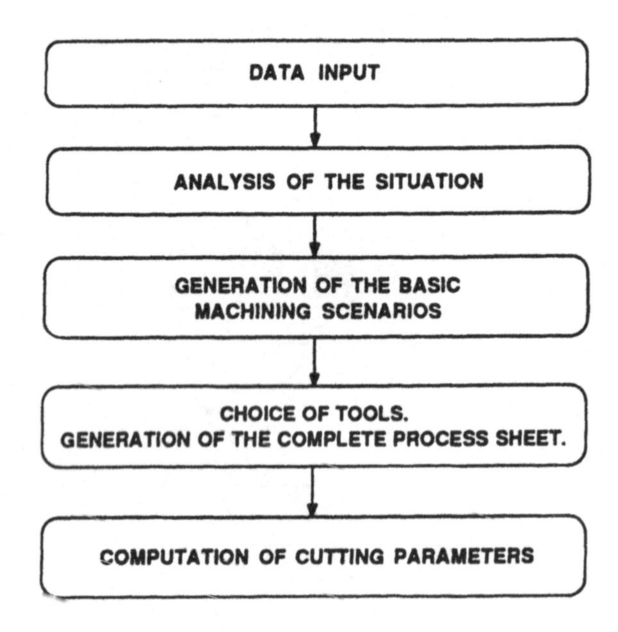

GENERAL STRUCTURE OF THE SYSTEM
FIGURE 2

4.1 Data input

The user must first input all the data required by the system. This data applied to the NC machine, the part and the operation to be performed.

The machine is described by various parameters (number and types of axes, dimensions of the bed-plate, etc.) which are used to check that the machine is capable of performing the operation required (stage 2).

The part is described by parameters such as dimensions, strength, type of material, etc.

The operation is the precise description of all the holes to be machined on the part (numerical values for their size and location on the part). In fact, the various holes are divided into sets with the same geometrical model and those sets are put together in different working sessions called operations or OP (see **figure 3**).

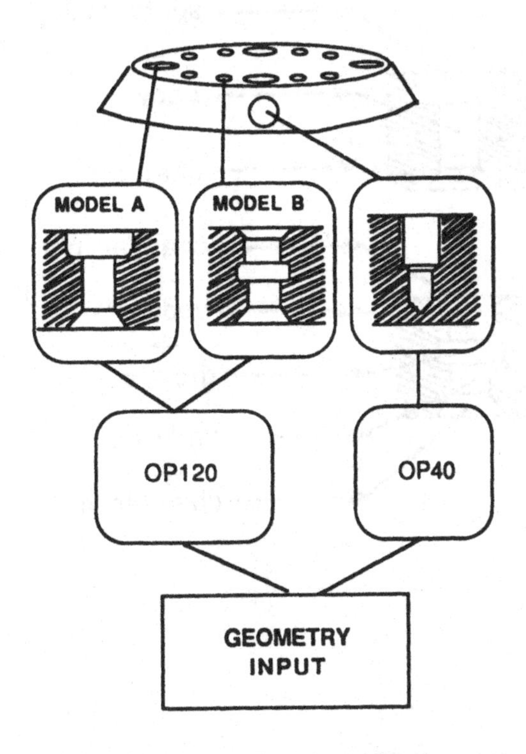

DEFINITION OF THE OPERATIONS (OP)
FIGURE 3

The most tedious task is describing the different geometries of the holes in the operation (model A and B for OP120 in **figure 3**). The large number of possible geometries for holes and the associated machining techniques led us to break each complex model down into a series of simple predefined geometries (basic models). Each complex model can then be broken down from top to bottom as a set of basic geometric models (see **figure 4**)

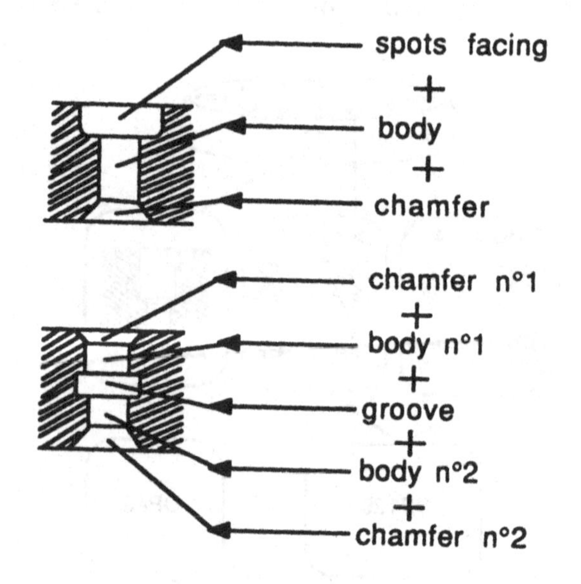

BREAKDOWN OF COMPLEX MODELS
INTO BASIC MODELS
FIGURE 4

All the possible geometries are shown in figure 5. Each of them can be described by their own set of parameters.

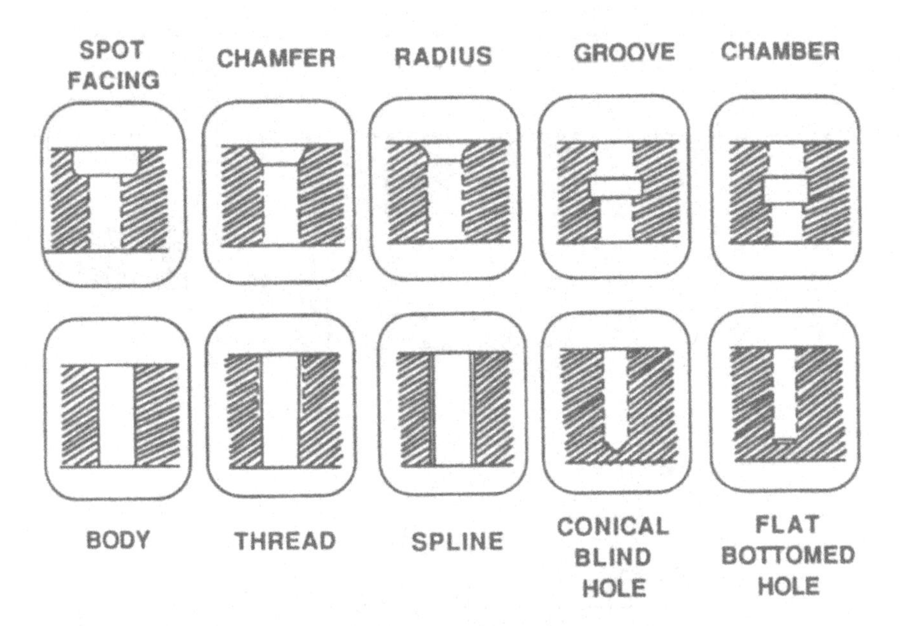

THE 10 BASIC MODELS
FIGURE 5

Due to the number of parameters to be entered, describing the geometry to be machined is always a long, tedious process. It is during this stage that it is most important to facilitate the work of the technicians. The system is connected to DB2 database which store descriptions of the NC machines, the parts and the operations defined during previous sessions. It is also possible to recall a similar geometry and modify the parameters which need to be changed. Finally, links with one of the CAD systems used at SNECMA (**CATIA**) are being tested to improve input of geometric data.

The system is designed to simplify and accelerate data input. This is achieved by using interactive graphic software where the technicians choose the different basic models required for the complex set (see **figure 6**)

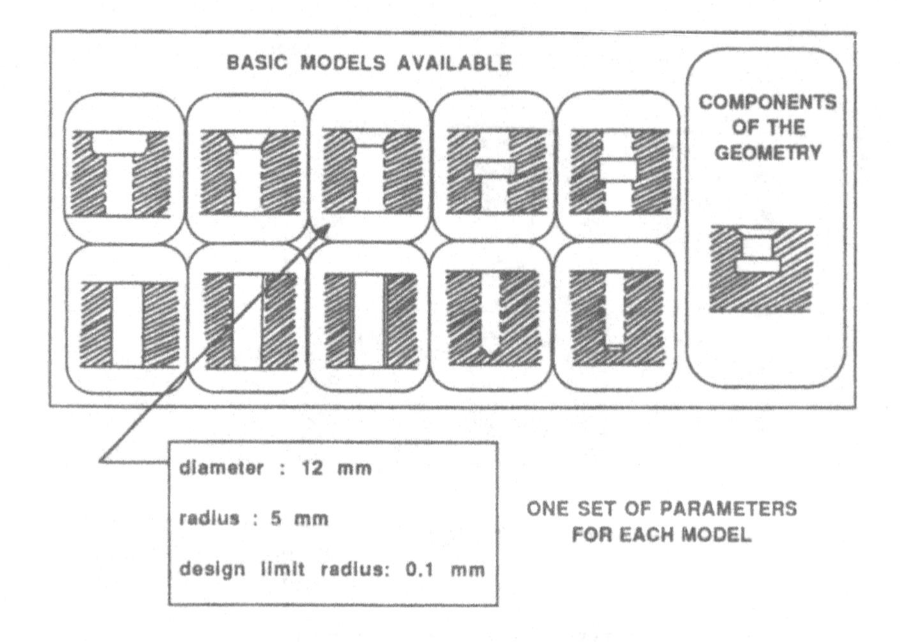

INTERACTIVE CHOICE OF COMPONENTS FOR A COMPLEX MODEL
FIGURE 6

4.2 Analysis of the situation

This is a short but very important step before the generation of the final machining process. During this step, the system checks whether it is possible to define the drilling operation. Two important questions must be asked :

* Is the numerical controlled machine suitable for the operation?

* Is another manufacturing process possible or mandatory (Electro-erosion, ...) ?

4.3 Generation of the basic machining scenarios

This is the most important stage in generating the process sheet (in terms of complexity and time). A simplified sheet (scenario) is derived for each model in the operation (see **figure 7**). At the next stage, all these individual sheets will be grouped together to form the final process sheet . At this point, there is still no reference to the tools to be used. The different basic operations available on the system are listed by type (punching, drilling, reaming,etc.) and possibly by basic machining model (drilling of the nth body , etc.). Furthermore, each of the elementary operations has its own set of parameters (length, drilling or reaming diameter, etc.). At this point, some of the parameters may already be fixed, while others may change during the following stages (ie: diameter of a pilot hole, etc.). In chapter 4.4, we will show why it is important to maintain this degree of freedom.

It is also important to note that these scenarios are independent, i.e. the geometrical components of other holes are not taken into account. The possible links between machining two complex models are studied in the next stage.

The various possible scenarios are displayed at the end of this stage (see **figure 7**). The user can then validate or reject the results before continuing with the next stage. He can modify the order of the operations, add new machining operations or delete useless ones. The availability of different choices in these scenarios gives the technician a chance to intervene early in the generation of the process sheet and customize at least a part of the final result (the complete sheet).

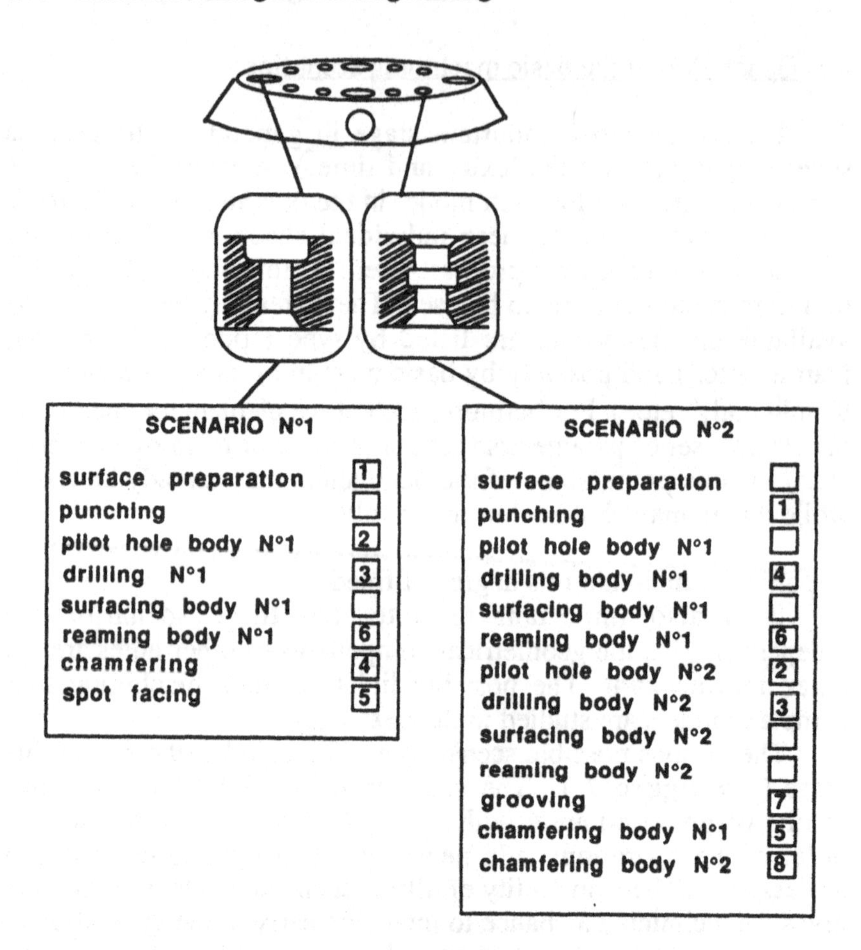

GENERATION OF BASIC SCENARIOS
FIGURE 7

4.4 <u>Choice of tools and generation of the complete process sheet</u>.

The choice of the tools necessary for the manufacturing process defined in the previous step is made based upon the type of machining (punching, drilling, reaming, etc.), the known parameters (fixed parameters) and the general characteristics of the part (material, stiffness, etc).

The unknown parameters (degrees of freedom) can be computed following the parameters for similar machining operations on other complex models of the same operation. For example, in **figure 3**, the diameter of a pilot hole on model A can be equal to the drilling diameter of model B. Hence, the pilot hole of A and the body of B can be drilled using the same tool.

The choice of the tool is made among the tools used at SNECMA (**existing** tools) and the tools offered by our suppliers (**theoretical** tools). We might therefore have to design a new SNECMA tool to perform an operation if it is not yet available in our shop.

4.5 Computation of cutting.

This last module computes the different parameters used for machining (cutting speed, rate of feed, etc.).

5. TECHNICAL ASPECTS OF THE SOFTWARE.

5.1 Modular software

In **figure 8**, we describe the modular structure of the system. There are several advantages of such a system :

5.1.1 Technical advantages.

Breaking the system down into its main functions facilitates software maintenance. At any stage in the procedure, only useful elements (rules and objects) are used. The low number of active rules loaded in the system increases the speed of each process.

5.1.2 Functional advantages.

The system can operate in two different modes.

In automatic mode, the system loads, in a preset order, the various modules required for complete process sheet generation: geometry input, scenario generation, tool choice, parameter computation.

In manual mode, the technician chooses the next module to be executed. The different entry points (see **figure 8**) allow the technician to use the system to solve partial problems without having to generate the whole manufacturing process (ie: choice of an NC machine, computation of cutting parameters, etc.). The technician may only be required to give just part of the geometry. For a given sub-problem, a large number of geometric data is rarely required. A pilot module controls the software and is in charge of managing the calls to the different modules for both modes (see **figure 8**).

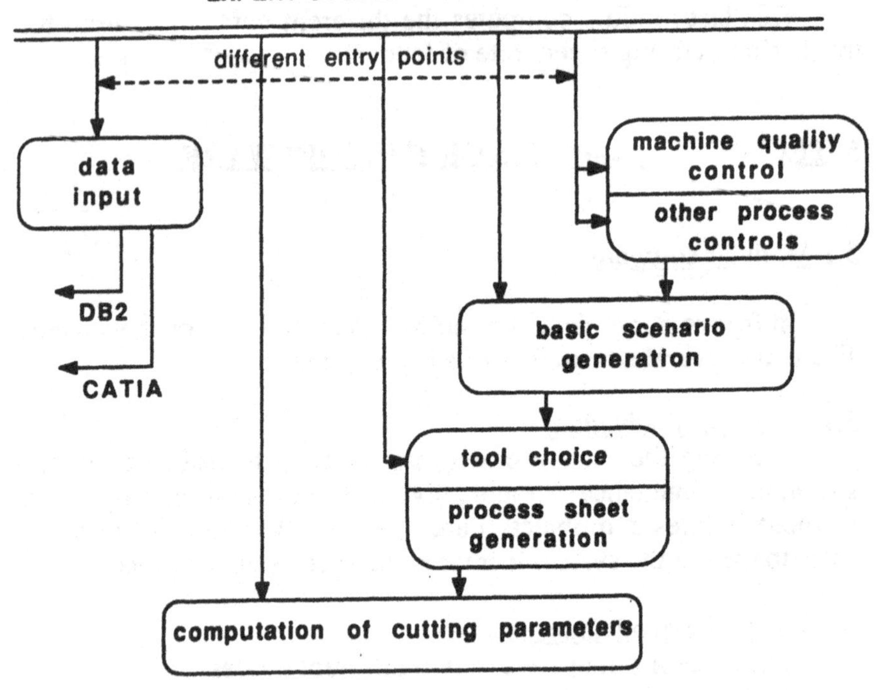

MODULAR STRUCTURE OF THE SYSTEM
FIGURE 8

5.2 Interactive software.

The need for interactive software and its functions have been described in 4.3. The expert system shell (**NEXPERT OBJECT**) is linked with external software and any values given by the user are checked after each external interactive routine has been called.

5.3 Functional breackdown.

Different objects (dynamic objects) of unknown type and number should be able to call the same set of rules for a specific sub-problem. For example, the question " IS THERE ANOTHER POSSIBLE MACHINING PROCESS AVAILABLE ?" could be asked at any time for each complex model. The problem "SHOULD A PILOT HOLE BE USED ?" can be studied for different drilling at the most crucial point. All these facts provide us with the following functional breakdown of the knowledge base:

5.3.1 Basic functions.

At the lowest level, the rules are organized by functions in a library: each set corresponds to a very precise sub-problem (see **figure 9**).

The basic functions and the objects using them are completely independent of each other :

5.3.1.1 structural independence

The organization and techniques of each set of rules is independent from the calling structure.

5.3.1.2 independence timing

The function and the corresponding set of rules can be called at any time during a session.

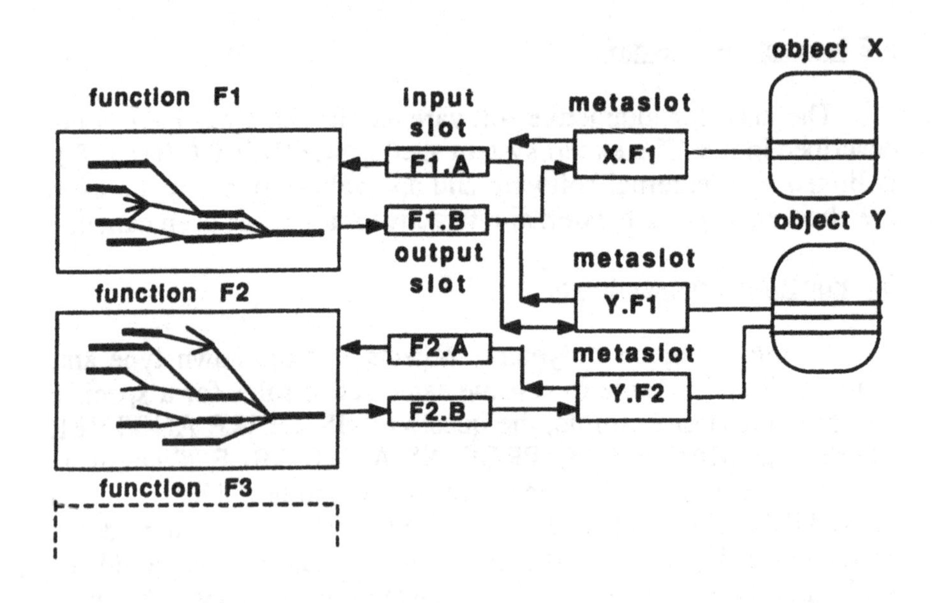

LIBRAIRIES OF SETS OF RULES
FIGURE 9

This independence is based upon simple communication techniques between the calling object and the set of rules called. The function is viewed only by a set of input and output parameters (slots F1.A and F1.B on **figure 9**). For objects, the function is called via one of its parameters (slots X.F1, X.F2 and Y.F2 on **figure 9**). The management of the input parameters (write A.F1) and of the output results (read B.F1) is individually handled for each object by the metaslot of the calling properties (metaslots of X.F1, X.F2 and Y.F2 on **figure 9**).

5.3.2 Complex functions.

More advanced functions have been built on top of the basic functions described above. These functions are organized according to their level of complexity (see **figure 10**).

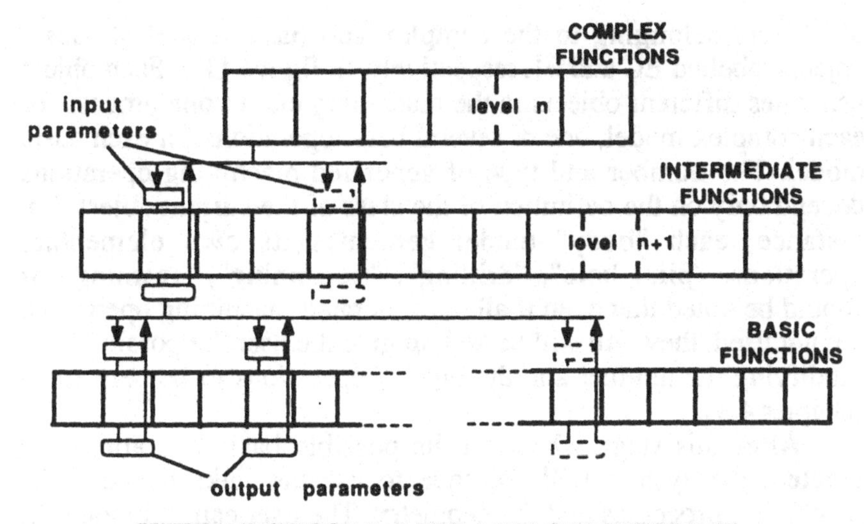

HIERARCHICAL LIBRARY STRUCTURE
FIGURE 10

Every function at level n uses one or several functions at level n+1. Lower level functions can also be called by several functions at a higher level. Communication between two different levels is ensured by fixed input and output parameters. The hierarchical structure offers flexibility in terms of structure, development and maintenance of complex functions for the generation of a process sheet.

6. REPRESENTATION AND GENERATION OF THE PROCESSES.

In this chapter, we will describe how the expert system links the structure of the geometry (chapter 4.3) and the possible machining operations. The number, type and parameters of the possible drilling operations are defined by the breakdown of complex models into a set of basic models. **Figure 11** shows the objects and the different links used during a complete session. To start with, we have four parallel class structures: the process and the complex model classes, basic operation (machining) and basic model classes. The description of the complex geometry gives a set of objects belonging to the complex and basic model classes (objects labelled EC and EL respectively on **figure 11**). Each object generates different objects in the machining class : one process for each complex model, one or several basic operations for each basic model. The number and type of generated machining operations depend only on the definition of the class of the current object. For instance, each "body" model generates its own elementary operations: "pilot hole", "drilling", "trepanning", "reaming". It should be noted that even if all the generated machining operations are not used, they will still be kept to give the user the possibility of modifying the number and the type of operations to be performed (chapter 4.3).

After this stage where all the possible basic operations are created, the system will continue to use the links between the machining processes and the geometry. The user can, for example, use the global parameters of the complex model and the specific parameters of each model to check the possibility of introducing a new operation or compute associated parameters.

7. SIMPLIFIED COMPUTATION OF THE PROCESS SHEET.

The final process sheet will be determined by a analyzing of the various basic operations generated according to the given geometry (chapter 6). The analysis of each machining process must answer the two following questions :

7.1 Need for an operation in a process sheet.

For some operations, a new step in the manufacturing process always needs to be created. For example, chamfering operations are always needed to create a geometry with chamfers. Other operations are optional (pilot hole, punching, etc.) and have to be studied more carefully. These operations must be analyzed according to the final geometry and the operations already defined in the manufacturing process. At this stage of the analysis, for example, a punching operation may need to be added depending on the other surface operations.

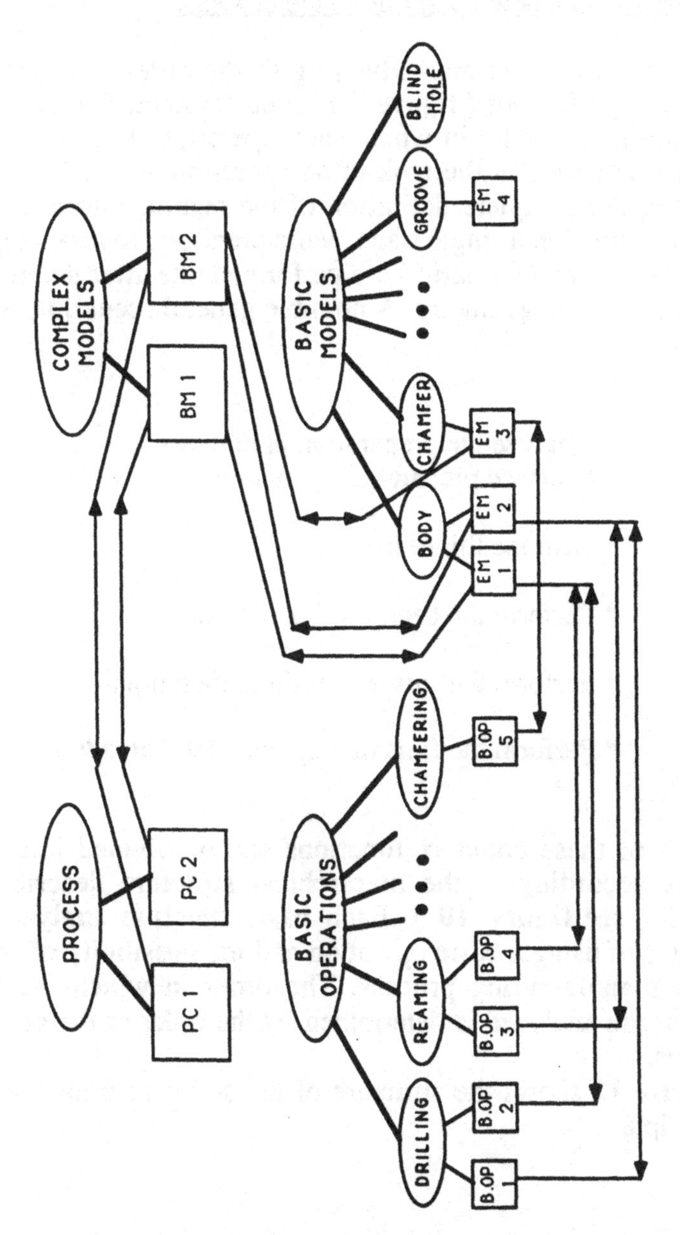

LINKS BETWEEN GEOMETRY AND MACHINING OPERATIONS
FIGURE 11

7.2 Where to put a new operation in the process.

The process sheet must always give the order in which each operation is performed (**figure 7**). In our system, this order will correspond to the order in which each operation is studied. Since both the existence and the rank of an operation are defined at the same time, the complete definition of the manufacturing process can be performed in a single stage. More precisely, several complex operations (general scenario) are performed one after the other to generate the whole geometry. A possible general scenario could as follows :

* Analyze the preparation operations (surface preparation, punching)
* Drill the thinnest body
* Perform the back milling operations
* Perform the forward milling operations
* Perform the finishing operations(Threads)

Each of these complex functions can be divided into basic functions according to the hierarchical structure described in chapter 5 (see **figure 10**). Each basic function analyzes the advantages of using a basic operation and the possibility of adding it to the manufacturing process. The order in which the basic operations are analyzed is determined by the order of the complex operations.

Figure 12 shows the structure of the software implementing this principle.

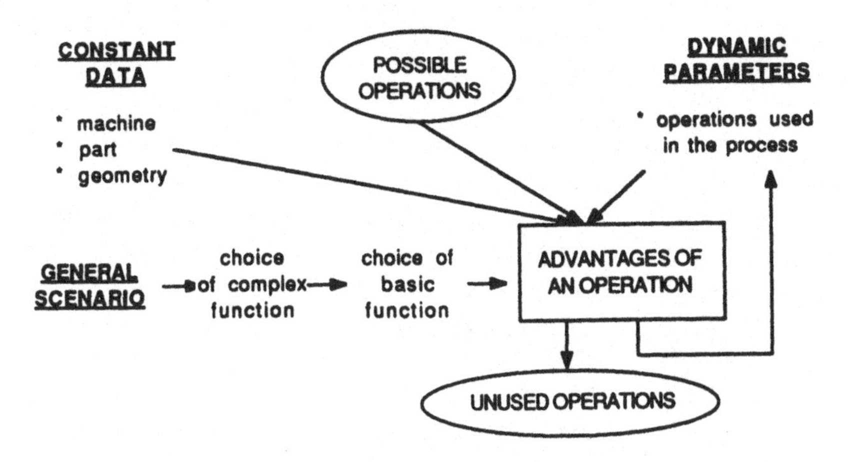

SIMPLIFIED DESCRIPTION OF THE SOFTWARE
FIGURE 12

8. CONCLUSION.

In this paper, we have described why a drilling tool selection aid system is useful for a company like **SNECMA**. The technical solutions proposed must fulfill and anticipate the needs of our users. The software developed will be in the line of the future technician's workstation (assistance software, tool databases, CAD/CAM, expert systems on the same hardware).

REFERENCES.

M. SCHERER, ' Une nouvelle conception du tournage', Industries et techniques,12 november 1990.

R. HICKMAN ' Analysis for knowledge-based systems ', Ellis Horwood Edition, 1989.

A. ROLLING, M. HITTINGER, S. SERFATY ' Optimizing the choice of a cutting tool for turning operation', Nexpert European User group and Distributor's, MUNICH, GERMANY, 26-28 november 1990.

FIGURE 12

CONCLUSION

[illegible] have [illegible] by a [illegible] and [illegible] [illegible] for [illegible] only the SYRCOM[illegible]. The [illegible] [illegible] [illegible] [illegible] [illegible] of the [illegible] [illegible] on the same [illegible]

REFERENCES

[illegible] SYRCOM [illegible]

[illegible] Analyse [illegible] systems [illegible]

[illegible] European [illegible] and [illegible] 24-28 November [illegible]

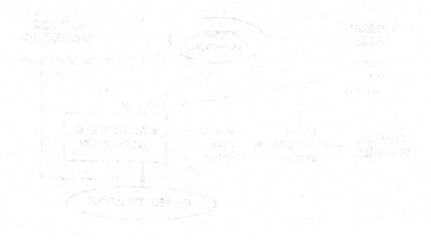

Structured Selection Problem in Pavement Rehabilitation

J.J. Hajek

Research and Development Branch, Ontario Ministry of Transportation, Downsview, Ontario, M3M 1J8, Canada

ABSTRACT

This paper examines the application and major features of a *structured selection problem* – a knowledgeable selection from a finite set of possible solutions using a reasoning process – in the context of selecting pavement rehabilitation treatments. The general problem of structured selection, with its main modules, solution sets, condition sets and search strategy, is described and applied to the task of selecting rehabilitation treatments for asphalt concrete overlays. The developed system, a knowledge-based expert system called OVERLAY, can be used to assist pavement engineers in designing cost-effective pavement rehabilitation treatments. The knowledge base for OVERLAY was obtained by studying relevant documents and interviewing 17 pavement design and evaluation professionals who are responsible for the majority of the Ministry's pavement rehabilitation designs. The results show that by using the framework of the structured selection problem, it is possible to organize and manipulate the knowledge base such that the recommendations for pavement rehabilitation treatments reflect the practices of a large agency. To lessen the development effort required, a modification of the solution paradigm should be further considered.

INTRODUCTION

Pavement rehabilitation is a group of pavement preservation treatments which significantly improves pavement structure and/or pavement performance. Unlike pavement maintenance, which is often annual, rehabilitation is usually done on a 10- to 20-year cycle and plays a key part in preserving the largest portion of the transportation infrastructure – pavements.

Despite an increasing knowledge of pavement material properties and the availability of several analytical design methods and non-destructive pavement testing procedures, the selection of rehabilitation treatments for asphalt concrete pavements is still based mainly on personal experience and judgement. This situation prevails at the Ministry of Transportation, Ontario, (MTO). Reliable pavement rehabilitation design models are scarce and field observations show that designs based on experience fit local conditions and yield the best results. At MTO, about 20 individuals, with considerable knowledge and experience in the pavement evaluation and design field, are responsible for the majority of the pavement rehabilitation designs.

Acquiring, analyzing and encoding this knowledge and experience were the basic steps towards developing a knowledge-based expert system for the design of rehabilitation treatments for asphalt concrete pavements. The knowledge-based expert system, OVERLAY, that resulted from this process was developed to (a) assist pavement engineers in designing cost-effective rehabilitation treatments, and, (b) enable the generation of practical rehabilitation alternatives for programs optimizing allocation of pavement preservation funds (e.g. Hajek and Phang [1]).

The purpose of this paper is to describe not only the development, structure, and function of OVERLAY, but also the application of expert system technology to the general problem of selecting the most suitable solution from a list of solutions, referred to as the *structured selection problem.*

The application of expert system technology to the field of pavement preservation is not new; Hendrickson and Janson [2], Ritchie et al [3], Aougab et al [4], Hall et al [5], Haas and Shen [6]. Indeed, because of the size of highway networks and the prevalent use of engineering judgement in the management of pavement preservation, the field can greatly benefit from the application of AI techniques [7]. However, based on an extensive literature search, it appears that the structured selection problem has not been systematically evaluated before in AI literature.

The research work described in this paper is part of a long-term effort by the MTO to apply knowledge-based expert system technology to pavement management [7, 8].

CHARACTERISTICS OF THE STRUCTURED SELECTION PROBLEM

The task of recommending a pavement rehabilitation treatment can be viewed as the general case of a structured selection problem; that is, a knowledgeable selection from a finite set of possible solutions using a reasoning process. A similar task is faced by a tradesperson selecting a tool from his or her toolbox or by a personnel manager selecting the best candidate for the job. The structured selection problem can also be part of other, more complex systems. For example, in the case of a real-time (industrial process) control system, structured selection can be used to recommend the best course of action given time-dependent feedback data and data describing operational and other parameters of the process.

In general, the structured selection problem can be organized into three modules: solution set, condition set and a search strategy (Table 1). The solution set contains a list of all possible or feasible problem solutions. The condition set contains all possible combinations of (selection) conditions which may occur at the same time and which are necessary for solving the problem. The search strategy (reasoning process) provides the interaction between the two sets. The complexity of the reasoning process depends on the complexity of the solution and condition sets. For OVERLAY, the solution set contains a list of all pavement rehabilitation treatments which are used by the MTO, and the condition set contains all pavement condition scenarios which may occur in the field, particularly those which may require a rehabilitation treatment.

To gain a better understanding of the mechanism of the structured selection process, several main variations of the solution and condition sets and interactions

Table 1 / Solution and Condition Sets for Structural Selection Problems and their Linkage

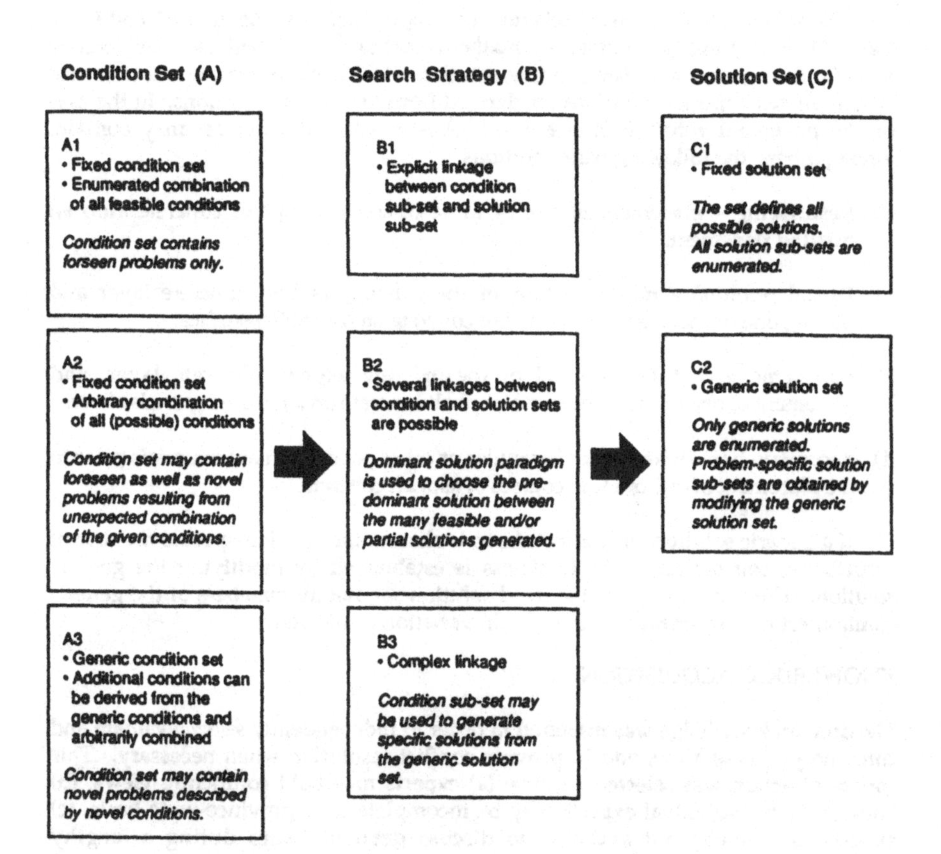

Condition Set (A)	Search Strategy (B)	Solution Set (C)
A1 • Fixed condition set • Enumerated combination of all feasible conditions *Condition set contains forseen problems only.*	B1 • Explicit linkage between condition sub-set and solution sub-set	C1 • Fixed solution set *The set defines all possible solutions. All solution sub-sets are enumerated.*
A2 • Fixed condition set • Arbitrary combination of all (possible) conditions *Condition set may contain foreseen as well as novel problems resulting from unexpected combination of the given conditions.*	B2 • Several linkages between condition and solution sets are possible *Dominant solution paradigm is used to choose the predominant solution between the many feasible and/or partial solutions generated.*	C2 • Generic solution set *Only generic solutions are enumerated. Problem-specific solution sub-sets are obtained by modifying the generic solution set.*
A3 • Generic condition set • Additional conditions can be derived from the generic conditions and arbitrarily combined *Condition set may contain novel problems described by novel conditions.*	B3 • Complex linkage *Condition sub-set may be used to generate specific solutions from the generic solution set.*	

Note
Condition sub-set - defines a set of conditions for a specific problem
Solution sub-set - defines a solution(s) applicable for a condition sub-set

between them should be considered. The basic variations of the structured selection problem modules are summarized in Table 1 in order of increasing complexity.

The solution and condition sets may be simple fixed lists (i.e. sets A1 and C1 in Table 1) or they may be defined by mathematical or other functions which can be used for their generation. For example, solution set C2 contains a generic solution set where problem specific solutions are derived from the generic solutions. In the case of the pavement rehabilitation example, the generic solution set may contain, among others, the following basic solutions:

a) Resurfacing – placement of one or more courses of asphalt concrete atop an existing pavement.

b) Partial removal – milling a part of the existing asphalt concrete layer and placing one or more courses of asphalt concrete on the milled surface.

c) Full removal – total removal of the existing asphalt concrete layer, and placement of one or more courses of asphalt concrete on a restored granular base.

d) Reconstruction – placement of granular or other pavement base materials prior to the placement of one or more courses of asphalt concrete.

If a generic solution such as (a) above is chosen, the location-specific number of resurfacing courses and their thickness is established by modifying the generic solution. Thus, in contrast to the fixed solution set, not all members of the generic solution set are enumerated and a certain variation is allowed.

KNOWLEDGE ACQUISITION

The domain knowledge was mastered in order to independently select, evaluate and combine points of view, and to provide conflict resolution when necessary. This course of action was selected because (a) experts may hold conflicting views, (b) knowledge of individual experts may be incomplete on a province-wide basis, (c) experts are usually not available to discuss detailed issues during a lengthy calibration process, and (d) knowledge from different sources must be evaluated and combined.

Knowledge for OVERLAY has been acquired from written sources and interviews with pavement experts. It was a substantial and time consuming iterative process. Knowledge acquired from written sources was discussed during the interviews; knowledge obtained during the interviews was compared and verified with other information.

Written sources

Information was extracted from the written sources by:

a) studying available documents where it already existed in an organized form (explicit knowledge), and
b) extracting and organizing information from pavement rehabilitation design documentation, such as contract drawings (implicit knowledge).

The sources of explicit knowledge included (a) engineering textbooks and literature, (b) official, jurisdiction specific, guidelines and directives, (c) nonbinding recommendations contained in internal reports, and (d) pronouncements and recommendations from various sources, such as minutes of committees coordinating geotechnical activities within the MTO, staff trip reports and laboratory reports.

The most important sources of implicit knowledge were recently prepared pavement rehabilitation designs. These documents provided detailed descriptions of the pavement rehabilitation work to be done, as well as pavement deterioration before rehabilitation. Consequently, they were invaluable in establishing links between the pavement condition and the corresponding actual rehabilitation designs. Altogether, 55 recent pavement rehabilitation designs were studied in detail. Data collected from each design included historical traffic data and traffic projections, subgrade type, thickness and composition of pavement layers before rehabilitation, pavement maintenance history and pavement age, history of pavement performance before rehabilitation (including severity and density of 15 pavement surface distresses [9]), environmental data, and a detailed description of the rehabilitation design (amount and type of all materials removed and added).

Interviews with pavement experts

Seventeen MTO experts, representing all 5 regional geotechnical offices responsible for designing rehabilitation treatments, were interviewed. The following interview format was used:

a) Introduction

The reasons for and the objectives of the interview were explained. It was emphasized that the goal of the project was to improve pavement management processes by developing a decision support system, not to dehumanize engineering decision making.

b) Unstructured part

This was an open discussion. Its main objectives were to establish the expert's basic design approach or philosophy and to resolve or understand any conflicting information obtained from other sources.

c) Structured part

The experts were asked to discuss in detail several rehabilitation designs which they had worked on recently or which had been completed recently in their region. The purpose was to learn how they utilize available data to select rehabilitation treatments.

The designs discussed were selected from the list of 55 recent rehabilitation designs. The selection attempted to obtain at least one representative design for each of the four generic designs described previously.

The use of the actual designs, rather then hypothetical rehabilitation projects, was instrumental in making the interview process interesting for the experts and proved to be an effective and productive method of knowledge acquisition. It provided a solid base for discussion and review of data requirements, design objectives, reasoning and, in a few cases, implementation results.

Information obtained during the structured part of the interview was summarized in terms of specific reasons offered by experts for selecting a given design. It became evident that experts based their design selection on a number of basic indicators, such as pavement structural adequacy, thickness of asphalt concrete layer, and the presence of key pavement surface distresses (e.g. rutting, flushing, and alligator cracking). These indicators can be arranged in a hierarchical order and represent a set of the main conditions governing the treatment selection.

ORGANIZATION OF KNOWLEDGE AND REASONING

The task of recommending pavement rehabilitation treatments was modeled, referring to the definitions given in Table 1, as a condition set A1, solution set C1 and search strategy B1. There were several reasons for the selection of this model:

a) The number of alternatives used by the MTO for rehabilitation of asphalt concrete pavements is relatively small (about 40). For example, it is impractical to construct an overlay course that is less than 40 mm thick and each additional course usually increases the thickness by about 40 mm. Thus, the overlay thickness is not a continuous variable.

b) The selection of the predominant alternative (search strategy B2 in Table 1) is not usually done by experts. They tend to concentrate on what they perceive to be the critical conditions governing the treatment selection and then select the treatment addressing them.

c) The arbitrary combinations of the selection conditions (included in condition set A2), while being theoretically possible, do not occur in practice. For example, the combination of a high Pavement Condition Index [9] (which is associated with a high quality pavement) and the simultaneous presence of extensive alligator cracking (associated with severely damaged pavements) do not occur. Thus, it is not necessary to anticipate all theoretically possible combinations of the condition set members.

d) Commercially available expert system development shells require an explicit organization of domain knowledge and an exact linguistic match. They do not readily support any linguistic approximations for the generation of additional conditions from the generic conditions.

The structured selection problem, defined by modules A1, B1 and C1 (Table 1), can be represented by several knowledge representation schemata such as decision tables, semantic networks (solution graphs) and decision trees. The last representation schema, decision trees, was adapted for OVERLAY because it reflects the hierarchy of different members of the condition set and provides transparent and convenient representation of the domain knowledge and reasoning. Decision trees can identify unique solutions for each branch. This reflects the expectations of the intended users, is easy to modify, and enhances the ability of the system to explain a line of reasoning. The main limitation of decision trees is that it may become quite cumbersome to display, modify and code them for large problems.

DECISION LOGIC

The main characteristics of the decision tree used to represent the domain knowledge and reasoning of OVERLAY are summarized in Table 2. An overall schema of the decision tree below level 1 (PCI level) is illustrated in Figure 1. The tree has nine levels and each level has two or more branches. The nine levels represent the basic indicators or decision criteria which govern the selection of the

Table 2/ List of Decision Criteria Arranged in Hierarchical Order

Decision Criterion		Branching Levels		Decision Criterion Variables		Decision Tree Schema
No.	Name	Names	No. of Lev.	Names	No. of Var.	
1	Pavement Condition Index (PCI)	Increments of 5 from PCI ≤ 35 to PCI > 75	10	PCI	1	1 2 2 2 2 2 2 2 2 2 2 Level 1, Branching 10
2	Structural Adequacy	Adequate Inadequate	2	See Figure 2	4	2 3 3 Level 2, Branching 2
3	Adequacy of Granular Base Equivalent Thickness	Adequate Inadequate	2	See Figure 3	14	3 4 4 Level 3, Branching 2
4	Uniformity of Pavement Deterioration	Uniform Non-uniform	2	• 4 Types of alligator cracking • Distortion • Rutting	6	4 5 5 Level 4, Branching 2
5	Critical Distress Combinations	Not Present Type 1 Type 2	3	See Figure 4	3	5 6 6 6 Level 5, Branching 3
6	Location	Rural Urban	2	% Urban Location	1	6 7 7 Level 6, Branching 2
7	Condition of Single and Multiple Transverse Cracks	Severe Not Severe	2	Severity and Density of Transverse Cracking	2	7 8 8 Level 7, Branching 2
8	Traffic Category	T1: AADT≤2000 T2: 2000<AADT≤ 4000 T3: AADT>4000	3	AADT Truck % Truck Growth	3	8 9 9 9 Level 8, Branching 3
9	Thickness of Hot Mix	(1) ≤ 80 mm (2) 80 –130 mm (3) > 130 mm	3	Hot Mix Thickness	1	9 solution list solution list solution list Level 9, Branching 3

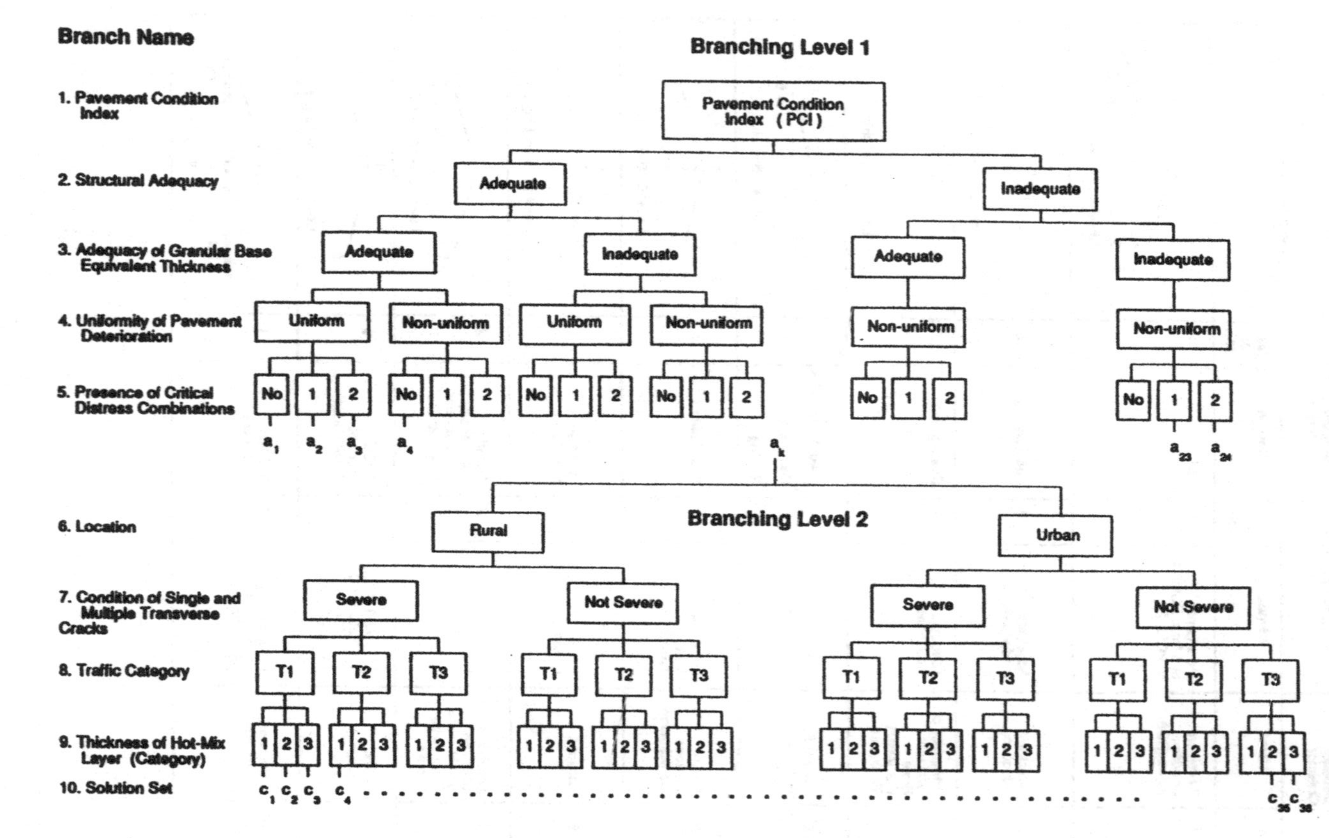

Figure 1 Full Decision Tree for Overlay

rehabilitation treatments identified during the knowledge acquisition phase. The full decision tree would require 7876 individual branches (the members of the condition set can be arranged in 7876 possible combinations). However, the combinations which were not considered feasible were pruned, and the combinations for which the solutions differed only slightly were combined. Altogether, only 785 branches were required.

The nine levels are outlined below. The outline uses a practical example to show how a multifaceted problem, depending on about 40 variables and factors, can be solved using structured selection. However, it may not be detailed enough for the reader to fully understand the decision process involved.

Level 1. Pavement condition index (PCI)
PCI is an overall measure of pavement condition which provides a basic indication of rehabilitation needs. Ten PCI branching levels, defined in Table 2, were used.

Level 2. Structural adequacy
A structural adequacy criterion was used to determine whether the existing pavement was structurally adequate based on its past performance. The structural adequacy was determined by evaluating the condition of alligator cracking using four diagrams as shown in Figure 2. The diagrams are based on pavement surface distress data which are available from the MTO pavement management data bank. These distresses are evaluated in terms of 15 distress types, including ravelling, flushing, rutting, distortion, and different types of cracking such as pavement edge, transverse and alligator. Each distress is evaluated separately both on a severity scale and a density scale, with each scale ranging from 0 to 5. The evaluation is done annually or biennially.

In Figure 2, the condition states of all 4 recorded types of alligator cracking distresses are represented by an area obtained by orthogonally plotting the possible combinations of severity and density. These combinations, considered to indicate structural inadequacy, are shown as shaded areas. Thus, each diagram represents a heuristic rule regarding structural adequacy. For example, the rule based on the first diagram states that if a pavement has wheel track alligator cracking with a severity and density greater than 2 (corresponding to slight severity and intermittent density measured on the scale from 0 to 5), the pavement may be structurally inadequate.

In general, using detailed, routinely available data, the basic decision criterion of pavement structural adequacy was modeled by (a) identifying all pertinent data, (b) considering all possible combinations of the data, (c) creating a decision criterion, and (d) including the decision criterion at the appropriate hierarchical level of the decision tree. The process enables one to synthesize extensive detailed information and use it in the selection process. A similar modeling technique of establishing decision levels was used for other levels of the decision tree (Levels 3, 4 and 5).

Level 3. Adequacy of granular base equivalent thickness
The granular base equivalent thickness was calculated using the procedure developed by Kher and Phang [10]. The estimation of granular base equivalency factors for existing, possibly deteriorated asphalt concrete layers, was done using the decision logic documented in Figure 3. It again utilizes all available pertinent data.

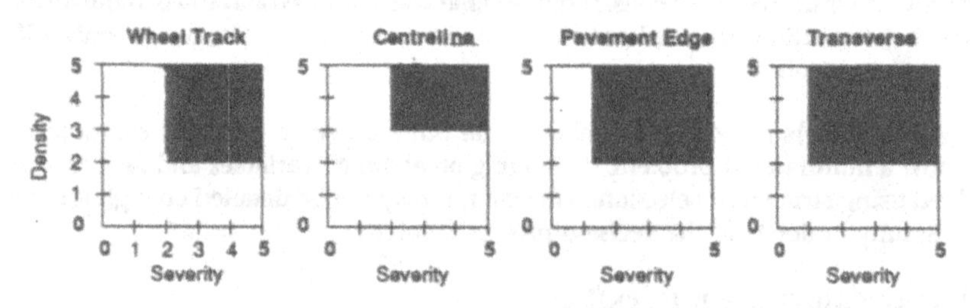

If any distress falls into the shaded area, the pavement is structurally inadequate

Figure 2/ Decision Criterion for Structural Adequacy

Condition 1: If pavement is structurally inadequate, GBE factor is 1.3 or 2.0 depending on the thickness of the asphalt concrete

Condition 2: If any combination below exists, (a combination of density and severity falls into the hatched area), GBE factor is 1.6 or 2.4 depending on the thickness of the asphalt concrete

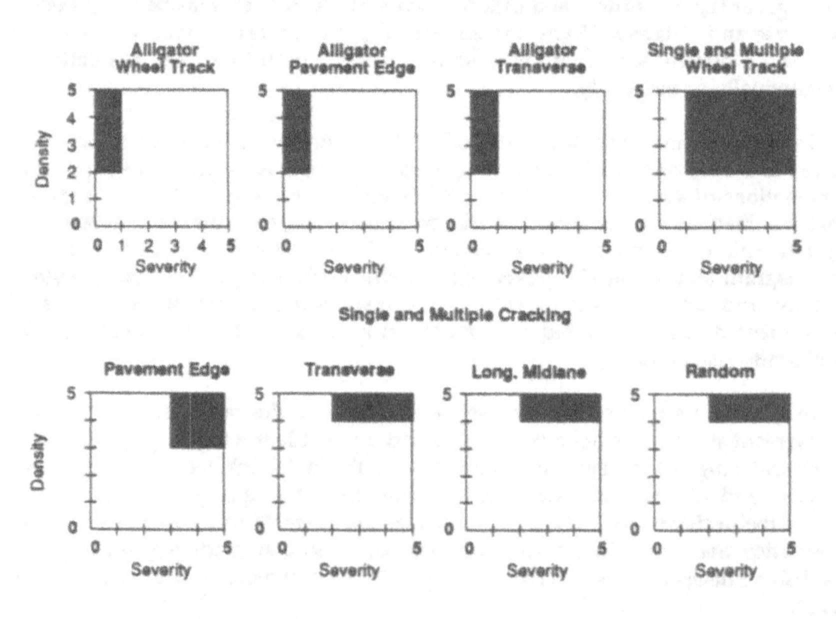

Condition 3: If Condition 1 or Condition 2 does not exist, GBE factor is 2.0 or 3.0 depending on the thickness of the asphalt concrete

Figure 3/ Determination of Granular Base Equivalency Factors (GBE) for Existing Asphalt Concrete

Level 4. Uniformity of pavement deterioration
This decision criterion was used to estimate the need for padding or extra milling before overlay placement. The uniformity of pavement deterioration was estimated using the severity and density values of all four types of alligator cracking (Figure 3), distortion and wheel track rutting.

Level 5. Presence of critical distress combinations
The presence of some distresses, namely flushing, ravelling and rutting, occurring at specific levels of severity, constitutes a safety hazard and may require a rehabilitation treatment regardless of the condition of other pavement deterioration characteristics. These three distresses are referred to as the critical distresses, and certain combinations of these distresses, which require a rehabilitation action, are referred to as critical distress combinations. The presence of the critical distress combinations is determined by heuristic rules summarized in Figure 4.

Figure 4 systematically lists and evaluates all possible relevant states of the three critical distresses to determine which of them results in the critical distress combinations. For example, the presence of Critical Distress Combination 1 (shown as a shaded square under column "Result, Crit. 1", in Figure 4) indicates the existence of ravelling or flushing requiring a rehabilitation treatment; the presence of Critical Distress Combination 2 (shown as a shaded square under column "Result, Crit. 2") indicates the existence of rutting requiring a treatment. (Two blank squares under "Result" in Figure 4 indicate that the critical distress combinations do not exist).

Level 6. Location
The location was either rural or urban. It was assumed that urban locations have, and would be required to have, pavements bounded by a concrete curb and gutter.

Level 7. Condition of single and multiple transverse cracks
The presence of severe or very severe transverse single and multiple cracks occurring on a substantial portion of the pavement section, was considered to require different, often stronger rehabilitation treatments, than those required by less severe transverse cracks.

Level 8. Traffic category
Three branching levels were used for the traffic category (Table 2). These levels are mainly based on Annual Average Daily Traffic (AADT) volumes, although a higher traffic category than that indicated by the AADT volumes alone was used for sections with above average heavy truck traffic.

Level 9. Thickness of hot mix layer
Three branching levels were used as shown in Table 2 and Figure 1.

SYSTEM DESIGN AND PROGRAMMING

System programming was done using the EXSYS expert system development package [11]. The nine decision levels were established by simple rules which utilized all pertinent data. Once the nine decision levels were established, meta rules were used to link all feasible states of the decision levels (representing the condition set) with recommended rehabilitation treatments (representing the solution set). The

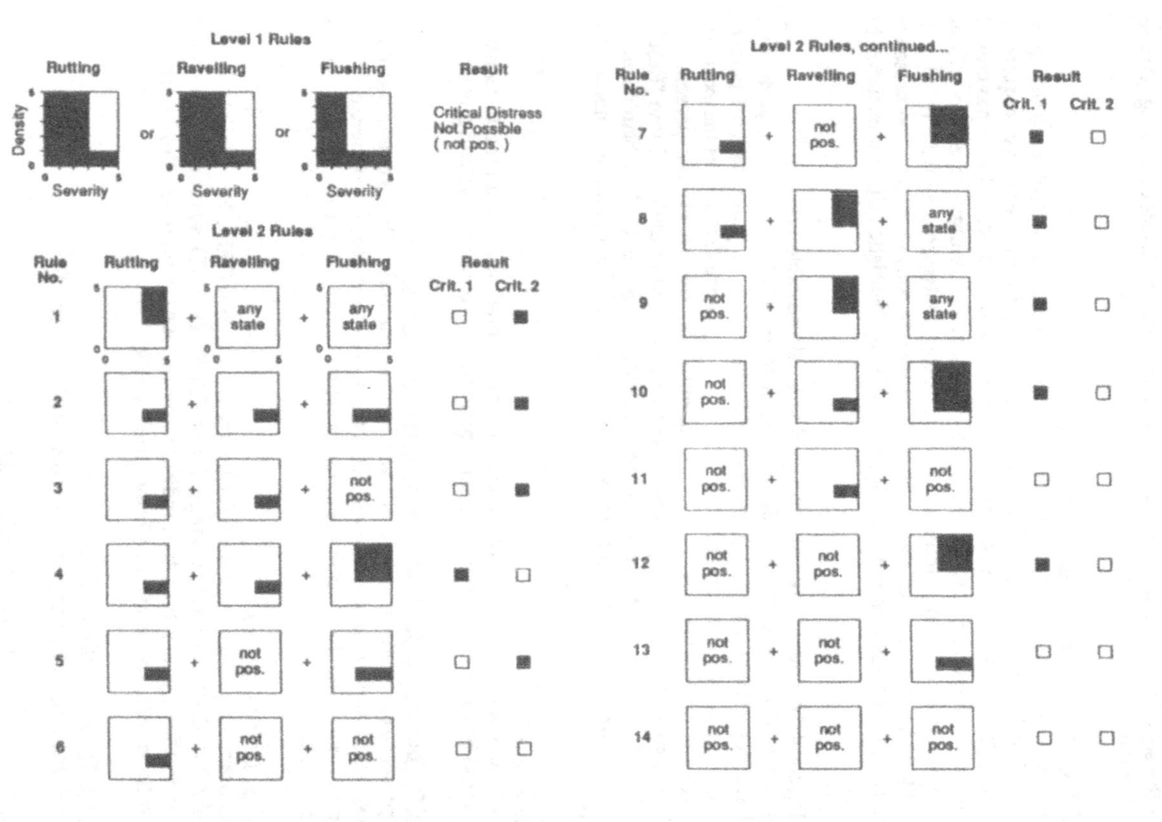

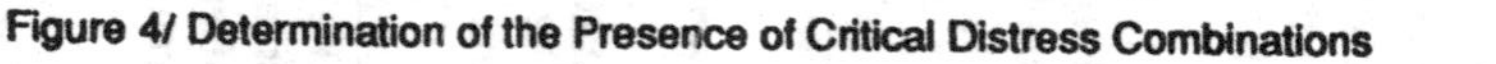
Figure 4/ Determination of the Presence of Critical Distress Combinations

solutions recommended by OVERLAY also include a degree of certainty that is expressed on a scale of 0/10, 1/10, 2/10, ... 10/10. Altogether, OVERLAY contains about 90 simple rules and 785 meta rules. The number of meta rules equals the number of feasible combinations of decision levels for which a solution is required.

ALTERNATIVE FORMULATION OF THE STRUCTURED SELECTION PROBLEM

The most difficult and time consuming task in developing OVERLAY was to recommend, in a consistent and logical manner, one or more rehabilitation treatments for all 785 feasible combinations of the decision levels and their values. Another difficulty encountered was the lack of flexibility in changing the knowledge base. For example, if an additional decision level was added, a substantial reconstruction of the knowledge base would be required. Similarly, an addition to the solution set may necessitate the review of many of the 785 combinations. It should be noted that it is the modification of the knowledge base, and not the computerized manipulation of the knowledge base that is at issue.

The experience gained in developing OVERLAY provided an incentive to return to the question of problem formulation and to explore the usage of an alternative formulation of solution sets, condition sets and search strategies which would simplify the selection process. The following two alternative formulations were explored but were not implemented.

a) Generic strategy model
A model using a generic strategy was formulated in terms of the fixed condition set A1, the generic solution set C2 which contained seven generic solutions, and the search strategy B3. The model required eight separate sets of rules, one set to select a generic solution and the seven remaining sets to determine the corresponding variable part of the generic solutions. It was concluded that this model provided no practical advantages for this application.

b) Dominant strategy model
There are two stages to a model of this type. The first stage creates a number of partial solutions (solution subsets) whose members solve specific subproblems (condition subsets). The idea is to simplify the problem by breaking it into a number of subproblems. The second stage considers all partial solutions and determines the dominant solution which encompasses all solution subsets. The solution subsets were formulated as pavement deterioration scenarios (e.g. rutting or fatigue cracking) and scenario-specific solutions were assigned to them.

The main problem encountered was that the deterioration scenarios were not always independent. A simultaneous presence of several deterioration scenarios may require a different solution than that generated for any single one. The model was considered unsuitable for this application. It appears that alternative approaches to the structured selection problem should be investigated, perhaps using fuzzy logic.

CONCLUSIONS AND RECOMMENDATIONS

1. The structured selection problem -- a judicious selection from a list of possible alternatives -- is one of the most common engineering problems and is well suited to the application of expert system technology.

2. It is feasible to develop a knowledge-based expert system formulated as the structured selection problem for recommending pavement rehabilitation treatments that reflect the practices of a large agency.

3. Recording, organizing and encoding of knowledge, including the reasoning process, is interrelated and requires a systematic approach.

4. Decision trees are cumbersome to use, but are transparent and user understandable, and can be easily coded using if – then rules. While it is also possible to solve decision trees using conventional programming techniques, the use of expert system development shells such as EXSYS provides productivity advantages by enabling a mechanical interpretation of the knowledge base, supplying editing programs and user interface, and facilitating rapid program modification and testing through built-in inference mechanisms.

5. At this stage, OVERLAY can be used to guide pavement rehabilitation engineers by generating basic rehabilitation concepts. These concepts can be subsequently refined by considering additional factors not included in OVERLAY, such as the type of asphalt concrete to be used for different overlay layers, and the exact location of padding.

6. OVERLAY's knowledge base should be enhanced using a cyclic process of field testing by experts followed by code modifications by the system developer.

7. To lessen the programming effort, a modification of the solution paradigm should be further considered. OVERLAY contains 785 condition sets, and each condition set has its corresponding solution subset. This approach requires a considerable data base development effort. It is recommended that the possibility of modeling the treatment selection problem using alternative approaches be fully explored.

ACKNOWLEDGEMENTS

The significant assistance of Mr. E.P. Hayden, systems engineering student, University of Waterloo, and Mr. Klaus Bodker of the Engineering Academy of Denmark, who did most of the OVERLAY programming work, is gratefully acknowledged.

REFERENCES

1. Hajek, J. J. and Phang, W.A. Prioritization and Optimization of Pavement Preservation Treatments, Transportation Research Record 1216, TRB, National Research Council, Washington, D.C., pp. 58-68, 1989.

2. Hendrickson, C.T., and Janson, B.N. Expert Systems and Pavement Management, Proceedings, 2nd North American Conference on Managing Pavements, Toronto, November 1987, pp. 2.255-2.266.

3. Ritchie, S.G., Che-I Yeh, Mahoney, J.P., and Jackson, C. Development of an Expert System for Pavement Rehabilitation Decision Making, Transportation Research Record 1070, TRB, National Research Council, Washington, D.C., pp. 96-103, 1988.

4. Aougab, H., Schwartz, C.W., and Wentworth, J.A. Expert System for Pavement Maintenance Management, Public Roads, Vol. 53, No. 1, pp. 17-23, June 1989.

5. Hall, K.T., Connor, J.M., Darter, M.I., and Carpenter, S.H. Expert System May Aid CPR Field Work, Planning, Roads and Bridges, pp. 35-39, April 1988.

6. Haas, C., and Shen, H. Preserver: A Knowledge Based Pavement Maintenance Consulting Program, Proceedings, 2nd North American Conference on Managing Pavements, Toronto, pp. 2.327-2.338, November 1987.

7. Hajek, J. J. and Haas, R.C.G. Applications of Artificial Intelligence in Highway Pavement Maintenance, Artificial Intelligence in Engineering: Diagnosis and Learning, Computational Mechanics Publications, Southampton, pp. 175-195, 1988.

8. Hajek, J. J., Chong, G.J., Haas, R.C.G., and Phang, W.A. Can Knowledge-Based Expert System Technology Benefit Pavement Maintenance?, Transportation Research Record 1145, TRB, National Research Council, Washington, D.C., pp. 37-47, 1988.

9. Hajek, J. J., Phang, W.A., Wrong, G.A., Prakash, A., and Stott, G.M. Pavement Condition Index (PCI) for Flexible Pavements, Report #PAV-86-02, Ontario Ministry of Transportation and Communications, Downsview, August 1986.

10. Kher, R., and Phang, W.A. OPAC Design System (Ontario Pavement Analysis of Costs), Proceedings of the 4th International Conference on Structural Design of Asphalt Pavements, University of Michigan, pp. 841-854, August 1977.

11. EXSYS - Expert System Development Package, EXSYS, Inc., Albuquerque, NM 87194, 1986.

HyperQ/Process: An Expert System for Manufactuing Process Selection

K. Ishii, S. Krizan, C.H. Lee, R.A. Miller

Engineering Research Center, College of Engineering, The Ohio State University, Columbus, Ohio, USA

ABSTRACT

This paper describes an expert system that helps designers select a manufacturing process in the earliest stage of product design. First, the paper focuses on net-shape manufacturing processes and identifies the major factors that affect the selection of an appropriate process. A versatile methodology should consider all the factors *simultaneously* in assessing the suitability of the candidate processes. The proposed system uses the concept of design compatibility analysis to represent the suitability of candidate processes with respect to the given product specifications. The expert system uses this knowledge to eliminate incompatible candidates and rank the compatible set of processes. A prototype system called HyperQ/Process uses HyperCard and Prolog to implement the proposed methodology. HyperQ/Process also contains information related to each process.

1. Introduction

1.1 Background

In recent years, concurrent engineering has emerged as a key practice in enhancing the competitiveness of a product. Most people agree that the cost and quality of a product are "locked" into the layout design. Many companies are actively pursuing means to integrate the life-cycle values of the product early in its development. In particular, design for manufacturability (DFM) has provided engineers a systematic methodology to reduce development time, cut production cost, and reduce defects. DFM typically focuses on the particular manufacturing process, e.g., machining, stamping, injection molding, assembly, etc., and seeks to incorporate into the early product design measures that can prevent manufacturing problems and significantly simplify the production process.

While this type of activity certainly enhances product competitiveness, it usually applies to a specific process. What precedes DFM is a very important decision, selection of the material and manufacturing process. Frequently encountered process selection targets include 1) electronics housing: sheet metal forming or injection molding, 2) automotive parts: machining or die casting or investment casting. These decisions not only affect the DFM methods that follow, but also the product's overall market competitiveness. A variety of factors influence this decision, many of which cannot be estimated accurately, e.g., volume of sales. While there are many handbooks for qualitative guidance in selecting a process, they do not provide a quantitative means to compare the suitability of each process to a given part. Today, most engineers select a process based on their experience and intuition in addition to "guesstimation" (estimation based on educated guesses) of many of the influencing factors. Engineers can greatly benefit from a design tool that allows them to compare different processes in a more rational, systematic manner, utilizing as much quantitative information as possible.

This paper reports on our research to develop a systematic methodology for process selection. We focus on net-shape processes such as injection molding, die casting, and forging. This paper also identifies the factors influencing process selection and points out the iterative nature of some of the decision variables. This information is currently compiled as HyperCard stacks intended for documentation as well as designer training. Then, we describe an expert system that utilizes qualitative information on compatibility of candidate processes to various product specifications. A designer would use this system in the preliminary stages of design to screen through the large set of possible processes and derive a small set of suitable processes.

1.2 Related Work

The past five years have seen a surge of research and development work involving DFM. Perhaps the most notable work was in design for assembly (DFA) pioneered by Boothroyde and Dewhurst (1983). Their work on assembly has an indirect influence in process selection. DFA recommends separate parts to be integrated into one unless there is a compelling reason not to. Integration of parts usually leads to a different process, typically a near net-shape process like die casting or injection molding. Yet, DFA only focuses on assembly cost and does not take into account possible increases in part cost. There are studies of similar nature addressing different processes. Lai et. al. (1985) assume that the part cost is given. Cutkosky (1989), Shah (1990), and many others address machining process. Our own work focuses on design for net-shape manufacturing. Ishii, et. al. (1989b) look at design for injection molding, while Liou and Miller, et. al. (1991) focus on design for die casting. In Maloney et. al. (1989), we focused on the compatibility between forging designs and the proposed process and equipment. Each work cited above concentrates on a single process and deals with more or less the detailed design that is suitable for the process in question. Ishii and Nekkanti (1989a) pursued a general framework for representing knowledge about design for net-shape manufacturing, but the paper still concentrated on injection molding. Work that addresses the comparison between more than two processes is not generally available.

There are many textbooks and handbooks that describe different manufacturing processes, their pros and cons, etc. Some handbooks even identify major factors that influence process selection and give qualitative guidance. Of the many sources available in print, perhaps the most comprehensive is by Bralla (1986). He gives an excellent coverage of major manufacturing processes, and comments on their suitability with respect to materials, mechanical properties, general shape and size, production volume, etc. While it is an outstanding handbook, he does not deal with the iterative nature of some of the decision factors such as production volume and cost (the more you make, the lower the price, thus more sales, etc.). Also the book documents the decision variables and the influencing factors in a "free format" i.e., not completely uniform across different processes. This format sometimes makes it difficult to compare the suitability of one process to another. There are many other books that provide similar information with a focus on different processes. Ludema et.al. (1987) look at the economic aspect of process selection, while Bolz (1974) combines mechanical requirements with cost issues. Eary and Johnson (1962) give a comprehensive coverage of a variety of manufacturing processes, but this information is now slightly dated.

Despite the abundance of literature on manufacturing processes and DFM methodologies for individual process, very little work has been done in developing a computer-aid that accommodates information about different (old and new) processes, evaluates the suitability of each process with the designers' needs, and assists in selecting the most appropriate process.

Process-based Group Technology represents the most notable attempt at guiding designers in process selection. Niebel (1966) devised a group technology system for a wide range of manufacturing processes. He also proposed a decision equation that approximates the cost per part of the primary operation. While his method gives a good "first cut" comparison of different processes, the system only addresses a relatively rough geometry classification (9 classes), materials, and lot size. Group technology normally addresses one classification factor, e.g., shape. Extending beyond one classification factor is not trivial. Therefore, group technology works well when it addresses a single process or processes that lead to similar geometry classifications.

1.3 Our approach.

Our research focuses on single parts that are to be net or near net-shape manufactured, although our discussion may compare a full net-shape plastic part with a sheet metal product that consists of a folded sheet with auxiliary components (Ishii, Ho, and Miller, 1990). We first seek to identify the major factors that affect process selection at the early stages of design. Particular attention goes to decision variables that, in turn, affect original factors (e.g., materials affecting a process and process influencing detailed classification of materials, etc.).

The next step is to develop a representation scheme for knowledge about process selection. At different stages in product development, and depending on the amount of quantitative data available, engineers use a different type of knowledge in the selection process. At the very preliminary stage of design, engineers are likely to use more qualitative knowledge on compatibility between design specifications and the process than quantitative knowledge on life-cycle cost. The compatibility knowledge could be looked upon as good or bad templates of the design and process, i.e., case-based design rules. We call this approach "compatibility-based" or "case-based" because of the example cases of good or bad compatibility form the Knowledge Base. The framework of design compatibility analysis (DCA; Ishii et. al. 1988) adopts primarily this type of knowledge representation.

As the design progresses and more data become available, we can utilize a more quantitative form of compatibility information. One candidate form of compatibility representation is the **interval** information of various costs that leads to a ranking of different processes. The most quantitative compatibility measure comes in the form of the life-cycle cost estimate.

Our eventual goal is to combine these forms of compatibility information so that designers can utilize the most appropriate data at various stages of product development. This paper focuses on the first type of knowledge, case-based compatibility knowledge, and describes an expert system that uses qualitative information to deduce a compatible set of manufacturing processes and provide a ranking among them.

2. The Decision Variables in Process Selection.

2.1 The processes under consideration.

The current investigation covers the following net-shape processes:

1) Hot Forging
2) Cold Forging
3) Powder Metals
4) Hot Extrusion
5) Sand Casting
6) Investment Casting
7) Die Casting
8) Injection Molding
9) Sheet Forming

2.2 Factors that affect the process selection

The major factors that influence the process selection are:

1) Mechanical Properties (stiffness, hardness, etc.)
2) Part Shape
3) Part Size
4) Tolerances and Surface Finish
5) Materials
6) Time to Market
7) Production Quantity
8) Production Rate

Factors 1 to 4 of process selection are what people normally refer to as the functional requirement. Factor 5, materials, is tricky. The functional requirement and other environmental constraints (temperature, etc.) affect material selection while the choice of the process also imposes a constraint. Factors 6, 7, and 8 are related to the production requirements. Again, we see some interdependencies in these parameters. The production rate and volume depend on the product sales and the market life (i.e., how long the product remains competitive on the market). The sales depend on cost which in turn affects production volume (sales). Figure 1 shows the dependency between the factors related to product design and process selection.

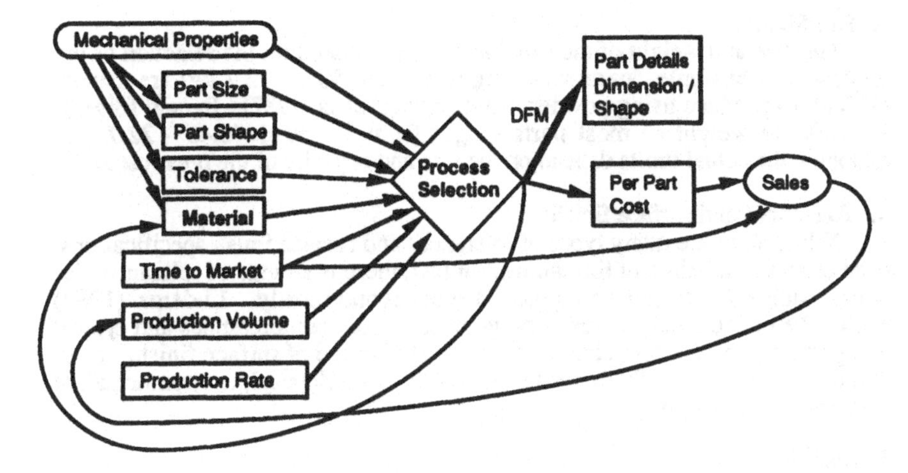

Figure 1. Design dependency diagram
This diagram views process selection as the main decision item, with arrows indicating dependencies in the decision process. Note the cyclic dependency between 1) process and material, and 2) sales and production volume.

The dependency between factors differs from product to product. This difference determines the sequence in which a designer would decide on the values of decision variables. In some cases, such as an electronics housing, the mechanical and environmental requirements give the designers a wide range of materials. Hence, the designers are likely to determine the process (injection molding or sheet metal forming) before deciding on the material. Naturally, the detailed design of the part and the determination of process parameters such as machine size and process conditions come after both the material and process selection. In essence, designers must resolve the eight major factors and select the appropriate process *simultaneously*. Let us briefly summarize how each major factor affects process selection.

1. Mechanical properties:
Mechanical properties such as stiffness (in one axis or several axes) and hardness have the biggest influence on part size and shape but they also have bearings on the process you chose. For instance, cold forging can have better mechanical properties than other processes because it can force the grain structure to follow the contour of the part.

2. Part Shape:
For geometrically simple parts such as bolts or straight shafts, the most economical method of manufacture is relatively apparent. As the shape of the part becomes more complex, selection of a suitable process becomes important. For example, cold forging is generally limited to cylindrical, square, hexagonal, or similar symmetrical shapes having solid or hollow cross sections. Therefore, if the part has an intricate and nonsymmetric shape, a casting or molding process may be considered more suitable.

3. Part Size:

The size and weight of the candidate designs also limit the selection of the process. For example, parts weighing more than 50 lb (22.7 kg) are usually difficult to produce using powder metallurgy (Trucks, 1987). In cold forging, the suitable weight of most parts ranges from 1 - 50 lb (0.45 - 22.7 kg), although the actual limits depend on the size and capacity of the press used.

4. Tolerances and surface finish:

While there are many types of tolerances and surface finish specifications, each process has inherent limitations. In fact, there is a range of tolerances for which each process can be employed most economically. Ludema (1987) shows the economical tolerance range in terms of size tolerances and typical component size. He also shows an economical range of surface finish. Note that tolerances are usually derived from assembly mechanical, environmental, or aesthetic requirements.

5. Materials:

Material selection is perhaps the single most important factor in both part design and process selection. The material is primarily dependent on the physical and mechanical properties required. In actual practice, the following properties are considered: strength (tensile, compressive, shear, creep), hardness, corrosion resistance, thermal conductivity, stiffness, weight, melting temperature, etc. These material properties directly influence the production methods by which the material is worked. Each net-shape process is limited by the suitable materials. For example, in the die casting process, only low melting point metals such as zinc, aluminum, magnesium, brass, lead and tin alloys can be used. Also, in die casting, metals with higher melting points still have problems even though they may become more and more economical as die materials are improved. On the other hand, there are cases in which the product drives the selection of materials, e.g., very thin walled die castings are typically manufactured from magnesium.

6. Time to market:

People use time to market as an indicator of success. The shorter the time from concept to market, the more competitive the product. Nevertheless, the time to market is driven by 1) the time the product is expected to be competitive on the market, and 2) the existence of competition to develop similar products. Time to market may affect process selection since the short market life of the product may not warrant lengthy design and fabrication of complex tools. A short time to market may rule out injection molding over sheet metal forming despite a possibly large expected volume due to the long lead time for mold development.

7. Production quantity:

The production volume affects process selection to a considerable extent (Ludema, 1987). The cost of a process has break-even points over the economic production quantities. Figure 2 illustrates an example of how the production amounts would influence the method of process. In net-shape manufacturing, tool (die and mold) design and fabrication costs take a significant percentage of the production cost of the part. The percentage differs from process to process. For example, in the design of electronics housings, tooling for sheet forming (bending, folding, and stamping) will be significantly less expensive and less time consuming than tooling for molding.

8. Production Rate:

Each process has its own possible production rate or an economical range of production rates although individual rates will differ depending on the machine capability. For example, the metal stamping process can produce parts at a rate of thousands per hour while the cycle time for injection molding is typically close to a minute.

3. Compatibility Knowledge Representation for Process Selection

For complex parts, there is generally more than one acceptable way of deploying a net-shape process to produce the part. Variations arise from the number and type of secondary operations required in the process plan. Additional machining or assembly operations may be necessary to add features which are difficult to form, to bring dimensional tolerances within limits, or to improve the microstructural properties of the part. In the case of tolerances or additional features, it is sometimes possible to bring the part closer to a final shape with more complex and elaborate tooling or tighter process control. The use of more sophisticated tooling reduces or eliminates secondary operations but it entails tighter tooling costs and may involve higher operating and maintenance costs. The hypothetical cost vs. volume curve shown in Figure 2 illustrates the points at which the best processing choice will change. The issue is to identify the basic factors which control the location of such breakpoints and to establish methods to estimate the incremental tooling costs which determine them. Naturally, we must also incorporate into our compatibility consideration other factors such as mechanical properties and time to market.

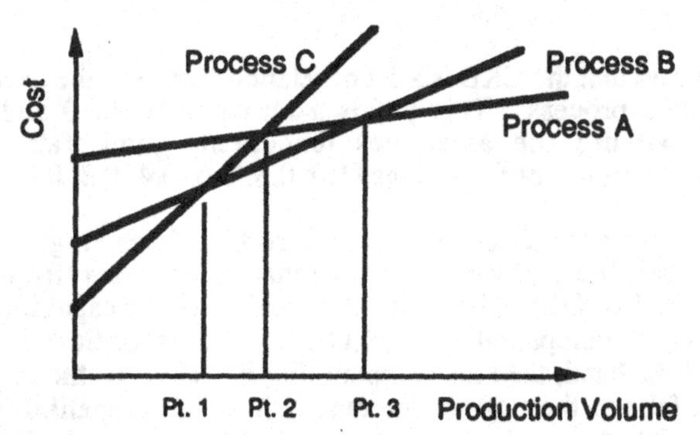

Figure 2. Incremental tooling cost breakpoints.

Most existing process selection procedures depend to a large degree on historical cost data. Such procedures have an inherent problem with maintaining current and accurate data. They are generally not useful for new processes or technologies which have no historical base. Further, accurate cost data only apply to relatively detailed stages of design. The main objectives of our research are to develop a representation scheme for the compatibility measure of a candidate process to the various specifications and to construct a methodology to evaluate compatibility early in the design stage. The information is rather uncertain at the early stages of design. Thus, our program must utilize qualitative, case-based knowledge that address the compatibility of each process with the product specifications.

In the following sections, we will be using the following notations:

X_i = Universe of discourse of the decision factor i
P = Universe of discourse of the process

$$XF = \prod_{i=1}^{n} X_i = X_1 \times X_2 \times \ldots \times X_n \qquad (1)$$

: decision factor space

X = Subset of XF, i.e., $X \subset XF$

3.1 Case-based compatibility knowledge.

The first type of knowledge representation we propose is the case-based compatibility representation, i.e., good, poor, and bad examples of concept geometry and the selected process. An earlier paper on design compatibility analysis (Ishii, et. al, 1988) focused mainly on the qualitative design rules compiled as good and bad templates of design. Each template, called a c-data, has a qualitative rating (good, poor, bad, etc.), justification for the rating, and suggestions for improvement. The qualitative rating is later mapped to a number between [0,1]. The template is grouped by the factor it addresses. The c-data comprise a set of data called the compatibility knowledge-base (CKB).

$$CKB = \{ \text{c-data} \mid \text{c-data} \subset X \times P \times [0,1] \} \qquad (2)$$

Equation 2 shows that the CKB is a set of relations between the decision factors and a candidate process. This yields a rating between 0 and 1. In our application, we use the adjectives [excellent, good, fair, poor, bad, incompatible] to represent the ratings [1.0, 0.8, 0.6, 0.4, 0.2, 0.0].

Let us give an example c-data related to surface finish. Figure 3 shows the surface finish capability of various manufacturing processes (Bralla, 1986). If the user specified surface finish falls in the middle of the capability range of a certain process, the compatibility is "excellent." If the specification corresponds to the edge of the band, then the compatibility is "fair." If the requirement is rougher, i.e., falls to the right of the range, then the compatibility is "poor", since the candidate process has redundant capability. Obviously, if the specified surface finish is finer than the capability band, i.e, falls on the right of the band, the rating will be "incompatible."

This qualitative information yields a set of c-data. Figure 4 illustrates one such c-data that indicates incompatibility between sand casting and surface finish specification of less than 5.0 micro meters. The figure is a screen dump from a knowledge-base maintenance tool in HyperQ/Process.

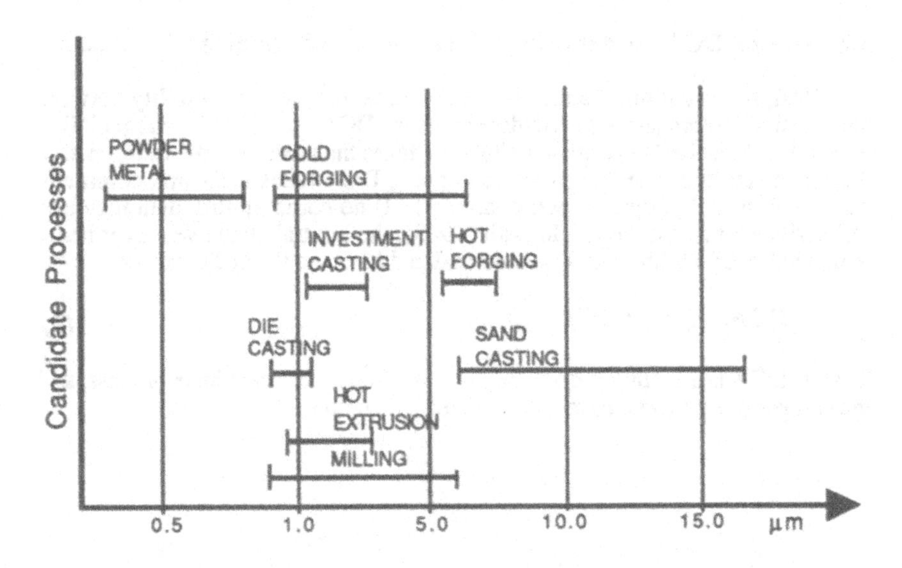

Figure 3. Surface finish capability of manufacturing processes (c.f. Bralla)

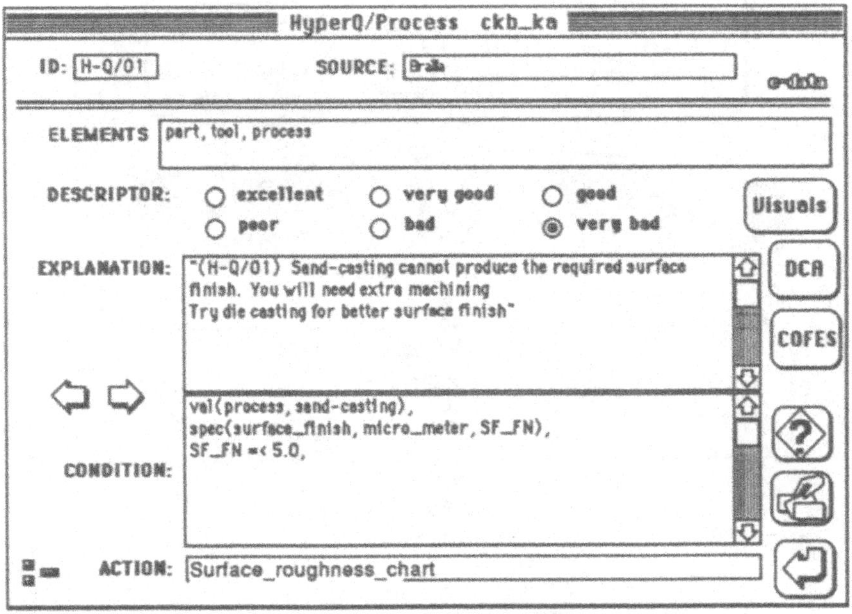

HyperQ/Process ckb_ka

ID: H-Q/01 SOURCE: Bralla

ELEMENTS part, tool, process

DESCRIPTOR: ○ excellent ○ very good ○ good ○ poor ○ bad ◉ very bad

Visuals

EXPLANATION: "(H-Q/01) Sand-casting cannot produce the required surface finish. You will need extra machining
Try die casting for better surface finish"

DCA

COFES

CONDITION: val(process, sand-casting),
spec(surface_finish, micro_meter, SF_FN),
SF_FN =< 5.0,

ACTION: Surface_roughness_chart

Figure 4. Example c-data in a compatibility knowledge-base.

3.2 Use of DCA in deducing a ranked set of candidate processes

Design compatibility analysis (DCA) measures the compatibility between the decision factors and the candidate process. DCA compiles the compatibility object that "matches" a particular situation, takes the most extreme rating (i.e., if there is more than one "negative" comment, DCA takes the worst comment; otherwise, it will adopt the best comment). If no compatibility data matches, DCA gives a neutral value. In Ishii [1988], the neutral status was assigned a match index of 0.5 (the index was normalized between 0 and 1). Hence,

$$\text{DCA: } X \times P \times CKB \rightarrow [0,1] \tag{3}$$

That is, DCA is a mapping from the decision factors, the candidate process, and the compatibility templates to a normalized evaluation.

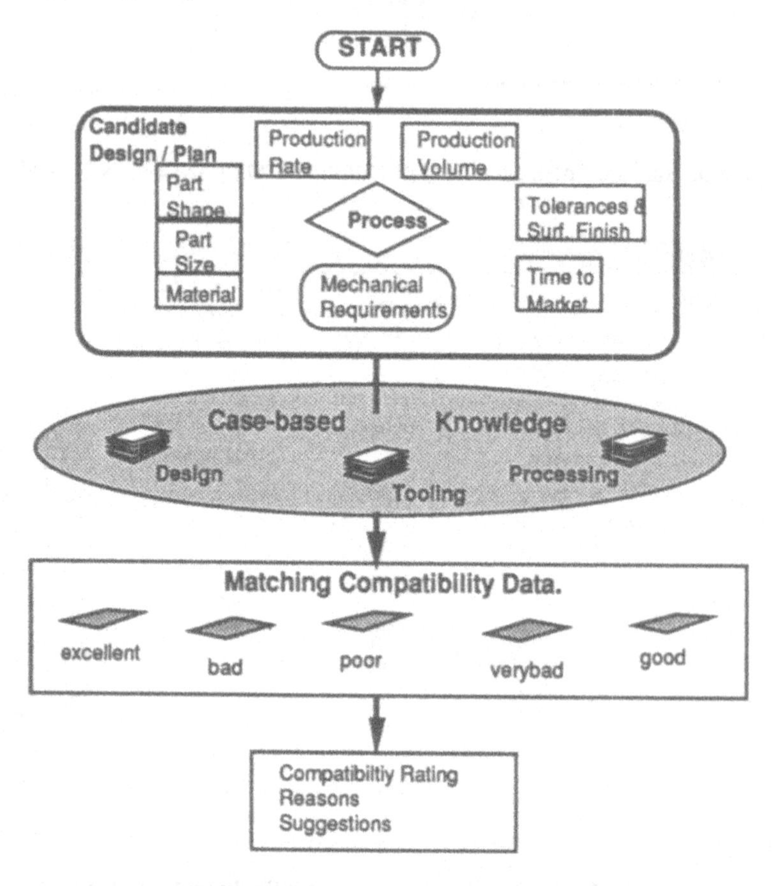

Figure 5. Schematics of Case-based DCA

Figure 5 shows the schematic of DCA using case-based compatibility rules. The figure illustrates the "simultaneous" nature of the evaluation process. Once the user specifies the values of various decision factors such as production volume and surface roughness, DCA scans through the compatibility knowledge-base for each candidate process and determines which c-data matches. In some cases, only positive comments match, while for others, you may find several totally incompatible c-data to apply. Based on the number and the nature of applicable c-data, DCA computes an overall rating. In short, the ratings will be:

1) 0.5 if no c-data matches
2) The rating that corresponds to the worst adjective, if there is at least one negative comment
3) The rating that corresponds to the best adjective, if there is no negative comment.

This method has proven its effectiveness in the very early stages of design, when there is little quantitative cost data available. We view this overall rating to be a normalized estimate of production cost per part. For details of DCA, refer to Ishii et.al. (1988) and Ishii and Nekkanti (1989a).

The idea behind our expert system for the screening of manufacturing processes is to use DCA and rank the candidates. Here are the proposed criteria in using DCA.

1) Eliminate any candidate process that matches with at least one "incompatible" c-data.
2) Compute the DCA compatibility rating of the remaining candidates.
3) Rank the candidates according to the ratings.

This process provides a set of candidate processes. Designers may want to focus on the top two or three candidates and proceed with detailed design based on these processes. As more detailed information becomes available (part geometry, production volume, etc.), designers may wish to employ a more quantitative measure of process compatibility such as cost estimate intervals (Ishii, Ho, and Miller, 1990).

Nevertheless, the screening procedure based on case-based knowledge will provide the designer with a focused view on possible manufacturing processes and encourage him to consider tailoring part designs for the candidate processes. Such consideration will greatly enhance design for manufacturability and reduce life-cycle cost of products under development.

4. IMPLEMENTATION OF HyperQ/Process

We have implemented our proposed procedure using HyperCard and Prolog. The prototype program, called HyperQ/Process takes a structure as illustrated in figure 6.

The current version utilizes Logic Manager, an implementation of Prolog developed by Apple Computer, and has over 100 c-data, in addition to over 30 inference rules related to characteristics of each manufacturing processes.

In addition to the selection feature using DCA, the program incorporates a library of process information for net-shape manufacturing. The information is organized into a HyperCard information stack. Figure 7 shows an example card out of the information stack.

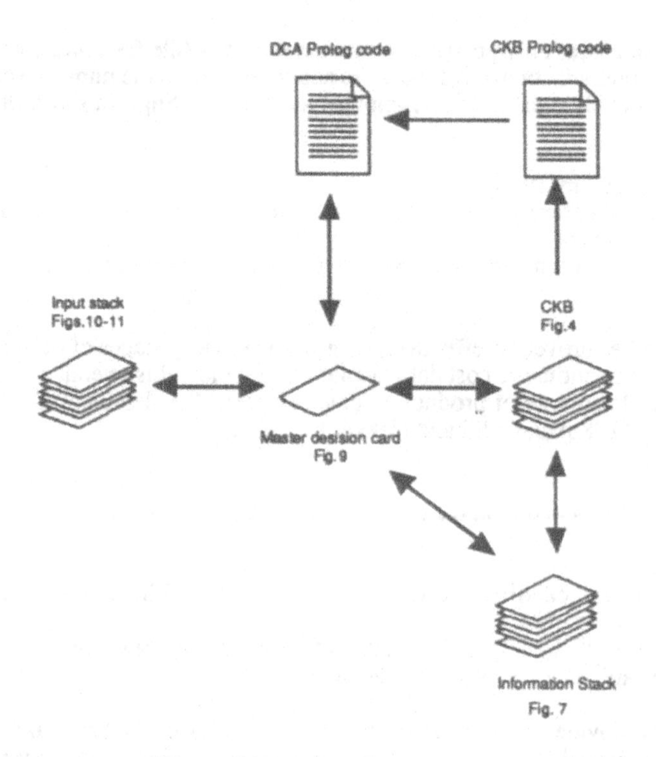

Figure 6. Program Structure of HyperQ/Process

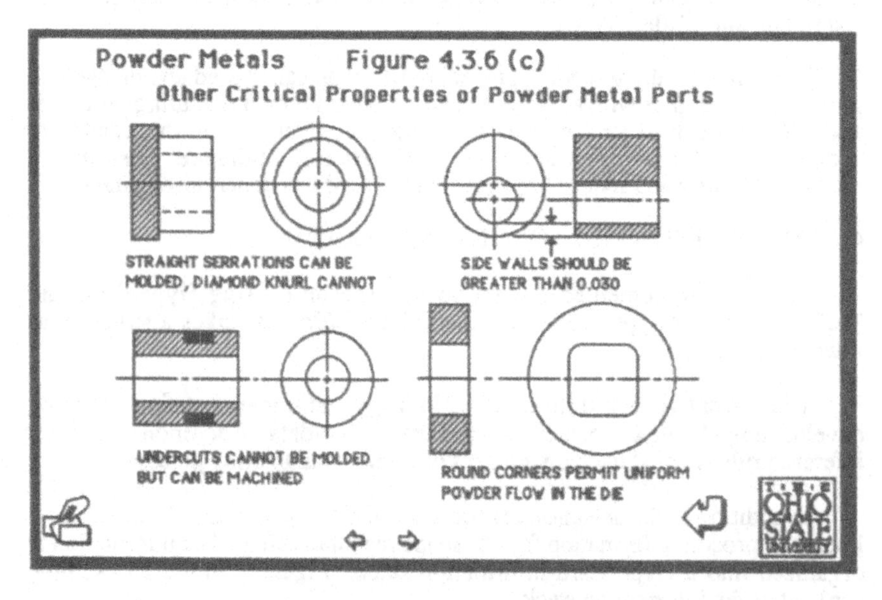

Figure 7. An example card from HyperQ information stack

5. Example run of the program

This section gives an example case of how HyperQ/Process can be used to find an appropriate process for a given part design. Figure 8 shows a proposed design of a part. The specified material is aluminum alloy, the minimum dimensional tolerances is of ± 0.5 mm, and a surface finish specification is 0.8 μm (R_a).

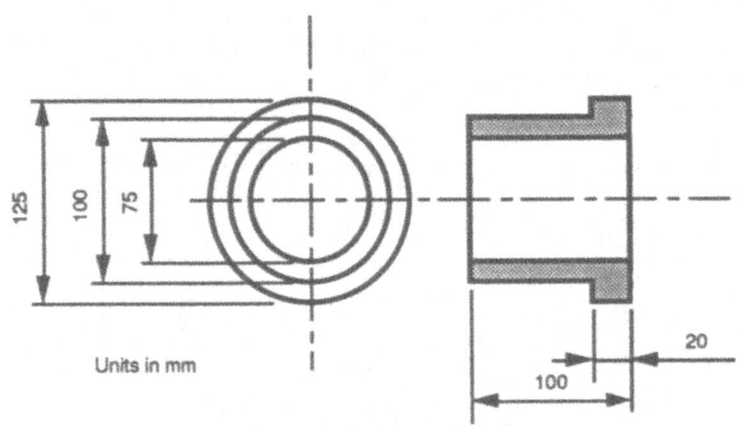

Fig 8. Example Part

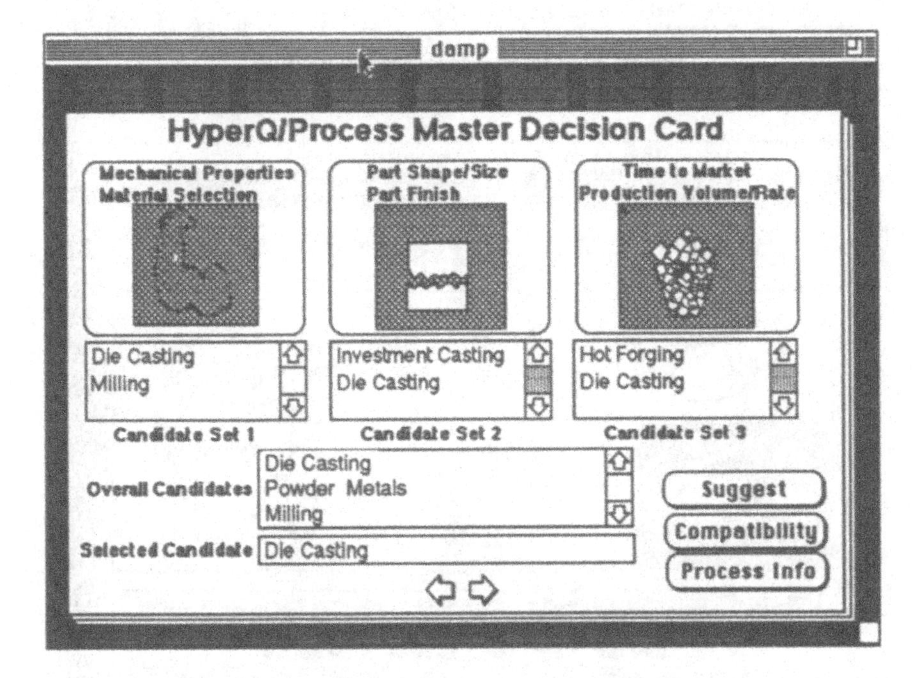

Figure 9. Master Decision Card in HyperQ/Process

Figure 9 shows the master decision card in HyperQ/Process which navigates the user to various specification input modules and help files. This card accepts the user inputs in three modules: 1) material selection and mechanical characteristics, 2) part characterization, and 3) production volume/rate and time to market. HyperQ/Process accepts these inputs through user interaction cards. Examples of these cards are shown in figures 10 and 11.

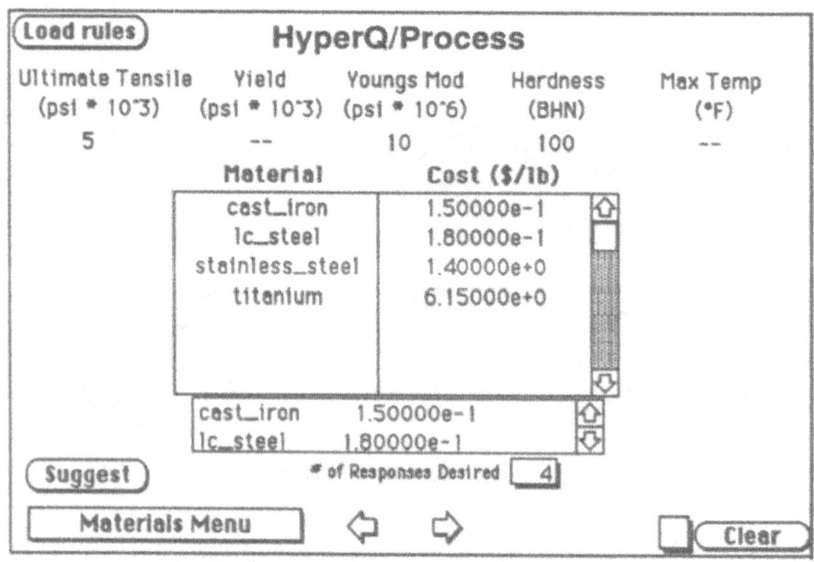

Figure 10. Material specification module

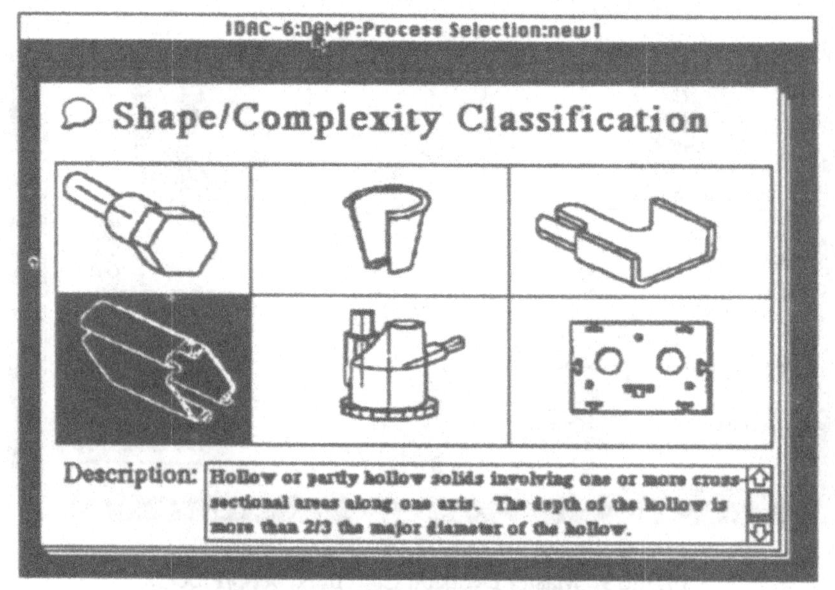

Figure 11. Part and shape classification

After the user specifies these input variables, HyperQ uses the case-based knowledge and DCA to screen each process and suggest alternative processes based on the design factors previously mentioned in section 3.2. Die casting, powder metallurgy, and machining satisfy the dimensional tolerances and the surface finish requirements. DCA deduces three suggested processes and displays the results in an output card as illustrated in figure 12. The information is sent back to the master decision card.

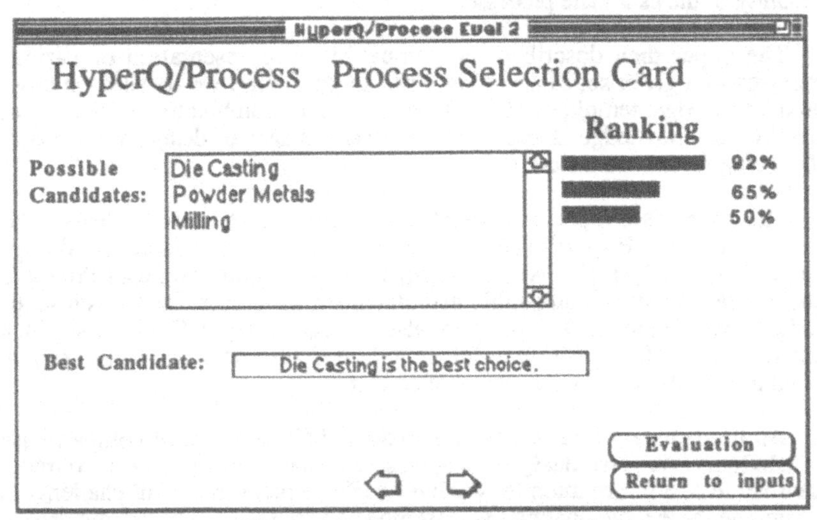

Figure 12. Process selection card

This example illustrates the trade-off between production cost and dimensional accuracy in die casting, powder metallurgy, and machining. Machining can achieve tighter tolerance than die casting but the cost will be much higher for the estimated production quantity. This is because the total tooling cost of the die casting is amortized over a large number of parts even though the initial investment for tooling is large. The machining process has tooling costs proportional to the production quantity even if the initial cost is much lower than for die casting.

This kind of case-based knowledge allows the designer to carefully look back at the design specifications and reconsider whether some of the design factors can be changed without sacrificing any functions of the part. For example, if dimensional tolerance can be relaxed to ± 0.8 mm, then hot forging can be added to the list of candidate processes. Now, if the tolerances were tightened to ± 0.25 mm, die casting would no longer be acceptable. However, a two step process plan with die casting followed by machining may be cost effective. The use of die casting as the primary process would minimize the amount of material to be removed resulting in lower total cost than machining from raw stock. Similar arguments may apply for forging and other processes.

6. CONCLUSION AND FUTURE WORK

This paper described our proposed methodology for process selection that applies to early stages of product design. We focused on net-shape manufacturing processes and identified the major factors that affect the selection of an appropriate process. Some of these decisions are iterative. The sequence in which designers typically base decisions depends largely on the nature of the product and the development environment. Thus, we concluded that our methodology should consider all the factors *simultaneously* in assessing the suitability of the candidate processes.

The paper then described the compatibility representation of various processes to a given set of specifications. In particular, we focused on case-based knowledge: templates of good, bad, and poor combination of decisions. This type of knowledge is appropriate for early stages of design when many influencing factors are uncertain.

This case-based representation of process selection knowledge lead to the development of HyperQ/Process, an expert system based on design compatibility analysis (DCA). This HyperCard-based program uses Prolog to deduce a ranked set of compatible manufacturing processes for a given set of product specifications. The program also includes a HyperCard stack which stores process information in an object-oriented fashion and allows users to search though the stack using a navigation map.

An immediate future task is investigate a different form of compatibility knowledge. As part designs progress, engineers have access to more quantitative cost information for various candidate processes. Our challenge is to combine these more quantitative cost models with the case-based, qualitative compatibility measures as designs become more detailed and quantitative measures become available. Ishii, Ho, and Miller (1990) outlines our program of research in this direction.

Another aspect we must consider is the modularity of parts. A proposed part may at times be more efficiently produced if it were broken up into two parts and manufactured separately. This question of modularity is a tradeoff between assembly cost and production cost per part. Yet another factor is the combination of processes. Rather than producing a working component in one process, a combination of net-shape process and a finishing process may yield more efficient overall production while maintaining the specified quality. The key here is the overall optimization of production cost that takes into account not only the component manufacturing cost but also assembly, service, and perhaps even recycling cost.

Our long range goal is to develop an integrated design assessment tool for net-shape manufacturing. In the early stages of design, the approach outlined here only requires classification of part geometry. As the design progresses and as designers seek a more accurate cost estimate, our tool needs to account for the shape complexities and detailed geometry of the part. The level of details will require us to integrate our method with geometry modeling environments.

Acknowledgement

This research was support by the US Army Material Command through the NSF / ERC at Ohio State University. Partial funding also came from the NSF DTM 888810824. The compilation into HyperCard of the process information is due to Chris Stevens and George Nicoloulias. Many thanks to Steve Weyer and Ruben Klienman at Apple's Advanced Technology Group who helped us in implementing our system with Logic Manager.

References

Barkan, P. et.al.. (1990) Competitive Product Design for Manufacturability. Book in preparation, to be published by MacMillan Publishers.

Boothroyd, J. and Dewhurst, P. (1983) Design for Assembly: A Designers Handbook. Boothroyd Dewhurst Inc., Wakerfield, Rhode Island.

Bralla, J.G. (1986) Handbook of Product Design for Manufacturing. McGraw-Hill Book Company.

Bolz, R.W. (1974) Production Processes; The Productivity Handbook. Conquest Publications, Novelty, OH 44072.

Cutkosky, M.R., D.R. Brown, and Tanenbaum, J.M. (1989) Extending Concurrent Product and Process Design Toward Earlier Design Stages. Proceedings of the Concurrent Product and Process Design Symposium, ASME Winter Annual Meeting, December, 1989, San Francisco, CA. pp. 65-73.

Eary, D. and Johnson, G.E. (1962) Process Engineering for Manufacturing. Prentice-Hall.

Ishii, K., Adler, R, and Barkan, P. (1988) Application of Design Compatibility Analysis To Simultaneous Engineering. *Artificial Intelligence in Engineering Design and Manufacturing (AI EDAM).* Vol. 2, No.1

Ishii, K. and Nekkanti, R. (1989a) Compatibility Representation of Knowledge About Net-Shape Manufacturing. Proc. of the ASME Design Theory and Methodology Conference.

Ishii, K., Hornberger, L, and Liou, M. (1989b) Compatibility-based Design for Injection Molding. Proceedings of the Concurrent Product and Process Design Symposium, ASME Winter Annual Meeting, December, 1989, San Francisco, CA. pp. 153-160.

Lai, K., and Wilson, W.R.D. (1985) Computer-Aided Material Selection and Process Planning, Proceeding of North America Manufacturing Research Conference, S.M.E., May 19-22, 1985, U.C. Berkeley, CA., 505-508

Ludema, K.C., Caddel, R.M., and Atkins, A.G. (1987) Manufacturing Engineering; Economics and Processes. Prentice-Hall.

Maloney, L.M., Ishii, K., and Miller, R.A. (1989) Compatibility-based Selection of Forging Machines and Processes. Proceedings of the Concurrent Product and Process Design Symposium, ASME Winter Annual Meeting, December, 1989, San Francisco, CA. pp. 161-167.

Liou, S.Y. and Miller, R.A. (1991) Design for Die Casting. *International Journal of Computer Integrated Manufacturing*, to appear.

Niebel, B.W. (1966) An Analytical Technique for the Selection of Manufacturing Operations. *The Journal of Industrial Engineering*. Vol. 9, No. 11. pp. 598 - 603.

Poli, C, Graves, J. and Sunderland, J.E. (1988a) Computer-aided Product Design for Economical Manufacture. Proc. of the ASME Computers in Engineering Conference, Vol.1, pp 23-27, San Francisco, July, 1988.

Poli, C and Fernandez, R. (1988b) How Part Design Affects Injection Molding Tool Costs. *Machine Design*, November, 24. pp. 101-104.

Shah, J., Hsiao, D., Robinson, R. (1990) A Framework for Manufacturability Evaluation in a Feature Based CAD System. Proc. of NSF Design and Manufacturing Systems Conference, Tempe, Arizona, Jan. 1990. pp. 61-66.

Trucks, H.E. (1987) Designing for Economical Production. SME Marketing Division, Dearborn, Michigan.

Ishii, K., Lee, C. H., and Miller,R. A. (1990) Methods for Process Selection in Design. Proceedings of the ASME Design Theory and Methodology, September,.1990,. Conference, Chicago,.Illinois.

Neural Networks in the Colour Industry

J.M. Bishop (*), M.J. Bushnell (*), A. Usher (**), S. Westland (***)

(*) Department of Cybernetics, University of Reading, Berkshire, UK

(**) Courtaulds Research, Spondon, Derby, UK

(***) Department of Communication & Neuroscience, University of Keele, Staffs., UK

ABSTRACT

In the past ten years there has been an explosion of academic interest in Neural Network research, yet the techniques are still viewed with some suspicion by many engineers faced with real world problems. The purpose of this paper is to illustrate how a simple neural network is being used to help solve a difficult physical problem. The work, sponsored by Courtaulds Research, involves colour recipe prediction. It is a difficult problem to solve using conventional computer techniques as the model that is most widely used (Kubelka-Munk theory) breaks down under a variety of conditions. The paper will discuss several of the design decisions, common to many neural network applications, that have been made in the process of developing the Courtaulds Recipe Prediction System.

INTRODUCTION

Colour control systems based on spectrophotometers and microprocessors are finding increased use in production environments. One of the most important aspects of the quality control of manufacturing processes is the maintenance of colour of the product. The use of colour measurement for production and quality control is widespread in the paint, plastic and dyed textile industry but is also prevalent in many other areas including food stuffs. An industrial colour control system will typically perform two primary functions relating to

the problems encountered by the manufacturer of a coloured product. Firstly the manufacturer needs to find a means of producing a particular colour. This involves selecting a recipe of appropriate dyes or pigments which when applied at a specific concentration to the product in a particular way, will render the required colour. This process is known as recipe prediction and is traditionally carried out by trained colourists who achieve a colour match via a combination of experience and trial-and-error. Instrumental recipe prediction was introduced commercially in the 1960's and has become one of the most important industrial applications of colorimetry. The second function of a colour control system is the evaluation of colour difference between a batch of the coloured product and the standard on a pass/fail basis.

The first commercial computer for recipe prediction[1] was an analog device known as the COMIC (COlorant MIxture Computer) but all colour systems on the market today employ digital computers. A typical colour control system consists of a reflectance spectrophotometer connected to a PC-based machine with various peripherals and costs in the region of £20,000 - £50,000. All computer recipe prediction systems developed commercially to date are based on an optical model that relates the concentrations of individual colorants to some measurable property of the colorant in use (e.g. reflectance). The model must also describe how the colorants behave when used in mixtures with each other.

The model that is almost exclusively used is known as the Kubelka-Munk theory[2]. It relates measured reflectance values to colorant concentrations via two terms K and S, which are the Kubelka-Munk version of the absorption and scattering coefficients of the colorant. The Kubelka-Munk theory is a highly simplified version of rigorous radiative-transfer theory[3] whereby only two fluxes of radiation are considered. Attempts have been made to introduce more complex theories by allowing the use of three or more fluxes[4], but the application of these more complex theories is generally not practical[5]. The use of the exact theory of radiation transfer is not of practical interest to the coloration industry.

The use of the conventional two-flux Kubelka-Munk theory has attracted criticism[6]. The popularity of the Kubelka-Munk equations is undoubtedly due to their simplicity and ease of use. The equations give insight and can be used to predict recipes with reasonable accuracy in many cases. In addition the simple principles involved in the theory are easily understood by the non-specialist. However in order for the Kubelka-Munk approximation to be valid a number of restrictions are assumed[6].

There are many applications of the Kubelka-Munk theory in the coloration industry, where these assumptions are known to be false. In particular,the applications to thin layers of colorants, for example, lithographic printing inks[7] and fluorescent dyestuffs[8,9] have generally yielded poor results.

NEURAL NETWORKS AND RECIPE PREDICTION

The performance of the Kubelka-Munk theory in certain areas of coloration is such as to warrant an alternative approach. An empirical colorant mixture model has been suggested[10] using high-order polynomial models but the accuracy of such a model will clearly be influenced by the exact choice of polynomial. The trained colourist accumulates experience of the behaviour of the colorants and is able to extrapolate and interpolate from this experience to predict recipes for new shades without the use of Kubelka-Munk theory or any other algorithmic model. Artificial Intelligence techniques are now beginning to be able to emulate the performance of human operators or experts in many areas of science and engineering and the use of one such technology, Neural Networks, is being investigated on the recipe prediction task.

It was expected that the Neural Network approach would provide a novel and profitable new solution to the recipe prediction problem. It was hoped that a suitable network system would be able to automatically learn relationships between colorants and colour, and hence learn to predict which colorants, and at which concentrations, need to be applied to a particular substrate in order to produce a specified colour. The preliminary results of this work have been published elsewhere[11,12]. In the course of this evaluation, several problems, common to many potential network applications, were met. This paper will highlight some of these and the steps that were taken to overcome them.

NEURAL NETWORK METHODS

Neural Networks consist of collections of connected processing elements that are not individually programmable. Each element usually computes a simple non linear function f on the weighted sum of its input (Figure 1). The output of this function is defined as the *activation* of the cell. Long term knowledge is stored in the network in form of interconnection weights linking a network of these cells. Unlike a conventional computer solutions to specific problem, a neural network is not explicitly programmed to complete a given task, instead it acquires knowledge over time by adapting the strengths of the interconnections between cells. It is this ability to learn the solution to a problem that enables the use of network models on computationally ill-defined systems.

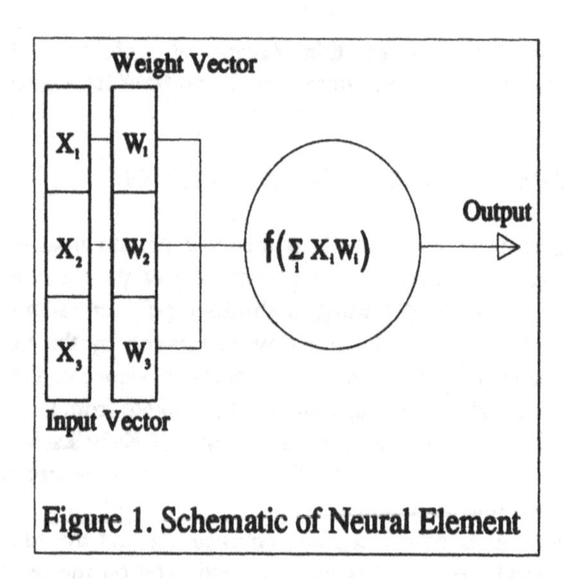

Figure 1. Schematic of Neural Element

The network model that has been used with this research is a multi-layer feed forward network architecture, taught using the Generalised Delta rule[13]. This works by performing gradient descent in Error/Weight space. That is, after each pattern has been presented, the resulting error on that pattern is computed, by comparing the actual output with the desired output, and each weight in the network modified by moving down the error gradient towards its minimum for that input/output pattern pair. Gradient descent involves changing each weight in proportion to the negative of the derivative of the error, defined by this pair.

Information enters the network in the form of a vector of real values, the CIELAB colour coordinates, applied to the input layer of the network. The network output is the vector of real values, the dye concentrations, defined by the activation values of the output layer. The network is trained by the repeated presentation of a list of such input/output vector pairs. The task of the Network is to learn a set of mappings between its input and output, such that the sum of the output errors squared, is minimised. The error, for each pattern in the training set, is defined as the difference between the actual network output and the desired output.

The Network architecture used in these experiments consisted of an input layer where cell inputs are clamped to external values (scaled CIELAB values), a number of hidden layers, where cell inputs are defined by the weighted activation values from the cells in the previous layer that they are connected to, and an output layer connected to the last hidden layer (Figure 2).

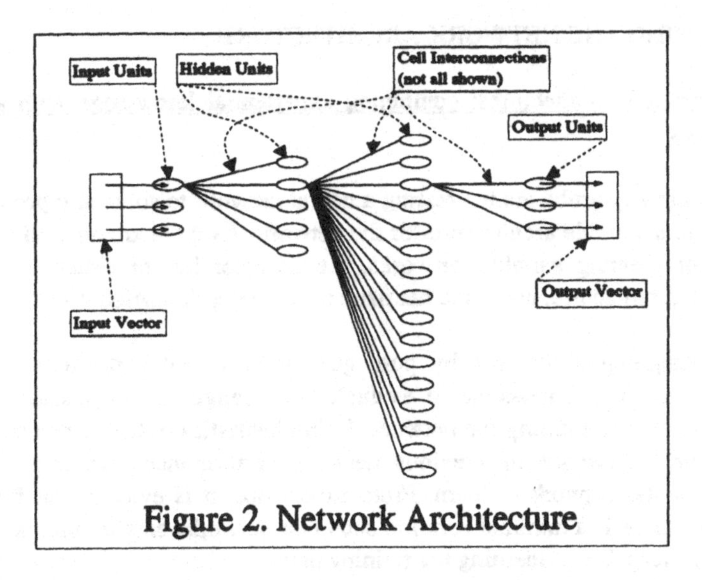

Figure 2. Network Architecture

In one learning sweep with the Generalised Delta Rule, the following sequence of events occur;

i: The network is presented with an input pattern (all the input units of the network are set to the required values).

ii: This input vector is used to compute the output values by feeding forward through the net and computing activation values for all the other units according to the weight values.

iii: The output vector for this pattern is compared to the required, or target, pattern and the error term is calculated for every output unit.

iv: Error terms are recursively propagated backwards through the net to the other units in proportion to the connection strengths between the units.

v: The weights are then adjusted in such a way as to reduce the error terms, by performing gradient descent in weight error space.

vi: The process is repeated for all the input/target pattern pairs in the training set.

The process of presenting to the network all the patterns over which it is to be trained is defined as an epoch. Training continues for as many epochs as are necessary to reduce the overall error to an acceptably low value.

DESIGNING THE NETWORK ARCHITECTURE

Hand crafted architectures: combining experiential knowledge with simple heuristics

One of the key problems in creating a neural network to solve any problem is to design a suitable architecture for the network. As there are no well defined rules for creating suitable networks, the designer has to either "guess" a network architecture, use some heuristic or use an optimisation process.

Designing a network by pure guesswork is not a profitable design strategy, as it is impossible to evaluate the strength of the chosen design without actually training the network. Using heuristics in the design strategy, based on experience of previous network performance, can improve the ability of the network to learn. From experience, it is evident that for each problem there is a minimal solution such that networks smaller than a certain size are incapable of learning the training data.

It has been shown that any classification task can be solved by a two layer back propagation network using a single layer of hidden units[14]. However, if the number of hidden units exceeds the number of training patterns then the network may store knowledge in a very localised manner, with each hidden unit responding strongly to one training pattern. Clearly, this type of network is impractical when there were many training patterns. In addition, such a network is less able to perform useful generalistion as each unit becomes highly tuned to one training pattern. What is required is a multi-layered network which distributes knowledge throughout the net, such that the training patterns are represented by a pattern of activation across the network. Using such an architecture, the network should be capable of generalising over unseen inputs to produce meaningful outputs.

A useful heuristic in the case where the network is required to generalise is that networks with multiple layers and fewer units in the first layer generalise better than shallow networks with many units in each layer[15]. A further heuristic is that the network should be larger than the minimum necessary to perform the task (ie. larger than that required to just learn the training data).

Even with the use of such prior knowledge, the eventual network design is still largely a product of "inspired guesswork". It may be necessary to design many such networks by hand in order to generate one useful net. What is required is an automated method of network design which is capable of optimising network parameters such as the number of layers and the number of neurons needed in each layer.

Problems with generalisation. Using the design heuristics described above, it is possible to produce networks that can learn the training data reasonably well, (total summed squared error (TSS) of the order 10^{-3} to 10^{-4}). However, such networks often failed to generalise well for unseen data - in the case of the recipe prediction system, failing to accurately predict recipes that they have not been taught. Clearly, a network that can only predict recipes which are already known is not of practical value!

Genetic optimisation of network architecture

A programming strategy known as Genetic Algorithms has been used to optimise the design of network architectures[16]. Genetic algorithms use evolutionary techniques to perform an optimising search whose search space is defined by a binary coding of the parameters to be optimised. In the case of neural networks, the genes can represent the numbers of neurons in each layer of the network. An initial population of networks is generated randomly and tested to obtain a measure of the merit of each initial net. The test involves training each network for a short period, in relation to the usual total training time, to obtain a measure of the total error over all the training patterns. This error is then used to calculate a merit function for the net. In the case of the recipe prediction problem, the training period chosen was 20000 epochs, which is short compared to the usual training period of 200000 epochs but long enough for the network to begin to converge if a solution can be found.

Once the initial population has been created and tested, evolutionary techniques are used to improve the population. At each generation, a parent network is chosen from the population using a *weighted roulette wheel* such that each net has a probability of being chosen proportional to its merit after training. The genetic string is then manipulated by one of a number of operations (also chosen on a weighted random basis) to produce a child. The offspring is tested in the same way as the initial population to determine its merit. If the new network is better than the worst network in the population then it survives and the old network is removed from the population.

The possible operations performed on the parent string to produce a daughter string are as follows:

a) **Inversion:** Two sites are chosen randomly within the string to determine how much of the string is to be inverted. Then, the portion of the string defined by these random sites is simply reversed. For example, if the parent string was 0110010110 and the sites chosen were 3 and 7 the offspring created by inversion would be 0101001110 after inverting the 10010 portion of the string.

b) **Crossover:** This operation involves two parent strings, so a second parent is chosen in the same manner as the first, using the weighted roulette wheel approach. A crossing site is selected and the offspring is created by splicing the beginning of the first parents string, up to the crossing site, with the end of the second parents string from the crossing site. For example, if the first parents gene string was 111111 and the second parents string was 000000, using a crossing site of 4, the new offspring would be 111100.

b) **Mutation:** This operation has a very low probability, as it does in nature, but is the simplest of the three operations. A mutation site is chosen at random the the bit value at that site is simply changed: a 1 to a 0 or vice versa.

Although all these operations and choosing mechanisms rely on random numbers, a population acted upon in this way soon improves. This is due to the fact the the parent genes are more likely to be those who gave rise to better networks and that poor genes are killed off by new offspring, where such offspring produce improved networks. Thus the entire population tends to improve.

In this way, after a number of generations the population will consist almost entirely of very similar networks which are all capable of solving the problem. This is a direct result of the optimisation strategy which ensures survival of the fittest and also propagates the best characteristics of those fittest into new generations.

THE EFFECTIVENESS OF A NEURAL NETWORK RECIPE PREDICTION SYSTEM

Performance using hand crafted neural network architectures

The first results to suggest that neural networks might be capable of performing recipe prediction came from a set of training data synthesised by an ICS-Texicon Colour system. The colour coordinates of a selection of recipes were calculated according to standard Kubelka-Munk theory using the ICS-Texicon colour measurement system. The sets of dye concentrations and colour coordinates were then treated as real world data. About half the data obtained in this way was used to train a network whose hidden layers comprised of 8 units followed by 16 units. The input data to the network consisted of CIELAB colour co-ordinates expressed in a three dimensional cartesian format. Here, the L value refers to the lightness of the colour, A to the redness/greeness of the colour (positive A refers to redness and negative A to greeness), and B refers to the yellowness or blueness of the colour. The outputs from the network corresponded to the concentrations of the three dyes used in the experiment. All the numbers involved were scaled to the range 0.00 to 1.00, limited by the range of the sigmoid activation function used in the network.

The results obtained in this way were reasonable, with approximately 60% of all the predictions having a ΔE value (a measure of colour difference) of less than one (0.8 is the figure normally used in making pass/fail decisions on colour samples). In most of the cases where ΔE was large, it was observed that the prediction required one or more of the dyes to have zero concentration. There is an inherent difficulty when using the network to learn mappings which require some of the output units to achieve zero activation, since the sigmoid activation function needs extremely large negative inputs to achieve zero output.

To reduce this problem, all further experimentation involved scaling the output data to fit the interval 0.1 to 0.9, reducing the need to learn values at the extremities of the activation function. Experiments performed using this scaling mechanism showed a significant improvement, with approximately 80% of all predictions having ΔE values of less then 0.8. This order of performance is similar to that obtained by conventional recipe prediction systems.

It is interesting to note that even with the improved scaling mechanism, the network still predicted multi-dye recipes much better than single or two dye recipes, having a failure rate of only 6.5% for three dye recipes. This situation is the reverse of that which occurs in conventional instrumental colour systems, in that the approximated Kubelka-Munk theory works better when few dyes are involved.

<u>Performance using a genetically optimised network architecture</u>

An optimised network was created by running the genetic algorithm program under the following conditions. The initial population consisted of ten genetic strings which were used to create the networks. These networks were each tested by running under the neural network program for 20000 cycles. The genetic algorithm then ran for 20 generations, with each generation creating 2 new offspring and removing the two worst networks from the population. Finally, the best network found by the genetic algorithm was trained for a further 60000 cycles and compared with the best hand-crated network trained for the same total number of cycles (80000).

Although in the initial stages of training the hand-crafted network had a smaller error the the genetically optimised one by the end of the 80000 cycles, the G.A. produced network performed much better than the hand-crafted one. The G.A. network had a final error of 0.000427 while the hand crafted network had an error of 0.003931, almost 10 times worse. Unfortunetely neither network was effective enough when predicting untaught recipes (most ΔE values being significantly greater than 0.8). Although networks have been able to predict synthesised recipes well[11,12], generalisation still remains a significant problem when dealing with real data. Current research involving adding gaussian noise to the training data[17], is aimed at overcoming this problem.

CONCLUSION

Neural Networks could potentially be a very useful tool for the engineer faced with a problem that is not computationally well defined. Results using a simple back propagation network on one such problem from the Colour Industry, have demonstrated that neural network techniques can potentially be used to solve recipe prediction problems. It has been shown that the Kubelka-Munk model has been approximated without any a priori knowledge about the system. There is no reason to believe that similar neural networks cannot learn the relationship between colorant concentrations and colour coordinates for practical coloration systems, and preliminary results from ongoing research indicate that this is indeed the case[11,12].

In the field of Colour Recipe Prediction, the use of neural networks offers several potential advantages over the conventional Kubelka-Munk approach.

i: The network can be trained on real production samples. Most dye-houses, for example, maintain historical shade data and this would be most suitable for training. The conventional Kubelka-Munk systems necessitate the preparation of special data base samples and up to ten samples per colorant is not unusual.

ii: By allowing the network to continue to learn after the initial training period, it will have the potential to adapt to changes in the production process, in a similar manner to the way that a colourist would adapt to such changes over time.

iii: The neural network approach may be able to learn the behaviour of colorants for coloration systems for which the mathematical descriptions are complex. For example, fluorescent dyes and metallic paint systems, are currently difficult to treat using standard Kubelka-Munk theory.

However the use of Neural Networks is not without problems. At present, there is no well defined set of heuristics that enable the engineer to specify an optimal network architecture and it may be difficult or impossible to learn the solution to a particular problem using an unsuitable architecture. Initial research involved the use of hand designed network architectures. Many of the designs did not learn the colour data at all, and the performance of those that did was not perfect. Recent research has involved the use of Genetic Algorithms to optimise network They have several advantages and disadvantages;

i. They are very computer intensive, taking many days of CPU time on a SUN Sparc Station to converge.

ii. They can produce networks that over learn the data - that is they are able to reproduce the training data very accurately but are unable to generalise over unseen data.

iii. Networks designed by GA's are inherently very stable. That is, learning is insensitive to the initial random weights used in a given training run.

Other recent developments that have been reported in network design include the use of networks that dynamically change their topology[18] and the use of Gaussian noise to improve generalisation performance[17]. The use of both these techniques on the recipe prediction problem is ongoing.

REFERENCES

1. Davidson, H.R., Hemmendinger, H. & Landry, J.L.R. A System of Instrumental Colour Control for the Textile Industry, Journal of the Society of Dyers and Colourists, Vol.79, pp. 577, 1963.

2. Judd, D.B. & Wyszecki, G. Color in Business, Science and Industry. 3rd ed., Wiley, New York, 1975, pp. 438-461, 1975.

3. Chandrasekhar, S. Radiative Transfer. Clarendon Press, Oxford, 1950.

4. Mudgett, P.S. & Richards, L.W. Multiple Scattering Calculations for Technology. Applied Optics, Vol.0, pp. 1485-1502, 1971.

5. Mehta, K.T. & Shah, H.S. Simplified Equations to Calculate MIE-Theory Parameters for use in Many-Flux Calculation for Predicting the Reflectance of Paint Films. Color Research and Application, Vol.12, pp. 147-153, 1987.

6. Nobbs, J.H. Review of Progress in Coloration. The Society of Dyers and Colourists, Bradford, 1986.

7. Westland, S. The Optical Properties of Printing Inks. PhD Thesis, University of Leeds, (UK), 1988.

8. Ganz, E. Problems of Fluorescence in Colorant Formulation. Colour Research and Application, Vol.2, pp. 81, 1977.

9. McKay, D.B. Practical Recipe Prediction Procedures including the use of Fluorescent Dyes. PhD Thesis, University of Bradford (U.K), 1976.

10. Alman, D.H. & Pfeifer, C.G. Empirical Research and Application. Vol.12, pp. 210-222, 1987.

11. Bishop, J.M., Bushnell, M.J. & Westland, S. The Application of Neural Networks to Computer Recipe Prediction. Color, Vol.16, No.1, pp.3-9, (USA), 1991.

12: Bishop, J.M., Bushnell, M.J. & Westland, S. Computer Recipe Prediction Using Neural Networks. Proc. Expert Systems '90. (London), 1990.

13. Rumelhart, D.E., Hinton, G.E. & Williams, R.J. Learning Internal Representations by Error Propagation. in D.E.Rumelhart, J.L.McClelland and the PDP Research Group (Eds), Parallel distributed processing: Explorations in the microstructure of cognition: Vol.1, Foundations. pp.318-362. MA: Bradford Books/MIT Press.

14. Funahashi, K. On the approximate realization of continuous mappings by neural networks. Neural Networks, Vol.2, No.3, pp.183-192, 1989.

15. Rumelhart, D.E. Parallel Distributed Processing. Plenary Session, IEEE Int. Conf. Neural Networks, San Diego, CA., 1988.

16. Dodd, N. Optimisation of Network Structure using Genetic Techniques. Proc. INNC. '90. Paris, 1990.

17. Sietsma, J & Dow, R.J.F. Creating Artificial Neural Networks That Generalize. Neural Networks, Vol.4, No.1, pp.67-79, 1991.

18. Hirose, Y., Yamashita, K & Hijiya, S. Back-Propagation Algorithm Which Varies the Number of Hidden Units. Neural Networks, Vol.4, No.1, pp.61-66, 1991.

The Design of a Simulator System for Educating Engineers

M. Quafafou (*), O. Dubant (*), J.P. Rolley (**), P. Prévot (***)
(*) EMA/CSP, 6, Av. de clavières 30100, Alès Cedex, France
(**) EMA/IMGM, 6, Av. de clavières 30100, Alès Cedex, France
(***) INSA de Lyon, 502, Informatique, 69621 Villeurbanne Cedex, France

ABSTRACT

This paper describes the design of a system used to assist Engineering Schools in learning mining prospecting techniques. The instruction goal is accomplished by simulation of exploration drillings. Different aspects of this simulation process are separated into specialized components. Benefiting from the development of the use of computers in education, the design of our system, which is currently developed in C++, is based on the traditional architecture of Intelligent Tutoring Systems (ITS). We exploit the mechanisms of classes for coping with the complexity of programming and for using inheritance in description of geological knowledge. The discussion centres around shape modelling, representation of geological knowledge, specification of student requests and his principal tasks. Before concluding we discuss the possibility of evolving towards an intelligent system.

INTRODUCTION

The use of computers in education has made possible the development of systems for Computer Aided Instruction (CAI). A distinguishing feature of these systems is a one-way teaching interaction; the system presents a question or a problem to the student and gets his answer. Then the system assesses and comments upon it. Other studies have taken an interest in the problem of having a computer which understands its user and makes intelligent decisions during their interaction [20]. So, from research on Artificial Intelligence (AI) and Cognitive Psychology, new systems have been developed, called Intelligent Computer Aided Instruction (ICAI). These systems are characterised by a significant application of AI methodologies and techniques, called more recently Intelligent Tutoring Systems (ITS).

On the other hand, the teaching of mining prospecting techniques has been studied since the seventies [19, 12, 8, 3]. For the most part, those examples entail establishing links between theorical training and concrete application. The mining prospecting process has become more expensive and its elaboration implies choices which are often irreversible. Hence, the interest in familiarizing the student with the different phases of deposit exploration.

Some software tools has resulted from the extensive cooperation between the Mining School of Paris and Lausanne University. At present, they are also used at the Mining School of Alès. The instruction goal is to make the student aware of the phenomena which are involved in the estimation of deposit reserves and feasibility problems.

Benefiting from the development of the use of computers in education, we are studying the problem of teaching the mining prospecting process using simulation techniques. So, the instructor constructs a synthetic geological site, and the student tries to discover deposits which are economically exploitable by analysing data which results from exploration drillings. The student interacts with an environment where the conditions resemble practical ones, i.e. natural conditions of the geological site or socio-economic conditions. Doing this, we hope to familiarise the student with both deposit research strategies and decision methods.

In order to conceive a flexible system for instruction by simulation we have to take into account both geological knowledge and complexity reasoning. In fact, the geological knowledge used by the system is complex and presents different aspects :

• *geometric* : mathematical models for representing different geological shapes or the distribution functions of mineral.

• *semantic* : different varieties of rocks, mineral and natural phenomena.

• *space* : both the description of a geological site and the exploitation drillings handle information which describes spacial properties.

In addition, the diversity and the uncertainty of the interpretations of given geological situations (results of drillings) make deposit research strategies very complex.

The goal of our project is the development of tools for learning mining prospecting techniques. We have used the classical architecture of ITS which contains four traditional components : a domain expert, a teaching expert, a student model and an interactive component [4]. We exploit the mechanisms of classes for coping with the complexity of programming and for using inheritance in the description of geological knowledge. So, different hierarchies are used for representing the aspects of geological data and their relationship.

The efficacity of the simulation is influenced by the capacity of the system to exploit geological objects and to deal with data, hence the interest of the geometrical component. So, we have built an experimental modeling package based on generic procedural models.

In this paper, we firstly sketch the design of the system then, after a quick overview of the geometrical problem, we discuss the specificity of the geological knowledge. We propose an architecture of the system taking into account both the different components of ITS and the simulation of exploration drillings.

We have studied (and continue to expand) the problem of the student model. At present, the criticism of the student strategies is carried out by the diagnosis commitee which consists of the instructor, the geologist and other experts. In the last section we analyse the different actions of the student. We analyse also the possible evolution towards an intelligent system by giving hints on further developments of the student model. We have developed our prototype in C++.

THE GENERAL ARCHITECTURE

In this section we briefly overview the traditional architecture of ITS and the different phases of the simulation of the mining prospecting process. This succinct introduction may be useful in clarifying some of the issues involved in the construction of our system. Then, we expose the analytical we are following for the development of our system.

The traditional components of an ITS

The classical architecture of ITS is described in figure 1. It consists of :

- *an expert module* : which contains the field of knowledge and the reasoning mechanisms;

- *a student model* : which contains a description of the learning status of the student. It is a set of information about the student;

- *pedagogical module* : which essentially embodies the teaching expertise of the system. This module uses and updates the student model;

- *a user interface.*

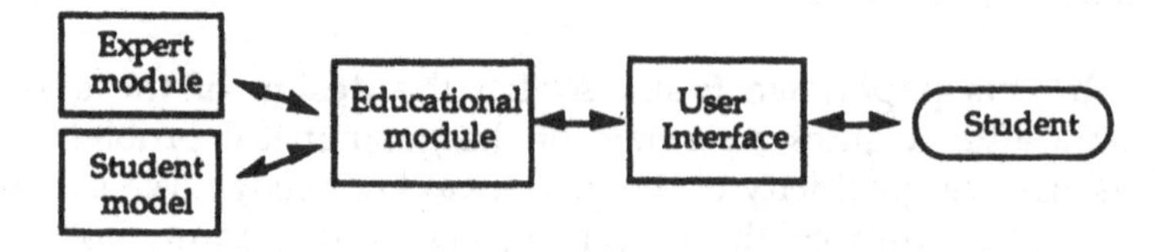

Figure 1. Classical architecture of ITS

The diagrams of ITS presented in literature are sometimes different from these shown in figure 1. In fact, the components can have other names or other functions [1, 2, 11, 17]

<u>The simulation principle</u>
Let us now illustrate the different aspects of the simulation process, see figure 2. Firstly, the geologist and the instructor have to define the teaching objectives and to construct interesting geological situations. Next, the simulation of the mining prospecting session can begin. The student explores the synthetic geological site and interprets data from exploration drillings. The teaching activity can be viewed as a process divided into four phases :

1 - *conception phase* : the simulation goal is to put the student in a position to research one or more deposits which are economically exploitable and to evaluate their mineral content. So, the instructor and the geologist stress the conception of geological situations and their educational interests.

2 - *preparation phase* : during this phase the instructor describes the concepts defined during the preceding phase. The realisation of a geological model requires some tests in order to validate it. This model must be sufficiently complex to reproduce a realistic impression but not so complex that the problem is reduced to geometrical recognition, or a random exploration.

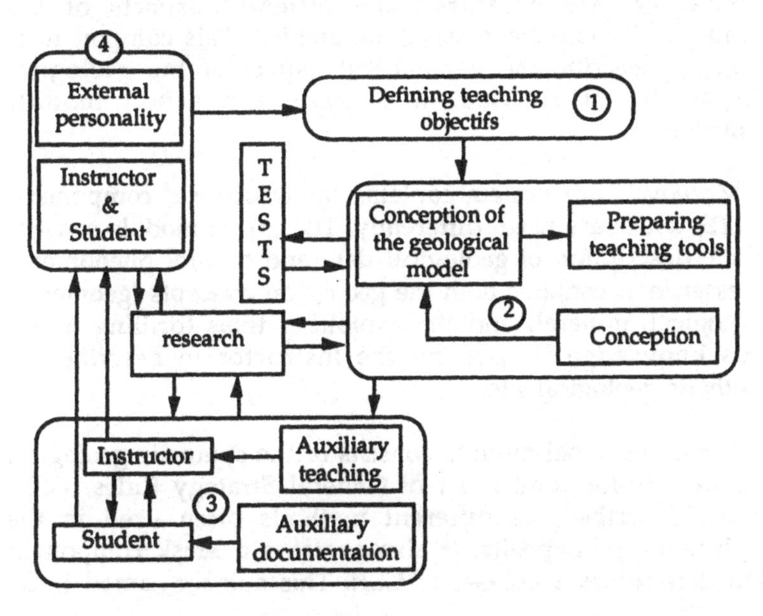

Figure 2. Different phases of the simulation

3 - *deposit research phase* : the student is the principal agent in this phase. He explores the geological site and interprets the different drilling results. Each drilling cost is deducted from the budget allocated to the student for his mining prospecting. In addition, he can have access to the auxiliary documents and discuss with the instructor during the exploration.

4 - *diagnostic and critical phase* : Finally, the instructor and other personalities discuss and criticise the different aspects of the student strategies.

The system design

The architecture of our system is described in figure 3. The various components of this architecture will be illustrated in the following sections and more information will be given afterwards.

The external environment consists of the various seminars which the student attends, auxiliary documents that he can collect from different experts and other useful external modules of data processing.

The geometrical group is a geometrical modeling framework. We organized the different aspects of the geometrical component using hierarchies. This component is used to describe the geometrical aspect of the geological shapes. So, it is used in defining the expert module component.

Moreover, our system contains the traditional components of ITS, which are used differently. The expert module consists of the description of geological data and natural phenomena. In essence, it contains both the geological concepts (geometric, geological, mineral) and the exploiting tools (drilling tools). This knowledge is used by the instructor to describe the synthetic geological site.

The educational module consists of the synthetic geological site description, and a set of General Strategy Rules (GSR) which describes the different methods often used in the exploration of deposits. It also contains a Mask component that determines a sub-set of GSR. This sub-set named Local

Strategy Rules (LSR) embodies only the rules suitable to a given synthetic geological site. The Mask is used as soon as the geological site is created or modified. The educational module is also used to compute drilling results.

The student model is a module which contains a set of mining prospecting strategies (LSR), suitable for the synthetic geological site. Each student request is analyzed by the Analyser of strategy using LSR. The student history represents all his interaction with the system.

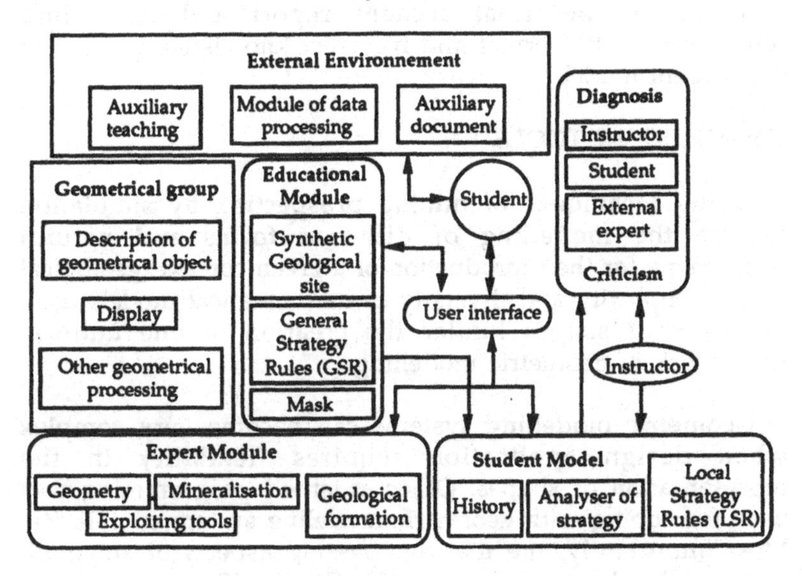

Figure 3. The design of the system

When starting a simulation session with a student, the system proposes a synthetic geological site which must be explored and the student is given a limited budget. Before the beginning of the simulation the geological site has already been constructed by the instructor and the LSR has also been defined by the Mask component.

During the simulation, the student asks the system questions. Each request is described by three parametrs <#DrillingPosition,#DrillingDepth, #DrillingTool>. The system checks whether it is an admissible request. If it is correct the system requires of the student a justification of his request. It is also possible that the student comments upon his requests. Then the system adds this information to the student model, evaluates the request and comments it. In addition, the system computes the answer using the pedagogical model, and returns it to the student. Naturally, this answer is added to the student model.

The student can ask the system for information about general geological knowledge. So, the expert module is then used to satisfy the student requests. Finally, the student strategies are analyzied by the analysis commitee. The analysis is based on the final student report and takes into consideration the correct and incorrect knowledge present in the student model.

GEOMETRICAL TOOLS

Teaching techniques of mining prospecting by simulation require the modelling of different forms and natural phenomena for the reproduction of a given context (geological site). Complexity and diversity of mathematical models used to represent shapes hinder the creation of one uniform framework in geometric modelling.

Geometric modelling systems can become very complex when design application requires flexibility in the representation of shapes. Object-oriented programming was used for coping with geometric modeling system complexity. Most importantly, the message-passing aspects of smalltalk language has been emphasized [7]. The specification of classes has also been taken into account [5], and the concept of class has been exploited in modelling and display software. The object-oriented organisation gives a programmer the benefits of encapsulation and inheritence.

In our system, the geometrical knowledge has been represented by means of different hierarchies of classes. We share this goal with several projects [5]. Each of the hierarchies describes a component of the geometric modelling system, as *geometry class hierarchy* (taxonomy of diverse geometric representation), *display class* (represents a display package) and *transformations* (to deform any basic shape). Going into detail would bring us out of the scope of the present paper, for more information see [14].

The conception and the realisation of the geometrical component are separate from its application. Our geometrical environment can therefore be used in modelling of 3D shapes, in animation process, etc...

EXPERT MODULE

The expert module contains the geological knowledge useful for the construction of a synthetic geological site. Geological objects are represented by three concepts : morphology, geological formation and mineralisation. As already said, each of those concepts is represented via a hierarchy. For example, the root of the morphology hierarchy contains some attributes and some behaviour inherited by all the shapes (all the shapes have to be drawn). We have also defined some scenarios for collecting geological objects to construct a geological site.

Geometrical aspect

To create a system which manipulates geological knowledge and integrates geological reasoning, we have firstly to represent and to understand geological objects. Several approaches have been developed for handling different aspects of geological objects.

A formalism was developed to represent geological objects and more particularly any geological section [6]. Computer generated image techniques were also exploited in modelling geological deformations [13]. In order to avoid geometrical computing, a relational data base was used to represent geological objects, so spacial queries are solved by relational algebra [9].

On the other hand, Interesting Situation Generator is a hierachical clustering program [18]. It discovers equivalence classes of situations (e.g. types of geological formation) that give rise to qualitatively distinct manifestations (e.g. different patterns of geophysical measurements corresponding to different types of geological formation). The frame concept was also exploited to represent geological knowledge [15]. So, the diversity of techniques used prove the complexity of the geological site modelisation.

In the expert module, the morphological hierarchy describes geometrical aspects of different rocks. For example we present, in figure 4, a family of folds which have the same description and different constraints.

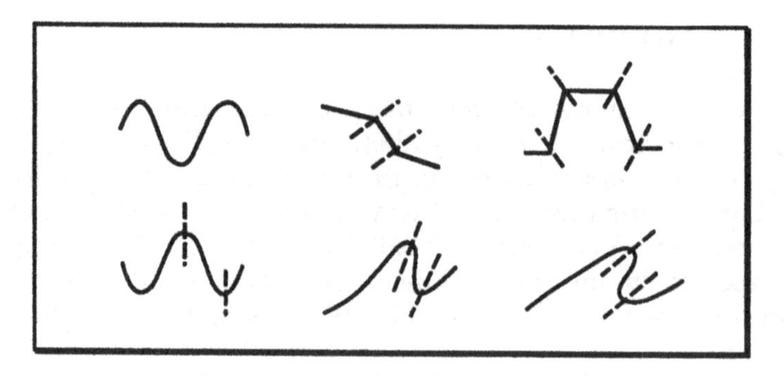

Figure 4. Example of folds family

So, knowledge about the geometrical aspects of rocks used in the construction of a geological site has been presented via a hierarchy of classes. These different classes are described by a geology expert using the above geometrical components.

Types of geological formation and mineralisation

Rocks are extremely varied, and they can be described by complex attributes like structure, texture and grain. The interaction between the different rock families also creates various and complex phenomena. However, this aspect has been represented by a hierarchy of class, in order to study the principal features and properties of rocks. This taxonomy takes into consideration the origin of rocks and their possible transformation.

Apart from aspects of rock composition, we considered some of their behaviour such as :

- *Impact* : it represents the capacity of a class of rocks to impair other rocks by modifying their mineral aspect, or generally their composition and consequently their nature.

- *Reception* : this is the reciprocal of the above behaviour.

- *Mineralisation* : Enables one to know if a given rock can contain a mineral or not.

The first two behaviours define the interaction phenomena between two families of rocks. These behaviours are integrated into the different classes of the hierarchy which describe rocks.

The mineral problem is not only the description of different kinds of type of minerals, but we must also model their diffusion functions in space. In order to simulate this distribution function many mathematical methods have been exploited and especially statistical ones. For example Krigeage theory, which was originally invented for mineral estimation.

A geological site is a collection of geological objects organised according to given scenarios. Firstly, we describe the geometrical and the types of geological formations of these objects, then we add the mineral description and its position in space.

Scenario

Now, we illustrate how to use the above concepts for constructing a synthetic geological site. We describe the geometrical aspect by collecting different shapes using Constituent-Element (CE). Some of them assure the geometrical continuity when others express discontinuous phenomenon, such as a fault system, for example.

In order to illustrate the continuous CE, we refer to the example of the construction of a geometrical object composed by two basic shapes. Each one of the basic shapes is constructed using both a surface element P and a set of transformations T. Let us note F1[P1,T1], F2[P2,T2], the basic shapes, where ∂[P1] and ∂[P2] are the borders of P1 and P2 that will be connected after applying T1 and T2 transformations. In order to connect F1 and F2, we consider the surface element A, described by the two borders ∂[P1] and ∂[P2] and we try to deform it by an automatic deformation T, computed from T1 and T2. The two conditions necessary to retain geometrical continuity are :

$$\left| \begin{array}{l} T/\partial[P1] = T1 \\ \\ T/\partial[P2] = T2 \end{array} \right. \qquad (1)$$

where T/∂[Pi] is the restriction of the T transfomation to the border ∂[Pi]. So, we define the CE with a primitive A and the deformation T. For example we can define T as following :

$$T(p) = \Psi(p)^{*}T1(p) + \Gamma(p)^{*}T2(p)$$

where p is a dot defined by (x,y,z). In order to satisfy the above constraints system (1), we consider the following constraints :

$$\left| \begin{array}{l} \Psi/\partial[P1] = 1 \\ \Psi/\partial[P2] = 0 \\ \Gamma/\partial[P1] = 0 \\ \Gamma/\partial[P2] = 1 \end{array} \right. \qquad (2)$$

So, we have to define the fonctions Ψ and Γ considering the constraint system (2).

$$\Psi(p) = \frac{d(p, \partial[P2])}{d(p, \partial[P2]) + d(p, \partial[P1])}$$

$$\Gamma(p) = \frac{d(p, \partial[P1])}{d(p, \partial[P1]) + d(p, \partial[P2])}$$

where d is the distance between a dot and a border. For simplicity's sake, we give a 2D example to illustrate the preceding interpolation method.

We define the two primitives F1 and F2 by an interval and a function and we note F1[I1, f1] and F2[I2, f2], where

$I1 = [-3\frac{\pi}{2}, -\frac{\pi}{2}]$ $\quad \forall x \in I1 \quad f1(x) = \cos(x)$

$I2 = [\frac{\pi}{2}, 3\frac{\pi}{2}]$ $\quad \forall x \in I2 \quad f2(x) = \sin(x)$

$I = [-\frac{\pi}{2}, \frac{\pi}{2}]$ $\quad \forall x \in I \quad h(x) = \Psi(x)^{*}\cos(x) + \Gamma(x)^{*}\mathrm{sinx}(x)$

Where $\quad \Psi(x) = \dfrac{|x - \frac{\pi}{2}|}{|x - \pi/2| + |x + \frac{\pi}{2}|}$

$$\Gamma(x) = \frac{|x + \frac{\pi}{2}|}{|x + \frac{\pi}{2}| + |x - \frac{\pi}{2}|}$$

So, we have the geometrical continuity propriety :

$$h(-\frac{\pi}{2}) = f1(-\frac{\pi}{2}) = 0$$

$$h(\frac{\pi}{2}) = f2(\frac{\pi}{2}) = 1$$

The preceding example deals with the connection of two geometrical objects retaining geometrical continuity. The scenario is more complex when we consider the composition of geological objects taking into consideration both their geometry and their type of rocks, see figure 5. Firstly, we define a shape from a surface element to which we associate a type of rock. A surface element divides space into two parts. All the space under the surface element is occupied by a given type of rock. So, we use this geological basic object to construct extremely complex geological objects.

The composition consists of two levels. The first one is a sample composition of the basic geological objects by defining their mutual positions. This level implies geological conflict. In fact, in figure 5, each rock is represented by one of the following symbols : {▥, ▤, ▧}. Each zone which contains more than one symbol is a conflict zone. The second level resolves the conflict problem and obtains a coherent geological object. In fact, we associate a scalar value with each geological basic primitive representing its priority in the composition process. Additionally, if two surface elements intersect, the one having priority (its value being greater) blocks off the other. So, for computing a log result we calculate all its intersections with the composed object considering the individual surface elements. After that, we correct this primary result using the priority system. Figure 5 shows an example of a log result. The primary result is defined by two intersections and three types of rocks, but the final result is characterised only by one intersection and two types of rocks. The other results were blocked off.

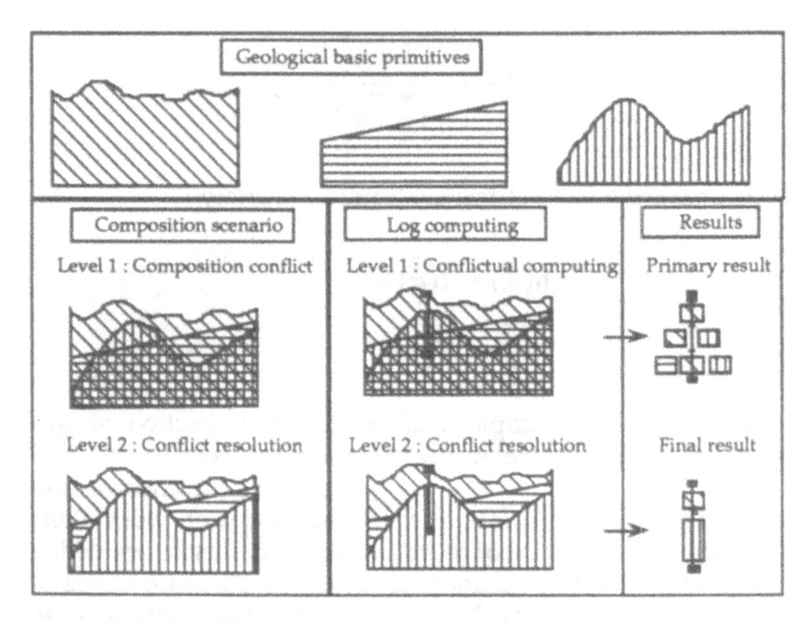

Figure 5. Composition scenario and log computing

THE STUDENT MODEL

In a geological environment, the student must research one or more deposits which are economically exploitable and estimate their reserves. Initially, the student has some information about the synthetic geological site at his disposal, and explores the site using the drilling tools. As he tries to discover the mineralisation and the nature of the terrain, he interacts with the system and requires results of some drillings which are defined by their origins and their depths. Then he interprets this data through some hypothesis. He continues his exploration by asking other questions and making other hypotheses. So, two working spaces are distinguished : the Request Space (RS) and the Hypothesis Space (HS).

We classify the student tasks during the exploration into ten actions represented by the following interaction rules. Those rules specify the relationships between RS and HS.

1/		⇒	Request
2/	Request	⇒	Request
3/	Hypothesis	⇒	Request

4/ Request $\Rightarrow$ Hypothesis
5/ Hypothesis $\Rightarrow$ Hypothesis
6/ Request $\wedge$ Hypothesis $\Rightarrow$ Request
7/ Request $\wedge$ Hypothesis $\Rightarrow$ Hypothesis
8/ Request $\wedge$ Hypothesis $\Rightarrow$ Request $\wedge$ Hypothesis
9/ Request $\Rightarrow$
10/ Hypothesis $\Rightarrow$

These rules represent the primary elements of the student reasoning. The left side of each rule represents inputs (results of previous requests or hypothesis which motivate the student action), the right side describes the new student actions. In fact, the interpretation of the results of the previous student actions can generate other requests or hypotheses. For example, the second rule means that the student uses only the results of some drillings to require other requests. The two last rules mean that there are some requests (respectively some hypotheses) that the student does not know how to use.

At present, the student must gives a justification for each request and he can comment on it. So, we describe the student's action by :

<request, justification, comment>

The different justications inform the teacher about the student actions. The basic justifications have been represented by :

<verb, subject, attributes>

The verb embodies the student action as for example Verify or Confirm. The subject is one of the three geological concepts : geometry, types of geological formations and mineralisation. The attributes characterise the position in space of the concepts as for example an angle or a depth. The following two justifications state that the student wants to verify that there is a fault orientated by 10 degrees, he also wants to confirm that there is schist at a depth of 30 metres :

Verify fault 10
Confirm schist 30

Generally, we represent a justification as a logical formula using the logical operators : ∨, ∧, ¬. We rewrite the formula into Kowalski notation using the following rewrite rules :

1/	P ∧ Q ∧ R ...	=>	P, Q, R ...
2/	P	=>	-> P
3/	P ∨ Q ∨ R ...	=>	-> P, Q, R ...
4/	P -> Q, ¬R, S	=>	P, R -> Q, S

For example, the next justification states that the student wants to verify that there is a fault orientated by 10 degrees and to confirm that there is schist limited between 10 and 30 metres deeper or there is no clay :

(Verify fault 10 ∧ Confirm schist <10,30>) ∨ (¬Exist clay))

The preceding logical formula can be rewrited also as :

(Verify fault 10 ∨ (¬Exist clay)) ∧
(Confirm schist <10,30> ∨ (¬Exist clay))

rule 1 : (Verify fault 10 ∨ (¬Exist clay))
(Confirm schist <10,30> ∨ (¬Exist clay))

rule 3 : -> Verify fault 10, (¬Exist clay)
-> Confirm schist <10,30>, (¬Exist clay)

rule 4 : Exist clay -> Verify fault 10
Exist clay -> Confirm schist <10,30>

So that, the general form of the justification is :

[{<#DrillingPosition,#DrillingDepth, #DrillingTool>},
{ {<verb, concept, attributes>} -> {<verb, concept, attributes>} },
{comment}]

Usually, the instruction systems use the student response to choose from among facts and rules considered as known by the student [10]. Then the diagnosis module adds these facts and rules to the student model which will be examined by the pedagogical module to determine the best "teaching strategy" to apply. The criteria used to determine what the student knows and to guide the teaching dialogue are complex to express. In our case, the justification and the comment represent the request context and they express the student's intentions. So, the student model is an exact image of the evolution of the student strategies.

DIAGNOSIS AND CRITIRIA ANALYSIS

At present, the diagnosis process is not automatic. It is based on the final student report given to the instructor. The diagnosis commitee (instructor and other experts) analyse both the comments of the system and the student's justifications. This commitee discusses with the student his different strategies. We believe that this experimental phase may be useful in specifying teaching rules and clarifying some of the issues involved in the construction of an intelligent diagnosis module.

For the development of an intelligent system we have to take into consideration the following questions :

1. How to guide the teaching dialogue ?

2. What are the roles of the student errors and misconceptions in guiding the system to choose the teaching strategy, and to correct the possible errors ?

The diagnosis component can be viewed as an expert system specialised in the interpretation of geological data, especially in the interpretation of results of exploration drillings. We are in a situation where there is an initial state (drilling results) and several acceptable states (the possible interpretations of drilling results). In general, a forward chaining production is well suited to this situation. Such activity requires expertise on the interpretation of drillings data, on the teaching methodology and on the ways to personalise the dialogue.

It is true that the expression of the preceding expertise is collectively complex. A first step for approaching our goal is to develope the interpeter model. It will be based on the interpolation of factual data resulting from the exploration drillings. Therefore, the interpreter's results must be matched against the student's propositions (Pattern matching phase). At the moment, we are focusing our attention on the analysis and diagnosis of the student hypothesis and the semantic validitiy of his requests.

CONCLUSIONS

From the description presented so far, it is evident that there are two major problems :

- the modelling of different aspects of a geological site, conditional upon simulation performances and the system's capacity to allow the construction of an interesting geological situation. Consequently, it is conditional upon the instruction interest;

- the student modelling and the interpretation of geological data in order to evolve towards an intelligent system.

At present, different hierarchies are exploited to represent the geological knowledge which is used by the other components of the system. The student analyses drilling data, resulting from his requests, and tries to discover deposits.

The diagnosis process is not yet automatic. We expand and specify the teaching rules and the possible evolution towards an intelligent diagnosis. Also, we are intersted in the possibility of using other exploitation tools, such as for example seismic processes.

REFERENCES

1. Anderson, J.R., Boyle, C.F., Yost, G. The geometry tutor, Proceedings of the Ninth International joint conference on artificial intelligence, IJCAI9, August 1985.
2. Anderson, J.R., Reiser, B.J. The LISP Tutor, Byte, Vol. 10, N° 4, 1985.
3. Brüggemann, H., Kade, M. Use of 'Artificial intelligence' in open-pit mining, OPCOM 87, Proceedings of the Twentieth International Symposium on the Application of Computers and Mathematics in the Mineral Industries, Vol. 3 : Geostatistics, pp. 115-120 Johannesburg, SAIMM, 1987.

4. Cerri, S.A., Leoncini, M. Conceptual Modelling Systems for the Design of Tutorial Dialogues, in Proc. of IFIP Working conference "AI Tools in Education", Frascati (Italy), 1987.
5. Grant, E., Amburn, P., Whitted, T. Exploiting class in modeling and Display Software, IEEE CG&A, pp. 13-20, Nov 1986.
6. Hamburger, J. FROG : Formalisme de representation d'objets géologiques, huitièmes journées internationales Les systemes experts et leurs applications, Vol. 2, pp. 53-68, Avignon, 1988.
7. Hedelman, H. A data flow approach to procedural Modeling, IEEE CG&A, Vol. 4, N° 1, pp. 16-26, Jan. 1984.
8. Lanöe, S. Modélisation sur ordinateur d'un compartiment géologique et applications pédagogiques, Thèse de docteur Ingénieur, Ecole Nationale Superieur des Mines de Paris, 1981.
9. Laurini, R., Milleret, F. Solving Spatial Queries by relational algebra, Proceedings of the 9th Symposium on Computer-Assisted cartography AUTO-CARTO-9, pp. 426-435, Baltimore, April 2-7, 1989.
10. Luigia Aiello, Massimo Carosio, Alessandro Micarelli, The design of an intelligent tutoring system in mathematics : The SEDAF Project, huitièmes journées internationales Les systemes experts et leurs applications, Vol. 2, pp. 271-296, Avignon, 1988.
11. O'Shea, T., Bornat, R., Du Boulay, B., Eisenstad, M., Page, I. Tools for creating intelligent computer tutors. In Human and Artificial intelligence. Eds. Elithor & Banerjii, North Holland, 1984.
12. Pélisonnier, H., Woodtli, R. La formation à la recherche minière, Annales des Mines, pp. 1-12, Paris, 1977.
13. Perrin, M., Oltra, P.H., Rommel, E., Peroche, B. Modélisation de structures géologiques par images de synthèse, Exemple : déformation d'une stratification par failles et plis semblables, Colloque FI3G, pp.330-339, Lyon, 10-13 Juin, 1987.
14. Quafafou, M., Dubant, O., Prevot, P. L'approche objet dans la conception des systemes graphiques. Revue Internationale de CFAO et d'Inforgraphie, Vol 5, N° 4, pp. 55-72, Eds. Hermès, 1990.
15. Quafafou, M., Rolley, J.P., Dubant, O., Prevot P. Le concept objet et la modélisation géométrique : Application à la

modélisation d'environnement géologique. In Bull.Soc.Vaud.Sc.Nat. 80.1, pp. 83-98, 1990.
16. Simmons R.G. The use of qualitative and quantitative simulations. AAAI-83, Whashington, August, 1983.
17. Sleeman, D., Brown, J.S. Intelligent tutoring systems. Academic Press, London, 1982.
18. Wisniewski Edward, Howard Winston, Reid Smith, Michael Kleyn, A conceptual clustring program for rule generation, International journal of Man-Machine Studies, Academic Press, Vol 27, N°3, pp. 295-313, Sep. 1987.
19. Woodtli, R., Vannier, M., Toros, M. Présentation d'une carte géologique obtenue par simulation à l'ordinateur. Bull. Suisse Mineral. Pétrogr. 55/3 : pp. 583-585, 1975.
20. Zheng-Yang Liu, Diagnosing and modeling in intelligent teaching systems, huitièmes journées internationales Les systemes experts et leurs applications, Vol. 2, pp. 87-106, Avignon, 1988.

SECTION 3: PLANNING AND SCHEDULING

Process Parameter Origination using a Combination of Knowledge Based Paradigms

W.A. Taylor

Department of Mechanical and Manufacturing Engineering, The Queen's University of Belfast, BT7 1NN, UK

ABSTRACT

This paper discusses the origination of process parameters for planning activities prior to industrial arc welding using automatic & semi-automatic equipment. The planning process involves the reconciliation of several factors including product design requirements, particularly geometry and mechanical properties, contextual factors such as physical location and orientation, and economic considerations such as welding speed, volume of deposited metal and procedure qualification time. The key issues in this planning process are highlighted, particularly those connected with the design of the edge preparation of the welded joint. Moreover, the nature and representation of the knowledge used for this purpose is detailed and the working system described. This system has been developed in close collaboration with a large industrial fabricator.

INTRODUCTION

In order to understand the design and implementation of this knowledge based system, it is important to circumscribe the domain and its position within its business environment.

The challenge of the market

Large scale fabrication of offshore platforms, pressure vessels and ships is a specialised engineering sector. It is increasingly being characterised by the growing diversity of materials being used and by the widening range and complexity of joining configurations. When coupled with a fall in demand for conventional products, this has resulted in a migration to new markets involving less traditional areas of expertise. Consequently the amount and variety of design and planning work has increased considerably. Furthermore the greater competition for contracts represented by this market shift has focused more attention on product lead times and on the quality of pre-production expertise.

Choice of problem domain

One of the critical success factors for such organisations appears to be the planning function relating design requirements to operational data. The decision to address this function was largely prompted by the recognition that it was analagous to what is usually referred to as process planning in other metalworking sectors [1]. It was only upon subsequent analysis that the wider strategic implications became apparent.

In other words, the application was not initially driven by the needs of the business but rather by technical curiosity and a desire to explore the potential which knowledge based systems claimed to offer. In the context of large scale fabrication this planning function is normally the responsibility of the welding engineering department, (Figure 1).

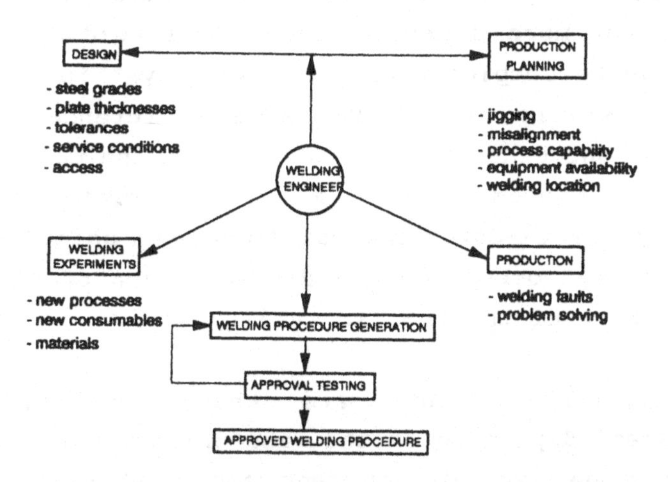

FIGURE 1 ACTIVITIES OF WELDING ENGINEERING DEPARTMENT

<u>Outline of problem domain</u>

The "process plan" which is generated is regulated in terms of its format and content by contractual standards such as BS 4870 Part 1 1981. Such plans, referred to as welding procedure specifications (WPS), are ultimately approved by a third party such as Lloyd's Register of Shipping and become significant legal documents in the event of loss or damage. They can also be used to assure clients, in advance of construction that there is a high probability of the welded structures meeting the specified design requirements. Moreover, in the event of a joint failure the WPS can be used for fault diagnosis investigations.

Overview of welding processes

The specific processes considered were the Submerged Arc process and the Self-Shielded Flux-Cored process. Both processes are widely used in heavy fabrication, the former solely for horizontal welding and the latter for more difficult positional welding. Thus they can be complementary processes, yet in terms of their operation there are some differences. For example, Submerged Arc welding shields the molten weld pool from contaminants by a layer of granular flux, whilst the Flux-Cored process depends on shielding gas produced by combustion and decomposition of the flux in the core of the tubular welding wire.

Fundamentally, both are true arc generating processes, depending on the same primary variables of arc voltage, welding current and welding speed to achieve the specified design requirements. With reference to knowledge about welding, one of the objectives of this work was to note the extent to which rules could be identified which would be common to both.

KNOWLEDGE ENGINEERING

Knowledge identification

Prior to any structured attempts to elicit knowledge it was necessary to identify the approximate scope of the planning function and the types of knowledge used in producing a WPS (Figure 2). The objectives of the planning function were threefold, namely,

(i) the production of defect free welded joints,
(ii) the minimisation of risk of subsequent cracking,
(iii) the achievement of requisite mechanical properties

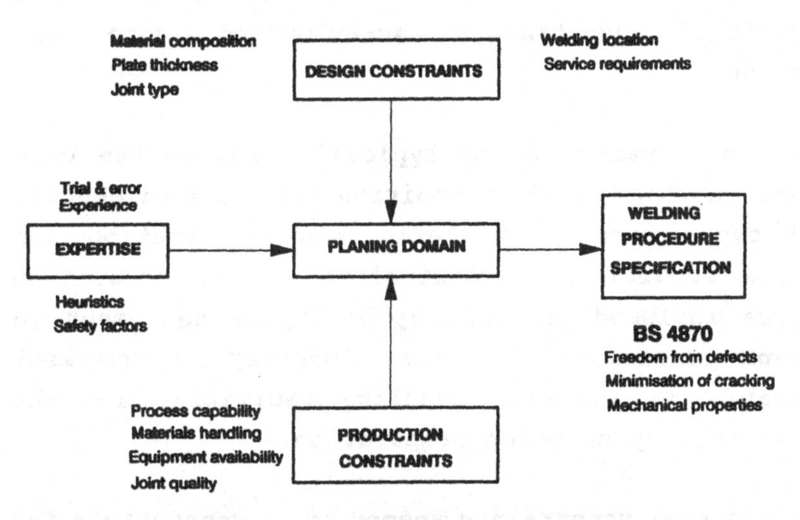

FIGURE 2 PROCESS PLANNING IN THE WELDING DOMAIN

Each of these three objectives was further subdivided into more specific criteria, e.g.
Freedom from defects:- Porosity, Undercut, Geometric irregularity, Lack of fusion.
Cracking risks:- Solidification cracking, Stress corrosion cracking, Hydrogen induced heat affected zone cracking.
Mechanical properties:- Fracture toughness, Impact strength, Tensile strength, Heat affected zone hardness.

Most of these criteria have an established body of theoretical knowledge associated with them based on engineering principles. Welding Engineers would have been conversant with these concepts as part of their education and training - yet there was little evidence from our investigations that such knowledge was formally and systematically used in practice. This appeared to be due to the large number of design and operational constraints superimposed upon any solution.

Design constraints would typically include the pre-selection of material composition and plate thickness, weld type (butt or fillet), welding location and service conditions. Constraints due to production factors included process capability of equipment (a concept uncommon in this industry), equipment availability, material handling facilities and the likley quality of joint preparation.

Such external constraints seemed to be responsible for the lack of causal reasoning by the welding engineers, since it would appear to be relatively difficult to quantify these contextual factors. Instead they tended to rely on well proven personal solutions based on experience and a considerable measure of trial and error. Of course, when moving into new business markets, previous experience is not necessarily relevant. Moreover, trial and error tends to introduce unnecessarily large factors of safety.

Knowledge elicitation and prototyping

Initially, three types of interview were used - unstructured, free-flowing discussion interviews, structured interviews focusing on specific aspects of the problem, and finally interviews where the agenda and questions were supplied well in advance. This phase extended over a period of about five months, at the end of which a prototype system was available, along with the knowledge base code which was supposed to be readable and verifiable by the experts, according to received wisdom.

The software did not facilitate this last objective due to its methods of representation (Production Rules and Bayesian Inference). For example, a rule for selection of a type of joint edge preparation could be represented by Bayes Theorem as:

```
EDGE PREP "Square Butt, No Root Gap"

ANTECEDENTS:

    Accessibility     LS 1.0   LN  0.0001
    Corrosion         LS 2.5   LN  0.33
    Distortion        LS 5.0   LN  0.22
    Weld volume       LS 2.2   LN  0.44
    Penetration       LS 3.7   LN  0.33

PRIOR PROBABILITY     0.5
```

Those familiar with this kind of representation can interpret the above coding in terms of the influence of each of the five antecedents on the likelihood of choosing the specific edge preparation - the weightings of the factors not being equal. However, if a welding engineer wished to add another antecedent he would have great difficulty deciding what weighting factors to assign.

Similarly in assessing the material composition in terms of its carbon equivalent (C.E.) value, the pertinent factors are the percentages of the constituent elements in the steel, the likelihood of a specific amount of hydrogen being entrapped in the welding flux, and the joint type, butt or fillet. Such rules might appear as:-

```
C.E. SCALE A

IF  (hydrogen level = 2
AND Joint type = Butt
AND Butt type = 1 OR Butt type = 2)
OR (Hydrogen level = 2
AND Joint type = Fillet
AND Fillet type = 1)
ELSE C.E. SCALE B
```

This is by no means the complete logic concerned with this matter but it should demonstrate another kind of complexity in programming which can give the user and sometimes the programmer a problem.

The meaning of weighting factors or degrees of belief is difficult to assimilate simply by inspection of the knowledge base. Further inspection of the prototype in operation helped but not in a rigorous way. More tangible validation of this knowledge was effected by conducting specimen welding tests based on the system's advice. This revealed further inconsistencies and areas of insufficient understanding.

Knowledge rationalisation

In parallel with these interactive sessions with the welding engineers, the wider subject literature was examined. This revealed a set of detailed approaches to the solution of the problem, none of which was actually employed in practice either in this company or by other market leaders.

These approaches are comprehensively discussed

elsewhere; they are summarised as follows:-

* Derivation of mathematical relationships between welding procedure variables and outcomes (design criteria for specific welded joints) [2]. Extensive work has been done but few of the equations have been used in practice due to their complexity [3].
* Graphical representation of process boundary conditions. Sometimes called the "tolerance box" method [4], it has been limited because of its inability to show more than three variables at one time.
* Studies of heat flow from the welding arc [5]. This has been limited by the inability to relate heat flow to welding procedure variables, weld bead geometry or microstructure.
* Development of guidelines for the avoidance of cracking. Nomograms have been produced relating joint dimensions, welding arc energy and material composition to preheating temperatures to avoid hydrogen induced cracking [6]. These have been widely used but they do not give any help with the choice of primary welding variables, edge preparation of joint, electrode diameter, polarity or wire/flux combination.

Each of the above methods has limitations but it was thought that if they could be combined in some way they might complement each other sufficiently to move closer to the solution of this problem. Furthermore they might be more acceptable in practice if embodied within a suitable computer environment.

Knowledge validation

For the Submerged Arc process, eighteen sets of published equations were identified in the literature, which could link design requirements to operational parameters. However, upon inspection they differed in so many small ways that overall they could not be regarded as consistent. For example there were variations in material types, plate thickness ranges, electrode wire diameters, fluxes used and electrode polarities.

An experimental programme was commissioned to evaluate the equations with regard to their suitability for incorporation within an expert system [7]. Of all the equations considered, only those of one author, McGlone [8], were found to be suitable for the proposed application. Conversely for the Flux-Cored process there was an almost complete absence of equivalent equations, the closest study being for gas-shielded flux-cored welding. Thus a separate experimental programme was initiated and a set of predictive equations produced [9]. A typical equation is shown in Figure 3.

MODIFICATION TO INITIAL SYSTEM SPECIFICATION

In view of the paucity of welding knowledge derived from the domain experts and with an enhanced understanding of the domain itself, it became clear that the use of rules and heuristics was only a small, albeit necessary part of the solution to this problem of process parameter origination (Figure 4). It was decided to use a more eclectic approach to the design of this knowledge based system, in recognition of the need to combine sources of knowledge.

BEAD WIDTH

$$W = 11.86 + 0.097V - 4.804F - 0.025S + 0.049E + 0.079G - 0.002VS + 0.278VF + 0.002VE + 0.002VG - 0.02FE + 0.001SE + 0.094EG$$

WELDING PARAMETERS	FIRST ORDER INTERACTIONS	
V - Voltage	VS	FE
F - Wire feed speed	VF	SE
S - Travel speed	VE	EG
E - Electrode stickout	VG	
G - Gun angle		

FIGURE 3 SAMPLE BEAD SHAPE EQUATION

Thus quantitative methods were used to produce base line values of the welding variables, either from a database if a matching solution was available, or from the predictive equations discussed earlier.

These base values would then be adjusted by heuristics to take account of contextual constraints. Moreover, in cases where the predictive equations were not valid, rules would be developed to make an estimate of the necessary parameter changes The estimates would be based on explicit qualitative knowledge about the physical effects of the welding variables on weld bead outcomes, without recourse to further experimentation which is costly.

Cost of experimentation is one of the prime reasons why progress in this field had been impeded thus far.

It was therefore imperative that further progress was not dependent on extensive experimentation.

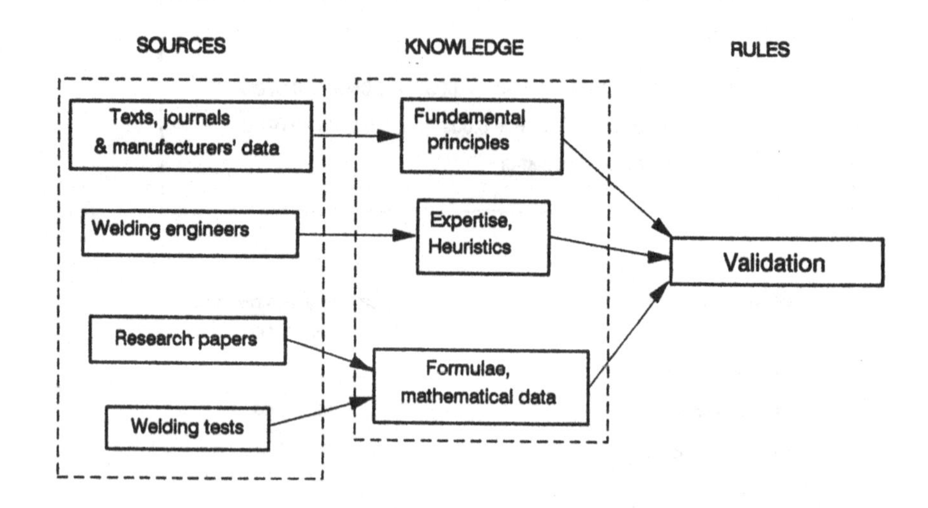

FIGURE 4 SOURCES OF KNOWLEDGE

Nevertheless, it was believed that some statistically based experimental work was necessary to provide a key role in knowledge validation - sometimes neglected in the expert systems development process.

SYSTEM PLANNING AND IMPLEMENTATION

The knowledge base for this system has three distinct aspects, viz;

(i) a database of previous solutions

(ii) a rule base capable of generating new solutions within the design scope of the system

(iii) a rule base to deal with situations outside the intended scope of the system

It is this latter category which is the most novel due to the need to assess the degree of variation and the extent of its implications.

General design principles

The system was designed on the principle of functional modularisation, each module addressing a specific aspect of the problem, e.g. choice of flux, selection of welding current and so on, (Figure 5).

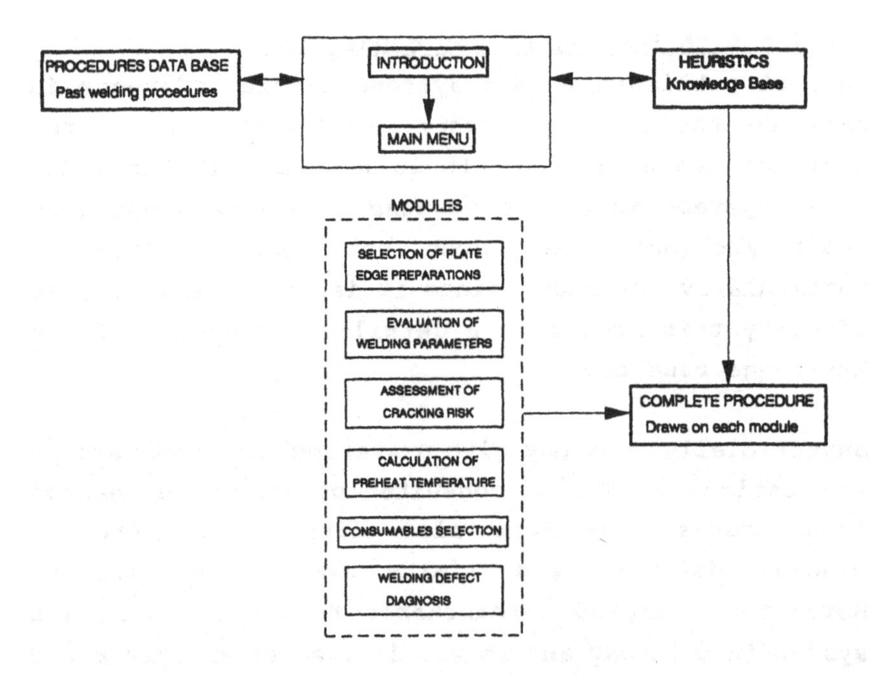

FIGURE 5 FUNCTIONAL MODULARISATION

In terms of knowledge representation, it was thought necessary to use a combination of several approaches, including objects, frames, rules and perhaps bayesian inference due to the diversity of the knowledge brought to bear on the problem. For example, the choice of edge preparations could be more succinctly expressed by assigning each option as an object with its attributes and values assigned specifically to it, i.e.

OBJECT: Edge preparation 1

Plate thickness range	10-17mm
Root face	6mm
Root gap	0.5mm max
Bevel angle	60 degrees
Backing Plate	Not required
Expansion Text	Call Graphics Proc(paras)

Dealing with "out of scope" conditions

In most published expert systems little reference is made to the specific scope and boundaries of the systems' capabilities. It is argued that knowledge based systems must also possess knowledge about when their knowledge base does not apply. This is particularly important since it is not at all easy to identify this from even a detailed inspection of the knowledge base code.

Superficially, one may give a statement to the user at the beginning of the consultation which delineates these issues. One might also address this matter by placing defaults and limits on allowed inputs. However, it seemed unreasonable to constrain such a system in this way and it was decided to explore areas in which the knowledge base could be extended by heuristic rules to cater for out of scope situations.

For example, the system was intended to deal with certain types of edge preparation and joint design. It was postulated that if another edge preparation was produced, the same rules could apply concerning choice of arc voltage, welding current and welding speed since they all determine bead geometry and microstructure according to fundamental principles.

Most welding knowledge derives from experiments on Bead-on-Plate test pieces. This is a low cost approach due to minimal preparation of the plates and the avoidance of alignment problems. Results from such tests, although widespread in research and industry, have no value in themselves unless they can be related to actual welding configurations, i.e. to joints with bevelled edge preparations (Figure 6).

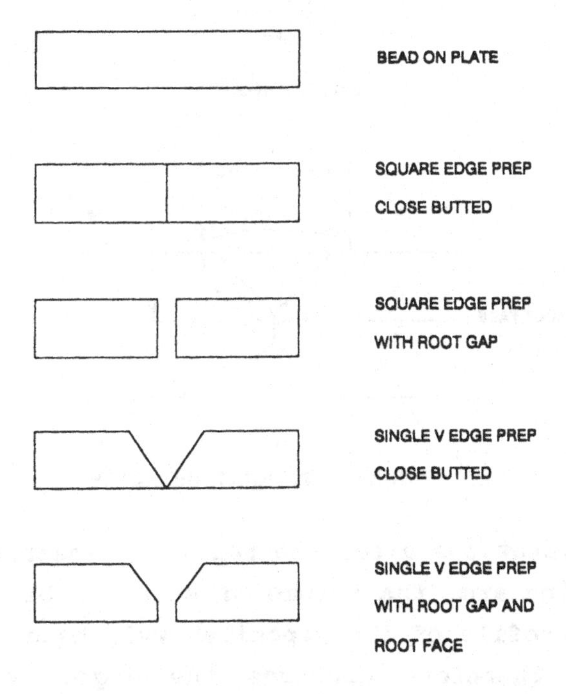

FIGURE 6 CONFIGURATIONS OF EDGE PREPARATION

Thus in Submerged Arc welding, any plate thicker than 17mm will require a single or double V groove. Essentially this is to avoid the excessively high currents which would be required to penetrate the plates from the surface. The high current and hence high heat input would have adverse metallurgical implications.

A standardised double V edge preparation was chosen, commonly referred to as a B1 preparation (Figure 7).

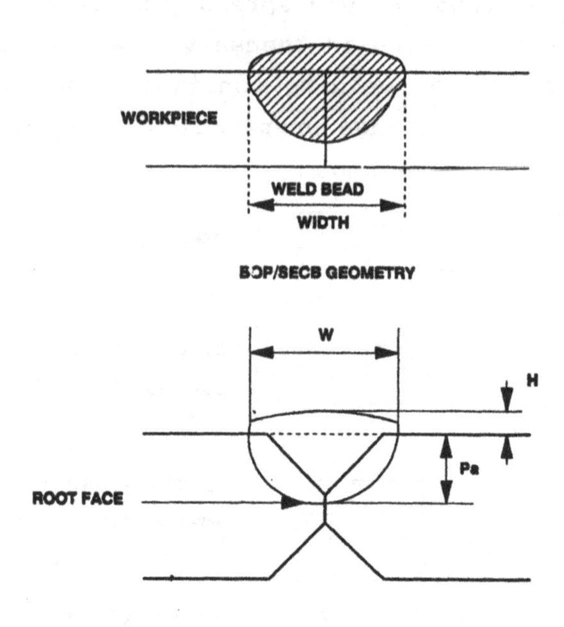

FIGURE 7 "B1" JOINT GEOMETRY

Such preparations alter the required penetration from the welding arc, the volume of metal to be deposited and the profile of the deposited weld bead. The key question therefore concerns the degree of change required to each of the process variables and indeed, the order in which to alter them, since they are interdependent.

The following rules are typical of those which were

found to apply to the relationships between BOP and B1 prepared joints.

```
IF B1 joint
AND BOP penetration is known
THEN B1 penetration = BOP penetration + 2mm

IF B1 joint
AND BOP bead width is known
THEN B1 bead width = BOP bead width - 2mm

IF B1 joint
AND BOP bead height is known
THEN B1 bead height = BOP bead height - 1.5mm
```

Such fine degrees of variation were obtained from experimental data and validated before being added to the system's knowledge base. These modified values can subsequently be used to determine arc voltage, welding current and welding speed as before - the fact that they are now required for a different edge preparation is not significant.

It should also be mentioned that the apparent simplicity of these additions belies the amount of effort involved in obtaining them and more particularly in ensuring that they were accurate. This also involved a detailed process capability study of the welding processes to quantify the amount of variation in penetration, bead width and height due to the natural variations in the equipment itself whenever the process conditions were supposed to be held constant [10]. This knowledge would also often appear to be omitted from engineering expert systems. It was knowledge which surprisingly was not available either from the industrial collaborators or from the suppliers of the equipment.

SYSTEM REVIEW

This system could justifiably claim to be truly knowledge based - indeed it has been acknowledged as such by practitioners in the field. Furthermore it would claim to be a novel system in terms of:

(a) the use of a wide range of approaches to the solution of this problem;
(b) the emphasis given to supplementary validation of the knowledge base;
(c) the attempt to incorporate knowledge which aids gradual degradation at the systems boundaries.

Whilst the knowledge itself is in demand from industrial users, its embodiment within the computer environment has not yet reached a commercially presentable level. There was also the view that had our knowledge elicitation methods been different, there might not have been the same need to resort to practical experimentation. To some extent that may have been valid, yet what we encountered was not so much a knowledge bottleneck, rather a knowledge "dearth" in the sense that no-one seemed to know the answers we were seeking.

It is also possible that our interviewing techniques were poor or inappropriate. Nevertheless, it still seems likely that some experimental validation of knowledge will be advisable, particularly when it is drawn from a domain heavily dependent upon physical principles.

Such a detailed examination of the domain literature revealed a surprising and disappointingly low quality of useable "knowledge". Certainly the reporting of investigative research should be aware of the need to clarify the scope of each programme and include all salient details.

In terms of knowledge base design, it is only after some three to four years of investigation that the domain is sufficiently understood to comment on this topic. Certainly it was possible to construct modules which reasoned about the problem but the key issue here is to match the methods of representation on to the different facets of the domain. This is still a part of the craft of software development; hopefully it will improve in the short term.

Overall, the success of this system has hinged upon the involvement and commitment of the user group rather than technical or knowledge based issues. There is certainly much more work to be done in this domain particularly with respect to qualitative modelling and the investigation of parallel processing.

REFERENCES

1. Taylor, W.A. Development of a knowledge based system for process planning in arc welding. Proceedings of the 1st Int. Conf. on Applications of A.I. to Engineering Problems. Southampton. April 1986.

2. Raveendra, J and Parmar, R.S. Mathematical models to predict weld bead geometry for flux-cored arc welding. Metal Construction. Jan 1987, pp31R-35R.

3. McGlone, J.C. Procedure prediction for single wire submerged arc welding - symmetrical joint geometry. Welding Institute Research Bulletin. July 1980. pp189-192.

4. Jones, S.B. Process tolerance in submerged arc welding: initial report. Welding Institute Members Report No 1/1976/PE. The Welding Institute 1976.

5. Kohno, R. and Jones, S.B. An initial study of arc energy and thermal cycles in submerged arc welding of steels. Welding Institute Report 81/1878/PE.

6. Bailey, N. The establishment of safe welding procedures for steels. Welding Research Supplement. April 1972. pp169-177.

7. Weimann, D.H; Taylor, W.A. and Bahrani, A.S. Review and evaluation of four decades of predictive equations for submerged arc welding. The Queen's University of Belfast. Dept of Mech Eng. Report No 1939. January 1991. 54pp.

8. McGlone, J.C. Weld Bead Geometry Prediction - a review. Metal Construction. July 1982. pp378-384.

9. Martin, P.J. Taylor, W.A. and Bahrani, A.S. Weld bead geometry prediction for self-shielded flux-cored arc welding. To be published in Int J. Joining Sciences. Summer 1991.

10. Weimann, D.H. Taylor, W.A. and Bahrani, A.S. Repeatability test results on the ESAB A6B-UP Submerged arc welding set. The Queen's University of Belfast. Dept of Mech Eng. Report No 1938. January 1991. 18pp.

Real-Time Means Planning Ahead to Look Back

R. Milne, E. Bain, M. Drummond

Intelligent Applications Ltd, Kirkton Business Centre, Kirk Lane, Livingston Village, West Lothian, EH54 7AY, UK

INTRODUCTION

Most plants need data interpretation and diagnosis. Significant benefits in terms of reliability and quality can be achieved if this is performed automatically [1]. Expert systems provide a very powerful means for achieving this [3]. They provide the appropriate mechanism for capturing the skill of an experienced engineer and turning that into an automatic computer program. However, traditional expert system packages require humans to input all the parameters.

In many situations this is not effective. The data may appear in high volumes or at high speed, or there is not always a human available, hence the need for on-line expert systems [6].

These systems are able to acquire data continuously from the plant but need to pre-process the data to a form suitable for the expert system diagnosis. The expert system then checks for a large set of possible faults and can provide extensive diagnostic information [4]. Traditional expert system shells only provide a means of implementing the diagnostic rules; they do not provide the extra functionality needed to develop a real time system and interface to the plant.

In this paper we discuss a strategy for implementing real-time expert systems, when the expert system hardware/software combination is apparently too slow. We outline the demand for a real-time system and how real-time analysis has been provided in spite of the fact that the expert system processing is slower than the data input. The heart of the real-time capability is a result of planning ahead for various types of data analysis and providing adequate history management so that the necessary data is available at the appropriate time. The techniques described in this paper have been fully implemented on a large scale real-time expert system for British Steel at Ravenscraig in Scotland [2]. In this paper we briefly outline the application, real-time systems and how they fit together..

REAL-TIME

The precise definition of real-time always causes a debate. There are a number of aspects to real-time systems. Different applications demand different aspects of the definition. In general, a real-time system is one that provides timely output without losing any necessary input data. For many applications, particularly in the expert system world, real-time simply means running fast enough.

The time constraint of a real-time system potentially has two major implications. The first area is the speed with which the answer is produced e.g. a safety monitoring system on a high speed turbine must be able to detect a fault and shut the turbine down within milliseconds. Some expert system strategies adopt the view that it is necessary for the system to always have an answer ready. If time for processing runs out, an answer is available, otherwise further time allows further refinement.

The other aspect of real-time is the speed at which data is acquired and analysed. This is the more commonly used aspect of the definition. It is important that no data is lost and that the expert system analysis is able to work with all the relevant data which has been acquired by the system. This has the important implication that if the expert system diagnosis is taking longer than the data acquisition then data may be lost. However, not all of the incoming data may be required. Full diagnosis can be achieved using only the important features and events. In general, it is necessary for the expert system to run faster than the data acquisition system.

Being able to provide output in spite of high speed data acquisition is a considerable problem when standard hardware/software is used. In most situations, the standard hardware/software combination runs much slower than the data acquisition system and yet it is too expensive to purchase hardware/software just to run faster. As a result, another solution is needed.

In the traditional world of real-time systems, much emphasis is placed on determining how long each particular part of the software will require. Related to that is the ability to interrupt the system at any stage to react to an incoming signal. Dealing with guaranteed response times in this interrupt environment, is a major concern of real-time system development. However, to date standard expert system tool kits do not provide support for these aspects. As a result, an alternative strategy such as the one proposed here is required.

It should be noted that another common interpretation of a real-time system is that it provide facilities for the time management and analysis of data. This should be considered a temporal, but not a real-time system. Just being able to reason about time does not guarantee that you would respond in real-time.

THE APPLICATION

We have developed a VAX based, large scale real-time expert system at British Steel Ravenscraig. It monitors the steel making process both to identify the faults they have now, and to predict faults that will occur in the future.

A ladle of molten iron (over 300 tons in this case) from the blast furnace is poured into a convertion vessel. The iron is then blasted with oxygen from a water cooled lance. Oxygen combines with the carbon impurities in the iron, producing carbon monoxide waste gas, leaving liquid steel. This process is called Basic Oxygen Steelmaking (BOS). It fits between the operations of the blast furnace, producing the molten iron and the mills where they roll it out into tubes, plates and a variety of beams.

The diagnostic system monitors the waste gas extraction and the primary process variables for the engineering (as distinct from production) areas. Any fault requires attention by an engineer and may also affect production quantity, quality and safety.

Primarily, what we are trying to identify are the faults that need to be fixed now and those that will need attention in the near future. The engineering staff want to know "What do I fix now, what do I fix in the future and what should I do to prevent a predicted fault occurring?". The system provides various degrees of information. The senior managers are able to look and see high level information and make strategic decisions about what has to be fixed and when, for example if there is a short shut-down. On the other hand, junior staff want to be able to look at a display, get all the relevant information about that fault and repair it without involving more senior staff.

This is not an experimental system, but is currently one of the largest real time expert systems in daily use in the UK. It is "state of the art", in that there are very few systems like this installed and working.

Such a large system is inevitably complex, comprising several cooperating sub-systems, including data acquisition and processing, expert system based fault diagnosis and prediction, fault management and display combined with extensive reporting facilities from the system databases. The data acquisition system was designed to interface with the existing British Steel data sources.

Annie [5] provides the capability to couple the expert system with this data acquisition allowing real time diagnosis. In this paper we are only concerned with the data processing portions.

The primary output of the system is a list of plant faults. These faults are divided into two categories. Those which have been identified as already existing and those that are predicted to happen. The plant engineer is able to look at a summary of the number of faults in each sub-area of the plant. He is then able to examine for each fault, the trigger that caused the system to investigate it, and an explanation of the fault including the past values related to the particular control items. For example, the controller output time history would be included with a controller output fault. This provides for a more comprehensive explanation and background behind the fault.

Another important aspect is a measure of how this fault has affected the production. Some faults will occur continuously after their first occurrence, others may occur in an intermittent fashion, depending on the actual parameters and state of the steel making process. The expert system output includes; a summary of each fault, of how long since it first occurred and how many steel batches were affected by this fault. This is particularly useful in spotting intermittent faults. It is also possible for the engineer to look at the history of faults in the database to see for example, what other occurrences of a controller valve being stuck have happened over the past several months.

A major aspect of the system is it's ability to examine long term trends and predict faults. For example, consider the control cone. To control the air flow a large cone is moved up and down. One thing that we can look at over a period of time is the build-up of particles on the outside of the cone. It gradually gets materials sticking to it and that clogs it up. What we can do is trend over time how high the cone is and look at its average position. As it gets built up we have to keep it further and further open each time. We can then trend on how long until the cone is fully open, at which point it has lost control and can only close in the one direction. At that point it has to be repaired and the system can now forecast how many days until they must be accomplished.

The outputs of the expert system analysis are directed to a number of places. In the first instance they are stored in a database on the primary diagnostic VAX, so that past histories and reports can be produced. There are a variety of reports available including; end of shift, end of week and end of campaign reports. The screen fault displays are available on a number of terminals throughout the engineering portion of the steel works.

Finally, it is hoped in the future to link the output to the plant maintenance system, rather than have maintenance actions triggered directly from the expert system. It is proposed to develop further links to integrate the output into the existing plant wide system for the generation of work orders. In this way, the output from the expert system is closely integrated with the way in which they currently deal with engineering problems.

THE REAL-TIME STRATEGY

The system being described is implemented on a VAX cluster under VMS. This is a multi process computer and as a result, is able to perform several tasks simultaneously. The data acquisition is primarily performed by an attached special processor. This processor acquires data from throughout the plant and makes it available over the VAX network to the expert system VAX. A high priority process runs continuously acquiring data over the network and placing it into the memory of the VAX. Currently, the data acquisition is cycling approximately every three seconds. The expert system itself however, is only cycling every ten to fifteen seconds. Without special actions, this implies that only twenty percent of the data would actually be processed, possibly missing important events.

The heart of the real-time support is the *Annie* software from Intelligent Applications. It takes care of the management and access to the incoming data. The *Annie* data acquisition process places the incoming data in a large circular buffer. This buffer is configurable, but holds approximately two minutes worth of data. The buffer cycles continuously, inputting new data with the appropriate time stamps. The expert system is then at its leisure to remove data from this buffer.

Because the data required may be spread out through the two minute history, it requires special processing functions to properly access the input. In support of the time processing, there is a database description for each individual channel. It contains information with regard to how fast it is updated, how long values must be at certain levels to be considered consistent, as well as thresholds for normal, high and low values. One of the difficulties of any plant is filtering out short term unimportant changes or random variations in this signal level. As part of this database description, the time limit over which a channel is considered stable is identified.

This data acquisition strategy represents a realistic industrial situation. The data acquisition was specified as standard within the steel plant. It would not have been appropriate for the expert system software to attempt to take control of data acquisition and specify when values should be measured. It is important to recognise that the data acquisition goes to a number of supporting computers and not just the expert system. It would have been totally inappropriate to have a specific scheduler to acquire data when the expert system wanted it or on the expert system schedule. That would not have been consistent with the rest of the plant wide data acquisition systems.

LOOKING BACK

The expert system itself is implemented in the Nexpert system shell. The current rulebase contains 1,500 rules covering 450 faults, cycling every 15 seconds providing fault output to a system of databases and fault displays. The *Annie* software is responsible for the interface between Nexpert and the data acquisition system. For the purposes of this paper we will only discuss the aspects of the expert system related to data acquisition.

The simplest and a very common form of access for incoming data from the expert system is to ask for its current value. The separate *Annie* processes have already taken care of matching the value to levels normal, high or low, depending on the time values needed. This pre-processing greatly reduces the amount of work needed at the expert system end.

A key observation is that only some of the data is actually used by the rules. The pre-processing step filters some of the data which would never be used, and abstracts it to a more useful form. The rulebase is also structured so that a trigger event will begin a more indepth analysis of a particular set of problems. Many of the more complex functions are only used when such a trigger event has occurred. This results in an efficient processing of the incoming data and reduces un-needed analysis.

Many of the rules are oriented around a time history of the values. Particular events such as a flow setpoint altering significantly, trigger the expert system to check a set of possible faults. In this case, *Annie* provides extensive functionality to provide access to the history of values. Because the system has planned ahead to acquire the data in a suitable form, these functions are able to look back and determine the pattern at the time the rulebase runs.

As an example rule, consider the gas analyser on the flare stack. For safety reasons, it is important to detect whether the gas analyser is faulty. In a normal situation the mixture of a gas such as CO is constantly varying. The *Annie* time history function **percent variation less than for** is used to determine whether the CO level is constantly changing. If the gas analyser is changing by less than a certain percentage, then we can deduce that it has failed. This can also be confirmed by comparing the percent variation of other gas analyses in other parts of the system. The percent variation function examines the two minute time history of the data. It looks at the way it has been changing over that period and the amount of variation which has occurred. To implement this looking back is relatively straight forward, however, to have implemented this as a constant analysis as data was arriving, would be extremely difficult. Table 1 provides a list of further examples of the *Annie* time history functions. Multiple signal versions of many of these are also available.

Table 1
Annie Historical Data Access Functions

equal_for
constant_for
less_than_for
greater_than_for
less_equal_for
greater_equal_for
channel_increased
channel_decreased
value_between_for
number_outside_limit
number_above_limit
number_below_limit
percent_variation_lt_for
max
min
gradient
mean
integrate
last_change_of
at_status_for
last_time_to
response_time
read_time
read_system_time
read_rules_loop_time
start_timer
read_timer
timer_running
timer_status

EXAMPLE DIAGNOSTIC RULE

The following is an example of the type of rule which can be implemented; Imagine the fault to be detected is a control valve, being stuck or too slow to respond. If a flow rate increases significantly above the set point, we should see the controller output decrease and as a result the valve position decrease.

Figure 1 shows an example rule to investigate this as well as the time history of the data. In this case the dashed lines indicate correct operation of the central system while the solid branch illustrates this particular fault. Once the expert system has found that the flow rate is too high and the set point is OK then it checks to see if the controller output to altered to attempt to reduce the flow. We would say if the flow rate is too high and the controller output has decreased to close the valve and the valve position is not closing, then the control valve is stuck.

In order to determine whether the control output has decreased, *Annie* looks back over the last two minutes of data, and through one of its access functions, is able to determine that the control output is gradually decreasing. It is then able to look at the valve position over the last two minutes and determine by how much it has decreased. If the valve position has not decreased significantly after being instructed to close then we would say that it is stuck or has a slow response. Elaboration of the rules in this area would allow us to differentiate between these cases. It should be noted that by looking at the time history, this is easy to detect. To try and determine from the current values could be difficult and unreliable.

SUMMARY

In this paper we have described a large scale real-time application. The critical aspect of real-time is that none of the required data is lost during the analysis process. The existing hardware/software combination runs slower than the data acquisition system. We have implemented a software strategy that involves planning ahead to look back. With this strategy, all the information required for real time diagnosis is available on demand. The data is pre-processed and stored in special history areas. The diagnostic rules are then able to examine this history with powerful feature extraction capabilities in order to determine how the system has behaved over time. This provides an effective strategy for a large number of real-time systems.

REFERENCES

1. Milne, R.W., Artificial Intelligence for Online Diagnosis, IEE Proceedings, Vol. 134, Pt. D, No. 4th, July 1987

2. Milne, R.W., Case Studies in Condition Monitoring. Knowledge-Based Systems for Industrial Control, IEE Control Engineering Series 44. (J. McGhee, M.J. Grimble and P. Mowforth, Editors), p255-266. Peter Peregrinus Ltd Publications, 1990

3. Milne, R.W., Diagnostic Strategies. Systems & Control Encyclopedia Supplementary Volume 1, Madan G. Singh, Editor in Chief, p155-160. Pergamon Press Publications, 1990

4. Milne, R.W., Expert Systems On-line. Systems & Control Encyclopedia Supplementary Volume 1, Madan G. Singh, Editor in Chief, p243-251. Pergamon Press Publications, 1990

5. Milne, R.W., Monitoring Process Control Systems. Advances in Engineering Software, Computational Mechanics Publications, vol 12, no. 3, p129-132, July 1990.

6. Milne, R.W., On-Line Artificial Intelligence, The 7th International Workshop on Expert Systems & Their Applications, Avignon, France, 13th-15th May 1987.

FIGURE 1: EXAMPLE DIAGNOSTIC RULE

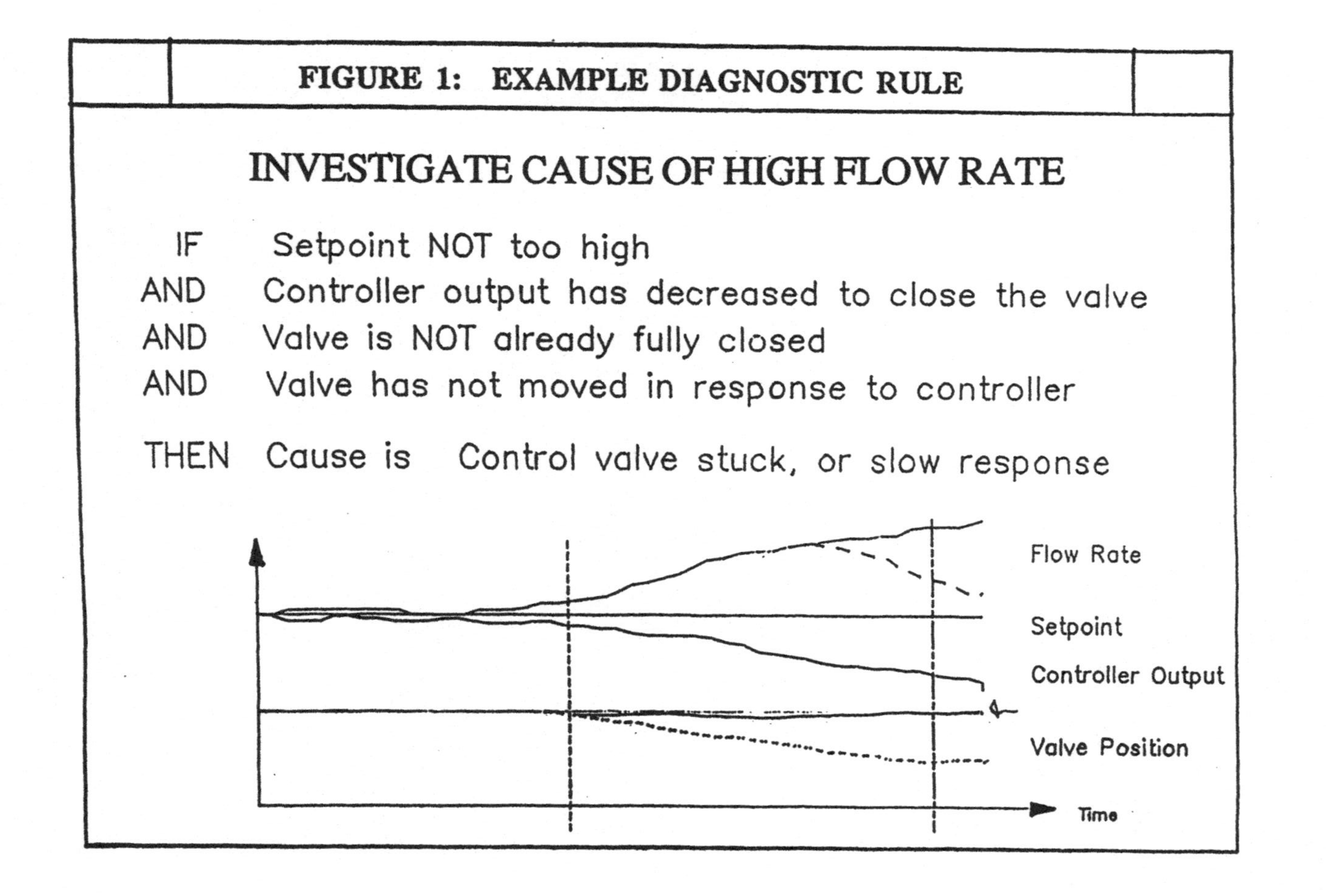

INVESTIGATE CAUSE OF HIGH FLOW RATE
IF Setpoint NOT too high
AND Controller output has decreased to close the valve
AND Valve is NOT already fully closed
AND Valve has not moved in response to controller
THEN Cause is Control valve stuck, or slow response
Flow Rate
Setpoint
Controller Output
Valve Position
Time

Resource Leveling in PERT by Neural Network

T. Shimazaki, K. Sano, Y. Tuchiya

Department of Civil Engineering, The University of Tokyo, 7-3-1, Hongo, Bunkyo-ku, Tokyo 113, Japan

INTRODUCTION

Recently, the application of neural network theory are tried in various fields, and are successful. In the field of civil engineering, Akamatsu et al.[1] and Nakatuji[2] showed that the possibility of the application of the theory, for example.

Major fields of the application of the neural network theory are pattern identification, optimization and so on. This paper examines the applicability of the theory to resource leveling problems in PERT analysis for construction projects. The problem is practically impossible to solve analytically. Accordingly, various heuristic solution procedures were proposed, but they cannot necessarily assure the global optimum solution.

First, we briefly review the neural network theory from the view point of applications to optimization problems. Then, we formulate resource leveling problems in PERT construction scheduling in the adequate form for the neural network theory in two way. We apply the formulated problems to simple examples to investigate convergency, computation time when the von Neumann type computer is used, and so on. Finally, we discuss the applicability of the theory, especially when the parallel type computer is developed.

OUTLINE OF NEURAL NETWORK THEORY

Principle of Neural Network Theory

The definition of information processing by neural networks is not established yet. Conceptually, the neural network theory is explained as information processing by network mechanism consisting of huge number of the simple information processing units which are connected each other and can exchange signals[3]. The theory is simulating the function of human brain.

Various types of the simple information processing unit are proposed. Most general one is the unit which accepts several inputs, changes the internal state of units according to inputs, and output a signal according to the internal state. This type of unit is corresponds to neurons in human brain, and can be shown as figure 1 schematically. Hopfield model is one type of transform from input signal to output, and can be expressed by the following equations[4].

$$du_i / dt = - u_i + \Sigma T_{ij} V_j + I_i \quad (1)$$

$$V_i(t) = g(u_i(t)) \quad (2)$$

where, u_i, v_i, and I_i are internal state, output, and the eigen state of the i-th neuron respectively, Tij is the connection matrix which express the relation among neurons, and g(*) is the sigmoid function(see figure 2). Most frequently used sigmoid function is the following logit type function.

$$g(u) = (1 + \tanh(u/\theta)) / 2 \quad (3)$$

where, θ is the parameter which represents the sensitivity of the neuron, and is called as temperature parameter in analogy with the statistical dynamics. Equations (1) - (3) are named as the equations of motions of neuron.

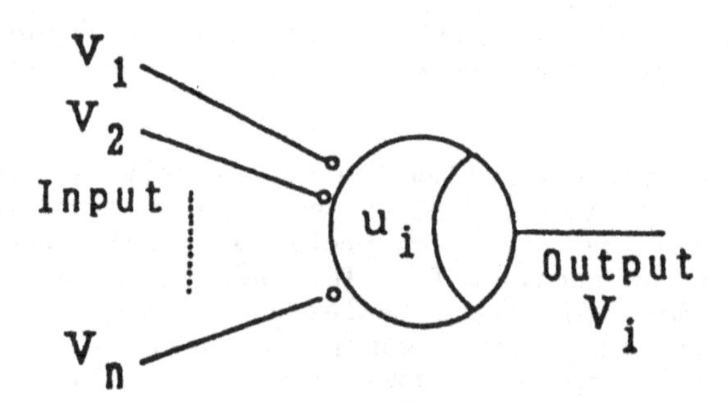

Figure 1. Schematic Diagram of Neuron

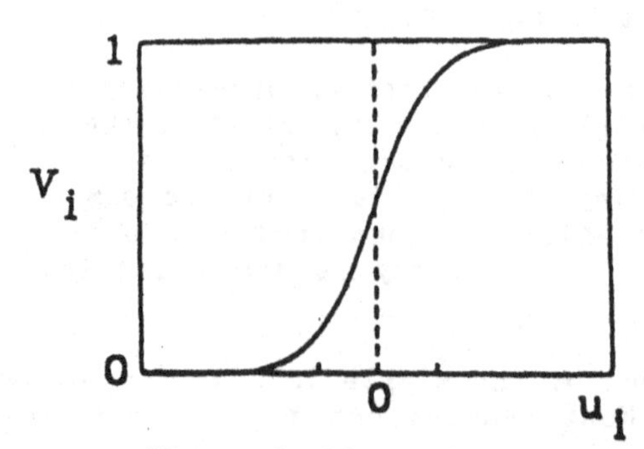

Figure 2. Sigmoid Function

Typical models which express the connections of neurons include hierarchical network, interconnected network and so on. According to the neural network theory, neurons change their state in parallel so as to minimize Lyapunov function defined on the whole neural network. Based on this characteristics, optimization problems can be solved by the neural network theory when an objective function and variables of optimization problems can be transformed to Lyapunov function and internal state of neuron, respectively.

Hopfield showed that the following function (4) becomes Lyapunov function in case that the equations of motion of neuron are expressed in the form of equations (1) - (3), and that the connection matrix T_{ij} is symmetrical. It means that internal states of neurons change so as to minimize the value of function (4), and the minimum point of E is stable. The function (4) is called as the energy of network.

$$E = -\frac{1}{2}\sum_i\sum_j T_{ij} V_i V_j - \sum_i I_i V_i + \sum_i \int_0^{V_i} g^{-1}(\omega)\, d\omega \qquad (4)$$

In function (4), the third term of the right hand side becomes 0 when variable V_i takes only 0 or 1. This fact can be shown by directly integrating the term. When temperature parameter θ is large, V_i can take the real number between 0 and 1, and the value of the term becomes large. When the temperature parameter is very small V_{ij} takes only 0 or 1, and the value of the term becomes 0. Accordingly, it is possible to solve the discrete variable problem using continuous variables by decreasing the temperature parameter. Based on this, Hopfield and Tank solved the traveling salesman problem by neural network theory[5]. However, it should be noted that the procedure may converge to local optimum in case of an inadequate initial solution.

FORMULATION OF RESOURCE LEVELING PROBLEM

Assumptions for Formulation

In formulating resource leveling problems in the form of neural network theory, we made the following general assumptions.

(1) Activity cannot be divided.

We assume that an activity is minimum unit of job and cannot be divided anymore. When an activity can be dividable, divided activity should be considered as the unit activity. So, this assumption is not restrictive.

(2) Project duration is given.

A project duration is given, typically, the duration when all the activities are started on the earliest start time by the usual PERT calculation.

(3) Number of resource is one.

In general, several resources are considered in resource leveling problems. However, all the resources are not optimized at once even in such cases, but the optimization is done with priority. This assumption can be generalized by defining problems to optimize the weighted average of several resources.

(4) Resource requirement per day for an activity is constant during a duration.

Resource requirement may varies day by day in general. It is assumed that resource requirement is constant here. This assumption can be generalized by dividing an activity into unit activities which has the constant resource requirement.

Formulation of Model

To solve optimization problems by neural network method, it is required to find out the correspondence between internal states of neuron and the variables of a given problem first. Here, we used two types of formulation. In model 1, a state of neuron is defined by whether some activity is being executed on some day or not. In model 2, a state of neuron is defined by whether some day is the start day of an activity or not.

Model 1 is formulated as follows. The internal state of neuron is defined as equation (5).

$$V_{ij} = \begin{cases} 1 & \text{the i-th activity is being executed on the j-th day,} \\ 0 & \text{the i-th activity is not being executed on the j-th day.} \end{cases} \quad (5)$$

The variable V_{ij} may take 1 between the earliest start time of an activity and the sum of the latest start time, the duration and the total float. So, following constraints should be satisfied.

$$S_i \leqq j \leqq S_i + d_i - 1 + f_i \quad (6)$$

where, S_i is the earliest start time of the i-th activity,
d_i is the duration of the i-th activity, and
f_i is the total float of the i-th activity.

In relation to the duration of an activity, following constraints should be satisfied.

$$\sum_j V_{ij} = d_i \quad \text{for all } i \quad (7)$$

where, the summation on j is taken over the range where the constraints (6) is satisfied.

From assumption (1), V_{ij} whose value is 1 should be continuous in regard to j, following constraints should be hold.

$$\sum_{j} V_{ij-1} \cdot V_{ij} = d_i - 1 \quad \text{for all } i \tag{8}$$

Lastly, following constraints should be satisfied in relation to the order of the activities.

$$\sum_{(i,i')\in P} \sum_{j'} \sum_{j=j'}^{S_i+d_i-1+f_i} V_{ij} \cdot V_{i'j'} = 0 \tag{9}$$

where, P is a set of activity pairs (i, i') in which the i-th activity is the predecessor activity of the i'-th activity. The summation on j' is taken over the range where the constraints (6) for the i'-th activity.

Objective function can be expressed as the minimization of the deviation of resource requirement for each day, as follows.

$$\min \sum_{j} \left(\sum_{i} V_{ij} \cdot r_i\right)^2 \tag{10}$$

where, r_i is the resource requirement per day for the i-th activity.

As a result, the resource leveling problem can be formulated to obtain the value of V_{ij} which minimize the objective function (10) under the constraints (6) - (9) in model 1.

Similarly, model 2 can be formulated as follows:

Objective function:

$$\min \sum_{j} \left(\sum_{i} \left(r_i \sum_{j'=j-d_i+1}^{j} V_{ij'}\right)\right)^2 \tag{11}$$

where, the summation is taken over the range between 1 and (project duration - activity duration + 1). And, r_i is resource requirement of the i-th activity, di is duration of the i-th activity, and
V_{ij} = 1 the i-th activity is started on j-th day,
0 the i-th activity is not started on j-th day.

Constraints that starting day is only one:

$$\sum_{j} V_{ij} = 1 \quad \text{for all } i \tag{12}$$

Constraints on order of activities:

$$\frac{\partial E}{\partial V_{ij}} = r_i \sum_{j'} r_{i'} V_{i'j} - \mu_i$$

$$- R_1 d_i + R_1 \sum_{i} V_{ij}$$

$$- \lambda_i (V_{ij-1} + V_{ij+1})$$

$$- R_2 (d_i - 1) (V_{j+1i} + V_{j-1i})$$

$$+ R_2 (V_{j+1i} + V_{j-1i}) \sum_{j} (V_{ji} V_{j+1i})$$

$$+ \sum_{i_1} \delta ii1 \ \nu (i_1, i_2) \sum_{j'} V_{i_2 j'}$$

$$+ \sum_{i_2} \delta ii2 \ \nu (i_1, i_2) \sum_{j'} V_{i_1 j'}$$

$$+R_3 \sum_{(i,i') \in P} (\delta_{jj1} \sum_{j'} V_{i'j2}$$

$$+ \delta_{ii2} \sum_{j'} V_{i'j1}) \cdot (\sum_{j} \sum_{j'} V_{ij1} V_{i'j2}) \qquad (15)$$

In equation (15), the second term of the right hand side should taken as 0 when the corresponding variables are not existing. Ri's are the adequate parameters of the penalty term of second order. In solving this, the revision of the coefficients of Lagrange can be done by following procedure;

$$\mu'_i = \mu_i + R_1(d_i - \sum_{i} V_{ij}) \qquad (16)$$

$$\lambda'_j = \lambda_j + R_2(d_{i-1} - \sum_{i} V_{ji} V_{ji} + 1) \qquad (17)$$

$$\nu'(j_1, j_2) = \nu (j_1, j_2) + R_3 \sum_{i} \sum_{i'} V_{j_1 i} V_{j_2 i'} \qquad (18)$$

$$V_{i^n j} \sum_{j'=1}^{j+d_{i^n}-1} V_{i'_n j'} = 0 \qquad \text{for all } n \tag{13}$$

where, in is precedence activity of the n-th activity pair which has precedence-following relation, and i'n is following activity of the n-th activity pair which has precedence-following relation.

Calculation Procedure of the Problem

The resource leveling problem above formulated is expressed as the minimization problem with constraints. To solve the problem by the neural network method, it is required to transform to the minimization problem without constraints. The problem was transformed by extended Lagurangian Method of multiplier method[6]. That is, the problem was transformed to the optimization of the following objective function without constraints.

$$E = \sum_j \left(\sum_i V_{ij} \cdot r_i\right)^2 + \sum_i \lambda_i \left(d_i - \sum_j V_{ij} \right)$$

$$+ \frac{1}{2} R_1 \sum_i \left(d_i - \sum_j V_{ij}\right)^2$$

$$+ \sum_i \mu_i \left(d_i - 1 - \sum_j V_{ij-1} \cdot V_{ij}\right)$$

$$+ \frac{1}{2} R_2 \sum_i \left(d_i - 1 - \sum_j V_{ij-1} \cdot V_{ij}\right)^2$$

$$+ \sum_{(i,i')\in P} \nu_{(i,i')} \sum_{j'} \sum_{j=j'}^{S_i+d_i-1+f_i} V_{ij} \cdot V_{j'i'}$$

$$+ \frac{1}{2} R_3 \sum_{(i,i')\in P} \left(\sum_{j'} \sum_{j=j'}^{S_i+d_i-1+f_i} V_{ij} \cdot V_{i'j'}\right)^2 \tag{14}$$

where, λ, μ, and ν are the coefficients of Lagrange, and R_i's are the adequate parameters.

Sum of the second and the third term of the right hand side of equations of motion of the neural network (1) is coincides with the derivative of the energy of the network (4). As mentioned before the third term of the right hand side is negligible if the term is small in regard of V_i. The derivative can be expressed by following equation in this case.

The whole calculation procedure is as follows;

(1) Using the arbitrary initial solution of V_{ij}, calculate the output by equation (2),
(2) Calculate the amount of revision of V_{ij} by equation (1), considering equation (15),
(3) Revise the values of λ, μ, and ν based on equations (16) - (18), and
(4) Repeat these steps until converges.

Similarly, model 2 can be transformed to minimization problem without constraints in the following form, and the revision procedure for Lagrange's coefficient as follows.

Objective function:

$$E = \sum_j \left(\sum_i \left(r_i \sum_{j'} V_{ij'} \right) \right)^2$$

$$+ \sum_i \mu_i \left(1 - \sum_j V_{ij}\right)$$

$$+ \sum_n \nu_n \sum_j \left(V_{inj} \sum_{j'} V_{i'nj'} \right)$$

$$+ \frac{R_1}{2} \sum_i \left(1 - \sum_j V_{ij} \right)^2$$

$$+ \frac{R_2}{2} \sum_n \sum_j \left(V_{inj} \sum_{j'} V_{i'nj'} \right)^2 \qquad (19)$$

Revision procedure of Lagrange multiplier:

$$\mu'_i = \mu_i + R_1 \left(1 - \sum_j V_{ij} \right) \qquad (20)$$

$$\nu'_n = \nu_n + R_2 \sum_{j'} V_{inj} \; V_{i'nj'} \qquad (21)$$

In this procedure, it is important to select the value of temperature parameter θ adequately. It is supposed that the third term of equation (4) is very small and is neglected in the process of induction of the calculation procedure, so the temperature parameter should be small in the final state of convergence. However, when the value of temperature parameter θ takes small value from the beginning, neurons behave as if they have the deterministic characteristics, and they tend to converge to local minimum. So here, we used the annealing method which keeps the temperature parameter large at the first time, and gradually decreasing it. There are no general rule to decide the initial temperature parameter, and the rate to decrease it. The most general equation (22) was used in this attempt.

$$\theta = \theta max/\log(t+1) \quad t>0 \qquad (22)$$

where, t is the number of iteration, and
θ max is the initial temperature parameter.

EXAMPLE CALCULATION

To check the performance of the neural network method for resource leveling problems, we applied the method to simple examples. The applied problem is as shown in figure 3 and table 1 for model 1.

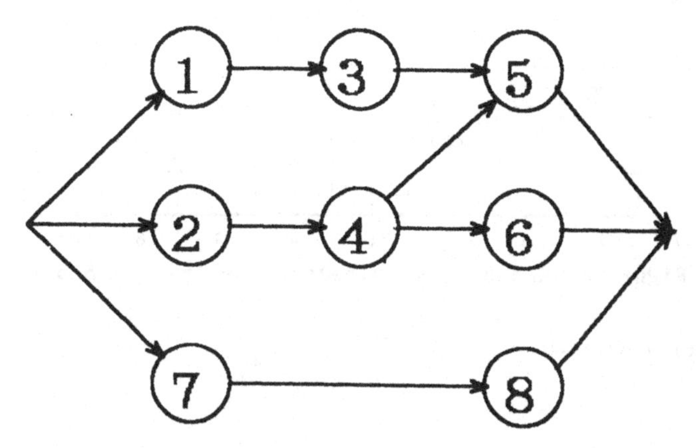

Figure 3. Example of Applied PERT Problem

Activity	Duration	Preceding Activities	Resource Requirement
1	2	—	4
2	1	—	3
3	3	1	9
4	3	2	6
5	4	3, 4	8
6	4	4	8
7	3	—	3
8	1	7	3

Table 1. Conditions of Applied PERT Problem

Resource requirement when all the activities were started on the earliest start time is shown in figure 4, and the optimized result by model 1 is shown in figure 5, which coincides with the real optimum obtained by enumeration method in this case. Sum of squares of resource requirement was reduced 10 % in this case. Similar examples were optimized by model 2, and compared with the solution by enumeration method. The result shows that real optimum solutions were obtained in 3 cases out of 10. In other 6 cases, the difference of value of objective function is less than 5 %. The largest difference was about 7 %.

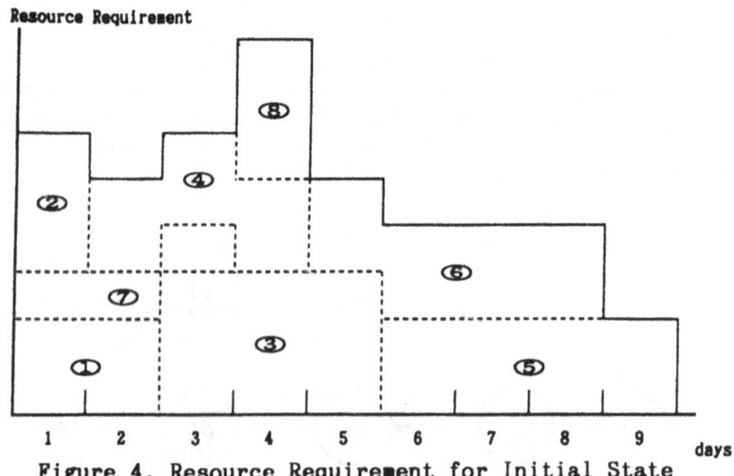

Figure 4. Resource Requirement for Initial State

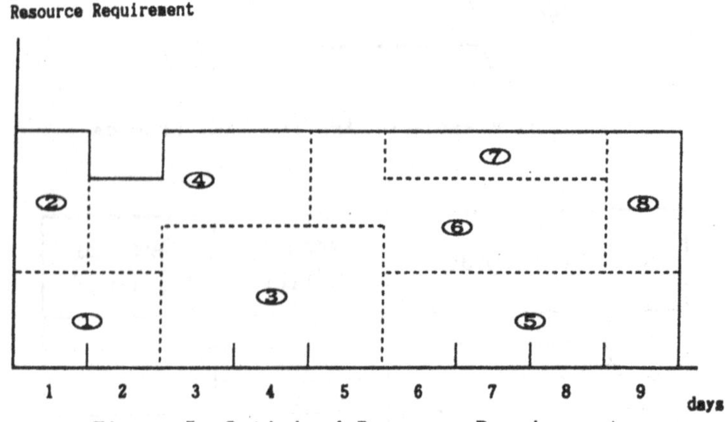

Figure 5. Optimized Resource Requirement

Number of iteration to obtain the result of figure 5 was about 3000 times by model 1, and the computation time was about 2 minutes on small personal computer(PC9801VM). The required number of neuron is the order of number of activities times the project duration. In the example, the number of neurons is 72.

When the parallel neural processor is developed, the required time can be reduced to 1/72. This effect becomes larger when the scale of problems become large. The CPU times for model 2 were similar level with model 1. Required number of neuron for model 2 is the order of sum of number of activities times its total float.

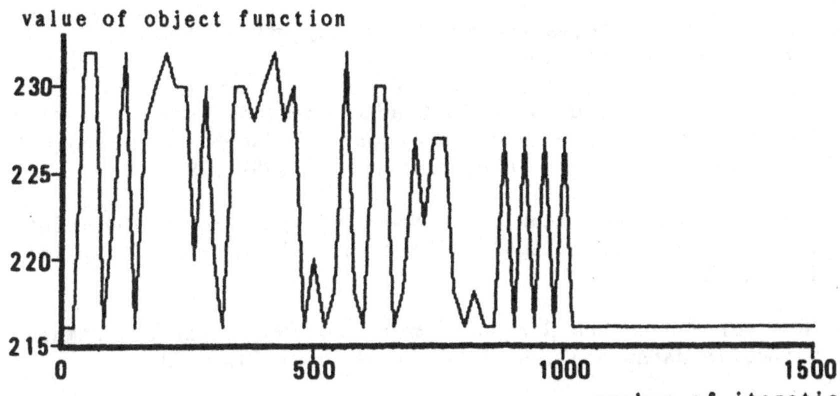

Figure 6. Convergence of the Model

Figure 6 shows the state of convergence. Vertical axis is the value of objective function, and the horizontal axis is the number of iteration. The figure is plotted for the solution which satisfies the constraints within 0.5 % in terms of error. In addition, the convergence is checked by the extent how the constraints are satisfied. In general, constraints on continuity is tend to converge fast, but the other constraints converge slowly. It looks better to check the level of convergence and output the result which has the some level of convergence. These can be said also for model 2. As convergence is not necessarily complete, the value of neuron Vij takes the real number between 0 and 1 in this formulation, which should be discrete value of 0 or 1. The result of the figure 5 is the solution obtained by rounding the value.

CONCLUDING REMARKS

Theoretically, the resource leveling problem can be solved by neural network theory. However, it is difficult to solve the problem by the sequential type computer from the view point of computation time. When the neural computer is developed, this procedure may be practical.

REFERENCES

1. Takashi Akamatsu, Yuuji Tuchiya, and Toshikazu Shimazaki,

"Application of Neural Network Theory to Urban and Transportation Planning(In Japanese)", Proc. of 44th Annual Meeting of Japan Society of Civil Engineers, pp.160-161, 1989.

2. Takashi Nakatuji and Terutoshi Kaku, "Application of Neural Network Model to Traffic Engineering(In Japanese)", Proc. of Infrastructure Planning, JSCE, No.12, pp.297-304, Dec. 1989.

3. Hideki Aso, "Information Processing by Neural Network", p.4, Sangyou Tosho, June 1988.

4. Hopfield,J.J.,"Neurons with Graded Response have Collective Computational Abilities like those of Two-State Neurons," Proc. Natl. Acad. Sci. USA, 81, pp.3088-3092, 1984.

5. Hopfield,J.J. and Tank,D.W.,"Neural Computation of Decisions in Optimization Problems," Biological Cybernetics.52, pp.141-152, 1985.

6. Hiroshi Konnno and Hiroshi Yamashita, "Nonlinear Programming Method(In Japanese)", Nikkagiren, March 1978, p.23.

CONTRALTO: Constraint Reasoning Applied to Logistics for Transport Organisations

C. Guimaraes (*), J.-M. Le Dizes (**)

(*) Ed. Nùcleo des Transports, SAN Quadra 3 Lotes O e N, 70.040 BRASILIA-Distrito federal, Brazil

(**) Consultant, Les Figons 13510 Eguilles, France

Abstract :

This paper presents an attempt to apply the constraint reasoning paradigm to a real life problem. It has involved three types of skill: that of road freight transport operators, the transport logicist, and the Artificial Intelligence specialist.
It has resulted in a tool designed to assist freight carriers in finding the best satisfactory solution for matching freight transport demand with carrier supply.
Empirical knowledge has been gained from operators. The transport model has been constructed by a logicist. The prototype software has been coded with Prolog III™ which embeds constraint handling in its core solving process.
In conclusion, we state (we suspected it before) that there is no 'declarative miracle' insofar as declaration is never sufficient...how the problem should be solved must be specified as well. It seems that the art of constraint reasoning is the art of adjusting constraints in order to keep within overconstrained and underconstrained situations.An art which is not the least part of the application expertise.

Keywords : Road Freight Transport, Logistics, Constraint Satisfaction Problem, Constraint Programming, PROLOG III™(*).

(*) PROLOG III is a registered trademark of Prologia Company (Marseille FRANCE)

1- Introduction

Determining the optimal solution for reasonably large problems in a reasonable amount of time has always been the ultimate goal of Operations Research. However, real life problems are so complex that they cannot be handled as a whole; a usual way to tackle such problems is to split them into 'quasi' independant subproblems according to the everlasting reductionist methodology.

Yet, it is highly desirable to find new techniques which might allow us to solve a problem in a reasonable time while keeping its globality, even though it means giving up any optimum. That might also mean that a set of satisfactory candidate solutions is acceptable. This is precisely the reason for the constraint reasoning paradigm gaining momentum. It meets the needs of some decision makers involved in planning, scheduling or assignment tasks.

Freight Transport Logistics aims at meeting the transport demand with the most economical use of resources. It involves distinct resources, conflictual constraints and various options.

Resources are vehicles, load units, drivers; Constraints are mainly demand related (e.g. type of cargo, pickup & delivery place and date,..). Options are supply related (route, schedule, vehicle/driver/load unit assignment, ..).

So far, constraint reasoning has not been attempted in this area; A study (Guimaraès [15]) has been conducted among actual Freight Transport Companies. It points out that the real life assignment problems , with which planners and schedulers are daily faced, are quite amenable to constraint reasoning.

Recent implementations of this emergent paradigm into so-called constraint programming languages, have allowed us to experience with it on a real concrete scale to meet real needs.

The needs expressed by operators, the willingness of specialists in modelling
(in both areas:Transport Modelling and Artificial Intelligence), the availability of adequate tools, have been the three major preconditions which made the experiment, a success.

2- Road Freight Transport : The new challenges of a pan-European deregulated market.

2-1 A Growing Activity :

Over 70 % of freight (measured in ton/kms) in the EEC countries (European Economic Community) is moved by road.

The volume has increased about 25 % over the past five years, while the total volume of international transport, as a whole, regardless of the mode, has increased about 15 % and the rail transports has marked a zero increase (Eurofret [1]).

European deregulation in freight transport is viewed by transport professionals as an opportunity to rationalise their logistics systems, such as gaining economy of scale, increasing flexibility and mobility , seeking cooperation between companies to cut operating costs and improve customer service.

In particular, cooperation might be an opportunity to find **return shipments** (backhauling), a big concern for transport suppliers.
An approximation says that some 30 or 40 % of the vehicles are **running empty** in one of the directions in international traffic (Eurofret [1]). The reasons are strict regulations, imbalances in trade or transport between areas or countries , imbalances in demand with special transport features, as well as ineffective information systems.
Furthermore, a major change for transport companies on a deregulated European market is the allowance to pick up cargo anywhere.

Strong competition is likely to restructure the market segments; recent surveys (Delphi poll, quoted in [1]) have shown that some market segments are expected to grow dramatically (dedicated contract transport, express transport and groupage). Hauliers will have to be able to meet the new market demands , such as "Just-In-Time" production with small shipments and precision (on-time delivery).

Tough competition should lead medium-sized operators to cooperate with each other in order to set up a network structure allowing a better utilization of fleet, vehicles, personnel and information systems.

To-day key problems in freight logistics are centered around four questions :

(i) How to meet the freight transport demand ?
(ii) How to minimise the economic operating constraints ?
(iii)How to match freight transport demand with carrier supply ?
(iv) How to communicate and cooperate with other operators ?

2-2 RTI-Technologies in Road Freight Operations: an overall assessment

2-2.1 -Current use and expectations

Recent Surveys (Fleet [2]) have shown that RTI (Road Transport Informatics) Technologies have been little used so far.
Traditionally, freight transport involves three management levels :
Freight Management
Fleet Management
Vehicle Management
Apart from the conventional managerial tasks, all three levels suffer from a lack of computerised processing.

An inventory (Fleet [3]) shows that there exist a number of systems to help in carrying out tasks such as transport order processing and even tour planning, but few of them are in routine use. Some simulation logistics tools are available (Strada [16]) .
Resource allocations are mostly processed manually (driver duty, freight and vehicle assignment, scheduling, routing); transport documents are mostly handled manually or mailed, the use of telephone, telex and telefax are the dominant communication means for cargo tracking, notification of shipment status and estimated time of arrival.

However strong needs have been expressed, especially needs for an integration of communication and monitoring facilities within logistics systems. The greatest expectations are put on EDI (Electronic Data Interchange) communications in order to create an integrated information system. It is viewed as a prerequisite to actually take advantage of advanced techniques such as artificial intelligence.

The most widely expected benefits concern two specific criteria : **"improved planning capability"** and **"increased fleet utilization"** (Fleet [2]) within the framework of a cooperative policy of fleet management.
These criteria are central to the capability of transport companies to meet the "just-in-time" production and eventually survive the oncoming deregulated pan-European freight transport market.
Meeting these criteria implies more care in the solving of resource allocation problems.
Some basic questions are :

- How to ensure that resource allocation reflects the most satisfactory solution regarding the conflictual demand-offer constraints ?
- How to increase the vehicle load and control the total amount of empty movements ?

Most research dealing with freight transportation has focused exclusively on loaded vehicles and very little work has been specifically done regarding empty freight returns(Attain [4], Hutson-Wood [7]) .
However, empty trips generate cost but no revenue, consequently, empty movements reduce the global revenue.

2.2-2 State-of-the-Art in similar areas

Logistics problems belong to the realm of problems which are characterised by the risk of being computationally untractable. Many of them are NP-complete problems. Operations Research usually tackles such complex problems by splitting them into pieces which make them tractable with less effort. As an example, the tactical planning activities are separated from each other: timetabling, vehicle assignment, crew scheduling.

A great deal of research has been carried out in this area: An entire generation of researchers has focused on the best algorithms or heuristics to apply to specific sub-problems such as activity planning, route finding, task scheduling, resource assignment, ... Many of them have been implemented in application specific software packages. In the public road transport area (for mass transit), an area which is close to Freight transport, Operations Research based tools are routinely in use for selecting transit routes, assigning vehicles to service lines, assigning crews to vehicles, ..
A complete review has been made in (Rousseau [6]) and more recently within the framework of (Cassiope [5] , Le Dizès [17]).
Attempts have been made also in the space planning area ; in particular in cargo stowage planning (ship or railways) (Sansen [8], Beurrier [9]). Adopted approaches range from blackboard based expert systems to rule based constraints.
Constraint satisfaction approaches have been experienced for assignment problems such as gate assignment for railways and airlines(Brazile & al. [10]).

3- Constraint Reasoning: an ideally suited paradigm for logistics problems

Constraint reasoning is an emergent paradigm for formulating knowledge in terms of a set of constraints, without specifying methods to satisfy them. Constraints are usually applied to some entities or relationships between these entities which make up the universe of discourse (or the problem space).
More theoretical definitions can be found in (Jaffar & al.[12] and Van Hentenryck & al. [14]).
What makes this paradigm work is the powerful underlying search strategy which is based on the "constrain & generate" principle rather than on the conventional "generate & test" principle.This feature allows a dramatic decrease in the search for solutions.

As stated earlier, many logistics problems are not easily tractable by Operations Research optimisation techniques, no optimum being attainable within a reasonable amount of time. In practice, constraints are so conflictual that it requires compromises between constraints to get workable solutions which are acceptable.

Recent programming languages, such as CHIP (Dincbas [11]), CLP($\mathcal{R}$) (Jaffar & al.[12]), and PROLOG III (Colmerauer [13]) have embedded features which enable the programmer to implement the constraint reasoning paradigm.All these languages are soundly based on formal semantics. Furthermore, they take advantage of the powerful expressiveness of declarative logic .

4-What does CONTRALTO consist of ?

CONTRALTO is a tool to assist freight transport companies in finding out the best satisfactory solution for matching freight transport demand with carrier offers.

4.1- The Freight Transport Model

The freight transport model is demand driven; i.e. its first objective is to meet the demand needs as far as it is compatible with operation capabilities.
It is based on the concept of 'circuits' with special emphasis on the empty return loads. More details can be found in (Guimaraès [15]).
Built-in knowledge is real human practitioners' expertise, captured by the transport specialist.

CONTRALTO suggests solutions which best satisfy two conflictual sets of constraints :
- The constraints of freight transport demand.
- The constraints of the carrier.

The major demand constraints are (Fig. 1) :

1- Physical freight characteristics:
Volume(or bulk) and Weight
Type : foodstuffs, chemicals, machinery, metal products, general, dangerous, fertilisers, ..
2- Availability and Delivery dates.
3- Origin and Destination.
4- Cost and special service packages, ..

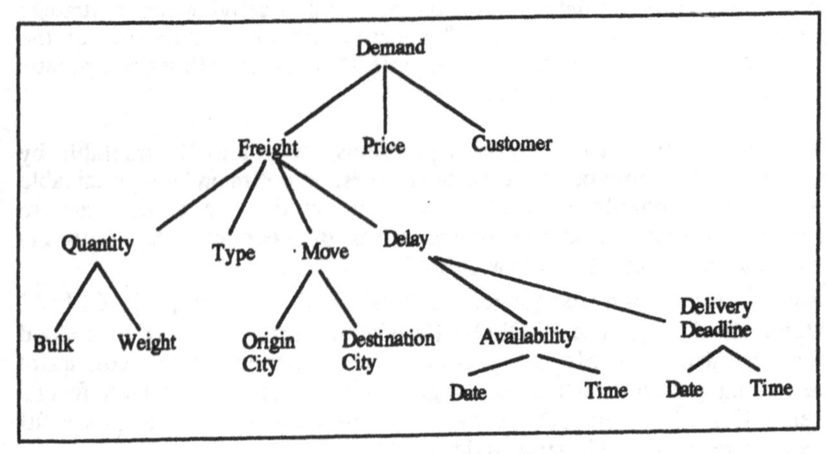

- Fig. 1 - Demand Structure

The major supply constraints/alternatives are (Fig 2) :

1- Fleet size and loading capacity(maximum size and weight).
2- Type of operation (intermodal, own account, hire & reward).
3- Type of load units(full/semi trailers, swap bodies, ISO containers, refrigerated units, ..).
4- Type of service(regular line haulage, tramp haulage, regular dely/pickup, parcel/courier service, "Just-in-Time" systems,..).
5- Cargo Compatibility.
6- Availability(places, dates) of load units, vehicles and drivers.
7- Limitations regarding the total mileage of vehicles between two maintenance inspections.
8- Limitations regarding transport regulations and labour rules.

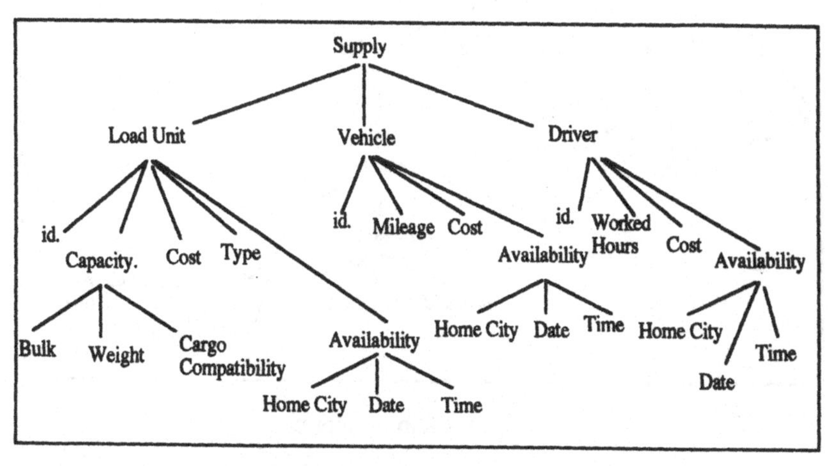

- Fig. 2 - Supply Structure (id. stands for identification)

The main tasks are :

(1) to rank-order the freight units according to their origin and destination.
(2) to fill the available load units taking account of the loading capacity and the cargo compatibility.
(3) to determine routes for each trip which minimise the empty returns.
(4) to determine the time schedule for each trip taking into account the demand delivery requirements.
(5) to assign vehicles and drivers in compliance with enforced regulations and possibly with driver preferences as well as in coordination with vehicle maintenance needs.

The system searches for all solutions which are compatible with constraints.
Each solution consists of a detailed description of each trip (Fig 3) :

- Vehicle identification (+ total mileage)
- Load unit identification.
- Driver identification. (+ total number of worked hours).
- Total cost.
- The move list :
  For each move :
    - Link: Departure & Arrival places.
    - Actual cargo (List of cargo identifications, volume, weight, departure & arrival times).

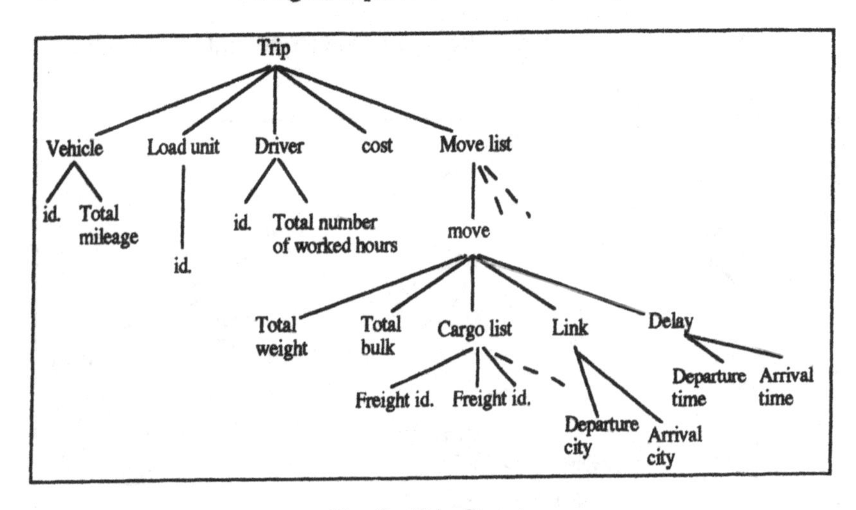

- Fig. 3 - Trip Structure

Some of the above-listed constraints are flexible; it is up to the user to adjust them and to rank them until a reasonable number of solutions is yielded.
Constraints can be reinforced in order to reduce the number of candidate solutions or to provide first the most satisfactory. e.g. taking advantage of a provisional excess in supply to make the decision of delaying some departure or suggesting that drivers take time off.
Likewise, if no solution can satisfy the set of constraints, constraints can be gradually relaxed. e.g.: in case of supply deficit, a certain tolerance can be allowed for some temporary driver overtime, extra delays before vehicle technical inspection.

4-2 The Constraint Programming based Implementation.

CONTRALTO has been coded with PROLOG III ™.
The demand and supply data as well as the geographical network are implemented as Prolog facts (tail free Horn clauses with bound variables).

Each arc of the geographical network(whose nodes are stop places) is labelled with compound cost, distance and travelling time.
(In fact compound cost should be calculated for each type of load unit, vehicle and driver).
Constraints are embedded in Prolog III clauses (this is an extended feature of Prolog III which allows the expression of numeric/boolean/tree constraints between local variables).

The solving process is guided by so-called meta-constraints. i.e. constraints to express the chronological order of constraint evaluations, to force the system to provide first the most satisfactory solutions.

<u>A case study</u>

Figures 4, 5, 6 describe a simple case study which involves the following elements :

12 freight of various types, origins, destinations, volume and weight.
4 vehicles; 4 drivers, 4 load units.
A geographic network linking 8 cities.

Some permanent constraints are :

1- For any move, the actual volume and weight should not excess the maximum volume and weight of the load unit.
2- For any move, the fret compatibility must be checked.
3- Demand satisfaction is Priority N#1. All must be done to ensure that goods are delivered on time(see meta-constraints).

Some flexible constraints are :

1- Vehicle mileage may temporarily excess the company regulations.
2- Driver overtime may be allowed.

Some meta-constraints are :

1- Try to load and deliver all goods(and in due time)=>Do not wait until a load unit is full before starting a move.
2- Load all available compatible goods in load units which are being loaded rather than beginning to load another empty load unit.
3- Plan loading goods in expected incoming load units rather than assigning another empty one.
3- If necessary, to ensure that all goods are handled, plan empty movements from the nearest place.

```
DATA BASE
/* Freight structure : ft(id,<weight,bulk,type>,<origin,destination>,<date1,date2>) */
/* date1 et date2 : available date & Latest delivery date expected ( hourly increment) */

ft(<1,<5,3,general>,<marseille,lyon>,<0,72>>)->;
ft(<2,<10,6,chemicals>,<marseille,avignon>,<0,96>>)->;
ft(<3,<4,6,general>,<lyon,arles>,<0,24>>)->;
ft(<4,<1,24,foodstuff>,<marseille,lyon>,<0,24>>)->;
ft(<5,<1,1,general>,<marseille,lyon>,<0,96>>)->;
ft(<6,<2,3,general>,<marseille,avignon>,<0,72>>)->;
ft(<7,<20,10,fertilisers>,<avignon,lyon>,<0,96>>)->;
ft(<8,<1,2,metal_products>,<lyon,dijon>,<0,96>>)->;
ft(<9,<10,28,machinery>,<lyon,grenoble>,<0,24>>)->;
ft(<10,<5,7,general>,<dijon,lyon>,<0,24>>)->;
ft(<11,<10,8,general>,<lyon,arles>,<0,24>>)->;
ft(<12,<10,30,general>,<avignon,lyon>,<0,48>>)->;

/* Vehicle Structure : veh(<id, mileage,Home_city,Availability_date>) */

veh(<BG20,4000,marseille,0>)->;
veh(<CE30,2000,avignon,0>)->;
veh(<DF88,4500,marseille,0>)->;
veh(<EZ43,4500,marseille,0>)->;

/* Driver Structure : dr(<id,worked_hours,Home_city,Availability date>) */

dr(<Simon,10,marseille,0>)->;
dr(<Brian,12,marseille,0>)->;
dr(<Dan,15,avignon,0>)->;
dr(<Carlo,8,marseille,0>)->;

/* Load unit Structure */
/* lu(id,type,<Max_weight,Max_Volume,Compatibility>,Home_city,Availability_date) */

lu(<100,tilt,<20,28,"A">,marseille,0>)->;
lu(<101,tank,<18,24,"Z">,arles,0>)->;
lu(<102,tank,<20,30,"Z">,marseille,0>)->;
lu(<103,tilt,<20,35,"A">,marseille,12>)->;

/* Freight compatibility */
ft_compa(racked,[general,foodstuff])->;
ft_compa(tilt,[general,foodstuff,metal_products,machinery])->;
ft_compa(tank,[chemicals])->;

END OF DATA BASE
```

- Fig 4 : A simple case study : The data base

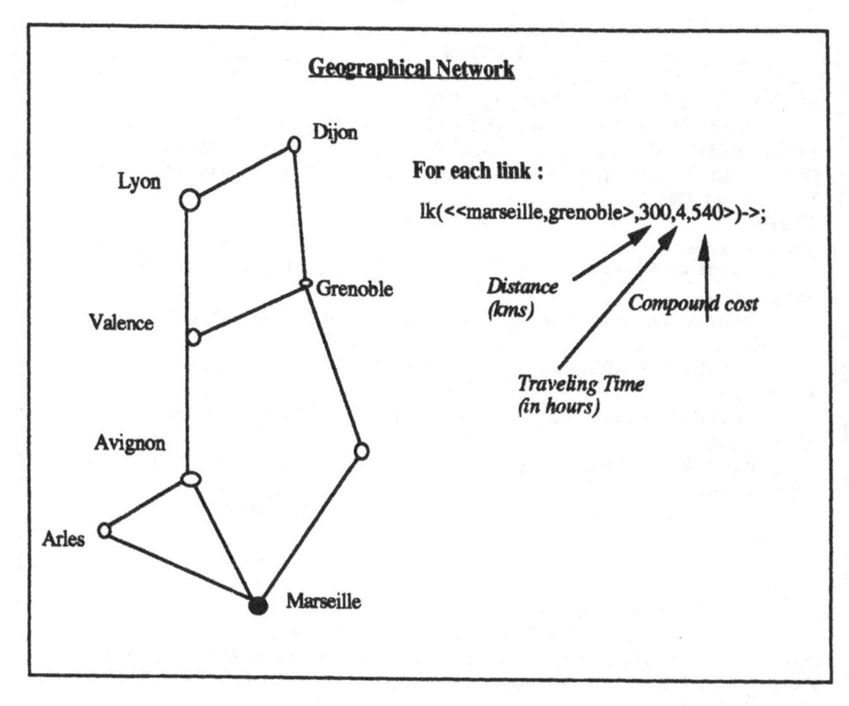

Fig 5 : A simple case study : The geographical network

The fig 6 (see below) describes one solution of the case study.(many solutions satisfy the set of constraints).
It consists of 3 trips.
Each trip involves a load unit, a vehicle and a driver.
The routes are described, move by move. Each move consists of a geographic arc, departure and arrival times,a list of cargo identifications and actual volume and weight.
Time unit is hour. Initial time is 0.
Trip 1 starts only at 12 (due to availability time of load unit 103), but all goods are delivered in due time.
All vehicles are back to the home depot. (This constraint should be relaxed within the framework of a cooperative "just-in-time" production).
Trip 3 handles 5 cargos on 3 arcs.
Freight 7 is not loaded (It concerns fertilisers, for which nothing has been said about freight compatibility (see fig. 4).

```
_____________________________ SOLUTION 1 _________________________________
---- TRIP : 1 ----
• LOAD UNIT:103  DEPOT:marseille     TYPE:tilt
          MAX. WEIGHT:20      MAX. VOLUME:35    AVAILABLE:12
• VEHICLE:EZ43  DEPOT:marseille     MILEAGE:[ Total:5430, Previous:4500, Max:10000]
• DRIVER :Carlo  HOME CITY:marseille  WORKED HOURS:[ Total:20, Previous:8, Max:40]
• TOTAL COST:1385
• ROUTE :
   <marseille,avignon>
      DEPARTURE:12  ARRIVAL:13        CARGO:[6]   VOLUME:3   WEIGHT:2
   <avignon,lyon>
      DEPARTURE:13  ARRIVAL:15        CARGO:[12]  VOLUME:30  WEIGHT:10
   <lyon,dijon>
      DEPARTURE:15  ARRIVAL:17        CARGO:[8]   VOLUME:2   WEIGHT:1
   <dijon,lyon>
      DEPARTURE:17  ARRIVAL:19        CARGO:[10]  VOLUME:7   WEIGHT:5
   <lyon,grenoble>
      DEPARTURE:19  ARRIVAL:20        CARGO:[9]   VOLUME:28  WEIGHT:10
   <grenoble,marseille>
      DEPARTURE:20  ARRIVAL:24        CARGO:Empty VOLUME:0   WEIGHT:0
---- END OF TRIP 1 -----

----  TRIP : 2   -----
• LOAD UNIT:102  DEPOT:marseille     TYPE:tank
          MAX. WEIGHT:20      MAX. VOLUME:30   AVAILABLE:0
• VEHICLE:DF88  DEPOT:marseille     MILEAGE:[ Total:4660, Previous:4500, Max:10000]
• DRIVER :Brian  HOME CITY:marseille  WORKED HOURS:[Total:14, Previous:12, Max:40]
• TOTAL COST:200
• ROUTE :
   <marseille,avignon>
      DEPARTURE:0  ARRIVAL:1        CARGO:[2]   VOLUME:6   WEIGHT:10
   <avignon,marseille>
      DEPARTURE:1  ARRIVAL:2        CARGO:Empty VOLUME:0   WEIGHT:0
---- END OF TRIP 2 -----

----  TRIP : 3   -----
• LOAD UNIT:100  DEPOT:marseille     TYPE:tilt
          MAX. WEIGHT:20      MAX. VOLUME:28  AVAILABLE:0
• VEHICLE:BG20  DEPOT:marseille     MILEAGE:[ Total:4670, Previous:4000, Max:10000]
• DRIVER :Simon  HOME CITY:marseille  WORKED HOURS:[ Total:17, Previous:10, Max:40]
• TOTAL COST:695
• ROUTE :
   <marseille,lyon>
      DEPARTURE:0  ARRIVAL:3        CARGO:[5,4,1] VOLUME :28  WEIGHT:7
   <lyon,arles>
      DEPARTURE:3  ARRIVAL:6        CARGO:[11,3]  VOLUME:14   WEIGHT:14
   <arles,marseille>
      DEPARTURE:6  ARRIVAL:7        CARGO:Empty   VOLUME :0   WEIGHT:0
---- END OF TRIP 3 -----

          **** Number of trips : 3
          **** UNLOADED FRET --> 7
          **** AVAILABLE LOAD UNIT ---> 101
          **** AVAILABLE VEHICLE ---> CE30
          **** AVAILABLE DRIVER ---> Dan
___________________________ END OF SOLUTION 1 ______________________________
```

5- Conclusions

CONTRALTO is an attempt to tackle a complex assignment problem with global scope; i.e. without splitting it according to a reductionist view.
Constraint reasoning has proven to be workable for freight transport, albeit on a small scale. It could be extended to various similar assignment/scheduling tasks or spatio-temporal problems. a first extension could be the management of boxes and containers, at a European scale. Another extension could be the intermodal freight transport (road, rail, air, ship) in any area.

Nevertheless, two conclusions should be pointed out :

(i) From the transport viewpoint, significant improvements in the real operations are closely related to further advances in computerised communications between vehicles and operation rooms, as well as intercommunications between transport companies. Work done within the European DRIVE programme should contribute to such advances, this is a prerequisite for taking advantage of promising techniques such as constraint reasoning.

(ii) Constraint reasoning is much more than constraint declaration. Iterative and interactive constraint adjustments are very necessary to obtain satisfactory results. The main point is to find out how constraints (which ones ?, to what extent ?, in which directions ?) should be modified to rapidly converge towards the best satisfactory solutions.
In fact, just like a navigation between Charybde and Scylla, the Art of Constraint Reasoning is to beat to windward between an overconstrained and underconstrained situation.

6- References

[1] EUROFRET : European system for international road freight transport operations (DRIVE Project V1027). -1990

[2] FLEET : Freight and Logistics Efforts for European Traffic DRIVE Project V1044. -1990

[3] FLEET : Freight and Logistics Efforts for European Traffic (DRIVE Project V1044 WorkPackage 2200.) -1990

[4] ATTAIN: Applicability to Trafic and Transport of Artificial Intelligence (DRIVE Project V1039.) -1990

[5] CASSIOPE : Computer Aided System for Scheduling, Information and Operation of Public Transport in Europe. (DRIVE Project V1019). 1990

[6] JM Rousseau
Strategic Planning in Public Transport. Montreal Canada. - 1988

[7] P. Hutson & C.L. Wood
"TSA: A trunker scheduler advisor"
in Expert systems conference and their applications - Avignon 1990
Specialized conference : Artificial Intelligence & Transportation

[8] H. Sansen
SHIP-PLANNER Navale-Delmas-Systemia
in Expert systems conference and their applications - Avignon 1990
Specialized conference : Artificial Intelligence & Transportation

[9] Beurrier & al.
Empty freight railcar assignment by expert system
in CCCT'89 Paris 1989

[10] Brazile, Swigger & al.
"An expert system for the airline gate scheduling problem"
in Expert systems conference and their applications - Avignon 1988)

[11] M.Dincbas,P. Van Hentenryck, H. Simonsis, & Al.
"The constraint logic progamming language CHIP"
In Proceedings of the International Conference on Fifth Generation Computer Systems '88, 1988

[12] J.Jaffar & JL Lassez
"Constraint Logic Programming" in
Proceedings Conference on Principle of Programming Languages. Munchen,1987

[13] A. Colmerauer
"An introduction to PROLOG III" in
10 th International Workshop "Expert systems & their applications"
Avignon '90,1990

[14] P. Van Hentenryck and M. Dincbas
"Domains in logic programming" in
Proc. of the AAAI 87.

[15] Célia Maria Guimaraès Anchieta
"Gestion de déplacements à vide d'une flotte de véhicules: une approche basée sur la connaissance heuristique".
CRET. Centre de Recherche d'Economie des Transports- Aix-en-Provence 1991
[16] STRADA. A computer based simulation model for strategic transport planning
Swedish Transport Research Institute, Stockholm. Sweden 1990
[17] JM Le dizès & al.
Identification of Key Areas in the field of road traffic, transport and safety engineering
Report M3.1 ATTAIN DRIVE Project V1039 - 1989

RAPS - A Rule-based Language for Specifying Resource Allocation and Time-tabling Problems

G. Solotorevsky, E. Gudes, A. Meisels (*)

Department of Mathematics and Computer Science, Ben-Gurion University of the Negev, Beer-Sheva, 84-105, Israel

Abstract

A general language for specifying resource allocation and time-tabling problems is presented. The language is based on the expert-system paradigm which was developed previously by the authors and which enables the solution of resource allocation problems by using experts' knowledge and heuristics. The language enables the specification of a problem in terms of: resources, activities, allocation rules and constraints. The language syntax is powerful and allows the specification of rules and constraints which are very difficult to formulate with the more traditional approaches. The language is independent of the inference engine that performs the allocation and can therefore be used to evaluate different allocation strategies.

1 Introduction

Some version of the generic resource allocation problem exists in almost any organization. Typical resource allocation problems include the allocation of flight crews to missions in air-lines or in an air-force, the assignment of classes and teachers in high schools and universities, and the allocation of machines for the production of parts or products in industrial environments. The name timetabling problems or scheduling problems is sometimes used in referring to these problems. Typical engineering applications in which these problems occur are: project management, manufacturing planning, job-shop scheduling and others.

These problems are very difficult to solve exactly or optimally. Most organizations solve their specific problems by using human experts who have gained much experience and intuition into their own specific problem. Furthermore, there are cases that require minute changes to the general allocation plan, as for example in the case of a sick pilot who is scheduled for a flight on a specific morning. In current real world systems, a needed (last moment) change of pilot is met by some specific expert actions which result, usually, with just a small change of the overall allocation plan.

The above state of affairs led many workers in this field to choose the approach of creating a *tool to help a human* in allocating resources, rather than trying to achieve a program that solves the problem completely. Such tools become more and more similar to Expert Systems (ES) [2]. The similarity to expert systems

(*) This research was partially supported by the Frankel Center for Computer Science and the Iwaair Center for Production Management, Robotics and AI.

arises from two main features of these programs. The first feature is the coding of human expertise and knowledge into Rules, and the second is the use of a search mechanism which is controlled by a Ruled-based type of control, [5,1].

The advantages of rule-based expert systems (RBS) have been documented in the literature [2]. Here we like to emphasize their advantages with regards to resource allocation problems:

- Rules are very convenient for specifying both the constraints on the problem and the heuristics used by the expert to solve it.
- A RBS can generate explanations easily, by following the path of rules that fired in the reasoning process. This is of particular importance in solving an RA problem because the expert has to be convinced that the proposed solution is plausible and in case the expert system calls for help from the user. The reason for failure, generated by the explanation mechanism, affects the user response and improves it.
- The ES approach has the ability to specify constraints more naturally, and the ability to pursue multiple objectives rather than a single objective as is traditionally the case with the analytic approches (see [6]).

In [3] we presented a general methodology for solving Resource Allocation (RA) problems using an expert system, and described the major components of an expert system shell for this purpose. In [4] we presented an evaluation of the methodology based on the implementation of two applications: the first one assigns crew-members to flight missions, the second one schedules classes of Mathematics in our department. An important component of a generic expert system is a language for the specification of an RA problem, including its con-constraints and allocation strategies. Such a language is presented in this paper. Examples demonstrating the syntax of this language are taken from the above two applications.

2 RAPS - A methodology for RA and the RAPS Language

2.1 The General Framework

Our framework has three major components:

1. A common set of concepts and terms, by which most resource allocation and timetabling problems can be defined. The important ones are:

   Activities - a list of tasks which must be performed. Activities of the crew assignment problem are the different tasks that are part of flight missions.

   Resources - a list from which allocations must be made in order to perform activities. The Resources of the crew-assignment problem are entries in the list of crew-members of different qualifications.

Priorities - both activities and resources are assigned priorities, which direct the allocation procedure. In the crew-assignment problem a higher priority is assigned to activities (missions) in a "dense" day of the week (in terms of activities).

2. A set of rules which the allocation procedures should follow. The rules are divided into two parts:

   - **Restricting rules** - these are constraints which the allocation procedure must follow. In the crew assignment case, an example of a restricting rule is that a crew member has to have a certain "relaxation" time-slice between two assignments.
   - **Recommending rules** - these are rules which the allocation procedure tries to follow, in order to either prune the search space, or get a "better" solution. For example, a recommending rule chooses to allocate crew-members with lower degree of availability first. (these rules are called also "assignment rules)

3. A strategy for resource allocation, which can be tuned by the user. This strategy, which was described in [3], has three components.

   (a) Forward allocation using priorities and recommending rules, one time-period at a time.

   (b) Consistency checking, i.e checking the current allocation against the relevant constraints.

   (c) Backtracking by the use of a special strategy, that replaces the depth-first policy, called the *local_change policy*. The local_change policy uses special kind of rules, local_change rules which specify what to do in case no resource can be assigned to the current activity. A typical rule in the crew-assignment problem will try to release an assigned crew-member and allocate him for the problematic activity.

2.2 Introducing RAPS

RAPS - Resource Allocation Problem Specification language is a major component in the generic expert system shell for resource allocation. The representation of a problem in RAPS permits the use of an allocation engine to solve the problem. The representation in RAPS isn't a computer program, in its regular sense; it doesn't describe an algorithm to find a solution and it doesn't necessarily lead to a certain solution, i.e., different allocation engines may reach different solutions based on the same RAPS specification.

On the other hand, the specification of a problem in RAPS includes not only a description of the resources and activities that are involved, and the constraints that apply to them, but also a specification of the way in which an expert would try to solve the problem; this specification is in terms of recommending rules which are followed in a particular order. Also, the set of recommending rules (and therefore the allocation strategy) can change by changing the "context" of the problem state. Thus, the expert's knowledge of finding a solution can be encoded quite closely.

2.3 An example

The following is an example of a RAPS specification of the activities and resources of the class-scheduling example. There is one activity called "class", and two resources called "room" and "time". The definition of activity also includes the specification of the resources it needs.

```
class is activity.
class attributes :
     key num is integer,
     participants is someof [math, physics, philosophy],
     name is alfanumeric,
     students_num is integer.
class needs resources named place,hour.

room is resource.
room attributes:
     key building is oneof [b, c, s],
     key number integer,
     seats_num is integer.

time is resource.
time attributes:
day_of_week is oneof [sun,mon,tue,wed,thu,fri],
hour is integer,
     math_classes_at_time is counter of
     (assigned to activity class where :
          1 | 'maths' member_ in class.participants.
     ).
```

An example of a recommending rule for the same problem is the following assignment that specifies that any room can be assigned to attribute "place" in any class.

```
assignment 1 (start):
select class.
     assign to class.place room.
```

Finally, there is a constraint which specifies that at least one class taken by Math students, should be on Sunday.

```
constraint 1:
```

```
must succeed.
    select class where :
        1 | 'math' member in participants.
        2 | (select time where :
            1 | time was assigned to class.
            2 | day_of_week equal 'sun'.
        ).
```

An important characteristic of RAPS is its independence from the allocation engine. This has the following advantages:

- We may use the same RAPS description to test several engines.
- It is enough for the user to know the syntax and semantics of RAPS in order to describe a RA problem and he doesn't need to know the mechanism of the allocation engine.
- The improvement of an allocation engine won't demand a change in the RAPS description.

2.4 RAPS as a Knowledge Representation and Acquisition Tool

As designers of RAPS, we had three major goals:

- Design a language with enough power and flexibility to describe a large variety of RA and timetabling problems.
- Design a language which would enable users to define and tune allocation strategies that will achieve high quality solutions fast.
- Design a language and an interface which will enable convenient knowledge acquisition and easy interaction with the experts.

Clearly, it is very difficult to achieve the three goals concurrently, but we hope that we found the right compromise. To provide enough power and flexibility we looked at several applications that we had solved previously [4]. We found that the rules we used in these applications were quite complex. They involved, in general, multiple entities, Boolean operators, functions (max, avg, etc.), set operators ("member in"), and others. Simple IF-THEN rules like in MYCIN [2] were not sufficient. We realized that the closest analogy in capabilities to our requirements is Database query languages such as SQL [8], and specifically the rules syntax is quite similar to a typical syntax of a database query language. Although, a language such as SQL may not be as friendly as our third goal demands, the use of Macros as demonstrated in Sections 3 and 4 can make the language simpler for naive users.

The second goal ties into what Marcus calls "Strategic knowledge" [7]. Marcus describes a tool for knowledge acquisition in constraints-based applications called: SALT. She then demonstrated how SALT can be used to define the order of tasks and sub-tasks, and the order in which various constraints will be checked. Similar capabilities exist in RAPS. First, by defining priorities we can order activities, resources assignment rules and constraint rules. For example, it is easy to cause the allocation engine to check the most difficult constraint first, by defining it with the highest priority. A second mechanism is the mechanism of *contexts* (see Section 3 and 4). This mechanism allows the user control on the order of allocation. In particular, using contexts, one can define different sets of assignment rules and priority rules in each context, (including local_change rules) and thus tune the allocation strategy to the state of the allocation. Contexts and priority rules are powerful mechanisms to define strategic knowledge.

To achieve the third goal we provided the language with a lot of redundancy which makes it more readable, and provided the Macro facility mentioned earlier. Furthermore, in our expert system shell - ESRA (see Section 4), we are implementing a friendly interface using EMACS [9] which creates a syntax-directed editor environment. This knowledge editor helps users define RAPS clauses by completing phrases, providing menus and help screens.

3 The RAPS Language

A RAPS description of a RA problem is constituted of: definitions, initiation of values, contexts declaration, and rules. We will detailed each below.

3.1 Definitions and Syntactic Conventions.

We define the entity types which affect the problem, i.e., the activity and resource types that participate in it. An entity type is defined by its name and its attributes, which may be static or dynamic. A static attribute is one with a fixed value during the whole allocation process. In RAPS, the dynamic attributes are called counters; their value may vary throughout the process due to the shell operations or due to explicit actions in the rules. At least one of the static attributes is used as a key. An attribute which is used to assign values of a resource is also called a "role". A role has a name which is used in assigning resources in the assignment rules (note, the role name and the resource name may be different.)

Three definitions of types of entities appeared in the example presented in section 2. The first one defines an activity named "class" that has four attributes, the first attribute, named "num", serves as the key of "class"; at the end of the definition of "class" there is a declaration of the resources needed by "class": a "place" and an "hour". The second one is a definition of a resource named "room", its first and second fields serve as the key of this resource. The attribute, "math-classes-at-time", of the third declared entity - "time" - is an automatic counter which counts how many Math classes (i.e of which there are Math participants) the resource was assigned to.

Most of the rules (vide infra) and the automatic counters are based on the select operation. The select (or exist) operation consists on selecting an entity that satisfies a condition list. A condition list is a list of expressions which are "anded" or "ored" with the usual meaning. An expression may contain nested select statements (similar to SQL), with appropriate scoping rules.

The precise syntax and semantics of RAPS can be found in [13]. In the following example we just illustrate the syntax and scoping rule of a "select" statement

using a rule taken from the air-crew application. It is a "select" operation that succeeds if there is a mission that fulfills the following requirements: (1) either (1.1) its name is 'patroll' and the active context (stored in a system's variable) is 'start' or (1.2) the mission name is 'patrol2'and (3) there are at least fifty other missions with (3.1) keys different from the key of the mission selected before and which (3.2) take place in the same day. The assignment of "mission" into the variable "Mission_var" in (2) allows us to refer to the mission chosen by the select operation during the nested exist operation.

```
(select mission where :
     1 | (or
          1 | (and
               1 | name = 'patroll'.
               2 | active_context equal 'start'.
          )
          2 | mission.name = 'patrol2'.
     )
     2 | Mission_var = mission.
     3 | (exists 50 mission where :
          1 | not (mission.name equal Mission_var.name).
          2 | day equal Mission_var.day.
     ).
).
```

3.2 Initiation of values.

Entities are created by initiation of values to the previously declared entity types. The next example shows the creation of six entities: two "classes", two "rooms", and two "times".

```
class values :
     {201189,[math,philosophy],AI,120} {201234,[math],'DB',89}.
room values :
     {b,201,60} {s,209,60}.
time values :
     {sun,16,0} {mon,12,0}.
```

Another type of initiation is a set of assignments which is demanded by the user as a fixed requirement. Below is a demand to assign the room whose key is b-209 to role "place" in class number 20189.

```
assign to class.place 201189 room (s,209).
```

3.3 Context declaration.

A context is a work environment in which certain rules are active and where the priorities of entities and the active rules get re-computed. The motivations for the use of the context methodology are:

- The possibility of dividing a large problem into sub-tasks i.e **Modularity**. This enables the clarity and speed that may be achieved by having a lower number of active rules in each stage.
- The capacity of selecting different approaches to the same task according to the problem state.
- In particular, the local_changes policy can be implemented as rules in a separate context.

Any rule which has no context list attached to it, is by default, active in the "start" context. The allocation session always begins in the "start" context. The following is a declaration of two contexts: problematic and hyper-problematic.

% context list problematic, hyperproblematic.

3.4 Rules.

Several kinds of rules exist in RAPS; they are all based on the "select" operation, but each has a distinct heading. The different types of rules are: assignment rules, constraint rules, context rules, priority rules, local change rules.

Each rule may appear as an explicit rule or as a **rule template**. A rule template is a model of a rule that has some "template variables"; the user may create several explicit rules from a template rule by instantiating its "template variables" to diverse values. Templates are especially useful when we want to solve the same process with slightly different rules, e.g., a template of a constraint which states that a certain class can not take place in a specific hour, is handy if this hour changes every semester.

It is possible to add to any rule a list of the contexts and the respective priorities in which the rule is active. The default context is "start" and the default priority is "0".

Assignment rules. An assignment rule is used to assign a resource to a certain role in an activity. The rule is constituted of two parts, the first one is used to select the activity and the second one, to select the desired resource.

The structure of an assignment rule is:

a) Select an activity that needs a resource for a given role, and satisfies certain demands.

b) Select a resource that satisfies some other demands.

c) If a) and b) succeed, then the resource selected in b) is assigned to the given role of the activity selected in a).

Following are three examples of assignment rules: The first selects an activity of type "class" and assigns to the role "place" a resource "room"; there are no further restrictions on the activity or the resource. The second assignment rule selects a class which has "math" students among its participants and assigns to that class an hour on Sunday, which has been assigned to less than two classes (note, this rule uses the automatic counter: classes_at_time). The third rule is an assignment template with two "template-variables": "Class" and "Room"; this rule assigns a room with a key equal to "Room" value given by the user, to a class with a key equal to "Class" value given by the user. The first rule is active only in context "start" with the default priority; the second and third rules are active in contexts "start" and "problematic" with priorities 5 and 7 correspondingly.

```
assignment 1 {start}:
    select class.
    assign to class.place room.

assignment 2 {start 5, problematic 7}:
    select class where :
        1 | 'math' member in participants.
    assign to class.time time where :
        1 | classes_at_time < 2.
        2 | day_of_week equal sun.

template assignment 3 {start 5, problematic 7}:
    templates vars
        Class is class.
        Room is room.
    select class where :
    1 | class equal Class.
    assign to class.place room
        1 | room equal Room.
```

Constraint rules. There are two kinds of constraints: positive and negative ones. A negative constraint (this is actually a **constraint**) is a condition that must fail in every context of the allocation process where the constraint is active. A positive constraint, is actually an **objective** which the allocation program should attempt to satisfy at the end of the allocation, but it may be not true while the allocationls are taking place (although it may be checked earlier). Neither a negative constraint nor a positive one, tell the system what assignment rule is recommended; their task is to check the assignment proposed by an assignment rule and to accept or reject it.

The structure of a constraint rule is:

a) Declare that the selection in b) must fail or succeed.

b) Try to do a certain selection.

c) If the selection in b) succeeded and in a) was declared "must succeed" or if the selection in b) failed and in a) was declared "must fail", then the rule succeeds.

The first example states that no class may receive an hour on Friday. The second example is a positive constraint which says that at least one class of math should get an hour on Sunday.

```
constraint 1:
```

```
must fail.
    select class where :
        1 | (select time where :
            1 | time was assigned to class.
            2 | time.day_of_week equal 'fri'.
        ).

constraint 2:
must succeed.
    select class where :
        1 | 'math' member in participants.
        2 | (select time where :
            1 | time was assigned to class.
            2 | day_of_week equal 'sun'.
        ).
```

To ensure the satisfaction of the objective expressed by the second constraint it will be better to add an assignment rule which recommends to assign a time on Sunday to class which has participants from the Math department. An example of a more powerful positive constraint would be a rule specifying that the number of classes in each day should be approximately the same. This can also be specified in RAPS.

Local change rules. Local change rules recommend local changes required when there are activities that still need resources, but there is no active assignment rule able to recommend a legitimate assignment. If these rules fail to solve the problem, then the context rules are used to change to the context in which the local change rules are active.

The structure of local change rules is

a) Select an activity which satisfies certain requirements.

b) Select a resource, that was assigned to the previous selected activity, which fulfills other requirements.

c) Select a resource, different from the one selected in b), which fulfills some conditions.

d) If a), b), and c), have succeeded, then assign the resource selected in c) to the activity selected in a) instead of the resource selected in b).

The following example is a local change rule which selects a class to which a room was assigned, whose number of places is bigger than the number of participants + 10; this room will be swapped with a room which has allocations for a number of participants lower than the number of participants + 10 but higher than the number of participants in the class.

```
local change 1:
    select class.
    select room assigned to activity class where :
        1 | room.seats_num > class.students_num + 10.
```

```
exchange it with room where :
    1 | seats_num > class.students_num.
    2 | seats_num < class.students_num + 10.
```

Context rules. These are the meta-rules which select the active context; using these rules, the set of assignment rules and their priorities is selected according to the state of the system. For description and examples of context rules see [13].

Priority rules. These rules determine the priorities of the resources and the activities in each context. The priority of an entity in a context is the maximal priority assigned to it by a priority rule in that context. The priorities calculation takes place whenever a context is entered.

The structure of these rules is: give a priority according to the value written in the rule to all the entities that fulfill the "select" condition in the rule.

The following priority rule gives priority "2" in context "start" to all the classes that have students of math as participants.

```
priority rule 1:
    set class priority 2
    of class where :
        1 | 'math' member in participants.
```

4 An expert system shell based on RAPS

As mentioned in the introduction, we are currently developing an expert system shell for solving RA problems. A first prototype of this shell called ESRA - Expert System for Resource Allocation - has already been developed. ESRA provides an integrated environment for developing and running RAPS' programs. ESRA includes six main components: an inference engine, a RAPS compiler, a graphical interactive display interface, an user interface, a knowledge-base editor, and a data base (see figure 1).

The inference engine is implemented in Quintus Prolog [12]. It has two main components: the allocation engine and the condition evaluator. The allocation engine is responsible for the allocation policy; it uses a generate and test mechanism which can allocate in a forward manner, and can interpret and follow recommending rules during allocation. The inference engine also checks the satisfaction of local and global constraints. The inference engine is responsible of backtracking and of using the meta-rules both of the local change policy and of the change of contexts. The allocation engine uses the condition evaluator in order to solve queries about the data base. The separation of the inference engine into these two modules allows a clear distinction between using different allocation strategies and using different strategies of calculating database queries.This separation also allows easy expansion of the power of "select" with new functions or operators.

The RAPS compiler compiles a problem specification written in RAPS into a lower level language which is understood by the inference engine. The compiler

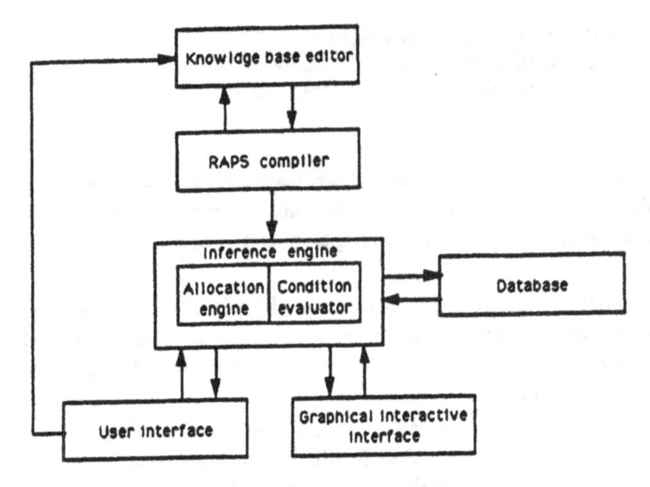

Figure 1: The architecture of the ESRA system

takes care of the type checking that may be needed, doing so it makes sure that the inference engine won't be slowed by performing type checking. The compiler is implemented using Yacc [10] and Lex [11].

The graphical interactive intereface provides a mean of following and intervening in the allocation process. Its principal task is to provide tabular information about the activities, resources, and done allocations ; it is able to represent the information in various forms according to the user's requests.

The default representation (see figure 2.) is a table in whose vertical axis are the activities, and in whose horizontal axis are the roles. Each entry in the table contains the resource which was assigned to the corresponding role in the horizontal axis of the corresponding activity in the vertical axis. The user can use this tables to interact with the allocation engine in order to delete/add allocations or to swap between two resources allocated to different activities/roles. After each request of an allocation modification done by the user, the graphical interface asks the inference engine if it is legal, if so the request is performed in the data-base and updated in the table, otherwise the user is notified about the reason of the rejection of its request (note that swapping can not always be implemented by a sequence of additions and deletions, since the status of the data-base may be inconsistent with the restrictions during these steps, but consistent with the state of the data-base before and after the swapping transaction). The user can ask from the graphical interface to display different amounts of information for each entry in the table; he can also choose different axis and entries in the table.

Other types of interaction between the user and the inference engine are provided by the user-interface which enables the allocation to run in "batch" or "interactive" modes. The user-interface can be used to set different debugging and tracing options, and to relax constraints when backtracking fails. We plan to move these options, in the future, into the graphical-interface.

The knowledge-base editor was built to support RAPS' syntax and semantics; the editor was built on top of an Emacs [9] editor; thanks to this strategy we didn't have to worry about the implementation of low level editing functions

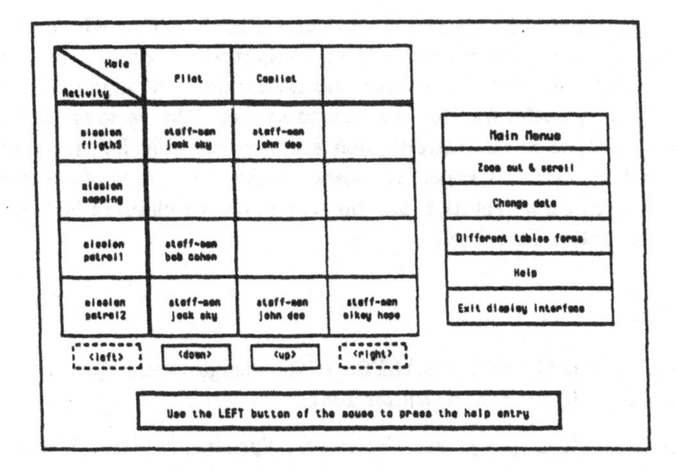

Figure 2: A partially allocated table obtained by the graphical interactive intereface

and of machine dependent qualities. Emacs was easily transformed into a RAPS editor by adding Lisp functions to Emacs extensible code.

The knowledge based editor provides several levels of syntax and semantics checking. The editor can also give the user help about possible completions of a RAPS sentence, and give explanations about what is expected from the user at a certain place in a RAPS description; this allows different types of users to be comfortable with it, e.g., a new user can ask the editor to report each suspected error in the code being edited and he will often ask for possible completions; meanwhile a more experienced user can set the editor to report only severe errors and to ignore missing parts of a RAPS representation, which he plans to fill later.

The editor also provides other facilities such as placing the cursor on errors found by the RAPS compiler, automatic saving , etc.

The ESRA system will be described with more details in a future paper.

5 Conclusions.

As we have seen, RAPS is a flexible language with a structure which enables the representation of various kinds of RA problems. We have used RAPS and the ESRA shell to solve several resource allocation problems which include staff allocation for air force missions, and class scheduling in our Department. These problems have been solved by us before by expert systems specially tailored for each of them. The specification of these problems in RAPS was easy and natural, and even though we used ESRA, which is a shell and not a tailored expert system, we obtained results which are comparable to those attained before, both in their allocation quality and in the time this allocations require.

Clearly, not all RA problems can be mapped easily into RAPS. Some problems are basically optimization problems such as "critical-path" problems and are better solved by graph searching algorithms. These problems may not use any specific expert heuristics. RAPS is most useful in situations where expertise and heuristics for solving them exist, but no computerized solution is available. Furthermore, RAPS interactive interface and its ability to change data dynamically make it a very useful tool for the human expert. The various engineering applications mentioned in the introduction are very suitable for mapping into RAPS, since a large body of expertise exists usually for this applications, and the stress is on engineering solution and on accomodating changes fast, and not on obtaining the optimal solution.

References

[1] D. Anderdon and C. Ortiz, "AALPS: A Knowledge Based System for Aircraft Loading", *IEEE Expert*, winter 1987.

[2] F. Hayes-Roth et. al. (ed.), *Building Expert Systems*, Addison-Wesley Inc., London, 1983.

[3] T. Kuflik, E, Gudes and A. Meisels, "An Expert System Based Methodology for Solving Resource Allocation Problems", *Proc. 3rd. IEA/AI*, Charleston, July 1990.

[4] A. Meisels, E. Gudes and T. Kuflik, " Limited-Resource Timetabling by a Generalized Expert System" FC-TR-029, Frankel Center for Computer Science, Ben-Gurion University, Beer-Sheva, April 1990.

[5] J. R. Slagle and H. Hamburger, "An Expert System for a Resource Allocation Problem", *Comm. of the ACM*, **28**, no. 9, 1985.

[6] Dhar V., and Ranganathan N., "Integer Programming vs. Expert Systems: An Experimental comparison," *Comm. of the ACM*, Vol 33, No. 3, March, 1990, pp. 338-348.

[7] Marcus S. "Understanding decision ordering from a piecemeal collection of knowldge", *Knowledge Acquisition* Vol. 1, 1989, pp. 279-298.

[8] Ullman J. D. *"Principles of Database Systems"*, second edition, Computer Science Press inc., Mariland, 1982.

[9] Stallman M. S., "EMACS The Extensible, Customizable Self-Documenting Display Editor" *ACM SIGPLAN Notice* Vol. 16 no. 6 June 1981, pp. 147-156.

[10] Johnson, S. C. "Yacc - yet another compiler compiler", Computing Science Technical Report 32, AT&T Bell Laboratories, Murray Hill, N.J., 1975.

[11] Lesk, M.E. "Lex - alexical analyzer generator", Computing Science Technical Report 39, AT&T Bell Laboratories, Murray Hill, N.J., 1975.

[12] *"Quintus Prolog User's Guide version 10"*, Quintus computer systems, Montain View, California, March 1987.

[13] Solotorevsky G., Gudes E., and A. Meisels, "RAPS - rule-based language for specifying resource allocation and time-tabling problems," Technical Report FC-032, Dept of Mathematics and Computer Science, Ben-Gurion U., 1990.

KBS in Marine Collision Avoidance

F. Coenen, P. Smeaton

Department of Computer Science, Liverpool University, Chadwick Building, P.O. Box 147, Liverpool L69 3BX, UK

ABSTRACT

A rule based system to provide decision support for the navigators of commercial, ocean going, vessels is described. The decision support is supplied in the form of advice on how best to avoid collision with both other vessels and land based objects. At the same time the system acknowledges the need for commercial vessels to proceed on a predefined passage plan.

The system is founded on a number of KBs representing knowledge about the collision avoidance domain, obtained from international regulations and expert mariners. Important attributes of the system are its real time mode of operation and its ability to take into consideration geographic constraints using a Knowledge Based approach to Geographic Information System (GIS) techniques. Further the system does not require any user interaction, all raw data is supplied by external sensors and from a chart data base.

The current status of the system is in the form of a proto-type. However it is considered that this proto-type is the first in its field. Initial testing in a simulated environment using experienced master mariners has produced encouraging results indicating the potential for the system to become a major aid to navigation and hence a commercially viable product.

INTRODUCTION

In this paper a rule based system to provide decision support for navigators of ocean going vessel is described. The idea of providing computer support for the mariner when carrying out the task of collision avoidance is not new, and in the last two decades many systems have been proposed (Hollingdale [1], Cannell [2]). However to date none of these systems have gained any acceptance within the maritime industry. This is largely due to the mathematical approach, based on plane trigonometry, adopted by these systems. An approach which produces good results in the case of open water two-ship encounters, but can not be applied to more complex encounters because of the many additional considerations and constraints imposed. Further this mathematical approach, other than for some initial analysis, bears very little relation to the problem solving processes applied by the navigator in practice.

In the light of the failure of these earlier systems a research team at Liverpool was set up to investigate the possibility of applying KBS techniques to the marine collision avoidance problem. This decision was supported by the large number of collisions, groundings and "near misses" that are still an almost daily occurrence through out the world and the costs involved, in the event of a collision, not only in human and material terms, but also in ecological terms.

This work has now culminated in a proto-type system designed to provide the mariner with navigational decision support. This is supplied in the form of advice on how best to avoid collisions with other vessels and land masses. The proto-type is designed to operate in real time without the need for any user interaction, all raw data being supplied to the system by external sensors connected to navigational instruments such as radars and gyro compasses. The system is currently mounted on an IBM AT. However it is envisaged that it will eventually be integrated with the hardware that forms an essential part of modern marine radars and thus be carried on board the navigating bridges of ocean going vessels.

The system has been tested in simulated conditions using experienced master mariners and has been well received. Further it has demonstrated that it can successfully resolve multi-ship encounters in a professional manner taking into consideration additional factors such as:-

1. The predicted movement of all vessels in the vicinity.

2. Chart constraints such as land masses and shallow water areas.

3. The commercial need for merchant ships to proceed on their voyages as efficiently and economically as possible.

It is felt that this proto-type is the first in its field, although it is acknowledged that there are other research teams through out the world working on similar systems (Grabowski [3], Sugisaki [4]).

In the rest of this paper the system is described in further detail. An overview of the operating domain is given in Section 2 and the system itself in Section 3. The discussion then focuses on a number of aspects of the system, including raw data processing and analysis, advice generation, prediction of expected manoeuvres on behalf of detected vessels and chart constraints (Sections 4, 5, 6 and 7). The operation of the system is illustrated using an example in Section 8. Finally some conclusions are drawn in Section 9.

THE OPERATING AND SYSTEM DOMAIN

The principle aim of any commercial marine navigation exercise is to proceed from A to B as efficiently and as safely as possible. Essentially this involves not colliding with with other vessels or land masses. The system's domain can thus be considered to consist of ownship, on which the system is fitted, a theoretically unlimited number of other vessels referred to, for historical reasons but perhaps unfortunately, as targets and a number of land and shallow water areas.

The navigation process can be broken down into two stages (a) planning and (b) execution. The first is carried out prior to the commencement of the voyage and results in a passage plan drawn up by members of the ship's staff. During the execution stage the plan is implemented. This involves a number of "watch keeping officers" working round the clock in shifts, sometimes for more than ten days in succession, ensuring that the passage plan is executed without involving collision. It is this second process with which the system described here is concerned.

Collision avoidance is carried out according to a set of regulations called the Collision Avoidance Regulations (IMO [5]) and what is termed the practice of good seamanship. The existence of a set of regulations lends itself immediately to a KBS application. However the nature of the watch keepers task, which involves keeping a constant lookout, means that a traditional consultative KBS, where the user answers questions or fills in forms when prompted by the host machine, is rendered unsuitable. The system must gather all necessary information concerning its domain from external sensors and relate this to a chart database without the assistance of the user. Further the time constraints involved, a collision can occur in a matter of minutes, means that the system has to operate in real time. A user independent, real time KBS to provide decision support for watch keeping officers when executing a passage plan was therefore proposed.

SYSTEM OVERVIEW

The system is founded on a number of rule bases. These are designed to reflect the accumulated experience and practical know how, acquired over many years, of the expert mariner. The navigational advice produced by the system is generated from these rule bases in such a manner that land constraints and the location of other vessels in the vicinity and their intended manoeuvres are taken into consideration. The advice generation process can be considered to consist of four stages:-

1. Data Acquisition, Processing and Analysis. The raw data received is processed and analysed. The result is a complete situation description at given instant in time. The generation of this description is described further in Section 4.

2. Advice Generation. Appropriate navigational advice is generated according to the nature of the analysed data and by consulting a

Collision Avoidance KB. A set of Collision Avoidance Heuristics, which take into account the predicted movement of detected targets, serves to limit the search through the KB. The structure and content of the Collision Avoidance KB is described in Section 5, and the method of predicting and describing the expected movement of targets in Section 6.

3. Testing. The efficacy of the proposed advice generated in stage 2 is tested. If unsuccessful the system returns to stage 2 to produce alternative advice and then repeats stage 3. A feature of this testing process is the systems ability to take into consideration chart constraints. How this is achieved is described in Sections 7.

4. Output. The generated advice is output to the user together with a computerised representation of the chart area surrounding ownship and information about other vessels in the vicinity.

A block diagram illustrating the above process is given in Figure 1.

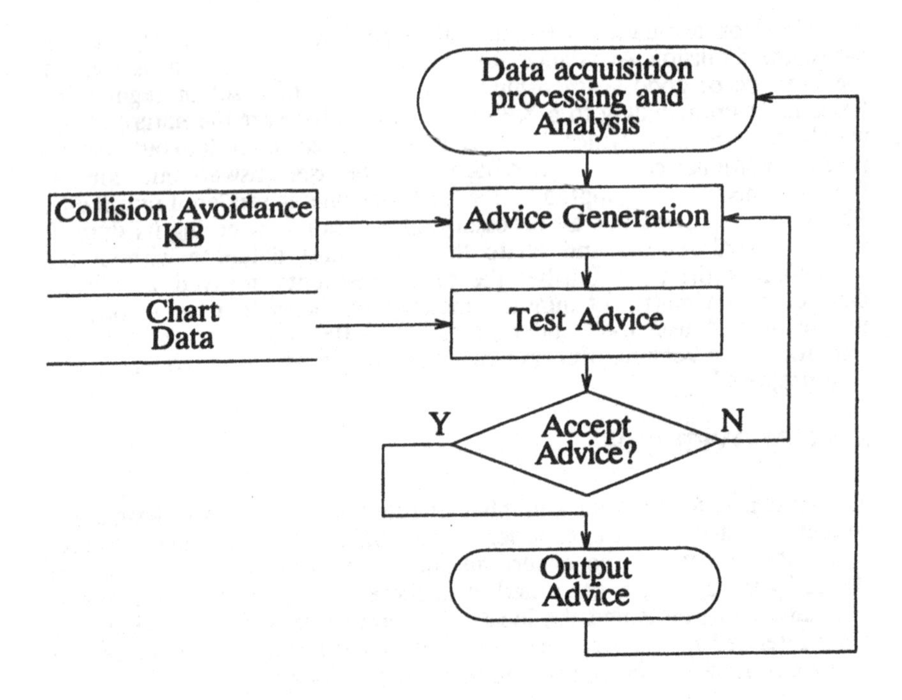

Figure 1: Schematic of Advice Generation Process

DATA PROCESSING AND ANALYSIS.

Knowledge of the movement of other vessels in the vicinity of ownship is made available to the system from radar information. This information is processed and a description of the movement of each target generated in the form of pair of descriptors referred to as the Primary and Secondary Status Descriptors (PSD and SSD). An example of a Status Descriptor pair is:-

targetCrossingStbdToPort, passingAstern

The degree of risk associated with a target is ascertained with heuristic guidance based on well documented marine domain theory (Goodwin [6], Davis [7]). Domain theory in this context is essentially concerned with the theoretical area about ownship which the mariner wishes to keep clear of all other vessels. The degree of risk associated with each target, according to this analysis, is described by a further descriptor pair referred to as the Primary and Secondary Risk Status Descriptors (PRSD and SRSD). An example of a Risk Status Descriptor pair is:-

closeQuartersSituation, unsafe

Targets whose SRSD is set to "unsafe" are considered to represent a risk to ownship.

The use of descriptors allows the system to encapsulate the movement of detected vessels, which would otherwise be defined by large amounts of numeric data, in a simple and concise manner.

ADVICE GENERATION.

Navigational advice is generated following a generate and test paradigm. This involves the generation of an initial collision avoidance strategy for the most hazardous target considered to represent a risk of collision to ownship, which is then tested against all other detected vessels and chart constraints. If suitable the strategy is output to the user in the form of advice. Otherwise an alternative strategy is generated. This process continuous until either a suitable strategy is found at or a default strategy is arrived at. The latter usually consists of an emergency action and is only proposed in very complex situations when ownship is "boxed in".

The advice is generated by reference to a Collision Avoidance KB. This is entered with the Status Descriptors for the most hazardous target, and exited with a suitable strategy. A simplified example of a Collision Avoidance KB rule would be:-

crossingStbdPort, onCollisionCourse -> stbdAlteration.

The knowledge base is organised in such a manner that each PSD type has as many as eight alternative strategies, ordered according to desirability, associated with it. In some cases the number of alternatives may be refined according to the SSD. There are three main factors affecting the desirability of a strategy. The first is the content of the Collision Avoidance Regulations, some Rules in the Regulations express a preferred action, for example Rule 15 states that:-

> *When two power-driven vessels are crossing so as to involve risk of collision, the vessel which has the other on her own starboard side shall keep out of the way and shall, if the circumstances of the case admit, avoid crossing ahead of the other vessel.*

Thus the rule expresses a preference that a vessel taking avoiding action, in a crossing situation, should refrain from crossing ahead of the other vessel if at all possible. The second factor affecting the desirability of a collision avoidance strategy is the manoeuvrability of the vessel. It is often difficult for merchant ships to reduce speed without prior warning and even then it may take some 20 minutes for a large vessel to come to rest. The third factor is the commercial expedience of the voyage. We want to avoid straying to far off our course line or carrying out complicated manoeuvres, such as time consuming round turns, if it can possibly be avoided.

The most hazardous target is selected according to a set of heuristics based on a number of numeric factors such as the relative speed of the target, angle of approach and location. The establishment of risk levels is of course very subjective. However the identification of a most hazardous target here serves only to provide the system with a suitable starting point for the generate and test sequence and to reduce the solution generation time required. An alternative target could just as well be used as a start point, but the generation time may be substantially extended.

To prevent the selection of strategies that will obviously be rejected some heuristic guidance is provided during the selection process by a set of Collision Avoidance Heuristics. These heuristics take into account the current movement, location and predicted movement of other vessels in the vicinity. For example a starboard alteration strategy will obviously be unsuitable if there is another vessel located close by on ownships starboard side. Alternatively a speed reduction strategy is undesirable if another target is expected to take some collision avoidance strategy of her own by altering course to pass round ownship's stern. In this latter case a speed reduction on behalf of ownship would cause it to reduce speed into the clear water astern required by this other vessel. The method where by the system can take into account the expected movement of other targets is described in the following Sections.

When a suitable strategy has been generated the efficacy of this strategy is tested. A proposed strategy is considered to be unsuitable if (a) it does not resolve the current collision situation by an acceptable miss distance or (b) it successfully resolves the current collision situation but it creates another one either with another vessel or a chart feature. What constitutes an acceptable miss distance is governed by the domain dimensions as discussed in the previous Section. If a strategy is considered to be unsuitable a return to the generation process and the

Collision Avoidance KB is implemented. Otherwise the strategy is output to the user together with a suitable "Return Action". Return actions are generated by reference to another Knowledge Base, the Return Action KB.

EXPECTED TARGET MANOEUVRES.

The predicted movement of targets is arrived at by repeating the analysis described in Section 4 while viewing the situation from each detected target in turn. An expected target manoeuvre KB is then consulted with the resulting PSD and SSD. The structure and content of the expected target manoeuvre KB is similar to the Collision Avoidance KB described earlier except that a descriptor representing a most likely manoeuvre, if any, is returned. This descriptor is referred to as the Expected Target Manoeuvre Status Descriptor (ETMSD) and plays an important role in the Collision Avoidance Heuristics used to refine the search through the Collision Avoidance KB and hence make it more efficient.

This alternative view approach marks a major departure from all previous work. Another departure from previous work is the consideration of chart constraints. This is described in the following Section.

CHART CONSTRAINTS.

The system takes chart constraints into consideration by applying a KBS approach to well tried Geographic Information System (GIS) Techniques (Samet [8], Gehagen [9]). In principal this involves tessellating a two-dimensional area of interest down to a minimum square size and storing the information discovered at the leaf nodes in a quad tree data structure. The primary advantage of this data structure is that it facilitates spatial operations such as area searches.

In the proto-type system under discussion here three types of terrain were identified as appropriate primary geographic constraints on the advice generation process:-

(a) Land.
(b) Sallow Water.
(c) Clear Water.

The first two were considered "unsafe" areas and the second "safe". Shallow water was defined as water with a maximum depth of less than 15m below chart datum because of the type of vessel used as a test vessel during system development. This had a draft of 9.14m. The handling of point features such as buoys and isolated rocks are still the subject of further research work.

The chart covering Lands End in Cornwall, England, was selected as an appropriate test area. The tessellating process involved quartering this chart area into smaller and smaller squares until either each square was completely filled with an area of a uniform characteristic or a minimum square size was reached. Where a minimum sized square was filled with more than one terrain type the predominant terrain type, with a bias towards safety, was selected.

The tessellated chart as described above was then mapped into a quadtree data storage structure. Each node in the quad tree represents a division of the chart. The branches in turn may then either consist of another node, indicating that the quadtree has been tessellated further, or a leaf node representing a single tessellation or tile.

A number of rule based routines were developed to allow interaction with this quad tree represented chart (Coenen [10]). For the purposes of this paper the Track Test Routines is of the most interest. This is used to check that any proposed course line, or set of course lines, which deviated from the proposed voyage plan, do not (a) pass through any unsafe area and (b) pass all unsafe areas by a suitable margin. What constitutes a suitable margin is again subjective. However the concept of a "land domain", based on empirical data, was developed i.e. an area about ownship similar to the ship domain described earlier, but which the navigator will wish to keep clear of all geographic constraints. Thus any course alteration suggested by the system will be considered safe if no unsafe tessellation are discovered along its length or within half the land domain width on either side. This requirement is implemented within the track test routine by testing a number of parallel tracks, spaced at a distance equal to the minimum tessellation size.

OPERATION.

The system's operation can best be illustrated by an example. A typical scenario involving a multi-ship encounter and including geographic constraints is given in Figure 2. The scenario is located north of Lands End and consists of four ships, ownship proceeding in a north-easterly direction, a fishing vessel proceeding towards the Bann Shoal and crossing ownship's course line from starboard to port so as to involve risk of collision, and two other ships, both being overtaken, one of which may cause a problem at a latter date. In the Figure the graduated arrows represent the predicted movement of each vessel over the next 36 minute period, each graduation representing 12 minutes (0.2 hours). The ship shapes are not drawn to scale, they simply give an indication of each vessel's true speed, the speed being directly proportional to the size of the ship shape.

In this scenario Target 1 is considered to be the most hazardous target and will be the target with which the Collision Avoidance RB is entered. Rule 15 of the Collision Avoidance Regulations states that in a crossing situation "the vessel which has the other on her own starboard side shall keep out of the way and shall, if the circumstances of the case admit, avoid crossing ahead of the other vessel". The initial advice generated by the system would therefore be for ownship to alter course to starboard so as to pass round the stern of the fishing boat. This will then be tested against against each additional detected target. In this particular case the collision avoidance strategy will be accepted. However when

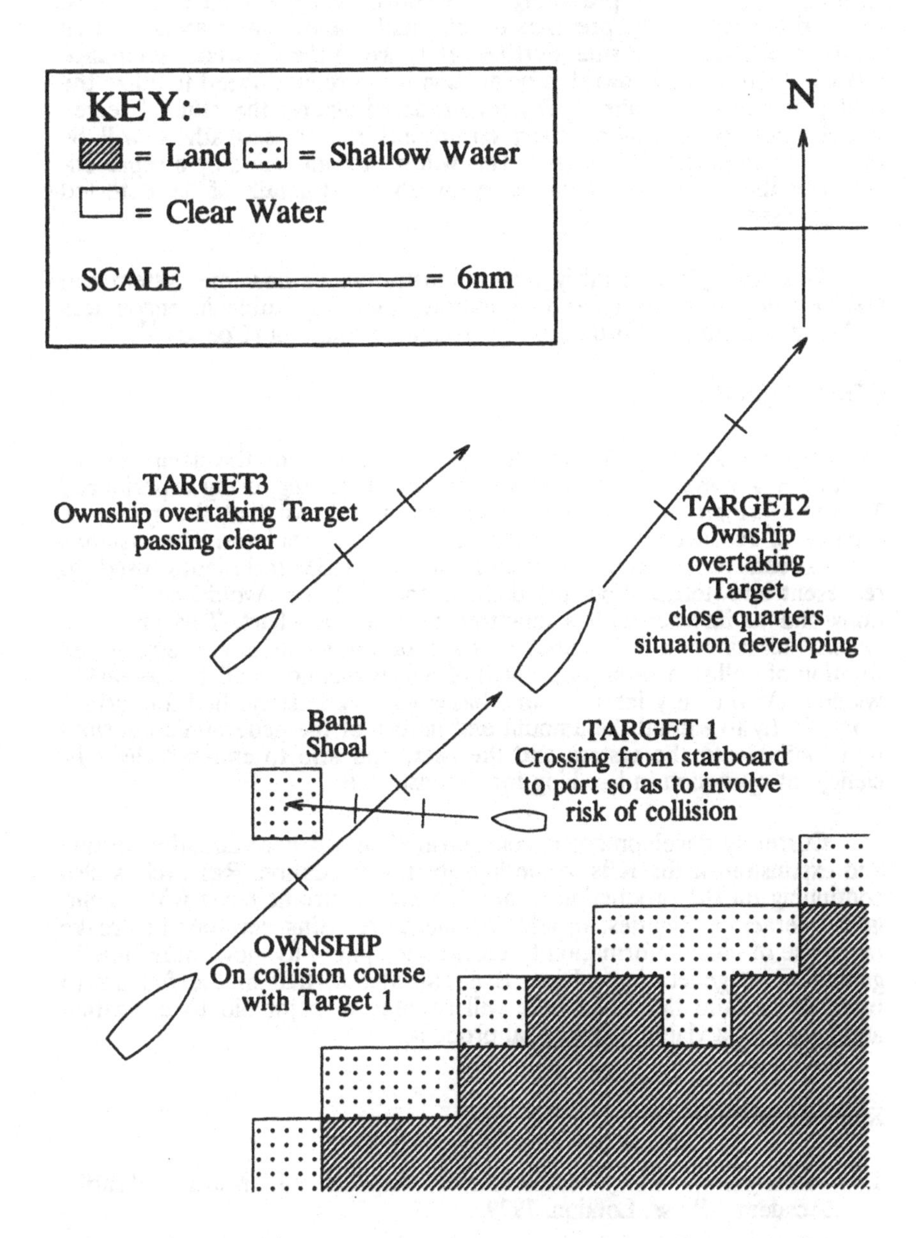

Figure 2: Example of Collision Scenario

considered against the prevailing chart constraints the advice will be be rejected because of the presence of land and shallow water areas located on ownship's starboard side. Further reference to the Collision Avoidance RB will then result in ownship being advised to reduce speed to allow the fishing vessel to pass ahead. When considered against the other detected vessels and the prevailing chart constraints the revised advice will be accepted and output to the user. This will be accompanied by a representation of the chart area surrounding ownship and details of the detected fishing vessel.

This example is a fairly typical illustration of how the system can resolve fairly complex collision situations involving multiship encounters and land constraints. Further examples can be found in (Coenen [11]).

CONCLUSIONS.

In this paper a system to provide decision support for the navigators of ocean going vessels has been described. Initial testing, using experienced Master Mariners, in simulated conditions, has shown that the system is capable of resolving complex collision avoidance problems. The system's success can be attributed primarily to the KBS techniques used to represent knowledge about the domain, the Collision Avoidance Regulations and to interact with a quadtree represented chart. This gives the system access to a substantial amount of information concerning the solution of collision situations, much of which was not available to earlier systems. At the very least a sound bases has been established for further work, firstly to extend the amount and nature of the geographic information available to the system and the user, and also to enhance the efficiency of operation in handling these constraints.

Currently development is concentrated on further validation testing and expansion of the KBs in the light of this validation. Research is also continuing on the production of an effective electronic chart with intelligent interactive facilities. In addition ideas are being developed to make more use of chart information to enable the system to show more "intelligence" when generating advice. It is considered, that in the fullness of time, the system described here will become a major aid to navigation and hence be a viable commercial product.

REFERENCES.

1. Hollingdale, S.H. Mathematical Aspects of Marine Traffic. Academic Press, London, 1979.

2. Cannell, W.P. Collision Avoidance as a Game of Co-ordination. Journal of Navigation, Vol 34, 1981.

3. Grabowski, M. and Wallace, W. An Expert System for Maritime Pilots, its Design and Assessment using Gaming. Report for U.S. Department of Transport, Maritime Administration 1987.

4. Sugisaki, A.M., Imazu, H., Tsuruta, S., Inaishi, M. and Matsumura, H. Expert Systems for Coping with Collision Avoidance and the "Ex Post facto" at Sea. Proc. International Navigation Conference, Sydney, 1988.

5. International Maritime Organisation. The International Regulations For Preventing Collision At Sea. HMSO, 1972.

6. Goodwin, E. M. A Statistical Study of Ship Domains. Unpublished CNAA Ph.D. Thesis, City of London Polytechnic, 1975.

7. Davis, P.V., Dove, M.J. and Stockel, C.T. A Computer Simulation of Marine Traffic Using Domains and Arenas. Journal of Navigation, Vol 33, 1980

8. Samet, H. The Quadtree and Hierarchical Data Structures. Computer Surveys Vol 16, No 2, 1984.

9. Gehagen, M. and Hogg, J. A Pilot Geographic Information System Based on Linear Quadtrees and a Relational Database for Regional Analysis in Spatial Data Processing using Tesseral Methods (Ed. Diaz, B. and Bell, S.). NERC, 1986.

10. Coenen, F. and Smeaton, P. Rule Based Algorithms for Geographic Constraints in a Marine KBS. In press, Knowledge Based Systems, 1991.

11. Coenen, F. and Smeaton, P. A Real Time KBS in Marine Navigation in Research and Development in Expert Systems VII (Ed. Addis, T.R. and Muir, R.M.). Cambridge University Press, 1990.

[illegible], A.M., [illegible] M., [illegible] Matsumura, [illegible] Expert Systems [illegible] Quality [illegible] and the [illegible] Fourteenth International [illegible] Conference, Sydney, 1988.

5. International Masonry [illegible] The [illegible] for Elevating [illegible] HMSO [illegible]

6. [illegible] M. A. [illegible] CMAA [illegible]

7. [illegible] M. [illegible] T. A. [illegible] of Masonry [illegible] Journal of [illegible] Vol. [illegible]

8. [illegible] Handbook and [illegible] Data [illegible]

9. [illegible] Pilot [illegible] Nuclear [illegible] Research [illegible] for Regional [illegible] Method, [illegible]

10. [illegible] Based [illegible] for Categorical [illegible] In press, Knowledge Based Systems.

11. [illegible] KBS [illegible] Cambridge University Press, [illegible]

An Expert System Tool to Aid Production Schedulers

C. Ready

Sira Ltd, South Hill, Chislehurst, Kent, UK

INTRODUCTION

Information Technology at the shop floor level in the process industry is now well developed. Compared with the manufacturing industry the process industry has been a regular user of instrumentation and control systems for at least two decades. Computer-based control systems and computer-literate personnel in this industry are commonplace. With the lower level operations well under control, attention has recently being focussed onto the top level of the 'control' hierarchy, ie that of overall production control.

Systems are now coming into use that deal with overall factory planning. Based on the capacities of the production facilities these systems are used to decide what the plant should produce and when. However, these systems typically treat the plant as a "black box" specifying only input and output materials and quantities over extended time periods.

The area where little attention has been focussed to date is that of the short-term scheduling of the operations within the plant. This is one of the areas of operation least suited to conventional computer-based solutions. It is however a very important area since wrong decisions, breakdowns, mistakes at this level can wreck weekly production schedules and cause product quality deteriorations, wasted materials, unnecessary plant idle time, late deliveries, etc.

In the process industry in particular with wide ranges of products; alternative methods and recipes for production; the need to deal with by-products, etc, the complexity is such that the scheduling is typically manual and in the hands of a supervisor or manager who has the experience to deal with the problems and appreciate the consequences of actions.

In contrast the manufacturing and automotive industry a considerable amount of work has been done in recent years in developing computer-based scheduling systems to deal with assembly and discrete part manufacture.

Automated process industry shop floor scheduling is therefore an area which is difficult and least advanced. It is also an area where Sira has paid considerable attention over the last 4 years resulting in the development of an Expert System based scheduling package.

Schedulers experience and the scheduling task

Although some computer aids may be available to help the scheduler, the process of production scheduling process still tends to be mainly manual (pencil, calculator, and planning board), and is heavily reliant on the past experience of the scheduler, and on his 'rules of thumb' about how to produce acceptable schedules.

In practice, the schedule is likely to be reviewed by production management, but assuming no major conflicts are found, and that the requirements of key customers are dealt with satisfactorily, it is likely to be accepted. There is often no real way to assess the quality of the schedule produced, and the scheduling process itself can be very time consuming. Of course, should any sudden changes in requirement occur, or if a key plant item breaks down, the whole process must be repeated.

The scheduling task is simple to understand as an intellectual activity, but in most industries is exceedingly difficult to perform as there are an abundance of possible 'solutions', most of which are in some way inefficient. The skill of the schedulers is in producing a schedule with the following characteristics:

* Meets all of the key demands.
* Meets most of the other demands.
* Does not impose massive stock variations.
* Limits inefficiencies such as change overs.
* Is 'cost effective'.

In creating these schedules, schedulers must have:

An understanding of the gross behaviour of the production plant. This includes the manufacturing times for various production processes, knowledge of economic batch quantities, the capacities of production plant and storage, the number and location of items of special equipment, routing options etc.

A knowledge of the characteristics of materials that affect production, eg colour, sweetness, shelf life etc.

A detailed knowledge of past schedules that have worked, and characteristics of schedules that have not.

The major requirements of the schedule. Key customers that have to be satisfied and those which may, if absolutely necessary, have to wait.

As detailed above, there may be a number of solutions to the scheduling problem; however most of them will be sub-optimal. In working towards a good solution to the scheduling problem, the scheduler must firstly take note of 'hard constraints' such as plant capacities, maximum production rates etc; the difference between 'acceptable' and 'good' however is much more a matter of using his skill to satisfy large numbers of 'soft constraints' which he knows produces 'the best' schedules.

Soft constraints - the key to a good schedule

Soft constraints are restrictions not based upon physical limitations but are more to do with preferences based on experience. For example a plant may have three identical packing machines, but the scheduler knows that there are problems with one machine. Machines 1 and 2 both pack single and multi-packs equally well, but that for some reason machine 3 has problems with multi-packs, often tearing the wrapping paper. There is no real difference between the machines, and machine 3 could be made to pack multi-packs if no other machine were available, but the scheduler would prefer not to risk a disruption in production if possible.

Other soft constraints might relate to material colour, or other attributes of materials such as acidity, salinity etc. For example, do not follow dark material with light material as this causes contamination. This constraint could be ignored and a clean performed, but in creating a production schedule the scheduler prefers to make light products first followed by medium and then dark.

The key to effective scheduling aids is the production of a system that will allow the specification and use of these soft constraints, in addition to the normal hard constraints.

CONVENTIONAL SCHEDULING APPROACHES

There are a number of scheduling approaches in current use ranging considerably in their level of sophistication. Below is a brief review of some of the key ones.

Wall Charts

Typically the weekly scheduling activity could occupy up to a couple of man-days using a wall chart or pinboard and calculator to develop the schedule. This approach is practical but far from optimum as typically significant approximations are made and only a fraction of the total possibilities are examined due to the effort required.

Spreadsheets
Spreadsheets are increasingly being used to good effect to assist with the simulation aspects of scheduling for predicting how a manually devised schedule will perform by calculating stock profiles, etc.

These spreadsheets are well suited to straight-forward situations where the production route and recipes are well defined. Again, however, they rely on the actual scheduling being done manually with the spreadsheet used to simulate the flows and inventories as a check.

Electronic Planning Boards
Electronic planning boards provide an interactive means of manipulating schedules on a computer screen. Such systems can deal with a number of the complexities of the process domain but are typically configured at a very low level requiring a large amount of customisation. Again these systems do not perform the scheduling task but assist the manual scheduler to check that his devised schedule is feasible .

Linear Programming
Linear programming is a well established means of optimising complex operations based on cost functions. Unlike the systems described above it does generate schedules. These are based on minimising accumulated cost measures associated with various aspects of running the plant. Unit costs are associated with late delivery, idle time, product changeovers, etc and a schedule is developed which minimises these costs.

The linear programming type of solution does, however, have its problems in that the scheduling is purely algorithmic and the best schedule based on the cost functions is generated. Because of the difficulties of associating costs with differing aspects of the production operation it may be that the resultant schedule does not look or feel right to the expert scheduler since expected sequences etc may not necessarily be produced. With such an approach attempts are often made to adjust cost functions to produce a schedule that the experienced scheduler does feel comfortable with.

Linear programming solutions tend not to be popular with the shop floor staff who have to use them due to the feeling of 'groping in the dark'. If the system fails to schedule an order for some reason then it is impossible to discover the reason why this has happened and there is a general lack of visibility.

THE REQUIREMENTS OF AN INTELLIGENT SCHEDULING SYSTEM

In producing any form of automated scheduling system that aims to solve real problems, two components have to be present, a representation of the production facility (a simulation or model) and a means of capturing the reasoning process of the scheduler. Almost without exception, the available tools concentrate (albeit to a limited extent) on the modeling and simulation aspects; few, until now effectively none, provide any means for capturing and using the experienced scheduler's knowledge and skill in dealing with the soft constraints.

Despite the general lack of suitable software there is a substantial requirement for such tools. Recent advances in software, specifically those relating to the area known as Expert Systems, mean that a more realistic production scheduling tools (able to deal with both hard and soft constraints) are now available.

Expert systems

Expert systems are computer systems that emulate the reasoning of a human expert. Expert system differ from conventional systems in a number of ways, the most important ones being:

* The expert knowledge is separate from the computer program that manipulates it.
* The expertise is in an intelligible form, not encoded in a non-understandable computer language.
* The operation of expert systems should be understandable at each stage of its reasoning.
* Expert systems can explain their reasoning and therefore can justify their conclusions.

Significant work has been done in applying expert systems to industrial problems over the past few years, and the technology has now gained widespread acceptance and is relatively mature. Most expert system applications are diagnostic or monitoring systems that help an inexperienced operator to diagnose process faults. Often off-line in the past, on-line systems are now available and gaining acceptance readily.

Sira pioneered the application of expert systems techniques to production scheduling in the process industries in 1987 and, following collaborative work with a number of key clients, proved that the approach provided a very effective solution.

Following on from the early prototyping work, we have just released PROSPEX, a sophisticated expert system-based production scheduling tool packaged so that it may be used by the production scheduler. PROSPEX is one of a new breed of expert system tools that aim at tackling a particular application (production scheduling in this case) rather than being general purpose tools.

PROSPEX

PROSPEX is a production scheduling tool aimed specifically at the production scheduler. PROSPEX is designed to produce schedules from a specified set of orders or demands, and takes into account not only the physical limitations of plant etc., but more importantly, can generate schedules using the same knowledge as the experienced scheduler; PROSPEX differs significantly in this respect from other scheduling tools. In any PROSPEX application there are two key components, a model of the production process <u>and</u> a set of rules that guide the scheduling.

<u>PROSPEX models</u>
The model of the production process includes the following descriptions:

The physical plant items and the connections between them.

The materials that flow through the plant, the raw materials, and intermediate and final products.

A specification of the production processes including recipes, machine capabilities and personnel requirements.

Timing information including shift times, planned maintenance time, unscheduled down-time etc.

A key facility of PROSPEX is that it comes with an extensive library of plant items etc. that the user can tailor to fit the particular requirements of their specific plant by adding extra user defined attributes. These user defined attributes may then be referred to by the rules, allowing the schedules to be generated on the basis of information specific to the particular plant, rather than simply relying on pre-defined attributes such as capacity and throughput which are common to all plant.

<u>PROSPEX rules</u>
The scheduling rules express two type of knowledge, soft constraints and sequencing information. A constraint would be expressed in a rule such as:

Do not load an order
if the order has the attribute CUSTOMER = CUSTOMER-A
and PACKER-37 last processed an order with the attribute CUSTOMER = CUSTOMER-B

The rule above is taken directly from a PROSPEX system and expresses a soft constraint about the destination of an order. The attribute CUSTOMER has been created and added to orders by the user and the rules can now refer to this attribute and use it in scheduling.

Sequencing information is important in scheduling, as the scheduler often has to express ideas such as

Start with CUSTOMER-A orders (the van has to leave early), then do any orders for plain cakes, then chocolate products (so we can limit clean times).

Make sure all of the products that need the decorator are done together (so it can be moved to another part of the plant).

In PROSPEX rules are provided that 'decide' what to do next at any point in the production, these have priorities and can be used to ensure scheduling follows the ideas expressed above.

What PROSPEX does
Once a PROSPEX system has been configured the scheduler can enter a production requirement. PROSPEX will then produce an appropriate production schedule taking into consideration all of the hard constraints held in the plant model and, more importantly, the soft constraints and sequencing information expressed in the rules.

Once a PROSPEX model has been run, the scheduler can maintain the schedule, updating the model based on actual production data. In this way, the model can always be kept in step with the plant, and if unexpected events occur, the scheduler can immediately reschedule from any point in time to take account of the changing circumstances.

As PROSPEX works with a knowledge of all of the things that affect production, from stock availability through shift times to personnel requirements, PROSPEX only ever generates schedules that may actually be achieved. In this, it differs from simpler capacity planning systems, and details exactly when and how the production takes place.

The system produces output directly for the scheduler. Gantt charts show the operation of production units or the sequence of an order through the plant. Graphs of stock profiles are provided for storage vessels showing the changing stock levels throughout the production period. Additionally a worksheet for any particular plant item can be printed, detailing what should be done and when. Statistical information on machine utilisation, materials consumed, products produced and their associated costs are also available.

Should any production requirement not be satisfied PROSPEX will given an exact reason why the failure occurred in an easily understood set of 'excuses', a significant improvement over conventional systems. Excuses are text messages generated whilst PROSPEX is scheduling that detail any points where problems arise. These are collected together, sorted and presented to the user through a special interface.

BENEFITS OF THIS APPROACH

PROSPEX enables production schedulers to use a computer system to automate for the first time what has in the past been a complex and time consuming task. Encapsulating the expertise of the scheduler provides direct benefits to the scheduler and to his company because:

* Expertise is available 24 hours a day.
* The strategies developed can be applied consistently.
* The scheduler may be freed to undertake other tasks.

These benefits are the same as would be derived from automating any manual process. However as PROSPEX is an expert system, it has significant additional advantages. These all spring from encapsulating in a computer system what was previously intangible, namely the expertise itself. This leads to the following:

* Expertise may be distributed and made more widely available.
* Expertise can be analysed.
* Expertise can be "understood".
* Expertise may be questioned.

and ultimately

* Expertise may be altered and tested.
* Expertise may be improved.

This may at first seem somewhat far fetched, but there is a significant history of expert systems that have outperformed the experts they once emulated. Improving production scheduling has substantial pay-backs in two areas: increased throughput and improved customer satisfaction.

REFERENCES

1. Warwick, A.M. and Walters, H.M.J. A rule based planning and scheduling system for manufacturing industries in Expert Planning Systems, IEE Conference Publication 322, pp. 104-109.

2. Braun, F. and Lebsanft, E. System d'ordonnancement expert de lignes robotisees d'empaquetage pour l'industrie alimentaire in Avignon 90 Special Conference AI, food processing, biotechnological, chemical and pharmaceutical industries, pp. 263-274.

3. Wild, R. Production and Operation Management, Cassell, London, 1989.

REFERENCES

1. [illegible]

2. [illegible]

3. [illegible]

SECTION 4: MONITORING AND CONTROL

A Real Time Interpretation Model

M. Blaquiere (*), F. Evrard (**), A. Awada (**)
(*) TOTAL NORGE, Haakon VII's gate 1, Oslo 1, Norway
(**) ENSEEIHT, 2 rue Camichel, 31071 Toulouse cedex, France

I- ABSTRACT

The monitoring of a real time process generally implies a reasoning on temporal events where time, duration, order, long term dependancies etc... have to be modelled.[4] Monitoring a process does not only mean to reason on what has been collected in a facts database, but also to follow dynamically what is happening through a set of axioms. Indeed it should be possible to describe what is in action, as well as what could possibly occur in the future resulting from these actions.

For a given axiomatic, including inference rules like modus ponens and necessitation, we think that the time dependent management of these inference rules, applied on formulas, is not neutral for a demonstration result. This means that this management will have an effect on the validity of the formula which is being evaluated.

We first show how a certain reasoning on temporal events can be more or less handled by different demonstration methods. Grammars, classical, temporal or default logics demonstrators are reviewed to point out the importance of the demonstration itself towards the logic used.

Based on the aspect of controlling a certain order of evaluation, we propose a formalism in which the inference control is clearly syntaxically expressed in rules. In consequence, contrary to usual resolution, the plan of a formula demonstration is specifically constructed from the formula itself. This planning approach explicitly gives the method to assemble elementary stages of demonstration.

II- RULES INDUCED BY MONITORING A PROCESS

One particularity of real time interpretation is that the flow of data on which the reasoning is handled is endless. From a set of sensors, one tries to gather values and associate them with symbolic states in order to catch and recognize from time to time pertinent situations. Interpreting such situations, derived from the analysis of sensor signals, can be seen as a problem of pattern recognition where two main methods are conventionally used: the statistical approach and the structural one. Each of these methods integrate a time representation to express the dynamics of analysed events.

The statistical approach considers time as an additional dimension in the representation space. The structural approach apprehends time by a reasoning on events or states, based on temporal operators or sequential analysis. These states can thus form part of the time signature characterizing a certain process operation, onto which other anomaly characteristics can be superimposed.

In [4] we are motivated to represent a state by a starting time, an ending time and a truth value at a given moment. The goal is to deal with durations and time relations in terms of interval based calculus. The temporal constraints are explicitly expressed in the rules. In this work, the rules may be of various types. Let recall some examples of these rules concern:

three rules that define the expertise and which can be seen as a part of the axioms of a particular theory

```
P(t,v) <- V(t,v1)&T(t,v2)&(v0=n*r*v2/v1)&(v=f(v0,t))
```

where v,v0,v1,... are numerical values, V,T,P represent sensors or analog variables and t is a discreet variable representing time, this rule is a computation formula which dynamically links an output according to several inputs, and where there is an implicit time constraint: a single instant, this kind of rule gives the possibility in calculating any formula from sensors values at a given instant,

```
I.A(t) <- P(t,v1)&Q(t,v2)&(v1≤a*v2)
```

this rule generates an instantaneous symbolic state "I.A(t)" from sensor values, it is an elementary state which corresponds to a pertinent recognized situation,

```
F(t1,t3) <- A(t1,t2)&B(t2,t3)&(t2-t1=1)&(0≤t3-t2≤3)
```

this rule defines a durative states sequence where tj-ti = n means that there are n sampling units between ti and tj, this rule combines durative states together generating more abstract state in a grammar manner,

three rules which seem to play the role of axioms schemata

```
D.X(t1,t2) <- I.X(t1)&I.X(t2)&(t1<t2)&(t2-t1<Σ)
```

this rule generates a durative state "D.X(t1,t2)" from several identical instantaneous states, where X is a predicate variable, this durative state represents the continuity principle of events,

```
D.X(t0,t3) <- D.X(t0,t2)&D.X(t1,t3)&(t0≤t1≤t2≤t3)
```

this "agglutination" rule governs the generation of persistant facts, as the time is discreet and counted in terms of sampling units, this rule is related to the previous one.

```
D.X(t1,t2) <- D.X(t0,t3)&(t0<t1<t2<t3)
```

this rule expresses a temporal inclusion

At this point of course the problem is to interpret correctly real facts with such rules. It is necessary to have a good demonstration model. To reach such a goal let us specify what are the roles of the main components of a demonstrator.

The three aspects of a demonstrator:

An axiomatic system consists of a set of axioms (meaning formulas that are considered valid), and a set of inference rules enabling the generation of new valid formulas from the existing ones. The resolution's principle or any set of inference rules define the way the rewriting has to be done. Associated with this rewriting mode, it is necessary to define an algorithm or mechanism that will apply the resolution principle (or the inference rules) on the set of axioms.

The demonstration of a theorem is founded upon three components:

- a set of axioms (eventually ordered),
- a set of inference rules,
- an algorithm (mechanism) that couples rules and axioms[9][10]

The formulas obtained by the demonstration does not result from a succession of logical consequences, but are constructed by syntactical substitutions authorized by the rewriting mode given by the inference rules. We are going to show that this mechanism has the main importance in the fact that a demonstration does classical, default or temporal demonstrators.

CLASSICAL LOGIC DEMONSTRATOR:

Classical tools of demonstration suggest running an evaluation on the states history kept in a database from the beginning of the situations analysis. Thus, the control of executions is summarized by an unending loop where, for each iteration, the entire set of axioms would be matched with the past events stored in a database. In such demonstration, time can be expressed as one of a predicate variable componant:

Example V(t,x) (cf. previous rules)

The checking of orders, durations, long tem dependancies, or all kinds of temporal links expressed between states, is done by filtering all the elements every sampling time of the states database to see if the time values are in accordance with the temporal constraints. Finally, the evaluation order of premises of a rule is generally not important and it is often done from left to right (in PROLOG for example). [8]

But in the rule

```
C(f(t,t')) <- A(t)&B(t')&[(t'-t)(t-t0)>0]
```

where t0 is a constant, t and t' are variables, verifying A(t) can inhibit the triggering of the rule. If t'<t<t0 the first condition to verify is B(t') and not A(t) of which the ocurence will arrive latter. In this case a classical demonstrator can loop arbitrarily a number of times on A until B occurs.

In addition when the evaluation order of the premisses is staticly determined it is not possible to use a partial triggering of a rule. Nor is it possible trying a kind of default reasoning which would point C(f(t,t')) as a hypothetical forecast, as soon as a premisse occurs.

About the following rules,

F(t,t+9) <- G(t,t+9) & C(t+3,t+6)
G(t,t+9) <- A(t,t+3) & B(t+6,t+9)
which correspond to the events sequence,

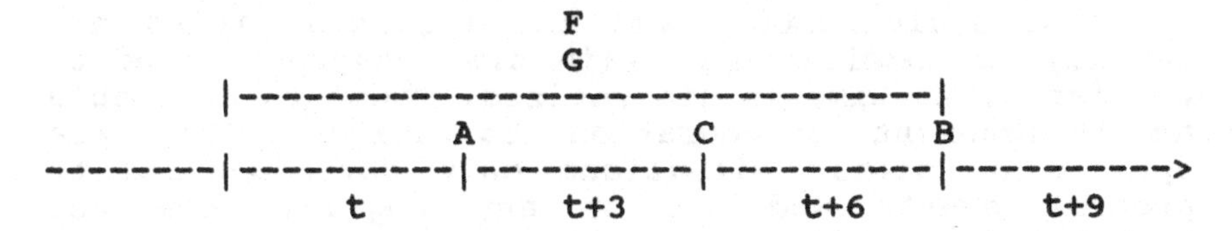

a classical demonstrator will have to wait or worse to loop until t+9, at earliest, to conclude F. It results from the fact that things must occur integraly and exactly to be considered as true. What is the part of responsability of the mechanism in this conclusion ?

Could others mechanisms give different conclusions with the same axioms and inference rules ?

Considering the agglutination and the inclusion rules (cf previous example), lets see that their application within a classical demonstrator can modify the conclusions and the proceeding of demonstrations.

Supposing that:
- during [t,t+3], A has occured in conformity with the rule
- during [t+3,t+6], C has samely occured
- At t+6, B has started,

at a time t0 between t+6 and t+9, we can conclude thanks to the agglutination that G(t,t0) is true. Same with t0' between t+3 and t+6, G(t,t0') is true. Without this rule, neither G nor F will be true for t0 or t0'.

If another rule where F is part of the premisses:

H(t,t+6) <- F(t,t+6) &...,

H(t,t+6) is true and can only be true thanks to the inclusion rule because [t,t+6] is included in [t,t0].

In a classical demonstrator even if F(t,t+9) is true, F(t,t+6) is false without agglutination and consequently so does H(t,t+6). And completing classical logic with agglutination, the demonstrator has unfortunatly to wait t+9 to conclude H.

We claim that a classical demonstration mechanism uses implicit rules of time manipulation: an integrality and exactitude principles, which results of the fact time is treated like simple variable with a rigid numerical value semantics on equality.

The agglutination and the inclusion rules are the way to modelize a certain time concept linked to the fact that time is discretized. These rules enable an assignement of duration to states. They are applied on atomic situations only corresponding to process events and not on any logical formulas. Though they do not correspond to axioms schemata.

The following question arises: is it sufficient to add rules like agglutination and inclusion ones to deal with some concept of time ?

Indeed the result of a demonstration (success or failure) in PROLOG, for example, is determined by applying the resolution principle that does not use in its working mechanism, the scheduling aspect of a time variable.

To simulate such logic, it is necessary in addition to use control metapredicates such as repeat until and cut, and other variables constraints facilities.

GRAMMARS

In process control it is possible to define a set of grammar rules which will recognize "sentences" formed by elementary tokens from the unending flow of data. Each of these tokens can represent a vector characterizing a particular process situation, where each components could correspond to a range of numerical values on such and such sensor. The sequentiality of events can be represented by the concatenation order of symbols, when the interpretation is done with a grammar. The concept of temporal order can be modelled by acting on the precedence of the symbols when writing the rules. The rules have a role of temporal control on the temporal order of events. One can even consider counting the time by the number of symbols and allowing for a time delay. It is the usual implicit and unique way to manipulate time and duration.

But grammar is a normative approach of events and unrecognized situations block the analysis in the sense that a rule like:

F -> A.A.A.B.B.C.D

which means 3 A-steps followed by 2 B-steps,..., would be transformed in an other set of rules containing an arbitrarily number of ?-symbols in place of some constituents of F. The purpose of this symbol would be to filter any arbitrarily long sequence of unrecognized steps. In such a case these rule can filter nearly anything. A proper analysis of F is avoided.

Likewise, describable abnormal situations require a parallel execution of complementary rules. This complementary grammatical analysis could be done for the recognition of abnomalies during a sequence of normal situations.

Example

```
F -> A.I.A.B.J.K.D
I -> A-anomaly-1 / A-anomaly-2 /...
J -> B-anomaly-1 /...
K -> C and ...
```

It is impracticable to pursue manually such a description of a process and especially to express temporal constraints like long term dependancies while a lot of events would happen in the meantime.

On the contrary an exact sequence of such elementary steps constitutes a program which is obviously related to the algorithmic aspect of a demonstrator. Grammar rules correspond to the rewriting control and they are to be considered as a program and not as axioms, or traditionaly expert system rules, like in the logical approach. This right control mechanism is what a demonstrator should perform.

TEMPORAL LOGICS

Temporal logic,[2] approached by modalities, enriches classical logic with temporal operators such as Next, Nextn (n times), Always, Until and so on. Defining the semantics of such operators assumes that it is possible to access states, operating on an accessibility relation R on them. To characterize a state "s" by "next A", it is necessary to be able to characterize the next state t of s (if: s R t) by the property A. Formally:
for all s,t

$$s \vdash \text{Next } A \quad \text{iff} \quad [s \; R \; t \Rightarrow t \vdash A].$$

It is obvious that the next operator is strongly linked to the symbols sequencing operator of a grammar.

We can associate a set of grammar rules to an automaton whose purpose is to recognize a sequence of states described by a temporal logic formula. For example: ababbababbbb....is a sequence generated by the grammar:

S -> a.B / a.B.S
B -> b / b.B

Each sequence, belonging to the language of this grammar, is characterized by the fact that: it is always true that b occurs just after a (there is not a second a after a). It is expressed by the temporal formula:

ALWAYS(a => NEXT (b)).

The necessity modality (i.e. ALWAYS) can be handled by two ways: either with the necessitation inference rule

if ⊢ X then ⊢ ALWAYS(X)

or with an equivalent formula defining ALWAYS(X)

ALWAYS(X) <=> X & NEXT(ALWAYS(X)).

In the first case, the interpretation of an event in terms of X implies that it will always be interpreted like that in every accessible future. This monotoneous result avoids to re-interprete differently a temporal fact from the past, in the light of new data. In the second case, it will be necessary to re-evaluate X at each iteration indefinitly.

Both cases are not acceptable for process control. For example it is possible that a peak, neglected in the past, becomes significant due to another correlated one in the present. Such past has to be re-interpreted.

DEFAULT LOGIC

When recognition concerns alarm handling, it is interesting to refer to the conclusion before all the premisses have completely happened.

These conclusions can themselves be part of other premisses refering other level of conclusions. This triggering property of rules expresses a non monotoneous principle. Certain conclusions can be destroyed as the time passes, or if certain new data occurs.

When monotoring a process with alarm handling, it is attractive to reason with a logic based on R. Reiter's [11] default logic. The defaults are considered as inference rules, possibly represented as follows:

$$(A \wedge MB) \supset C \quad \text{or} \quad \frac{A:B}{C}$$

where MB signifies B is consistent, compatible...In other words, if one cannot deduce ¬B, C is then concluded by default.

The demonstrations consist of upgrading an extension graph by looking at incoming flow of facts. This graph represents the different hypothesis which can be handled, for example, with a logic containing hypothetical truth values [12]. Every sampling, the demonstrator generates the new hypothesis and tries to supress the branches where the facts are surely not demonstrable any more.

The visualisation of such a graph is relevant when monotoring a process for alarm handling. Indeed, it is possible to show when alarms could hypotheticaly occur, by associating temporal constraints with the hypothetical graph.

However the number of hypothesis with such a demonstration process is in practise quickly enormous. It is difficult to share a constant monitoring with this method.

III- TEMPORAL MECHANISM OF DEMONSTRATION / DYNAMIC LOGIC

In some temporal logic applications [5][4] and in dynamic logic,[1] a formula such as $(\alpha)P$ means that after the execution of the program α the property P is true. The modality which appears within the brackets, allows to write programs using a programming language.

A process control system has to know at any moment what it has to do. Depending upon the way the demonstration is driven, the system can or not recognize real situations. Logically speaking, it means that the truth value of a formula is not always the same and depends upon the demonstration.

It is necessary not only to pursue a process, but also to make forecasts in order to answer about the truth of any property in the future. More, it is necessary to know how it can be achieved. Then the programs, within the modality we have to deal with, are particular ones: demonstrations.

We are no more interested with provability but with proves, and we propose a proof-logic to avoid defaults of the previous ones we have presented. Concerning either default logic or tree temporal logic, they have to manage future possible words.

In compensation our logic has to manipulate programs. And the main fundamental question we want to answer is: how to build a program, the execution of which leads to the truth of a property. Forecasts are based now on the existence of such programs concerning a formula. If it is possible to build them, it will mean that a situation can occur, on the contary if it is not there is no way to reach such a situation.
Any process expertise rises a critical point: the completeness of the associated theory is required to conclude in the possibility or impossibility of a situation. Of course it is difficult to have a complete set of representations of normal-considered situations , and it is still worse with abnormal and unpredictable situations. It remains that just with what is known we are still able to recognize that a normal situation can no more occur.

To illustrate this idea, taking the two rules seen previously:

```
F(t,t+9) <- G(t,t+9) & C(t+3,t+6)
G(t,t+9) <- A(t,t+3) & B(t+6,t+9)
```

lets try to find the program P allowing to demonstrate F(t,t+9).

```
              MP(α,β,X,Y)
[P]F(t,t+9)  ----------->   [α]X & [β](X ⊃ F(t,t+9)
              P=α.β
              Y=F(t,t+9)

X=G(t,t+9)&
-------------> [α](G(t,t+9) & C(t+3,t+6)) &
C(t+3,t+6)       [β](G(t,t+9) & C(t+3,t+6) ⊃ F(t,t+9))

β= φ
------------> [α](G(t,t+9) & C(t+3,t+6)) & [φ] R
G(t,t+9) &
C(t+3,t+2) ⊃ F(t,t+9) = R
```

R is used here for comodity.

```
Dec(γ,δ,X,Y)
-------------> [γ]G(t,t+9) & [δ]C(t+3,t+6) & [φ]R
X=G(t,t+9)
Y=C(t+3,t+6)
α= γ.δ
```

```
MP(ε,η,Z,Y)
-------------> [ε]Z & [η](Z ⊃G(t,t+9)) &
γ=G(t,t+9)         [δ]C(t+3,t+6) & [φ]R
γ=ε.η


Z=A(t,t+3) &
-------------> [ε](A(t,t+3) & B(t+6,t+9)) &
B(t+6,t+9)     [η](A(t,t+3) & B(t+6,t+9)⊃ G(t,t+9))
               & [δ]C(t+3,t+6)  & [φ]R


  η = φ
 --------------->    [ε](A(t,t+3) & B(t+6,t+9))
A(t,t+3)&B(t+6,t+9)⊃   & [φ]S & [δ]C(t+3,t+6)
G(t,t+9) = S           & [φ]R

                     S is used for comodity.

Dec(u,v,X,Y)
------------> [u]A(t,t+3) & [v]B(t+6,t+9) & [φ]S &
X=A(t,t+3)                [δ]C(t+3,t+6) & [φ]R
Y=B(t+6,t+9)
ε=u.v

[ ⊢X]X
---------->u =  ⊢A(t,t+3)  ,v=  ⊢B(t+6,t+9)  ,
           δ =  ⊢C(t+3,t+6)

sumarizing
                                               rule 3
P=α.β=(γ.δ).φ=((ε.η).δ).φ=(((u.v).φ).δ).φ =    (u.v).δ

P = ( ⊢A(t,t+3). ⊢B(t+6,t+9)). ⊢C(t+3,t+6)
  rule 6
P   =    ( ⊢A(t,t+3);wait(t+3,t+6); ⊢B(t+6,t+9)).
            ⊢C(t+3,t+6)

P  =  ⊢A(t,t+3); ⊢C(t+3,t+6); ⊢B(t+6,t+9)
```

The rules used are:

1) MP(α,β,X,Y) : $\dfrac{[\alpha]X \ \&\ [\beta]X \supset Y}{[\alpha.\beta]Y}$ is the modus ponens

2) Dec(α,β,X,Y) : ∃α,β, $\dfrac{[\alpha]X \ \&\ [\beta]Y}{[\alpha.\beta]X\&Y}$

3) [α.φ]X = [α]X   is the neutral element for .

4) [⊢X]X (after execution of "demonstrate X", X is true)

5) [α][β]X   [α;β]X

```
6) [ ⊢X(t1,t2). ⊢Y(t3,t4)] =                   (time model)
 1  (t2<=t3)    ( ⊢X(t1,t2);wait(t2,t3); ⊢Y(t3,t4));
 2  ((t2<t4) & (t1<t3))   ( ⊢X(t1,t3); ⊢X(t3,t2) //
                   ⊢Y(t3,t2); ⊢Y(t2,t4));
 3  ((t2<=t4) & (t3<=t1))   ( ⊢Y(t3,t1);
                              ⊢X(t1,t2) //
                            ⊢Y(t1,t2);⊢Y(t2,t4);
 4  ((t4<t2) or (t3<t2))   ( ⊢Y(t3,t4). ⊢X(t1,t2))
```

The notation // means that X and Y are executed in parallel.

The rule 6) corresponds to the composition of two programs which have only one instruction each.

6 cases are possible; they are represented by 4 clauses.

* The first case is without any recovery between the two intervals

```
          X           Y                    X       Y
     |--------|   |-----|    or    |--------|-----|
     t1       t2  t3    t4         t1       t2    t4
                                           t3
```

* The second case is with a recovery

```
          X
    |-----------|  Y
            |----------|
```

* The third one is an inclusion of the first in the second

```
   |-----|        |-----|        |-----|       |----|
 |---------|      |---------|  |--------|      |----|
```

* The fourth clause treats same cases with a communication of intervals.

Like grammars with rules, we have through program with our logic what has exactly the demonstration to do. As classical logic we can dispose of a set of rules which give a model of time we need. These rules link a particular structure over time (for example the ordered natural numbers structure), a particular one over programs defined in terms of a programming langage based on an adequat set of primitives instructions and operators (ex: the number 6 rule). The interest of temporal or default logics, which consists in the forecasting abilities, is now garanted here by the ability of the effective construction of a program. An other interest of temporal and dynamic logic is to speak about properties of programs. Our work re-assume it, using modalities to explicit the control to which the demonstrator is submited.

What has not appeared in the previous example session is that the program obtained is only an intermediate programming form. Indeed elements like - A(t,t+3) can be still decomposed to return to instantaneous instructions discretized according to the sampling units ot time.

BIBLIOGRAPHE

(1) D.Harel, Dynamic logic, D Gabbay and F.Guenthner (eds.), Handbook of Philosophical Logic, Vol.II, ch 10, 497-604, 1984, D.Reidel Publishing Company.

(2) J.P.Burgess, Basic tense logic, D.Gabbay and F.Guenthner (eds.), Handbook of Philosophical Logic, Vol.II, ch 2, 89-133, 1984, D.Reidel Publishing Company.

(3) R.Bull and K.Segerberg, Basic modal logic, D.Gabbay and F.Guenthner (eds.), Handbook of Philosophical Logic, Vol III, ch 1, 1-88, 1984, D.Reidel Publishing Company.

(4) E.Audureau, P.Enjalbert, L.Farinas del Cerro, Logique temporelle semantique et validation de programmes parallèles, Etudes et Recherches en Informatique, 1990 Masson.

(5) B.Moszkowski, Executing temporal logic programs, 1986, Cambridge University Press.

(6) A.Thayse et alii, Approche logique de l'Intelligence Artificielle, Tome 1 et 2, Dund Informatique, 1988-89, Bordas.

(7) J.Lambek and P.J.Scott, Introduction to higherorder categorical logic, Cambridge studies in advanced mathematics 7, 1986, Cambridge University Press.

(8) D.Maier and D.S.Warren, Computing with logic, logic programmings with Prolog, 1988, The Benjamin/Cummings Publishing Company Inc.

(9) C-L.Chang and R.C-T.lee, Symbolic logic and mechanical teorem proving, 1973 Academic Press.

(10) D.W.Loveland, Automated theorem proving: a logical basis, fundamental studies in computer science, Vol 6, 1978, Noth Holland Publishing Company.

(11) Reiter, R. 1980. Logic for default reasoning. Artificial Intelligence 13:pp 81-132

(12) Mc Dermott, D. & Doyle, J. 1980. Non monotonic logic I. Artificial Intelligence 13:pp 41-72

(13) Blaquière, M. & Evrard, F. 1987. Recognition of temporal phenomena in petroleum processes. Knowledge Based Expert System in Engineering, classification education and control. Editors D. Sriram & R.A. Adey. Computational Mechanics publications. pp 255-270

(14) Blaquière, M. & Evrard, F. 18, 19 Juin 1987. Simulation of processes for risk previsions. ORIA 87. Artificial Intelligence and Sea Marseille France

(15) Fox, M.S. & Kleinosky, P. August 1983. Technics for sensor-based diagnosis. VIII. IJCAI, Karlsruhe. pp 158-163

A New Method to Obtain the Relation Matrix for Fuzzy Logic Controllers

D.T. Pham, D. Karaboga

Intelligent Systems Research Laboratory, School of Electrical , Electronic and Systems Engineering, University of Wales, P.O. Box 904, Cardiff, CF1 3YH, UK

ABSTRACT

This paper introduces a new method for designing Fuzzy Logic Controllers (FLC). In order to find the optimal relation matrix which represents the rule-base of a FLC, the Genetic Algorithm, a directed random search procedure, is used. The paper presents simulation results obtained for a FLC designed by this technique to control a time-delayed second-order system.

Keywords: Fuzzy control; genetic algorithms; expert control; optimal controller design.

1. INTRODUCTION

It is well-known that human operators are often more successful than conventional controllers at controlling complex or ill-defined processes for which mathematical models do not exist. Where the control knowledge of these operators can be captured in the form of production rules or where the operators' control actions can be modelled using such rules, it is possible to design controllers which implement the rules and replace the operators in their control tasks. There are techniques based on fuzzy logic [1,2] for synthesising controllers from sets of rules. These controllers, first proposed by Mamdani [3], are known as Fuzzy Logic Controllers (FLC). The key element in a FLC is the relation matrix which represents the original set of rules. This matrix defines a mapping from the space of errors between the desired and actual process outputs to the space of control actions. The performance of the controller thus depends on the relation matrix which, in turn, depends on the rules from which it was constructed.

This paper describes a method of synthesising relation matrices which does not require the use of rules. With this method, the optimal coefficients of the matrix are obtained either during trial on-line control of an experimental plant or, preferably, during simulated control of a plant's input-output behavioural model. Such a model would have been built after identifying the dynamics of the actual plant, using, for example, neural network techniques [4].

The main body of the paper comprises four sections. Section 2 reviews some of the previous work on the design of FLCs. Sections 3 and 4 give brief introductions to fuzzy control and genetic algorithms (GAs). The latter are a form of directed random search procedures [5-6] used in this work to locate the optimal relation matrix. Section 5 presents simulation results obtained for a FLC designed by the genetic algorithm technique to control a time-delayed second-order plant.

2. PREVIOUS WORK ON FLC DESIGN

Researchers have concentrated on the problem of extracting control rules for FLCs. Manual extraction of rules has two major difficulties. First, experienced operators from whom rules can be acquired may not be readily available. Second, such operators may not be able to represent their process control knowledge as accurate and consistent rules. Thus, efforts have been devoted to finding methods to extract rules automatically. For example, Procyk and Mamdani [7] have described a self-organizing FLC capable of this. Lee and Berenji [8] have reported a self-learning FLC employing reinforcement techniques to learn the required rules. Automatic rule learning has also been achieved in the work by Patrikar and Provence [9] who have used neural networks for this purpose.

The automatic generation of control rules based on a model of the behaviour of the controlled process rather than the operator's actions or experience has also been attempted. Work in this area includes that by Czogala and Pedrycz [10] and Shen [11]. The latter author has employed a method known as the cell-state-space method for constructing the rule base. Peng [12] has described a parametric function optimisation method for deriving control rules for systems with known mathematical models. FLCs designed in this way have been shown by the author to possess a better performance than the conventional PID controllers available for such systems.

3.FUZZY LOGIC CONTROL

The basic structure of a FLC is conceptually shown in Figure 1. The knowledge base of the FLC is its relation matrix R which, as previously mentioned, represents the rules for controlling the plant. Using R, the computation unit produces a fuzzy output B′ from a fuzzy input A′ . B′ is defuzzified by the defuzzification unit to give the output v to control the plant. A′ is obtained from the fuzzification unit. The input to the latter is the error between the desired plant output r and the actual plant output y.

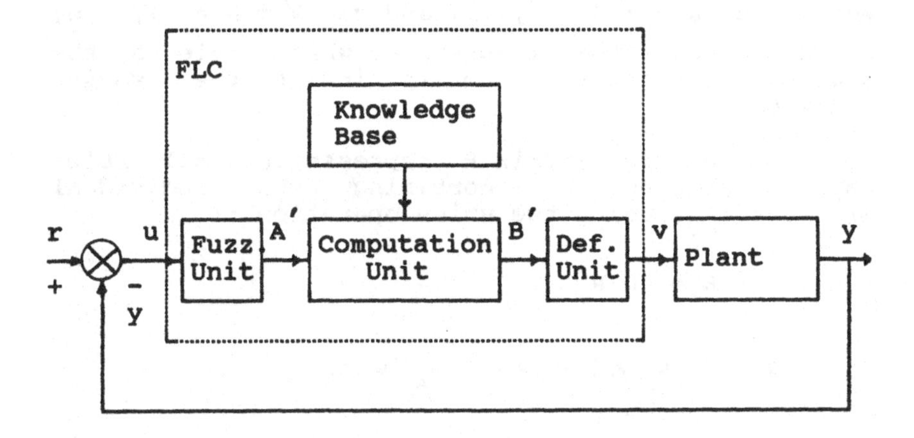

Figur 1. Basic structure of a fuzzy control system

3.1 Relation matrix

In this section, the derivation of R from control rules will be explained to show the physical meaning of R, although this derivation has not been used in the current work.

Consider a control rule expressed in the following form :-

if A_i then B_i

where A_i and B_i are fuzzy input and output variables defined in the universes, or sets of all possible inputs, U and, outputs, V, respectively. Corresponding to each rule, an individual relation matrix R_i could be obtained using the cartesian product :-

$$R_i = A_i \times B_i$$

When A_i and B_i are defined by membership functions μ_{A_i} and μ_{B_i} on finite discrete universes U and V, the elements of R_i can be found using :-

$$\mu_{R_i}(u_j, v_k) = \text{Min}\ [\ \mu_{A_i}(u_j)\ ,\ \mu_{B_i}(v_k)\ ]$$

where ($u_j \in U : j = 1,..,m$ and $v \in V : k = 1,..,n$) and Min denotes the Min-operator which selects the smaller of the two numbers inside the square brackets.

The relation matrix R representing all rules may be obtained by combining the individual matrices, R_i, using the union operation :-

$$R = \cup R_i$$

$$\text{or}\quad \mu_R(u_j, v_k) = \underset{r}{\text{Max}}\ [\ \mu_{R_i}(u_j, v_k)\ ]$$

where r is the total number of rules and Max denotes the Max-operator which selects the largest $\mu_{R_i}(u_j, v_k)$ from amongst all the R_i's. The dimension of R is the same as that of R_i and is equal to m×n.

3.2 Computation unit

During operation of the controller, if the fuzzified

observed plant error is A' , the controller's fuzzy output B' will be produced by the computation unit according to the compositional rule of inference [1] as follows :-

$$B' = A' \circ R$$

$A' \circ R$ is the sub-star composition of A' and R defined by using either the Max-Min operator or the Max-Product operator. In the former case, the membership function of B' is given by :-

$$\mu_{B'}(v_k) = \underset{j}{\mathrm{Max}}[\mathrm{Min}[\mu_{A'}(u_j), \mu_R(j,k)]$$

In the latter case, the membership function of B' is :-

$$\mu_{B'}(v_k) = \underset{j}{\mathrm{Max}}[\mu_{A'}(u_j) . \mu_R(j,k)]$$

The Max-Product operator has been adopted in this study as it has been shown to produce smoother control actions [13].

3.3 Fuzzification unit

This unit converts a "crisp" error value u ($-5 \leq u \leq +5$) into a fuzzy set A' the elements of which are linguistic variables. The following seven linguistic variables were used in this work : Negative Large (NL) or u_1, Negative Medium (NM) or u_2, Negative Small (NS) or u_3, Zero (ZE) or u_4, Positive Small (PS) or u_5, Positive Medium (PM) or u_6, Positive Large (PL) or u_7. These linguistic variables are themselves fuzzy sets defined by the membership functions shown in Table 1. Note that as A comprises up to seven elements, the number of rows, m, of R is equal to seven.

3.4 Defuzzification unit

The output B' of the computation unit is a fuzzy set. The maximum number of elements of B' is equal to the number of columns, n, of R, which has been chosen as eleven in this work. That is, the output space of the FLC has been quantized into eleven different levels. These are defined in Table 2.

$$\mu_{NL}(u) = \begin{cases} 1 & \text{if} \quad u < -4 \\ \exp(-\mathrm{abs}(u+4)) & \text{if} \quad -4 \le u < 0 \\ 0 & \text{if} \quad u \ge 0 \end{cases}$$

$$\mu_{NM}(u) = \begin{cases} \exp(-\mathrm{abs}(u+2.5)) & \text{if} \quad u < 0 \\ 0 & \text{if} \quad u \ge 0 \end{cases}$$

$$\mu_{NS}(u) = \begin{cases} \exp(-\mathrm{abs}(u+1)) & \text{if} \quad u < 0 \\ 0 & \text{if} \quad u \ge 0 \end{cases}$$

$$\mu_{ZE}(u) = \begin{cases} \exp(-\mathrm{abs}(u)) & \text{if} \quad -1 \le u \le 1 \\ 0 & \text{if} \quad -1 > u \text{ or } u > 1 \end{cases}$$

$$\mu_{PS}(u) = \mu_{NS}(-u)$$

$$\mu_{PM}(u) = \mu_{NM}(-u)$$

$$\mu_{PL}(u) = \mu_{NL}(-u)$$

Table 1. Definition of membership functions

v_1 = -5
v_2 = -4
v_3 = -3
v_4 = -2
v_5 = -1
v_6 = 0
v_7 = 1
v_8 = 2
v_9 = 3
v_{10} = 4
v_{11} = 5

Table 2. Definition of output variable levels

The defuzzification unit converts B' into a crisp value for controlling the plant. The centre-of-area method has been selected to implement the defuzzification. The crisp output v is thus produced from B' using the following equation :-

$$v = \frac{\sum_{k=1}^{n} \mu_{B'}(v_k) * v_k}{\sum_{k=1}^{n} \mu_{B'}(v_k)}$$

4. GENETIC ALGORITHMS

This section outlines the operation of a basic

genetic algorithm(GA) and presents the GA adopted in this study.

4.1. Basic structure of a GA

A basic GA consists of five components. These are a random number generator, a "fitness" evaluation unit and genetic operators for "reproduction", "crossover" and "mutation" operations. The algorithm is summarised in Figure 2.

The initial population required at the start of the algorithm, is a set of number strings generated by the random generator. Each string is a representation of a solution to the optimisation problem being addressed. Binary strings are commonly employed. Associated with each string is a fitness value as computed by the evaluation unit. A fitness value is a measure of the goodness of the solution that it represents. The aim of the genetic operators is to transform this set of strings into sets with higher fitness values.

The reproduction operator performs a natural selection function known as "seeded selection". Individual strings are copied from one set (representing a generation of solutions) to the next according to their fitness values, the higher the fitness value, the greater the probability of a string being selected for the next generation.

The crossover operator chooses pairs of strings at random and produces new pairs. The simplest crossover operation is to cut the original "parent" strings at a randomly selected point and exchange their tails. The number of crossover operations is governed by a crossover rate. This operation is shown in Figure 3.

The mutation operator, illustrated in Figure 4, randomly mutates or reverses the values of bits in a string. The number of mutation operations is determined by a mutation rate.

A phase of the algorithm consists of applying the evaluation, reproduction, crossover and mutation operations. A new generation of solutions is produced with each phase of the algorithm.

For more details of the different versions of GAs, see [5-6].

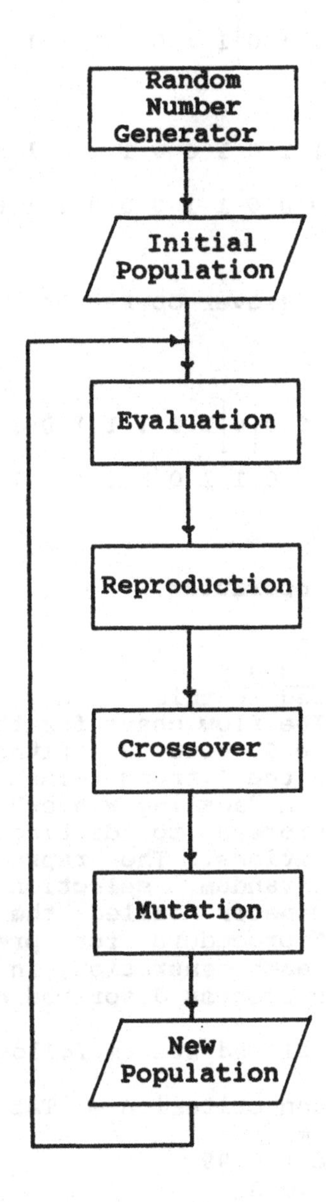

Figure 2. Flowchart of simple Genetic Algorithm

```
Parent 1        1 0 0 0|1 0 0 1 1 1 1 0   ← tail
Parent 2        0 1 1 0|1 1 0 0 0 1 1 0   ← tail


New string 2    0 1 1 0 1 0 0 1 1 1 1 0

New string 1    1 0 0 0 1 1 0 0 0 1 1 0
```

Figure 3. Simple crossover operation

```
Old string    1 1 0 0|0|1 0 1 1 1 0 1

New string    1 1 0 0 1 1 0 1 1 1 0 1
```

Figure 4. Mutation operation

4.2. GA used in this study

The GA version adopted in this study was that described in [14]. The flow chart for this algorithm is depicted in Figure 5. Note a "fitness scaling" unit for normalizing the fitness values computed by the evaluation unit. A "scaling window" is used in this normalization process to distinguish between good and better solutions. The reproduction unit also implements a random selection procedure, controlled by a parameter called the "generation gap", and an "elite" procedure for preserving the fittest solution in each generation, in addition to the seeded selection process described earlier.

The GA parameters employed are as follows :-

(i) Fitness evaluation criterion = ITAE (Note a)
(ii) Scaling Window = 1
(iii) Crossover rate = 0.99
(iv) Mutation rate = 0.01
(v) Generation gap = 1.0
(vi) Length of a solution string = 616 bits (Note b)
(vii) Number of solution strings in each generation=100
(viii) Maximum number of generations = 30

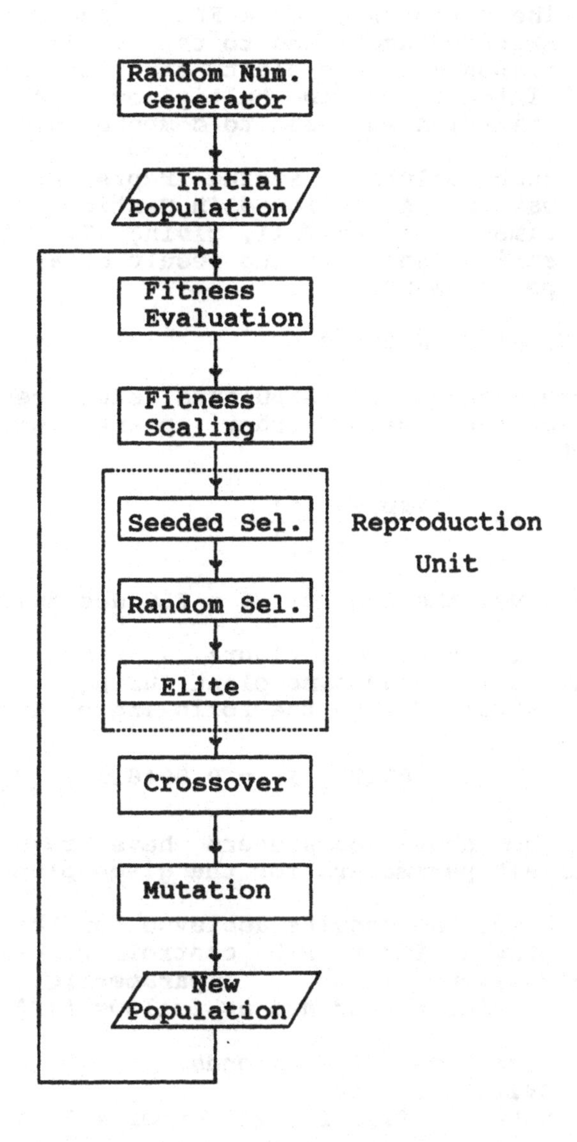

Figure 5. Flowchart of Genetic Algorithm used

Notes:

a- The performance of a FLC produced by the GA was measured according to the error in the step response of the plant that it controls. The "Integral of Time Multiplied by Absolute Error" criterion was used to compute this error [15].

b- Each solution string represents a relation matrix. As mentioned in Section 3 the matrix dimensions are 7×11, giving 77 elements. The string length is the result of assigning 8 bits per element.

5.SIMULATION RESULTS

Figure 6 shows the simulated step response of a time-delayed second-order plant with transfer function

$$G(s) = \frac{\exp(-0.4s)}{(0.3s+1)^2}$$

when under the control of a FLC designed by the GA.

For comparison, Figure 7 shows the response obtained from the same plant using a conventional PID controller with the following parameters :-

$$K_p = 0.630517\ ,\ T_i = 0.594813\ ,\ T_d = 0.237036$$

The above parameters have been adopted as "optimal" parameters for the given plant [16].

Also, the results achieved for this plant using an optimal Fuzzy PID controller based on the previously-mentioned parametric function optimization method are as follows [12] :-

Rise Time = 2.8 seconds
Overshoot = 4%
Settling Time for ±3% error = 5 seconds

6.CONCLUSION

This paper has described a new method for designing fuzzy logic controllers. The method is based on using a genetic algorithm to optimise relation matrices. Results obtained have clearly demonstrated the superior performance of controllers designed with the proposed method.

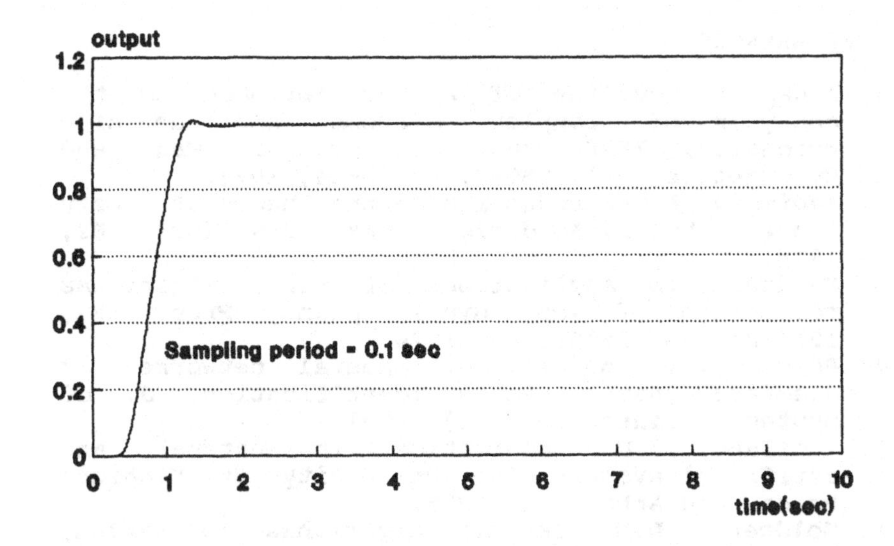

Figure 6. Step response obtained with FLC designed by GA

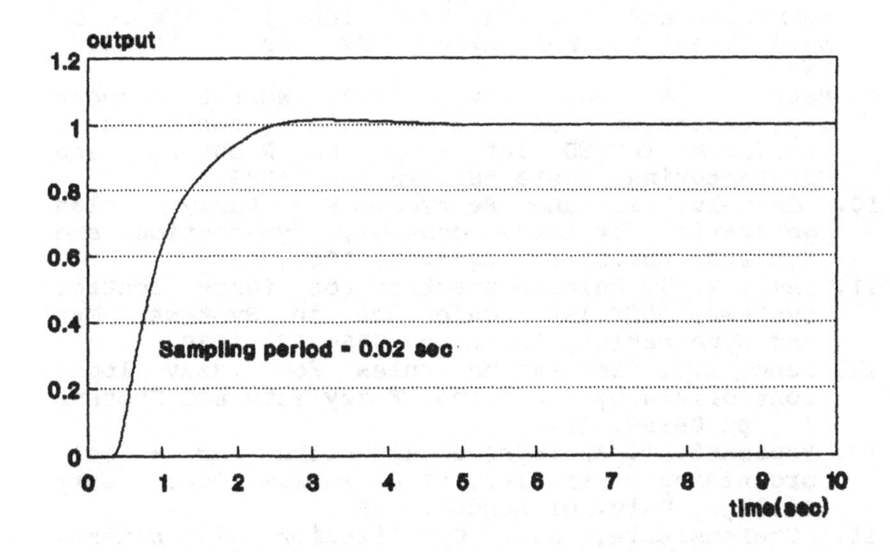

Figure 7. Step response obtained using conventional digital PID control

7. REFERENCES

1. Zadeh, L. Outline of a new approach to the analysis of complex systems and decision processes, IEEE Trans. on Systems, Man and Cybernetics, Vol. SMC-3, pp.28-44, 1973.
2. Kaufmann, A. Introduction to the theory of fuzzy subsets, Vol.1, Academic Press, New York, NY, 1975.
3. Mamdani, E.H. Applications of fuzzy algorithms for control of simple dynamic plant, Proc. IEE, 121(12), pp. 1585-1588, 1974.
4. Pham, D.T. and Liu, X. Neural networks for discrete dynamic system identification, J. of Systems Engineering, 1(1), 1991.
5. Holland, J.H. Adaptation in natural and artificial systems, The University of Michigan Press, Ann Arbor, MI, 1975.
6. Goldberg, D.E. Genetic algorithms in search, optimization, and machine learning, Addison-Wesley, Reading, MA, 1989.
7. Procyk, T.J. and Mamdani, E.H. A self-organizing linguistic process controller ,Automatica, 15, pp.15-30, 1979.
8. Lee, C.C. and Berenji, H.R. An intelligent controller based on approximate reasoning and reinforcement learning, Proc. IEEE Int. Symp. on Intelligent Control, Albany, NY, pp. 200-205, 1989.
9. Patrikar, A. and Provence, J. Neural network implementation of linguistic controllers, Proc.12th IASTED Int Symp. on Robotics and Manufacturing, Santa Barbara, CA, 1989.
10. Czogala, E. and Pedrycz, W. Fuzzy rules generation for fuzzy control, Cybernetics and Systems, Vol.13 pp. 275-293, 1982.
11. Chen, Y. Y. Rules extraction for fuzzy control systems, IEEE Int. Conference on Systems, Man and Cybernetics, Vol.2 pp. 526-527, 1989.
12. Peng, X.T. Generating rules for fuzzy logic controllers by functions, Fuzzy Sets and Systems 36, pp.85-89, 1990.
13. Yamazaki, T. An improved algorithm for a self organising controller, Ph.D. Thesis, Queen Mary College, Univ. of London, 1982.
14. Grefenstette, J.J. Optimization of control parameters for genetic algorithms, IEEE Trans. on Systems, Man and Cybernetics, vol. SMC-16, No. 1, pp.122-128, 1986.
15. Ogata, K. Modern control engineering, Prentice-Hall, Englewood Cliffs, NJ, 1970.

16. Peng, X.T., Liu, S.M., Yamakawa, T., Wang, P. and Liu, X. Self-regulating PID controllers and its applications to a temperature controlling process, in Fuzzy Computing, M.M. Gupta and T. Yamakawa (eds.), Elsevier, Amsterdam, pp.355-364, 1988.

Using Process Knowledge for Adaptive User Interfaces

R. Denzer (*), H. Hagen (**), G. Kira (**),
F. Koob (**)

(*) Kernforschungszentrum Karlsruhe, Postfach 3640, D-7500 Karlsruhe, Germany

(**) University of Kaiserslautern, Postfach 3049, D-6750 Kaiserslautern, Germany

ABSTRACT

In 1989 we started to develop an architecture for a more "intelligent" human-computer interface for the process control room. Having implemented a process control system for the garbage burning plant TAMARA [1,2], a test facility operated by the Kernforschungszentrum Karlsruhe (KfK), we had experiences about the special difficulties the operators have running and optimizing the plant. One of the main experiences was, that there is a lack of online-knowledge in the user interfaces currently used, and that this knowledge must not necessarily come from sophisticated diagnose systems. We can do a lot (and it is much easier than implementing a whole diagnose system) if we integrate the a-priori-knowledge we have about the plant - knowledge which is available at the time the plant is built and which describes not all but an important part of the plants behaviour.

The user interface always has at least two faces: one for the end-user and one for the designer of the user interface. Making the design effective lead us to an object-oriented architecture. As for process control, it is always necessary to visualize data from within a distributed system, we have implemented our object system as a distributed system [2,3,4]. We use a distributed object interface to connect applications (diagnose system, process interface, user interface,...).

This paper shall focus on the use of process knowledge to build adaptive user interfaces. The main goal is to optimize the dialogue at runtime according to a given process context. We use rules to make these optimizations. Rules control the dialogue and even modify the dialogue itself at runtime.

INTRODUCTION

The task of technical plant operation is to manage a complex dynamical system in an optimal way. Most of the difficulties involved in this task are coming from the mass of data and from the causal and temporal relationships within the plant, which are often not exactly known. In critical situations operators have to decide in short time how to react on unforeseen events. The decision process is rather difficult because on one hand there is too much "low level" information (up to some hundred alarms within a few minutes), but on the other hand there is a lack of "high-level" information, namely what to do next. This situation has been described by Sachs with the term *cognitive overload* [5]. Therefore, a kind of information filtering is necessary for the user interface.

A lot of work has been done in the real time community to develop systems supporting operators with diagnostic knowledge, to filter information at runtime or to predict critical situations [6-9]. User interfaces have been improved using high resolution graphics and knowledge based techniques [6]. Many of the approaches are based on diagnose systems.

We think, that one of the drawbacks of these techniques is, that you need a sophisticated process model to implement such a diagnose system, because if you do not have a good model, what is the diagnose worth? Sometimes such a process model is too expensive to build or the process has a stochastic behaviour. Then you must work without the model.

Another drawback we see is the fact, that often the distributed nature of technical systems is not taken into concern. We still see a problem in the integration of knowledge based systems into process control systems. Also, sometimes the diagnose system is a separate system with a separate user interface in parallel to the process control user interface. If the two user interfaces are not consistent, this can make work harder. Our approach is to integrate knowledge based components (and other components) of a process control system in a highly distributed system with one consistent user interface for process operation.

AVAILABLE PROCESS KNOWLEDGE

From our point of view, there is an aspect which has frequently been overlooked: it is, that we already have a lot of useful knowledge about the technical process - without a backward chaining diagnose system:

- There is knowledge about the control relationships in the plant, which can easily be used to filter information in the user interface using forward-chaining rules.

- One of the experiences in the TAMARA project was, that the complex relationships within the plant have often been expressed by the operators in a form like "if the temperature x is ... and the pressure y is ..., then I need information about z1 and z2 and not about z3". This knowledge is also very useful to adapt the user interface.

- A lot of important informations are in the technical plans of the plant. Nevertheless they are not available in the user interface using the currently available process control systems. Making those informations available in the user interface in form of hypertext connected to the technical plant objects makes the user interface much more efficient and helps the operators finding the reason for wrong plant operation.

These are examples for available knowledge about the plant at the time the plant is built or after a short time running it. All this knowledge is much easier to get than to build a process model which is capable of diagnosing a complex chemical plant.

We would like to focus on the fact, that from our experience this kind of knowledge is especially useful to build user interfaces for the process control room. We would also like to point out, that all the knowledge we mentioned above can be expressed by graphical means using forward chaining rules for the dialogue.

SYSTEM OVERVIEW

We have implemented a distributed object system (fig.1) which consists of a design environment and a runtime system. Our main goal for the user interface was to combine object-oriented techniques with rules, as they express very well the behaviour of the plant objects. The main design decisions of the overall system were:

- The system had to be distributed.

- The system had to be open. That means, that for any task in the process control system, the best suited programming language or environment should be used, e.g. a knowledge based tool for diagnosis and C-programs for observers.

- We wanted to provide a user interface tool, which makes it easy to integrate the informations from the different subsystems at runtime. The integration should work by declaration and not by programming.

- In the user interface tool, the plant objects had to be modeled as a whole, combining technical and graphical entities and their behaviour, thus making it easy to reuse them.

The goals mentioned above lead us to a multi-layer approach. On the basic layer we have implemented a language and network independent communication model with an extendable message protocol to exchange state-change messages asynchronously. This layer is implemented in C. On top of this layer, we use a distributed object interface, which makes it possible to provide links between remote objects. Thus an instance variable of an object can share an instance variable of another object in a remote application. This is useful to integrate process variables directly into the user interface. The runtime component of the user interface tool is a distributed object system written in OPS5. It uses DECwindows/Motif as graphical system.

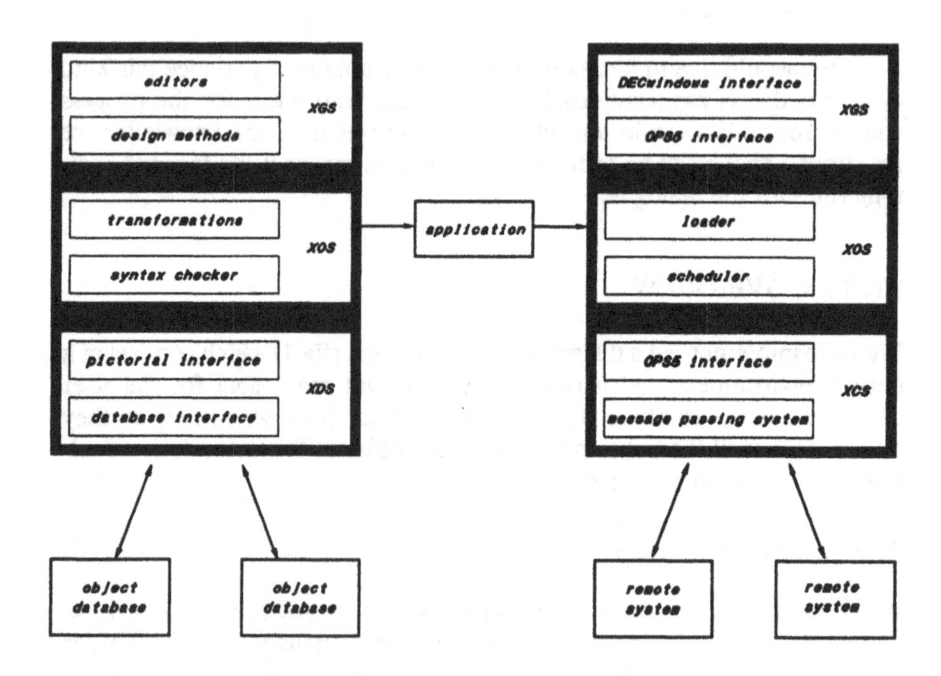

Fig. 1: System overview

There are four main modules of the XAVIA system: XCS (XAVIA communication system), XGS (XAVIA graphics system), XOS (XAVIA object system) and XDS (XAVIA database system), which are used in the design environment and/or the runtime environment.

The state of the implementation is as follows: the runtime system is mostly implemented and we can run user interface and process components in parallel using the communication mechanism. We load applications from a file, which describes the objects and rules of an application. Currently we are working on two fields: we design and implement the parts of the design environment which are not yet built. In parallel we try to get experiences, how we can use process knowledge for adaptive and in some way "intelligent" user interfaces.

FILTERING TECHNIQUES

Our first goal was to embed all kinds of available knowledge into the user interface, e.g. electrical plans. This makes the amount of information from which the operator can choose grow a lot. Therefore the user interface must be designed carefully to make the informations accessible in an easy and rapid way. Some informations may be necessary to visualize permanently, others may not.

The second goal was to filter informations in case of faults in the plant. This is made by the way objects look like at run time. The appearance of objects (usually the colour) change according to the context of the process, e.g. whether an object itself has a fault or whether another object has caused this fault. The conclusion is not drawn by the system. The user interface only changes the overall look and helps the operator to draw conclusions. It serves for quick understanding of what is happening.

The third goal was to modify the dialogue itself at run time according to the process context. We want to avoid unnecessary or impossible interaction, e.g. switching a pump which cannot be switched because it is blocked for any reason. This shall help making the dialogue as efficient as possible.

We use two concepts called *dialogue filtering* and *optical filtering*, which will be introduced and which we will show in the examples. The intention of both is rapid understanding and rapid access. Before explaining the use of both, we will introduce some terms which will be used in the sequel.

The term *primary information* denotes an information (or an object like a pump) which is of primary interest for controlling the overall behaviour of the plant. In contrary, the term *secondary information* denotes an information (or an object) which does not influence the overall behaviour of the plant as much, which might be hidden behind a part of the user interface and which might only be accessible by certain user interaction.

A *primary fault* is a fault of an object **O1**, for which the reason has been found within the object **O1**, e.g. a faulty fuse of a pump. Primary faults are important, because they may be the source of many other faults. A *secondary fault* denotes a fault of an object **O1**, for which the reason can be found in another object **O2**, e.g. a pump which has been switched off because of a pressure alarm in a boiler. A *bolt* is a state of an object **O1**, for which the reason is found in an object **O2** whereby access (e.g. switching on or off) is inhibited as long as the primary fault exists.

Dialogue Filtering

Dialogue filtering is a technique to modify the dialogue of the operator with the system at run time according to a given process context. The aim is to make the dialogue as quick as possible, especially in case of wrong operation of the plant. We use knowledge about the control relationships in the plant to do so. Fig. 2 shows an example of dialogue modification which is taken from the boiler of the TAMARA plant. See fig. 3 for the boiler window.

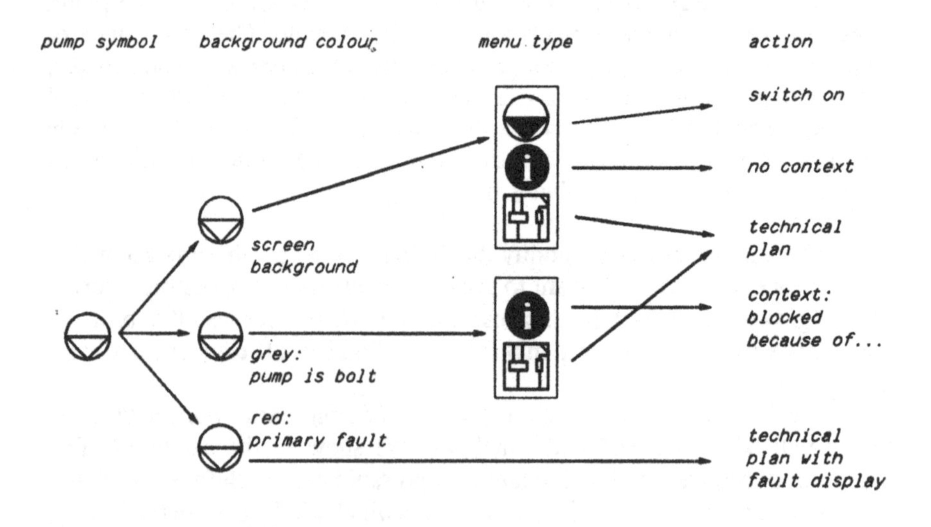

Fig. 2: Dialogue modification according to process context

In case of normal operation of the pump, the menu shows three entries: a switch entry, a context information entry and an entry for technical informations (fig. 4). Switching the pump changes the way it is displayed. Choosing the technical informations entry of the menu, we display the electrical plans of the unit including the process control system interface (fig.5). These plans may be scanned or produced by documentation software.

If the pump is bolt, switching makes no sense. Therefore switching is inhibited and the information symbol displays information about the process context which caused blocking the unit. In the following case, the pump **Pu2** is bolt, because the tank **B10** is empty. The menu is displayed before displaying the context information, because the technical plans must be accessible (fig. 6). Choosing the information symbol shows the context information (the reason) of the bolt (fig.7).

In case of a primary fault of the pump, there are several possibilities: there may be a number of possible reasons or a single reason which could be detected. Selecting the pump symbol in the following case (fig. 8) now inhibits the menu, because the reason could be found in a fuse. We directly display the electrical plan with the reason for the fault, because the context information is associated with the plan (fig.9). As you can see, we try to display faults the way they appear. If they appear in the electrical system, we display them in the documentation of the electrical system, because this is the most natural way to think about them.

Optical Filtering

Dialogue filtering mainly has to do with informations which are directly connected to one object or its environment controlling it. On the opposite, optical filtering is used for a set of objects, where the relationships between the whole set are too expensive to integrate into the user interface. Optical filtering may also be used to hide informations, which are not as important as others.

Fig. 10 shows a user interface for the whole incineration part of the plant. The operator can choose several views from the menus. In fig. 10, all the values from all measuring instruments are displayed (primary and secondary informations). For normal operation this is not necessary, because many of them are not of primary interest. Choosing to display only the primary informations reduces the amount of information (fig. 11).

An important application of optical filtering is the colour coding of faults if there are many of them. Coding primary faults in red, secondary faults in light red and bolts in grey is a possibility to focus the interest on the primary faults.

The final example shows optical filtering for a whole part of the plant. This part is usually hidden behind a symbol which denotes the start of a new dialogue (cursor position in fig. 11), because this part has been considered as secondary information. Selecting the symbol makes this part appear on the screen (fig. 12).

CONCLUSIONS

Using the control knowledge about a plant makes it possible to adapt a user interface at run time. The goal is to reduce the amount of information and to focus on important informations according to a given process context.

Our experiences building the system and designing the examples showed, that rules are a powerful tool to model the plant and the dialogue. The next interesting step is to use the experiences of operators.

ACKNOWLEDGEMENTS

Aknowledgements are due to Dr. Jaeschke and Prof. Trauboth from KfK who made the cooperation possible.

REFERENCES

[1] Dittrich G., Kerpe R., TAMARA - Versuchsanlage zur schadstoffarmen Müllverbrennung, GVC-Kongress, Baden-Baden, December 1989

[2] Denzer R., User Interface Management in Distributed Systems, in: Denzer R., Hagen H., Kutschke K.H. (eds.), Visualization of Environmental Data, Proc. of the GI-workshop, Rostock, November 1990, Springer-Verlag

[3] Denzer R., Hagen H., An Object-Oriented Architecture for User Interface Management in Distributed Applications, internal report, January 1991, to be published

[4] Denzer R., User Interfaces für die Visualisierung von Umweltdaten - Anforderungen und Architektur, in: Pillmann W., Jaeschke A. (eds.), proc. of the 5. Symposium "Informatik im Umweltschutz", Vienna, September 1990, Springer-Verlag

[5] Sachs P., ESCORT - an Expert System for Complex Operations in Real Time, proc. of the Alvey workshop on deep knowledge, IEE, London, 1985

[6] Alty J.L., Mullin J., Dialogue Specification in the GRADIENT Dialogue System, in: Sutcliffe A., Macaulay L. (eds.), People and Computers V, proc. of the 5. Conference of the British Computer Society HCI Specialist Group, Nottingham, September 1989, Cambridge University Press

[7] Tzafestas G. (ed.), Knowledge-Based System Diagnosis, Supervision and Control, Plenum Press, New York, 1989

[8] Khanna R., Moore R.L., Expert Systems Involving Dynamic Data for Decisions, proc. of the International Expert Systems Conference, Oxford, 1986

[9] IEE Colloquium on The Use of Expert Systems in Control Engineering, London, IEE Digest No. 1987/27, 1989

SCREEN DUMPS

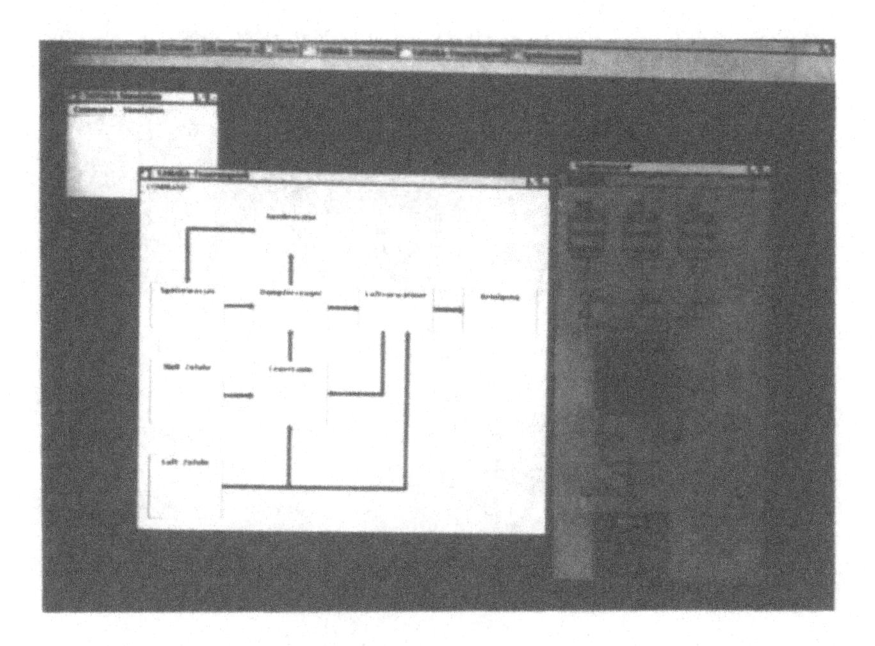

Fig. 3: Boiler window

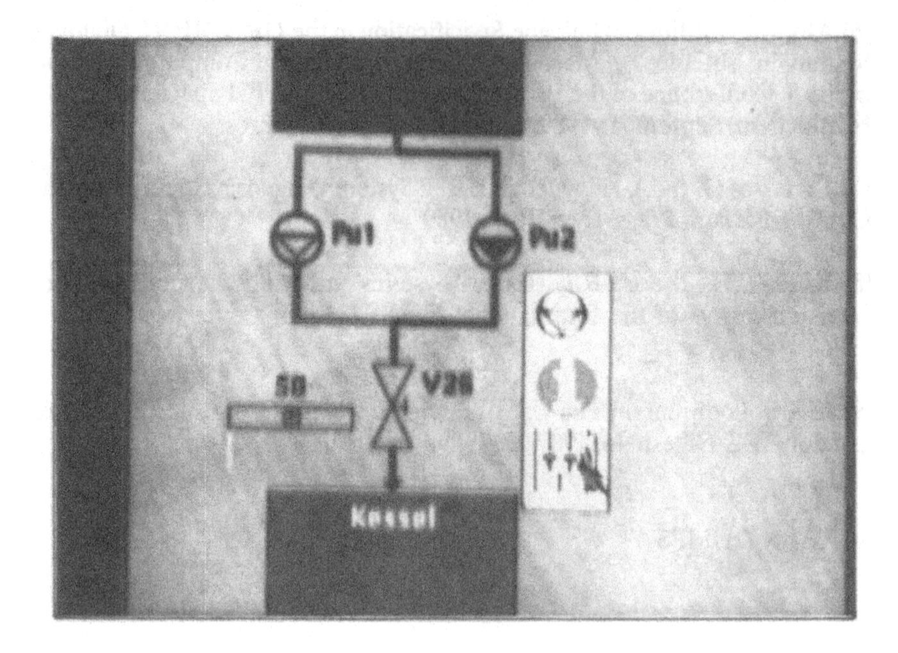

Fig. 4: Complete menu - selection of technical documentation

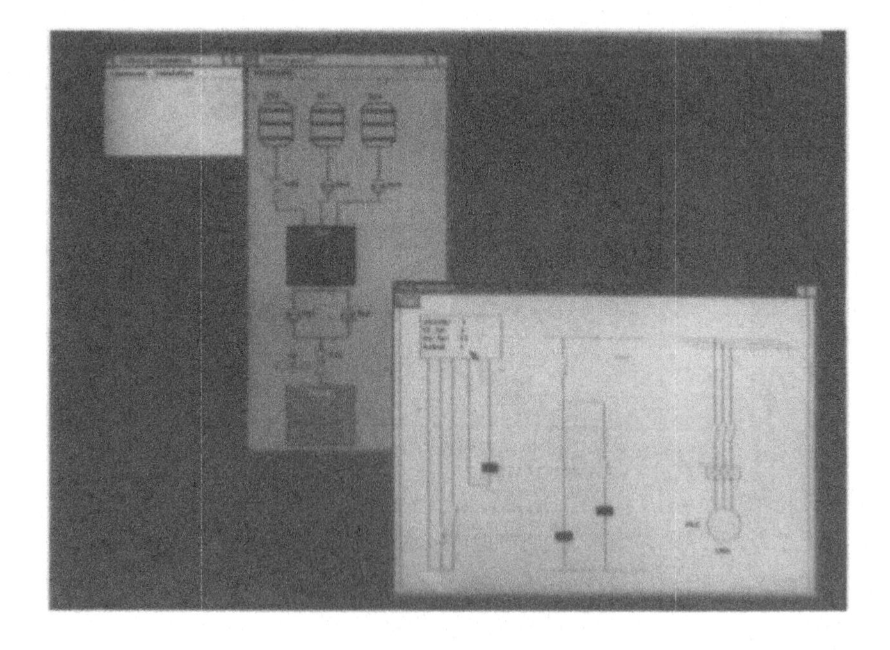

Fig. 5: Technical documentation

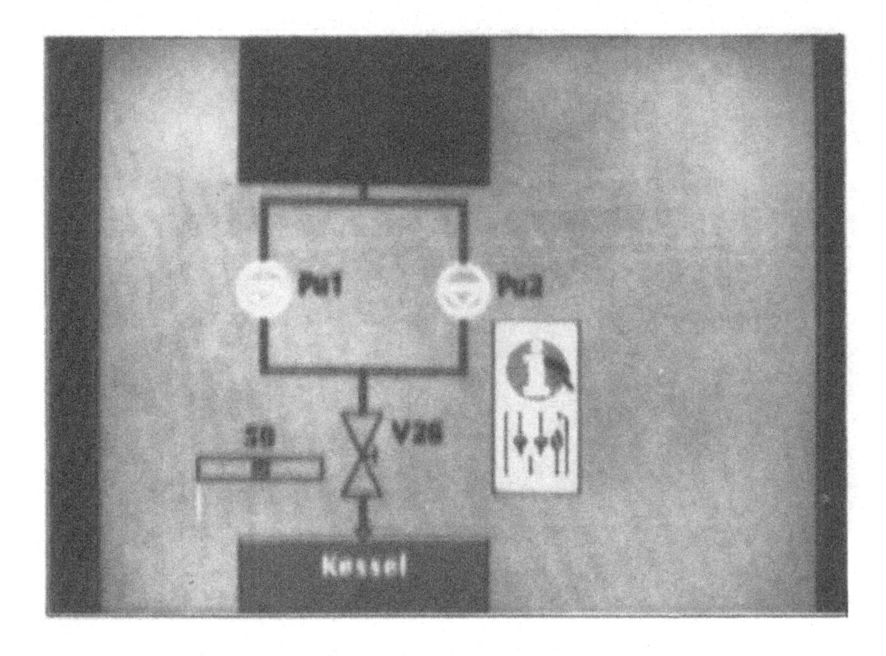

Fig. 6: Modification of symbol and menu in bolt state

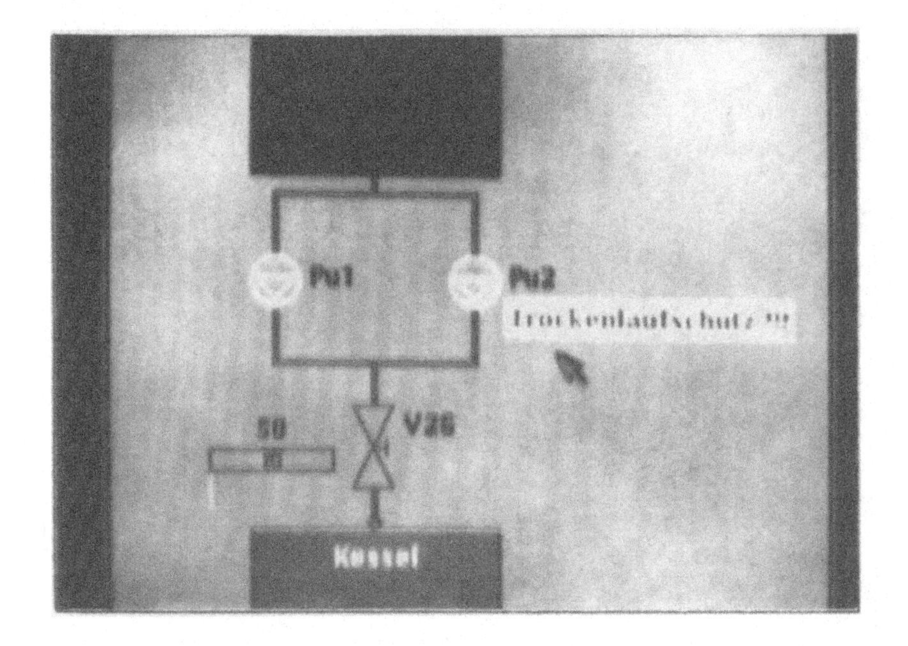

Fig. 7: Display of context information

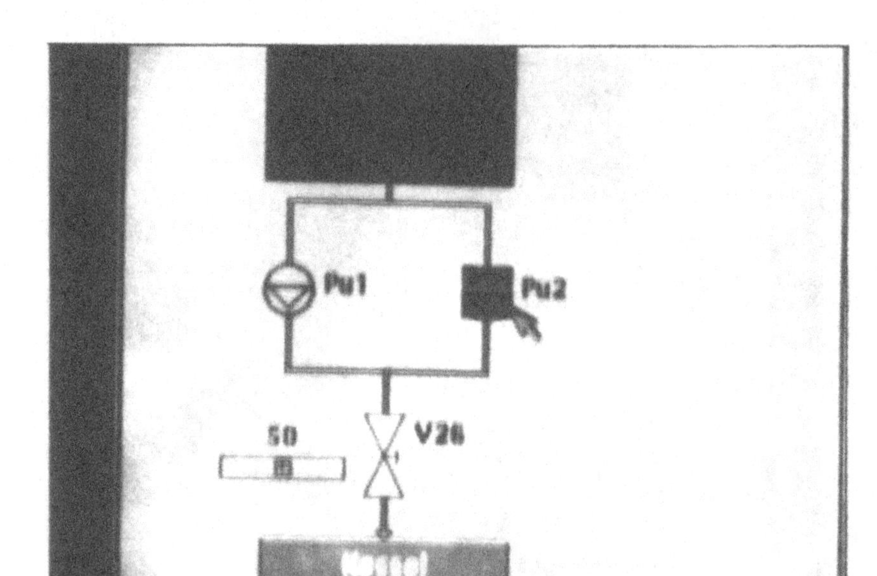

Fig. 8: Primary fault at unit Pu2

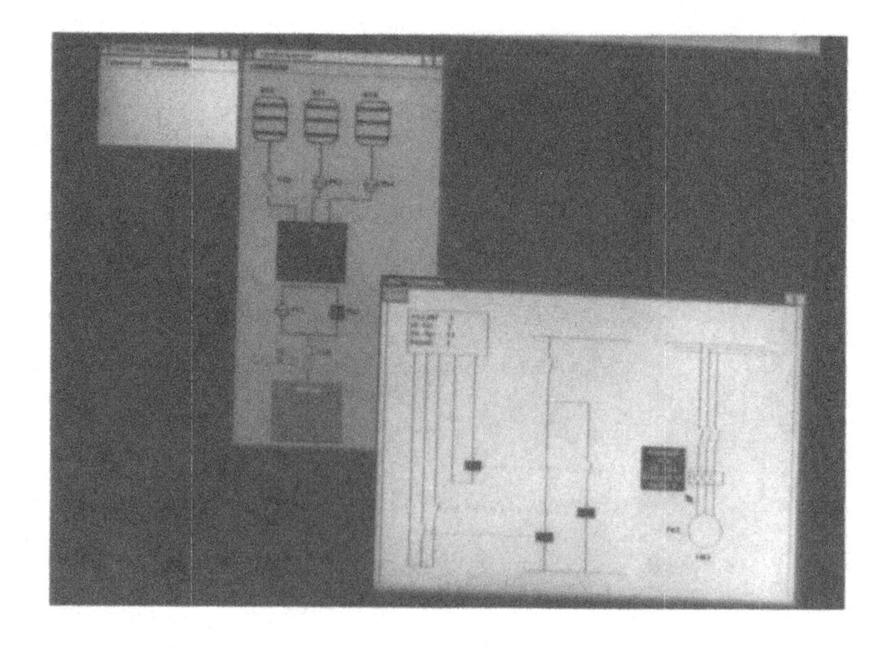

Fig. 9: Fault display

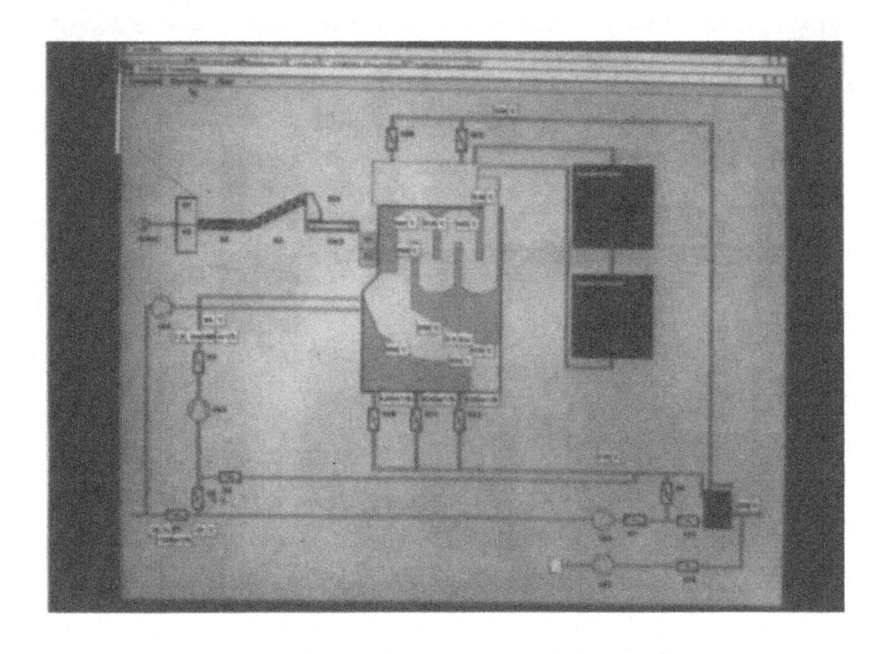

Fig. 10: Display of primary and secondary informations

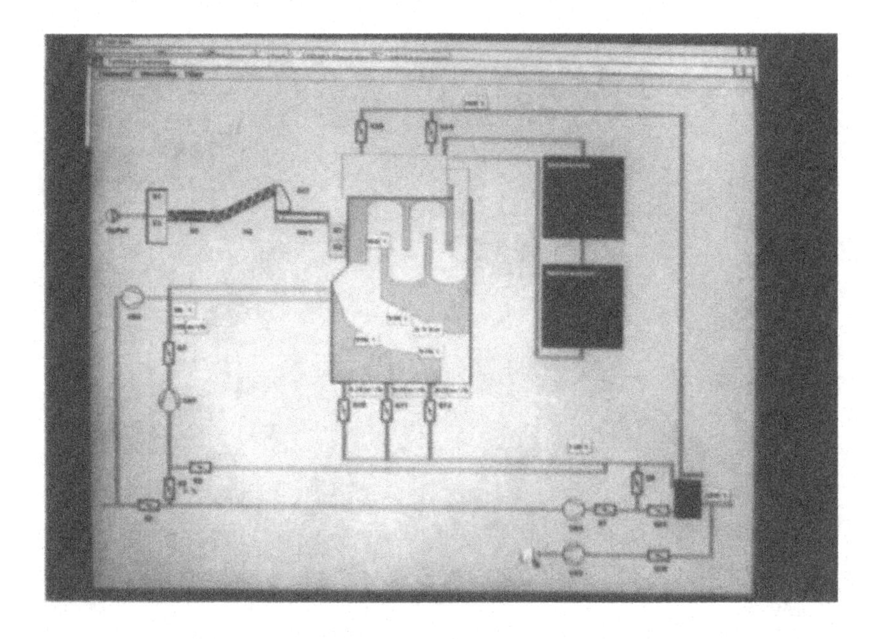

Fig. 11: Display of primary informations

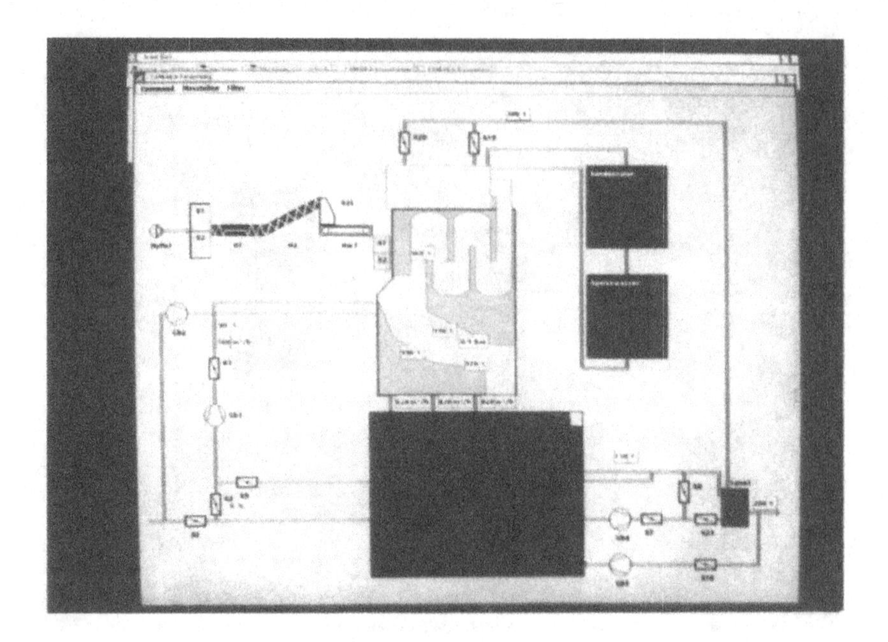

Fig. 12: Display of secondary information

Intelligent Machine Tools: An Application of Neural Networks to the Control of Cutting Tool Performance

H.A. Epstein, P.K. Wright
Robotics and Manufacturing Research Laboratory, Courant Institute of Mathematical Sciences, New York, NY 10003, USA

ABSTRACT

An Open-Architecture Machine Tool, based on a Sun/VMEbus/C/Real-Time Unix operating system has been constructed to provide a "machining research control platform" for the execution, sensing and gauging of precision machining. The real-time control of cutting tool performance is being monitored with dynamometers and thermocouples in order to monitor the stress and temperature acting on the tool's cutting edge. Initially, a closed-form engineering relationship between stress-temperature and speed-feed can be developed and used to adjust feed and speed so as to keep stress and temperature within safe but productive bounds. However, the control of the system, especially with a deteriorating tool due to wear, benefits from the application of a neural network. This approach "learns and updates" the relationship between speed-feed and stress-temperature over a broad range of operating conditions. Results for the cutting of steel with carbide tools are described.

1. INTRODUCTION: BACKGROUND AND SIGNIFICANCE

Recent publications have described an Open-Architecture Manufacturing System that demonstrates a prototype of a next generation controller for intelligent machine control. We have utilized components that are *de facto* standards in the computer culture. [1,2]

- **a general purpose workstation (a Sun 3/160 M)**
- **a standard unit of hardware (the VMEbus)**

- a standard operating system (a real-time version of UNIX)

- a standard language (the C language)

This environment provides a flexibility and openness as yet undemonstrated in today's Flexible Manufacturing Systems and individual machine tools on the factory floor. Today's controllers are more typical of programmable controllers of the 1970's; and even those that do exhibit greater flexibility are usually "closed" to outside users and developers so that the original manufacturer corners the market on sensor developments and system expansions. The true intelligence of today's systems is therefore limited: sensors, CAD systems, expert systems and other machines may be **interfaced** but never **integrated**. Today, one-bit handshakes pass between black-boxes that may be able to stop cutting if a tool breaks. But at present there is no suggestion of a shared goal, where if one sub-component of the system changes, all the other sub-components adjust in sympathy. However, the new Open-Architecture Machine tool does demonstrate the required synergy for sensor-developments and intelligent control strategies. As a result, the sub-components of the system can act both autonomously or collectively depending on the specific needs of the moment. The motion control boards can constantly be "open" to updates from probe-data, dynamometer data or recommended changes in plan from the supervisory level. Some examples of these interactions are now discussed in the context of a closed-loop control scheme that also utilizes a neural network for system learning and stabilzation.

2. CLOSED-LOOP CONTROL AND NEURAL NETWORKS

The advent of the open-architecture manufacturing system calls for the development of a practical closed-loop control scheme for the machining process. The system being implemented will use dynamometers and thermocouples to monitor the cutting tool's stress-temperature state and adjust feed rates and, or, cutting speeds to regulate that state and, consequently, regulate tool performance and life. Similar to an open-loop control system, the closed-loop control system will **initially** select the machining parameters such that the stress and temperature acting on the tool's edge lie in a region of the tool material's deformation map [see 3] which allows high cutting efficiency and minimum tool wear; however, the closed-loop control, enabled by the open-architecture, will **subsequently** allow adjustments to feed rates and/or speeds during the course of machining to reflect changing conditions occasioned by tool deterioration or changes in workmaterial.

While a servomechanism model can often serve as a practical means for achieving automated closed-loop control, it is uni-dimensional and can deal with only one setpoint, one command, and one feedback variable at a time. An intelligent control methodology requires a "multi-variable" servomechanism model. In a manned machining environment, an experienced craftsman provides this type of control by continuously monitoring several factors at once, sound and chip color for example, and translating these inputs into subtle changes in machining parameters. Thus, the automation of such intelligent control will require the utilization of a model capable of reacting to several input stimuli and calculating one or more output values on a real-time basis.

The neural network model, a simulation of the microstructure of the human brain, can serve as a tool to implement such a multi-variable servomechanism needed for closed-loop control in machining. [4] The neural network model can map a vector of original setpoints combined with a feedback vector into an output vector. Figure 1 illustrates how a neural network may fit into a closed-loop control system for machining. The set points are the desired feed rate and cutting speed found in a reference, such as the Machinablility Data Handbook, for the specific operation, work material and tool type. Additional input to the neural network includes data from the process itself: current feed and speed, and data from process sensors: temperature and stress.

The development of training data for such a neural network is based upon chip formation theories [5,6] that relate the physical state variables of stress and temperature at the cutting edge to the various types of tool wear and the control input variables of feed rate and cutting speed. These relationships permit an a priori calculation of expected stress and temperature based upon the chosen value of initial feed and speed. Changing conditions due to tool wear may cause changes in the relationships between the stress-temperature state and the feed-speed parameters. Such events are likely to require adjustments in process control parameters in order to maintain the machining process in a state that maximizes cutting efficiency while minimizing tool wear. Postulating the maintenance of open-loop stress and temperature values as our closed-loop system objective, if measured stress or temperature is greater than a value predicted by application of a machining theory relationship formula, then it can be assummed that the original relationship has shifted. New values of feed or speed that are requisite to maintain originally targeted temperature or stress can then be calculated based on the shifted function. The neural network multi-variable servomechanism for closed-loop control is trained to maintain the *a priori* values by recognizing how the relationships have to shift in order to yield measured values for temperature and stress relative to current values for feed and speed. [The pragmatic arrangements for using embedded thermocouples and piezoelectric dynamometers, and consequently calculating in real-time the precise temperature and stress acting on the tool's edge, are described in reference 2]

3. METHODOLOGY

For the purposes of this paper, we have limited our investigations to the single-point, semi-orthogonal cutting of steel on a lathe with a carbide tool. Figure 2 shows the hot compressive yield strength deformation map for various tool types.

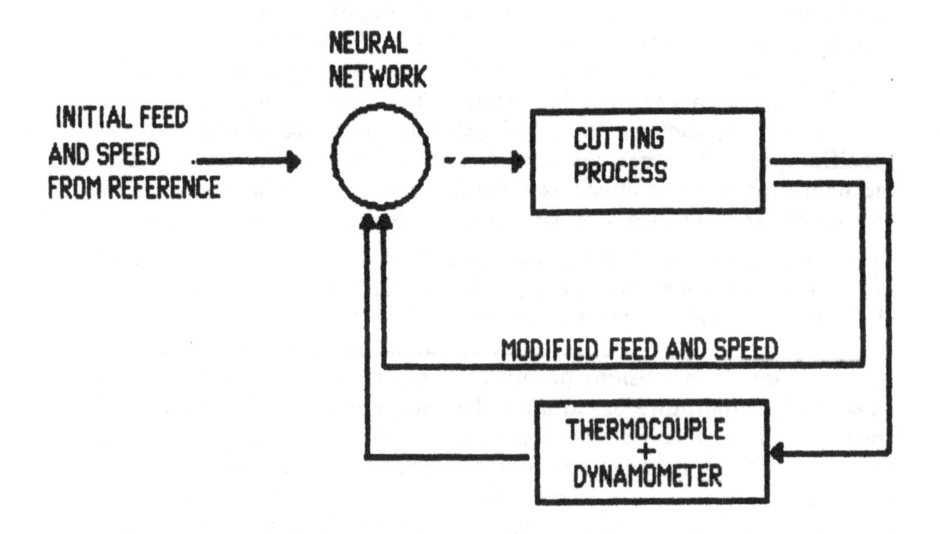

Figure 1

For different combinations of temperature and stress, different forms of tool wear will dominate. The shaded parallelogram represents "safe but productive" combinations of stress and temperature. The initial values for speed, v, and feed, f, should be chosen so that the resultant tool stress and temperature will lie somewhere within this defined area. The following formula attributable to Yen and Wright, [7] but based on standard theories [5,6] was employed to first ascertain the expected temperature rise, ΔT_p at the midpoint of the primary shear zone in semi-orthogonal turning:

$$\Delta T_p = \frac{.51(\frac{fv}{\alpha}\tan\phi)^{(1/.216)}\tau_p\cos\gamma}{\rho c\cos(\phi-\gamma)\sin\phi} \quad (1)$$

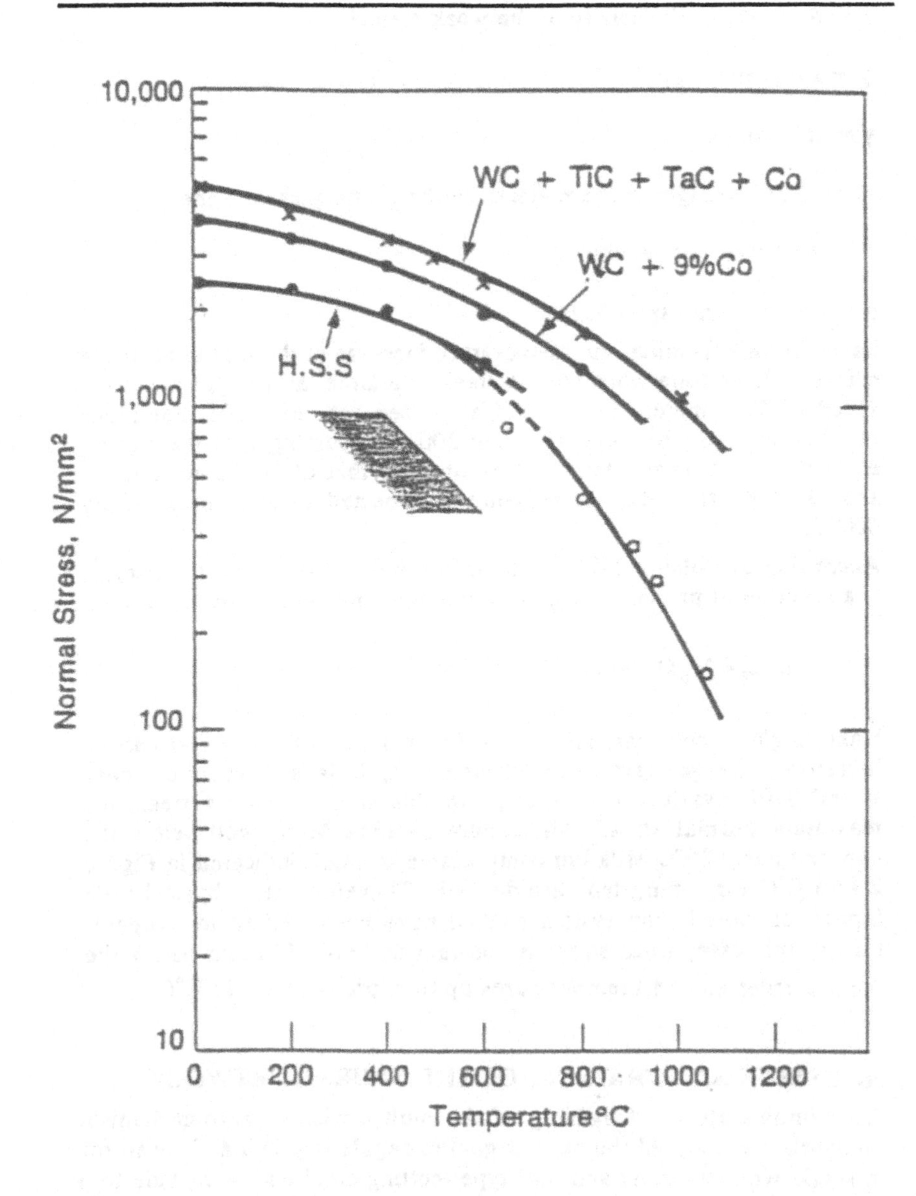

10,000
1,000
100
10
Normal Stress, N/mm²
WC + TiC + TaC + Co
WC + 9%Co
H.S.S
0
200
400
600
800
1000
1200
Temperature°C

Figure 2

where

α = the thermal diffusivity of the work material

ϕ = shear plane angle

γ = rake angle

τ_p = shear strength of the material in the primary shear zone

ρ = workmaterial density

c = workmaterial specific heat

Based on this formula, the temperature increase at the midpoint of the primary shear zone when cutting steel on a lathe at the typical cutting speed of 2.75 meters per second at a feed rate of .0001 meter per revolution would be approximately 300°C; assuming that the cutting edge of the tool is at a temperature of twice that of the primary shear zone [see 8], the edge temperature is expected to be approximately 600°C.

According to Loladze [9], maximum normal stress on the tool, σ_{max}, is a function of primary shear zone strength and rake angle as follows:

$$\sigma_{max} = 2\tau_p(1.3 - \gamma) \qquad (2)$$

Rake angle is constant, and as shown in much of the metal cutting literature, the primary zone shear strength is a "dynamic shear stress" [10] invariant with speed. In this case of cutting steel, the maximum normal stress will amount to 1344 MPa, well below the approximately 2000 MPa hot compressive strength indicated in Figure 2 at 600°C for a tungsten carbide tool. Therefore, the physical state inputs for closed loop system control have been limited to temperature in this case, since stress is constant at 1344 MPa and below the failure criterion at all temperatures up to approximately 1100°C.

4. DESIGN AND TRAINING OF THE NEURAL NETWORK

To demonstrate the feasibility of the multi-variable servomechanism network, we adopted the narrow goal of regulating feed and speed for a single work material and tool type: cutting steel with a carbide tool on a lathe. Feedback was limited to temperature alone based on equation (2) and typical working values of σ and T_p. The design of the neural network multi-variable servomechanism is shown in Figure 3. There are three input nodes: target feed rate, thermo-couple feedback and current feed rate. It was assumed that the cutting speed would remain constant, and only feed rate varied. There is a single output neuron to represent the new feed rate; the middle layer consists of five neurons. The network flows forward only and has no intra-layer connections. Backpropagation was employed for training. Each middle and output neuron receives a bias input in addition to inputs from the neurons in the layer above (the equivalent of an additional connection with a value equal to its weight). The target feed rate input to the middle layer may also be considered a bias since it remains constant for our single selected work material and tool combination. The network was trained to adjust feed rate in order to maintain the edge temperature at the approximate 600°C level originally indicated by the reference feed rate and speed. The following restatement of the Yen and Wright formula relates the feed/speed product to temperature change in the primary shear zone and was employed to plot the edge temperature curves shown in Figure 4, again assuming that $T_e = 2T_p$.

$$fv = \frac{\alpha}{\tan\phi}\left(\frac{\rho c\cos(\phi-\gamma)\sin\phi\Delta T_p}{0.51\tau_p\cos\gamma}\right)^{(1/.216)} \tag{3}$$

If process anomalies cause the measured edge temperature to be different from expectations, then it can be assumed that a shift in this curve has occured. For example, the curve labelled "shifted" in Figure 4 depicts how the relationship would change if edge temperature were measured at 640°*C* while the f * v product remains at .000275. Based on this new relationship, to attain the desired edge temperature of 600°, the f * v product would need to be reduced to approximately .0002. The neural network was trained to detect how the curve shifts, and to calculate process changes needed to achieve the basic edge temperature.

The shifting in the curve and calculation of the new target feed * speed product is accomplished by isolating the temperature component from the Yen and Wright equation as follows:

$$fv = \frac{\alpha}{\tan\phi}\left(\rho c\cos(\phi-\gamma)\sin\phi\Delta T_p\left(\frac{1}{0.51\tau_p\cos\gamma}\right)\right)^{(1/.216)} \Delta T_p^{(1/.216)} \tag{4}$$

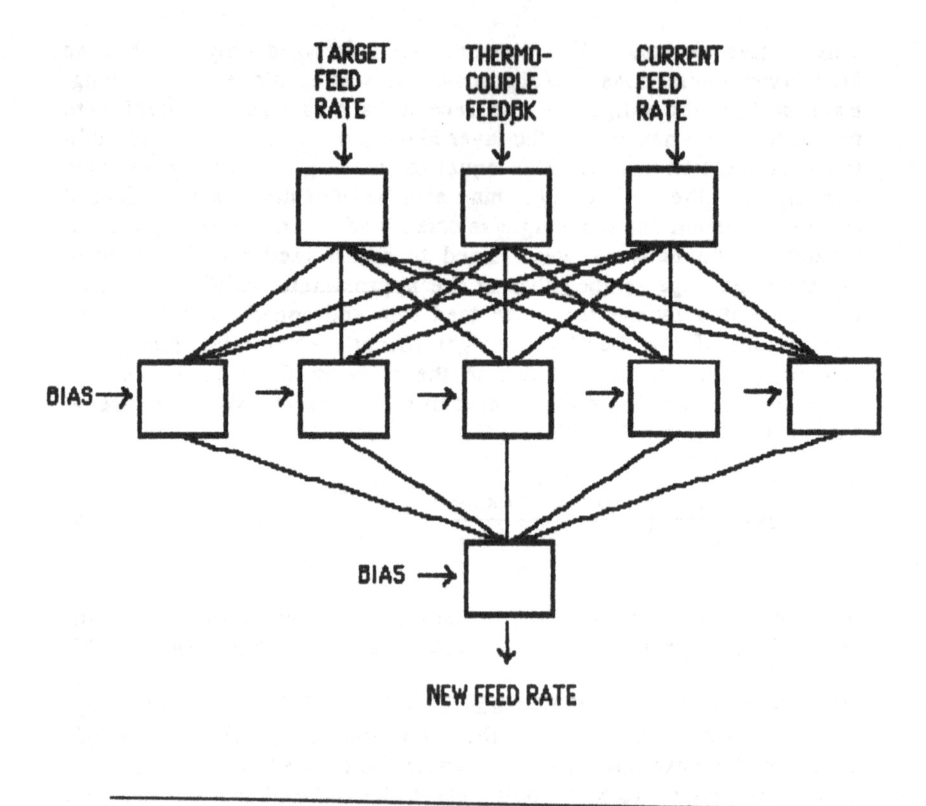

Figure 3

The shift in the non-temperature component, X', is calculated by dividing the feedback f * v product by the feedback temperature component; the new target f * v is then the product of the target temperature component and X'.

$$X' = \text{current fv/current } \Delta T_p^{(1/.216)}$$

$$\text{new fv} = (\text{target } \Delta T_p^{(1/.216)})\ (X')$$

The training set developed using this relationship is shown in Table 1. The target feed rate was maintained at .0001 for all cases (representing the reference value for this particular combination of work material and tool).

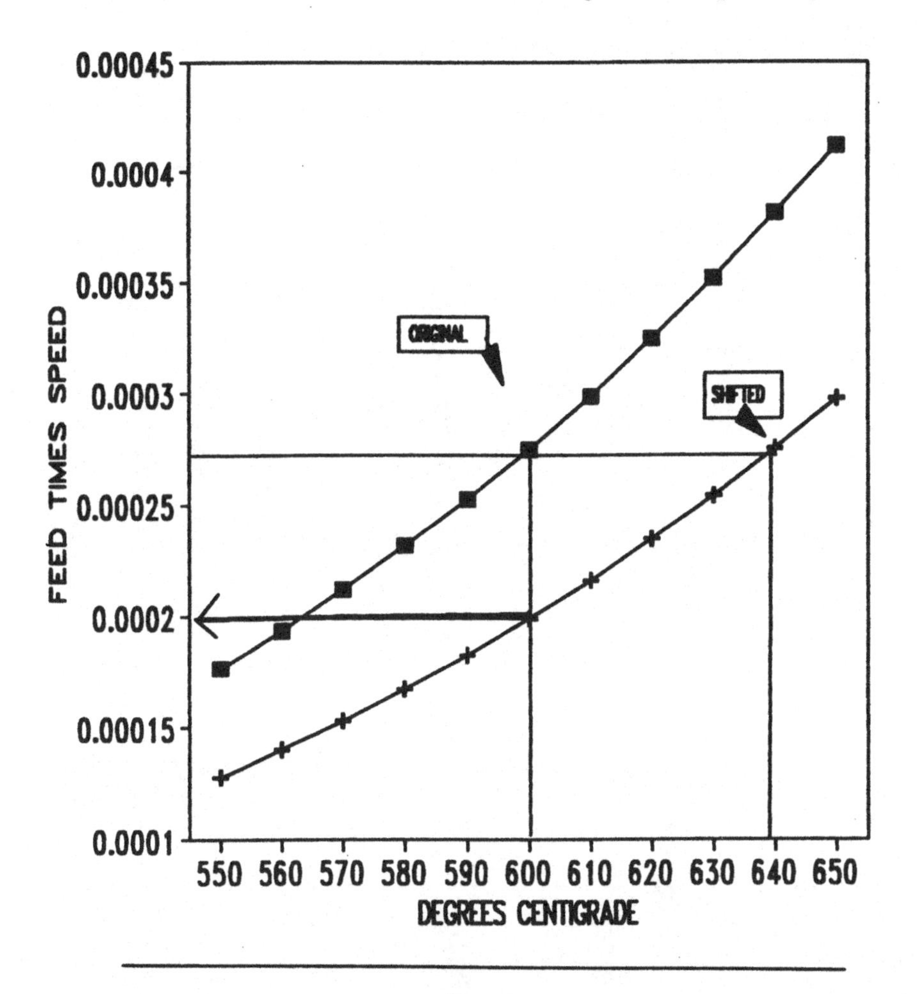

Figure 4

Temperature was varied in 10° increments by plus and minus 50 ° from the indicated 600°. The feedback feedrate parameter was allowed to assume only three values, the original target, 50% less and 50% more. While the feedback feedrate parameter is sparsely represented relative to the number of actual values that it may assume during machining, the span of values used for the training is sufficient to permit the network to adequately generalize the effect of this input variable.

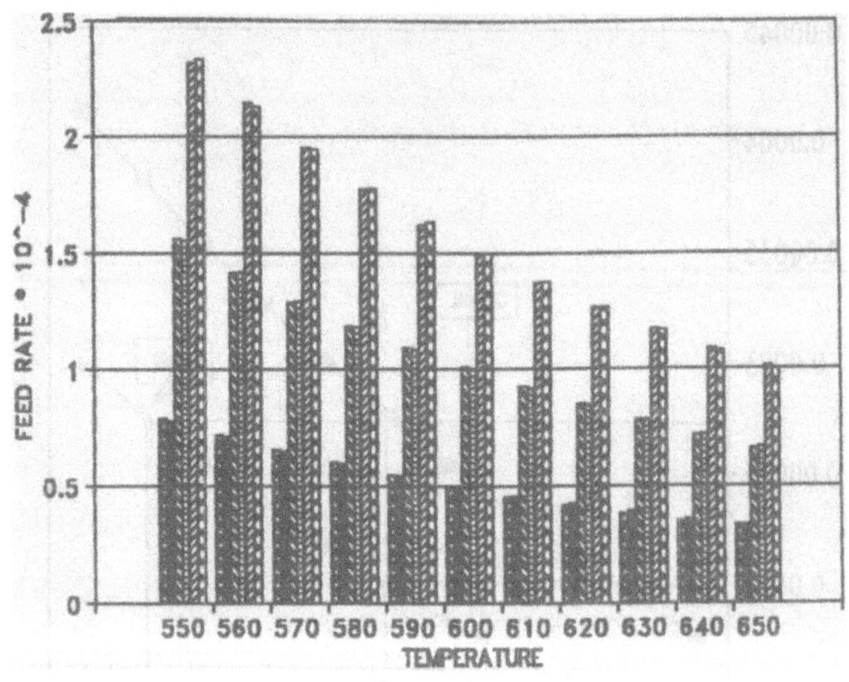

Figure 5

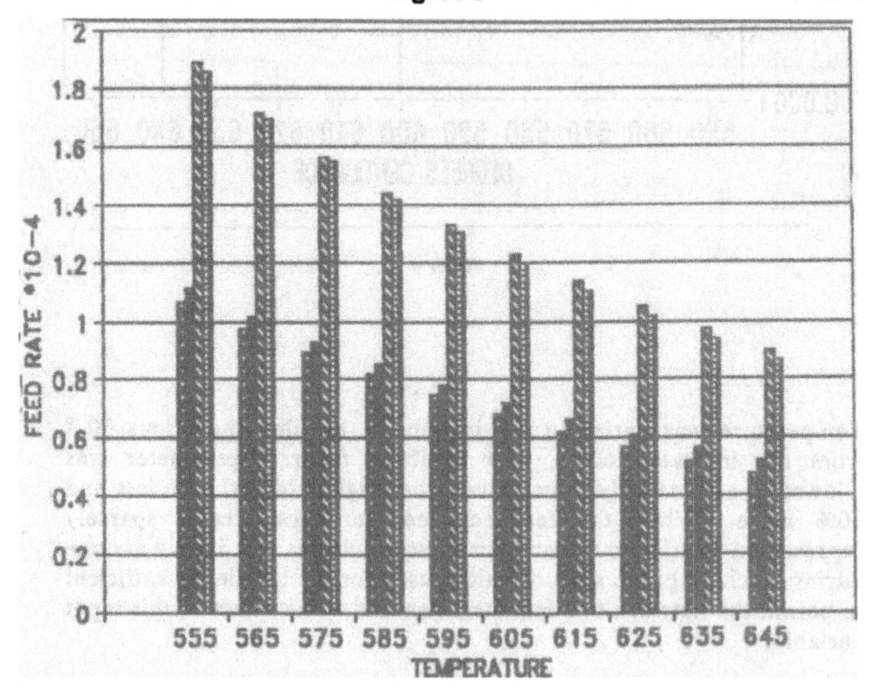

Figure 6

TG FEED	TEMP	CUR FEED	NEW
1	650	1.00	0.668334
1	640	1.00	0.722495
1	630	1.00	0.782087
1	620	1.00	0.847761
1	610	1.00	0.920262
1	600	1.00	1.000442
1	590	1.00	1.089277
1	580	1.00	1.187887
1	570	1.00	1.297564
1	560	1.00	1.419801
1	550	1.00	1.556325
1	650	0.50	0.334167
1	640	0.50	0.361248
1	630	0.50	0.391043
1	620	0.50	0.423880
1	610	0.50	0.460131
1	600	0.50	0.500221
1	590	0.50	0.544638
1	580	0.50	0.593944
1	570	0.50	0.648782
1	560	0.50	0.709901
1	550	0.50	0.778163
1	650	1.50	1.002501
1	640	1.50	1.083743
1	630	1.50	1.173130
1	620	1.50	1.271641
1	610	1.50	1.380393
1	600	1.50	1.500664
1	590	1.50	1.633915
1	580	1.50	1.781831
1	570	1.50	1.946347
1	560	1.50	2.129702
1	550	1.50	2.334488

Table 1

5. RESULTS

Figure 5 shows the network's success at learning the training set for the servomechanism demonstration. There are six sets of bars for each temperature classification; each adjacent pair within a set depicts the network transformation and the results as calculated by formula, respectively, for each of the three feedback feedrates used in training. The network was able to learn to reproduce the training set outputs with a high degree of accuracy. After 164,405 learning events, the amount of error was below .00001 for all of the training sets. In addition to this network error, there is also some inaccuracy introduced due to scaling (all network input and output must be mapped within a range of 0 to 1).

A better test of the viability of the neural network based multi-variable servomechanism is to determine its performance for combinations of feedback for which it was not specifically trained. Figure 6 shows the performance of the network compared to the formula for a range of feedback edge temperature from 555 to 645 degrees in ten degree increments. The pairs of bars within each temperature classification show network and formula calculated feed rates for feedback feed rates of .75 * 10^-4 and 1.25 * 10^-4, respectively. While the variation between network and formula is more significant than for the training set, the stability in amount and direction of variation suggest that the performance of the network could be improved with better defined training set.

6. CONCLUSION

The results of this project demonstrate the feasibility of a nueral network based servomechanism model for closed loop control in machining operations. The neural network is, of course, less accurate than any formula based control system. More important than the degree of arithmetic accuracy is the question of the validity of the model's premise. Is the process most efficient if the originally indicated edge temperature is maintained? Further study and refinement of the model is necessary in order to ensure that the servomechanism functions to maximize cutting efficiency while minimizing the various forms of tool wear.

It is also conceivable that the neural network servomechanism can be equipped with some of the non-quantifiable expertise of an experienced craftsman. For example, the activities of skilled craftsmen can be studied to determine the relationships between the changes they make in process variables and measurable process data such as temperature.

Regardless of whether they are trained to appreciate the behavior of engineering formulas or experienced craftsmen, neural networks evidence the characteristics necessary to implement real-time closed-loop control for the Open-Architecture Machine Tool. Firstly, they can serve as a platform for the integration of a great many sensory inputs without the need for complicated interrupt driven schemes or programming. As a state device, a neural network can react to all inputs at once to map out a set of process directives. Also, small simulated neural networks which synergistically incorporate knowledge as well as a system for accessing that knowledge in a relatively small set of interconnection weights, can operate at the speeds needed for efficient real-time control.

7. REFERENCES

(1) P.K. Wright and I. Greenfeld. Rapid Prototyping in an Open-Architecture Manufacturing System, *Conference on Applications of Artificial Intelligence in Manufacturing*, Boston, MA, July 1990, pp.3-28.

(2) P.K. Wright, I. Greenfeld and E. Pavlakos. Tool Wear and Failure Monitoring on an Open-Architecture Machine Tool, *American Society of Mechanical Engineers*, Winter Annual Meeting, PED-Vol. 43 - Fundamental Issues in Machining, pp. 211-228, 1990.

(3) C. Ghandi and M.F. Ashby. Fracture Mechanism Maps for Materials which cleave: FCC, BCC and HCP Metals and Ceramics, *Acta Met.*, Vol. 27, 1979, pp. 1565-1602.

(4) James S. Albus. Brains, Behavior, and Robotics. *McGraw Hill*, Peterborough, New Hampshire, 1981.

(5) G. Boothroyd. Temperatures in Metal Cutting, *Fundamentals of Metal Machining*, Arnold, London 1965, pp. 39-53.

(6) E.M. Trent. *Metal Cutting*, Butterworth, London, 1977.

(7) D.W. Yen and P.K. Wright. Adaptive Control in Machining: A New Approach Based on the Physical Constraints of Tool Wear Mechanisms. *ASME's Journal of Engineering for Industry*, Vol. 105, 1983, pp. 31-38.

(8) P.K. Wright. Physical Models of Tool Wear for Adaptive Control in Flexible Machining Cells. *Computer Integrated Manufacturing*, ASME Special Bound Volume at the Boston Winter Annual Meeting, Vol. 105, pp. 31-38.

(9) T.N. Loladze. Nature of Brittle Failure of Cutting Tool. *Annals of the CIRP*, Vol. 24, No. 1, 1975, pp. 13-16.

(10) S. Kobayashi and E.G. Thomsen. Some observations on the Shearing Process in Metal Cutting. *ASME's Journal of Engineering for Industry*, Vol. 81, 1959, pp.251-259.

Studies in A.I. Augmented Control Systems using the BOXES Methodology

D.W. Russell

Department of Electrical & Computer Engineering, Penn State Great Valley, Malvern PA 19355, USA

ABSTRACT

The BOXES paradigm has been successfully applied to control systems that are mechanically unstable. The primary application is the "Trolley and Pole" in which a freely hinged pole is balanced by rapidly reversing the direction of a guided trolley. This paper takes the BOXES (Michie [1]) methodology into the realm of continuous control systems, in which success is not measured by time to failure, but rather by the form factors of response to stimuli. To illustrate the method a classic second order, damped harmonic system is modified to include an AI contribution to its control parameters. The paper establishes that such contributions are both significant and desirable. The paper concludes with some encouragements for further study of systems that are poorly or partially defined.

INTRODUCTION and the BOXES METHODOLOGY

This paper describes studies of a control system application of the BOXES methodology. This method enforces an A.I. paradigm that causes continuous intelligent decisions to be generated without a-prior knowledge of the dynamics of the system being observed. A favorite application of the method is the TROLLEY and POLE (Russell [2], Russell & Rees [3], Russell, Rees & Boyes [4]. This application uses the BOXES learning algorithm to seek for a control matrix that keeps a freely hinged pole in controlled motion for sustained periods of time. The control matrix gives decisions in response to state variables read from the system.

In the case of the trolley and pole, the system state number, which is derived from the trolley position (x) and velocity (dx/dt) and the pole angle (θ) and velocity ($d\theta/dt$) indexes the matrix for LEFT, RIGHT (or NEUTRAL) motor switching. The state number is calculated by marking the integer regions that the system variables (e.g. x, dx/dt, θ, $d\theta/dt$) fall into at any sample time. These region numbers are then combined into a unique state number. It is this state number that indexes the control matrix. Each matrix cell contains a control decision for that state, for that run. When the system utilizes a state, an accumulation of usage and time spent within that state is kept by the automaton. When the run is over a statistical method examines the length of the run as compared to recent history and numerically rewards or punishes the decision making tuples.

Once complete, the automaton re-examines the values in the decision matrix, and recalculates the appropriate control values. The algorithm can be adjusted to give long runs, but with wildly oscillating transients or to give highly stable low motion solutions of short length.

Three areas of concern in the BOXES method are: the selection of appropriate boundaries, initial conditions and the fact that updates to the control matrix only occur after a *failure*. The selection of boundaries is difficult if true impartiality is to be maintained in order that absolutely no a-priori conditioning is imposed. The initial conditions of both the control matrix at the start of a study and the system variables for each run must be random if learning is to follow a true unbiased pedagology. It is however also apparent that to initiate a learning run from a totally uncontrollable position can not be conducive to the process of learning. The last restriction of the three is much more severe, because if the automaton is controlling a real world plant, then failure is not allowable for obvious safety reasons, under any circumstances.

It is because of this last fact that the notion of using a BOXES type paradigm to *augment* a stable controller with sporadic, risky but productive AI controlled interjections was considered feasible.

A.I. AUGMENTATION

A schematic of the proposed system is shown as figure 1.

The AI augmented control process falls naturally into five phases.

Conventional Control

In this mode, the system is controlled by a conventional method, such as PID (Proportional, Integral and Derivative action) or 100% feedback. The AI controller is merely an observer of the plant and essentially just passes the system variables into the state number allocator and waits. The conventional control algorithm keeps the system in stable control, handling transient demand changes, suppressing oscillations and generally running the system.

Transition into AI Control

As stated previously the AI processor is constantly monitoring the system and waiting for an opportunity to take over or supplement control when a suitable state number sequence occurs. For example, the transition algorithm might hone in on the system remaining in one state for some predetermined time interval, or if a preferred state number sequence occurs or repeats. At this juncture, the AI algorithm injects control signals to increase productivity using the BOXES methodology, by augmenting the conventional control actuators.

AI Control

Once in the AI control phase the BOXES paradigm plays out an intelligent *pseudo game* trying to optimize yield, or follow some other reward structured strategy. When the game is over, recalling the trolley and pole analogy, the *fail* condition is a signal to pass control back to the safe algorithm and to post-process data from the run for learning purposes in anticipation of another game in the near future.

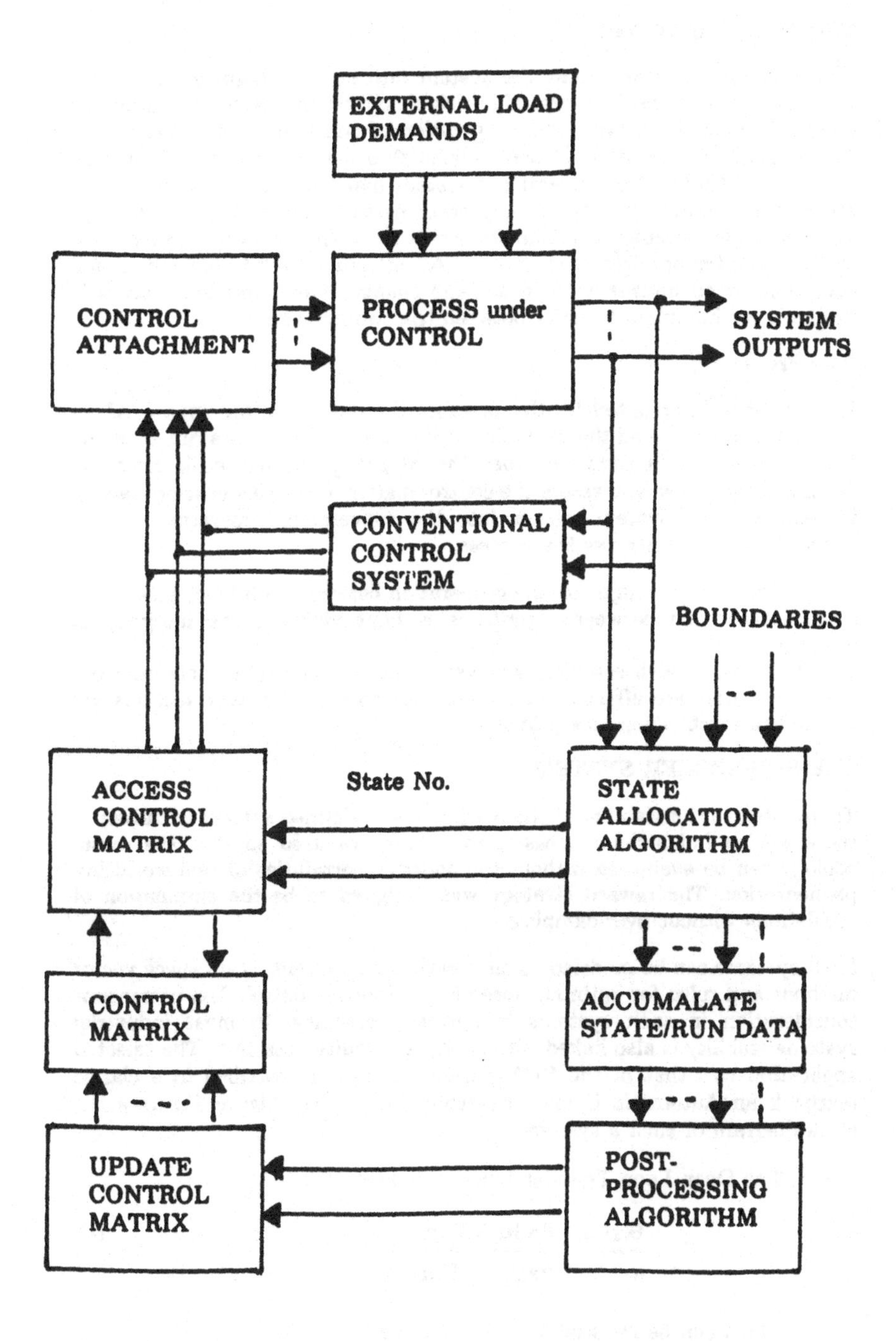

Figure 1. A.I. Augmented Control System Schematic.

Transition from AI Control

The learning algorithm takes the system into possible high-risk zones for short periods of time. The term high-risk should be understood as regions of control beyond the normal stable operation of the system, but well within safety margins. The AI controller relinquishes control to the conventional system gracefully. The conventional system now views the new state variables in an identical manner to any other perturbation and dispatches signals for control restoration. The transition from AI occurs at the end of a run or on interrupt from the system. An interrupt may be a fault or fail condition, or an unusually large demand change. The transition process is designed for bumpless transfer between the two systems.

Post Processing

If a run was interrupted by the process, the data collected in the AI algorithm is ignored, and the game declared *null and void.* If the run ends due to some out of limit condition, then the AI post-processor recalculates the statistical variables, and resets the control matrix. Once this post-processing is complete, the system relaxes into the conventional operation defined above. Figure 1 illustrates the process.

The system design hinges on the cooperation between both ideologies and a smooth transition between algorithms without violating the integrity of either process.
The first step in such research is to verify that the control signals from the BOXES matrix are effective in the control process. The next sections are devoted to establishing this principle.

THE SUN-SEEKER SYSTEM

To investigate the viability of the method a well defined system is chosen to represent an industrial process. Simulations are used so that the methodology can be evaluated without the added uncertainties of real world implementation. The reward strategy was designed to be the elimination of oscillations without over-damping.

Most systems can be modeled using describing equations, state space vector methods and rules for optimal, closed loop adaptive control. The paramount consideration in such systems is normally stability. In most industrial systems stability is also linked with safety and faulty detection. The selected application was that of the SUN SEEKER system described in a classic textbook an Automatic Control Systems, (Kuo, [3]). Figure 2 shows the block diagram of such a system.

The Open Loop Transfer function is given by:

$$\frac{\Theta(s)}{e} = \frac{Ks.Rf.K.Ki/n}{Ra.J.s2+Ki.Kb.s} \qquad (1)$$

which can be reduced to

$$\frac{\Theta(s)}{e} = \frac{K1}{s(s+\beta)} \qquad (2)$$

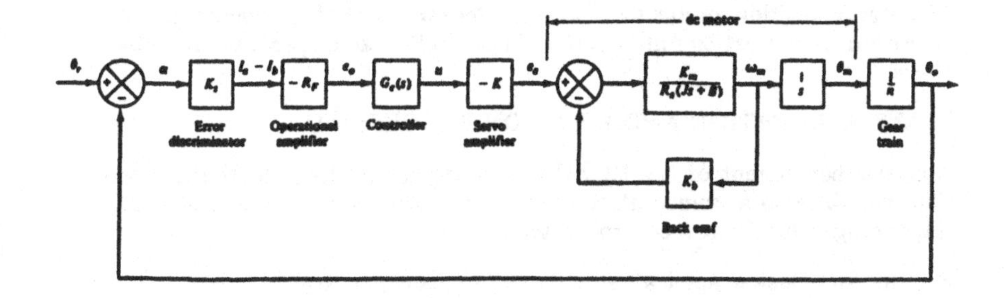

Figure 2. Block Diagram of a Sun-seeker Control System

Benjamin C Kuo. AUTOMATIC CONTROL SYSTEMS ,6e, c 1991, p.842.Reprinted with permission of Prentice Hall, Englewood Cliffs, New Jersey.

The Closed Loop Transfer function, for 100% feedback is given by

$$\frac{\Theta(s)}{\Theta r(s)} = \frac{K1}{s2 + \beta.s + K1} \qquad (3)$$

for the reduced system. Figure 3 shows a typical transient response to a unit step generated from the simulation to be used throughout this study.

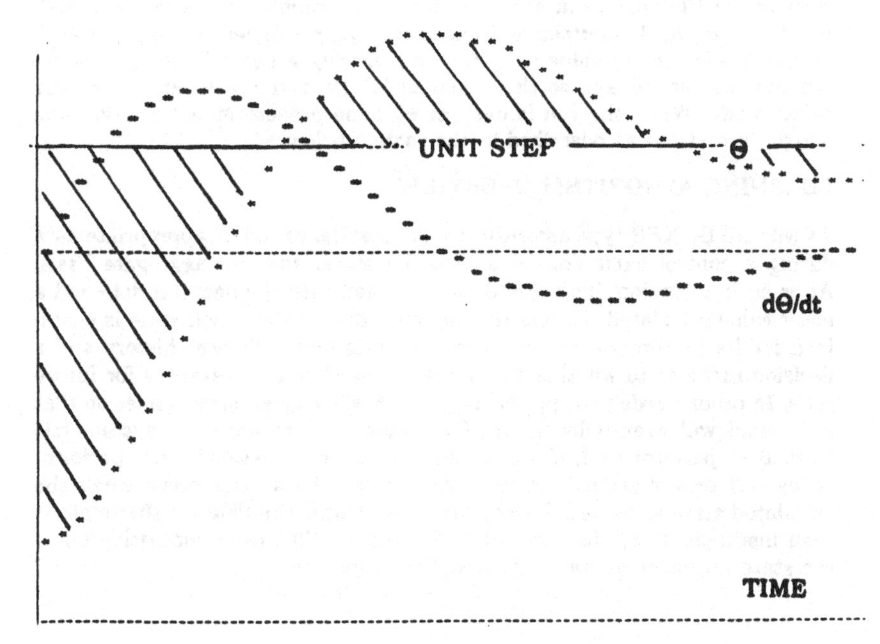

Figure 3. Conventional Control System Response.

To defer transition problems, the Sun Seeker model implements the AI algorithm from start to finish, rather than the Sporadic operation described above.

STATE ALLOCATION AND DECISION PROCESSING

A classic component of the BOXES methodology is the reduction of real-time variables to a unique state number, and the rapid retrieval of a corresponding value from a control matrix.

For the Sun Seeker application, only two indexes are required

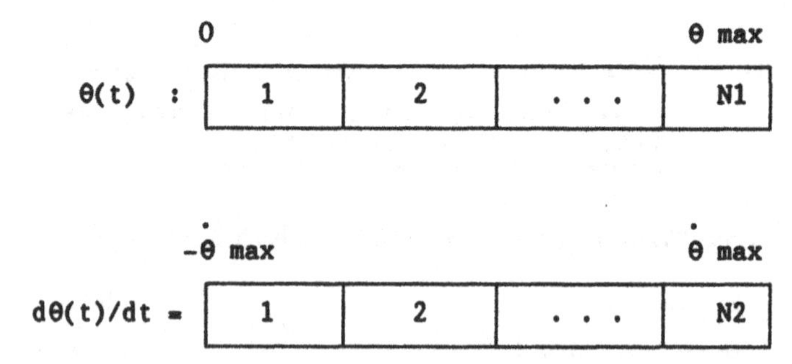

For any θ(t) and dθ(t)/dt range numbers i in Nl and j in N2 exist. The state number is therefore given by

$$m = (i\text{-}1) \times N1 + j \qquad (4)$$

If there are 10 divisions in each variable, m is a number between 1 and 100. If θ(t) or dθ(t)/dt fall outside of their range, the run is terminated. For each m, there is a matrix value u=UMAT(m). During a control run the matrix remains unchanged and so the values of 'u' are taken in a rapid read-only reflex mode. When the run is over, or at some predetermined interval, the matrix is updated as described in the section below.

LEARNING ALGORITHM OVERVIEW

As with all BOXES type algorithms it is essential to collect appropriate data during a control excursion on a state by state, and run aggregate basis. After each excursion, historical data is merged with the new run data and a *merit* value calculated. During the historical data update each state is examined for its performance in the last run, weighted with past history and a decision made as to whether the matrix value should be reversed for future runs. In other words poor performance runs will weaken state values so that a reversal will eventually occur. The notion is that once the system has learned to perform well, it will do so repeatedly. Occasional poor performances will only marginally affect the process. These can occur when the calculated state value falls into a naive, rare state. The data for that state is then insufficient for the recalculation process. With each successive entry the state improves as its decision statistics mature.

In the Sun Seeker system described above, run time is inconsequential in the learning process. It is the shapes of transient responses to disturbances that are all important. Classical methods such as P.I.D. attempt to shape the response to set stimuli. If the perturbation is in the form of a unit step, the ideal, if unrealistic response would follow the same form. The actual response is the well known second order harmonic waveform shown in Figure 3 above.

In the Sun-Seeker learning algorithm the divergence from the ideal waveform is used as the penalty/reward parameter. To avoid canceling out of plus and minus values the absolute value is used. The shaded area of Figure 3 shows the divergence factor, which gives an indicator of the poorness of the response as compared with the input step.

In the simulation, the value of angle $\Theta(t)$ is computed via a numerical integration. If the step length of the method is 'h' then the individual elements of the divergence can be calculated from $[\Theta(t)-\Theta r]\cdot h]$ where Θr is the input stimulus. This is the divergence contribution over any small time interval. If $\Theta(t)$ mirrors Θr, exactly, the divergence would be zero, the target for the algorithm.

As explained above, the values of $\Theta(t)$ and its derivative are condensed into a unique state number (m). In the active control mode, values for the divergence are accumulated on a state by state basis, as well as an incremental count of entries into the state, distinguished by whether the current control value is a 'zero' or a 'one'. A running grand total of the divergence is also kept for the duration of the run.

After the time designated for the run has elapsed, the off-line historical data process is performed. The algorithm is in two stages: Overall and Individual. The Overall performance calculation accumulates a weighted grand total of the divergences. The notion of merit is expressed by dividing the number of runs to date by this overall divergence. As the divergence decreases so the merit increases, thus indicating performance improvement or deterioration. Other merit algorthims are under consideration that more accurately represent the waveform shape and incorporate steady state errors etc.

The Individual performance values are handled in an identical way to the Overall merit but on a state by state basis. By keeping two counters for divergence and entry popularity the sub-merit figures can be calculated for each state and the control matrix values altered purely on the basis of a numerical determination on the relative strengths of the 'zero' or 'one' value.

Using this algorithm, small divergence generating states are encouraged to remain at the values that contributed to the low transient high merit run, and high divergence generating state values tend to be altered.

AI AUGMENTATION IMPLEMENTATION

Several experiments were tried to evaluate possible augmentation implementations of the control function of the system. The Sun Seeker system described above has the following transfer function when operating in closed loop mode with a feedback coefficient of c.

$$\frac{\Theta(s)}{\Theta r(s)} = \frac{K1}{s2+\beta s+K1.c} \qquad (5)$$

If c = 1, then the system behaves with normal damped harmonic motion. If c is zero, the system races into an unstable open loop mode. The value of c can be augmented by the u-value from the control matrix. For example

$$c^1 = 0.5\ (1 + u) \qquad (6)$$

so when u = 0, c^1 is 0.5, or, when u = 1, c^1 = 1

A bolder algorithm might equate c to u, in the belief that open loop transients when u = o can be controlled by heavy closed loop damping if the value of β is large enough whenever u = c = 1.

A second method is to attach the AI matrix value (u) to β itself. The rationale is that the velocity damping be reduced for speed (u=0) in some states, and increased (u=1) in others to catch the overshoots.

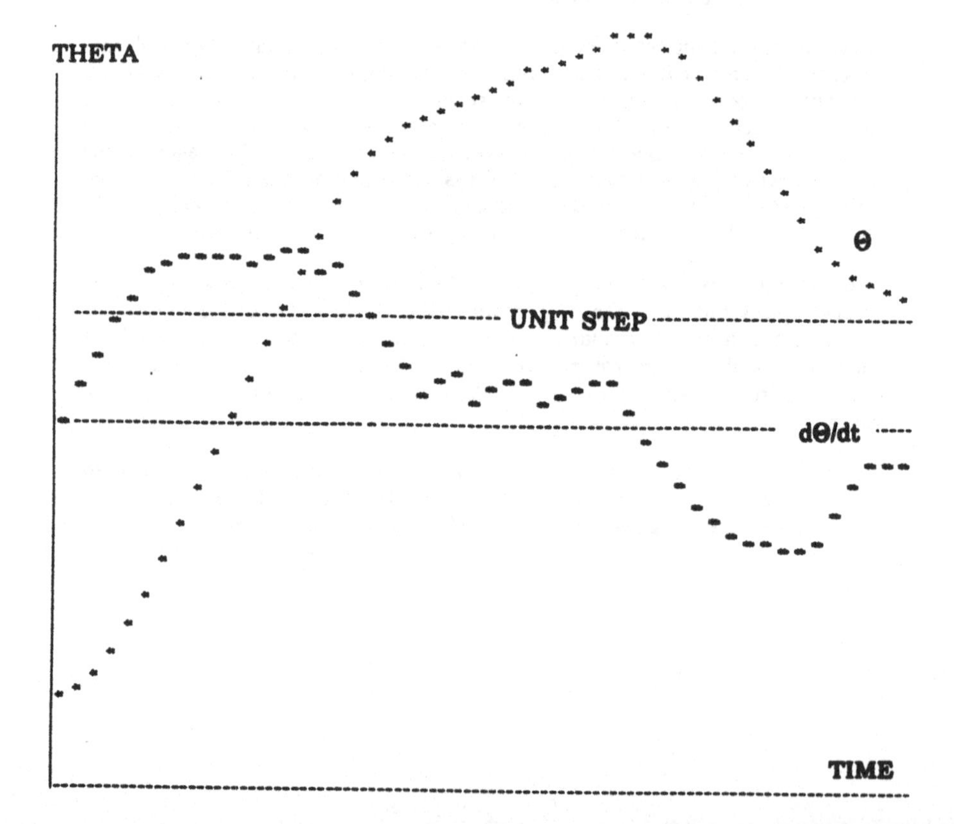

DYNAMIC SIGNIFICANCE OF THE AI ALGORITHM

A program was written that generates on demand plots of the dynamic response of the solar panel to input stimuli which in this study are step functions. Figures 4 and Figure 5 illustrate typical distortions that occur when control matrices are allowed to augment the feedback by 50%, as compared to the pure 100% feedback operation of Figure 3.

The total divergence for Figures 3, 4 and 5 are given for comparison

	Divergence Total
Figure 3. No AI Contribution	0.104
Figure 4. AI Causing Oscillation	0.28
Figure 5. AI Suppressing Oscillation	0.20

The above figures, taken after many exercises of the program lead to two major factors that are encouragements in the research. First, the u-matrix contribution to feedback is highly significant. Second, the divergence factor does reflect the form factor of the harmonic shape and provides a basis for a penalty and reward structure.

The program was is written to operate in learning mode, in which a varying step was repeatedly applied to the system. A run, or BOXES game is considered over after one half second. If the system is forced out of bounds, the BOXES method ignores the run. The output, can be displayed on a single screen showing a run count and merit figure. Merit is a measure of non divergence from the input step.

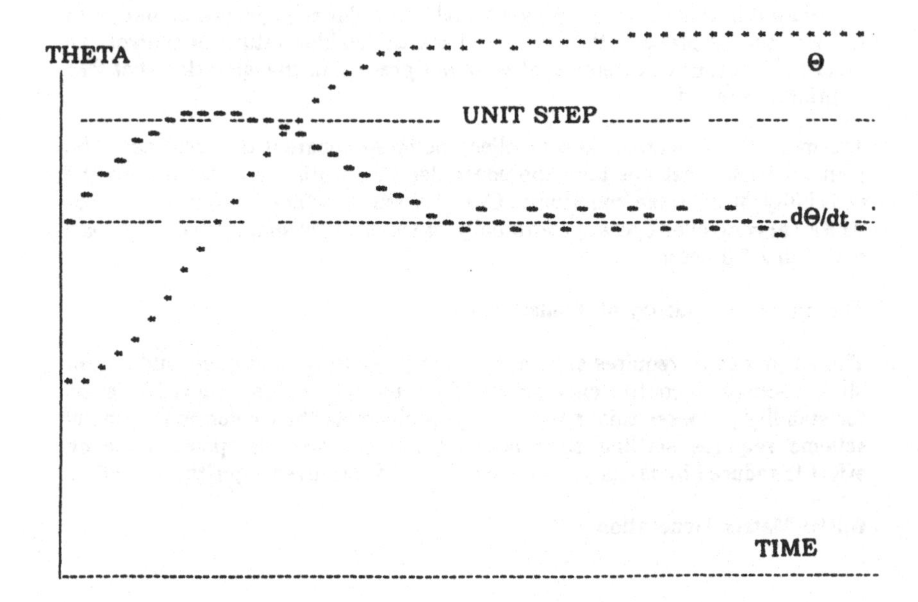

Figure 5. Response with A.I. Contribution Supressing Oscillation.

CONCLUSIONS AND CHALLENGES

The study of the Sun-Seeker system continues at the time of writing. Refinements in the process include

- Learning Run Evaluations
- Variation of State Boundary Values
- Reduction in the Number of State Boundaries (Bain, Michie & Sammut [6])
- Variation in the BOXES Contribution to the Controller
- Discovery of other Attachments to the Process

The original intent to attach a BOXES type of controller to an industrial process is still a high priority item. The idea of increasing yield by the application of positive control steps and by reduction of negative oscillations to those steps is still under consideration, but some limitations are obvious and inevitable.

Any cooperative control system has limitations; and this augmentation of a conventional system by an AI algorithm is no exception. The major limitations in the design that have surfaced in the research so far are as follows:

Transient Induction

For the pass-back to the conventional system from the AI controller to be manageable, the magnitude of the worst case transient must be limited to a percentage of the maximum step-change that the system can withstand. The reason is that after the conventional controller relinquishes actuation in favor of the AI process, then on pass-back the sudden return of control can induce high order transients that were not present in the operation that was in prior command.

The more sophisticated the controller, the more apparent this becomes. The transfer back must not take the controller by surprise, and force even the possibility of unstable conditions. One method of solution is to let the conventional controller operate continually in the background, exchanging roles with the AI process.

Timing and Frequency of Transitions

The AI processor requires substantial post-processing interludes, and so can allow re-entry to control excursions with some limited frequency. However, for stability purposes this is not really a problem as the conventional control scheme requires settling time between *games*. Also the post-processing effort is reduced by faster microprocessors and improved algorithmic design.

Initial Matrix Generation

The control matrix that is used in the AI controller must be initialized in a reasonable fashion or else the training effort is so tedious that practical effectiveness may never occur. One approach is to build the initial knowledge base from data taken in a pre-operation period from the conventional system. In this way, only reasonable values will be injected into the matrix and continued operation will give the statistical variables some credibility in early runs. The traditional methodology has dictated random values, but in the control situation some expert knowledge from the plant under control is considered advantageous and necessary.

SUMMARY and ENCOURAGEMENTS

These limitations are currently under study and do not appear to be insurmountable obstacles to the system veracity and future applications.

A.I. applications to real world systems are particularly susceptible to suspicion due to the *unknown* factors that are always present. These factors are also part of the human inference, deduction and reasoning processes. Knowledge based systems are only as expert as their imparted cognisance and learn only according to finite pedagology. The prospect of an unintelligible situation, leading to a nonsensical or null decision looms over the horizon in most systems. The system described in this paper seeks to balance the possibilities of failure and risk with improved performance over the short time periods that the AI system cooperates. A good analogy is the use of different driving skills in overtaking slower vehicles or avoiding critical situations as compared to simply "getting from A to B" on a safe road.

Boden [7] aptly notes, speaking of the mind

....failure is not the prime trigger of cognitive change, and adaptive control cannot be explained only in differential response to distinct classes of failure..

Failure plus adaption is the rationale of the process of this AI attachment to a real system.

The paper describes some encouraging results in the development of an AI augmented control system. The transition between the control layers in both directions is the crucial part and the subject of continuing research. The detection of an out of control region state due to unknown factors is handled by a simple passback to safe control. Unknown factors include unmanageable control demands, faults and accidents and poorly defined or overly approximated control equations.

The work at Liverpool in 1975-77 (Rees [8], Boyes [9]) indicated that the BOXES methodology could produce optimal seeking results and the corollary to reflux switching in the distillation column opened the door to the sun seeker applications. Other industrial application under consideration are a Trash to Steam Plant in Baltimore Md. (DiAngelo [10]) and a Penicillin Fermentation Process (Brownlowe [11]. The work of Barto, Sutton and Anderson [12] focussed on a neural network approach to the Trolley & Pole in which some element of scheduling (the *adaptive critic element*) affects the learning and operating process (the *associative search element*). The research currently underway at Penn State Great Valley deliberately avoids scheduling in the algorithms, although look ahead (feed-forward) in control systems is obviously respectable and may be incorporated in the future.

ACKNOWLEDGMENT

This work was sponsored in part by a Research Development Grant from Penn State University.

BIBLIOGRAPHY

[1] Michie, D. & Chambers, R.A. *Machine Intelligence 2...BOXES: An experiment in Adaptive Control.* Oliver & Boyd

[2] Russell, D.W. *The Trolley & Pole Revisited: Further Studies in AI Control of a Mechanically Unstable System.* AIENG '90. Boston, May 1990.

[3] Russell, D.W. & Rees, S. J. *System Control: A Case Study of a Statistical Learning Automaton.* Proc. 2nd European Meeting on Cybernetics & Systems Research. Vienna, 1974.

[4] Russell, D.W., Rees, S.J. & Boyes, J.A. *Microsystem for Control by Automata of Real Life Situations.* Proc. C.I.S.S. Baltimore, 1977.

[5] Kuo, Benjamin C. *Automatic Control Systems* 6 ed Chapter 8.3. Time Domain Design of Control Systems. pp. 480 ff. Prentice Hall. Englewood Cliffs, N.J. 1991.

[6] Bain, M. Michie D., & Sammut, C. *Experiments with the Pole Balancing System: A Suitable Case for Genetic Treatment.* Proc. ISSEK, Udine, Italy, 1988.

[7] Boden, M.A.. *Failure is not the Spur* Adaptive Control of Defined Systems. Proc. NATO Adv. Research Institute, June 1981, Moretonhampstead, Devon. Plenum Press, N.Y. 1984.

[8] Rees, S.J. *An Investigation Into Possible Applications of Learning Control Systems.* M. Phil thesis. Liverpool Polytechnic, 1978.

[9] Boyes. J. *A Man-machine Interface for Training an Automation for a Real World Situation.* M. Phil thesis. Liverpool Polytechnic, 1977.

[10] DiAngelo, J. *Trash to Steam Optimization Using the BOXES Methodology*, Unpublished Masters Paper. Penn State Great Valley, 1991.

[11] Brownlowe, W. *Control Strategies for Microbial Fermentation*, Unpublished Master's Paper. Penn State Great Valley, 1991.

[12] Barto, A.G., Sutton R.S. & Anderson C.W. *Neuronlike Adaptive Elements That Can Solve Difficult Learning Control Problems.* IEEE Transactions on Systems, Man & Cybernetics. SMC - 13: 834-846. 1983.

Artificial Neural Network for Alarm-State Monitoring

N. Dodd

Neural Solutions, 15 Celandine Bank, Woodmancote, Cheltenham GL52 4HZ, UK

Abstract

A device is described capable of signalling novelty in the state of a system monitored by an arbitrary number of instruments. The device learns by being shown examples only of "healthy" signals from the system and infers the class of "alarm" signals by default. The system is demonstrated on real data acquired from critically ill patients in an intensive care ward and is shown to provide a useful degree of alarm state detection.

1 Introduction

In numerous situations it is the case that a system of some kind requires monitoring for an unusual state. This may be a state of alarm. By the nature of the system it may not be possible to have access to examples of these unusual states in order to train a conventional learning system with examples of both healthy and alarm states. The technique of "default training" described in this paper allows the alarm state to be inferred by default without access to actual examples.

The application domains for such a technique include an industrial plant, environmental monitoring and the intensive care ward. The latter has been chosen for a major experimental study and results are presented for the performance of the technique on real data acquired from critically ill patients.

Some medical instruments, designed for the intensive care ward environment, are now being equipped with primitive monitors which sound an alarm when the reading of the instrument strays outside pre-determined limits. However, the bounds of acceptability are to some extent arbitrary and require setting by someone with prior knowledge of the patient's condition. Additionally, the individual single instrument monitor has no information regarding the other parameters of the patient's body that are being instrumented. By treating each bodily parameter as distinct and unrelated to any other bodily parameter we discard much useful information regarding the interdependence of these measurements. The human body is, after all, a unified whole whose parts must function in concord. The Neural Network Monitor learns the interrelationships between the samples, made by the instruments, of the bodily state. Any departure from normality of the entire system is signalled as an alarm.

The Neural Network Monitor receives output from any number of instruments and provides an output indicating a condition requiring attention. The Neural Network Monitor does not require the explicit statement of *rules*, as would a conventional expert system, but may be taught by example. After a period of supervision, during which it is given experience of the normal, or healthy, response to be expected from the instruments it is monitoring, the Neural Network Monitor is left to receive input without guidance.

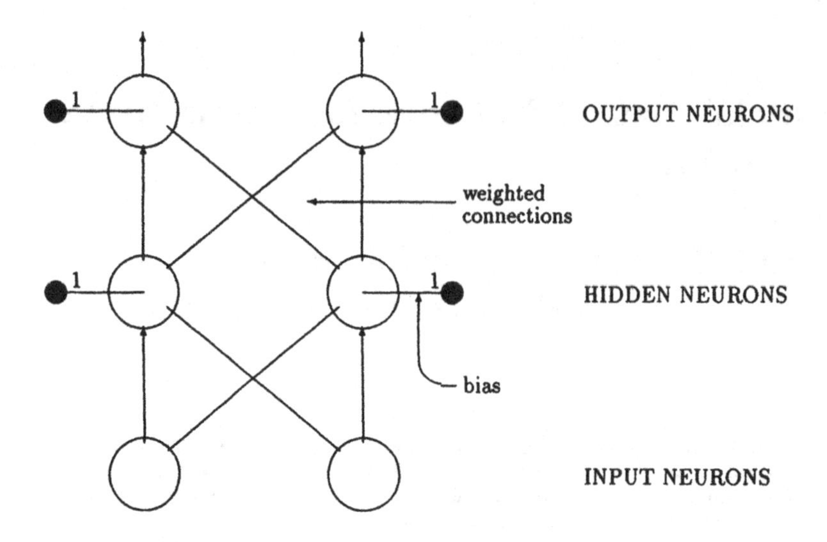

Figure 1: A Simple Three Layer Neural Network

At first the behaviour is somewhat "nervous" and the device will sound its alarm bell at any possibly "unhealthy" response from the instruments. The (human) supervisor may then indicate to the Neural Network Monitor that the response from the instruments that caused the alarm should be included in its concept of a normal or healthy response. After repeated reassurances, the Neural Network Monitor settles down to give infrequent false alarms.

A neural network consists of a number of simple processors, or *neurons*, linked together as in figure 1. The neurons combine their inputs and subsequent produce an output which is passed to other neurons. The links between neurons contain *weights* which control the amplitude of the signal passing through. In addition, each neuron has an associated *bias*, which is effectively a connection to a neuron which is always in the *on*-, or 1-, state. It is the weights and biases that embody the information required to classify the input signals, just as in the mammalian brain it is the links *between* the neurons that determine its function. In the training stage, the weights and biases are iteratively improved by applying input and output pairs to the network.

One way of viewing the operation of the Neural Network is as *interpolation* in a space defined by its parameters. The training patterns are the representative examples of classes. After training, previously unseen patterns are classified according to an interpolation between the training examples. The number of neurons in the network determine the complexity of the space in which interpolation is done. In this way, with enough neurons, a division of feature-space by arbitrary complex boundaries can be made, assimilating the fine distinguishing features of the input patterns. Alternatively, by providing only a few neurons, the network is forced to generalise and the outliers in its training set will be effectively ignored.

2 Properties

The major advantages of the artificial neural network for equipment monitoring include:

- learn by example[2.1]
- generalise from a representative training-set[2.2]
- tolerant of noise[2.3]
- incorporation of prior knowledge[2.4]
- adaptive to any combination of inputs[2.5]
- extremely simple to operate — no special skill required[2.6]
- small network implementable on standard PC AT + interface[2.7]
- no technology-limited upper bound to potential size of network[2.7]

These claims will be dealt with one by one.

2.1 Learning by example

The knowledge of the network is contained in the values of the weights between neurons. Initially these are set to small random values. The process by which the values of the weights are refined to represent better the mapping of the network's input to the required output is known as "error-backpropagation", the mathematical details of which are given in [1].

2.1.1 Default classification

If no examples of one class of output are available, the complementary class must be inferred by default. For example, if a particular system is being monitored for some dangerous condition, and if that dangerous condition cannot be produced at will to train the network explicitly, then the only examples available for training will be consistent with the healthy, non-dangerous state of the system. Training a network only on one class of inputs, with no counter-examples, causes the network to classify everything as the only class it has been shown. However, by training the network on examples of the "healthy" class but also on random inputs for the "dangerous" class, any input which occurs after training which does not resemble one of the previously encountered "healthy" inputs will automatically be classified as "dangerous". The network effectively behaves as a novelty detector.

To illustrate this, a network was trained as follows. It had five inputs and one output. The "healthy", class 1, input vectors consisted of elements a, b, c, d, e such that

$$b < c,\ d < c,\ a < b,\ e < d, \tag{1}$$

as illustrated in table 1. 50 training examples satisfying condition 1 were generated. A network with 5 inputs, 3 hidden neurons and one output was trained to output class 1 for this data. Additionally random data with the same first-order statistics as the data for output class 1 was synthesized for which the network was trained to output class 0. When tested on a new set of 100 inputs, half of which were produced by explicitly following condition 1, and half of which were generated randomly, the performance was 93% correct. For the randomly generated data there is a probability of $0.5^4 = 6.25\%$ of fulfilling condition 1 and so the network is in fact performing to within 1% of the inherent upper limit of performance.

A more rigours justification for synthesising the unavailable data with random numbers follows from the fact that training seeks to minimise the sum squared error over the training set. Consider a binary classification network with a single input v producing an output $f(v)$. The required outputs are 0 if the input is a member of class A and 1 if the

		input			*output*
a	b	c	d	e	*class*
⋮	⋮	⋮	⋮	⋮	⋮
0.150160	0.241971	0.496722	0.338752	0.163327	1.0
0.752625	-0.258011	0.050505	-0.144331	0.085486	0.0
0.390102	0.582408	0.667979	0.252589	0.037113	1.0
-0.894841	0.933459	-0.331432	-0.835807	-0.459371	0.0
-0.406147	-0.263851	0.074262	-0.330817	-0.446870	1.0
0.024406	0.173021	0.477517	0.743378	0.155935	0.0
0.769972	0.797939	0.964704	0.768290	0.724058	1.0
0.705362	0.622344	0.909775	0.808566	-0.722170	0.0
0.574456	0.622694	0.686187	0.684142	0.684017	1.0
0.173560	-0.250082	-0.946428	-0.070469	0.570686	0.0
⋮	⋮	⋮	⋮	⋮	⋮

Table 1: Output class 1 vectors are obtained using condition 1. Output class 0 vectors are synthesized from random numbers having the same mean as class 1.

input is a member of class B. If the prior probability of any data being a member of class A is P_A, and the prior probability of any data being a member of class B is P_B; and if the probability distribution functions of the two classes as functions of the input v are $p_A(v)$ and $p_B(v)$, then the sum squared error, E, over the whole training set is given by:

$$E = \int_{-\infty}^{\infty} P_A p_A(v)[f(v) - 0]^2 + P_B p_B(v)[f(v) - 1]^2 dv \tag{2}$$

Differentiating this with respect to the function f:

$$\frac{\partial E}{\partial f} = 2p_A(v)P_A f(v) + 2p_B(v)P_B[f(v) - 1] \tag{3}$$

and equating this to zero

$$f(v) = \frac{p_B(v)P_B}{p_B(v)P_B + p_A(v)P_A} \tag{4}$$

which is exactly the probability of the correct classification being B given that the input was v. So by training for the minimisation of sum squared error, and using as targets 0 for class A and 1 for class B, the output from the network assumes a value equal to the probability of class B. Substituting the $f(v)$ corresponding to a trained network back into 2 we get

$$E = \int_{-\infty}^{\infty} \frac{P_A p_A(v) P_B p_B(v)}{P_A p_A(v) + P_B p_B(v)} dv \tag{5}$$

and so the minimum attainable error is when there is zero overlap between the distributions:

$$\int_{-\infty}^{\infty} P_A p_A(v) P_B p_B(v) dv = 0. \tag{6}$$

In this way it is possible to model the default class to produce an error less than the error to be expected from using uniformaly distributed random variates.

2.2 Generalisation

In a situation where a network is apprenticed to a human to learn to distinguish a healthy class of signal from anything else that might come along, it is important that the

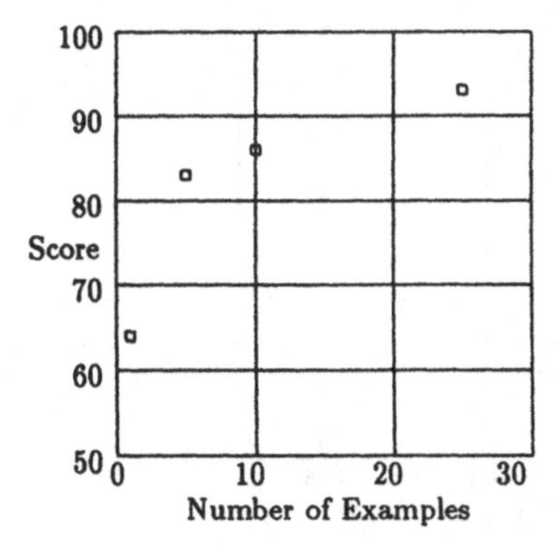

Figure 2: Illustration of the network's ability to generalise from only a few training examples: Score on unseen data versus number of training examples.

network should reach a state of learning where it can be left alone, as soon as possible. The training stage should be as brief as possible, leaving only the occasional false alarm to recall the human operator to give the benefit of his judgement. Reaching a useful level of performance with only a few examples is only possible if those examples are representative of all members of the class. To demonstrate that networks are able to generalise with relatively few examples, the training-set described above was used with different numbers of training examples. With only 5 unique examples of class 1 the performance is above 80% (see figure 2). As before 50 random vectors were used to synthesise class 0. In this example the network has 5 inputs, 3 hidden neurons and 1 output. There is evidence to suggest that as the hidden layer of neurons is made smaller, so the network is obliged to generalise better. This generalisation is at the expense of being able to classify correctly the outliers of the training set.

2.3 Tolerance to noise

The ability of a network with few hidden neurons to generalise suggests that a compact representation of the data is being made. If the data is corrupted by zero-mean noise, the essence of the data is retained from sample to sample while the noise is changing. Within limits this has little effect on the ability of the network to learn classifications. Figure 3 shows the effect on classification performance of unseen data with an increasing amount of noise present on both the training and test data. The score is 74% even with a signal to noise ratio of 1:1.

2.4 Incorporation of prior knowledge

For scalar inputs which vary slowly with time and have no syntax, an artificial neural network which is layered and has total connectivity between layers is just as good as any. However certain types of input may have an underlying generator which undergoes well defined state transitions. For these types of input the ideal network should contain the hardware (written in terms of the network formalism) able to exploit the regularity of the data and to extract parameters from it to be fed to the rest of the network. Further discussion of this subject, still under research, is outside the context of this summary.

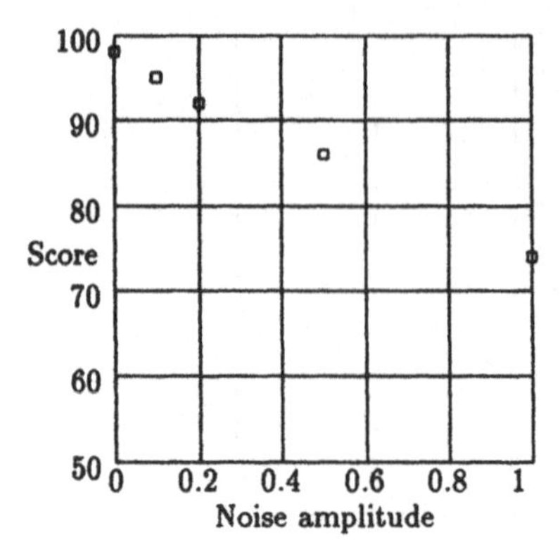

Figure 3: Illustration of insensitivity of network to noise in training and test inputs. Noise amplitudes are relative to mean signal level of 1.0.

2.5 Allocation of inputs

Since the network learns from example, the allocation of signals to inputs of the network is arbitrary. This is certainly the case for a homogeneous network consisting, for instance, of totally interconnected layers, however particular applications may require a structured network predisposed to address the characteristic variation in certain types of input. In this case inputs to the network will favour certain types of signal and must be allocated accordingly.

2.6 Operator skills

The envisaged method of operation is very simple and consists of

Connect it up: Connect the various instruments whose output requires monitoring to the Neural Network Equipment Monitor. If the instruments do not provide a line-output then it is usually a simple matter to provide one. If, however, breaking into the circuitry of the instrument is not allowed, then a pick-up coil mounted on the surface of the instrument will pick up any high-frequency signal, such as a video signal, which can be de-modulated and sent to the Neural Network Equipment Monitor. As discussed above, the Neural Network Equipment Monitor will adapt to virtually any type of signal and is tolerant of noise. There is an increasing tendency to equip intensive care wards with data collection centres which serve to collect the vital function data of a ward of patients. An instrumentation such as this would be an ideal platform for incorporating a Neural Network Equipment Monitor.

Teach it about "healthy" signals: Once the instruments are turned on and registering signals characteristic of a "healthy" state, press the OK button and keep it pressed (it can be equipped with a latch) for several minutes while the Neural Network Equipment Monitor learns the concept of a healthy signal, making sure, during this time that the signals are characteristically healthy.

Let it work alone: Release the OK button. If the concept of a healthy signal has been well represented by the training examples given so far, the false alarm rate will be low. If, however, a signal is produced which do not fit the networks concept of a healthy signal, the alarm will sound requiring the human operator to press the

OK button *if indeed the signal is a healthy one.* If false alarms are too frequent, a subsequent period of training may be required. Otherwise the Neural Network Equipment Monitor can be left to monitor the instruments unattended.

2.7 Implementation

The simulations used as examples in previous sections were implemented on a Sun 3[1] and took 75 seconds for 500 updates, each having calculated the error derivatives (see [1]) over the entire 100 training patterns. Once the network has learned, the input patterns can be processed at an approximate rate of 500 patterns per second. Even an 8-bit processor running 100 times slower than the Sun 3 would therefore be able to cope adequately with real-time input at a rate of 5 patterns per second, though the learning time of two hours might be impractical. Of course the learning can take place on a powerful machine leaving a much slower processor in charge of the monitoring once the network has learned.

If real-time learning is required, or if a much larger network is needed, there is no technology-limited upper limit to performance if the algorithm is implemented using parallel processors. A practical design using Transputers[2] allows the use of the language Occam[2] which addresses the parallelism in a program. Using this formalism a forwards pass through a network that has already learned the correct weight values would be

```
PROC forward.pass
  SEQ
    PAR
      calculate output for each neuron in input layer
    PAR
      calculate output for each neuron in next layer
    :
      :
    PAR
      calculate output for each neuron in output layer
```

A backward pass, wherein the weight updates are calculated, would have a similar, but inverted, structure. A complete learning cycle would consist of

```
PROC learn.cycle
  SEQ
    forward.pass
    backward.pass
```

The computing power of a T800 Transputer is very roughly equivalent to a Sun 3, and from this the number of Transputers required for a network of a given size to operate at a given speed can be calculated. For development, one of the many commercially available Transputer systems hosted by a PC would be used. However, for a conveniently packaged system suitable for use in a hospital, an expandable board containing 5 Transputers with memory, power-supply, etc., would fit into a box the size of a small briefcase. Its cost price would be around £5,000.

3 Trial with real data

A hospital intensive care ward was approached to obtain vital function data from a number of patients over a period of 24 hours. The data was scaled to lie between −1 and +1, and an artificial neural network trained to output 1 for the healthy data and 0

[1] Very approximately equivalent to a PC AT running in excess of 20 MHz
[2] Trademark of the Inmos group of companies

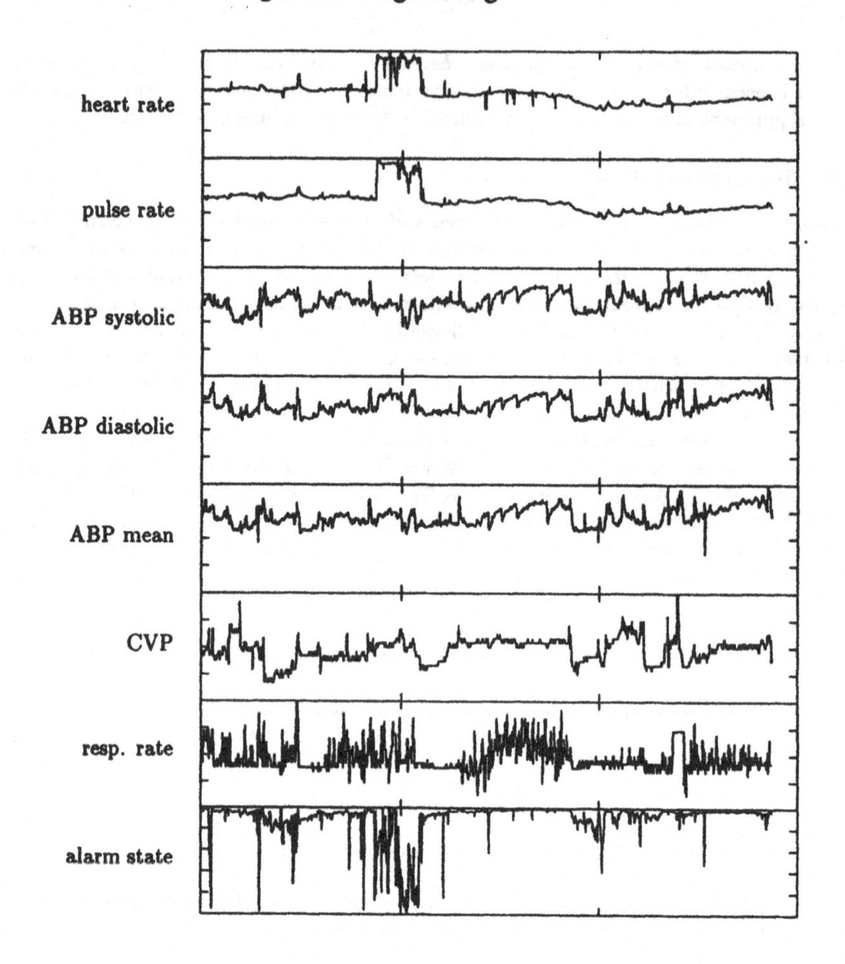

Figure 4: Traces of vital functions over 24 hours. Lowest trace is alarm state.

for uniformally distributed uncorrelated noise. Results are shown in figure 4. In figure 4 the output of the artificial neural network is given at the bottom, and is mostly high, indicating a healthy response from the patient. The low dips indicate alarm conditions and with each of these can be associated a departure from normal of one or more of the vital function traces above.

4 Conclusion

The technique of "Default Training" is a statistically proven method of supplying synthetic data for a class not available in the training of a multi-layer perceptron.

When used as an alarm monitor, such as in the intensive care ward scenario, the default trained multi-layer perceptron assesses correlates of the data as well as absolute levels and supplies a useful level of alarm state detection.

References

[1] David E. Rumelhart and James L. Mcclelland. Learning internal representations by error propagation. In *Parallel Distributed Processing: Explorations in the Microstructures of Cognition*, volume 1: Foundations, pages 318–362. MIT press, 1986.

Design and Experimentation of a Shell to Develop Knowledge Based Systems for Spacecraft Control

S. Gusmeroli, M. Monti

FIAR Space Division, via Montefeltro 8, 20156 Milano, Italy

ABSTRACT

Knowledge Acquisition in Expert Systems (ESs) development represents one of the major bottlenecks and obstacles to their effective operational delivery in the industry. Although many different tools exist at the moment on the market to ease and speed up knowledge acquisition, they often have been demonstrated too general and not tailored for particular restricted domains and classes of problems. Shells have been used for developing ESs addressing different kinds of problems (planning, scheduling, trouble-shooting, faults diagnosis, data interpretation, decision support) in different application domains (spacecraft control, industrial processes control, production scheduling, robotics, economic business), but have offered little guidance on which features to use in a specific application (a specific problem in a specific domain). Spacecraft control domain does not constitute an exception to this: space operations preparation, mission planning and scheduling, computer-aided operations execution, spacecraft monitoring systems and personnel training are different applications, which, however, share common data, procedures and rules. This paper presents the project SCS (Spacecraft Control Shell), started at FIAR to develop an intelligent tool to allow an easy and quick definition of the items needed by ESs for spacecraft control (spacecraft model, Flight Operations Plan, problem-solving knowledge, constraints) also by non-experts in AI. SCS is being tested on several ES projects Fiar Space Division is at present carrying out in Spacecraft Control domain.

1. INTRODUCTION

Developing Expert Systems in an industrial environment is a very difficult task and Knowledge Acquisition represents one of its major bottlenecks. In fact, Expert Systems (ESs) are usually applied to loosely structured domains where conventional software solutions cannot be applied. This requires the interaction between two specialists: the domain expert, who knows the characteristics, requirements and constraints of the specific problem domain, and the knowledge engineer, who is the expert in developing AI systems [BUC 88] [GRE 86]. The operational delivery of ESs in the industry, which is one of the major objectives in the 90's for Artificial Intelligence developers, has incremented the attention to the Knowledge Acquisition problem in expert systems studies.

Some of the main difficulties that can be found during the knowledge acquisition phase are the following [PEZ 90]:

1) Usually the domain expert and the knowledge engineer do not know anything of the other's field of work. The domain expert knows very little of AI, and the knowledge engineer ignores the domain specific problems and constraints. One problem is even to find a common language between the two.

2) The tools that are available on the market [TAU 89] to develop ES, also called shells, are very powerful environments that provide a variety of mechanisms to represent and handle knowledge. The problem is that little or no help is given on how to use these mechanisms, especially with relation to a specific application domain. Backward chaining, forward chaining, object-oriented features, frames, slots, etc. might all be available, but the knowledge engineer has to guess which one is best suited to represent the entities involved in the specific problem at hand, and the domain expert is of no help. The only source is the expertise of the knowledge engineer himself. But if two ES are developed by two different teams on two specific problems of the same domain (for instance two trouble-shooting ES applied to different radars), the probability that the resulting systems will have completely different architectures and representation mechanisms is almost 1.

3) The use of general tools makes it very difficult to re-use part of the software developed for a

specific application in another one of the same domain. As an example, if two planning expert systems are developed in the same domain, they might share a common requirement, for instance the need to explicitly represent and manipulate time intervals. It is however very difficult to use in both of the ESs the same temporal reasoning module.

4) Because of the large set of possible features and representation mechanisms available in a general tool, and the lack of explicit suggestions related to the specific domain, the initial phase of each project, in which the representation and architecture are designed, is replicated each time. This results in a scarce optimization of the effort, and in an unnecessary rediscovery of the same needs.

This is of course valid if the domain of application is sufficiently narrow to ensure that the same issues will recur. In other words, what presented applies for instance to the development of two expert systems for production planning in the industrial field, and not to the development of an expert system for production planning and one for trouble-shooting in the same factory. To address these problems, FIAR has adopted the concept of "domain-specific shell", and has started to develop tools to ease the generation of ES within specific domains. This paper presents this approach, and the tool which is being developed for generating expert systems for spacecraft control.

2. DOMAIN SPECIFIC SHELLS

The two main approaches available when developing an expert system shell can be called 'horizontal' and 'vertical'. The first consists of using one of the environments widely available on the market, such as KEE or ART. These tools are "horizontal", in the sense that they provide a large library of features and functions, which is not targeted to any specific domain or class of problems. The second option is to use domain-specific tools, which are not general, but are optimized for a specific kind of applications (for instance to produce expert systems for diagnostic problems in process plant control). Less expertise in AI is required, through the use of man-machine interfaces oriented to the specific domain. This is the "vertical" solution, which FIAR has adopted.

3. A SHELL FOR KBSs IN SPACECRAFT CONTROL

Spacecraft control involves several classes of problems like mission planning, operations preparation, planning and scheduling, spacecraft data monitoring, faults diagnosis and isolation, etc.. For the SCS project, FIAR has chosen planning and scheduling operations and experiments in Spacecraft control domain that includes different applications sharing, however, common features and needs. During the design of a Shell for developing KBSs in space domain, these common aspects have to be emphasized and put into a software architecture common to each application. Typical applications included in this domain are the following:

1) Producing a long term (not executable) schedule of experiments to be performed on the spacecraft starting from generic requests by the users of the mission. The purpose in question is to allocate a provisional time window for experiments, taking into account the characteristics of the spacecraft (some experiments can be executed only on certain payloads and by certain instruments) and the constraints expressed by experimenters (desired duration and frequency, relationships with other experiments, maximum delay allowed, visibility of stars, planets and other spacecrafts). The result is a proposal for an allocation of each payload of the spacecraft along a wide planning horizon generally lasting for a few months with the aim of optimizing on-board resources (payloads and instruments time, power, data recording tapes) and minimizing reconfiguration procedures.

2) Producing a short term (executable) schedule of space operations to be performed in order to satisfy the needs coming from a long term schedule of experiments and from platform operations to keep the spacecraft in the proper functioning state. Planning the schedule of operations for a specific period of time (say a week) for a given satellite is a complex activity, in which very limited or no support from automatic tools is at present given to the user. This because conventional software lacks the flexibility required to handle such a task. It is however of key importance to be able to develop such tools, in the view of increasing the degree of automatization, and provide more effective support to user operations. The task is very complex because the satellite might have different payloads, each with a list of potential users. The long term schedules for each payload might be incompatible each other; moreover, they have to be harmonized with the housekeeping

operations, that is, with the operations that have to be performed regularly to compensate for periodic errors, or in coincidence with specific events (like eclipses or space perturbations). Lastly, it has to be ensured that the different payloads don't interfere each other during operation, that the limited resource budget (for instance power, communication channels or data recording tape) is verified, and that no operational constraints of the satellite are violated.

3) Monitoring and controlling the execution of a plan of operations by coping with anomalies coming from the human controller (skip/abort of procedure steps, missing authorisation to proceed) and from the spacecraft or a simulator of it (delays, unsuccessful completion of procedures, faults, up-down link faults). Data interpretation tools and fault/recovery models are here used to produce new constraints and operations requests for a new planning session.

From an analysis of the applications outlined above, FIAR has designed and started developing a vertical domain-specific Shell [PEZ 90] for ES development in spacecraft control (see fig. 1) comprising two distinct knowledge sources (General Problem Solving Knowledge Source and Domain Specific Knowledge Source) cooperating to customize the three different kinds of knowledge (Factual, Procedural and Control Knowledge) needed by a KBS for the specific application.

General Problem Solving Knowledge Source comprises:

1) A Time Dependent Reasoner containing a set of modules able to represent and handle temporal knowledge expressed by absolute, relative or periodical temporal constraints. This module is based on Allen's interval algebra [ALL 84] and Rit's SOPO structures [RIT 86].

2) An inference engine derived from the default one included in OPS83 Run Time Support, but enriched in order to allow an easier debugging of programs and the loading of every kind of factual knowledge produced by other modules of the Shell.

Domain Specific Knowledge Source comprises:

1) A Spacecraft Model Editor to support the definition of different kinds of satellites at the level of abstraction required by the specific application. For instance, passing from application 1

to 3, the needed amount of structural information about the satellite increases. This impacts on factual knowledge.

2) A Flight Operations Plan (FOP) Editor useful for defining and implementing the possible operations to be performed on the units of the spacecraft. This also contributes to the construction of the application specific factual knowledge.

3) A Problem-solving Knowledge Editor to define and introduce knowledge about how to solve the class of problems in question (planning, diagnosis, trouble-shooting). This impacts on procedural knowledge as far as procedural problem solving rules and on control knowledge as far as problem solving control rules are concerned (e.g. rules implementing the sequence of activation of contexts).

4) A Constraints Editor able to handle particular limitations of the problem-solving knowledge specific of the application. This impacts on the structure of the control knowledge. Temporal constraints have to be linked to the correspondent modules of the Time Dependent Reasoner in order to build usable knowledge.

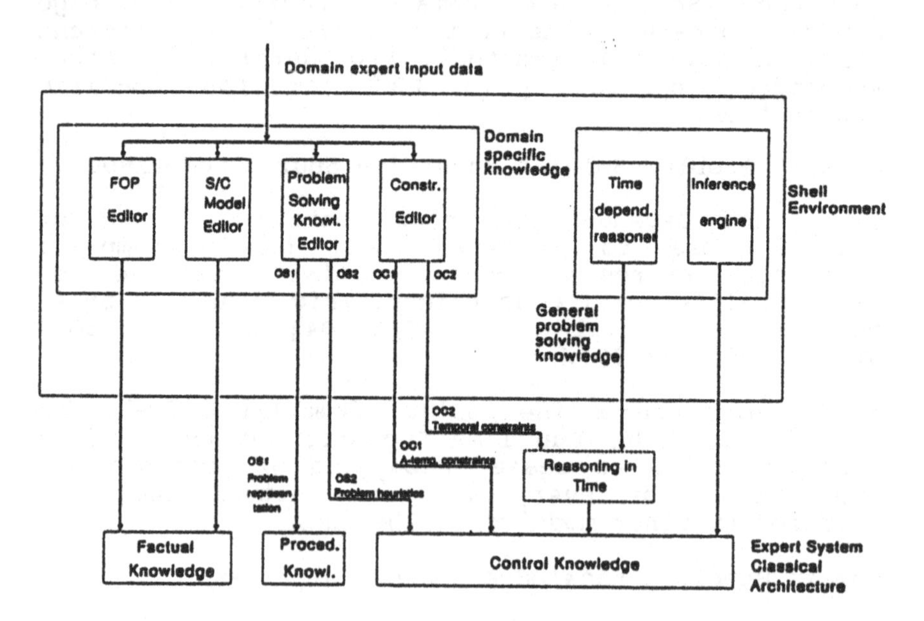

Fig 1. Shell architecture

For each of these items an interface has been defined to support an easy definition. These interfaces, integrated into a single environment, translate the users definitions into the internal representation formalism, which is then used by the inference engine and software architecture that have been developed.

The process of building an ES through the Shell can be decomposed into three steps (see fig. 2):

Step 1: Problem-solving Knowledge Editor defines the structure of the ES on the basis of the task it has to cope with and make the inference engine suitable to the specific class of problems (planning, monitoring, diagnosis, contingency recovering).

Step 2: Spacecraft Model and FOP Editors customize the ES skeleton developed in Step 1 on the specific spacecraft in question, while Constraints Editor introduces knowledge concerning the specific bounds to be observed.

Step 3: The ES developer realizes the interfaces bringing to end the development of the ES application.

Modules composing the development shell are described in the following sections.

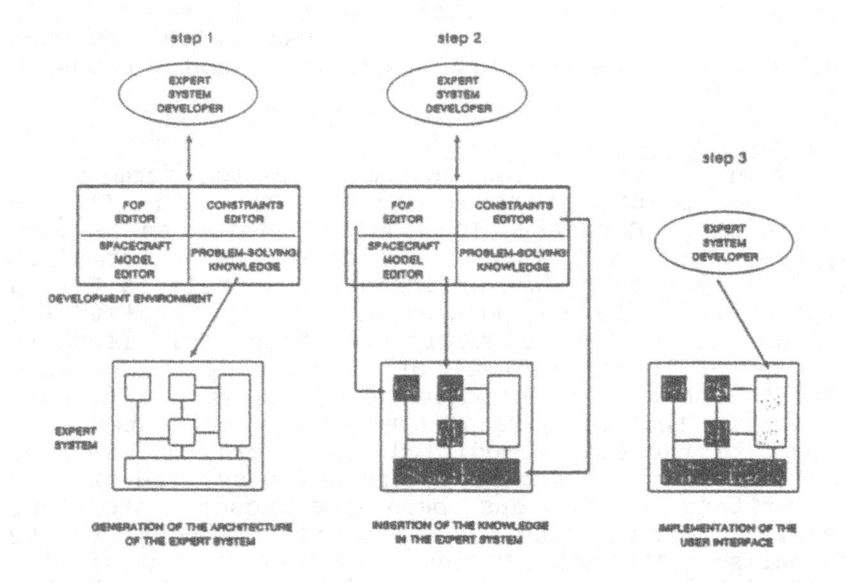

fig. 2 - generation of an expert system by means of a vertical shell

3.1 Spacecraft Model Editor

By an interactive session with this Editor, the user is able to define the spacecraft subject of the application at the required level of abstraction. For example, in planning applications like 1 and 2, the user can define the structure of the specific satellite on which planning has to be performed. This knowledge describes how the system is composed by subsystems, the subsystems by units, and the units by components. For each entity defined, the user inserts then the allowable states and the foreseen behaviour in presence of particular events like eclipses. It is important to note that the purpose is not to insert in the ES the complete knowledge of the satellite, down to every board or single component. This would be a very long process, which would only clutter the knowledge base with a lot of unnecessary details, not needed for the purpose of planning. The goal is to provide the ES with the information that is relevant to the planning problem. Therefore the definitions have to regard 'macro states' of payloads or units (such as unit-off unit-on, stand-by, pointing of an antenna, and similar), and not elementary states of individual electronic components. The source of the information is again the FOP and only marginally the ORH (Operations Requirements Handbook). In monitoring and control applications knowledge about the single components and the way they may break down is in addition required. The same data structures used in planning (say state-graphs) are useful even in this case, but have to be enriched with further information concerning anomalies and recovery techniques.

3.2 FOP Editor

In order to codify the procedures which form a FOP (FCP = Flight Control Procedures and CRP = Contingency Recovery Procedures) into an expert system, the first step is to determine a formalism able to represent them efficiently. Normally a FOP is written by different people, and it is conceived for manual rather than automatic execution. This leads to a scarce optimization of the code and many replications. As a consequence, it is very difficult to verify that no inconsistencies are generated when a part of the FOP is modified. In general maintenance and access to the knowledge are not easy. The process of writing a FOP has been decomposed into four levels, each with a specific representation formalism: FCP and CRP level, Flight Procedure Step (FPS) level, substep level and descriptive level. The FCP/CRP level definitions are stored into an explanation module which will be used by the target

ES for providing the user with rationales of operation (in fact what will actually be used by the ES for planning are the FPSs and not the FCPs). The substep level (telemetry and telecommands) are stored in an execution knowledge base, and the descriptive level forms an on-line manual for the human operator.

Three different types of FPSs can be defined:

* State-check FPS, which check the conditions defining each state of the graph and are drawn in the graph as labels of the nodes;

* State-to-state FPS, which realize a state transition in the graph and are drawn in the graph as arcs connecting nodes;

* In-state FPS, which change some parameters (gain levels of amplifiers, pointing coordinates of antennas) of the spacecraft without changing its state and are drawn in the graph as loop rings on nodes.

In the following, a description of a sample piece of FOP at the four levels of abstraction is outlined:

1) FOPxx composed by FCPaa FCPbb CRPcc

2) FCPaa composed by FPS11 FPS12 FPS13 FPS14 FPS19
FCPbb composed by FPS11 FPS15 FPS16
CRPcc composed by FPS17 FPS18 FPS11

3) FPS15 composed by

```
 send TCa1 and verify TM X008 is ON;
 check TM N089 is 2..8 into p4;
 if p4 = 2 goto label_x
   else wait TM X0011 is 3.5 time_out 30 exec FPS20
```

4) FCPbb realizes a state transition for payload XX from state checked by FPS11 to state checked by FPS16 by executing FPS15

CRPcc returns to normal state checked by FPS11, starting from an anomalous state checked by FPS17 by executing state-to-state step FPS18

FPS15 sends TeleCommand TCa1 and check its completion by verifying that the value of TM X008 is ON. Check the value of TM N089 within the interval 2..8 and put the returned value into p4 variable. If p4 is less than 2 go to label_x, otherwise wait for TM X0011 to become 3.5. If it does not happen within 30 sec., abort the FPS and call FPS20.

The integration of a set of FPSs and the model of the spacecraft leads to the construction of structures representing static and dynamic behaviour of the spacecraft in normal (planning application 2) and anomalous (application 3) conditions. These structures are called state-graphs, with bubbles indicating the states, directed arcs indicating the transitions (implemented by FPSs) and labels of the bubbles indicating the conditions to be verified in order to access the spacecraft in that state (also implemented by FPSs). An example of such a graph is reported below for payload "cmp" of a spacecraft, where dotted arcs and bubbles indicate information needed only by application 3, in order to diagnose and suggest the proper corrective actions.

3.3 Problem-Solving Knowledge Editor

Using the interfaces provided by the shell, the Expert System developer is able to customize the inference engine of the system being developed by introducing in it problem-solving knowledge [CLA 83]. This knowledge concerns the plan representation and construction, the methods to be used for solving conflicts arisen during planning and finally some spacecraft deep models (e.g. state-graphs, fault-trees, Petri-nets) developing facilities to be used in particular applications. Two levels of decision have in general to be taken about problem-solving knowledge definition [DRU 89]: how to represent the plan generated by the system and how to design the search space through which the system will find a path as a solution to its problem.

As far as plan representation, two approaches are the most promising ones:

State-Space approach: plans are represented by means of states as snapshots of the condition of the problem at each stage and by means of operators as means for transforming the problem from one state to another [BUC 88].

Action-Ordering approach: plans are represented as a set of actions with constraints on the order in which the actions have to be executed.

Moreover, two search space interpretations are possible:

States of the world: where the states of the search space are world states and the plan is constructed simultaneously with the space navigation as a collection of actions associated to each arc included in the solution path

Partial plans: where the states of the search space are partial plans obtained from a process of problem reduction into simpler subproblems

In spacecraft operations environment, the output of an ES devoted to develop a provisional schedule of experiments starting from generic experimenters requests (cfr. application 1) can be easily represented in terms of spacecraft states and the search space process can accordingly follow a world states approach: at each step of the search, the system generates all possible world states characterised each by the allocation of a different experiment in the next free space, evaluates them and chooses the most promising one for next expansion. On the other hand, the output of an ES for planning the operations to be performed on board (cfr. application 2) can instead be better represented as an order (total or partial) among actions and the correspondent search space as a set of partially completed plans. The Shell provides both kinds of problem-solving knowledge through its implementation in rule-based software modules. In developing Essope [GUS 90], an Expert System for Spacecraft Operations Planning and Execution, via the development shell, a partial plans search space has been chosen. Each complex problem to be solved (as the root of a problem decomposition tree) is expressed as a goal configuration of the satellite to be achieved and is derived either from the specification of some housekeeping periodical operations to be carried out on board or from the specification of space experiments to be performed. In both cases, the initial problem can be represented as a set of goals to be achieved as shown in Fig. 3.

A decomposition tree of the complex problem is then generated by handling each subgoal independently and following the below steps:

D0) Decompose root goal into subgoals supposed independently achievable;

D1) Search preconditions of each goal examining the history of the spacecraft already planned [CHA 85] (by considering a Closed World Assumption and a kind of Modal Truth Criterion): already achieved goals are cancelled;

D2) Decompose complex goals into simpler ones with the aid of deep models in the application domain (in this case state-graphs of the spacecraft components);

D3) Match the subgoals with the set of FPSs defined by FOP Editor and mark as completed the corresponding states;

D4) If some goals are still not reached, go to step 2.

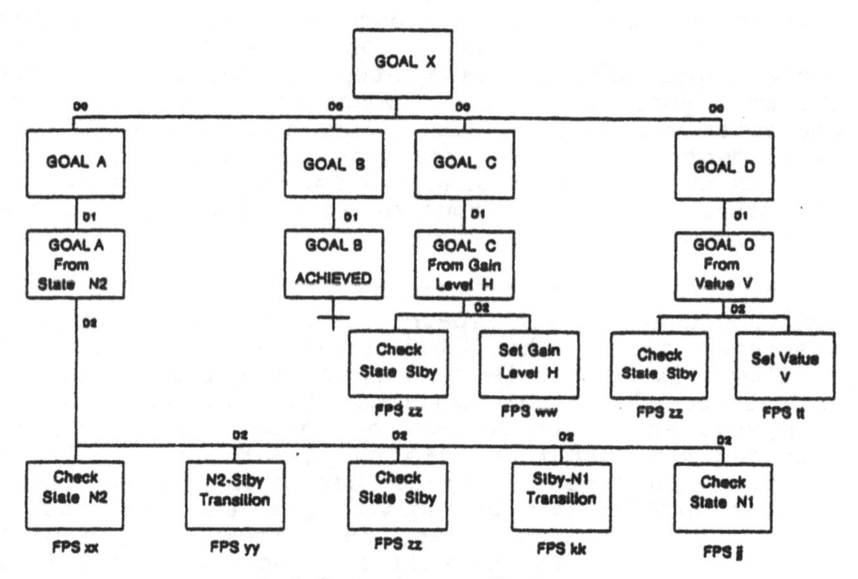

fig. 3 Goal x decomposition tree

At last a synthesis process is started to find a total order of application for the FPSs, following the below steps:

S1) Analyse FPS pre- and post-conditions, obtaining a partial order among steps;

S2) Solve the conflicts (steps executable in parallel) via heuristics (e.g. gain level FPSs should be performed before any other in-state FPS);

S3) Optimize the FPS sequence, by eliminating duplications.

In the example above, step S2 allows to find an order between achieving goal c or goal d first. Step S3 is finally used to cancel useless duplications of check steps.

The sequence of orders this way obtained is shown here below:

INITIAL: No Order

AFTER S1): xx,yy,zz,(zz and ww) or (zz and tt),kk,jj

AFTER S2): xx,yy,zz,zz,ww,zz,tt,kk,jj

AFTER S3) FINAL: xx,yy,zz,ww,tt,kk,jj

Once the FPS sequence achieving the goal configuration has been synthesized, conflicts can arise in putting it on the planning horizon with previously allocated experiments or procedure steps sequences. In this case, problem-solving method definition allows the user to customize the conflict solver to his needs and exigencies.

Three levels of methods are available:

- Criteria, given to decide the winner and the loser of the conflict (e.g. it is the shortest experiment or the experiment with the maximum priority value that wins);

- Policies, given to establish the actions to be performed (e.g. shortening an experiment, shifting a procedure);

- Tactics, given to determine the next policy to apply in case of failure (e.g. a tactic may be to apply all the available policies to the loser and only in case of failure of all of them to pass to the winner).

Fig. 4 shows examples of application of criteria and policies to a conflict arisen between the procedures of an experiment E2 and the previously allocated experiment E1. Tactics guide the conflict solver to perform its task: let us suppose to have given high priority to a shift policy (SHI) and low priority to a shorten (SHO) one and that E1 is the loser of the conflict identified by the chosen criteria. Two possible tactics (depth and breadth) giving respectively emphasis to criteria and to policies will try different orders of policies, as indicated here below in a Lisp-like language:

```
Depth: cond (SHI(E1) exit)
            (SHO(E1) exit)
            (SHI(E2) exit)
            (SHO(E2) exit)
            (T failure)

Breadth: cond (SHI(E1) exit)
              (SHI(E2) exit)
              (SHO(E1) exit)
              (SHO(E2) exit)
                 (T failure)
```

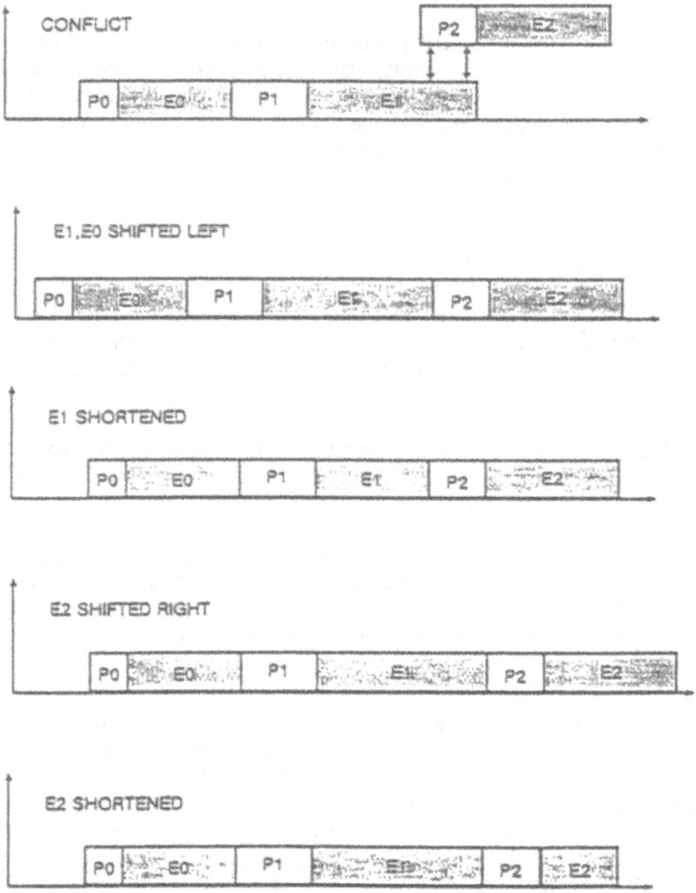

Fig. 4 Examples of conflicts solution

The result of this conflict solving activity is one total ordered plan for each payload/subsystem of the satellite. But the main goal of the Essope project and of other similar ones is to integrate in a unique knowledge-based environment planning and plan execution activities. To achieve this goal, that is to make the developed plan executable, constraints about the characteristics of the executor (a-temporal and temporal constraints) must be considered as shown in 3.4. The third main component of problem-solving knowledge is represented by deep models of the spacecraft editing facilities. The Shell provides a tool to ease the development of graphs, nets and trees to be used in many different applications. For instance, a KBS for space operations execution monitoring and control we are at present developing uses payload state graphs to plan recovery actions, restoring a nominal spacecraft functioning state after an anomaly has been detected. An example of such a state graph is reported in Fig. 5.

3.4 <u>Constraints Editor</u>

Many constraints bound problem-solving knowledge application for the specific application. Some of them are better introduced inside the spacecraft model or the FOP, some other need a specific representation language and definition tool. Constraints can be divided into a-temporal (1..4) and temporal (5) categories:

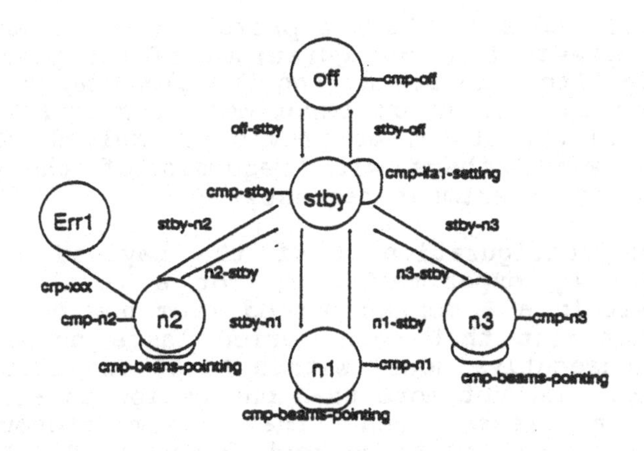

fig. 5 An example of a state-graph

1) Operating Constraints: express the behaviour of the satellite through the definition of the allowable states of the units for normal operation. They are codified in the Spacecraft Model

2) Platform Operations Constraints: express additional limitations valid during (before or after) particular operations needed by the platform of the spacecraft as for instance a reduction in the power available on board during eclipse periods. They too are included in the Spacecraft Model.

3) Cross Constraints: regard constraints about the simultaneous execution of space operations (for instance two specific FPSs cannot be executed in parallel on two different payloads because of interferences) or the simultaneous achievement of some states of the units (two states of two different units are not allowed simultaneously). These constraints are included in the FOP definition via the construction of a matrix of compatibility.

4) Execution Constraints: express limitations that have to be kept in order to make the schedule executable. Number of FPSs that can be executed in parallel or speed of the executor in answering information requests can be sample constraints of this kind and influence the nature itself of the plan. Such a constraints are specifically handled by Constraints Editor module.

5) Temporal constraints: express time dependent relations between the configurations of the payloads of the satellite required during the planning horizon by platform operations or experiment requests. Each configuration (or state, as previously called) might be related with others with requests of the type (derived by an experiment request):

> "Achieve configuration X of the payload W, as requested by experimenter Y, for a total of 60 hours within a 8 months period from now on, with the constraint that each period lasts between 2 and 5 consecutive hours within 05:00 and 12:00 of each day, that not more than one period is allowed within a week, and that, simultaneously, configuration Z has to be kept for payload K."

As an example of how to cope with such constraints, let us return to Essope to see how execution constraints have been taken into account. The executor of the plan generated by Essope interfaces with a so called Multi Satellite Support System

(MSSS), which allows operations on different payloads (not involving the platform) to be executed simultaneously. The Shell provides the way of specifying such a constraint and the rule-based modules able to achieve a total order in platform operations and a partial order in operations on different payloads. An example of an executable plan is reported in fig. 6. For instance, procedure P7 is considered as incompatible with any other procedure, on the basis of the cross constraints reported in the matrix of compatibility. The order in which operations B,C,D will be executed is not specified at planning time: only a time window associated to each of them specifies the admissibility intervals for their execution. The speed of MSSS answers allow the three requests to be satisfied "in parallel" as in a time-sharing computer system. A violation of an admissibility interval of one of them will cause a delay exception in execution and the need to analyze the present situation again, in order to decide whether or not to go on with replanning. In the Shell, temporal constraints are handled by a set of modules implementing a Time Dependent Reasoner (TDR) based on Allen's interval algebra [ALL 84] and Rit's SOPO structures [RIT 86]. A typical application of TDR is the problem of finding a provisional schedule of experiments on the basis of generic requests such as the one presented above: states generated by the expansion algorithm will be pruned by TDR if they violate temporal constraints. An example of formalisation of a temporal constraint is given below, according to a language derived from Allen's Interval Algebra:

```
REQUEST

name = req1
experimenter = Y;
payload = W;
horizon = 01-05-90 .. 31-12-90
amount = 60
t-constr =
 interval_x(req1) starts|during|ends|equals
           (05:00 - 12:00)                          AND
 duration(req1) > 2                                 AND
 duration(req1) < 5                                 AND
 ^((interval_x(req1) starts|during|ends week_k)     AND
  (interval_y(req1) starts|during|ends week_k))     AND
   interval_x(req1) equals interval_j(req2);
```

Such a specification is then translated into executable rule-based pruning modules.

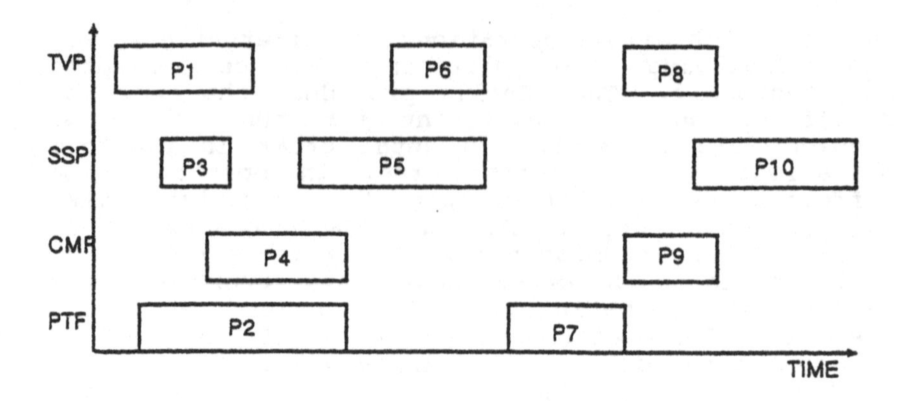

fig. 6 An executable schedule of operations

4. EXPERIMENTAL RESULTS

The SCS shell is at present being developed in FIAR's laboratories. Some modules (Spacecraft Model Editor, Time Dependent Reasoner, Inference Engine, part of the FOP Editor and of the Problem Solving Knowledge Editor) have already been implemented and demonstrated their usefulness in developing Essope, an expert system commissioned by ESA/ESOC (the Operations Center of ESA) for Olympus spacecraft operations planning and execution monitoring. This Expert System, connected on line with a Multi Satellite Support System and with an Olympus simulator, has been delivered to ESOC together with a part of the SCS used to develop it. The SCS shell is written in C and OPS83 programming languages, utilizes the OpenWindows graphical toolkit and runs on a Sun workstation.

5. CONCLUSIONS

Aim of this paper is to demonstrate the effectiveness of vertical approach to Expert Systems shell development in spacecraft control domain. Besides Essope, other applications in planning/scheduling of operations and in Flight Operations Plan construction areas are at present under development via the SCS shell and confirm the appropriateness of our approach. Next work to be done is to extend the capabilities of the shell in order to cope with diagnosis and contingency recovery problems which resulted of fundamental importance in Spacecraft control domain.

REFERENCES

[ALL 84] J.F.Allen 1984, Towards a general theory of Action and Time, Artificial Intelligence 23, 2.

[BUC 88] B.Buchanan & al 1988, Design of Knowledge-Based Systems with a Knowledge-Based Assistant, IEEE Trans. on Soft. Engineering, Vol. 14, No. 12.

[CHA 85] D.Chapman 1985, Nonlinear planning: a rigorous construction, Proc. IJCAI-9, pp 1022-1024

[CLA 83] W.J.Clancey 1983, The Advantages of Abstract Control Knowledge in Expert System Design, Proc. Third Nat. Conf. Artificial Intelligence, pp. 74-98

[DRU 89] M.Drummond 1989, AI Planning: A Tutorial and Review, Artificial Intelligence Applications Institute technical report AIAI-TR-30

[GRE 86] C.Green & al 1986, Report on a Knowledge-based Software Assistant, in Artificial Intelligence and Software Engineering, C.Rich and R.C.Waters Eds. Morgan Kaufmann 1986, ch. 23, pp. 377-428.

[GUS 90] S.Gusmeroli & al 1990, ESSOPE: Expert System for Spacecraft Operations Planning and Execution, 1st Int. Symposium on Ground Data Systems for Spacecraft Control, Darmstadt FRG

[PEZ 90] A.Pezzinga & al 1990, The shell concept to ease the development of knowledge-based systems for spacecraft control, 1st Int. Symposium on Ground Data Systems for Spacecraft Control, Darmstadt FRG

[RIT 86] J.F.Rit 1986, Propagating temporal constraints for scheduling, Proc. 5th Nat. Conf. on A.I. AAAI-86, Philadelphia USA, pp. 383-388.

[TAU 89] A.Taunton 1989, Tools for the Trade, Systems International, April 1989, pp. 25-29.

Coupling Expert Systems on Control Engineering Software

H. Hyötyniemi

Helsinki University of Technology, Control Engineering Laboratory, Otakaari 5 A, SF-02150 Espoo, Finland

E-mail: hhyotyniemi@sorvi.hut.fi

Abstract. Computer aided control systems design is an example of those real world application areas that offer many challenges to the artificial intelligence community. The basic approaches in these two disciplines are different: combining the worlds of qualitative and heuristic artificial intelligence with the quantitative and analytical control engineering, is not straightforward. This *coupling* has become an area of growing interest. Current trends and problems that are faced in coupled systems are discussed in this paper. Automation of the redundant phases of design is discussed as the next target of future design systems. It is concluded that some rethinking is needed to facilitate the fundamental enhancements in design environments. Developing an intermediate level interface between the mathematical programs and the conceptual AI tools is proposed as a first step towards the solution of the deep coupling problem. As an application example, guidelines for further development of an existing integrated control engineering CAD environment are presented.

1 Introduction

The advances in computer hardware and software have made it possible to solve traditional design problems using new and ambitious methods. Artificial intelligence is becoming a real tool in various fields of modern engineering. Being based on a combination of diverse more or less heuristic design procedures and a toolbox of numerous mathematical algorithms, control systems engineering is among the most challenging fields of applied AI. This paper will concentrate on design of control systems using a combination of traditional tools and AI-like approaches. The field of artificial intelligence is wide, and in this context AI is not only identified with the best known and commercially most successfull application areas, knowledge engineering and expert systems. Also some other AI-like ways of thinking may turn out to be fruitful in the near future.

In computer-aided control engineering (CACE) there is a large number of design methods, each of the methods having its own special features and optimal application areas. No systems designer can be acquainted with all of these paradigms and be aware of the possible deficiencies and special properties of all methods. The need of tools for mastering all this variety is evident, and naturally AI has been the magic buzzword. In computer-aided control systems design (CACSD) expert systems have been applied to help in the design process for quite a long time.

There are many challenging fields for applying AI methodologies within control engineering, but at the moment the emphasis is only on the controller or compensator design, supposing that a mathematical model has already been achieved somehow. This kind of a restriction makes the discussion more concise—it minimises the need to deal with some of the fussy problems that are always present in AI work, like those caused by the *common sense* presuppositions, etc. The tools proposed would be useful also on other branches of systems theory or control engineering.

In what follows, problems and possibilities of applying AI methods in CACSD systems are discussed, and the basic principles of the forthcoming project are outlined. The project starting is to some extent a preliminary study, and propositions presented here are still somewhat abstract.

2 Control systems design

Some aspects of control engineering that will probably be specially emphasized in future research work are briefly summarised, and problems with the state-of-the-art existing systems are pointed out. The field of computer-aided engineering is in turmoil and methodologies are changing fast now when applications of artificial intelligence are becoming commonplace.

2.1 CAD system trends

Control engineering has been an active application area of mathematical programming, and there are now plenty of program packages available for various steps of control systems analysis and design. Modern CAD/CAE tools for control systems design are surveyed in [3], and in [4] MacFarlane and Ackermann discuss their views of the future developments of control engineering CAD systems.

The trends within CAD in general are outlined by Ohsuga [14]. The main topic he discusses is intelligence being added to forthcoming systems in various forms. The point is that the design system has to be adaptable and extendable in a flexible way, and it must be able to use knowledge dynamically to achieve the user's goals. The object representation and manipulation are the key issues in CAD. Ohsuga concludes that it is necessary first to develop new efficient general-purpose information processing systems: the field of computer programming is not yet mature, either.

Denham [2] discusses the concept of a *CAD environment* for control systems consisting of a specialised design language, a design knowledge base, and a range of analysis and synthesis tools, together with advanced interface mechanisms. He emphasizes the need of the design language providing application oriented data structures, extensibility of the language, and reasonable efficiency. He also points out that user control and interaction must not be overlooked, and human factors should be taken into account when designing the user interface.

The role of AI in control systems CAD in general is described by Lamont and Schiller [11]; A fascinating description of the role of the computer science in control engineering is given by Wonham [24].

Almost all researchers emphasise the need for interactive design environments and sophisticated high-level user interfaces. However, very often concentrating on the outermost level only results in fancy user interfaces that are easy-to-use—and perhaps easy-to-sell. This pragmatical approach only hides the fundamental problems there exist: problems of software integration, and those caused by the variety of information and knowledge representations and transformations, to name only some. Making minor polishing on the products, never attacking the underlying problems, means bulky products and need for excessive amount of development work needed for each effort of enhancement. The truth is that as far as there are no commonly recognised standards—and most things can never be standardised—developing user interfaces actually always starts from scratch.

2.2 Characteristics of control systems design

Elaborating on CACSD, first we have to have some intuitive picture of that field—what is the current status, what kind of applications are commonplace, and what is needed in the future? Roughly, there are three ways of attacking the design problem, depending of the size of the plant:

- Small systems—simplified processes, usually for educational purposes. Flexible flow of control in the design system is needed, it has to allow experiments and 'what if' analyses
- Medium-sized design problems—for instance, compensator design for well-defined linear multivariable systems, where the design is based on well-known, iterative procedures. This kind of somewhat simplified design tasks are common in practice and of great practical value
- Large-scale plants—real life industrial processes. The design procedures developed for problems specific to individual plants may have no use when solving design problems for other processes. Heuristic design methods are used to avoid the explosion of complexity, and the expertise of the designer is essential.

The first two items offer a field of feasible research, and the benefits can be significant. Having contemporary machinery of computer hardware, the automation of large-scale plant design is not available. However, a homogeneous framework of basic utilities for systems research is valuable also in this case.

In Figure (1), outlook of a basic design process is given. Design consists of many tasks of first stating the targets, then applying the design methods and finally assessing the results. In control engineering design, these tasks could be as follows:

1. Defining, or later, modifying, specifications for the plant properties
2. Choosing an appropriate design method for the plant
3. Applying the design procedure, or finding a compensator
4. Simulating the behavior of the compensated system and assessing the result.

By definition, design is always a trial-and-error process, where successive steps often have to be repeated. Fast prototyping and prototype analysis is the basic cycle in the design process. A beginner who cannot really spot the most essential features of the structures, usually has to do a great deal of extra work, while an experienced designer can minimize the amount of iteration. However, why should it be the human that is always consulted in the iterative loop? During a complex design task, the user has to evaluate the same things over and over again. Essentially, all information needed for decisions is embedded in the data structures and given specifications. Automating the internal steps, or hiding superfluous iterative loops of the design process from the user, will be the focal point in the research project.

Put another way, in this study, the target of applying AI is to *facilitate automation in design*.

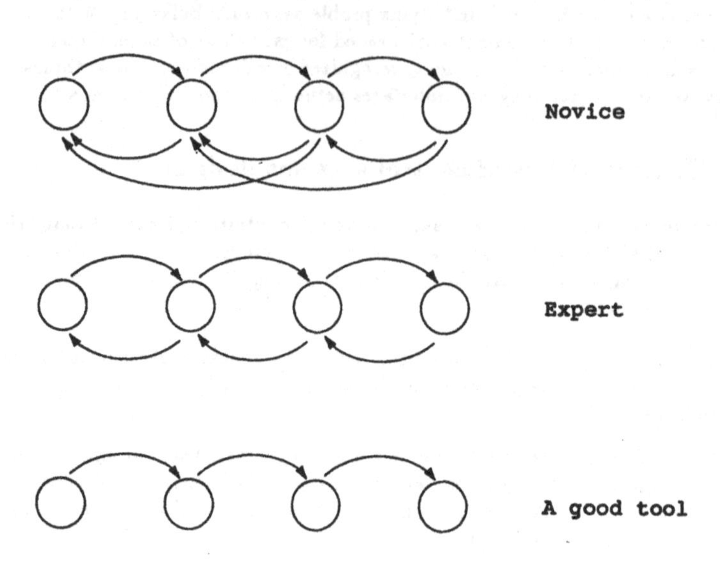

Figure 1. The process of design becomes more and more streamlined with experience, but with good tools it might even become straightforward

2.3 Programming paradigms

Object-oriented programming paradigm is becoming popular, and especially in systems modelling this new way of thinking is very welcome. Using object-oriented formalisms it is easier to model real life systems consisting of various subsystems with differing properties hierarchically, and large-scale model updating is easier in the objects world than in traditional procedural programming environments. The data structures themselves contain knowledge about the reasonable operations that are applicable to them, and changing these operations is much more straightforward. Object-oriented databases also promise to relieve the data abstraction problem that becomes acute when using traditional rigid database structures for information storing.

Advances in computer capacity has, of course, resulted in growing interest in new algorithms and design methods. However computationally expensive the general solutions based on symbolic calculation may be, in many special cases the new packages for doing symbolic manipulation of mathematical expressions promise to facilitate very interesting new ways of design: often the solution space can be parameterized using the adjustable design parameters. On the other hand, methods based purely on number crunching or simulation are becoming more and more commonplace. These methods usually offer little concise analytical information, but huge amounts of raw data. So, the ways of data representation are becoming scattered—extremely structural and extremely structureless to be handled in the same design environment.

When building a complex system it is reasonable to organise it hierarchically, so that the system can be defined, analysed, and constructed in a modular way, on different levels of abstraction. One of the main issues is a flexible and transparent interface between the conceptual level and the routines written in lower level programming languages.

Beforehand, it is not possible to define or express the personal needs of the designer exactly. The syntax of the intermediate level formalism must be easily modifiable to match optimally the design problem at hand. The needs of design system users are varying and changing, and there cannot be any closed general purpose systems for all of those inherently individual design tasks. By offering tools for flexibly modifying the design environment, the user can reach the reasonable outlook of the data access. The 'design system designer' becomes nearer to the end-user, the control system designer—a good design environment can be tailored hierarchically by the end-user, and the personal needs and desires of the user can be taken into account.

Because of the various program packages available for control systems analysis, it is reasonable to utilise ready-to-use building blocks in future design environments. However, the communication between programs and surrounding world is far from standardized, and there has to be some kind of *drivers* for adapting the inputs and outputs. Problems of combining several independently developed design tools in computer-aided control design system are discussed by Spang in [20]. He calls this kind of an integrated environment a *federated* system. His approach is based on some basic design tools for modeling, analysis, and simulation that together span the major part of the design process.

3 AI in control engineering

There is a difference between the worlds of traditional engineering and artificial intelligence. The school of engineering work relies on analytical solutions and numerical algorithms, while the artificial intelligence community uses heuristics and symbolical concepts. When AI has been applied to CACE, the chosen approaches have also been predictable and productive—this has traditionally meant mainly expert systems. However, to preserve its status as a refreshing paradigm, AI should not lose its ideals and touch of cognitive phenomena.

3.1 Expert systems in CACSD

Expert systems have been used in the field of control systems design for quite a long time, and there are numerous examples of expert system projects. Strategies of applying expert systems in control engineering are illustrated and research in that field is surveyed by Rychener [18]. In [16], Pang describes a systematic way of applying an expert system in multivariable feedback controller design. Usually, not very much criticism is targeted on the very narrow scope of the existing design support systems. Expert systems tend to be very specialized and application oriented—however, in [22], Taylor outlines his view of a general expert system architecture for CACE, and in [21] he discusses different expert system approaches.

Interest on expert systems has traditionally been focused mainly on the surface level, coding the knowledge using some of the many commercially available expert system building tools. Seen on this surface level, the problems of applying expert systemsthat are encountered are very similar in different application fields. These problems are usually connected to knowledge acquisition and are not considered in this paper [9]. Neither are the *embedded expert systems* interesting when aiming at general purpose tools and they are not elaborated on—those systems are tailor-made and application dependent, and extending their scope may be very difficult.

An AI example from another field, MYCIN, the expert system for making medical diagnoses, has been reported often to outwit its human competitors in practice—that is partly because it always extracts all information of the patient it can get [5]. A human

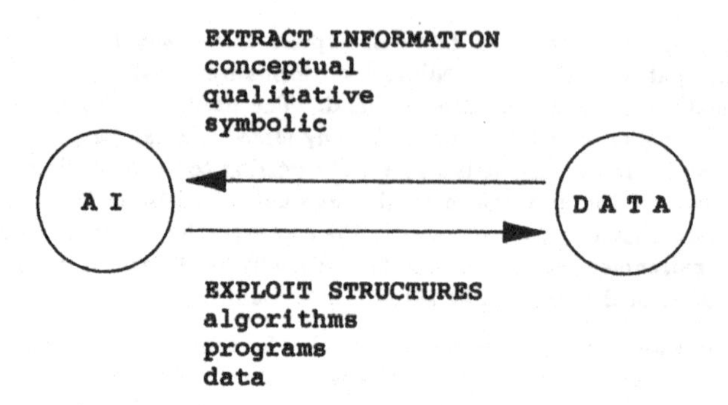

Figure 2. 'Deep coupling' of data structures and artificial intelligence tools as the key to connecting AI to real world

physician often discards the 'less probable' hypotheses, and sometimes makes a false decision based on the insufficient data. In control engineering design the situation is just the same: a more efficient design procedure or a more sophisticated controller structure can be chosen if more is known about the process. For MYCIN, however, the facts once given remain stable, but in a design environment each operation may change the facts. This makes the problem of maintaining consistency with a reasonable amount of user interaction much more difficult in this case. Because of the numerous aspects to be judged during the design, a tool for finding the information automatically is invaluable to reach optimal results. The human designer would soon be bored during the iterative design and analysis cycles if he were the only link between the process and the expert system to be consulted.

It has been said that a good design tool makes the *most probable* outcome of the design process near to the *optimal* one. So, the lessons learned should be taken seriously when planning future CACSD systems, the MYCIN experience being a good example.

3.2 Coupled expert systems

The view of expert systems as monolithic, independent black boxes is becoming oldfashioned as AI is being applied to real life environments. *Coupled expert systems* must, by definition, have some knowledge of the numerical processes embedded in them and be able to reason about the results of those underlying numerical processes [10]. The coupling is necessary when trying to assist the designer using complex numerical algorithms and programs, or when insight into the problem solving processes is needed to obtain the solution or to interpret the results.

Traditionally, control engineering tools offer a great deal of accuracy, not insight to the underlying procedures or data structures. Integrating formal mathematical methods with methods based on conceptual knowledge might be the key for making problem solving environments more robust, and better in handling problems involving ambiguous, contradictory, or imprecise data. In [7], a survey of the current status of coupled systems is given—it is concluded that the research work is scattered, and some common basis of methodology and terminology would now be needed.

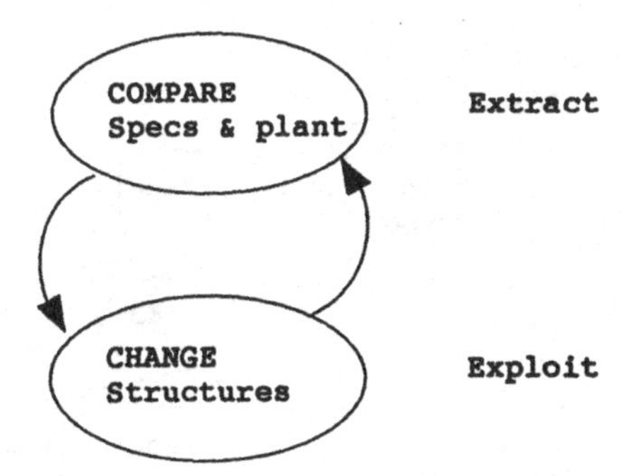

Figure 3. What deep coupling means in control systems design environments: successive steps of synthesis and analysis

It is useful to make the distinction between *shallow* and *deep* coupled systems. Deep coupling means ability of extensively utilising knowledge of the processes involved, and ability of exploiting the data structures (Fig. 2). Current expert systems that use mathematical models tend to treat the modelling component as a black box, in a shallow way, not being able to find out information about the underlying structures, and using production rules to fill in the numerical parameters required by the model. Some advantages of the deeper approach are summarised in [23]:

- **Domain independence** is achieved by building a separate knowledge-based system to serve as a modelling kernel; mapping modules define the interface to each domain. This makes the system easier to extend and maintain.
- **Robustness,** model structure flexibility, allows one to find the most appropriate model structure to match the situation at hand. The overgenerality of rigid templates planned for the worst-case problem, and the oversimplicity of models fitting the requirements of an algorithm can be avoided.
- **Explainability** of the expert system reasoning process can be enhanced.

Attempts to apply the coupled expert system paradigm to control engineering problems are surveyed in [8]. In control systems design environments, deep coupling means ability of comparing specifications and the existing plant, and ability of modifying data structures using predefined procedures (Fig. 3). Contemporary coupled systems usually seem to be only intelligent front ends between the user and the system. The steps of synthesis and analysis should be repeated automatically. As they put it in [25], several knowledge structures and computational skills new within AI programming paradigm are needed, when developing an interface with the deeper type process connection.

3.3 Combining AI and CAD

In the field of control engineering, the dichotomy between artificial intelligence and practical design is easy to spot: there are dozens of programs to do the number crunching for CACSD applications, and there are dozens of expert systems to help the designer in

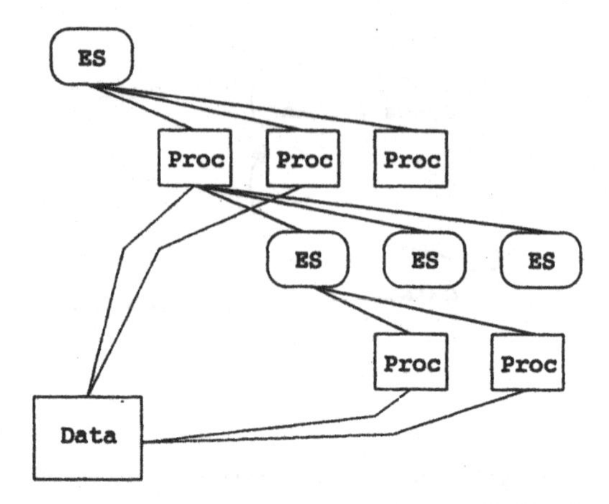

Figure 4. The ultimate goal: various hierarchical knowledge bases for different subtasks, and procedural data processing in between

choosing the right ways of attacking the facing problems. Even in the up-to-date, ambitious realisations combining these worlds do not really meet: it is always the *designer himself* who has to act as the server between the various parts of the design environment, extracting the needed properties of the plant and inputting them into the expert system. This cannot be the architecture in the long run. However, there are differing opinions—for example, Pang and MacFarlane, who have developed expert systems for multivariable feedback design [15], claim that the computer can only be used as an assistant expecting continuous user interference during the design.

In design environments of the future, there should be flexible interaction and data transfer between the different ways of modelling, numerical or conceptual, whichever is the most appropriate way of thinking of any particular aspect of the design. Figure 4 shows the structure of a future design environment, with many independent knowledge based subsystems for various steps in the design, separated by blocks of procedural control for data manipulation.

Summarizing, the AI tools should be implanted *between* the designer and the process, interpreting the commands and the results, and minimizing the unnecessary or redundant information flow through the hands of the human design process operator—relieving the burden of mechanical data processing (Fig. 5). Flexible communication between the worlds of data and AI would facilitate realizing independent and automatic loops of data processing and analysis. The human can never be substituted from the design system, but giving his judgements just once should be enough.

3.4 Special topics

There are three main aspects about applied AI that need special emphasis in this context. First of these is the *deterministic coupling*, or mapping between well-defined concepts and corresponding data structures. These conversions are standard operations and can be essentially transparent, because there is no need of human interference. The two other aspects are *pattern recognition* and *multiobjective optimization*.

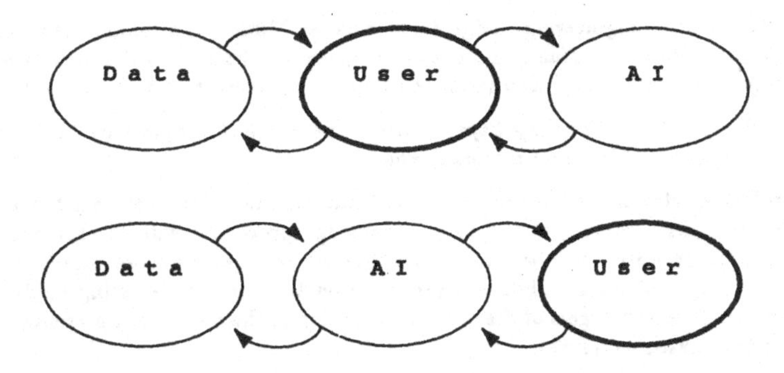

Figure 5. The role of the human in today's design systems and in those of tomorrow

Most control engineering design procedures are essentially planned for traditional design methods. In a design process, there usually exist so many adjustable design parameters that some kind of compression of information has to be used to uncover the features of interest in an optimal way. This optimal way may be very complicated if the computer is used only as a sophisticated calculator and the human is to make all the reasoning and decisions: human pattern recognition capability is superior compared to his analytical skills. The result is that most often design is based on figures and plots, for example. Some design methods rely heavily on these pattern recognition capabilities. Nyquist-like methods, for example, are in this class, while some methods, like LQG design, are somewhat more oriented on exact numeric measures and less dependent of heuristic valuation decisions. Of course, we cannot expect design methods to change, and a good environment for implementing these methods has to contain some approach to attack this feature extraction problem.

Having only some mathematical expressions to tackle with should be of orders of magnitude easier problem, compared to most other pattern recognition tasks. At least in some cases, features like stability margins and robustness properties could be expressed numerically and calculated explicitly. But even if some formulae for extracting features could be found, that is not the way how a human accomplishes the task, and experimenting with more ambitious approaches would be needed: When is the result 'good looking' and when it is not?

Further, finding a good compromise between many possibly contradictory optimality criteria can be difficult. The 'good looking' process response must not be oscillating, on the other hand, but must not be too slow either. Offering the user flexible tools for affecting the design quality assessing process presupposes ability of utilising some multi-objective optimization methods, like the *analytic hierarchy process*, AHP [19]. The AHP method can be used to weigh contradictory aspects two at a time, assessing their relative importance, an automatic procedure taking care of the combining of the comparations.

4 Framework for a research project

ISEE, the integrated systems engineering environment [13], is a control engineering design system developed for systems analysis, modelling, simulation, and controller design tasks. The ISEE approach is to integrate a collection of some independently developed and established control engineering design tools, such as Matlab and Simnon [17], under a uniform user interface. This basic control engineering design environment is already operational and can be used as a testbench for further research and experiments.

The **CES** project, or *Coupling Expert Systems on Control Engineering Software*, will focus on the aspects that were discussed above.

The project started in the beginning of 1991 and its first phase will last ten months. During this time a picture of the most promising approaches will be sketched and a simple prototype system will be constructed to study and test the proposed ideas. Later, further development work will then concentrate on these most interesting applications. The most emphasised targets of the project, according to the discussions presented above, can be paraphrased as follows:

1. First, define and implement an intermediate level formalism for making access of data structures and procedures easy and transparent [6]
2. Next, using that infrastructure, study the possibilities of deep coupling, ultimately aiming at automating the iterative control systems design process.

According to discussions with experts, reaching the Step 1 would already be a major achievement, facilitating unpredecented design methods. Attacking the problems of the Step 2 would be frontier research.

Acknowledgement

The ongoing project is financed by The Academy of Finland, and this support is gratefully acknowledged.

References

[1] Aho, A. V., Sethi, R., and Ullman, J. D.: *Compilers: Principles, Techniques, and Tools*. Addison-Wesley, Reading, Massachusetts, 1986.

[2] Denham, M. J.: *Design Issues for CACSD Systems*. Proceedings of the IEEE, vol. 72, no. 12, pp. 1714–1723, 1984.

[3] Downtain, D. C.: *Survey of CAD/CAE Tools for Control System Design*. Proceedings of the 1988 American Control Conference, Atlanta, Georgia, pp. 737–742.

[4] MacFarlane, A. G. J., Gruebel, G., and Ackermann, J.: *Future Design Environments for Control Engineering*. Proceedings of the 10th IFAC World Congress on Automatic Control, Munich, FRG, 1987, pp. 235–246.

[5] Harmon, P. and King, D.: *Expert Systems: Artificial Intelligence in Business*. John Wiley & Sons, New York, 1985.

[6] Hyötyniemi, H.: *Tools for Coupling AI and Control Systems CAD*. Proceedings of the Second International Workshop on Computer Aided Systems Theory (EUROCAST'91), Krems, Austria, 1991.

[7] Jacobstein, N., Kitsmiller, C. T., and Kowalik, J. S.: *Integrating Symbolic and Numeric Methods in Knowledge-Based Systems: Current Status, Future Prospects, Driving Events*. Coupling Symbolic and Numerical Computing in Expert Systems, II (eds. Kowalik, J. S. and Kitsmiller, C.T.), North-Holland, Amsterdam, 1988, pp. 3–11.

[8] James, J.R.: *Lessons Learned in Coordinating Symbolic and Numeric Computing in Knowledge-Based Systems for Control Design.* Coupling Symbolic and Numerical Computing in Expert Systems, II (eds. Kowalik, J. S. and Kitsmiller, C.T.), North-Holland, Amsterdam, 1988, pp. 207–220.

[9] Johannsen, G. and Alty, J. L.: *Knowledge Engineering for Industrial Expert Systems.* Automatica, Vol. 27, No. 1, pp. 97–114, 1991.

[10] Kitsmiller, C. T. and Kowalik, J. S.: *Symbolic and Numerical Computing in Knowledge-Based Systems.* Coupling Symbolic and Numerical Computing in Expert Systems (ed. Kowalik, J. S.), Elsevier Science Publishers, Amsterdam, 1986, pp. 3–17.

[11] Lamont, G. B. and Schiller, M. W.: *The Role of Artificial Intelligence in Computer-Aided Design of Control Systems.* Proceedings of the 26th Conference on Decision and Control, Los Angeles, California, 1987, pp. 1960–1965.

[12] Maciejowski, J.M.: *Multivariable Feedback Design.* Addison-Wesley, Wokingham, England, 1989.

[13] Marttinen, A. and Telkkä, T.: *A Hierarchical Process Modelling Environment.* Proceedings of the 11th IFAC World Conference, Tallinn, Estonia, 1990, Vol. 10, pp. 73–78.

[14] Ohsuga, S.: *Towards Intelligent CAD Systems.* Computer-Aided Design, vol. 21, no. 5, 1989, pp. 315–337.

[15] Pang, G. K. H. and MacFarlane, A. G. J.: *An Expert Systems Approach to Computer-Aided Design of Multivariable Systems.* Springer-Verlag, Berlin, 1987.

[16] Pang, G. K. H.: *An Expert System for CAD of Multivariable Control Systems Using a Systematic Design Approach.* Proceedings of the 1987 American Control Conference, Minneapolis, Minnesota, pp. 555–560.

[17] Rimvall, M.: *Interactive Environments for CACSD Software.* Preprints of the 4th IFAC Symposium on Computer Aided Design in Control Systems CADCS'88, Beijing, P.R.China, 1988.

[18] Rychener, M. D.: *Research in Expert Systems for Engineering Design.* Expert Systems for Engineering Design (ed. Rychener, M. D.), Academic Press, Boston, 1988, pp. 1–33.

[19] Saaty, T. L.: *The Analytic Hierarchy Process.* McGraw-Hill, New York, 1980.

[20] Spang, H. A.: *The Federated Computer-Aided Control Design System.* Proceedings of the IEEE, vol. 72, no. 12, 1984, pp. 1724–1731.

[21] Taylor, J. H.: *Expert-Aided Environments for CAE of Control Systems.* Preprints of the 4th IFAC Symposium on Computer Aided Design in Control Systems CADCS'88, Beijing, P.R.China, 1988.

[22] Taylor, J. H. and Frederick, D. K.: *An Expert System Architecture for Computer-Aided Control Engineering.* Proceedings of the IEEE, vol. 72, no. 12, 1984, pp. 1795–1805.

[23] Wellman, M. P.: *Reasoning About Assumptions Underlying Mathematical Models.* Coupling Symbolic and Numerical Computing in Expert Systems (ed. Kowalik, J. S.), Elsevier Science Publishers, Amsterdam, 1986, pp. 21–35.

[24] Wonham, W. M.: *Some Remarks on Control and Computer Science.* IEEE Control Systems Magazine, April 1987, pp. 9–10.

[25] Żytkow, J. M. and Żytkow, A. N.: *Numerical and Symbolic Computing in the Applications of Law-Like Knowledge.* Coupling Symbolic and Numerical Computing in Expert Systems (ed. Kowalik, J. S.), Elsevier Science Publishers, Amsterdam, 1986, pp. 147–159.

Real-Time Drought Control of Storage Reservoir by Combining Middle and Long-Term Weather Forecast and Fuzzy Inference

S. Ikebuchi (*), T. Kojiri (**)
(*) Water Resources Research Center, Disaster Prevention Research Institute of Kyoto University, Gokasho, Uji, Kyoto, Japan
(**) Department of Civil Engineering, University of Gifu, Japan

Abstract

In this paper, we will propose the long-term and real-time reservoir operation systems considering the middle and long-term weather forecast provided by the Meteorological Agency. Firstly, after expressing the forecast weather conditions with some numerical values, the amounts of weekly and monthly precipitation are predicted. Secondly, formulating the drought control structures, the control rules of release and storage volume are formulated by using fuzzy inference theory. Then the fuzzy membership functions for storage volume are changed depending on the previous operational performance and the prediction accuracy through adaptive fuzzy inference. Finally the release from reservoir is decided through the fuzzy control theory to consider the uncertainty of prediction and operation.

INTRODUCTION

Recently, we have been suffering from the serious drought situation because of the complexity of weather analysis and the rapid increase of water demand in the metropolitan area. Moreover, the water resources system has not found the effective solution for drought because the accuracy of long-term hyetograph and hydrograph prediction were too low.

Traditionally, two main approaches have been taken to the real-time drought control; namely (i) deterministic optimization estimating the future hydrograph with statistical methods (Takasao et al., 1982) and (ii) stochastic optimization extrapolating the hydrograph feature with prediction accuracy or residuals (Ikebuchi et al., 1981). In the former case, the operational result may not be optimized because the predicted hydrograph includes the uncertainty and can be called as the expected operation. In the latter case, as the uncertainty increases according to the predicted periods, the result depends on the maximum point of predicted residuals. However, from the view point of real-time operation, the hydrograph is succesively updated at prediction stage and the operational strategy is modified after getting the prediction of residuals and drought level by Kalman filter. So, we want to establish the long-term reservoir operation systems introducing the adaptive fuzzy inference to consider the above specific drought situation. Moreover, the information of middle and long-term weather forecast provided by Meteorological Agency should be handled to evaluate the hydrograph accuracy and the desirable level of membership functions in fuzzy set. The proposed systems will be applied into the real river basin for verfication of theories. In this paper the operationg time stage is five-days period.

PREDICTION OF LONG-TERM PRECIPITATION AND DISCHARGE

Middle and long-term weather forecast

From middle and long-term weather forecast in Japan, the weekly. monthly and three month informations are provided. The weekly forecast is announced at every day on weather condition, mean temperature with three levels and total amount of precipitation. The monthly forecast is done at the end of every month as weather condition with monthly and ten-days, mean temperature with three levels and total amount of ten-days precipitation. Three months forecast is done at every 20th on weather condition, monthly mean temperature with three levels and total amount of three months precipitation in future. Weather trends for next six months are announced at the beginning of hot and cold seasons. Table 1 shows the numerical values of the mean temperature and precipitation to the announced logical levels in these forecast information.

Table 1 Classification of middle and long-term weather forecast

Factor	Unit	Small	Normal	Big
Deviation of Temperature	month	−0.6	−0.5~0.5	0.5~
	ten days	−0.9	−0.8~0.8	0.9~
	five days	−1.1	−1.0~1.0	1.1~
Ratio of precipitation	month	~69	70~119	120~
	ten days	~39	40~139	140~
	five days	~19	20~119	120~
Occurence probability		30%	40%	30%

Prediction of long-term precipitation

The long-term discharge is predicted by using information of weather forecast as shown in Fig. 1. In the prediction system (RCONV I), probability of weekly precipitation is gained as mean value among precipitation rate (PPL) according to the weekly forecast level. Assuming that weather conditions is represented as numerical value such as 0=clear, 1=cloudy, 2=rainy and 3=stormy, weather condition for next seven days is expressed as (PW1, PW2, PW3, PW4, PW5, PW6, PW7) with the numerical number. The expected five-days precipitation is calculated as the averaged weekly precipitation RP multiplied by PPL. As the representative precipitation RRP(n) is gained as

$$RRP(n) = (RP \cdot PPL) \left(\sum_{n=1}^{5} PWn\right) / \left(\sum_{n=1}^{7} PWn\right), \tag{1}$$

and as the averaged precipitation AP(n) at time stage n(n-th five-days) is given, the precipitation ratio is formulated as follows;

$$PRF(n) = RRP(n) / AP(n). \tag{2}$$

The precipitation ratio is converted into logical symbols of big, normal or small. Then, in the second prediction system (RCONV II), the probability of monthly precipitation DF(n) is calculated as the mean value for precipitation rates PPM according to the monthly forecast level. The precipitation sequence for this month is predicted multiplying averaged precipitation at each five days by precipitation probability DF(n). At these five days, prediction is taken by using the calculated probability in RCONV I. Total amount of this predicted precipitation is converted into logical symbols, too. In the third prediction system RCONV III, the trend index of precipitation after this month is estimated through long-term forecast. Assumingg the forecast value "small" is -1, "normal" is 0 and "big" is 1, the weighting values of next month, the month after next month and the month after next two months are set as 1.0, 0.8 and 0.6, respectively. The announcement at last month is considered as the forecast data from the first five days to fourth five days. The information from the fifth five and sixth five days are used with the updated index announced at current month. Therefore, the summation of forecast values multiplied by weighting factors at each month is called as the trend index for long-term.

Prediction of long-term discharge

The minimum, mean and maximum precipitation sequences until the end of i-th month are predicted at the beginning of each time stage by the above three conversion systems. And the minimum, mean and maximum inflow discharge sequences are predicted with those precipitation sequences and the linear regression equation from precipitation to inflow discharge.

PROCEDURES OF REAL-TIME OPERATION

Hierarchical structure of drought control

Though the drought damage has been evaluated through the function of total amount of (shortage of water)^2/(target discharge), the real drought control is consisting of several control levels considering the water demand and basic maintenance discharge for river environment as follows;

(i) level 0 : normal release
$QO(n) > DD+MD$ and $RDD=DD$
(ii) level 1 : warning for drought
$QO(n) > DD+MD_{min}$ and $RDD=DD$
(iii) level 2 : first restriction and 10% reduction
$QO(n) > 0.9DD+MD_{min}$ and $RDD=0.9DD$
(iv) level 3 : second restriction and 20% reduction
$QO(n) > 0.8DD+MD_{min}$ and $RDD=0.8DD$
(v) level 4 : third restriction and 30% reduction
$QO(n) > 0.7DD+MD_{min}$ and $RDD=0.7DD$
(vi) level 5 : emergency and special restriction
$QO(n) > 0.0$, $RDD < 0.7DD$

where, QO(n) is the release including the water demand RDD and basic maintenance discharge for river environment. All levels are hierarchically linked with each other. On the other hand, storage volume at every control stage is expressed as logical symbols such as (i) enough volume (level 0), (ii) slightly small volume (level 1), (iii) small volume (level 2), (iv) rather small volume (level 3), (v) very small volume (level 4), and (v) almost empty volume (level 5). Fig. 1 shows also the control procedure, where OPT is the subsystem of release discharge and INF is the inference subsystem of drought control level. In order to introduce the uncertainty of prediction and anxiety of operator, the desirable drought level is identified through the adaptive fuzzy inference and the reasonable release is decided to prepare the coming drought. This system is operated iteratively for the year round except the flood control.

Caluculation of sequences of release and storage volume

When the drought level is 0, the release is taken as the normal discharge of the water demand and maintenance discharge. However, from the drought level 1 to 4, the release is optimized minimizing the drought damage. The restricted water demand RDD(n) is formulated as follows;

$$RDD(n) = (1.0 - (DL(n) - 1.0)/10.0)\,DD(n) \tag{3}$$

where, DL(n) denotes the drought level at time stage n. The basic maintenance discharge RMD(n) is optimized to minimize the summation of deviations of the storage volume and the basic maintenance discharge as follows;

$$J(n) = \frac{(S_0(n+1) - S(n+1))^2}{S_0(n+1)} + \frac{(MD(n) - RMD(n))^2}{MD(n)} T(n) \tag{4}$$

where, S_0 means the target storage volume at time stage n+1 and T(n) is the control periods in time stage, respectively. The continuous equation in reservoir is represented as

$$S(n+1) = S(n) + ((QI(n) - RDD(n) - RMD(n))T(n) \tag{5}$$

where, QI(n) is the averaged inflow at time stage n. On the condition that S(n)+((QI(n)-RDD(n)- RMD(n)) T(n) is greater than zero, J(n) is minimized with the following value of RMD(n).

$$RMD(D) = \frac{S(n) + ((QI(n) - RDD(n) - RMD(n))T(n)}{S_0(n+1) + MD(n)T(n)} MD(n) \tag{6}$$

If the ratio on storage volume in the right side of above equation becomes greater than one, RMD(n) is set as MD(n) from the drought condition that the exceed release is permitted in the case of predicting the overstorage in future. In the case of drought level 5, RMD(n) is equal to the minimum maintenance discharge as

$$RMD(n) = MD_{min}(n). \tag{7}$$

Moreover, the optimal release for water demand at the beginning of time stage is gained as follows;

$$RDD(n) = \frac{S(n) + ((QI(n) - RDD(n) - RMD(n))T(n)}{S_0(n+1) + 0.7DD(n)T(n)} 0.7DD(n) \tag{8}$$

If the ratio on storage volume in the right side of above equation (8) becomes greater than one, RMD(n) is set as 0.7DD(n). Adding the actual storage constraints such as the feasible storage volume and feasible release capacity, the sequences of release and storage volume until considered control periods are calculated successively.

Fuzzy set of the desirable drought level

Decision variables of storage volume are defined as following normalized values and logical value PSL;

$$PSR = S(n)/S_0(n) \qquad (0<PSR<1) \tag{9}$$
$$DSR = S(n)/S_0(n) - S(n-1)/S_0(n-1) \qquad (-1<DSR<1) \tag{10}$$
$$PSL = SL(n) \qquad (0<PSL<4) \tag{11}$$

Furthermore, the control rule is consisting of three cases comparing the control results and storage volume; namely the first case is that the current storage volume and discharge are close to the preferable situation. The second one is that the next drought control should be changed into more severe level because the current control result is not fully performed enough. The third one is that the next drought control should be taken as looser level because of enough storage volume. The storage and release sequences are calculated through the predicted inflow sequences to the minimum, mean and maximum sequences as follows;

$$(QO_{min}(n), QO_{min}(n+1), \ldots, QO_{min}(n+k))$$
$$(QO_{ave}(n), QO_{ave}(n+1), \ldots, QO_{ave}(n+k))$$
$$(QO_{max}(n), QO_{max}(n+1), \ldots, QO_{max}(n+k))$$
$$(S_{min}(n), S_{min}(n+1), \ldots, S_{min}(n+k))$$
$$(S_{ave}(n), S_{ave}(n+1), \ldots, S_{ave}(n+k))$$
$$(S_{max}(n), S_{max}(n+1), \ldots, S_{max}(n+k))$$

These storage volumes are handled as the antecedent conditions to be compared with above control rules. So, the wishes of changing the desirable drought control in these seven five days is represented as the numerical number DDLP(n) ($-1 < DDLP(n) < 1$). Concretely,
(i) When $DDLP(n) < -0.5$, the restriction is relaxed to one level until the end of current month.
(ii) When $DDLP(n) > 0.5$, the restriction is reinforced to one level until the end of current month.
(iii) When $0 < DDLP(n) < 0.5$, the restriction will be relaxed to one level at the changing stage BAk as follows;

$$BAk = 7 - [DDLP(n)\,(7-j)/0.5] \tag{12}$$

where, [ ] denotes the Gaussian integer function and j is the current time stage. If BAk is equal to one, the previous control level is kept until the end of current month.
(iv) When $-0.5 < DDLP(n) < 0$, the restriction will be reinforced to one level at the changing stage BBk except the current five days because of its reliability as follows;

$$BBk = 7 - [DDLP(n)\,(6-j)/0.5] \tag{13}$$

As the reservoir operator must be anxious about the predicted discharge and wants to take the safer control, the weighting factors for discharge sequences are given as w_{min}, w_{ave} and w_{max}, where the ratio of w_{min}, w_{ave}, and w_{max} is a:b:0 (a+b=1). Then the inflow sequence is assumed and the release sequence after next five days is calculated subjected to DDLP(n).

Fuzzy inference approach to storage volume and release
The storage volume space PRS and the difference storage ratio DSR are divided into six and three fuzzy spaces such as (i) Not S :not small storage volume, (ii) SS : slightly small storage volume, (iii) S : small storage volume, (iv) PS : pretty small storage volume, (v) VS : very small storage volume, (v) ES : extremely small storage volume, and (vi) Normal : normal storage volume,

and (i) N : negative difference, (ii) ZO : zero difference and (iii) P : positive difference. Table 2 shows the control rules of reservoir. Fig. 2 shows the membership functions in antecedent and consequent. p means the ratio of minimum release under the restriction of 30% to the target release as follows;

$$p=100.0((0.7DD(n)+MD_{min})T(n))/S_0(n) \quad (14)$$
$$p1=(100.0-p)/5.0 \quad (15)$$

Table 2 Control rules on reservoir operation

No.	Control rules
R_1	if PSR is Not S, then PSL is 0
R_2	if PSR is SS and DRS is P, then PSL is 0
R_3	if PSR is SS and DSR is ZO, then PSL is 1
R_4	if PSR is SS and DSR is N, then PSL is 2
R_5	if PSR is S and DSR is P, then PSL is 1
R_6	if PSR is S and DSR is ZO, then PSL is 2
R_7	if PSR is S and DSR is N, then PSL is 3
R_8	if PSR is PS and DSR is P, then PSL is 2
R_9	if PSR is PS and DSR is ZO, then PSL is 3
R_{10}	if PSR is PS and DSR is N, then PSL is 4
R_{11}	if PSR is VS and DSR is P, then PSL is 3
R_{12}	if PSR is VS and DSR is ZO, then PSL is 4
R_{13}	if PSR is VS and DSR is N, then PSL is 5
R_{14}	if PSR is ES, then PSL is 5
R_{15}	if PSR is Normal and DSR is N, then PSL is 1

Decision of desirable drought level

The desirable drought level DDLP(n) in consequent is also divided into seven fuzzy subspaces such as (i) NB : negative big, (ii) NM : negative medium, (iii) NS : negative small, (iv) ZO : zero or pretty small, (v) PS : positive small, (vi) PM : positive medium and (vii) PB : positive big. Defining the antecedent variable as the difference x1 between PSL and DL(n), the common control rule is consisting seven fuzzy subspaces such as (i) NB : negative big, (ii) NM : negative medium, (iii) NS : negative small, (iv) ZO : zero or pretty small, (v) PS : positive small, (vi) PM : positive medium and (vii) PB : positive big. In addition to the variable x1, other six variables to the control levels are defined, where x2 is the averaged storage ratio at the end of current month (AVSR), x3 is the minimum storage ratio at the end of current month (MISR), x4 is the maximum storage ratio at the end of current month (MASR), x5 is the trend parameter of future precipitation (FPTP), x6 is the averaged critical storage parameter in current month (AVCP) and x7 is the minimum critical storage parameter in current month. Fig. 3 shows the membership functions of x2 to x7. The parameter p' is modified through the adaptive fuzzy concept as follows;

$$p' = 100.0((0.7DD(6.0i+1.0)+MD_{min})T(6.0i+1.0))/S_0(6.0i+1.0) \quad (16)$$
$$p2 = (100.0-p')/M \quad (17)$$

Where, M is the control time stage with five days. The membership function of SB (the grade of smallness is big) and Not S in x2 to x4 are defined comparing the feature of SM (the grade of smallness is medium) to cover the wide fuzzy spaces of control levels. As the discharge sequence is predicted with fuzzy inference to consider the continuously less precipitation, the antecedent variables are defined as PSL and the trend index of future precipitation with long-term forecast (FPTP) and the consequent variables is defined as the minimum discharge ratio (RMIN). Fig. 4 shows these membership functions and p3 is calculated as follows;

$$p3 = 0.2(DL(n-1)+DDLP)/5.0+1.0 \quad (18)$$

By using gravity center method of fuzzy inference theory, the discharge is inferred and the release is gained through the decision process of storage and release.

APPLICATION OF METHODOLOGIES TO REAL RESERVOIR

Information and initial condition of reservoir operation

The storage capacity of applied reservoir is represented in Table 3. The total operational periods is 73 five days from April 1st to March 31st. Discharge is estimated by using following regression equation with the precipitation r and identified parameters shown in Table 4.

$$QI(n) = a1\ QI(n-1) + a2\ r(n-1) + a3\ r(n) \quad (19)$$

The upper part in Fig. 5 shows the averaged discharge with bar graphs, the design release sequence with the solid line and the target release sequence with dotted line under the condition of averaged precipitation. The lower part in same figure shows the target storage sequence with bold solid line, the discriminative storage capacities with narrow solid line for two control seasons and the storage sequence against standard drought event with broken line, which is defined under the return period of ten years on annual precipitation. As the initial hydrological conditions at the beginning of April, storage volume S(1) is 7.36 $\times 10^6$ m^3 ($=S_0(1)$) and at the previous time stage or the end of March, the precipitation r(0) is 21.9 mm, the discharge QI(0) is 13.3 m^3/s, the release level is normal (DL(0)=0), storage volume $S_0(0)$ is 6.17 $\times 10^6$ m^3 ($=S_0(0)$), respectively. The damage functions of water demand, the maintenance discharge and storage volume are defined as follows;

$$DDAM(n) = (DD(n)-RDD(n))^2\ T(n)\ /(DD(n)\ 86400) \quad (20)$$
$$MDAM(n) = (MD(n)-RMD(n))^2\ T(n)\ /(MD(n)\ 86400) \quad (21)$$
$$SDAM(n) = (S_0(n)-S(n))^2 T(n)\ 10^6/(S_0(n)\ 86400) \quad (22)$$

Table 3 Information on storage capacity

Available storage capacity	50.0 $\times 10_6$ m^3
Storage capacity for flood	30.0
Storage capacity for water demand	20.0
Municipal water demand	5.0 m^3/s from Oct. to June
	6.0 from July to Sept.
Irrigation demand	5.0 from the middle of
May	
	to the end of Sept.
Maintenance discharge	
Target discharge	6.0
Minimum discharge	4.0

Table 4 Coefficients of regression equation

Month	$QI(n-1)$ a_1	$r(n-1)$ a_2	$r(n)$ a_3
4	0.5	0.01	0.3
5	0.5	0.01	0.4
6	0.5	0.01	0.4
7	0.4	0.01	0.3
8	0.6	0.01	0.4
9	0.7	0.01	0.4
10	0.5	0.01	0.5
11	0.5	0.01	0.5
12	0.7	0.01	0.3
1	0.9	0.01	0.1
2	0.9	0.01	0.1
3	0.6	0.01	0.3

Control result in the standard hydrograph

The weather forecast in this application is given as the normal precipitation level. To investigate the proposed operation effect, the controlled results will be compared with other three operational strategies : namely (i) no restricted release I where the target release is taken only when reservoir has storage volume, (ii) no restricted release II though the release exceeding the current storage volume is prohibited, (iii) restricted release where release over the water demand is prohibited. Among these release strategies, the water demand is calculated as follows;

(i) When the target release is not satisfied and the maximum feasible release ABLD is greater than the minimum release in the drought control level ii,

$$RDD(n)=0.9DD \text{ and } RMD(n)=ABLD-RDD(n). \quad (23)$$

(ii) When ABLD is less than $0.9DD+MD_{min}$ and greater than the minimum release in the drought control level iii,

$$RDD(n)=0.8DD \text{ and } RMD(n)=ABLD-RDD(n). \quad (24)$$

(iii) When ABLD is less than $0.8DD+MD_{min}$ and grater than the minimun release in the drought control level iv,

$$RDD=0.7DD \text{ and } RMD(n)=ABLD-RDD(n). \quad (25)$$

(iv) When ABLD is in the drought control level v and the maximum feasible storage is greater than the summation of water demand and maintenance discharge in the restriction percentage 50%,

$$RDD(n)=0.5DD \text{ and } RMD(n)=ABLD-RDD(n). \quad (26)$$

(v) When ABLD is equal to or greater than MD_{min} and less than the volume of case (iv),

$$RMD(n)=MD_{min} \text{ and } RDD(n)=ABLD-RMD(n). \quad (27)$$

(vi) When ALBD is less than the maintenance discharge,

$$RMD(n)=RDD(n)=ABLD/2.0 . \quad (28)$$

Fig. 6 and 7 show release and storage sequences and drought damages, respectively. In May, from July to August, and major periods in winter season, the drought control level i is taken though the users of water does not suffer from damage. Moreover, there are not big differences between target storage and controlled storage volumes. In the case of no restricted release I, around at time stage 25, the storage volume goes down and in May it becomes almost empty. On the other hand, no restricted release II, the first level of drought control is suddenly taken in February and the second level is taken at time stage 67, though it is operated on the basis of saving the water. The restricted release gives the steady control after time stage 26 when the first level of drought control is started.

From the view point of accumulated damage, the amount damages of proposed and the restricted strategies are less than the damages of no restricted release, because the latter do not consider the forecast information. Thus, the storage volume is consumed according to the target release.

Control result in the representative drought event

In the case of 1978 drought as shown in Fig. 8, the storage volume is decreasing in April and May, the storage volume becomes minimum at September and the controlled storage volume is different from the target volume from January to March. The monthly weather forecast at September announced the big precipitation, though the actual one was normal. So, the system predicts to get the enough discharge until the end of this month. From the fuzzy inference process, considering the difference between the previous predicted strategy and actual storage level, the restriction was judged to be reinforced. However the controlled sequences denote the fine results to be compared with the known weather and discharge patterns since the fuzzy inference can handle the uncertainty of forecast, prediction and decision making.

CONCLUSIONS

In this paper, the drought control which has become the serious problem in the area with high population and scarce water resources was discussed to be reduced its damage and to perform the reservoir effects considering the weather forecast. To sum up. the following results are gained:

(i) Application of weather forecast : To handle the uncertainty of precipitation in future, the weather forecast was used with fuzzy set according to the forecast fidelity.

(ii) Formulation of operational rule of release : Rule base of release is formulated considering the actual decision making for drought.

(iii) Introduction of fuzzy inference process to get the release : The final decision of release is inferred through fuzzy control theory. To evaluate the uncertainty of prediction and anxiety of operator, the desirable drought levels and the membership functions are calculated with adaptive fuzzy inference process.

REFERENCES

Ikebuchi S., Takasao T. and Kojiri T., REAL-TIME OPERATION OF RESERVOIR SYSTEMS INCLUDING FLOOD, LOW FLOW AND TURBIDITY CONTROLS, Experience in Operation, Water Resources Publications, 1981, pp. 25-46.

Takasao T., Ikebuchi S. and Kojiri T., On-line Real Time Operation of Dam Reservoir Systems Including Low Flow and Turbidity Controls, Proc. of the 32nd Japanese Conference on Hydraulics, JSCE, 1988, pp. 379-385.

Sugeno M., Fuzzy control, Nikkan Kogyo Newspaper CO., LTD., 1988, pp. 67-109.

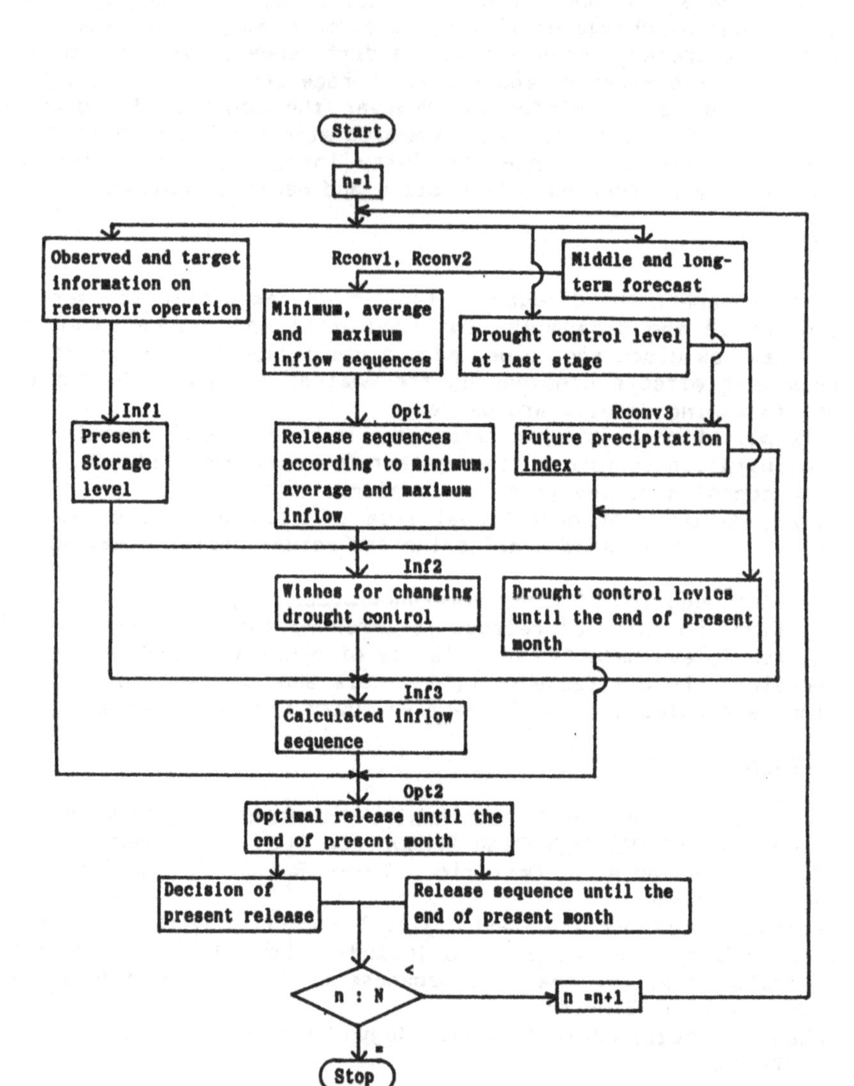

Fig. 1 Procedure of real-time operation

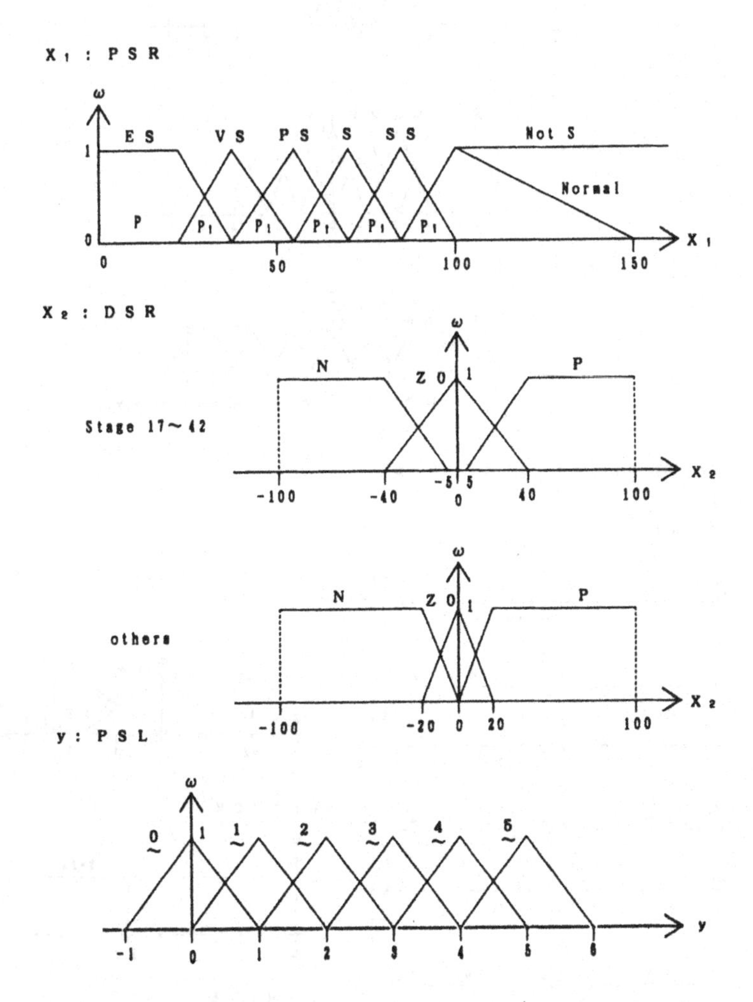

Fig. 2 Membership functions of storage ratio and release

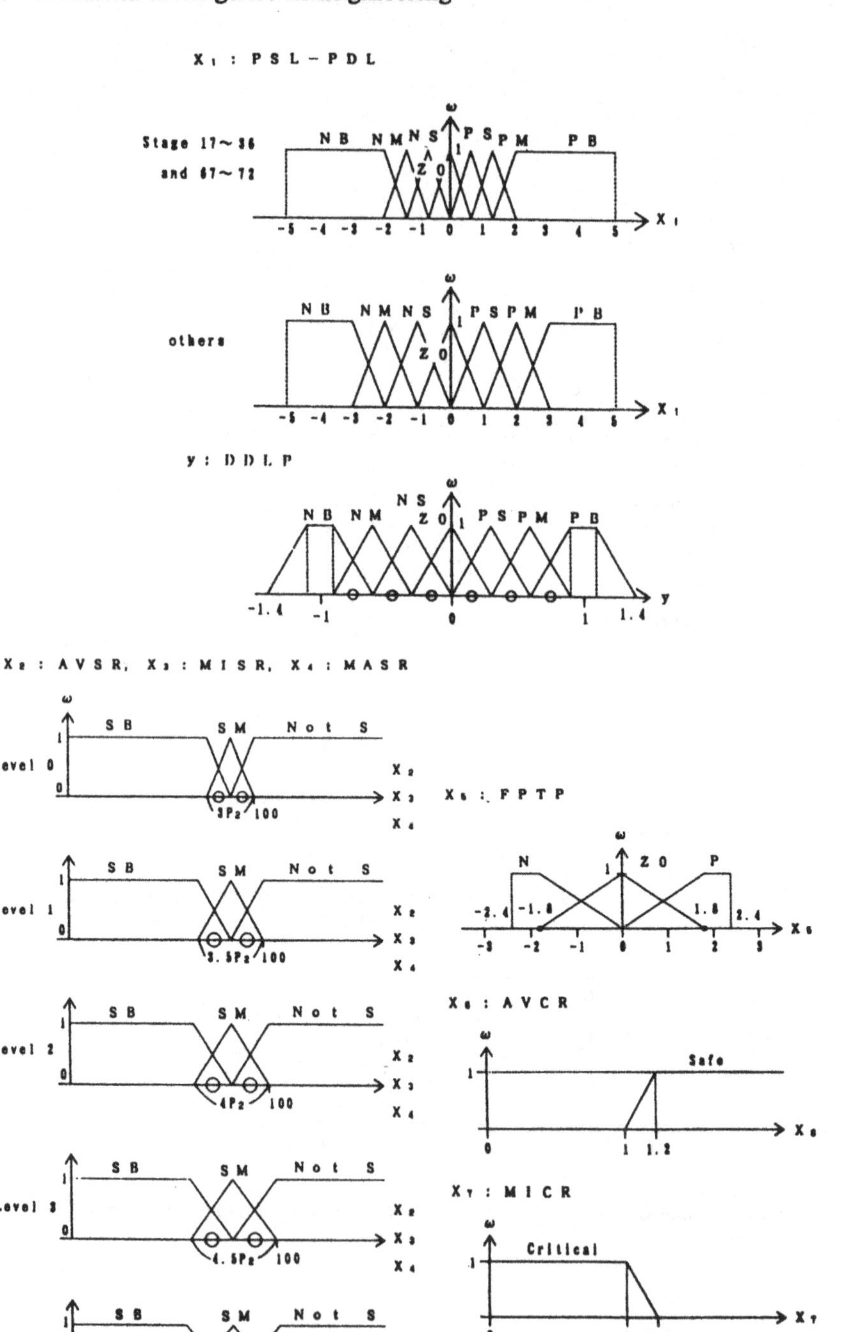

Fig. 3 Membership functions of desirable drought level

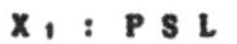

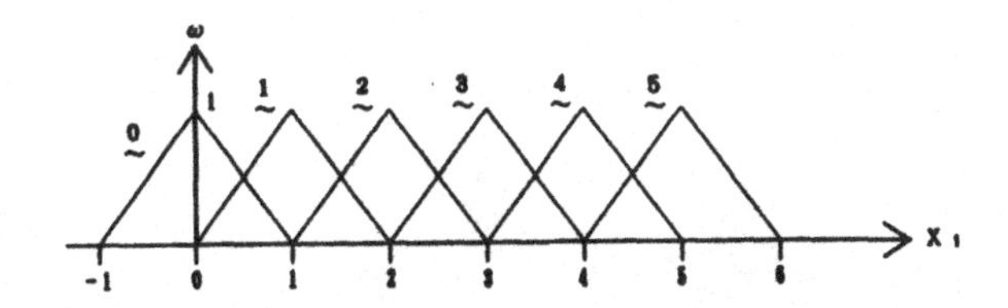

X2 : FPTP

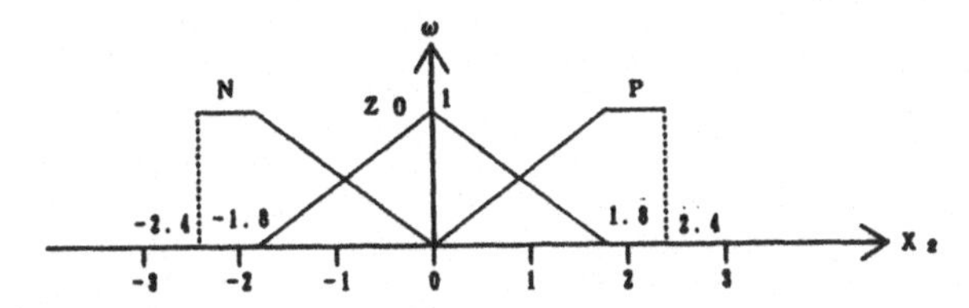

y : RMIN

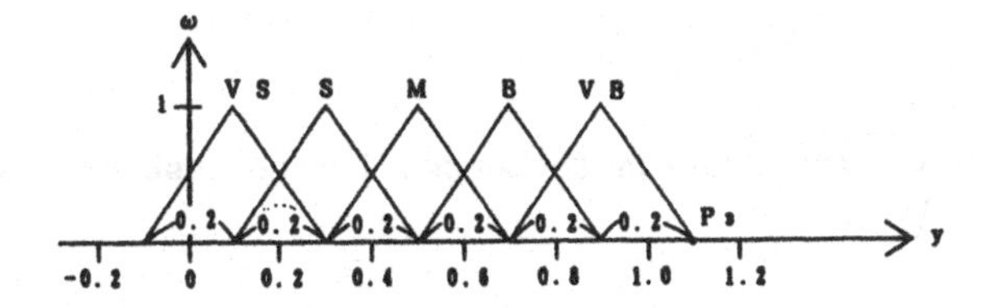

Fig. 4 Membership functions of precipitation and trend index

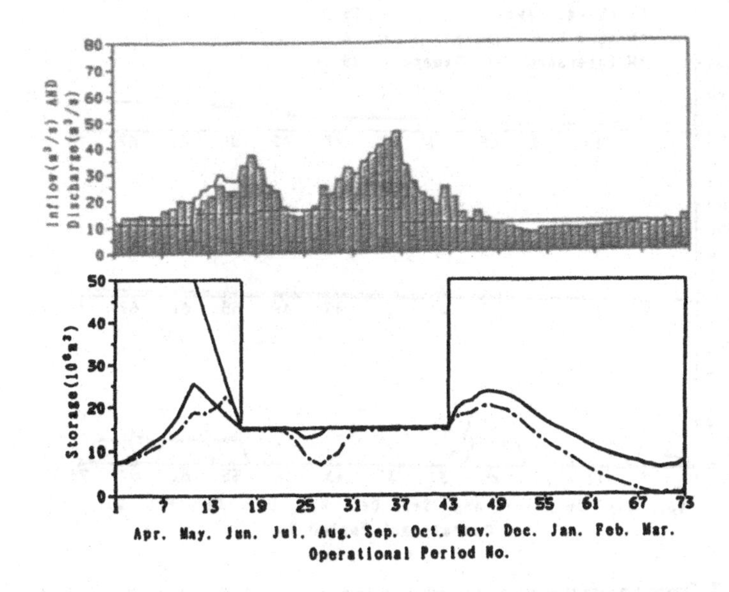

Fig. 5 Sequences of averaged discharge, design release and target release

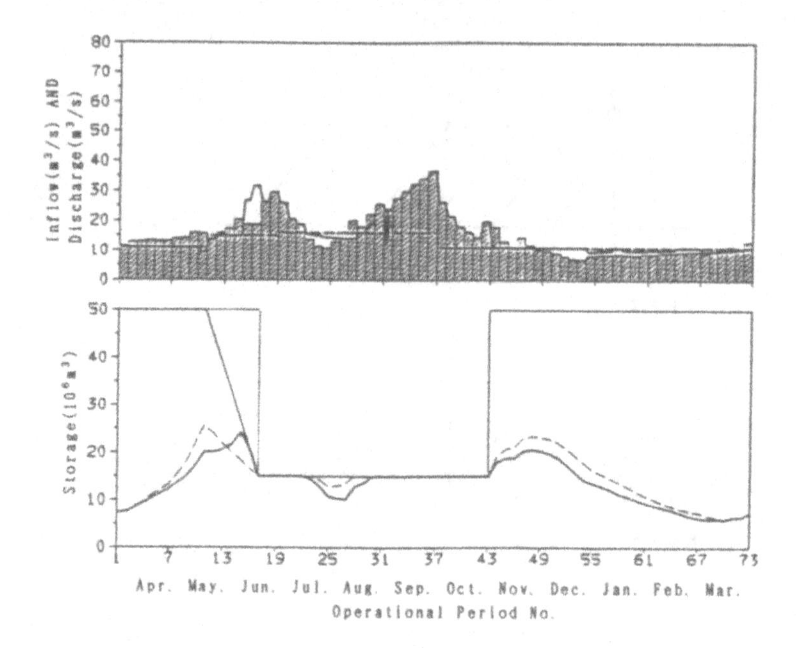

Fig. 6 Release and storage sequences in the case of standard hydrograph

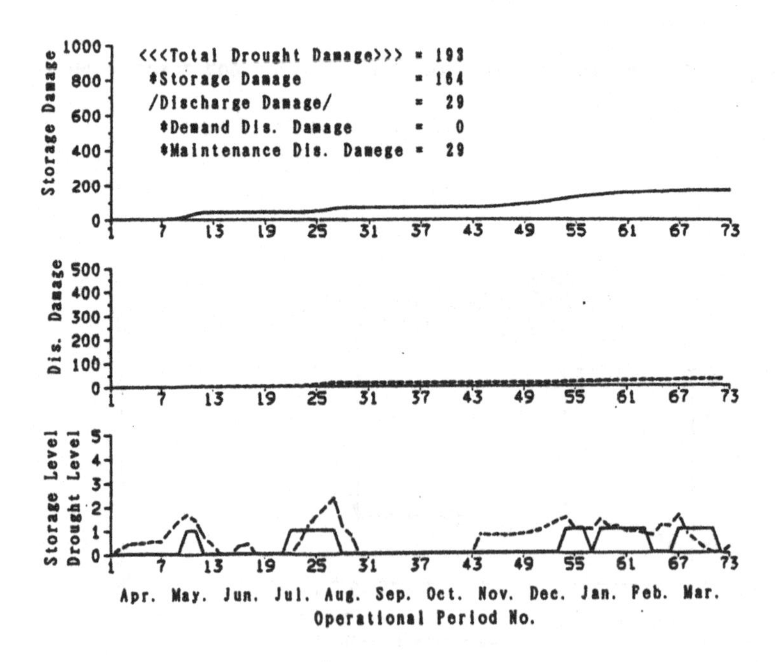

Fig. 7 Damage sequence in the case of standard hydrograph

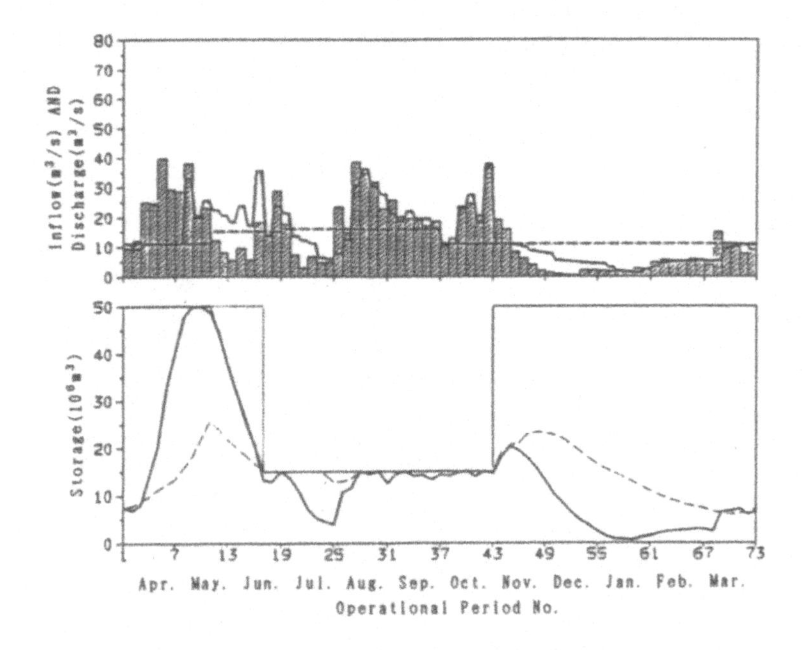

Fig. 8 Release sequence in the case of representative drought event

Vehicles Controlling: Representation of Knowledge and Algorithms of Multi-Agent Decision

P. Mourou, B. Fade

Institut de Recherche en Informatique de Toulouse, Université Paul Sabatier, 118 route de Narbonne, 31062 Toulouse Cedex, France

ABSTRACT

This paper deals with the problem of vehicle control in highway traffic. The vehicles are equipped with co-pilots which aid the human driver in two ways: either they drive the vehicle, in an automatic way, according to the instructions given by the driver or they act, in a supervising way, by warning the human pilot or generating avoidance trajectories when an accident is imminent. The vehicles are considered as agents in a multi-agent world and exchange information by means of a communication network. The use of co-pilots should increase security and make traffic easier. This paper describes the knowledge representation of an agent, the time management, and the structure of action plans generated by the planner. The notion of influence between two agents will lead to the planning method, first by supposition, then by verification: in the first phase we generate a plan supposing the other agents' behaviour, and in the second phase we adapt our behaviour according to what the other agents send. We present two planning algorithms used to generate a behaviour, and compare them.

INTRODUCTION

Controlling vehicles moving in highway traffic is a challenge for researchers in artificial intelligence. It extends to a non-structured world the difficulties of controlling robots in a factory, which is still a very open subject for researchers. It needs intelligent functions, adapted to perception, communication and action. The functions of perception will not be studied here. We actually suppose that our system is constantly supplied with information about the environment in which our vehicle is moving. Our main interest is in fact the communication between the vehicles and the generation of the actions which must be executed. Thus, we suppose that each vehicle is linked with another vehicle by a short distance communication network. The data and plan representation along with their use in action planning is studied in the multi-agent approach. This approach allows us to think about a control distributed among different agents, each agent intending to solve the conflicts between it and the agents it meets. However, the implementation of a centralized system is, in fact, very difficult because of the great number of agents and the complex work which must be done.

Plan generation in a multi-agent world is far from the classic problems of plan generation in which a goal is resolve into sub-goals. We must solve conflicts between agents in order to solve multi-agent problems. The conflicts appear when the common resources are shared (machines, space...) and can only be solved by a mutual knowledge and an exchange of arguments. Logical representations, based on the model of the human way of thinking, were used by Bessiere [1] and Wilks and Ballim [2] in order to model desires and beliefs, both of which represent mutual knowledge and are very useful to solve conflicts. Persuasion, e.g. Rosenschein [3] or Sycara [4], is the aim of the arguments exchange. The communication network allows the agents to exchange their intentions for the future and also their persuasive arguments. Some works, e.g. Morgenstern [5], which takes into account incomplete knowledge, will be fully used in our project when we will start the study of the second part of the development, namely by taking into account vehicles with no co-pilot. The only information we have on such vehicles is taken from the sensors of all the equipped vehicles surrounding it. These agents' intentions are unknown. We intend to make the co-pilot of our vehicle generate the hypothetical plan of the non-equipped vehicle to associate intentions with this agent as Konolige and Pollack [6] do. This is however not the subject of the present paper.

Planning our vehicle's behaviour is closely related with the allocation of space. To make this space division possible without using a traditional resources management, we propose a method: a code of good manners allows a rapid and sure space allocation, without any backtracking.

Then, we analyse the agents' influence between themselves and the suppositions done by an agent when generating a plan. Besides, we compare two general algorithms producing an action according to several agents' influences. The notion of selective backtracking will be pointed out and developed by using two algorithms.

DATA STRUCTURE

Wood [7] was interested in representing a multi-agent world to create a simulator for the testing of the automatic generation of vehicles' itineraries. Our approach is similar, and leads us to choose the data representing any multi-agent world (see [8] for more information).

General description

The data structure allows us to construct a model of a multi-agent world in which agents, similar or not, can meet in a common environment. This data structure is composed of a set of data bases, one for the environment, one representing the agent himself (named "us"), and a data base for any other agent. The data base of each agent intends to express its static characteristics as well as its history, actual state, and plan of action for the future.

Classification of variables

We can classify different variables in distinct categories; for example, a few variables which could be fully used in the representation in highway traffic of a given vehicle B are:

Length of vehicle B (1)
Speed of B (2)
Relative position of B (3)
What vehicle does B overtake? (4)

We can classify (1) in the category of a STATIC variable because it is unchangeable and as MONO-AGENT since it characterizes only one agent.

Variable (2) is DYNAMIC and MONO-AGENT.

Variable (3) depends on two vehicles. Moreover, in this case, one of the two vehicles is ours. So we say that (3) is INTER-AGENT. It is always DYNAMIC.

Variable (4) gives the identity of the vehicle overtaken by B. It is also DYNAMIC and we say that it is INTER-CO-AGENT because it links two agents different from us.

We notice that there is a difference between INTER-AGENT and INTER-CO-AGENT for the INTER-CO-AGENT variables are sometimes redundant when the INTER-AGENT variables are known. This happens when we have transitive data. For example: we only need to know the relative distances between us and B (INTER-AGENT) and between us and C (INTER-AGENT) to conclude by transitivity the relative distance existing from B to C. So, for transitive data, the variable can be INTER-AGENT. On the other hand, when we need the values of the variable existing between any two agents, we will classify it in the different INTER-CO-AGENT variables. However, this variable can take a value relative to us ("what vehicle does it overtake?" can take the value "us").

Furthermore, we notice that we have neither a static INTER-AGENT variable, nor an INTER-CO-AGENT one. In fact, the relations between agents are dynamic in general. However, we can find another example: in highway traffic, the variable "relative position" could be STATIC, and either INTER-AGENT or INTER-CO-AGENT, if we intend to take into account the towed vehicles.

Here is the summary of the different categories of the variables fully used to represent a multi-agent world:

Table 1. Categories of variables

	MONO-AGENT	INTER-AGENT	INTER-CO-AGENT
STATIC	•	•	•
DYNAMIC	•	•	•

<u>The distribution of information</u>
That is how the different variables classes are distributed on the different data bases.

The MONO-AGENT information is located on each agent (for example: the speed of a vehicle is in the data base of this vehicle) (cf Figure 1).

The INTER-AGENT information is memorised in the data base of the concerned agent (for example: the relative position of agent B is in B's data base).

The INTER-CO-AGENT information is memorised in one of the two data bases of the two agents.

There is no INTER-AGENT or INTER-CO-AGENT data in the data base of the agent called "us". It is true that a value such as our relative position is useless.

<u>Time representation</u>
Time must be taken into account as soon as we are interested in a multi-agent world, which is very dynamic in its essence. A planner must constantly verify that the resources which may be used by the agent will be free at the execution time. For example, in highway traffic, the agent must verify that the needed space will be free and if it is not, it must plan

actions which will allow it to go into other spaces. A policy of space representation and management must be settled down.

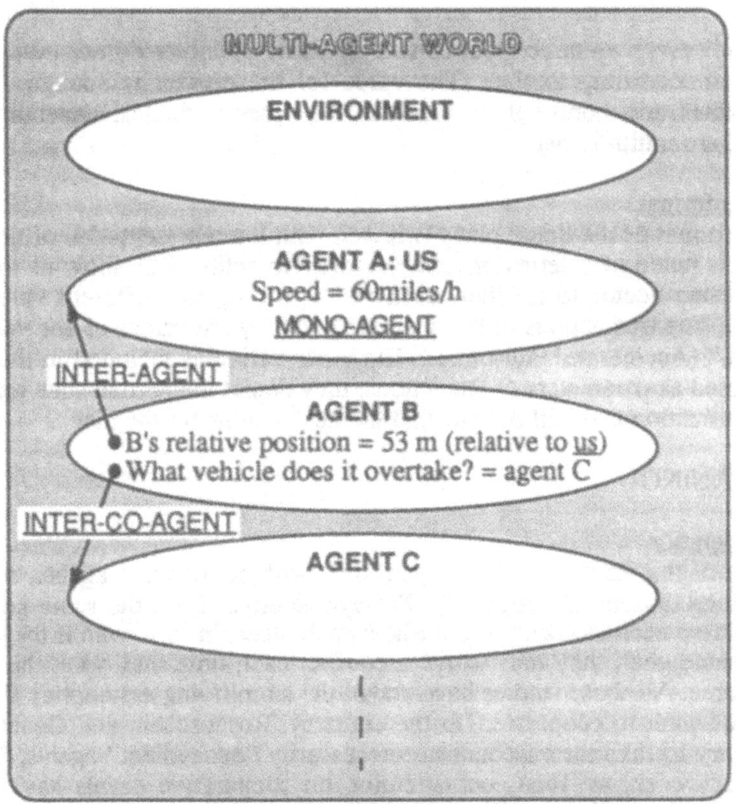

Figure 1. Data Structure of a Multi-Agent World

Data bases give a "photograph" of the state of the world at the present instant. This photograph is represented by all the values of the data base variables, it is composed of the history of the world (the past and the present which no longer change), and of suppositions about the future (supposed or expected plan), which still can change. In our case the plans are linear, each dynamic variable has a list of values connected with time.

A few values of the past, the value of the present and the values expected for the future, will form the variable value. In our representation, the list of values of a variable is connected with a list of temporal values in the time variable which is developed according to the system needs. The value for the T-instant of a variable can be found at the rank of the T-instant in the "Time" variable. For example, the value for 23-instant of Variable-X, which is represented by:

Time =(21 22 23 24 ... m)
Variable-X =(val21 val22 val23 val24 ... valm)

is val23, the value of rank 3 in Variable-X, rank of 23 in variable Time.

However, we notice that certain dynamic variables do not need to be known according to time. The value for the present is enough. These variables are mono-valued whereas the other variables, depending on time, are multi-valued.

Plans format

The format of the linear plans is linked with the representation of time. A plan is made of a series of state vectors. An action will allow us to pass from one vector to another, each vector giving the different values to reach. The true actions of this plan are given by the values of one variable "Plan" (Accelerate, Pull-out...). The other variables included in the plan are used as parameters of the actions, they depict the normal state reached by this action and will help to find the new actions for the plan.

PLANNING

Cooperation

Most of the studies on multi-agent deal with cooperative agents, see for instance Cammarata et al. [9]. The agents either have the same goal, or they have each its own goal, but help each other. In fact, even if they have the same goal, they must solve a conflict each time they want the same resource. All those studies have added the simplifying assumption that all agents want to cooperate. On the contrary, Rosenschein and Genesereth [10] try to take into account not-necessarily "benevolent" agents. In the present work, we focus our attention on cooperative agents having the same main goal: a traffic with highest efficiency and unprecedented safety. Each agent has nevertheless its own goal: it must try to maintain the speed wished by the driver. The agents communicate by a network, and conflicts are resolved by a common code of good manners. This code is the kernel of the strategy of cooperation of Cammarata et al. [9]. Here, this code allows us to resolve a conflict between two agents. In highway traffic, in most cases, an agent's decision can only depend on the past data of another agent. In other cases, when it needs forthcoming data, it will lead to a dialogue, until the two agents agree. In that case, the code of good manners does not prevent an exchange of agreements. The capacity of an agent in influencing the others' behaviour is necessary, so that an agent asserts its rights and negotiates a decision.

Allocation of space

The allocation of space is essential in highway traffic. In fact, the system intends to appreciate the geometrical situations of different agents, which change in time. It tries to project these geometrical situations in the future. To connect geometrical areas with different agents, we can use an automatic computing technique based on a division of space into areas, which are used as resources. However, this method does not fit highway traffic because the agents have their own changeable speed. The area appointments are done at different rhythms, thus complicating the management of these areas. The use of areas whose lengths are proportional to the agents' speed cannot solve anything. The length of these areas shall be reduced to appointed areas of different lengths made up of a large number of elementary areas. The management of those micro-areas is slow. Another difficulty in the use of such a method is that a conflict between two agents will only be seen at the last minute, when an area is asked at the same time by two agents; thus we must go backward until a choice can be changed. This technique, called backtrack, is well known and used in every planning system, but is rather expensive.

A code of good manners, the highway code and some definite safety rules, helps us to generate trajectories without backtracking. The rules of conduct ("No change lanes when a vehicle is on the left", "If the safety distance behind a vehicle is reached then pull out or slow down"...) help to find certainly good trajectories when all the parameters of choice are known. For example, the trajectory of an agent A around an agent B will be generated straightforward if it can be determined by only using the past of A and B. It would not always be the case with the automatic computing technique.

Influence of other plans

In our case, the agents constantly exchange the plans they have built (by the communication network). To determine the main element of the planning operation, we must wonder: which part of B's plan will agent A use in order to plan its own behaviour, i.e. what will be B's influence on agent A. To generate its action for the time T_i, A may use B's past, i.e. it uses every action of B during all the instants before T_i: from T_0 to T_i (cf Figure 2). It may need also to use B's choices for T_i, taking into account what B and A will do at the same time, to respect a constraint like "If it does this, I cannot do that". But an important constraint like "If it does this before three steps, I cannot do that" can be respected if A takes into account what B will do in the future (from T_{i+1} to T_{i+3} in this case). The choice of the influence of an agent's plan on other agents' plans has to be done by taking into account the needs of the context. In the highway traffic example we need actions for time T_i from the other agents to deal correctly with examples such as: three vehicles are following in order, the last two vehicles want to pass the first one, the action of the last one "I am going to pass" prevents the second from passing since the last vehicle has the highest priority in the highway code.

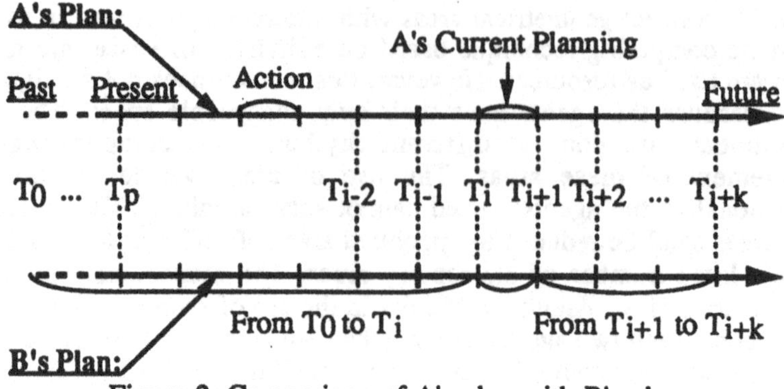

Figure 2. Comparison of A's plan with B's plan

Suppositions

We intend to let each agent be as autonomous as possible during the planning, and we do not let any agent slow down an other agent. Therefore, even when agent B's decisions for T_p are necessary, A must produce a temporary plan, which is built on suppositions on B's choices. This plan is useful for the rest of the community. Another agent C can thus use the temporary plan to generate its own plan. If A's suppositions are exact, after they have been confronted with B's decisions, a definite time will be saved. Otherwise, nothing will be lost, a new plan for A will be built after B has given its decisions. Moreover, if A's plan depends on the future of B's plan, and B's plan depends on the future of A's plan, then, in order to generate these plans, it is necessary for A and B to make successive tries without all the data. The implementation of different methods for the recognition of an agent's desires is necessary, but we must point out that the suppositions are used to save our time. As Durfee and Lesser [11] emphasize it, we must choose between a definite appreciation of an agent's desires and wasting our time. In the highway traffic case a very simple rule can be used: for example, an agent which has not given his decision for the time T_p will be considered as a passive agent, i.e. it stays in the same lane without accelerating. Despite the fact that this rule is simple, it is very efficient because it expresses a common behaviour. If the suppositions are false, the decisions we take are not automatically wrong. For example, an agent B may have accelerated instead of staying passive, but this attitude does not change at all our well-justified decisions. In that case also, we get much benefit.

We use specific tests, produced during the planning, to check that an action enters in the framework of our suppositions. For example, if we want to pass B, as long as B has not yet decided to move to another side, we admit that our suppositions are still good. Therefore a small gap can sometimes be absorbed, but this gap must not increase in the future (it is true that several successive gaps could be disastrous for our suppositions...). That is why all the tests produced since the first supposition was made must be checked when a new piece of information is given. If a new conflict between B and us arises, then re-planning or backtracking becomes necessary. In the general case, a common code does not immediately allow an agreement between the present agents, who must discuss T_p until they agree. We must point out that this backtrack is not the one mentioned in the paragraph "Allocation of space". The one used by the automatic computing technique was necessary to produce a plan even if the other agents' plans were known. On the contrary, the backtrack mentioned here is used to restore the condition in which a choice was made when there was a lack of information.

Production of an action

Introduction It is not easy to produce an agent's action depending on several other agents because there are numerous possible influences on our agent. Classic planning methods, by resolving a goal into sub-goals, cannot be used to solve this problem. We suppose that problem solving is not necessary or else that it has already been realized (cf Allocation of space), i.e. the confrontation of A with B has already been solved. The solution for A and B's confrontation is assumed to be given by a rule-base which adapts A's behaviour to B's one by choosing a behaviour (a rule) by means of a code of good manners (ex: the highway code). We call that-rule base the "bi-agent" one because it solves the conflict between two agents. We simply intend to find a method of confrontation when the present agents are numerous. Our concern is the confrontation of A's desired behaviour with many other agents' desired behaviours, the conclusion of one confrontation throwing back into question another one.

Different possible algorithms If we have no bi-agent decision rules, we could try to build a decision system approaching an agent's local world through successive refining processes. This system should recognize the number of agents and their different influences; therefore it should recognize every multi-agent situation which might be produced. That is an utopian solution and would certainly be very hard to build because of the combinatorial explosion. Let us come back to the bi-agent's decision system case. These bi-agent decision rules can be used in different ways to adapt A's behaviour to the behaviour of all N-1 other agents:

<Algorithm 1>. A system could take a decision by the successive confrontation of a "behaviour going on till now" with each co-agent behaviour. A first behaviour would be chosen by being compared to a first agent. Then the current behaviour should be accepted by the N-2 other co-agents. The "behaviour going on till now" could be associated with a rule which was fired by all the confronted agents. When all the agents would have been taken into account, the chosen behaviour could be introduced in the plan as an action.

```
Algorithm 1:
begin
--- Phase 1---
for every agent B do memorise that B does not agree.
for every non-agree agent B do
  if the current behaviour does not fit (same non-fired rule)
  then
    begin
    for every agent C do memorise that C does not agree.
    choose a possible behaviour for B
      (another possible fired rule).
    if there is no behaviour
      then exit backtrack
        (a precedent action must be changed).
    endif.
  memorise that B agrees.
  endfor.
--- Phase 2 ---
add the new action to the plan.
end.
```

<Algorithm 2>. This system is successively interested in N-1 co-agents, but it differs from Algorithm 1, for whenever a behaviour is found, it is memorised in a concise and non-redundant form in a structure called "results". At the end of the exploration of all interactions with co-agents, the desired behaviours are given by the results. In a second phase the convenient and desired behaviours would be chosen depending on the behaviours asked, and it would be added in the plan as an action.

```
Algorithm 2:
begin
--- Phase 1 ---
for every agent B do
  compare B's behaviour to A's state to find
    the desired behaviour for A.
  add this desired behaviour to the results.
  endfor.
--- Phase 2 ---
analyse these results to find the chosen behaviour.
add the new action to the plan.
end.
```

In the example (Figure 3), agent A wishes to pull out to overtake vehicle B, and also does not wish to change lanes because vehicle C is on the left side. During the phase of choice the algorithm will find the convenient behaviour "decelerated until it has reached B's speed so that it does not hit it".

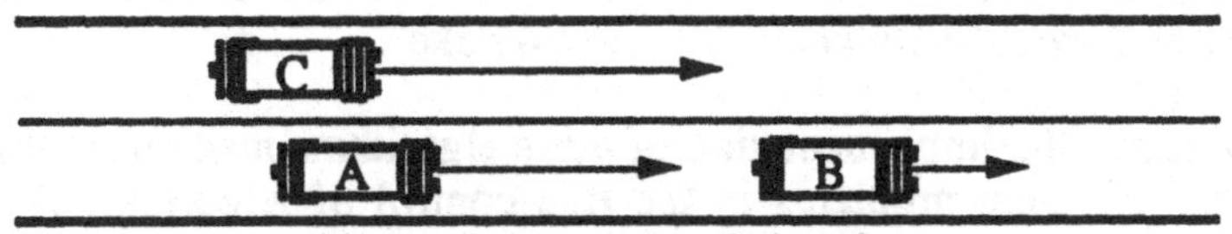

Figure 3. Example of situation

Now the "results" structure for this highway traffic example is given: it is a concise and non-redundant representation of the agent's wishes (see Figure 4).

No-change-lanes	t
No-cut-in	()
No-accelerate	()
Pull-out	t
List-which-vehicle-pull-out	(B)
Cut-in	()
Accelerate	()
Decelerate	()
List-speed-decelerate	()

Figure 4. Example of results

Criticism of both algorithms Algorithm 1 opposes A's state (the value of A's state-variables) and a current behaviour (the fired rule) to B's state and B's behaviour. Algorithm 2 only opposes A's state to B's state and B's behaviour, because no "current behaviour" has been chosen, but only some desires have been expressed. In Algorithm 1, a behaviour refused by an agent can no longer be considered during the confrontation with other agents. On the contrary, in Algorithm 1, the confrontation with one agent can be repeated, while in Algorithm 2, only one confrontation per agent is done.

Moreover, the fact that in Algorithm 2 the choice of the good behaviour is done only at the end enables us to take more intelligent decisions. In Algorithm 1, a given behaviour will not be taken if it is not useful (for example: it is useless to overtake as nobody is in front) or if it is refused (for example: one is not allowed overtake when there is an agent on the left side). Therefore, we cannot take into account all the refusals to influence the following confrontations (if the behaviour "overtake" is not taken we do not know if there was a vehicle in front or not). The lack of information can be noxious to a good choice. For example (see Figure 3), if A must overtake B and, if A can not change lanes because C is on the left, the intelligent decision, given by Algorithm 2, could be "overtake + do not overtake --> decelerate", whereas Algorithm 1 would only given "do not overtake".

Moreover, the implementation of these algorithms made us realize the existence of some problems in the rule control of Algorithm 1. These rules are more complex than for Algorithm 2 because they are fired in several different ways:

- actively, by at least one agent which produces the behaviour (for example: the vehicle that is in front provokes the overtaking);

- passively, by all other agents which are not opposed to this behaviour.

Moreover, when all agents have been submitted to this rule, at least an active fire must have been recorded, otherwise the rule is refused because this behaviour is useless.

Rules management is almost not natural and becomes complex if we try to solve the problem mentioned before, i.e. finding "overtake + do not overtake --> decelerate", instead of finding "do not overtake". A specific rule should be fired:

- actively by at least B, which produces the overtaking;

- actively by at least C, which produces the refusal of overtaking;

- passively, by all the other agents which are not opposed to this behaviour.

If we try to avoid these problems we can imagine to cut this rule into three and then come back to Algorithm 2, that seems apparently more adequate for the multi-agent planning. However, the existence of data structures which could be used as "results" for a given application has not been proved.

Selective backtrack B sends an action B_i (B's action for the time T_i), either because it has just planned this action, or because it has re-planned it after a backtrack, in which case it sends back a new action. All our planned actions produced by supposition on action B_i or using the action it had sent previously (the previous B_i) must be checked. If the new action is not similar to what was taken into account during the previous plan, these actions must be re-planned. Depending on the chosen algorithm, we can do a selective re-planning of an action:

Algorithm 1 will be started over by examining again the behaviours which had been refused by B (re-validate all the rules which had been refused by B), without taking into account the behaviours which had already been refused by other agents except B itself (leave aside all the rules which have been rejected by other agents except B). This process will lead to a new action A_i for our plan.

Algorithm 2 will enable us to calculate in a first phase, and only for B, our desired behaviour, in order to adapt it to B_i. We call B-Results-2 the results connected to B. We suppose that the results we found in the previous pass were memorised in B-Results-1, whereas the final results were memorised in Final-Results-1. If these new results are included in the first final results (i.e. B-Results-2 included in Final-Results-1), the wish expressed to satisfy the new action B_i, was already included in all agent's wishes for the previous pass. It is useless to enter the second phase, because the resulting action will be the same. We can notice that if B's new results are already included in the previous one (i.e. B-Results-2 included in B-Results-1 because the new action B_i has not influenced the behaviour we already wished), that implies its inclusion in the final results (i.e. B-Results-2 included in Final-Results-1). On the other hand, if we do not get one inclusion, it means that a new wish has appeared. The second phase of Algorithm 2 must be executed to find the new action.

At the end of selective re-planning (algorithm 1 or 2), if the new action is different from the old action, all the actions following it must entirely been re-planned. On the other hand, if this action is like the old one, the selective re-planning continues for the following action.

Both algorithms enable the selective re-planning. This possibility is fundamental for the agents who will often use it. Most agents will get ahead with their planning and will be obliged to check their fully justified suppositions on a large part of other agents.

CONCLUSION

Our aim is to realize a co-pilot which controls a vehicle, moving in highway traffic. In this paper, we point out the multi-agent aspect, characterizing this world and we also try to describe structures and methods which could be used in any multi-agent world. Besides, we analyse the knowledge representation of an agent, the time management, and the structure of action plans generated by the planner.

We point out the agents' influences on each other and show that conflicts can be solved by temporary suppositions, then by confrontations, as fast as the other agents' planning. At last, two algorithms of multi-agent planning are presented and compared. Their comparison has been based on the way they act on simple cases. Their qualities enable them to answer efficiently to selective re-planning problems. It has been stressed that the second algorithm is much more intelligent and easier to use than the first algorithm.

The concepts described in this paper have been programmed and tested with a multi-agent simulator within the context of highway traffic. The continuation of our work will concern the "multi-agent" execution monitoring of those plans, with a co-pilot which assist the human pilot (i.e. warning human pilot or generating avoidance trajectories when there is an imminent accident) while taking into account other vehicles having no co-pilot.

REFERENCES

1. Bessiere, P. Un Formalisme Simple Pour Representer La Connaissance Dans un Contexte de Planification Multi-Agent, in RFIA/84, tome 2, pp. 345-354, Actes du 4ième Congrès de Reconnaissance des Formes et Intelligence Artificielle (AFCET, Agence de l'Informatique, INRIA), Paris, France, 1984.

2. Wilks, Y. and Ballim, A. Multiple Agents and the Heuristic Ascription of Belief, in IJCAI/87, vol 1, pp. 118-124, Proceedings of the Tenth International Joint Conference on Artificial Intelligence, in two volumes, Milan, Italy, 1987.

3. Rosenschein, J. S. Synchronization of Multi-Agent Plans, in AAAI/82, pp. 115-119, Proceedings of the Second National Conference on Artificial Intelligence, Pittsburgh, Pennsylvania, USA, 1982.

4. Sycara, K. P. Argumentation: Planning Other Agents' Plans, in IJCAI/89, volume 1, pp. 517-523, Proceedings of the Eleventh International Joint Conference on Artificial Intelligence, in two volumes, Detroit, Michigan, USA, 1989.

5. Morgenstern, L. Knowledge Preconditions for Actions and Plans, in IJCAI/87, volume 2, pp. 687-874, Proceedings of the Tenth International Joint Conference on Artificial Intelligence, in two volumes, Milan, Italy, 1987.

6. Konolige, K. and Pollack, M. E. Ascribing Plans to Agents: Preliminary Report, in IJCAI/89, volume 2, pp. 924-930, Proceedings of the Eleventh International Joint Conference on Artificial Intelligence, in two volumes, Detroit, Michigan, USA, 1989.

7. Wood, S. Dynamic World Simulation for Planning With Multiple Agents, in IJCAI/83, volume 1, pp. 69-71, Proceedings of the Eighth International Joint Conference on Artificial Intelligence, in two volumes, Karlsruhe, Germany, 1983.

8. Mourou, P. Représentation et Simulation d'un Agent dans un Monde Multi-Agent: un Véhicule en Circulation Autoroutière, Rapport Interne N°IRIT/90-10, Février 1990.

9. Cammarata, S. McArthur, D. and Steeb, R. Strategies of Cooperation in Distributed Problem Solving, in IJCAI/83, volume 2, pp. 768-770, Proceedings of the Eighth International Joint Conference on Artificial Intelligence, in two volumes, Karlsruhe, Germany, 1983.

10. Rosenschein, J. S. and Genesereth M. R. Deals Among Rational Agents, in IJCAI/85, volume 1, pp. 91-99, Proceedings of the Ninth International Joint Conference on Artificial Intelligence, in two volumes, Los Angeles, California, USA, 1985.

11. Durfee, E. D. and Lesser, V. R. Incremental Planing to Control a Blackboard-based Problem multi-agent, in AAAI/86, volume 1, pp. 58-64, Proceedings of the National Conference on Artificial Intelligence, Philadelphia, Pennsylvania, USA, 1986.

SECTION 5: DIAGNOSIS, SAFETY AND RELIABILITY

DIPLOMA - The Seal of Approval

K. Oldham (*), R.P. Main (*), J.M. Cooper (*),
N.F. Doherty (**)
(*) Lucas Engineering & Systems Ltd., Solihull, B90 4JJ, UK
(**) Manufacturing Systems Research Group, University of Bradford, Bradford, BD7 1DP, UK

ABSTRACT

This paper describes the development of a knowledge-based system to assist in the quality assurance of complex electro-mechanical products during initial assembly and test, and subsequent overhaul and refurbishment.

INTRODUCTION

Consistent product quality is a key success factor for a manufacturing company in today's competitive world. Achieving this requires the use of well-organised business processes supported by the appropriate tools and techniques at all stages of product introduction and manufacture. Many of these rely on knowledge and experience to obtain the best results and so this is a fertile ground for the application of intelligent knowledge-based systems. Lucas Engineering & Systems have been working in this area for nearly seven years and have developed several systems and started work on others during that time.

One highly effective way of contributing to the quality of electro-mechanical and similar products is to reduce the number of components making up the product. Separate components are necessary only if there is relative movement between them, the application requires the use of different materials, or they have to be taken apart for maintenance purposes. The Design for Assembly system [1] provides a structured way of assessing this in terms of a design efficiency ratio as well as providing a measure of the ease with which the remaining components can be assembled. Reductions in the number of components through using this system in a team-based approach are typically between 15% and 50%. Designing the remaining components so that they are less susceptible to variations during manufacture and determining the optimum conditions for manufacture can both be facilitated by using the Taguchi system [2]. Finally, the most effective way of forming attachments between components, whether it be by using mechanical fasteners, welding or adhesives, can be determined with the help of FACES (Forming Attachments of Components Expert System) which incorporates the adhesive selection system Stick [3].

However, with complex products of this type, the performance is governed by the relative dimensions of combinations of components. In this case, ensuring that each component conforms to specification is necessary but not sufficient to ensure that the product meets the specified performance requirements while still allowing component interchange during product maintenance. Component conformance assurance during manufacture has to be followed by product acceptance testing.

The life-cycle of a typical product can be depicted as shown in Figure 1. A product specification is agreed and passed on to a project team who produce detailed component drawings, acceptance test procedures and manufacturing instructions. Components are then manufactured according to those instructions, ensuring that they match their detailed drawings. A complete set of components is formed into a kit which is assembled and tested in accordance with the acceptance test procedures to check that the product as a whole matches its original specification. The unit then goes out into service and may eventually be returned to the manufacturer for repair or refurbishment. Following a survey of the state of the unit on receipt, the unit is stripped, repaired, re-built and finally tested before it is sent back to the customer. In-service information, such as durability, gathered at this stage is used to update design/ manufacturing to improve product reliability.

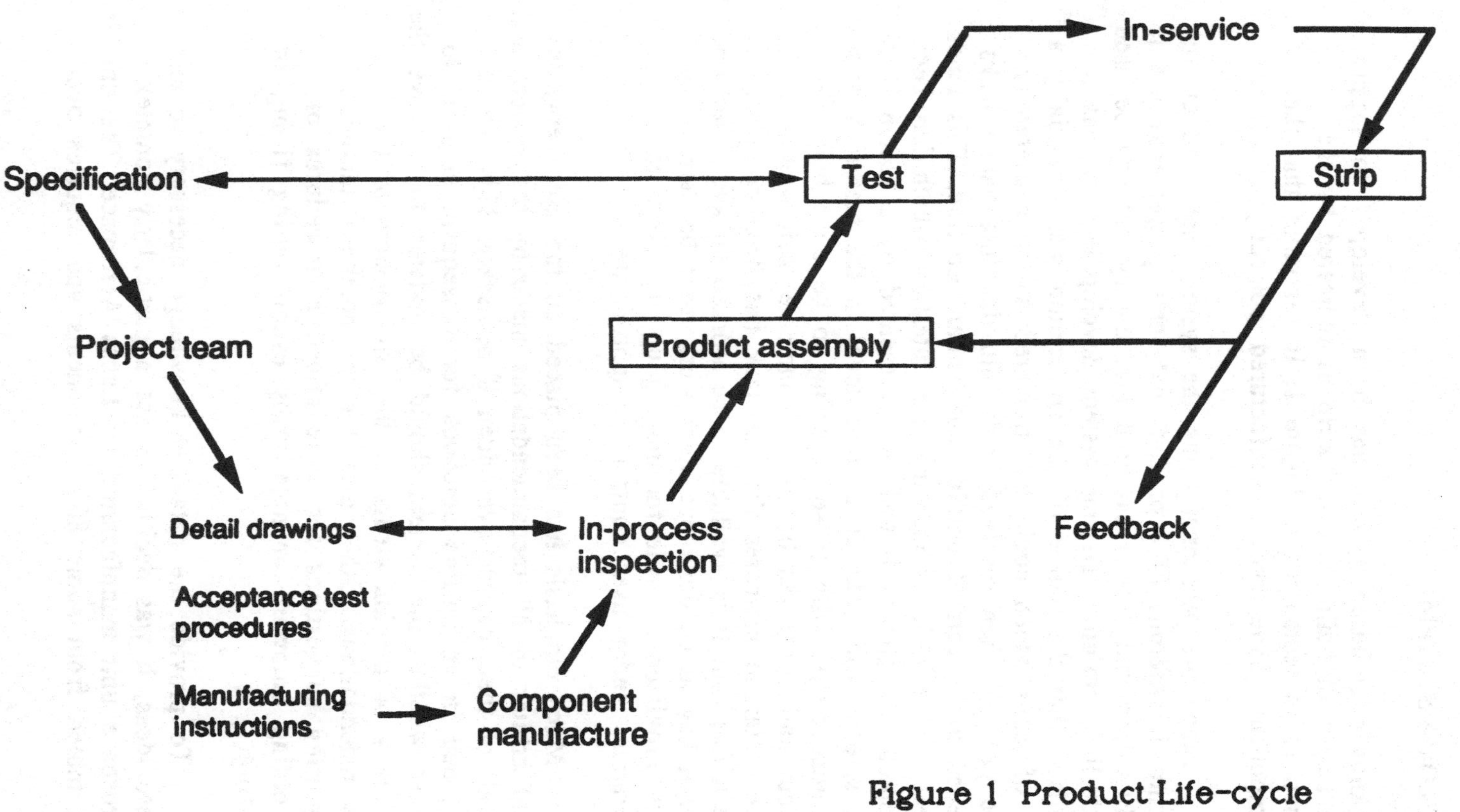

Figure 1 Product Life–cycle

DIPLOMA SYSTEM

A knowledge-based system has been developed to support the stages of build, test and strip as indicated by the functions in rectangles in Figure 1. It assumes that the components have been manufactured correctly.

Since the personnel using the system may well change during the lifetime of a product and some people may use it only infrequently, it is essential that the system can be used with little training. During system development, it was realised that the nature of the application was essentially a set of tasks which could be structured within a hierarchy. The problem then resolved itself into deciding which tasks needed to be carried out in a particular situation. This lends itself to the style of man-machine interface which has been adopted whereby each task is represented by a rectangle on the screen. The tasks at a given level in the hierarchy appear simultaneously, with those requiring to be done being in colour and connected by lines to indicate their sequence. A blue rectangle indicates that the task has been completed, a green one that it is available to be worked on and a red one that it has to be done but other work must be carried out first. Provision is made for moving up and down the hierarchy under user control as appropriate.

Much emphasis has been placed on the user being the final arbiter of all recommendations made by the system and so all of them can be overridden if necessary. This can be recorded in an on-line notebook for subsequent analysis to assess whether the system should be updated to improve the accuracy of the knowledge or the effectiveness of the recommendations. Other uses for the notebook include capturing suggestions for more effective procedures or recording untoward or noteworthy events during fitting or testing.

To provide the data and knowledge necessary to test these ideas, it was decided to use a particularly complex aerospace unit manufactured by Lucas Aerospace. The unit is assembled from some 800 components and requires over

200 tests to ensure that it is of serviceable quality. Building can take up to a week and testing, a similar length of time, provided that nothing is wrong. At the start of the project, this unit had already been in production for several years and so there existed a considerable body of knowledge about it. Since the units can remain in service for twenty or thirty years after regular production ceases, there is a strong incentive to capture this knowledge so that it is available for the lifetime of the unit. From time to time, engineering changes, either mandatory or optional, or to satisfy customer requirements, are made to the design of the unit. Where these are sufficiently significant, a unit incorporating them is deemed to be of a different 'build standard' and any support system must cater for units of differing build standards. These may reflect differences in reliability or release life and so it is very important to identify them clearly. Although the existing body of knowledge is specific to the target unit, the system should be applicable to many different products in a variety of industries, merely by adding the relevant data and knowledge.

The system that has been developed is called DIPLOMA (Design and Implementation of a Process capability and Life-cycle Optimisation Module - Alvey project). Its software architecture is shown in Figure 2. The objective was to create a quality assurance system consisting of a linked collection of robust intelligent knowledge-based systems. Adjoining rectangles indicate that the corresponding software modules can interact. The four main functional areas are those of strip, test, trend and database. The test and strip areas each consist of three modules since they include the associated knowledge base and maintenance modules.

The 'test module' guides the test engineer through the complex test schedule requiring some 230 individual measurements. The sequence in which the tests have to be performed is indicated by representing the test schedule graphically as a set of hierarchical precedence networks where the colour of each node indicates if the test is available to be carried out, partially done, passed, failed, etc. The hierarchical nature of the networks means that they can be viewed at the level appropriate to the user's expertise.

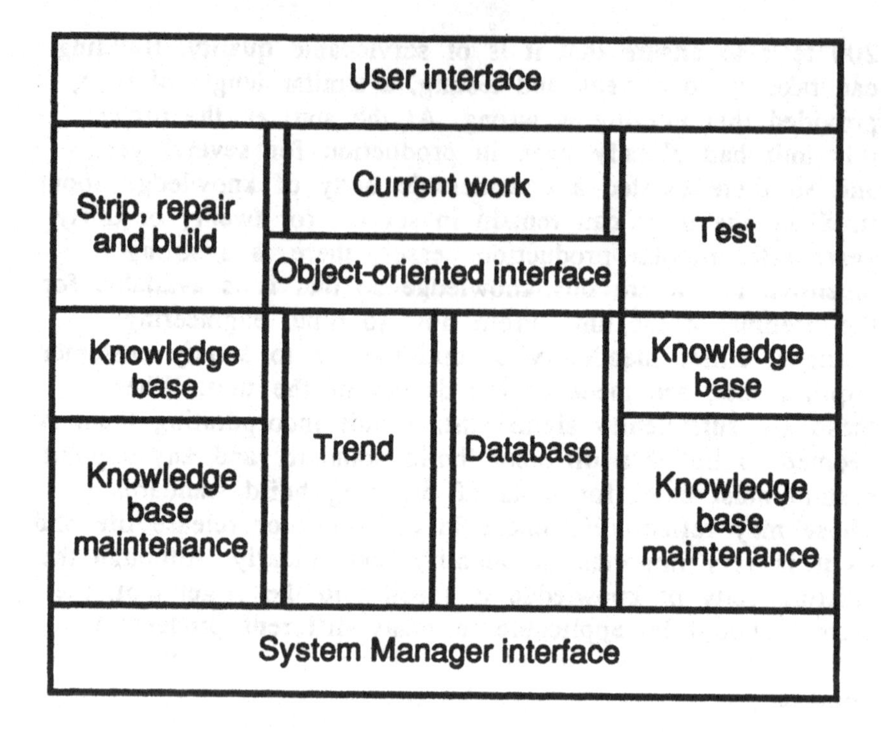

Figure 2 DIPLOMA Software Architecture

The system evaluates each test result as it is entered to determine whether any faults are present and, if so, which ones. If remedial work has to be done, this may invalidate some of the test results already obtained. In this case, the system will inform the tester as to which tests must be repeated. During all of these activities, supporting documentation including the test schedule, relevant diagrams and lists of tools required, is made available via a set of function buttons which co-exist with the networks on the display.

Figure 3 depicts a typical screen display schematically with the function buttons at the top and four nodes in the lower part. The absence of arrows to and from the Repair node indicate that, although it may be required in certain circumstances, it is not needed in the present case.

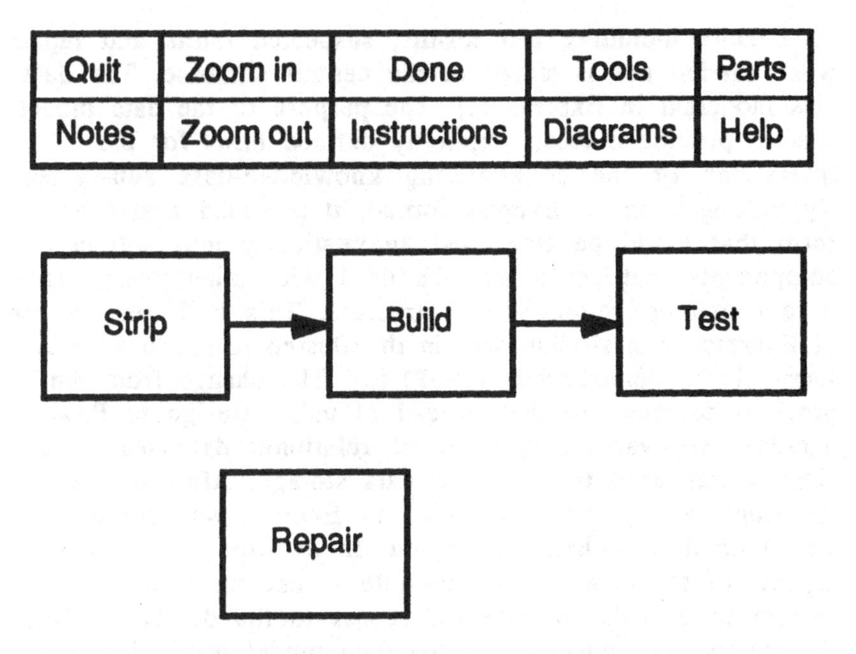

Figure 3 Schematic Diagram of Screen Layout

The 'strip module' instructs a fitter on how to disassemble, inspect and rebuild the unit in order to cure the faults found by the test module. The strip module can also give similar assistance for full overhaul, repair and original equipment build work. To do this, it has to select appropriate strategies for undertaking the original assembly of a unit, or repair/overhaul activities; incorporating appropriate engineering modifications and work resulting from the receipt survey of units returned from service. Using the selected strategies as the basis, it generates a precedence network specifically tailored to the work requirements of the unit and determines the parts required to complete the work successfully. Having ordered and kitted the parts, the fitter is guided through the necessary work by means of a graphical precedence network similar to those for testing. During the network session, the fitter is required to provide feedback on the progress of the work and any faults detected. The fitter is also able to review relevant sections of the parts list and the tool schedule using the function buttons on the screen.

Data including test results, suspected faults and repair work carried out is stored in the central database. The data was modelled in Express [4]. The purpose of the data model was to provide a clear, explicitly defined basis for the interaction of the collaborating knowledge-base sub-systems. By putting it in an Express format, it provided a standard form that could be translated automatically into software components that could then be used with other components in a distributed client-server structure. This built on the use of Express, in a similar role, in the Design to Product Alvey Large Scale Demonstrator (DtoP) [5]. The change from the previous project was that instead of using Design to Product database software, a commercial relational database system, Oracle, was used for the final data storage. After several iterations, an agreed data model in Express was produced and a database schema generated. In addition to the static aspects of the data, it was possible to use the translation system to provide routines for access to the database. Thus the statics and dynamics of the data model were all captured in a single Express file. This provides maintainability for the model that can be used as a firm basis for interaction between the individual knowledge-base sub-systems.

The 'trend module' applies statistical tests to the information in the database in order to determine if any fault trends have become apparent. This trend data is fed back to the test and strip modules to improve their future performance and also to the engineering and manufacturing departments of the business so that any appropriate re-design of the product or manufacturing processes may be considered. The trend module also has the capability of determining the most effective way of remedying a suspected fault. The updating of the knowledge-bases has to be achieved without reducing the stability of the system to the point where small changes in input data would result in dramatic changes in the recommendations to the user. Various approaches, such as neural networks and Bayesian theory, were investigated but a conventional statistical method has been adopted as it can cope better with noise in the data. It is lowly tuned to ensure that the module is not liable to instability or to jump to conclusions on the basis of only one or two items of data. This means that the error limits of conclusions are readily calculable. The knowledge-bases are only updated by the trend module when the error limits of a calculated result are lower than the error in existing information. Conservative error assessments are used throughout.

The other modules are the interfaces and current work selection. The end-user interface is the one that will be used by the fitters/inspectors and testers during normal operation. When deciding which unit should be worked on next, they will access the current work selection module which maintains a record of the status of each unit on site. They can then either choose a unit already known to the system or enter the details for an additional unit. The object-oriented interface has the effect of transforming the relational database into a database with object-oriented characteristics. Access via the system manager interface is restricted to nominated people who are authorised to maintain the knowledge in the knowledge bases. This will be necessary, for instance, if the build standard is changed. They can also determine whether the trend module has identified any patterns in the data and can query the database directly.

DEVELOPMENT METHOD

Typically, systems with a large knowledge-based content are developed incrementally, using the RUDE (Run - Understand - Debug - Edit) methodology [6]. Serious consideration was given to this approach with the suggestion that the test module should be developed as a stand-alone module in the first instance to demonstrate the ideas as early as possible to the potential users. However, the size and complexity of the problem and the geographical separation of the development team suggested the need for a more robust approach. It was therefore decided that the disciplines and control mechanisms offered by the structured approach for conventional software development, POLITE (Produce Objectives - Logical/physical design - Implement - Test - Edit), would be needed to ensure the consistency and compatibility of the software modules.

The chosen development strategy was primarily to use the tools and techniques of the structured methodology, bolstered where necessary by prototyping to gain greater understanding of the problem and obtain direct feedback from the user community.

Some difficulties were encountered in using the structured approach, but some of these were due to the nature of the project rather than the approach itself.

- Because the system was intended to be applicable to many different products in a variety of industries and not restricted to the target unit, it was difficult to identify potential users who could validate the design for all possible applications. Even for the target unit, it became apparent that there were many people who had a vested interest in the system, i.e. were stakeholders. Consequently, a full stakeholder analysis had to be carried out to capture all of their requirements and the key stakeholders had to be identified and involved in the development process.

- It was not clear where the task of eliciting and documenting the knowledge fitted into the general framework of the structured approach.

- Because the majority of end users were shopfloor personnel, whose experience of computers was limited, system specifications using data flow diagrams were of limited value as a means of promoting understanding and validating the system.

- The process of writing, checking and issuing formal documentation was exceedingly time consuming due to the size and scope of the system, and consequently insufficient time was left for detailed prototyping. This was exacerbated by different people on the various sites using different names for the same data item and the same names for different items even though regular meetings were held in an effort to avoid this happening.

- Maintenance of the documentation became impractical because of the magnitude of the changes required as the understanding of the true system requirements developed during the design stages. This was due in part to the changes in organisation at the target site with attendant changes in the roles of the people involved and in part to the limitations of the package used.

In using the structured approach, as well as reaping the usual benefits such as facilitating the design of the system architecture, the following lessons were learned.

- The approach allowed the project to be planned in a systematic way and the progress to be monitored and assessed at each stage. The plan had to be flexible as people and machines were not always available when it would have been most convenient.

- The graphical techniques and attendant documentation meant that changes in personnel working on the project could be accommodated with little detriment to the progress of the project.

- Experience demonstrated that the analysis of the procedures in use at the target site must be carried out in considerable detail and all the ramifications explored to form a sound basis for the design of the new system.

- The earlier the core data model can be defined, the better.

Controlled prototyping, as defined by Born [7], is a useful technique for resolving uncertainty when developing a system. The following experience was gained in its use on the project.

- When prototypes were shown to the stakeholders in an effort to clarify the user requirements, the stakeholders tended to assess the user interface rather than the functionality that the prototype was intended to demonstrate.

- Where there were several possible ways of meeting a user requirement, prototyping proved to be very effective in obtaining user preferences. An example of this is whether a set of tasks that had to be carried out were better represented by a network of boxes or a series of menus.

- Similarly, where there were several possible ways of implementing a function, prototyping was used to determine the best one. For example, different algorithms were prototyped for colouring the boxes to indicate whether tasks had been done or not.

KNOWLEDGE ENGINEERING

In developing knowledge-based systems, knowledge engineering warrants particular consideration and this was recognised from the outset. Whilst certain activities were easily identified, e.g. the identification of experts and the appropriate use of knowledge elicitation techniques, there were various issues which proved problematic. These are outlined below. In particular, their significance was difficult to assess in advance and therefore they could not be ignored.

As noted above, it was difficult to decide when the bulk of the knowledge engineering activities should be carried out.

- An investigation into the on-going research into expert systems development methodologies did not provide any help on this aspect.

- Data flow diagrams were used to model the design of the knowledge-based system. However, as far as the knowledge in the knowledge-based system is concerned, there is only one process - the inference engine - that uses it to produce conclusions. Hence progress in designing the knowledge components of the system could not be monitored by reviewing the progress in generating the data flow diagrams.

As part of the project planning, it was necessary to estimate how much time would be needed for knowledge engineering. It is not sufficient to make the estimates based only on the availability of the experts that were identified in the stakeholder analysis. A better understanding of the size of the problem was needed:

- In order to assess each expert's individual role in the existing system it was not sufficient simply to know who were the experts. The volume and depth of their knowledge, under what circumstances and in which parts of the existing system it is applied, had to be considered.

- Whilst it was useful to focus on each expert individually, their interaction could not be ignored. For example, there are occasions when the combined expertise of several experts is utilised in diagnosing faults.

- In order to validate knowledge using a prototype, the knowledge must first be elicited and analysed to determine a suitable representation before software can be developed to encapsulate that knowledge and display it for validation. However, the best representation for validation purposes may well differ from that for efficient use in the final system. If this is the case, a paper-based approach is preferable if a suitable representation can be devised as it avoids the development of additional software.

- Such knowledge will be incorporated in the knowledge-based system so that the system requirements will be met. It is important that the corresponding knowledge representations support efficient use of the knowledge. Consequently, any transformation from the form in which it was initially obtained from the experts needed consideration so that the size of this task could be assessed.

However, these aspects are very difficult to quantify before elicitation has been completed. In view of these issues, it was deemed wise to commence the knowledge engineering early so that these uncertainties could be resolved while there was still time to replan the project if necessary.

Initial discussions with the experts indicated that the highly technical engineering terms could make effective communication with the knowledge engineers difficult. In an effort to remove this language barrier, and also to optimise future discussions, the knowledge engineers obtained the supporting documentation for the unit and digested those documents which it was thought would be most useful. This took longer than expected as insufficient consideration had been given to their accessibility.

Particular consideration was given to the management of the knowledge elicitation sessions. This included:

The duration of the session
- care was taken not to overrun the allotted time.

The frequency of the sessions
- this was based on the availability of the experts and the length of time needed to both represent and analyse the elicited knowledge.

The concentration span of the expert
- if the session exceeded this limit the accuracy of the content could not be assured and his interest also decreased.

The level of difficulty
- if the expert was required to think very deeply it could not be sustained for long periods of time. A series of sessions proved to be a more effective way of tackling this problem.

The accuracy of the knowledge
- this was aided by making audio-tape recordings of interviews which were subsequently transcribed and analysed. The experts clarified the technical content

where required to remove ambiguity and any misunderstandings.

The environment

- an office was used so that the noise and distractions of the working environment would not impinge on the knowledge elicitation sessions.

Prior to this project, the experts had no experience of using knowledge-based systems, neither had they been involved in the development of a knowledge-based system. It was important therefore that they had a basic understanding of knowledge engineering. The initial discussion with each expert included a summary of the objectives of the knowledge engineering activities, an outline of the time constraints, and an overview of the activities in which they would be involved in order to emphasise the importance of their role. It also enabled them to talk more authoritatively to their managers and hence helped to maintain management commitment to this project.

There are many knowledge elicitation techniques available, each with its own strengths and weaknesses. Several were used in this project, the reasons being:

- It was important to establish good working relationships with the experts. Observing them in action and interviewing them were particularly helpful in this respect.

- The experts had acquired their expertise in different ways. For example, the testers had developed heuristics in the course of performing the same tests many times. They sometimes found it difficult to verbalise their knowledge as this was not required in the normal course of applying their expertise. In such cases, observing the testers on the job, in conjunction with interviews focussed on particular aspects of their job, was fruitful.

- It was important that the knowledge elicitation activities were effective. In deciding which technique was appropriate, efficiency was a factor; some techniques are more efficient than others, e.g. structured interviews are more efficient than unstructured interviews. In addition, where it was apparent that the expert had quite a depth of knowledge, the knowledge was accessed in a variety of ways by using different techniques, e.g. card sorts [8] and structured interviews, and hence it could be explored more fully.

It was obvious that the knowledge elicitation activities would span several months, and so it was crucial to maintain the experts' interest. Where the knowledge elicitation required a particular expert regularly over a long period of time, it was important that the activity was flexible. For example, an activity which comprised a series of structured interviews based around a card-sorting exercise was carried out. It was designed so that both the length and the frequency of the interviews could be varied. This was necessary as it was more difficult to ensure the availability of the expert as the frequency of the interviews increased.

Although a great deal of information was collected during the knowledge elicitation activities driven by the knowledge engineers, it was difficult to assess whether the resulting models of the knowledge were complete. Feedback from the experts was always encouraged. Their ideas indicated whether additional sources of information or other methods were needed in order to obtain models of the knowledge which were more complete.

Particular care was taken in representing the knowledge. Whilst the expert may be able to validate the knowledge in the form which represents what was elicited, subsequent representations must not transform the structure of the knowledge because the expert may not be able to validate it in its transformed state. The knowledge representations were frequently in tabular form - this did not destroy the underlying relationships in the knowledge and its simple structure facilitated validation.

PEOPLE ISSUES

Many computer systems, and particularly those developed as part of 'research projects', are rarely used to their full potential because they satisfy requirements that the user does not have and fail to tackle the real problems and requirements. The need to gain and maintain the commitment of staff at other levels in the company was essential. Accordingly, close contact was maintained with the demonstrator site to ensure their involvement during all stages of the project including the design, implementation and subsequent review of the system.

A detailed analysis was performed of stakeholders, i.e. people who were likely to have some information to contribute to the design of the system or could be affected by the system. This started with a few people who were likely to be most closely associated with the system but expanded over time to include one or more interviews with each of some fifty individuals at the factory. Having obtained this initial input, it was felt necessary to maintain the interest and involvement of managers of departments which were peripheral to, as well as those central to, the operation of the system. The main objective was to promote the feeling that the system was theirs, rather than that of the development team, and to provide a recognised way of participating in the development of the system. In addition, it was important that all concerned had an accurate and up to date knowledge of the progress of the system. This was achieved by the creation of the 'validation committee'. This included some fifteen departmental managers, all of whom had some interest in the system. Meetings were held at two monthly intervals where development progress was presented and any associated issues discussed. Additionally, each member of the committee was asked to declare interest in some detailed aspects of the system, such as reporting or fault diagnosis, so that sub-committee meetings could be called to discuss particular points as and when they arose.

Demonstrations were also given to shopfloor personnel, including trade union representatives, and they were involved in some of the knowledge elicitation and man-machine interface validation sessions to promote their acceptance of the final system.

PROJECT DETAILS

The work was carried out under a grant from the Department of Trade and Industry as part of the Alvey initiative (project number IKBS 114). The four collaborators on the project were Lucas Engineering & Systems Ltd., Lucas Aerospace Ltd., the University of Bradford and GEC Electrical Projects Ltd.. Lucas Engineering & Systems were responsible for overall project management and, together with Lucas Aerospace, contributed six and a quarter man-years of effort. The University contributed six man-years and GEC, two and three quarter man-years. The level of funding from the Alvey Directorate was 50% for the industrial partners and 100% of costs for the University.

Throughout the project, great emphasis was placed on achieving a working system within the project lifetime rather than pursuing interesting ideas until the money ran out. Sometimes this meant that compromises had to be made. For example, it was agreed, in view of the size of the problem and the limitations on funding and resources, that the quality assurance aspects of the work would be given full attention during the period of funding and that the practicality of linking the system to the test rig would be investigated fully as part of the on-going development during the evaluation period at the target site.

EXPECTED BENEFITS

The system will soon be installed at the target site and, after an evaluation period of several months, will be used in production on a full-time basis. It is expected that the system will yield considerable benefits. Following the example of the DTI Guidelines [9], these are listed below under the headings of infrastructure, people, efficiency/ productivity and external competitiveness.

Infrastructure:

- Knowledge created/preserved - the preservation of product knowledge for the lifetime of the unit, which is twenty to thirty years after regular production has ceased, is seen to be the major benefit of the system.

- Best expertise widely available - the system provides an ideal vehicle for capturing product knowledge, such as the interpretation of test results and the best ways of carrying out various strip and re-building tasks, while the unit is in production and making this knowledge available to the repair and overhaul staff who may be based on other sites.

People:

- Staffing situation improved/better use of scarce resource - capturing all the product-specific knowledge enables the tasks to be carried out efficiently by skilled fitters and testers who are unfamiliar with the product.

- Spin-off training/development - having all the product knowledge and information available at the touch of a button means that people should be able to learn about the unit even though the representation of the knowledge has not been chosen with this use in mind.

Efficiency/productivity:

- Improved throughput/timeliness - the use of the system will improve the accuracy of the prediction of the requirements for replacement parts, thereby developing spares provisioning into an MRP-style calculation.

- Better performance/decisions - capturing the expertise of the best fitters and testers and supplementing this automatically in light of practical experience will ensure that the quality of decisions will be enhanced and the best practice for performing tasks will be employed.

- Better equipment utilisation - the improvement in the interpretation of test results will ensure that the test rig will be used more effectively in that units will not be removed for repair work and replaced more than is absolutely necessary.

- Improved knowledge/understanding - the process of eliciting and analysing the product knowledge of itself improves understanding of the unit.

- Better data records/preservation - holding all the relevant information about the products in a unified database, rather than on separate paper records, means that preservation and retrieval of information will be that much easier. In addition, the records will be more complete, while being more compact, and the reports based on that data will be produced in a consistent manner.

- Better explanation/information - during the evaluation period, the need for explanation facilities will be explored more thoroughly.

External competitiveness:

- Better image/client service - the introduction of state-of-the-art techniques to support the later phases in the product life-cycle will complement the use of advanced technology elsewhere in the business for the earlier phases. This will further enhance the image of the company and improve the service to customers.

- More reliability, less risk - the improvement in feedback from service returns to product engineering and manufacture will result in more reliable products in the future.

- Enables new activities/products - the use of a computer-based system related to quality of manufacture and performance in this area of the business will provide an opportunity to link it electronically to existing systems in the company as part of a drive to become a Total Quality Organisation.

CONCLUSION

Since the system is generic in nature, it can be configured for any manufacturing organisation, in which testing and quality assurance of comparably complex products is important, simply by adding the relevant product-specific knowledge and data. Thus it is relevant to large sections of British manufacturing industry. Lucas Engineering and Systems intend to capitalize on this by exploiting the system both within and external to Lucas Industries plc.

The collaboration proved successful in that it brought the specialised and diverse skills of four organisations, industrial and academic, to bear on a single problem and the links formed during this partnership will be continued into the future. Considering the geographical locations of the organisations involved, communication and 'team spirit' were established and maintained remarkably well. The success of this project and its contribution to IKBS technology in British industry has proven the value of, and need for, collaborative projects between academic and industrial partners supported by government funds.

Acknowledgement

The author wishes to acknowledge the contributions of all concerned to the success of the project.

REFERENCES

1 Smith J.U.M. and Owen K. Engineering Design: Appraisal and Optimisation - Lucas Engineering & Systems Ltd DFA Application, Department of Trade and Industry Expert System Opportunities Case Study 9, HMSO, London, 1990.

2 Oldham, K. A knowledge-based system for product and process quality, Proc. Autotech 89, Institution of Mechanical Engineers, London, C399/39.

3 Teal, D.E., Oldham, K. and Eccleson, P. Developing systems in a manufacturing environment, the Lucas Industries experience, Proc. 1st Conf. Artificial Intelligence and Expert Systems in Manufacturing, IFS, Bedford, pp. 37-44, March 1990.

4 Schenck D. Express Language Reference Manual, ISO TC184/SC4 document number N496.

5 'Design to Product', An Alvey Programme Large-Scale Demonstrator Project, Final Report to Department of Trade and Industry and The Science and Engineering Research Council on Contract Number LD/004, The DtoP Consortium, 22nd January 1991.

6 Bader J., Edwards J., Harris-Jones C. and Hannaford D. Practical Engineering of Knowledge-based Systems, Information and Software Technology, Vol. 30, 5, pp. 266-277, June 1988.

7 Born G. Guidelines for Quality Assurance of Expert Systems, Computing Services Association, London, 1988.

8 Burton A.M., Shadbolt N.R., Rugg G. and Hedgecock A.P. The Efficacy of Knowledge Elicitation Techniques: a Comparison across Domains and Levels of Expertise, Knowledge Acquisition, Vol. 2, 2, pp. 167-178, 1990.

9 Shaw R. Guidelines for the Introduction of Expert Systems Technology, Department of Trade and Industry Expert System Opportunities, HMSO, London, 1990.

Knowledge Based System for Fault Diagnosis of a Hot Strip Mill Downcoiler

M.H. Littlejohn

Rolling Technology Group, BHP Research, Melbourne Laboratories, 245-273 Wellington Rd, Mulgrave Victoria, 3170, Australia

ABSTRACT

A computer based system has been developed to diagnose the probable causes of out of specification steel product in a hot strip mill downcoiler. The system uses a knowledge based system approach to analyse both operational heuristics and process data in a structured manner. Automatic interrupt driven data logging and analysis extends the scope of the diagnosis to include high speed process data. The development of a structured control hierarchy allows the system to conform closely with standard process operating practices, during diagnosis, to minimise disruption to the process. The general approach to fault diagnosis in an industrial environment is also discussed.

INTRODUCTION

The increased use of high speed automated equipment in steel processing mills has meant that the skills in operating and controlling such equipment can significantly influence the quality and manufacturing cost of the product. Any variation in the performance of the individual mill components or the coordination between these components, may adversely affect the quality of the product. This has resulted in the need to determine the factors affecting the performance of such equipment so they can be maintained in an optimum production condition. By observing tight process practices supported by regular maintenance programs, performance can be optimized while potential equipment problems are kept to a minimum. Also in the case of an equipment malfunction or drift in performance further diagnosis may be required to regain satisfactory performance.

The hot strip mill at a BHP steel plant rolls steel slabs through a series of rolling stands to produce thin flat strip. After rolling, the strip is wrapped into a coil by a downcoiler machine See Figure 1. There are two downcoilers which are used alternately to catch the strip by forcing it through a series of rolls and around a rotating mandrel assembly to form a coil See Figure 2. A downcoiler is set-up for each incoming coil by the mill process computer and the downcoiler operator. Set-up consists of a number of positional settings and timing sequences to coordinate the operation of the downcoiler for varying strip input. During the coiling operation, the downcoiler components must be positioned accurately and actuated at precise times to achieve a satisfactory result. The mill process computer controls the speed and timing of the downcoiler components and some manual operator intervention is required to fine tune the process or to correct major operational problems if they occur.

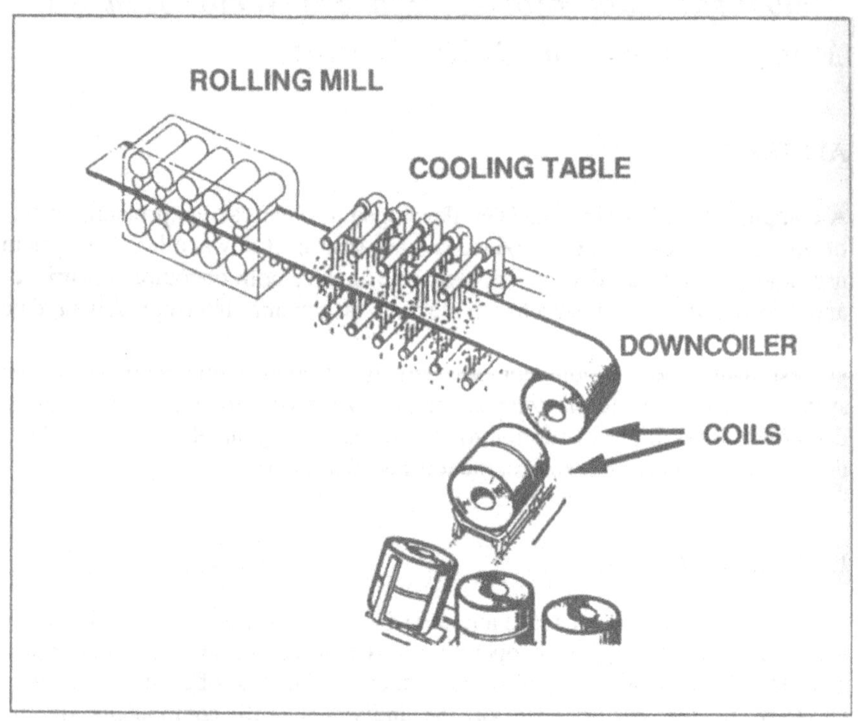

Figure 1 -Schematic of a Hot Rolling Mill showing the location of the Downcoiler [From "An Introduction to the Hot Strip Mill" John Lysaght (Australia) Ltd 1976.]

The overall ability of the downcoiler to produce coils within dimensional specification is determined by the accuracy of the component settings. Due to the complex interactions of the components, a high level of operational experience is required to determine the direct cause of any problem occurring. This paper describes a systematic method of diagnosing possible causes of substandard performance in the downcoiling operation using a Knowledge Based System and highlights the methodology required to successfully implement such a system.

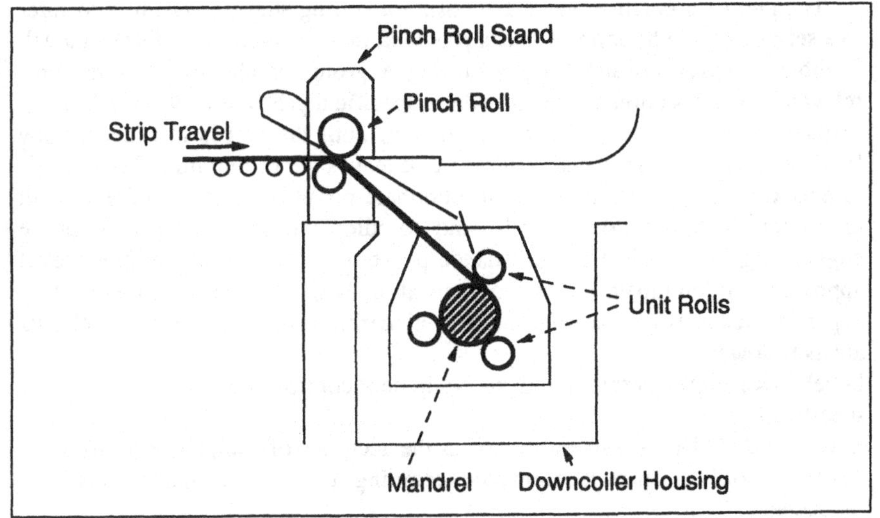

Figure 2 -Schematic of a Downcoiler showing the Strip threaded onto the Mandrel

DOWNCOILER PERFORMANCE DIAGNOSIS

The aim of the downcoiling operation is to produce tightly wrapped, straight sided coils while minimising delays to the overall process due to operational problems. Production quality of the downcoilers or coiling performance is largely determined by the physical appearance of the coil [or coil shape]. The coil shape must be within strict dimension limits, both to avoid damage during handling, and to conform with downstream processing requirements. Any coils not conforming to specifications are reprocessed resulting in increased costs. If the coil shape specification is not consistently met, operational and diagnostic checks are performed on the downcoiler to rectify the problem, often resulting in mill delays. Poor coil shape can result from a number of factors including inaccurate positional settings, wear or failure of components and component timing inaccuracies. Some examples of poor coil shape are shown in Figure 3.

To optimise coiling performance, the downcoilers are operated to a set of standard operating and maintenance procedures. When the shape of the coils drifts towards the specification limit for any reason, the operator attempts to correct any fault by adjusting machine settings. This approach is based largely on personal experience and the experience of others [through standard practice manuals and work orders etc] and so may vary with changes in personnel. The action of the operators is often successful in the short term however an overall drift in the machine settings and operational conditions can result, leading to further deterioration in performance with time.

The previous practice when diagnosing a coiling shape problem is to refer to a set of Trouble Shooting Checks provided in a standard procedures manual. Trouble shooting consists of performing a group of checks or procedures relevant to each section of the coil where specific a problem is observed. Each of three sections, head end [start of the coil until the coil is threaded], body [middle section], and tail end [after the coil leaves the rolling mill] has 4 levels of checking. These are designed to eliminate possible causes of defects with minimum disruption to the mill, and to allow scheduling of appropriate engineering tasks into the maintenance program. The aim of the multi-level approach is to eliminate minor problems at an early stage prior to committing expensive resources to perform major engineering tasks. The levels of checks are as follows:

Level 1 (Rolling) Operator confirms equipment controls are within specification.

Level 2. (Short Delay) Operator verifies the accuracy of equipment controls.

Level 3. (Rolling) Chart performance recordings compare "actual" versus "expected".

Level 4. (Long Delay) Major engineering checks to equipment.

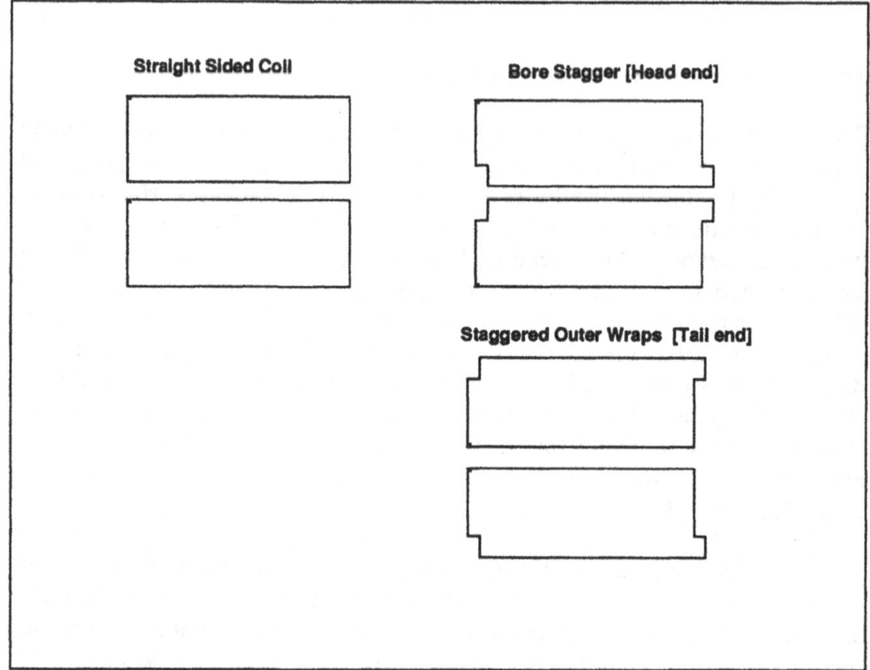

Figure 3 - Examples of Downcoiler Coil Shapes viewed in profile

DOWNCOILER DIAGNOSTIC SYSTEM

The previous trouble shooting aid provided the basic framework for effective trouble shooting. In order to improve diagnosis the trouble shooting aid was expanded to include the high level of process experience of the engineers and operators and relevant plant data. Several advantages were identified using this approach including increased general knowledge of the process, tighter implementation of the specifications and improved consistency at all times. The central role of the chart performance recordings for diagnosis of complex problems [Level 3] demanded that considerable effort was concentrated in this area. A scheme was included in the diagnostic system to record and analyse this data automatically.

Knowledge Based System

The technique chosen was to develop a computer based diagnostic system which incorporates the features of knowledged based systems [KBS]. A knowledge based system can capture the expert information and problem solving steps without extensive programming, and manage large amounts of information flexibly and efficiently. Use of a computer program automates problem diagnosis to ensure the correct order of events within the existing level structure is followed. The system uses "knowledge data" gathered from engineering and operator experience, technical knowledge in running the downcoiler, and existing standard practices.

Knowledge based systems are the collective name for a variety of methods for accumulating, specifying and presenting information about a problem or process. KBS's can solve problems that are difficult enough to require significant human expertise for their solution, but are seldom sophisticated enough to perform major functions without human control. They are however an efficient technique for providing assistance and improvement to process standardisation and capturing the expert decision making in a process. A central idea in KBS is expressing computer data or knowledge in the form of rules of thumb or good judgement [ie IF ...a set of conditions THEN ....conclusion is proved]. By combining the condition of one rule to the conclusion of another, a set of rules can be linked into a decision tree allowing apparently unrelated facts to be linked.

Data Acquisition

Effective process diagnosis by operators and engineers relies heavily on the analysis of critical timing parameters within the downcoiler. These high speed signals, of component load and timing, are normally recorded on a chart recorder set up manually as required. The recordings are used by engineering staff to diagnose complex timing interactions of the downcoiler components as the strip passes through. For example, when the strip head end is threading the downcoiler and forming a coil, the coordination of the coiler speed with the mill speed is critical, as is the timing of the withdrawal of the various guiding components when the strip has pulled tight. The chart recordings are used to compare current conditions with a standard set of conditions, highlighting any significant differences, which can be interpreted by the mill staff to deduce the probable cause.

Due to the importance of this data in complex situations, the diagnostic system structure incorporated modules to log the timing data automatically, and analyse and classify these signals for inclusion into the knowledge base. To achieve this within the restraints of the computer hardware available a novel approach was required. This is discussed below in section titled "System Software Structure".

Initially the data was logged automatically and displayed off-line to allow manual trouble shooting by the engineering staff. Subsequently the information from the recordings was classified automatically and processed for input to the diagnostic system. A description is given in section titled "Data Analysis". This has replaced some of the manual operator input required and allowed complex trouble shooting at an early stage.

System Software Structure

The diagnostic system is coded in the "C" language using a combination of proprietary NEXPERT OBJECT library functions and user written control and input/output functions. The NEXPERT OBJECT suite of functions provides a comprehensive range of KBS strategies for representing, displaying, and controlling the information data. An important feature of NEXPERT OBJECT is that it can be configured as a set of "C" language callable routines to perform complex mathematical and screen display tasks in addition to KBS functionality. Such a feature was necessary when logging data from the process computer under the fast real time conditions required in a rolling mill process.

The main executable "C" program is run in protected mode under the DOS operating system. The program is loaded into extended memory and during operation loads several types of files from disk to perform specific functions [sets of logic rules, database information, formatted input/output screens, data file sets]. The program controls program flow and provides input/output formatting via customised screens. The screens have "hotkeys" to allow display of current and previous data and context sensitive help.

Data acquisition is performed by a software task installed "memory resident" to achieve constant monitoring of the downcoiling process while the KBS is in operation. The "memory resident" program is attached to a DOS system interrupt with a timer and monitors a primary process flag at regular intervals. If a coil is approaching the coiler the program invokes a routine to log at high speed for the required duration, before returning to the KBS.

The program writes the data to disk and updates a status file to allow the KBS to load the latest data from disk as it is required. This configuration allows data logging activity to be performed in real time together with data display and analysis. The program structure is shown in Figure 4.

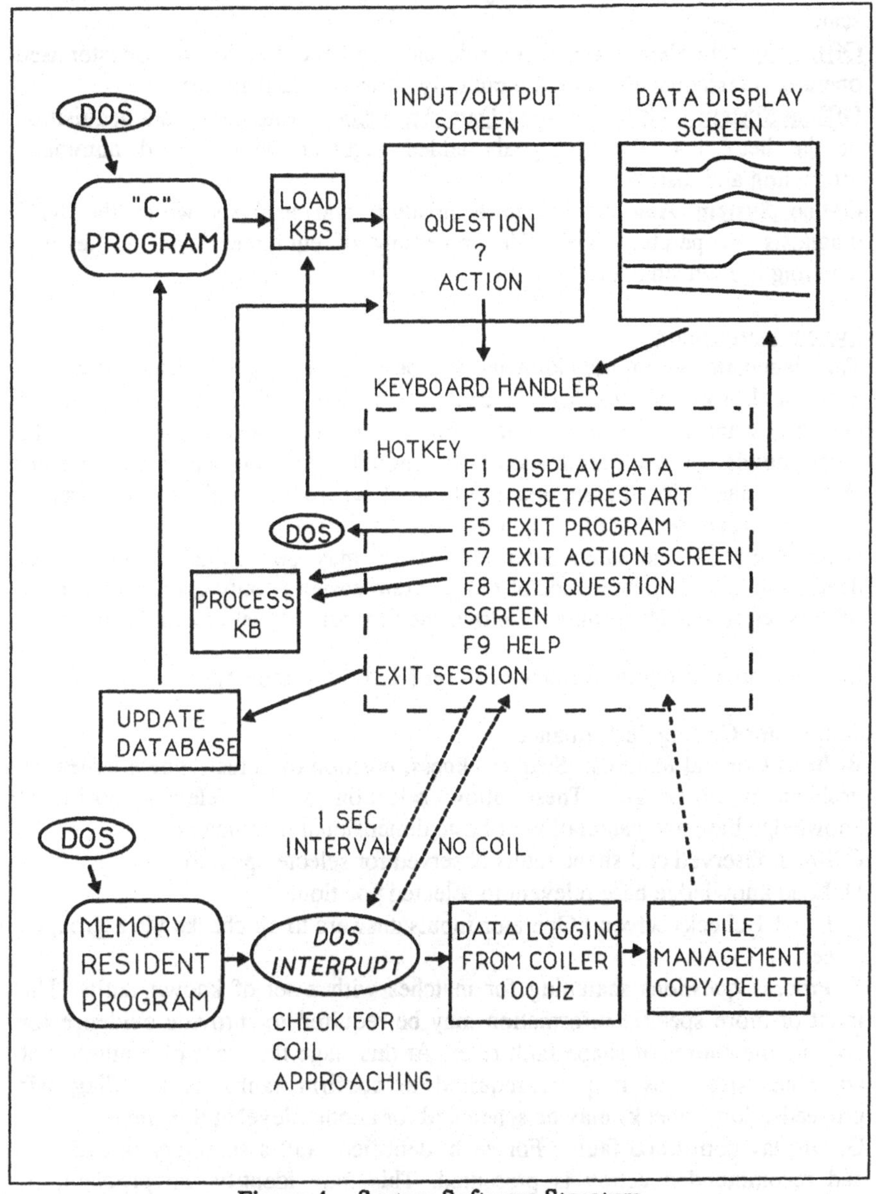

Figure 4 - System Software Structure

Development Stages

The system was developed in the four stages:

Prototype System Determine the level of knowledge available, build rule structures, construct candidate program formats, and demonstrate on a reduced scale.
Offline System Expansion of the rule base and the addition of a customised operator interface with manual input only [Question & Answer].
Offline System with background Data Acquisition Automatic data acquisition of the head end timing signals added together with limited automatic processing and display.
Online System Automatic data acquisition and analysis while the KBS diagnosis is operating. Data file processing of equipment data at Level 3 resulting in reduced operator input.

System Description

The diagnostic system uses knowledge bases to process questions/answers/data and search for conclusions. A central knowledge base containing all the definitions and control rules is in computer memory at all times, while the other knowledge base's are loaded as required, to focus the search and present relevant data for the section of the coil where the problem is currently observed [head, body, tail]. Each of these positional knowledge base's has the same format so the same control rules are may be applied. This makes development and maintenance of the program easier as only one set of control rules is required. The program follows the procedure illustrated in Figure 5.

Explanation of Program Control Features [Refer to Figure 5]

A. Monitor Coiling Performance
B. Input General features: Strip thickness, position of defect, which coiler the problem is observed. These allow selection of the relevant positional knowledge base and calculation of basic dimensional parameters.
C. Input observed coil shape faults observed for selected position:
D. Load knowledge base relevant to selected position:
E. Level 1 checks screen : Operator inputs answers to all checks via a grouped screen.
F. Process questions searching for matches with a set of known faults: The input of more specific information may be required to provide evidence for proving the causes of shape fault rules. At this stage the input of a number of additional questions may be required to confirm faults as the diagnosis proceeds. Some checks may be scheduled for another level at this time.
G. Display confirmed faults: For each identified fault a summary of problem and recommended action is presented. This may identify the problem, or suggest equipment adjustment or further checks to confirm the problem.
H. Roll another Coil: Test if problem still exists. If fault no longer observed, finish diagnosis, update database and finish diagnosis.

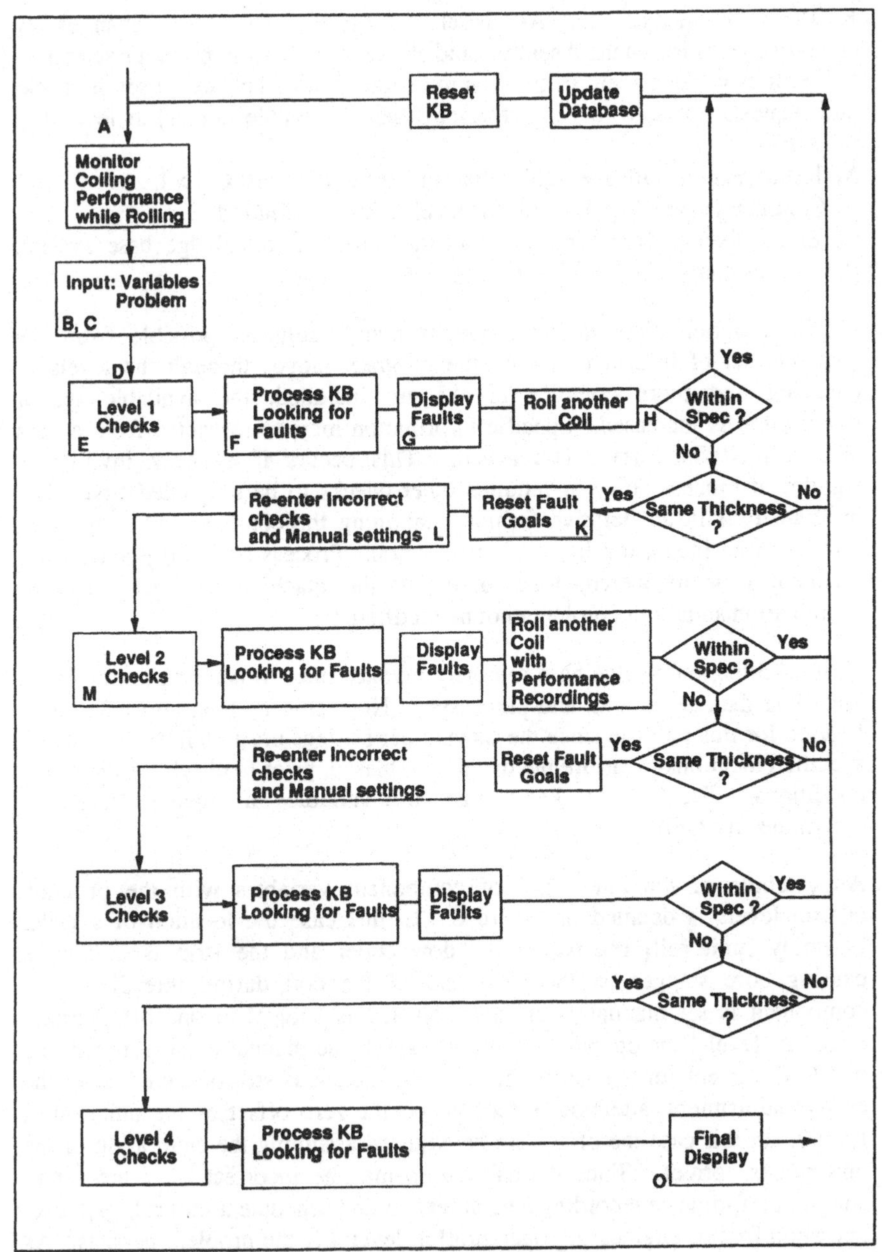

Figure 5 - Program Flow Chart

I. Check specification/dimensions: A significant portion of the data is specific to broad product thickness dimension. Data must be reentered if there is a change and dependent variables reset.

J. Increment Level of checking to level 2:

K. Reset knowledge base: Ask level 1 questions again [only changes are entered] to confirm manual settings and changes made from previous search.
L. Level 2 checks screen: In addition to further standard checks, a set of checks and requests not able to be performed [due to possible delays] at level 1 is presented.
M. Reprocess questions searching for matches with faults: G. -> I.
N. Continue processing J. -> M. for levels 3 &4 if required.
O. Final Display: Suggest action outside scope of knowledge base system. Present summary of relevant variables.

The program aims to determine as many faults as possible from the available set of information. As the program moves through the levels of checking and more complex checks are included, the available set of information is increased. Specific information may be restricted from access until a predefined level is reached. This occurs if a check involves a significant process delay or requires other checks to be scheduled first. The program runs in an iterative fashion examining the complete set of relevant rules with an increasing body of information. Process standard practices are maintained or recovered prior to setting the machine for more detailed examination should the problem not be rectified.

At level 3 the knowledge base routinely invokes rules to perform performance recording data extraction, and processing. Normally the next coil rolled will be used for this purpose since the downcoiler settings must be in their standard specified locations to ensure the data is not biased, prior to check for abnormal conditions. The resultant knowledge base variables are then processed to determine any faults.

An example of the interaction of component variables with the program control levels is detailed in Figure 6. In this case the location of a roller assembly [unit roll] relative to the downcoiler and the strip is critical in causing bore stagger on the head end of the coil during threading. The component is set manually for each coil and is subject to one off operator error. At level 1 the operator is simply asked if the manual is correct resulting in an adjustment for the next coil. If the problem is still observed, with the correct adjustment, attention is focused on the zero offset of the unit roll at level 2. In this case the offset can be adjusted assuming the calibration of the unit roll is correct. Since the calibration may be incorrect flags are set to analyse performance recording data at level 3 and schedule a unit roll gap reset operation [major engineering operation] at level 4 if the problem persists. As the program moves through the levels more complex information is scheduled for analysis by the rule set. At the same time the relevant search space is decreasing by elimination of minor problems in an iterative fashion. This method achieves efficient diagnosis within the strict frame work of production priorities.

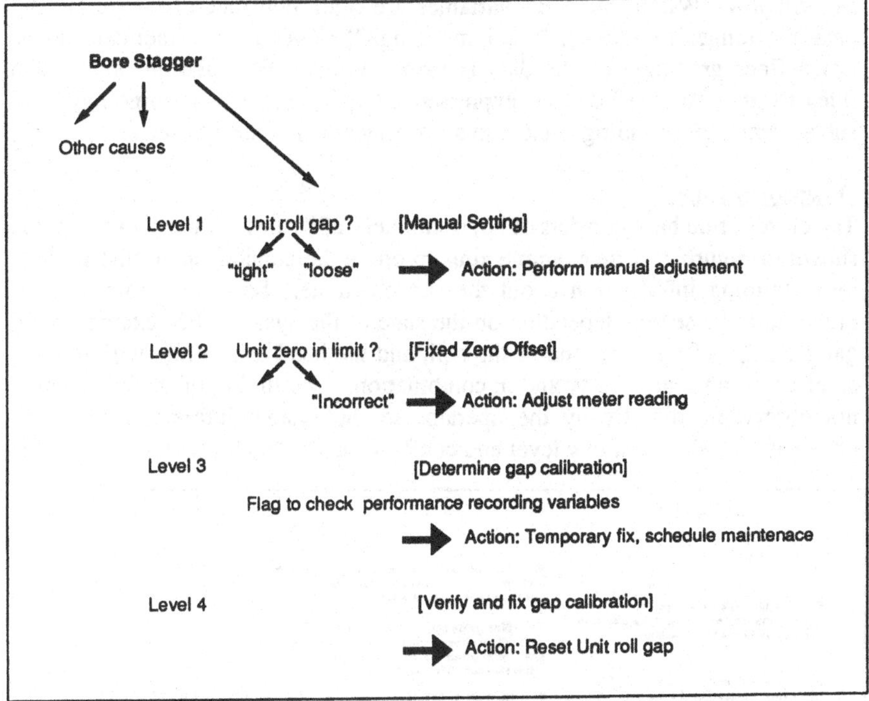

Figure 6 - Partial Decision Tree for a Roller Assembly Component with reference to the Program Control Levels

Data Analysis

Each data file consists of digital time series of the individual signals recorded at 1 msec intervals. These files are preprocessed at level 3 to identify critical "features" or variables for inclusion in the KBS rules search. The preprocessing routines are a sets of "C" functions which extract and classify the data as follows:

A set of basic events specific to each signal is extracted from the time series. These events include; a positive change, exceeding a threshold, a positive/negative value, change in slope, momentary flattening of a slope, attaining a constant level.

Up to 23 predefined "features" are identified or calculated from the basic events. This may involve a simple time difference, a series of events occurring in order, or a Boolean flag to signify if the event has occurred.

Specific features are normalised to allow comparison across a range of coiling conditions [eg mill speed, strip thickness]. Many features use the absolute or relative time of events and so are affected by the coiling conditions in operation, especially the speed of the operation. Range checking is performed to eliminate incorrectly measured data.

Feature values are compared with a database of heuristic "conditions" to determine a classification range where appropriate and transferred to the KBS as variables. Within the KBS variables are often conveniently described by heuristic ranges [eg "low", "medium", "high"] rather than numerical values. Often finer gradation of the data is inappropriate. Formulation of the data rules by the "experts" can be expressed using appropriate terminology, and subsequent understanding is clear to all operators using the system.

Downcoiler Rules

The current rule base consists of approximately 200 rules. A partial rule tree is shown in Figure 7. The example aims to prove the conclusion "found a fault" by examining information about the run out table. Several factors may be drawn into the search depending on the state of the system. For example strip gauge, shape of the head end of the coil, and the head end timing will all have an effect both individually and in combination. The timing of the head end is not observable directly by the operator so the system infers it from other observations while at a low level and confirms this at level 3 with process data.

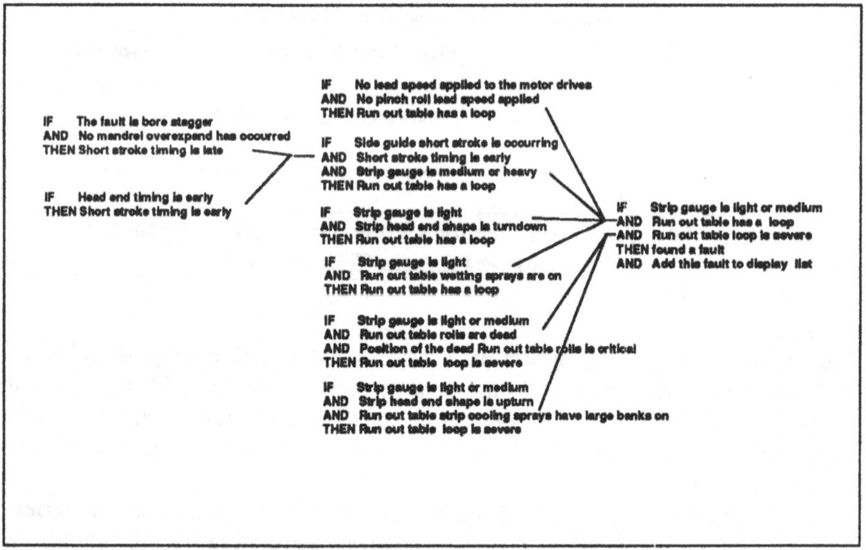

Figure 7 - Example of a Rule Decision Tree

DIAGNOSIS ISSUES

Large manufacturing processes often consist of many individual operations linked together to produce a final product. Each operation can be sufficiently complex to warrant the use of sophisticated procedures and maintenance and diagnosis strategies. To coordinate these activities for an individual operation and over the whole plant, detailed standard practices are developed to cope with the many problems which are likely to occur. Personnel working in these environments are trained to use these standard practices so that the process runs smoothly no matter who is in control.

Since any candidate diagnosis system must be fully compatible with the standard techniques and practices, it is important to consider implementation issues at an early stage to ensure that the final system gains total acceptance. There are several important areas which must be considered.

Customised User Interface

The computer interface must have a "look" and the "feel" that is compatible with other software systems in operation in the plant to be accepted and used by operators. The functionality of the keyboard and the screen formats must be correct to avoid confusion. Data input should be, where possible, via choice lists to avoid incorrect data entry. Consistent screen design must be maintained throughout, and context sensitive help available to explain input questions by referring to operational manuals and procedures. Where input of a large number of variables is required on a routine basis, repeditiveness can be reduced by presenting logically grouped questions on each screen with dynamic default values reflecting the current system status. Operator input is reduced to changing variables that are incorrect rather than inputing a complete set of information at each cycle. Such groupings may include information irrelevant to the current problem but this method ensures adherence to mill operational practices while performing problem diagnosis.

Other important features of a successful user interface are simple and concise presentation of diagnostic information, and careful attention to include terminology appropriate to the level of operator skill. Also the operator must have ability to override the system at any time to retain operational control.

Program control

Possibly the most important requirement for a diagnostic system in an industrial process environment is that the total system sequencing or event control conform closely to standard operational procedures or standard practices. If this is not the case then the scheduling of critical process and maintenance activities may be inaccurate and lead to delays in finding the root problem cause, thus increasing processing costs.

Achieving tight event control within a KBS program may require the use of a separate set of rules from the diagnosis rules. These rules perform overall control and set flags to restrict the diagnostic search to parameters relevant to the current available body of information. As diagnosis proceeds and more complex information is obtained from the process then the scope of the rule search is increased. In this way adherence to standard practice is maintained at all times. The knowledge base rules are configured as a combination of forward chaining strategy for the selection of the search scope, and backward chaining to search for specific problem goals within the current search scope.

The current system has four levels of search complexity designed to establish standard conditions and eliminate minor problems prior to committing substantial resources. At each level, revision of the input variables is required to check that standard machine settings have been established before continuing to perform checks of a more complex manner. For example a component setting may appear to be in error, so initially the program would check if the manual set-up input was inaccurate. If this was correct then the offset is checked and finally the zero of the instrument. Thus simple causes are eliminated before scheduling a check of the component's zero setting.

Data Acquisition /Classification

Real time process data is often required to enhance the diagnosis capability of a system. In fast processes visual observation of system behaviour is not possible. Engineers routinely use process variables to monitor the performance of the equipment. The data must be available to the KBS to provide accurate real time analysis at a high level. This helps to reduce the amount of information entered manually into the system which is subject to error and is often a time consuming exercise. Often simple logical computation or comparison with standards will expose many basic problems quickly and efficiently.

Operator/Engineer involvement

At a human level the involvement of the operators during the development phase is essential to completely represent all aspects of the problem domain and in gaining system acceptance. Operators will tend to trust their judgement over a machine if they have no input into the content of such a system.

A data base of the system usage and diagnosis results will highlight areas where improvements may be needed. Operator comments will provide feedback on system operation and possible new diagnoses.

Prototype Developments

Since knowledge is in the form of personal experience it is difficult to determine at the start how much knowledge exists about a particular topic and how to use it in the final system. One method used successfully elsewhere is to develop prototype systems. The objective of a prototype is to test the technical feasibility of the system, and establish the final requirements without committing substantial resources. It presents an opportunity to review the system function and to demonstrate it's use. Development of a prototype consists of building a simple system with a specific but reduced scope. The facts and basic elements of logic are included in detail to test both the available level of the "experts" knowledge and the functionality of the system. Little attempt is made to customise the "look" of the user interface of the program as this is often time intensive. Once the system format and functionality have been finalised it can be expanded to the full scale system adding the majority of rules and an appropriate user interface.

Knowledge Acquisition

The data and information which form the basis of the rules and decision structures in the system are developed from existing experience about the problem. The information is drawn from a series of discussion sessions with persons acknowledged as having an outstanding knowledge and/or experience about the area of interest ["experts"]. The discussions use available standard practices and problem scenarios to trace the steps involved in solving a problem in a logical manner, enabling descriptions of the inter relationship of the component parts of a problem. In the case of the downcoiler a small group of mill personnel were chosen to cover all relevant aspects of the downcoiling process including electrical, mechanical and operational areas.

CONCLUSIONS

A diagnostic system has been developed to improve the efficiency and reliability of the existing coiling performance trouble shooting process by providing a systematic diagnostic investigation of the downcoiling operation. Problems encountered during the development phase have demanded the final system rely on strict event control to accurately follow the standard mill practices.

Data logging of critical variables from the process has provided additional information to the KBS and significantly increased the usefulness of the system. Up to 23 data features are extracted and classified for inclusion as variables into the KBS system at level 3.

The system is currently installed in the downcoiler pulpit at the Westernport Hot Strip Mill of the Coated Products Division of BHP Australia. The impact of the system in reducing costs due to poor coiling performance is currently under evaluation.

ACKNOWLEDGEMENTS

The author wishes to thank The Broken Hill Pty Co Ltd for permission to publish this paper. The author also wishes to acknowledge the effort of the staff at the Coated Products Division Hot Strip Mill who were involved in discussion sessions and in particular Ed Neil who provided a major input into the project.

Towards an Automated FMEA Assistant

A.R.T. Ormsby, J.E. Hunt, M.H. Lee
Artificial Intelligence and Robotics Research Group, Department of Computer Science, University College of Wales, Aberystwyth, Dyfed SY23 3BZ, UK

Abstract

Failure Mode Effects Analysis (FMEA) is an important part of the design process for a large number of complex systems. However, current techniques for FMEA are largely manual, and tend to be slow and tedious. The FMEA procedure itself is usually carried out by experts whose time is expensive. For these reasons, any system which can help automate some aspects of the process are likely to be extremely valuable. This paper describes some work in progress to produce an expert system which can assist an engineer in the FMEA process by emulating some of the reasoning which an expert uses. The current application domain is automobile electrical systems, though the architecture of the system is extensible.

Introduction

This paper describes some work being carried out with the aim of producing a system which will assist engineers in performing failure mode effects analysis, (FMEA). We are currently investigating the analysis of electro-mechanical systems in automobiles. The first section (*The FMEA Context*) briefly describes the background to the project and the importance of FMEA.

In the following sections, two aspects of the work are described. The first of these sections, (*The Jacquard System*) describes the background to the diagnostic architecture which we are adapting to perform FMEA. The reasons for choosing a model-based approach and the decisions which have lead to this particular architecture are explained. Then, in the final section of the paper, one of the model simulators mentioned in the section on the Jacquard System is described. This simulator provides a means of modelling the structure and behaviour of electrical circuits in order to allow the effects of faults to be explored.

The FMEA Context

FMEA involves postulating faults in components of a system, examining the effects of such faults on the system as a whole and then analysing these effects to determine whether these faults are sufficiently probable or serious to merit redesign. FMEA is of growing importance in engineering as a means of both improving manufacturing processes and design of systems. It is of particular importance now that the responsibility of designers and manufacturers of products which may cause injury has been made much more explicit by the introduction of product liability legislation.

Coker [1] identifies two varieties of FMEA, *process* and *design* FMEA. The former is concerned with examining how the manufacturing process of an artifact may introduce faults, and what the effects of these faults may be. The latter is aimed at the indentification and assessment of all the potential faults in a device, with the aim of making changes to the design so as to eliminate the potential fault or mitigate its effects.

In order for design FMEA to be really succesful, it is important for it to be carried out as early as possible in the product lifecycle. This is emphasized in Dussault [5], together with the need for more efficient procedures. Only by performing FMEA as early as possible in the design process will the full benefits be realised and will the ability be gained to fold the knowledge obtained back into an improved design. The problem is that current manual FMEA techniques are slow and tedious to conduct. Not only that, but the process is particularly expensive as FMEA must be done by experts since the effects of a particular fault in a component may be wide-ranging, requiring considerable domain knowledge to predict and analyse.

As well as identifying potential faults, the engineer will typically also need to estimate the likelihood of each particular fault, its severity and the likelihood that the fault will be detected. (For example, the failure of the ignition system in a car would be obvious — the engine would not operate. But what are the effects of the failure of a warning light on the dashboard, a failure which may itself go unnoticed and which will potentially mask some other fault?) All of this requires a high degree of expertise and the exercise of judgement based on considerable experience. Given this, it is perhaps not surprising that to date, relatively little automated assistance is available to engineers involved in FMEA. Those computer systems which are available are mainly confined to assisting in the clerical work connected with the documentation of the process.

The Jacquard project is looking into the design of a expert system which will eventually act as an engineer's assistant in the FMEA process. This paper briefly describes the general outline of the project and the system that is being built, and then discusses some aspects of the electrical circuit simulator which is under development.

The Jacquard System

The impetus for the work described in this paper comes from the fundamental nature of the FMEA problem. That is, FMEA is the analysis of a system design in order to determine the potential faults so that the effects and possible consequences of these faults can be discovered. This may appear to suggest that heuristic-based expert systems are an appropriate technology to apply to this problem. However, as such first generation expert systems rely on the fact that the effects (and consequences) of potential faults are already known, these systems are inappropriate.

Although FMEA experts do refer to information which can be heuristic in nature, for example "I know that the failure rate of wires is low", the type of reasoning used to determine the behaviour of a system, when a particular fault mode is present, relies on the experts understanding of the domain. That is, the expert must resort to first principles in order to determine the effect of a particular fault mode. Very often, in man-made devices, the sort of knowledge that the human expert will use is related to the structure of physical objects in the domain plus relevant technical knowledge — in a chemical processing plant, the expert might be concerned with the structure of the plant (tanks, pipes, valves, etc) and the chemical reactions taking place in the plant.

This has lead us to investigate of how to apply model-based reasoning techniques in order to develop advanced artificial intelligence tools. In particular we have been examining the development of tools which will support the FMEA process for, and diagnosis of, vehicle control systems involving mechanical and electrical sub-systems.

Model-based reasoning

Model-based reasoning takes knowledge about the entities, structures and interactions in a particular domain and uses that knowledge as a foundation for problem solving. The key feature is that a model is maintained which *mirrors* the important structure and features of the domain. Many current model-based diagnostic systems generate fault hypotheses using only the information available within the model of the system under diagnosis. This is done by comparing predictions about the behaviour of modified versions of the model with the observed symptoms. Any modification that can generate all the symptoms is considered to be a possible explanation for those symptoms (for example [4], [2], [3] and [14]). However, such model-based systems ignore many of the processes, and much of the information, pertinent to the effective effective problem solving. For model-based reasoning systems to handle real applications, they must also treat problem solving as complex processes which require additional information not available within the model.

Augmentating model-based systems with such information can provide significant benefits, however it is not without its limitations. Its main drawback is that such additions tend to be incorporated directly into the model-based system in an ad hoc fashion:

- The additional knowledge required tends to be implicitly incorporated in the model-based system.
- Additional control structures must then be added to exploit the additional knowledge.
- Useful problem solving short cuts or tactics are often "hard coded" into the system until it achieves an acceptable level of performance.

Because these additions are added directly to the model-based system, it often becomes difficult to reuse parts of it in different domains. Worse, due to the "tweaked" nature of the system, which has implicit short cuts and domain information, it may become more fragile, failing to identify unlikely or novel faults. All this can lead to a system which is convoluted in structure, opaque in operation, specific to one application and may have lost any pretence of robustness.

For this reason, the augmentation of the model-based system must be carried out in a principled manner in order to benefit from the potential gains available from using relevant additional information while still preserving the advantages of model-based reasoning. One way to facilitate the easy construction of well–engineered model-based systems is to implement an architecture which encourages the system builder to explicitly distinguish between the model and the knowledge that is added to the system to improve problem solving performance. Such an architecture has been discussed in [9] and [10] and is outlined below:

A framework for the architecture

The first issue to consider is what type of framework is suitable for this architecture. An obvious candidate is a blackboard framework which is presented below. However, we will show that this is not an ideal solution to our problem and present a more suitable framework.

Blackboard architectures In a blackboard system (see [6]), a set of problem solving modules, typically called knowledge sources, share a common global database (called the blackboard). The contents of the blackboard are often called, and indeed denote, hypotheses and are often structured hierarchically. Knowledge sources respond to changes on the blackboard, and interrogate and subsequently directly modify the blackboard itself, by creating, modifying and solving hypotheses (in fact they are the only elements allowed to make changes to the blackboard). Each knowledge source only responds to a certain class or classes of hypotheses which often reflect the different levels in the blackboards' hierarchy.

BB1, described by Hayes-Roth in [8], extends the basic blackboard architecture described above, by adding a blackboard control architecture. There are now two blackboards; a domain blackboard and a control blackboard. The domain blackboard is just the same as the blackboard described above, the control blackboard however is concerned with generating a solution to the control problem. The control blackboard works in exactly the same way as a standard blackboard, however the knowledge sources are now control problem solvers and the solutions on the blackboard are solutions to the problem of which domain knowledge source should be activated next.

However, there are significant drawbacks associated with the use of a blackboard architecture as the framework within which to build out task specific architecture (for example, specific to diagnosis or FMEA). These include the following:

No communications language. The language for passing information between problem solvers is the language of the intermediate hypotheses on the blackboard. This means that rather than an explicit request for some task to be performed, such as the generation of a simulation, the request must be phrased in the terms used to describe partial solutions on the blackboard — this may or may not be appropriate.

Computational complexity of cooperation. The problem of determining which knowledge source, of the applicable set of knowledge sources, to invoke at any time has to be solved dynamically. In the case of BB1 this can be computationally very expensive, this is because BB1 faces the problem of the potential unbound nature of determining the next knowledge source to invoke. That is, if there is no control of the control blackboard, a great deal of control "problem solving" could occur before any real work is begun on solving the problem on the domain blackboard.

No guidance for task. Blackboard architectures do not give any guidance to a system builder on how to solve a particular problem or perform a particular task (such as design or diagnosis) within their framework.

No exploitation of task specific strategies. There is no explicit support for task specific and domain specific problem solving strategies, such as useful diagnostic short cuts. Such strategies would have to be implicitly added into the scheduling system of the architecture.

Global database. It is directly modified by the knowledge sources and assumes that knowledge sources will behave responsibly, if they do not chaos could ensue. This assumption goes against all the information hiding techniques which have been developed by software engineers to promote well engineered systems.

Difficult to incorporate existing facilities. It can be very difficult to incorporate existing resources such as existing databases etc. into the blackboard architecture.

Task specific architectures A task specific architecture provides structures for designing and building large grained tasks such as diagnosis or design. That is, the architecture is specific to a task such as diagnosis and can therefore provide facilities which will explicitly support that task (see [12] and [13] for a more detailed discussion of task specific architectures). The main benefits of task specific architectures are:

- that if knowledge of the tasks to be performed to solve the overall problem is available, then such knowledge can be exploited. This saves the dynamic generation of the task structure every time a problem is processed.
- that such an architecture provides a framework within which particular kinds of problem can be solved, and that such a framework provides a level of support for the knowledge engineer above that of a general problem solving architecture (cf. blackboard architectures).
- that the architecture can be tuned to the task being performed while possessing a simpler structure. This is because the additional flexibility of general problem solving architectures is not required.

In order to provide such benefits, task specific architectures must explicitly represent certain task specific information such as the task structure of the problem and potential task specific problem solving strategies which can guide the system to more effectively solve the problem. The architecture must also provide control strategies for, and support cooperation between, the problem solving modules which comprise the system. Thus the main features of such task specific architectures include:

- The integration of multiple problem solving elements. Each of which is expert at some task or operation, e.g. simulation.
- The provision of facilities for describing a complex problem in terms of the goals and subgoals required to solve that problem.
- The set of tasks required to process these goals and subgoals, called the task structure of the problem.

- The availablity of a set of problem solvers, selected on the basis of their ability to perform these tasks.
- Strategies for selecting which goal to satisfy next and subsequently which task to attempt next in order to satisfy the current goal.
- Strategies for selecting the next problem solver to allocate the current task to, based on dynamic case specific data (e.g. symptoms, requirements etc.) and problem solving specific data (e.g. problem state information).

A task specific framework

The manager–client framework The manager – client framework, presented in figure 1, provides the infrastructure for effecting control of, and cooperation between, the diverse problem solving modules of task specific systems built within the generic architecture. The manager represents the role of the managing director and the clients represent the directors on the board. A client is a resource, such as problem solvers, databases, programs etc. combined with the client interface module. The manager and all the clients are connected to a communications medium, represented by the circle in the center of the figure, through which information is exchanged.

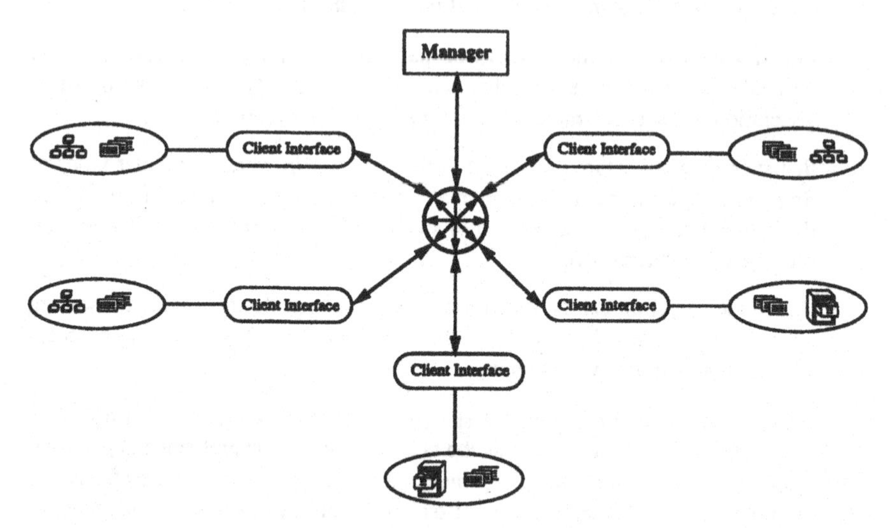

Figure 1: The manager-client approach

The manager monitors the progress of the problem solving process in order to maintain an overview of the current state of the problem. It uses this information to determine what goals to satisfy. These goals in turn dictate which tasks are selected. The manager also determines which clients to allocate the current task to. It is able to do this becuase it maintains a register of all the clients available to it as well as profiles containing any other relevant available information, such as the type of inference mechanism used (e.g model-based or heuristic-based reasoning). This information is made available to the manager by the clients either during an initial configuration phase or dynamically, during problem solving, when they become available. The manager also maintains the database of emerging solutions.

A client is the combination of some resource with the client interface. The client interface provides facilities for:

- integrating resources into the system. The client interface acts like a *plug in socket*, providing facilities for connecting to the communications system resources' existing interface.
- exchanging information with the manager and any other clients connected to the communications system.
- explicitly representing the problem solving role of the client and any other relevant information which is to be made generally available.
- accepting a request for information or problem solving activity which explicitly refers to the client.
- determining whether a general request for information or some task can be accepted, or not, based on the actual request and the role of the client.

This allows any problem solver, database, or tool to possess whatever representation and reasoning technique is appropriate. For example, a user interface written in pascal, or an Oracle database can be combined with a model-based reasoning system without compromising any one of the systems.

The third element of the diagram is the communications system, which is a fully connected star network. That is, it possesses a central node, which all clients and the manager are connected to. This means that all exchanges of information between diverse problem solvers are effected via the communications system. The communications system also provides a common language in which to express the domain information in order to promote the integration of that information. This common language is based on the language of the domain model, which provides the common reference for all clients. The language of the domain model refers to the terms used to represent the model of the system or device under investigation. For example, the terms used to represent an electrical circuit might include, resistor, wire, switch etc.

A diagnostic specific architecture A diagnostic specific architecture has been built using the design features described above. A diagnostic specific instance of the manager has been generated which possesses all the functionality of the manager. It also explicitly represents the goal structure of the diagnostic task. This includes; *Acquire symptoms*, *Identify candidate faults*, *Verify candidate faults* and *Return the diagnosis*. The tasks required to process these goals are also represented in the task structure of the manager. The tasks include; *Retrieve Symptoms*, *Generate candidate faults*, and *Verify selected faults*.

A diagnostic expert system has been implemented within the framework of this architecture and is illustrated in figure 2. It has been documented in greater detail in [11] and [10]. The clients available within the system include:

Model simulator This client animates a model returning a description of its behaviour.

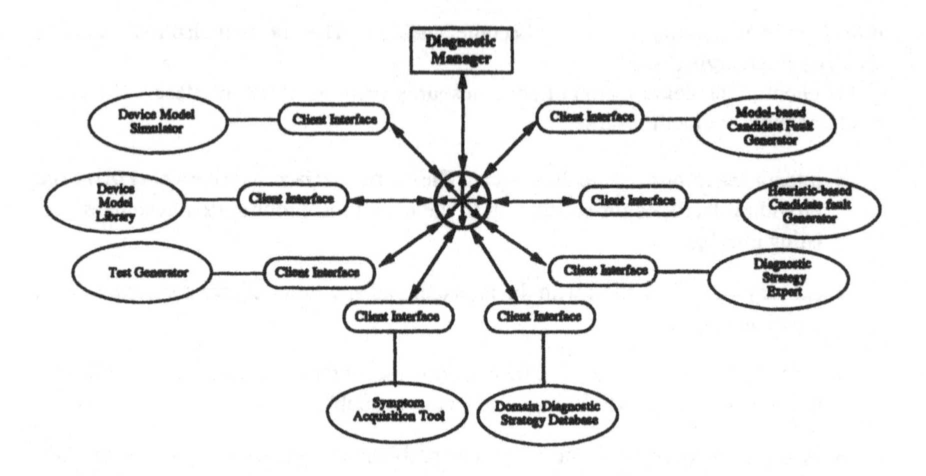

Figure 2: The Jacquard Diagnostic System

Model library manager This database contains libraries of model descriptions of real systems and their constituent parts.

Model-based candidate fault generator This problem solver generates a list of those modifications to the model which could potentially account for the symptoms.

Heuristic-based candidate fault generator This problem solver generates a list of possible candidate faults suggested by heuristic knowledge.

An FMEA specific architecture

At a simplistic level Failure Mode Effects Analysis (FMEA) is the analysis of possible faults and their likely effects. In this respect, FMEA is the reverse of diagnosis, i.e. FMEA considers possible faults and attempts to generate the possible associated behaviour, while diagnosis considers the behaviour and attempts to identify the possible faults which generated that behaviour. However, like diagnosis FMEA has significant problems in the number of possible faults and combinations of faults that must be examined.

We are currently working on an FMEA specific architecture built using the task specific framework used to implement the diagnostic specific architecture. The diagnostic manager is replaced by an FMEA manager. Both managers are built on top of the generic manager, but with different goal and task structures. The set of clients available within the FMEA architecture will also differ.

The Circuit Simulator

The particular FMEA domain which we have chosen to investigate is that of automotive electrical systems. Automotive electrical systems are rapidly increasing in complexity, with many more devices on a typical car now being electrically powered than just a few years ago. Some of these systems could be considered safety critical, with anti-lock braking systems and computer controlled engine management systems being two of the more obvious examples. This trend seems set to continue with "drive-by-wire" cars now likely to appear within the next few years.

For our current work, we have chosen to concentrate on the electrical systems after discussions with our industrial collaborators. We hope to examine non-electrical aspects later on in this project. As a step towards the FMEA for electrical systems, we have developed a qualitative circuit simulator which allows the exploration of the effects of certain types of failure on the rest of the circuit.

Why Qualitative?

When deciding to construct a simulator for electrical circuits, one of the first questions which was addressed was whether or not it was necessary to be able to produce precise numerical values for the various variables associated with the circuit — for instance, the voltage and current levels at each point in the circuit. Methods for doing this, most obviously using a package such as SPICE, already exist. However, this kind of solution would not accurately model the type of reasoning that seems to be employed by engineers when performing this kind of process. Presented with a simple circuit diagram, or even part of a more complex one, an electrical engineer will usually be able to say general things about the direction of current flow, the existence of short-circuits, and so on, without doing any precise numerical calculations at all. This "seat of the pants" reasoning is also usually very quick. As a result, we have chosen a *qualitative* as distinct from a *quantitative* approach.

Using a model-based approach, an object-oriented model of electrical circuits has been constructed. In this, the various components, such as connectors, wires, bulbs, and resistors are directly reprsented by instances of objects in the simulator. However, resistances are represented not by numerical values, but by one of three qualitative values: *zero*, *load* or *infinity*. Zero-load components represent items such as wires and connectors which have no significant resistance; load components represent those which do have a significant resistance, while the qualitative resistance value of infinity is used to represent components through which there is no significant current flow, such as at open switches or blown fuses.

The qualitative approach seems to be suitable for many of the kinds of faults which engineers are interested in when doing FMEA. "Stuck-at" faults are common, where a particular point on the circuit is "stuck" at the value of battery positive or negative. These failures represent changes in the topology of the circuit due to short- and open-circuit faults. Since the behaviour of the model is deduced in part directly from its underlying structure, structural changes of this type are not difficult to handle using the qualitative model.

A qualitative circuit algorithm

Having constructed a model of the structure of an electrical circuit, the next stage is to be able to make statements about the behaviour of that circuit. To do this, it is necessary to draw on some of the existing work in circuit theory and recast this in qualitative terms. Thévenin's theorm, for example, only shows that a change to a resistance at one point in the circuit makes all other points in the circuit candidates for change — not terribly useful if the only values are coarse symbolic ones as they are in this qualitative scheme. It therefore seems that a minimum of Kirchoff's and Ohm's laws are necessary to support reasoning about the behaviour of qualitative models of circuits.

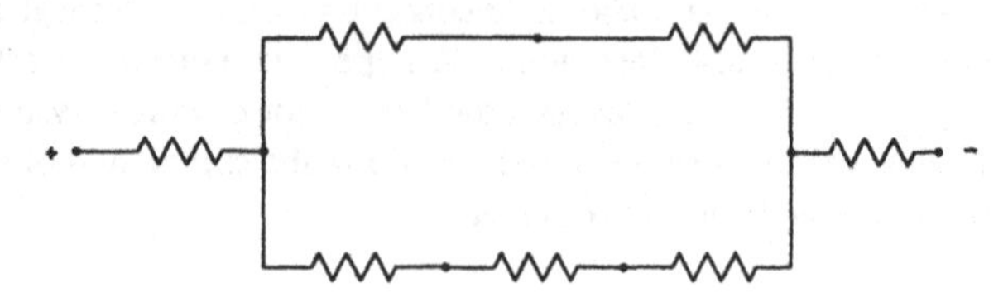

Figure 3: A simple circuit

In order to give some of the flavour of this technique, consider the very simple circuit in figure 3. Our algorithm works by treating the circuit as a graph of nodes, where electrical connectivity is represented by arcs in the graph. Using a variation on Dijkstra's shortest path algorithm, a form of best-first search, a procedure can calculate the shortest path between source and sink nodes, which initially might be the battery positive and negative terminals. A *load* qualitative resistance at any node in the circuit represents a cost, or distance increment in the path from the source to the sink; nodes with zero resistance add nothing to the distance. By performing this process from both positive and negative terminals through the circuit, it is quickly possible to identify short and open circuits. For what we have termed the "primary paths" in the network, this algorithm also unambiguously indentifies the direction of flow.

Figure 4 shows the example circuit as a graph. The circuit can be seen to have two paths. Tracing along each path from one of the terminal nodes to the other, it is clear that one path is shorter than the other. This is the "primary path". For each junction node on the path, two numbers in the form (f/r) give the distance from the positive battery terminal and the negative battery terminal (or source and sink) respectively. An interesting property is that the sum of these two values (the *total path resistance*, "*tpr*") remains constant along the primary path. From this, not only can the primary path be identified, but the direction of current flow along it can be determined using the convention that if the first of the two figures rises when the second decreases, a movement to a point "nearer" the sink node has taken place. (The primary path, and the direction of flow are denoted by solid arrows in figure 4). Where it is not possible to reach one of the source or sink nodes, that component of the *tpr* is marked infinite. No current can be flowing on that path.

If there are no non-zero values for the distances from the source and sink nodes on a path, then that path is a short circuit. This case is quickly found and can be flagged.

Extending this somewhat further, secondary paths (denoted by dashed arrows) can be identified as any in which the nodes have a total path resistance greater than that of nodes along the primary path. Careful case analysis of the transitions can allow current direction to be inferred in many cases. Methods for identifying potential future current flow (at an open switch, for example) have also been developed.

The example shown in figure 5 shows what might happen if a typical fault is introduced. In this case, a short to battery-negative is introduced at a point in the

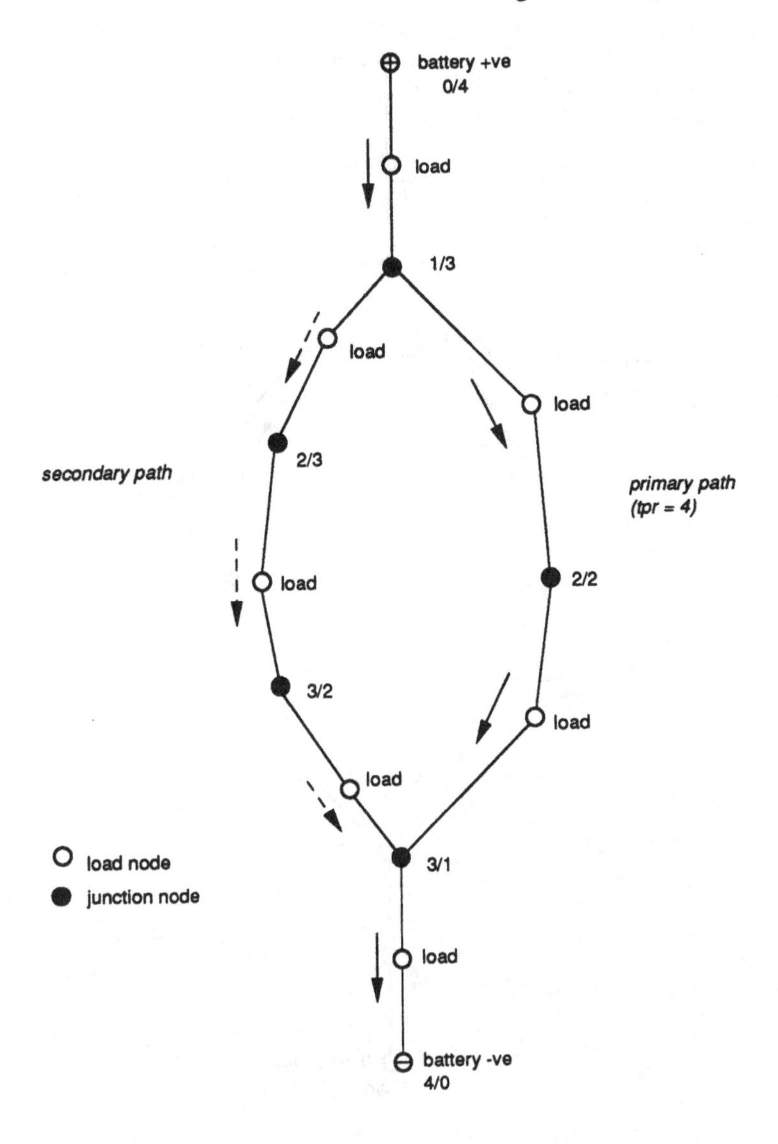

Figure 4: Representation of the simple circuit

cirucit. The new distances from the positive and negative terminals are calculated and the direction of flow is amended accordingly. It can be seen that the right-hand part of the circuit, which had perviously been the primary path, is now the secondary path. The direction of flow across one of the loads in the circuit is has also changed.

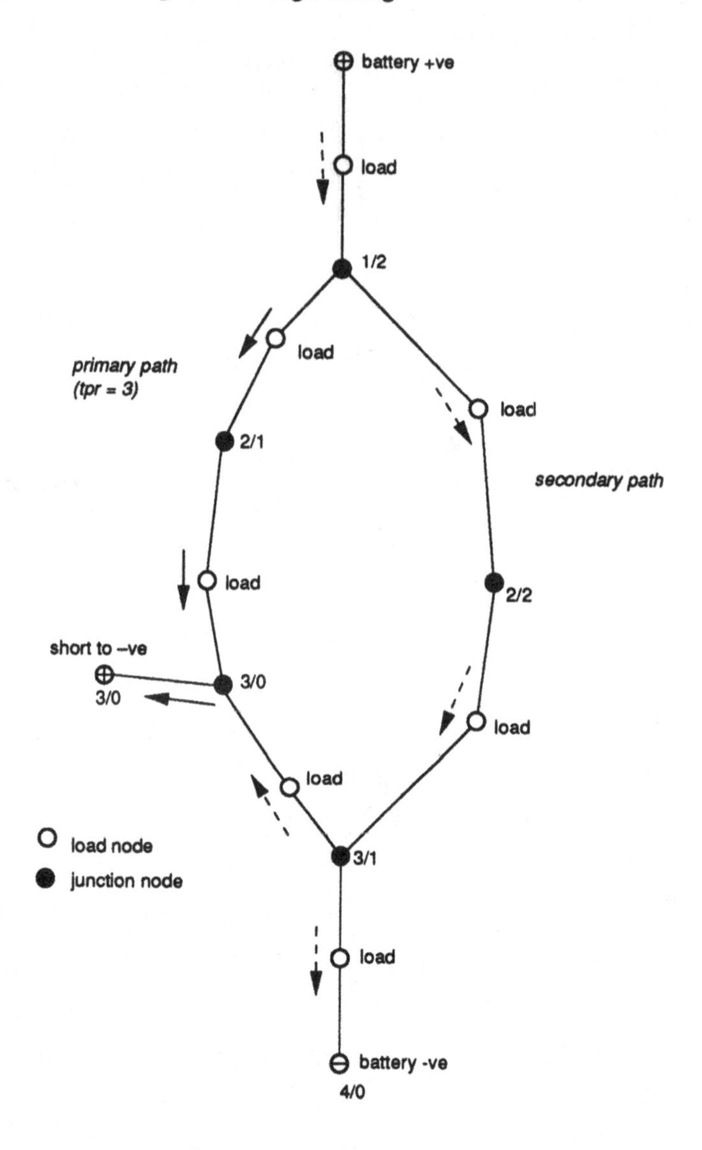

Figure 5: A short-circuit to positive and its effects

Further work

While the simulator forms the core of an intelligent FMEA tool, and our architecture provides the basis for its use, there will need to be extentions in order for the system to be used effectively. The reasoning which the simulator provides is low level and confined to the electrical domain. We are currently examining methods for linking different domains together and integrating the simulator with higher-level, more functional reasoning. In order to perform the final stage of the FMEA process, information about the likelihood of fault occurrence,detection rates and application specific criticality information will all need to be incorporated. We intend to report on the results of these developments in the future.

References

[1] A. O. Coker, J. A. Smith, S. Higgins and D. C. Cameron, *Computer-based Failure Mode and Effects Analysis for Quality Management — A Case Study*, Quality Assurance 15:3, p89–94, September 1989.

[2] R. Davis, *Diagnostic Reasoning based on Structure and Behaviour*, Artificial Intelligence 24, (1984).

[3] Davis. R. and Hamscher. W., *Model-based Reasoning: Troubleshooting*, in Exploring Artificial Intelligence, (ed) H. E. Shrobe, pub. Morgan Kaufmann, (1988).

[4] J. De Kleer, and B. Williams, *Diagnosing Multiple Faults*, Artificial Intelligence 32, p97–130, (1987).

[5] H. B. Dussault, *Automated FMEA — Status and Future*, Proceedings Annual Reliability and Maintainability Symposium, San Francisco, January 1984.

[6] L. D. Erman, F. Hayes-Roth, V. R. Lesser, and D. R. Reddy, *The Hearsay-II speech understanding system: integrating knowledge to resolve uncertainty*, ACM Computing Surveys, 12(2), pp 213-253, (1980).

[7] F. Hayes-Roth, D. Waterman and D. Lenat (eds), *Building Expert Systems*, Pub. Addison Wesley, (1983).

[8] B. Hayes-Roth, *A blackboard architecture for control*, AI, 26 pp 251–321, 1985.

[9] J. E. Hunt and C. J. Price, *Towards a Generic, Qualitative-based, Diagnostic Architecture*, Proceedings 9th International Workshop Expert Systems and their Applications, specialized conference on Second Generation Expert Systems, Avignon, France pp253–268, (1989).

[10] J. E. Hunt and C. J. Price, *Performing Augmented Model-based Diagnosis*, Presented at the International Symposium on Mathematical and Intelligent Models in System Simulation, Brussels, September, (1990).

[11] J. E. Hunt and C. J. Price, *A Model-Based Diagnostic Architecture*, Presented at the 7th UK Deep Knowledge-Based Systems Workshop, Gregynog, April (1990). Available from the University College of Wales, Department of Computer Science.

[12] W. F. Punch, *A Diagnosis System Using A Task Integrated Problem Solver Architecture (TIPS), Including Causal Reasoning*, Ph.D. Thesis, The Ohio State University, (1989).

[13] W. F. Punch and B. Chandrasekaran, *An investigation of the roles of problem-solving methods in diagnosis*, Proc. 10th International Workshop on Expert Systems and Their Applications, specialized conference on Second Generation Expert Systems, Avignon, France, (1990).

[14] O. Raiman, *Diagnosis as a Trial: The Alibi Principle*, A.I. Research Department, IBM Paris Scientific Center, 4 - 5, Place Vendome 75001 Paris, France, (1989).

Generic Diagnosis for Mechanical Devices

M.P. Feret, J.I. Glasgow
Department of Computing & Information Science,
Queen's University, Kingston, K7L 3N6, Canada

1 Introduction

This paper presents an approach to fault diagnosis based on hierarchical decomposition of mechanical devices. The diagnosis problem is reformulated as a problem of pruning a search tree corresponding to the structural decomposition of the monitored device. Thus, the knowledge acquisition phase for this approach consists of determining the hierarchical decomposition and the appropriate pruning rules for the particular application.

Many of the engineering devices used nowadays present resembling global characteristics. This leads to "families" of machines which often have similar components, are built along the same construction patterns and have parts which are bound by the same dependency relations. Due to these regularities, the reasoning involved in troubleshooting machines in the same "family" is often very similar from one machine to another. Thus developing a generic diagnosis algorithm for a specific family of machines appears to be a legitimate goal.

This paper presents the diagnosis component of a generic health monitoring system (the Automated Data Management System, or ADMS) which aims at providing a development framework for diagnosis systems (Féret [2]). Our ultimate goal is to apply our system to the Mobile Servicing System, Canada's contribution to the International Space Station. We have applied this paradigm to a robotic device called the Fairing Servicing Subsystem (FSS). The FSS is a robot placed at the rear of a boat to replace damaged fairings on a cable which drags an underwater detection system. The fairings are essential to prevent the detection system from drifting away from the axis of the boat. The FSS satisfies the requirements for the demonstration model subsystem since, in addition to a moderate level of sensor density, it comprises a complex arrangement of interactive modules. We are currently testing our architecture and diagnosis strategy on this

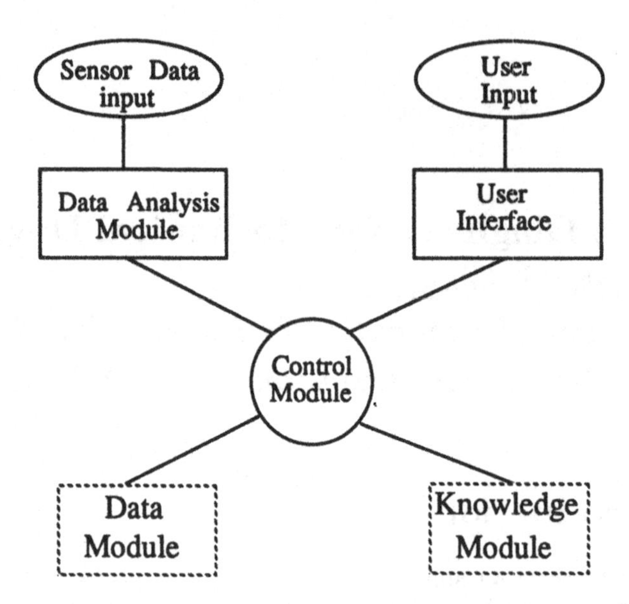

Figure 1: Architecture of a Real-Time Knowledge-Based System.

robot using a simulation program. Many of the examples provided in this paper come from this system (Spectrum [11, 12]).

Section 2 of the paper presents a brief overview of the architecture for the ADMS. We show how the different functionalities of the system are handled by independent communicating modules. We also argue that our architecture meets the generality requirements.

Section 3 of the paper describes the diagnosis module of the ADMS. This module can be decomposed into four submodules: the knowledge acquisition, knowledge representation, control and explanation submodules. The design and functionality of each of these submodules are also presented in Section 3.

Section 4 discusses the advantages of our approach and the problems which arise from applying it. It also suggests future directions of research, including the use of learning techniques to improve the diagnosis and the extension of the system to include a fault prediction capability.

2 The Automated Data Management System

Real-time systems generally refer to processes for which there exist time constraints on the processing of data (Sauers [14]). They typically contain several sensing devices that produce a steady stream of analog or digital signals. These systems are used to monitor and control processes where failure to interpret the current state of the process may result in damage, loss of production and/or loss of revenues (Seiler [13]).

The purpose of the ADMS is to manage large quantities of real-time sensor data, and use it to monitor the health of the device it is applied to. This includes performing fault diagnosis whenever a fault is detected in the device. The overall architecture of the ADMS is describes in Féret [2]. A review of the motivations and goals of such system can also be found in Lawson [7]. This section briefly summarizes these sources from a diagnosis point of view.

The architecture we propose is presented graphically in Figure 1. It contains four computational modules: Data Analysis, User Interface, Data Module and Knowledge Module. The Control Module provides a communication interface between each of these computational modules.

The *Control Module* is in charge of all the communications between the modules. It uses a simple blackboard paradigm. Each module picks up the tasks which are ready for it, in order of priority. The Control Module is the heart of our architecture and is fully documented in Féret [2]. The proposed architecture allows designers to accomplish several goals. It promotes the reuse of specific components of the system and allows for a clean separation of concerns. Each module performs very specific and clearly defined tasks. A centralized communication module allows for independent design and implementation of the computational modules.

The *Data Analysis Module* receives the real-time sensor data from the device. It performs simple analysis(e.g. range checking), samples the data according to rates which are adjustable by the operator and sends the data to the Data Module for storage. It also detects faults by comparing the readings to expected ranges of values. This module provides the data which will guide the diagnosis task. The sensor data analyzed by this module provides knowledge about the current state of the device. This state information is essential to the diagnosis process. It is stored in the Data Module.

The *Data Module* stores the sensor data from the Data Analysis Module. It can be further decomposed into a Data Management Module and a Database Module. We designed a set of generic relations covering all the information that is typically needed for diagnosis. The main relations include a library information relation (which contains static information about the components fo the device), an operation relation (which stores real-time readings from all the sensor on the device), a fault information relation (which stores information about the faults that occurred in the past). Other relations store the results of the diagnoses, the ranked potential diagnoses for each fault, maintenance history for components, etc. The important feature of the database is that it was designed to be generic.

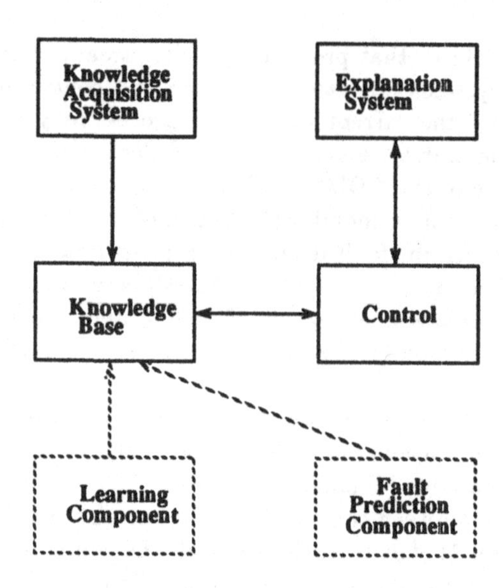

Figure 2: The Conceptual Decomposition of the Knowledge Module

Thus, it is adaptable to a variety of applications.

The *User Interface* provides access to the Database and to the Knowledge Base. It also presents the diagnosis process to the operator and provides for explanations of the decision making process; it is important that the operator understands how the diagnosis task works and why specific choices are made by the program at different points during the diagnosis. Section 3.4 gives more details about the explanation system for the proposed diagnosis module.

The primary task of the Knowledge Module is to perform diagnosis as accurately as possible. The goal is to provide a list of potential diagnoses, possibly representing multiple faults. This list is then ranked according to various criteria such as meantime between failures, frequency of failures, age, etc.

The next section presents how the diagnosis task can be seen as a generic task and how this can be implemented in a generic system.

3 Generic Fault Diagnosis

The diagnosis task takes a fault, a location (where the fault was detected), and a fault type as inputs. Based on this information, the current state of the device (sensor data) and knowledge of the structure of the device, the diagnosis process must determine what particular component (or components) have failed.

This fault diagnosis module can be conceptually divided in four separate subsystems as showed in Figure 2: the knowledge acquisition, knowledge representation, control and explanation subsystems (Figure 2 also shows the learning component and the fault prediction component which we discuss in Section 4). This section describes each of these subsystems.

3.1 Knowledge Acquisition

Our approach is based on considering the device as a hierarchy of components or groups of components. At the top of the hierarchy, the device itself serves as the start of the diagnosis process. All the recognizable components that can be diagnosed as sources of failure are situated at the bottom of the hierarchy. In between these top and bottom levels, the device is decomposed into substructures of the device.

From a structural point of view, each node in the hierarchy can be considered as a separate structural or functional entity in the device. From a diagnosis point of view, each node can be considered as a potential pruning step: one or more pruning rules are attached to each node in the hierarchy. A pruning rule is simply a description of a normal (or abnormal) behavior or context for each represented entity in the system. The association of nodes representing structural or functional entities in the device with the description of properties defining normal expected (or abnormal unexpected) behaviors or contexts can be considered as a semantic network. This semantic network has the shape of a tree representing the hierarchy of parts and subparts in the device.

In the context of our representation, performing diagnosis is no more than navigating in the semantic network according to the following overall strategy: if there is no evidence that the substructure can be faulty, then the whole substructure can be ruled out. If not, the part is examined further, i.e. each of its parts is applied some local diagnosis. Section 3.3 shows how this translates into a tractable algorithm.

3.2 Knowledge Base

This section describes how the structural knowledge of the device and of the pruning rules are represented in the knowledge base of the system. This representation preserves the system generality.

A common approach to representing a semantic network is to use a frame representation (Minsky [8]). Each level in the hierarchy is defined by a single frame holding all the information shared by the nodes of the same level in the tree shaped semantic network. All instances of such a generic frame contain information which is specific to the instance. The relation *IsPartOf* is used in the instance frames to represent the hierarchy of parts in the device. Generic frames can be used to represent generic substructures. Appendix 1 shows part of the frame structure we have implemented.

The pruning rules are considered properties of individual nodes. Some of these are as general as a type of part (for example, motors do not run without fuel or electricity); others are specific to a component (for example, a given motor might be working in a specific way due to the task it is performing, and might tend to break down because of overloading). Others can be specific to one level of substructure in the device. General rules (i.e. rules attached to a type of components) are stored in the corresponding generic frame, whereas specific rules are stored in the frame representing the component of which they are a property.

This frame representation can also contain pointers to the database where more information can be stored. For example, procedural attachment and inheritance can be used to query the database: queries can be stored at the appropriate level in the hierarchy and be evaluated when needed. This avoids storing information in the frames which fits more naturally in a standard database format.

3.3 Control

As mentioned earlier, decomposing the device into successive levels of complexity allows us to also partition the diagnosis process into successive steps corresponding to the structure of the device. Therefore, in our approach designing a diagnosis algorithm means following paths in the semantic network, according to a context which is partially defined by the real-time sensor data contained in the database.

As mentioned earlier, the diagnosis task can be considered as a top-down traversal of a tree-shaped search space which we are trying to optimize. Each step in the algorithm prunes branches at one level of the tree. The optimization problem is to try to eliminate branches as soon as possible in the diagnosis process, while preserving the validity of the final result. The design of the model representing the structure of the device is closely related to how the diagnosis and the pruning rules are understood: the different levels in the decomposition, as well as the number of parts at a given level of decomposition, depend on the availability of pruning heuristics. The more accurate the pruning rules are, the more precise and tight the diagnosis will be. The pruning rules are usually closely related to how human experts view and understand the device and how it can be diagnosed. They are often functions of the organization of the device as well. Thus defining and representing the model of the device and the pruning rules are tasks which need to be done in close association.

The diagnosis algorithm is defined by extracting the pruning rules from the semantic network, and building a series of pruning steps which correspond to levels and/or nodes in the hierarchical decomposition of the device. This extraction is done automatically. A rule can be more than just decomposing the current substructure into its subparts; it can involve complex queries for the next step of the algorithm, marking of other nodes in the semantic network, etc. Thus, pruning rules can be more complex than simple tests on substructures and/or components of the device.

The number of intermediate substructures between the top and the bottom of the hierarchy depends on the complexity of the device. It is also of interest to note that the more levels, and the more substructures at any given level of decomposition, the finer the decomposition will be. This does not imply that the pruning rules will be more specialized. Our experience shows that the same rule can be applied to many of the same substructures at a given level or even across different levels of decomposition. The efficiency of the diagnosis does not depend on the degree of decomposition of the device, but rather on the correspondence between the degree of the decomposition and the range of the pruning rules. Therefore a finer decomposition does not necessarily result in a more efficient search strategy.

In terms of complexity, the essential ingredient for an efficient strategy lies in the pruning rules. If the pruning rules are efficient (i.e. if they effectively prune subtrees which could not possibly contain faulty components, and if the remaining subtrees contain a minimal number of potentially faulty components), then the algorithm will be efficient as well. In fact, our paradigm reflects exactly how much is known about troubleshooting the monitored device, provided that the device fits into our hierarchical representation.

In the instance of the generic system we are currently working on, each test is executed in a constant time. Under this constraint, the algorithm will finish in a time proportional to the number of levels and to the average number of parts at the same level of decomposition. It also depends on the average number of tests needed at each level of decomposition. Experience shows that this number is quite low (below 10 evaluations of sensor functions, requiring only a few simple queries to the database).

Once the list of potential diagnoses has been produced, the diagnosis process ranks them by order of likelihood, according to criteria such as age, time remaining to life expectancy, meantime between failures etc. This final step of the the diagnosis process is important because it determines the order in which the final list of potential diagnoses will be presented to the operator.

3.4 Explanation Facility

Once the list of potential diagnoses is given to the operator, explanations, or documentary informations can be provided upon request. New hypotheses can also be explored, and new results can be compared with the results of the diagnosis algorithm. Finally one (or more) diagnosis is selected. All

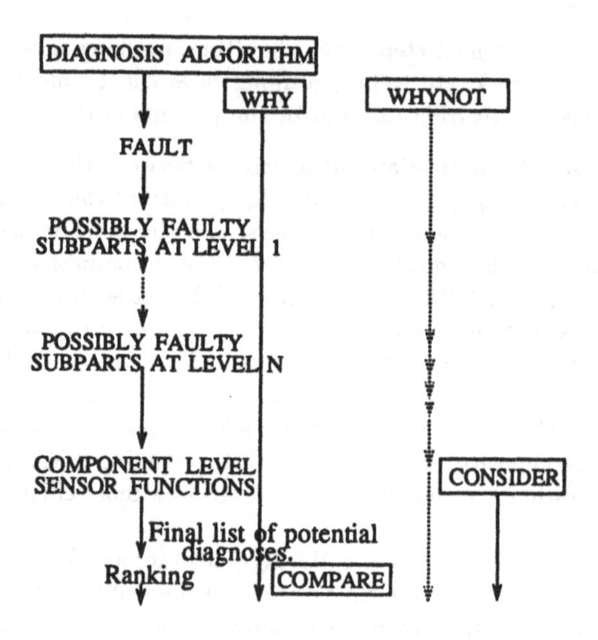

Figure 3: The Relation between the Diagnosis Algorithm and the Explanation Facility.

these options/actions are greatly eased by the structure of the diagnosis algorithm.

Several functionalities are available in the explanation system. These include the ability to:

- obtain explanations (*Explain*, *Compare*, *Why*, *Whynot*),
- ask the KEM to assess a component which has not been considered by the diagnosis algorithm (*Consider*),
- query either the database or the knowledge (*Query_db* and *Query_kb*),
- choose or postpone the final diagnosis (*Choose*), or
- ask for some *Help*.

Only the *Compare*, *Why*, *Whynot* and *Consider* functionalities are related to the design of the diagnosis algorithm. We now present each of these functionalities:

Compare lists all the possible diagnoses with the ranking factors and the corresponding values. This allows the operator to see why a diagnosis is preferred over another one.

Why explains, step by step, why a component was chosen as a potential diagnosis. This explanation procedure goes through the whole diagnosis algorithm and provides the operator with justifications for each step of the diagnosis. This corresponds to a standard trace facility of an expert system shell.

Whynot allows the user to ask why a component was not included in the final list. This is where the decomposition of the device into different levels is important: the diagnosis algorithm is run again for this specific component. An explanation is then provided which states where in the tree representing the search space (the structure of the device) the branch containing the component was ruled out. This gives a precise and accurate information about the decision process which lead to discarding this component as a potential diagnosis.

Following the *Whynot* function, the *Consider* function allows the user to ask the system to operate a local test on a specific component. This skips over the whole diagnosis algorithm and simulates the situation where the smallest subpart of the device containing the component is considered as possibly faulty.

The explanation facilities we provide in our system cover all the steps of the diagnosis algorithm, i.e. all the levels of decomposition of the device. Figure 3 also shows that the explanation facility, although built "around" the initial concept of structural decomposition of the device and of the diagnosis algorithm, is still generic: it depends neither on the numbers of levels, nor on each specific step of the diagnosis algorithm, nor on the actual structure of the device.

Appendice 2 presents a trace of a diagnosis: The algorithm starts by locating all possible modules. It then extracts all the current sensors, torque sensors and proximity detectors for these modules from the knowledge base. One of the sensors is giving incorrect readings. The stream to which the sensor belongs is then examined, component by component. A list of potential diagnoses is then produced (in a different window of the user interface, which we cannot reproduce here).

Appendice 3 illustrates a use of the *Why* facility on the first potential diagnosis of the list of Figure 4. It goes through the diagnosis process and explains each step carefully, especially the last one. At this point the operator can use the other explanation facilities to further understand the diagnosis.

4 Discussion

This paper presents a unique approach to diagnosis of mechanical devices, based on a knowledge representation perspective. It describes how considering a model of a device as a search space affects the way the diagnosis task can be designed and implemented. The decomposition paradigm on which our system is based allows for a comprehensive and flexibile in the specification of the diagnosis algorithm.

We illustrated how the diagnosis itself can be considered as a search problem which can be optimized given a specific state of the device (defined by the real-time sensor data). Traditional search optimization techniques as well as ad hoc heuristics can be applied to any instance of the system. The diagnosis is decomposed into a series of simple pruning steps, which are easy to understand and independent of one another. The algorithm is an offshoot of the model of the device: pruning heuristics are considered properties of defined substructures of the device. Therefore, constructing the model allows the knowledge engineers to simultaneously design the diagnosis algorithm. Our approach to knowledge representation allows for general rules, as well as ad hoc ones, to be represented at an adequate level of abstraction in the model. This makes the diagnosis algorithm a series of pruning steps, each of which prunes the search space of substructures and basic components, depending on the available knowledge of the device and of its troubleshooting procedures.

The diagnosis paradigm presented in this paper also facilitates a rich explanation facility. It allows for decomposiion of the explanations into very natural and comprehensible functionalities. Each of these functionalities is in charge of explaining or documenting one single aspect of the diagnosis process.

The diagnosis part of the current ADMS is implemented in Nial, a functional programming language designed at Queen's University (Jenkins [4, 5]). We have applied the ADMS to a few prototypical test cases with success. We are currently testing the system in combination with a simulator of the FSS that can generate 700 different faults. We are now tuning the sensor functions to allow for more accurate failure mechanism patterns. We are also planning to demonstrate generality and flexibility by applying the ADMS to a second device.

There are two additional areas where our paradigm can be applied within the framework of our project: fault prediction and learning.

In fault prediction, we attempt to to avoid faults before they occur instead of trying to diagnose and fix them afterwards. Fault prediction could be tailored to be pessimistic (the system will predict too many faults) or optimistic (the system will not predict some of the faults which will occur). Fault prediction can be incorporated to our system since it uses the same data as the fault diagnosis.

Machine learning aims at improving a system's overall performance by automatically integrating new facts or knowledge, sometimes provided by experience. It is very hard to have a complete and consistent set of pruning rules or of descriptions of (ab)normal states. It is thus likely that the diagnosis algorithm, as it is defined now, will not diagnose all the faults correctly. Machine learning could partially solve this problem. Learning techniques such as case-based reasoning (Kolodner [6], Riesbeck [10]) and general induction are applicable to our approach. The principle behind case-based reasoning is to store past cases indexed by features and to reason from and about this sum of past knowledge to try to improve performances over time and experience. Induction is the process of generalizing knowledge from examples which share some commonalities, possibly according to a set of constraints.

We feel that both learning and fault prediction can greatly improve the performance of our system, by preventing faults from occurring and by making the algorithm improve over time and experience. We also feel confident that the decomposition paradigm will be well suited for both of these extensions.

Acknowledgments

The application of our approach to Canada's contribution to the International Space Station is being sponsored by the Strategic Technology for Automation and Robotics (STEAR) program, in collaboration with the National Research Council of Canada. The application of the ADMS to the FSS was developed in collaboration with Spectrum Engineering Corporation Ltd., Peterborough, Ontario, Canada.

References

[1] Crowe, C., Douglas, P.L., and Glasgow, J.I. Development of an Expert System for Process Flowsheeting, Proceedings of Conference on Applications of Artificial Intelligence Chemical Engineering, Houston, April 1989.

[2] Féret, M.P., Glasgow, J.I., Lawson, D. and Jenkins, M.A. An Architecture for Real-Time Diagnosis Systems, in Proceedings of the 3rd International Conference on Industrial & Engineering Applications of Artificial Intelligence & Expert Systems, pp. 9-15, Charleston, South Carolina, July 1990. ACM Press, 1990.

[3] Glasgow, J.I., Jenkin, M.A., Féret, M.P. and Lawson, D. An Architecture for Fault Diagnosis, invited abstract, in Proceedings of the 40th Canadian Chemical Engineering Conference, Halifax, July 1990.

[4] Jenkins, M.A. and Jenkins, W.H. The Q'Nial Reference Manual, Nial Systems Limited, Kingston, Canada, 1985.

[5] Jenkins, M.A. and Jenkins, W.H. Artificial Intelligence Toolkit for Q'Nial, Nial Systems Limited, Kingston, Canada, 1987.

[6] Kolodner, J.L. Proceedings: Case-Based Reasoning Worshop (DARPA), Morgan Kauffman Publishers, Inc., San Mateo, CA, 1988.

[7] Lawson, D. A Design Architecture for Knowledge-Based Failure Diagnosis of Real Time Systems, Master's Thesis, Department of Computing and Information Science, Queen's University, Kingston, March 1990.

[8] Minsky, M. A Framework for Representing Knowledge, in The Psychology of Computer Vision (Ed. Winston, P.H.), McGraw-Hill, 1975.

[9] Rich, E. Artificial Intelligence, McGraw-Hill, 1983.

[10] Riesbeck, C.K. and Shank, R.S. Inside Case-Based Reasoning, Lawrence Erlbaum Assoc., Inc., Hillsdale, NJ.

[11] Automation of Operations Relating to an Automated Data Management System, Phase I, Final Report, Volume I, Spectrum Engineering Corporation Limited, Peterborough, January 1989.

[12] Automation of Operations Relating to an Automated Data Management System, Phase II, Technical Proposal, Volume II, Spectrum Engineering Corporation Limited, Peterborough, June 1989.

[13] Seiler, H.B. and Seiler, K.B. Process Control and Monitoring Using Micro-Computer Based Expert Systems, in Proceedings of the 2nd Annual AI and Advanced Computer Technology Conference, May 1986.

[14] Sauers, R. and Walsh, R. On the requirements of future expert systems, In Proceedings of the 8th International Joint Conference on Artificial Intelligence, 1983.

APPENDIX 1

An Example of the Structural Decomposition of a Device.

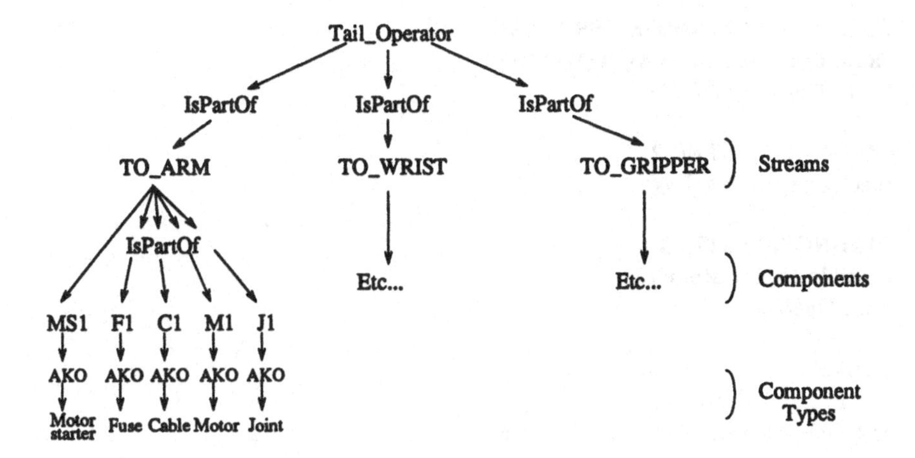

APPENDIX 2

The Trace for a Diagnosis

DIAGNOSIS : STEP 1
Fault detected by sensor: FSS-NO-SE2
Fault Type: MOTIONAL TIMEOUT
Fault Time: 929669782

DIAGNOSIS : STEP 2
checking system voltage

DIAGNOSIS : STEP3-4
Possibly Faulty Modules:
Nose Operator

Current Sensors, Torque Sensors
and Proximity Detectors:
FSS-NO-IS1 FSS-NO-IS2 FSS-NO-IS3

Out of range Sensors:
FSS-NO-IS2

Corresponding Streams:
FSS-NO-WRIST

Corresponding Components are:
FSS-NO-MS2 FSS-NO-F2 FSS-NO-C2 FSS-NO-M2 FSS-NO-J2 FSS-NO-G

Checking component: FSS-NO-MS2
Checking component: FSS-NO-F2
Checking component: FSS-NO-C2
etc ...
List Of Potential diagnoses:
Component : Nose Operator - Motor Starter #2:
Failure Mechanism #2: motor stater does not open when requested: 30.
Component : Nose Operator - Cable #2:
Failure Mechanism #1: conductor failure: 29.99
Component : Nose Operator - Motor #2:
Failure Mechanism #4: motor bearing stall: reduced speed: 27.5
etc ...

APPENDIX 3

An Example of the WHY Explanation Facility

WHY: Component : Nose Operator - Motor Starter #2
Failure Mechanism #2: motor stater does not open when requested

The detected fault was: MOTIONAL TIMEOUT
The fault was detected by the sensor: FSS-NO-SE2
which is in the module: NO

FSS-NO-MS2 is in stream: FSS-NO-WRIST where:
The sensor FSS-NO-IS2 is out of range

The sensor functions for FSS-NO-MS2 are: tf111 tf161
sf111: Does current sensor indicate 0?
s161: Was motor starter requested to close?

Their respective values are: oo
This triggered the Failure Mechanism #2 which has the following
condition pattern: tf111 tf161, yielding the values: ll

APPENDIX

An Overview of the WES [illegible] Facility

[illegible]

Fault Detection in Digital Filters for Satellite Systems

K. Raghunandan, F.P. Coakley

Centre for Satellite Engineering Research,

University of Surrey, Guildford, GU2 5XH, UK

ABSTRACT

This paper describes the application of database search techniques to simulate on-line fault detection in satellite communication systems. Conventional methods of fault detection by hardware is difficult in complex Digital Signal Processing (DSP) circuits which handle real-time data. Hence, a space search method using an Artificial Intelligence (AI) language - Prolog, is developed and it is shown that this approach complies with the requirement of on-board fault detection circuits. Application of the space search method to two key elements in DSP circuits - the multiplier and Adder/Subtractor is described. Implementation of such a novel fault detection method on-board a satellite, has constraints in terms of space and weight and the tradeoff needed is explained. Use of AI based search methods to analyse on-board satellite systems, are not only useful for fault detection but also for other diagnostic jobs in satellites.

INTRODUCTION

Circuits on-board a satellite have an important difference compared to most circuits on the ground in that they must work reliably in the space environment and if they fail due to any reason, repair is not possible. This situation calls for an analysis of the on-board systems with regard to their failure modes and possible methods of making them fail-safe. Traditionally, faults on-board satellites have been diagnosed by ground stations using the data received through a telemetry link, which informs ground controllers about health parameters like voltages, current, temperature etc., at different points in the satellite. Using the data received, a fault model is simulated on ground and a correspondence is established with regard to the problem faced. Corrective actions, like replacement of the suspected unit by a redundant unit, are then accomplished through a telecommand link. This method will not be sufficient in the forthcoming generation of satellites where the complexity is expected to increase considerably and experts systems like the PICON described by Leinweber [1] or an autonomous fault tolerance scheme, at least for some of the important sub-systems on-board, will be necessary to complement the work carried out by ground controllers. Amongst the on-board sub-systems proposed for future satellites, DSP circuits used for converting signals from the frequency domain to the time domain, will form one of the complex units of the communication sub-system, for which fault tolerance will be needed. A circuit used for such a signal transformation is known as a Transmultiplexer (TMUX) which consists of digital filters used to channelise signals into different frequency bands and to output these signals as time interleaved data. Bi and Coakley [2] describe a TMUX where the signals handled are bit serial in nature and many multiplication and addition operations are needed to separate the data into bandlimited signals. The multipliers and adders function in a bit-serial manner, making the circuit complex both in terms of the arithmetic operations as well as the timings for data selection. On-line fault detection in such circuit elements is the subject of the subsequent sections.

This paper is divided into five sections. The first section has introduced the application area, the second section briefly describes TMUXs. The third section deals with fault detection in multipliers using a space search of the database. The adder/subtractor chain with the multiplexers is considered in this section and a method to detect faults by applying space search is described. The fourth section considers the implementation of a space search system on-board satellites. The fifth section concludes this paper by showing the advantages of using heuristics with a database for diagnosis of faults on-board satellites.

TRANSMULTIPLIER - BASICS

This paper will not describe the theory of TMUXs (for details see [2]), but will illustrate the basic features of a TMUX and its function. There are several approaches to design the TMUX, of which the multistage tree approach, is one of the versions suitable for on-board applications due to the relatively high throughput as well as the limited number of components required, which offer weight and space savings needed for satellite applications.

The basic function of a TMUX tree structure shown in Fig 1(a) [2], is to separate the incoming serial data into a number of branches. The first stage separates the incoming signals into High and Low frequency branches, using operations like filtering and decimation. Each subsequent stage progressively separates the data into High and Low branches. At each stage the process of separation is accomplished by a Processing Element (PE), which consists of a band-splitting filter to separate the signals as well as buffers to store incoming and outgoing signals. This structure is used to separate the input signal into 2^n output channels, with n stages containing a total of $(2^n - 1)$ PEs. The operations of a tree structure can also be accomplished by configuring the PEs in a pipelined manner as shown in Fig 1(b) [2], where the operations of each stage of the tree structure, are performed by the PE of that particular stage in the pipelined version. In general, the 'i'th stage PE in the pipeline will have to be time shared by data belonging to 2^{i-1} branches, considerably increasing the control complexity. The output of the final stage PE consists of time interleaved data belonging to all the channels. An important feature of the pipelined version is the considerable reduction in the number of PEs required, since a total of just n PEs are required to demultiplex 2^n channels. In the tree structure, the input stage operates at the input sampling rate, while each PE in the succeeding stages operates at half the rate of the previous stage PE.

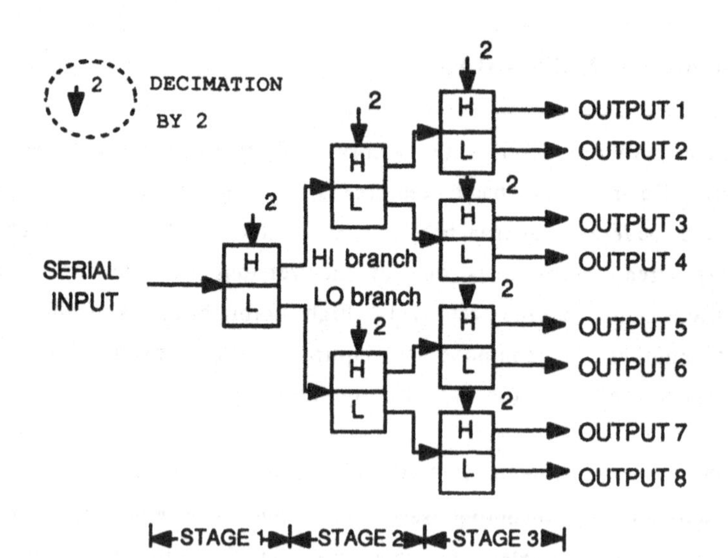

Fig.1(a) MULTISTAGE TRANSMULTIPLEXER - TREE STRUCTURE

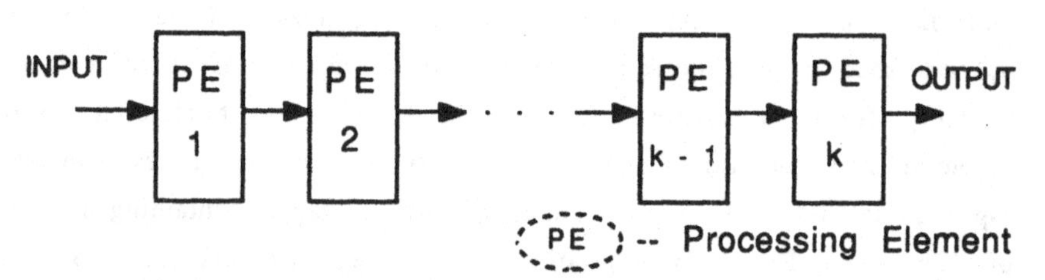

Fig.1(b) TIME MULTIPLEXED VERSION OF Fig.1(a)

But in the pipelined version all the stage PEs operate at the input sampling rate and output of each stage has to be stored in a buffer, while the succeeding stage is processing data from the other branch signals. The pipelined version also offers advantages in VLSI design, due to single wire connections and modular construction of each stage. From a fault tolerance point of view, the pipelined version has a limitation in that a fault at any stage will result in faulty outputs. Therefore, on-line fault detection at each stage is necessary and if a fault is detected, it must be possible to replace the faulty unit within a reasonable period of time (few seconds, at best). This paper limits its scope to the on-line detection of faults in a PE. Other aspects of fault tolerance like diagnostics using self-test, reconfiguration using redundant units, etc., are outlined by Raghunandan and Coakley [3]. Our study is restricted to faults in a typical filter section and the detection of faults, while normal data flow is taking place. The band-splitting filter section used in the PE is relatively complex although the basic operations are similar to a FIR (Finite Impulse Response) filter described in any standard text on digital filters. The complexity arises since the input data is in the form of a complex number and the timing circuits to select the appropriate data has a complex structure; further the buffers used in the iterative process add to the complexity, as described in [2].

FAULT DETECTION

Although the hardware circuit operations of a band-splitting filter occur in real-time, it is impossible to implement a fault detection scheme in real-time, due to the high-speed of operations. In the present case, fault latency due to a search routine occurs at the functional level since it informs the fault tolerance system that one part of a PE has been found faulty. At the next higher level, fault in a particular PE is reported and may be at a further higher level in the hierarchy, a fault in the sub-TMUX level is reported. All these transactions should be completed quickly so that the replacement of a faulty unit can be effected by reconfiguration within a short duration (typically a few seconds), so that a customer using the satellite does not notice it as a major breakdown in the service.

FAULT DETECTION IN MULTIPLIERS

A good survey of the different types of multipliers used in DSP circuits has been presented by Ma and Taylor [4]. Most multipliers use the technique of repeated addition to arrive at the final product. The filter section of a TMUX uses two's complement multipliers, with bit serial data inputs in real-time to produce one product in each interval of a word mark signal. The functional diagram of a bit-serial, real time multiplier which can be used in TMUX is shown in Fig 2, which handles 4 bit inputs (5 bits including the sign bit) and produces an eight bit output (nine bits including the sign bit). The product outputs emerge separately, with 4 LSB bits at the storage point shown below, while the other five MSB bits (including the sign bit) appear at the terminal y_{out}. These are later combined to produce the final product (not shown in Fig 2).

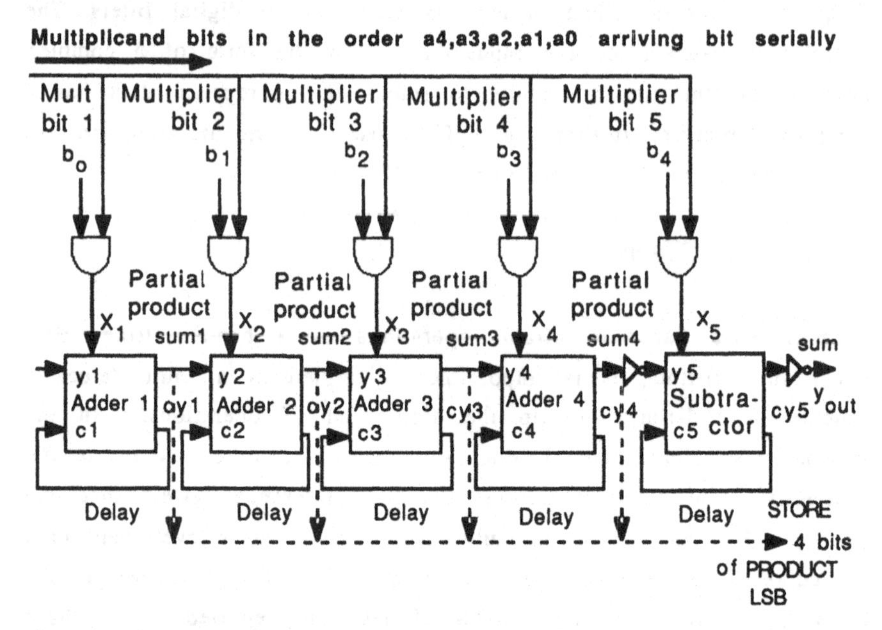

Fig 2. Two's Complement Multiplier, bit-serial scheme

Five adder stages are connected in a pipelined manner to handle the five input bits as shown in Fig 2, where the final stage is modified to act as a subtractor. While each of the multiplier bits are assigned to a particular adder stage, the multiplicand bits are sent into the AND gate in the order of their arrival. Details of timing delay and multiplier bit routing have not been shown in Fig 2 to retain simplicity. Inputs to the multiplier arrive bit-serially with the LSB arriving first. Each adder has an 'x' input which is an ANDed combination of the input multiplier bit (b_0 to b_4 depending on the stage) and the multiplicand bit (a_0 to a_4 which are used by all the stages at different time intervals). The other input to each adder is the partial product sum y, generated by the previous stage adder (except the first stage). Each adder has two outputs one for the 'sum' where the partial product sum is generated and the other for 'carry' which is used by the same adder stage during the next addition. The operations can be visualised by noting that in the first cycle b_0 and a_0 are ANDed by the first AND gate; in the next cycle b_0 and a_1 are ANDded by the first and b_1 and a_0 by the second gate and so on. The first four LS bits of the product (partial product sums sum1 to sum4) are routed into the LSB product storage, at the end of the first five cycles. When the fifth bit (the sign bit a_4) of the multiplicand input arrives, bit manipulation is effected (not shown in Fig 2) depending on the sign indicated by the fifth bit and all further product bits represent manipulated product bits. As a consequence, in the next four cycles the manipulated MS bits arrive serially, from the output y_{out} of final subtractor stage. In the mean time during the sixth cycle, the LSB of the next pair of inputs arrive and are processed by adder1, while the remaining 4 product MS bits of the previous input set, are still being processed as partial sums sum2 to sum4. This pipeline action continues till the end of the ninth cycle when final product bit of the previous input set emerges from the output y_{out}. The operations follow the well known Booth's algorithm [4], although certain implementation details differ.

Fault detection schemes for multipliers proposed by Wu and Wu [5] as well as Shen and Ferguson [6], have used hardware detection and their inherent limitation is that detection is based on certain preconditions. These preconditions include an initial state for each adder as well as a known set of inputs which must be loaded into each adder. In a real-time situation, a multiplier handles random inputs and the terminals of adders cannot be set to known conditions unless they are cut off from the actual circuit. Existing schemes aim at fault detection in the individual adder stages, which preclude faults due to timing delay, software (control signals) or breaks in line connections between adders. While a considerable increase in hardware is expected due to these fault detection circuits, they do not offer complete fault detection since the preconditions assumed are not likely to occur in practice. Since each filter section contains several multipliers, it is better to detect faults at the multiplier level rather than at individual adders levels.

Due to the nature of two's complement numbers, it is not possible to devise hardware schemes by which the two inputs ('a' and 'b') to the multiplier can be used to check the product, unless another circuit as complex as the multiplier itself, is used. We therefore propose a software based simulation scheme wherein each set of two inputs and their product are tabulated and the entire table is placed in a database. Using an AI language like Prolog, it becomes a simple task to search out the product, given any two input numbers (by monitoring the input), since Prolog allows facts and rules to be defined as clauses. The product searched out by the unification process, can thereafter be compared with the multiplier's output (also being monitored) to detect faults, if any. The problem therefore reduces itself to preparing a compact table so that the total set of input/output combinations to be stored in the database reduces to a reasonable size. To provide a reduction and yet maintain a good balance between database storage capacity and search time, it is necessary that the database structure be a combination of the bushy tree and skinny tree structure as described by Wong, Suda and Bic [7]. Therefore, it is better to divide the multiplier's output product (9 bits in this case) into the LSB (4 bits) and MSB parts (5 bits including the sign bit).

Fig 3 shows the search path, where the starting point is indicated by two separate points to denote the time delay which occurs while the fifth bit (sign) arrives. In terms of search completion time, there is little to choose from since we expect "only one answer for the product". Although more than one answer may be selected

during backtracking, only one correct answer must emerge from the search routine. For any multiplier handling two's complement number inputs four types of input combinations possible, viz.,
(i) both multiplicand and multiplier positive
(ii) multiplicand being positive and the negative
(iii) multiplicand negative and multiplier positive
(iv) both multiplicand and multiplier negative,
which makes up a total of 225 X 4 = 900 combinations, assuming 4 bit inputs (5 bits including the sign bit).

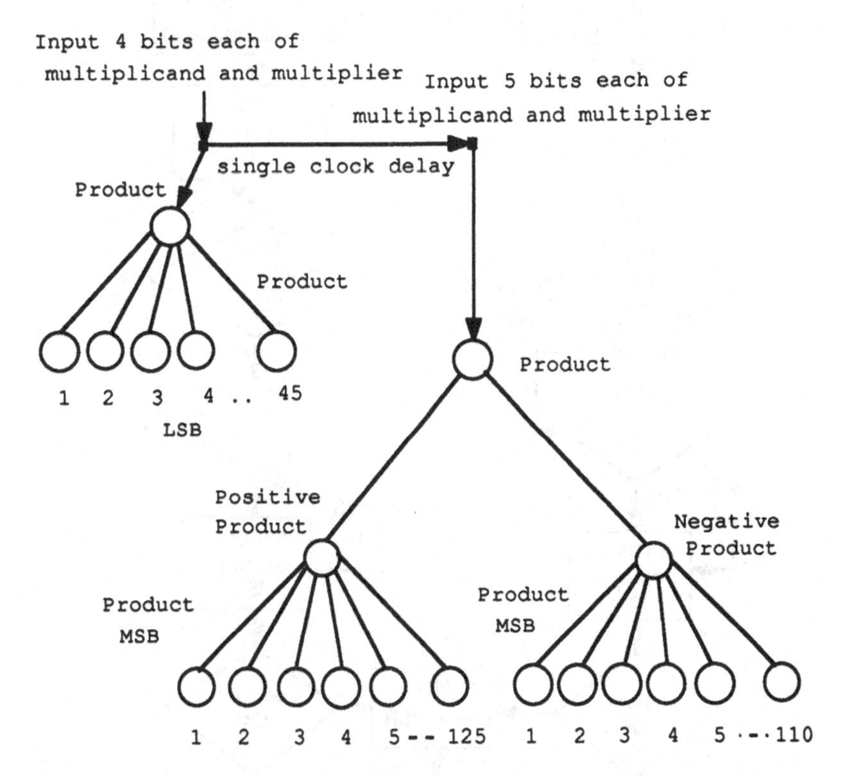

Fig 3. Search path for finding multiplier's "product"

Separate tables for the product LSB and product MSB bits can be constructed (900 combinations for each), which gives the complete group of tables to be stored in the database. Many of these combinations collapse into a much smaller group due to several factors some of which are: (a) the commutative law, (b) several combinations reducing to don't care bits etc. Just 45 such reduced combinations can be used to represent input sets for all the four product LSB bits.

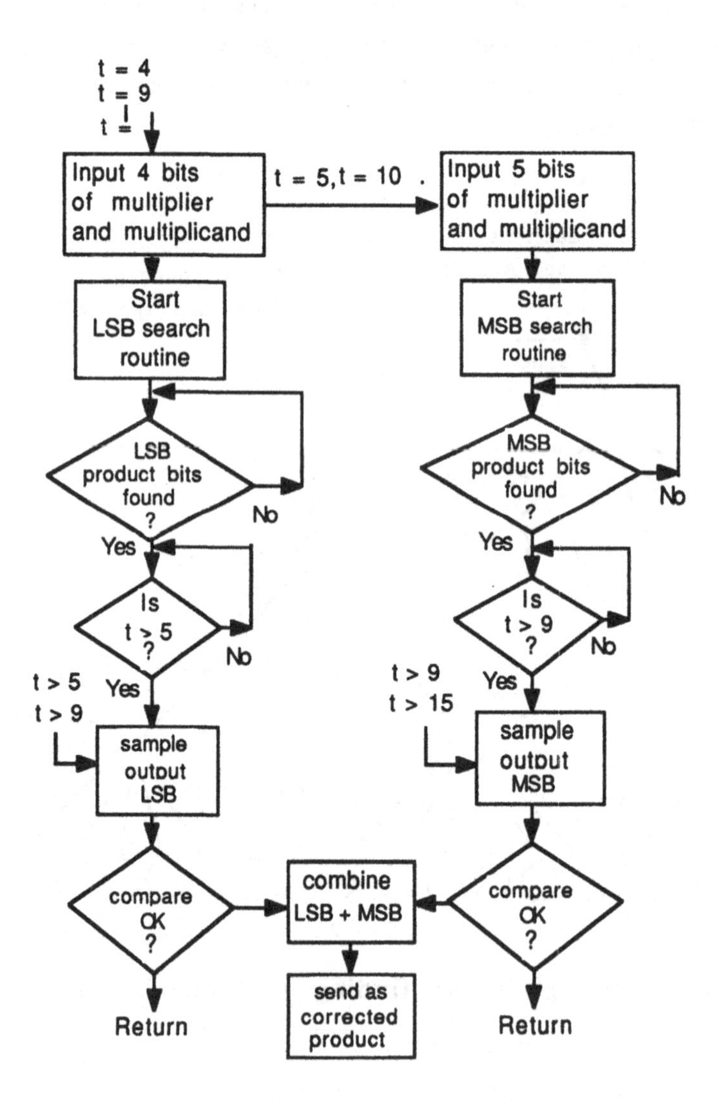

Fig 4. Software space search routine

Although further reductions are possible, it defeats the exercise since additional statements are necessary in the Prolog clauses and the overall detection time will not be improved because of the additional calls required when Prolog executes the search. The search sequence for a bit-serial multiplier is shown in Fig 4, where a search during each multiplication operation is indicated. But in practice it is rarely necessary to check a multiplier during each multiplication cycle. A routine check conducted over periodic time intervals, as decided by the system designer, would suffice. Arguments for and against fixing such time intervals are not within the purview of this paper. Prolog clauses for the product LSB and some of the combinations to be stored in database are shown below:

```
search(F, productl(A, B, P), Product):-
        setof(B, productl(A, B, Product), P).
search(F, productl(A, B, P), Product):-
        setof(A, productl(B, A, Product), P).
%       LSB product and input relationships
%               Multiplicand /Multiplier/ product LSB/
%                     /          //          //          /
        productl([X, Y, Z, 0], [1, 0, 0, 0], [0, 0, 0, 0]).
        productl([M, X, X, 1], [M, X, X, 1], [0, 0, 0, 1]).
                      /
        productl([X, 0, 1, 1], [X, 1, 0, 1], [1, 1, 1, 1]).% total of 45
combinations
```

The Prolog built-in predicate "setof" has been used here to advantage, in that it searches all the possible combinations and when a single answer for the product is obtained, it assures that one and only one product has satisfied the given input combination. For the MSB however, we need two such groupings, one for the positive product MSB (sign bit 0) and another for negative product MSB (sign bit 1). Prolog clauses and the combinations to be stored in database (125 for positive MSB and 110 for negative MSB) are shown below.

```
search(F, productmP(A, B, P), Product):-
        setof(B, productmP(A, B, Product), P),
        not(exclude(A, B, Sp)).
```

```
search(F, productmP(A, B, P), Product):-
        setof(A, productmP(B, A, Product), P),
        not(exclude(B, A, Product)).
exclude(A, B, Product):-
        (A, B, Product) = ([1, 1, 1, 1, 0], [1, 1, 0, 0, 0], [0, 0, 0, 0, 0]);
%                              /
        (A, B, Product) = (1, 1, 0, 1, 1], [1, 0, 0, 0, 0], [0, 0, 1, 0, 0]). % 13
cases of exception
                productmP([0, 0, 0, 0, M], [0, W, X, Y, Z], [0, 0, 0, 0, 0]).
%                    /
        productmP([0, 1, 1, 1, 1], [0, 1, 1, 1, 1], [0, 1, 1, 1, 0]). % total of 110
combinations
%       Clauses for the negative MSB products
 search(F, productmN(A, B, P), Product):-
        setof(B, productmN(A, B, Product), P),
        not(exclude(A, B, Sp)).
search(F, productmN(A, B, P), Product):-
        setof(A, productmN(B, A, Product), P),
        not(exclude(B, A, Product)).
exclude(A, B, Product):-
        (A, B, Product) = ([1, 1, 1, 0, 0], [0, 1, 0, 0, 0], [1, 1, 1, 0, 1]);
%            /
        (A, B, Product) = ([1, 1, 1, 1, 0], [0, 0, 0, 0, 0], [1, 1, 1, 1, 1]). % total
of 8 exceptions.
        productmN([1, 0, 0, 0, 0], [0, W, X, Y, Z], [1, 0, 0, 0, 0]).
%            /
        productmN([1, 1, 0, 1, 1], [0, 0, 0, 1, 1], [1, 1, 1, 1, 1]). % total of
125 combinations
```

Irrespective of bit-serial or bit-parallel multiplication schemes, there are limitations in conducting checks on every single product generated, both in terms of computational power as well as the time needed for a search. This aspect is important in fault detection and is discussed later, under the section on "on-board considerations".

FAULT DETECTION IN ADDITION/SUBTRACTION UNIT

Multiplexers are used to select appropriate data for the purpose of addition and subtraction.

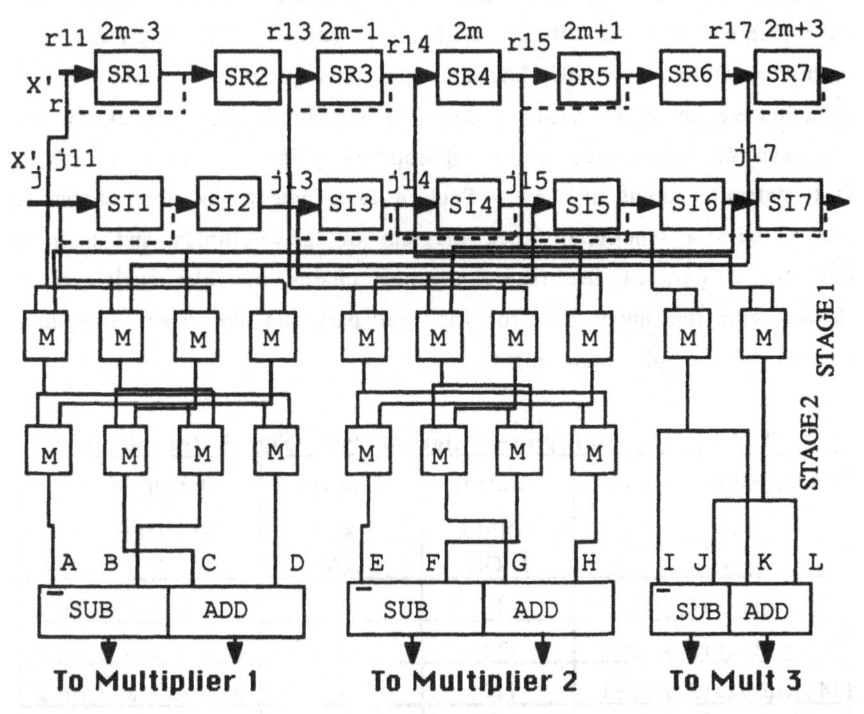

FUNCTIONAL BLOCK DIAGRAM OF MUX/ADDITION UNIT

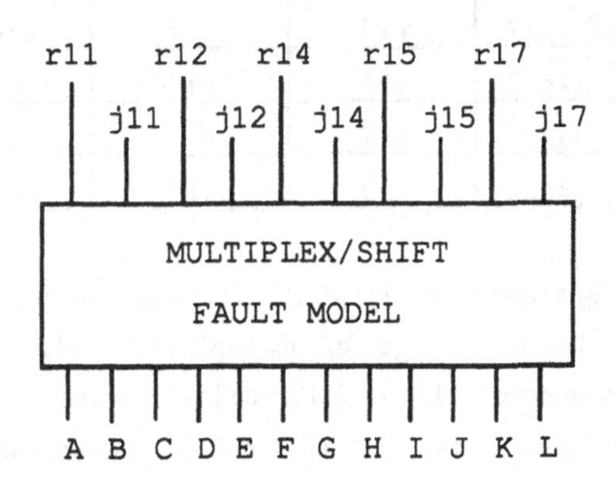

Fig 5. Simplified model for multiplexer/shift register

Multiplexers not only reduce the number of dedicated data connection paths but also provide a means to select the right data at the right time thus reducing the overall memory requirement. Therefore, it is better to consider the functions of Adders/Subtractors alongwith shift registers and multiplexers which select the appropriate data for these arithmetic units. Shift registers and multiplexers do not modify the data but only provide a means of data selection.We therefore need to derive a simplified fault model so that data can be represented at the appropriate points in the unit from a fault detection point of view. Such a fault model has been shown in Fig 5, with a functional block diagram of the multiplex and addition unit for a part of the filter; separate chains of shift registers are shown with the upper one for the real part and the lower imaginary part of the complex data input.

TABLE 1. Multiplexer signals (see Fig 5 for details)

Observation points	Control signal HIGH	Control signal LOW	Output of the multiplexer
Multiplexer1,Stage1	r11	j17	goes to 1 & 4 stg 2
Multiplexer2,Stage1	j11	r17	goes to 2 & 3 stg 2
Multiplexer3,Stage1	j17	r11	goes to 3 & 2 stg 2
Multiplexer4,Stage1	r17	j11	goes to 4 & 1 stg 2
Multiplexer1,Stage2	r11	j11	goes to input A
Multiplexer2,Stage2	j11	r11	goes to input C
Multiplexer3,Stage2	j17	r17	goes to input B
Multiplexer4,Stage2	r17	j17	goes to input D

The multiplexers used are of the 2:1 selection type and it can be observed that the first stage of multiplexers (left to right) select inputs in the sequence r11, j11, j17, and r17 when the control signal (not shown) is "high" and j17, r17, r11, j11 when the control signal is "low". The second stage multiplexers are used to swop data between the inputs A,B and C,D during alternate cycles of the control signal, as shown in Table 1. There will be similar multiplexer chains to select inputs for EFGH and IJKL, as shown in Fig 5. To detect faults in shift registers and multiplexers it is sufficient to compare the relevant input/output pairs shown in the fault model of Fig 5, using Table 1. Such a comparison can be avoided if we are not particularly interested in the exact location of the fault and it is sufficient to observe the outputs of the adder/subtractors (connections to the inputs of multipliers) since we know the data combinations expected. We could restrict our monitoring to the shift register points (r11, j11 etc) and the adder/subtractor outputs. The search routine would therefore take the inputs once and proceed with the search for both the "sum" (for adders) and "remainder" (for subtractors). One input set will be common to both routines but after the search, the first comparison is possible straight away while the second routine has to wait till the end of next cycle, for comparison.

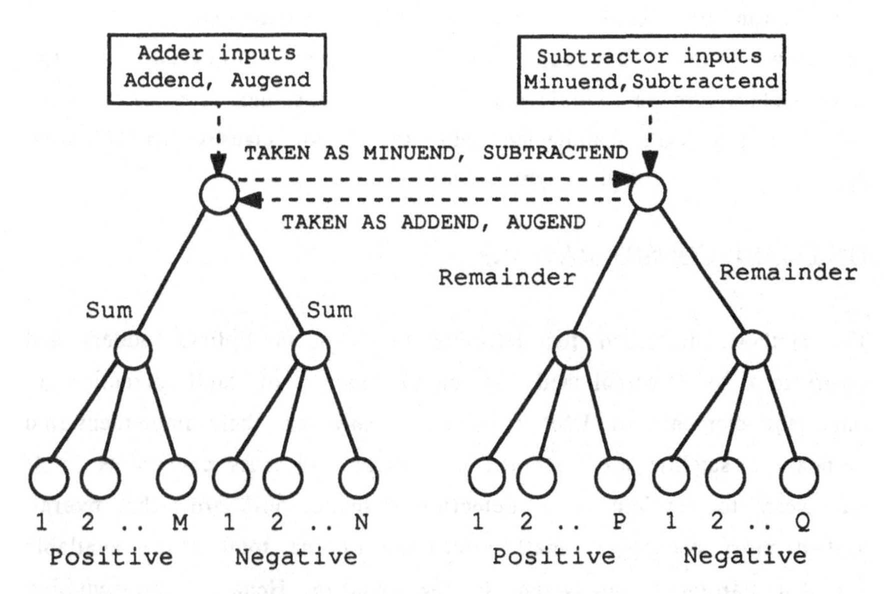

Fig 6. Adder/Subtractor search routines

If we start with the inputs to the adder as our reference, for example, then the search for the sum will be completed first and it can be compared with the adder output, while the search for remainder can begin simultaneously but the subtractor output is compared only at the end of the next cycle. The starting point can be the Adder or the Subtractor as indicated in Fig 6. In practice, the hardware based adder/subtractor units in a filter, execute the sequence much faster than our search routine and only the cycle timing is important for comparison. This method will not only ensure proper working of all the multiplexer and shift registers, but checks the adder/subtractor as well. To prepare database for adder and subtractor it is necessary to construct tables similar to the ones for multiplier. However, the difference is that there is no need to split the output sum into LSB and MSB parts, since both the input and output have the same number of bits. Instead, it would be advantageous to group the positive sum and negative sum (similarly positive and negative remainders) as shown in Fig 6. Such grouping ensures a shorter search time since the database will again have a combination of bushy and skinny structures indicated in [7]. A negation stage succeeds the multipliers connected to the adders/subtractors (not indicated in Fig 5); this unit can be checked by deriving a two's complement table and Prolog clauses can be used again to search for the right value.

ON-BOARD CONSIDERATIONS

The methods described for detecting faults in multipliers, adders and subtractors have established the unique nature of fault detection in important elements of DSP circuits. However, their implementation on-board a satellite are governed by volume and mass constraints. If we were to implement a detection scheme on-board, the overall system must occupy a small percentage of the total space available for that particular sub-system in the satellite. Hence, implementation of expert systems like PICON have remained in the exclusive realm of future space stations [1]. For satellites which are only a fraction of the space station in size, it is necessary to find implementations of either simple inference engines using fuzzy logic eg. Togai and Watanabe [8], or a list processing chip eg. Lee, Leung and Cheang [9], since the fault detection process described can be accomplished by list processing.

Assuming these limitations, the problem must be modified to represent the detection methods described earlier, into either fuzzy logic [8] or list processing [9]. To implement the detection process in terms of fuzzy logic, either previous recordings of the telemetry data or hypothetical data under fault conditions are necessary [1]. Since DSP circuits have not yet been implemented on-board for such communication systems, there is no possibility of obtaining the telemetry data and therefore hypothetical fault history could be built using simulations.

The reason for retaining 45 rules for the product LSB for example, was that we wanted to strike a balance between the Prolog search time (in terms of unification and list matching) and the memory. Therefore, a reduction in the number of rules will mean additional statements in the Prolog clauses. However, in translating the rules for multiplier products it may be possible to reduce the number of rules, since in fuzzy logic these rules can be represented in a compact form. To implement the rules as list processing [9] may be quite straight forward, but it is unlikely to yield substantial savings in terms of memory or operation time, hence the fuzzy logic approach seems to be a better option.

In terms of the chip area, the TMUX with a single channel can be built using a standard TMS32020 DSP chip which has an area of 1600 mm X 1500 mm. The inference engine described in [8] has an active chip area of 3mm X 3.5 mm to accommodate 32 rules. Since the inference engine can be designed in a rule-sliced manner, the number of rules accommodated can be increased by cascading so that many of them can be executed in parallel, with a linear increase in the chip area. To represent nearly the 300 rules for a 4 bit multiplier, an inference engine with a chip area of 30 mm X 35 mm may be needed. Further, the rules for adders/subtractors must also be incorporated, increasing chip area of the inference engine.

Despite these requirements, in terms chip area the percentage increase due to inference engine will still be less than 1% of TMS32020. This is quite small compared to the area needed by the hardware based fault detection circuits or self-test schemes, which is typically in the region of 5% to 20%. However, in terms of execution time there will be a considerable lag since each inference can be executed by the engine in 12 µSec, while every multiplication (using bit-parallel multipliers) in a TMS32020 needs two instruction cycles of 200 nS each (i.e 400nS per multiplication). This is to be expected since real-time detection can not be performed on any operation whose execution time is very small.

In signal processing operations, multiplication is accompanied by a summation (also subtraction) of the products and the fault due to a multiplication, for example, which contribute to a considerable time delay before its effect is felt at the TMUX operational level. The tradeoff between the additional chip area and the number of multiplications being checked per second is therefore quite important. The fact that most VLSI chips manufactured in the coming years will provide self-test features using boundary scan as prescribed in the new IEEE standard 1149.1-1990, will mean that fault detection becomes an essential pre-requisite before executing self-test on any VLSI chip. After all, if a self-test has to be performed, the chip in question has to be first cut off from the circuit and this should not be done unless there is some evidence to suspect the part (eg. the evidence provided by the fault detection scheme).

CONCLUSION

We have shown the application of database search as a useful means to achieve fault detection. Although the examples shown represent a small part of the on-board communication system, it is intended to reflect the special nature of the fault detection procedure offered by AI based search techniques in terms of a compact database and simple algorithms. Earlier, expert systems with a knowledge base have been applied at systems level for control and test purposes, but such systems can not be used on-board satellites, since they are large mainframe based systems [1]. Important advantages of using a software based fault detection scheme on-board are the flexibility offered by database and the possibility of using the same system for fault detection, diagnostics and reconfiguration management [3]. At a later stage, when sufficient data becomes available from operational satellites, it would be easier to implement expert systems on-board to diagnose faults, based on the "data trends". However, before considering the suitability for on-board use, one has to consider the increase in hardware/software as described in the previous section on on-board considerations.

Software reliability is an area of extensive study currently and it is possible to develop reliable software for fault detection purposes using known software schemes. For space applications, the data diversity scheme described by Amman and Knight [10] offers a good approach to attain software reliability, since data diversity requires some means of using the available data sources to attain better reliability. Application of data diversity scheme has been made in our adder/subtractor search routine where the source of data is the adder input as well as the subtractor input, during alternate cycles, as shown in Fig 6. Additionally, the flow of data through shift registers and multiplexers is tested by the different adder/subtractor routines at three adder/subtractor groups indicated in Fig 5. The levels at which detection should be made and how it affects the system performance, the number spares required and a systematic method to reconfigure the system in the presence of faults etc., are important areas which must be addressed to achieve overall fault tolerance of on-board systems.

ACKNOWLEDGEMENTS

The first author would like to thank the European Space Technology Research Centre of the European Space Agency (ESA), as well as the ORS awards committee of the Committee of Vice-Chancellors and Principals (CVCP), U.K, for the financial support rendered.

REFERENCES

1. Leinweber, D. Expert systems in Space, IEEE Expert, pp 26-36, Spring 1987.
2. Bi,G. and Coakley, F.P. The design of Transmultiplexers for on-board processing satellites using bit-serial processing technique, pp 613-622, Proceedings of the 13th AIAA Int. Conf. on Communication Satellite System, Los Angeles, California, March 11-15, 1990.
3. Raghunandan,K. and Coakley, F.P. Fault tolerance of on-board processors, Paper 3.5, ESA-WPP-019, DSP 90 (E.S.A), Proceedings of the 2nd Int. Workshop on Digital Signal Processing techniques applied to Space Communications, Turin, Italy, 24-25 September 1990.
4. Ma, G.K. and F.J.Taylor, Multiplier policies for digital signal processing, IEEE ASSP magazine, pp 6-20, January 1990.
5. Wu, C.C. and Wu, T.S. Concurrent error correction in unidirectional linear arithmetic arrays, Proceedings of the 17th Int. Symposium on Fault Tolerant Computing (FTCS -17), IEEE, pp 136 - 141, Pittsburgh, USA, July 6-8, 1987.
6. Shen, J.P. and Ferguson, F.J. The design of easily testable VLSI array multipliers, IEEE Transactions on Computers, Vol. C-33, No.6, pp 554-560, June 1984.
7. Wong,W.C., Suda, T. and Bic,L. Performance analysis of a message-oriented Knowledge-Base, IEEE Transactions on Computers, Vol.39, No.7, pp 951-957, July 1990.
8. Togai, M. and Watanabe, H. Expert system on a chip: An engine for real-time approximate reasoning, IEEE Expert, pp 55-62, Fall 1986.
9. Lee,K.H., Leung,K.S. and Cheang, S.M. A Microprogrammable list processor - for personal computers, IEEE Micro, pp 50-61, August 1990.
10. Ammann, P.E. and Knight, J.C. Data diversity: an approach to software fault tolerance, IEEE Transactions on Computers, Vol.37, No.4, pp 418-425, April 1988.

SECTION 6: ROBOTICS

The Co-operative Behavior of Multirobot Systems

A. Neki, K. Ouriachi, M. Bourton

L.I.R.R.F. U.A.CNRS 1118, Université de Valenciennes, 59326 Valenciennes Cedex, France

ABSTRACT

A co-operative behavior of Multirobot systems is described. The two main problems treated are the planning problem and the allocation problem. We used a temporal logic for the first problem, and a task allocation algorithm for the second problem. The main properties of temporal logic based on intervals are explicit and flexible time handling and symbolic constraints use. The principal characteristics of the task allocation algorithm are that it is fast and reliable.

INDEX TERMS- multirobot system, co-operation, plannning algorithm, temporal logic, allocation, unpreductible task, action, operation.

I. INTRODUCTION

Nowadays, the industrial world has recourse to robots to improve the throughput of production and to replace the human operator in monotonous tasks. However, if robotic integration leads to high flexibility and reliability of the manufacturing systems, the robotic tasks will be very complex. They need in this case a system of robots which must co-operate to achieve a given task. The co-operation of the different agents (robots, peri-robots) is expressed, for instance firstly by exploiting the inherent parallelism of the system, secondly: the absence of frequent tools exchanges for a product which requires several types of parts and thirdly by application on heavy and/or voluminous objects.

The faculty of processing such tasks involves a growth in complexity of the processing systems. Thus, it generates processes which progress concurrently. The presence of such processes have always posed problems that we classify in two categories.

-The co-operation of processes to achieve the same goal.
-The competition of processes sharing the same resources.

The approach of classical automation based on the functional tools, such as Petri-net, is particularly efficient to define the access by processes to numerous resources. Nevertheless, this efficiency is only proved in the limit of deterministic processes.

Since we want to execute, by several co-operative agents, "unpreductable" tasks (i.e. non periodic tasks which include uncertainty), the programming approach is well suited to formulate and to address the non deterministic behavior of co-operative processes. This is the object of LCOOP scheme in our laboratory L.I.R.R.F., where the aim is the design and the development of programming environment.

The contribution of this paper is to propose a formulation describing the co-operative behavior of a multirobot system.

In order to facilitate the work of the actual operator in manufacturing, the robotic task will be specified declaratively. But, it will not be executed as long as it is not factorized on executable actions. An analysis method is under investigation in our laboratory. It exploits the hierarchical knowledge representation to avoid combinatorial explosions. There are three levels of knowledge(1)(16).

-The most abstract one is the task level. It enables us to describe this task independently from the run time environment.

-The intermediate one is the operation level which takes into account the objects. Which are handled thanks to their description (shape, weight,..).

-The low level is the action level. The actions are directly executed by the agents, and their description depends on the environment behavior. Since the environment depends on the actions, we can say that, they are closely coupled. We can conclude that action execution depends on the environment historic.

The remainder of the paper is organized as follows. Section II describes our system model of multirobot systems and lists our assumptions. Section III presents the logical tool for temporal reasoning and its use in planning. Section IV describes the allocation algorithm. Section V summarizes the conclusions of the paper.

II. MULTIROBOT SYSTEMS

A- ARCHITECTURE OF THE SYSTEM

A multirobot system is composed of two or more robots and a set of peri-robotic elements (machinery, sensors). It is capable of executing operative process with effectiveness, flexibility and reliability. It may be considered as a distributed real time system(9). Each component of this system may act on its local environment. Actions are, at first sight, independent with each other. This independence is limited because robots have tendency to share common resource access, or to communicate informations to co-operate to achieve a common task.

This system is defined as a multi-agent system, where each agent is a physical or abstract entity capable of acting on its environment. This entity disposes of a partial knowledge of such environment, and can co-operate with other agents. Its behavior depends on its world perception, on its competence and on its interactions with the other agents(18).

Since a multirobot system is defined by a set of robots, it is important to define a robot. A robot is considered as a machine which is competent to achieve operative process as well as computing process. Indeed unlike a computer, it could, in addition, perform physical intervention on the work environment. Unlike a modular automatic machine, it performs different tasks.

From formal point of view, a robot can be represented only by its competence. This competence is defined by a set of entities which are actions. The property of these actions is that one single action is performed at once.

$$c=\{ a_i / \forall a_j \quad a_j \neq a_i \} \qquad (1)$$

Therefore, the multirobot system is defined formally as a set of parts of actions, where the reunion of intersections of parts is equal to the general competence of the system:

$$SMR = \{ Ri / \cup(\cap c_i) = C \} \qquad (2)$$

SMR: multirobot system
Ri : robot i
c_i: competence of robot i
C : general competence.

B- CHARACTERISTICS OF TASKS

a)-Action: an action is an activity caused by an agent. In robotic context, action make evolution of the robot operative space which executes it(12)(13).

b)-Operation: an operation is a relation applicable on two entities and which generates a result. In robotic, an operation is the result of the execution of one or more actions.

c)-Task: in general, a task is often defined as" the work which is giving to one or more persons and/ or machines". In robotic context, a task is a set of operations partially ordered working towards the same aim.

A task T is a set of operations,
T={Op1,........Opk} with card(T)=k

an operation Opi is a set of actions
Opi={Ai1,.....Ail} with card(Opi)=l

C- CO-OPERATIVE BEHAVIOR PROBLEM

The goal of this work is to develop a planning algorithm based on a logical tool and a dynamic allocation algorithm, where reliability and continued performance in the presence of failures is of an utmost importance.

The co-operative behavior problem is actually composed of multiple subproblems:

. flexibility in planning to take into account the unpreductibility of robotic tasks,

. dynamic allocation problem;

III. PLANNING OF ACTIONS

From a theoretical point of view, a multirobot system is considered as a problem resolution system. A real world inadequate modeling reduces significantly the abilities of such system. It is due to the fact that the time factor which is not considered in the proposed models.

Time may be handled by an algorithm as integer or real entities. It must have a well defined semantic since it participates to the algorithm semantic. At this level, we may consider time as a date or a duration. In both cases used measure is set up on the concept of time unit.

Beside this measurable time, there is another perception of time, which is essentially logic, where we are neither interested in duration, nor in absolute value of date, but we are interested in comparing two dates, that is the notions of " before" and " after".

Since the determination of durations and run-time values of actions is very difficult, it involves intuition and experience, an approach based on the logical perception of time enable us to overcome this difficulty.

1- TEMPORAL LOGIC

We suggest a model, that takes into account the past, present and future of this environment. It is based on the temporal logic of J.Allen(12)(17). The environment behavior is described as a set of assertions temporally specified. This enables us to describe the static aspects (properties) as well as dynamic aspects (occurrences). By exploiting the inference mechanism of temporal logic based on intervals, we can then schedule the resultant actions from the task factorization, different type of actions can involve different kind of resources (tools, sensors, particular abilities of the robots, tasks,...), but every action involves time.

Given this model, a schedule is a collection of assertions considered as an abstract partial simulation of the future, including actions the scheduler intends to take as well as other predicted actions, events, and states. In a coherent schedule, most of these events and states are causally related. A goal is a partial description of the environment desired. This description is not confined to a specific instant of time. It might consist of a sequence of states (e.g., grasp Part P1 and later Put P1 on P2), and constraints (e.g., clear P2 before Puting P1 on P2), or any other set of facts expressible in the temporal logic.

We will use an action formalism with temporal preconditions and effects. The temporal representation used is based on the J. Allen model, where the basic unit is a temporal interval, and intervals can be related by any thirteen primitive relations (see table no 1).

Table no1:Allen relationships

Relation	Symbol	Symbol of inverse	Example
X before Y	<	>	XXX YYY
X equal Y	=	=	XXXXXXX YYYYYYY
X meets Y	m	mi	XXXYYY
X overlaps Y	o	oi	XXXX YYYY
X during Y	d	di	XXX YYYYYYY
X starts Y	s	si	XXX YYYYYY
X finishes Y	e	ei	XXX YYYYYYY

Another complex relationship is one interval X wholly contains another interval Y.

Y IN X ----> X (di si fi) Y.

The inference procedure is based on the transitivity behavior of the relations. A typical example of such an inference rule is:

If X finishes Y and Y meets Z Then X meets Z

XXXX
YYYYYYY + YYYYYYZZZZZZ -----> XXXXXZZZZZZ

Our current concern is not suggesting any new methods for problem solving, but simply to investigate the consequences of the more general world model, because the current problem solving systems are constrained in their applicability by inadequate world models. Particularly, in most systems, the model of time is such that action must be considered to be instantaneous, and only one action can occur at a time. This is the case of systems based on the situation calculus.

The world is represented as a set of situations, each describing the world at single instant in time. An action is a function from one situation to another, and can be described by a set of prerequisites on the initial situation and a set of effects that will hold in the final situation. While this model has been extremely useful in modeling physical actions by a single agent in an other words static world, it cannot easily be extended to account for simultaneous actions and events as in multirobot systems. Furthermore, since an action in the situation calculus is equated with change, actions that involve no activity (e.g waiting for a signal from another robot or restore the world to its original state) cannot be modeled.

2- PRINCIPLE OF ACTION'S PLANNING

We intend to handle explicitly the temporal component of planning. From a plan of operations generated on an object level of robotic programming and a run time environment, we may specify and schedule different actions for each operation. We use the previous tool, which has many advantages; 1) handling explicitly and symbolically time, 2) existence of qualitative relationships, 3) flexibility of intervals. Nevertheless, it has a major drawback which is coherence verification of a set of interval relationships. It is a NP-complete problem. We first try to reduce it. We propose to verify coherence for each operation. This verification is made by the group of robots which will perform operation. This leads us to choice a decentralized control structure.

The planning diagram summarizes the method(see fig.1).

2.1. TRANSITION TABLE

The most important property of Allen relationships is the transition(8), because we may infer a relationship between two given intervals, since known two sets of relationships between three intervals.

For example, if A d B and B m C Then A < C

The infered relationship is deduced from Allen composition table(4).

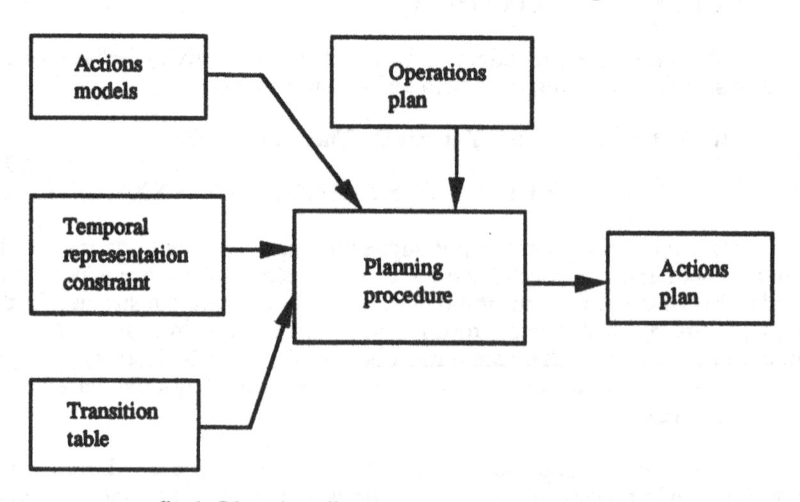

fig.1 Planning diagram

2.2. TEMPORAL REPRESENTATION CONSTRAINT

When we assert that a property is held over an interval t, this interval is assumed to be the most wide interval. This means that two intervals associated to the same property do not meet or overlap, otherwise equal.

2.3. ACTION MODELS

We will use a STRPs-like action formalism, except that the preconditions and effects of each action will be temporally qualified(10)(11)(3).

The preconditions and effects are considered as properties. The action model is:

If Occur ((A1, Ag1),t1) then
    Holds(P1,t2) such that
        t1 R1 t2
    And
    Holds(E1, t3) such that
        t1 R2 t3.

A1: action 1
Ag1: robot 1 which performs A1 over t1
P1: precondition during t2
E1: effect during t3
R1, R2 are two Allen relationships.

2.4. OPERATIONS PLAN

Operations plan generation was treated by (15). The purpose was to pass from a task static description (object, links, precedence) to task dynamic description (operation, precedence). This dynamic description is called operations plan.

Assuming that he has an operations plan and run-time environment, the programmer must draw up a program using a co-operative language. He must specify the set of actions of each operation without causal explanation. That being the case, the planning procedure will exploit one operation at a time, and use the transition table to generate links between actions of the same operation. We obtain a graph for each operation, the coherence verification problem will be reduced.

2.5. PLANNING ALGORITHM

For each operation of plan, we search for underlaying actions their models in model file. Place the index of these models in index file. To find relationships between the different actions, we exploit the temporal representation constraint, otherwise looking for each couple of models the existence of a temporal relationship.

To find all the possible relationships between actions of the same operation, we must exploit the transition table.

A- PSEUDO CODE

```
1-For each operation  do
        for each action  do
          if action    in model-file then end;
             else
                  index-file <---- index (action)
               forward  index-file
            end if
   end do
end do;
2- For each operation do
      for each index-couple do
         deduce a temporal relationship
         place binded couple indexes in relation-file
         end do
      forward relation-file
end do;
3-For each operation of relation-file do
      put existent relationships in result-file
      if number of couples >= 3 then
      deduce relationships between triplet
       put the relationship in result-file
      end if
end do.
```

B-COHERENCE VERIFICATION

This verification will be made locally, each group of robots will be engaged in checking the coherence of its proper graph. As soon as a new constraint arrives, it will be propagated on the graph.The coherence verification algorithm is proposed in (4).

IV. ALLOCATION OF ACTIONS

As a multirobot system is by "nature" distributed physically and / or logically and heterogeneous (operative processes, computing processes), these features lead to the same difficulties as in distributed systems, particularly the allocation problem.

In a multirobot system, each robot may have its own features, although they have some common features. For instance, two robots may share the same workspace, but have particular tools. There are two main methods proposed in the literature of distributed computing, the enumerative methods and heuristics.

a)- The enumerative methods

These algorithms are non determinist, enumerative. They give exact solutions but the complexity is raised by exponential temporal function. To avoid the complete enumeration, we use the following technics:

- non-linear programming(5),
- graph theory(6),
- bound and branching methods(7).

b)- The heuristic methods

The two main algorithm types are:

- iterative methods(8),
- greedy methods(14),

An heuristic algorithm is qualified to be "well" (14) if:

-the necessary calculation is realistic,
-the criterion value is near optimum,
-it must as simple as possible.

In spite of various methods of tasks allocation, few are these accounting the dynamic allocation Our approach to this problem is an heuristic method based on non-linear programming and Sceptre model.

Since previously a robotic task and a multirobot system are defined as two sets, allocate a task T consisted of "a" actions over a system R consisted of "r" robots will lead to seek all the application of T -----> R. There are r^a possible cases to study.

In previous work on the allocation problem, it was proved that it is an NP-complete problem(2). Our first aim is to reduce it, while accounting for the features of multirobot system.

We subdivide this problem on two subproblems; partitioning problem and dynamic allocation problem. The holding criterion is the task achievement in best conditions.

1. DYNAMIC ALLOCATION

1.1. MOTIVES

The motives which leads us to do a dynamic allocation for a robotic task over a multirobot system are:

- fault-tolerant,
-exploitation of the inherent parallelism,
-flexibility during the task execution.

1.2.. DIFFERENT PHASES OF ALLOCATION

1.2.1. PARTITIONNING OPERATIONS

We propose to affect statically each operation to a group of robots capable to execute it. Let have a multirobot system defined as a set of robots. The partitioning will be based on the competence of each robot and the robotic task. It will take into account the following constraints:

-the correspondence between an operation and a group of robots must be done such that each action of the operation could be executed at least by two robots (Fault tolerant constraint).

-The reunion of robot competences which fulfil the previous constraint must de great or equal to the operation (admissibility of operation).

-If several groups of robots can be allocated to an operation Oi and if it will be executed concurrently with another operation Oj, we affect Oi and Oj respectively to their groups which form a minimal intersection.

Example

Suppose that we have an operation O and four robots r1,r2, r3, r4. The processing of the partitioning is as follow:

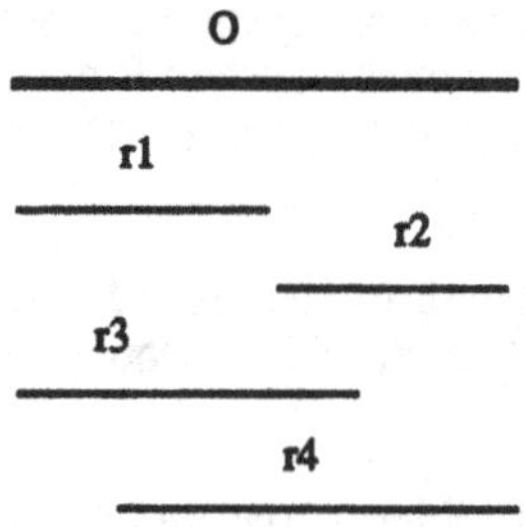

O

r1

r2

r3

r4

fig.2 Representation of operation and robots

$\cap 1 = r1 \cap r3 \neq \phi$

$\cap 2 = r1 \cap r2 = \phi$

$\cap 3 = r2 \cap r4 \neq \phi$

$\cap 4 = r1 \cap r4 \neq \phi$

$\cap 5 = r3 \cap r4 \neq \phi.$

leads to $\cup = \cap 1 \cup \cap 3 \cup \cap 4 \cup \cap 5 = O$

The group s consisted of r1, r2, r3 and r4 can be affected to the operation O

1.2.2. Dynamic allocation of actions

Since each operation is affected to a group of robots, we must allocate each action of the operation to a robot. Unlike the partitioning, the allocation will be done dynamically, in other words as soon as an action is ready, we proceed to do an election of a robot to execute it.

A-Model of the run-time environment

< As, Rs, Ps >

Where:
As is action state,
Rs is robot state,
Ps is part state.

1)The Action states

While refering to SCEPTRE (19), we can say that an action can be on one of the following states.

-locked,
-ready,
-busy,
-suspended.

2) The robot states

A robot may be in one of the three states:

-busy,
-ready,
-suspended.

3) The part states

A part may be present or absent.

-present,
-absent.

When an action is ready, we calculate the cost function for the robots capable of executing it. The cost function is presented as follows:

$$Fi=Ci*Di \tag{3}$$

where:
Fi is the cost function of the robot i,
Ci is the competence of the robot i,
Di is the availability of the robot i.

After the calculation of the different functions, we compare them and we assign the action to the robot with the highest cost function.

We illustrate this algorithm with this following example.

Given a multirobot system consisted of three robots R1, R2, R3, and an action schedule. When an action A1 is ready and the corresponding part P1 is present in the workspace, we calculate the three different cost functions of R1, R2 and R3.

Suppose that R1 and R2 are able to execute the action A1 and if R2 will execute it. R1 is not available at the moment.

$F1 = C1 * D1$
$F2 = C2 * D2$
$F3 = C3 * D3$.

R1 is:
- capable to execute A1 then $C1 = 1$,
- not ready then $D1 = 0$,
$F1 = 1*0 = 0$

R2 is:
capable to execute A1 then $C2 = 1$,
ready then $D2 = 1$,
$F2 = 1*1 = 1$.

R3 is
not capable to execute A1 then $C3 = 0$,
$F3 = 0$.

When comparing F1, F2 and F3, we conclude that

F1 is greater than F2 and F3, this yields to assign A1 to R1 and so on.

V. CONCLUSIONS

In this paper we only treated the co-operative problems, which is presented as a planning problem and dynamic allocation problem. We have just mentioned the competition problem. The latter will not be addressed here.

We have presented the relation between the flexibility in plan of actions and the unpreductibility of the executions of a given task.

REFERENCES

Book

2. P. Gaspart. Langages de programmation de la robotique, Hermes, Paris 1987.

18. J. Carlier, P Chretienne. Problemes d'ordonnancement, modelisation, complexité, algorithmes. Masson Paris 1988.

11. J.Nilson. Principle of Artificial Intelligence. Tioga Press 1980.

8. H. Bestougeff, G Ligozat. Outils logiques pour le traitement du temps, de la linguistique à l'intelligence artificielle. Masson, Paris, Fevrier 1988.

Journals

13. W. Chli, L. Hollway, M. Lan and K. Efe. Task allocation in distributed data processing. Computer, Vol 31, No 11, pp 54-69, November 1980.

14. H. S. Stone. Multiprocesseur scheduling with the aid of network flow algorithm. IEEE Trans - software Eng, Vol SE- 3, pp 85-93, January 1977.

15. O. I. El Dessouki, W. H. Huen. Distributed enumeration on computer. IEEE Trans - comp, Vol C-29, No 9, pp 818-825, Sept 1980.

16. S. H. Boukhari. On the mapping problem. IEEE Trans - comp, Vol C-30, No 3, pp 207-224, mar 1981.

3. K. G. Shin, M. E. Epstein. Intertask communications in an integrated system. IEEE Journal of robotics and automation, Vol R.A-3, No 2, Apr 1987.

9. R. E. Fikes and N. J. Nilson. STRIPS: a new approach to the application of theorem proving to problem solving. Artificial intelligence 2, (3), pp 189-208, 1971.

10. E. Sacerdoti. Planning in a hierarchy of abstraction spaces. Artificial intelligence, 5, (2), pp115-135, 1974.

5. J.F. Allen. Towards a general theory of action and time. Artificial intelligence, 23, pp123-154, 1984.

6. A. Galton. A critical examination of Allen's theory fo action and time. Artificial intelligence, 42, pp159-188, 1990.

Thesis

17. N. Chezal. Quelques méthodes d'ordonnancement de processus de traitement de signal sur un réseau de transputers. Thèse d'état, Université de Paris Sud Orsay, Oct 1988.

12. O. Kalafate. Planification des tâches opératoires robotiques basées sur le modèle d'acteur: aide à la construction de programmes d'un robot d'assemblage. Thèse de doctorat, Université de Valenciennes, Dec 1990.

Paper in Conference Proceeding

1. N. Benameur, A. Neki, K. Ouriachi. Système de factorisation de tâche robotique. Conference IA-90. Hermes, Paris, Dec 1990.

7. J. F. Allen. Maintaining knowledge about temporal intervals. Communication of the ACM, 26, 11, pp832-843, 1983.

Report

4. J. Ferber and M. Gallab. Problématique des univers multi-agents intelligents. LAFORIA, Université de Paris 6, 1989.

19. Browaeys et al . Proposition de standard de noyau d'exécution temps réel. Projet SCEPTRE, Rapport BNI, No 26/2, Sep 1982.

Piecewise Straight-Line Correlation Algorithm for Feedback Navigation Systems with Robotic Applications

A. Berman, J. Dayan

Faculty of Mechanical Engineering, Technion, Haifa 32000, Israel

ABSTRACT

Autonomous moving systems, such as free moving robots and "nursing" robots, accumulate errors during path tracking. To reduce these errors, it is required to measure the deviations from the path and to correct the system movement in order to find the correlation between the measurement of the pattern characterizing the path, and a previously recorded pattern stored in the tracking system computer. When the correlation process is performed, it is possible to evaluate the system deviations from the desired path and calculate the required corrections.

A novel correlation algorithm, based on a pattern of piecewise straight lines, is described. The pattern can be obtained by filtering the straight lines from a TV, IR, laser, radar or sound picture of the path vicinity. The Bayes Classifier or MAP (Maximum A - Posteriori) algorithm is the optimal correlation law which selects x_i as the location from which z_i was measured at instant i, with high probability for minimal error. This law is suitable due to the selected linear pattern of piecewise straight lines.

Two main issues in implementing the correlation process are discussed:

A. Developing a correlation algorithm that has near optimum performance along with simplicity allowing real-time calculations with minimum computational effort.

B. Evaluating the performance bound of the correlation process.

The correlation algorithm selected for this work is derived from MSD (Mean Square Difference), which has an advantage over other methods for path pattern correlations, when there are no scaling errors between the measurement and the pattern in memory. This claim is based on simulation, as well as analytical proof, showing that MSD converges to ML (Maximum Likelyhood) for errors which are small with respect to the pattern measurement.

It is shown that, with the proposed method, deterministic errors in measuring the distances to these lines do not affect the correlation in the longitudinal (x) direction. Thus, only random measurement errors among the lines will cause correlation errors in x. In the lateral (y) direction, however, the obtained error is identical in size to the bias error.

INTRODUCTION

For autonomous moving systems, such as robots, it is required to track a path using feedback measurement, in order to cancel possible errors accumulated during the movement. To close the feedback control loop, a suitable technique is needed in order to find the correlation between the measurement of the pattern characterizing the path and a previously recorded pattern stored in the computer of the tracking system.

Performing the correlation process makes it possible to evaluate the system deviations from the desired path and calculate the required corrections. The main problems in implementing the correlation process are:

1. Selecting the path pattern (and accordingly the measuring system to be used) having the suitable resolution required for the desired accuracy and, at the same time be, simple enough to allow real time correlation.
2. Selecting or developing a correlation algorithm that has near optimum performance along with simplicity allowing real-time computation with minimum computational effort.
3. Processing and recording of the selected pattern data in order to store it in the computer memory.
4. Evaluating the performance of the correlation process.
5. Integrating the servo system, that includes the measuring system as well as the computational algorithm and the correlation process, in the overall tracking system.

A novel correlation algorithm, based on section of nonparallel straight lines pattern, is described. The pattern is obtainable by:

1. Filtering the straight lines from a TV, IR, laser, radar or sound picture of the path vicinity
2. Marking the pattern along the path.

A typical schematic pattern of straight lines along a path is described in Fig. 1 along with the crossing measurement band.

Several correlation algorithms are possible candidates for this task as the fast development in LSI technology makes it possible to implement highly sophisticated correlation techniques in real-time applications. Clary and Russell [1] described a cross-correlation using FFT. Reed and Hogan [2] performed their algorithm according to:

$$\rho(r) \triangleq \left[N \sum_{n=1}^{N} z(n)d(n+r) - \sum_{n=1}^{N} z(n) \sum_{n=1}^{N} d(n+r) \right] / N^2 \sigma_z \sigma_d(r) \qquad (1)$$

where

- $\rho(r)$ - correlation index
- $z(n)$ - measurement components
- N - total number of measurement components
- $d(n)$ - pattern
- σ_z - variance of the measurement distribution
- σ_d - variance of the pattern
- r - distance from the robot to the reference pattern.

They also suggested other algorithms such as the EOR - Exclusive OR Algorithm,

$$EOR(r) \stackrel{\Delta}{=} \frac{1}{N} \sum_{n=1}^{N} \left[z_{EOR}(n+r) \oplus d_{EOR}(n) \right] \tag{2}$$

and the NPA - Normalized Product Algorithm,

$$NPROD(r) \underline{\underline{\Delta}} \sum_{n=1}^{N} z(n+r)d(n) \Big/ \sum_{n=1}^{N} z(n+r) \tag{3}$$

as well as the DGA - Direct Gradient Algorithm - which is based on a 1971 US patent (No. 3609762):

$$DGA(r) \stackrel{\Delta}{=} \frac{1}{N} \sum_{j=1}^{N} \left\{ z[S_2(n)+r] - z[S_1(n)+r] \right\} \tag{4}$$

where $S_1(n)$ - the beginning edge of the pattern
$S_2(n)$ - the ending edge of the pattern.

Another method, using moments (similar to FFT) has been described by Moskowitz [3] while Boland et al. [4] suggested the SSDA - Sequential Similarity Detection Algorithm - which is especially suitable for TV picture correlation. There is a whole class of correlations such as MAD (Mean Absolute Difference) and MSD (Mean Square Difference), which are claimed (Hinrichs, [5]; Wessely, [6]) to have advantage over other methods for area correlations when there are no scaling errors between the measurement and the pattern in the memory. This claim is based on simulation (Hinrichs [5]) and a proof showing that MSD converges to ML (Maximum Likelihood) for errors which are small with respect to the pattern measurement signals.

Since MSD approaches ML, which is likely to have minimal error (Young and Calvert [7]), and because of its simplicity, it has been selected as the correlation algorithm for the present work.

GENERAL ANALYSIS OF THE CORRELATION PROBLEM

Determination of the exact location by correlating a certain property of the path pattern (say, heights, colors, etc.) and the memory recorded picture of that pattern is not absolute, due to accompanying measurement noise and the errors between the picture in the memory and the real pattern. The relations between the measurement $\bar{z}$ and the pattern $\bar{d}$ is:

$$\bar{z}_i(x_i) = \bar{d}_i(x_i) + \bar{w}_i(x_i) \tag{5}$$

where the index i represents the sampling instant, and

$$\bar{z}_i(x_i) \underline{\underline{\Delta}} [z_1(x_i), z_2(x_i) \ldots\ldots z_n(x_i), \ldots z_N(x_i)]^T \in R^N \tag{6}$$

is the measurement obtained at location x_i. The index n denotes the nth -component of the measurement $\bar{z}_i$ corresponding to the nth - component of the path pattern.

The true pattern (without noise) at location x_i along the path is given by

$$\bar{d}_i(x_i) \underline{\Delta} \left[d_1(x_i), d_2(x_i), \ldots d_n(x_i) \ldots d_N(x_i)\right]^T \in R^N \tag{7}$$

and the overall noise accompanying the pattern $\bar{d}(x_i)$ is given by

$$\bar{w}_i(x_i) \underline{\Delta} \ [w_1(x_i), \ldots \ \ldots w_n(x_i), \ldots \ w_N(x_i)]^T \in R^N \tag{8}$$

The latter contains two factors,

$\bar{w}_v$ - noise of the measuring device

$\bar{w}_{map}$ - error of the memory stored map, used as base for the correlation, which includes quantization error relative to the real pattern.

Assuming that there is no problem of identifying the terms of the vectors $\bar{z}$ and $\bar{d}$, the indices n's are omitted from now on.

The correlation process goal is to evaluate $\hat{x}_i$ which minimizes (according to the selected criterion) the difference between the measured vector $\bar{z}_i$ and the selected terrain vector $\bar{d}(\hat{x}_i)$ of the reference map. Due to noise $\bar{w}_i$, an error between the real x_i of the measurement $\bar{z}_i$ and the evaluated $\hat{x}_i$, as observed by the correlation system, is likely to exist. This uncertainty in evaluating $\hat{x}_i$ is the main cause for inaccuracies in navigation by pattern correlation. Therefore, a criterion is needed that can assist in limiting the range of admissable correlation error.

The Bayes Classifier or MAP(Maximum A-Posteriori algorithm) is the correlation law which, according to Young and Calvert [7], selects $\hat{x}_i$ with minimal probability for errors, as the location from where $\bar{z}_i$ was measured. According to this decision law, $\hat{x}_i$ is selected in such a way that $P(\hat{x}_i / \bar{z}_i)$ is maximized, i.e., the a-posteriori probability that $\bar{z}_i$ was measured at $\hat{x}_i$ is maximum when

$$P(x_i/z_i) = f_{z/x}(z_i/x_i)P(x_i)/f_z(z_i) \tag{9}$$

where, $f_{z/x}(z_i/x_i)$ is the probability density function of obtaining measurement z_i at location x_i.

Our goals are:

a. To find a suitable algorithm for the Bayesian classifier (estimator) to work with the selected correlation method.

b. To find a sub-optimal correlation algorithm, allowing simple analysis, whose performance constitute a bound for the Bayesian algorithm. Obviously, it is desired that this bound will be "close", thus providing a measure for the correlation error.

It will later be shown that, due to the advantages of the proposed method, which is based only on a straight-lines pattern, deterministic errors of measuring the distances to these straight lines do not affect the correlation in the x-direction. Thus, only random measurement errors among the lines will cause correlation errors in x. In the y-direction, however, the correlation error is identical in size to the bias error.

Because of the cancellation of the bias errors by the correlation systems, only random errors will be considered in the analysis from now on.

SELECTING THE BEST MAP ESTIMATOR FOR THE LINE CORRELATION.

According to Sage and Melsa [8] the MAP estimate equations for measurements $\bar{z}$ with Gaussian noise $\bar{v}$ is independent of the measured variable $\bar{x}$ is:

$$\hat{\underline{x}}_{MAP} = \left(V_x^{-1} + H^T V_v^{-1} H\right)^{-1}\left(H^T V_v^{-1}\bar{z} + V_x^{-1}\bar{\mu}_x\right). \tag{10}$$

where

$$\bar{z} = H\bar{x} + \bar{v} \tag{11}$$

and

$$E(x) = \mu_x \qquad var(\bar{x}) = V_x \qquad \bar{x} \in R^2$$

$$E(\bar{v}) = 0 \qquad \sigma^2 = var(\bar{v}) = V_v \qquad \bar{v}, \bar{z} \in R^N, \quad N \geq 2$$

H is an Nx2 matrix, N is the number of measurements and 2 - the number of the state variables.

This problem is identical to the present estimation problem where the pattern $\bar{d}$ is defined by

$$d_{i,n}(x_i, y_i) = b_n x_i + a_n - y_i \tag{12}$$

where, $d_{i,n}(x_i, y_i)$ is the y distance from the nth line to the robot sensor location (x_i, y_i), i denotes the measurement at time i, and n - the component (out of N) of the pattern. Hereupon, index i is omitted. Then,

$$H = \begin{bmatrix} b_1, & -1 \\ b_2, & -1 \\ \vdots & \vdots \\ b_N, & -1 \end{bmatrix}; \qquad \bar{x} = \begin{bmatrix} x \\ \\ y \end{bmatrix}; \qquad var(\bar{x}) = \begin{bmatrix} \sigma_x^2 & 0 \\ & \\ 0 & \sigma_y^2 \end{bmatrix}$$

For continuous straight lines (with the exclusion of lines perpendicular to the axes), $\sigma_x^2 = \sigma_y^2 = \infty$.

Given the following equation for the probability density function of the measurements

$$f_z(\bar{z}) = k\exp\left[-\frac{1}{2}(\bar{z}-H\bar{\mu}_x)^T\left(HV_xH^T+V_v\right)^{-1}(\bar{z}-H\bar{\mu}_x)\right] \tag{13}$$

where

$$E(\bar{z}) = H\bar{\mu}_x \triangleq HE(x) = \bar{b}\,x + \bar{a} - \bar{T}y \tag{14}$$

$$\bar{T} = (1,1,.....1)^T \tag{15}$$

the variance σ_z^2 for the pattern $\bar{d}$ is obtained by

$$\sigma_z^2 = V_v + HV_xH^T = V_v + \bar{b}\,\bar{b}^T\sigma_x^2 + T\sigma_y^2 \tag{16}$$

where

$$\underline{T} \triangleq \bar{T}x\bar{T}^T \tag{17}$$

The probability density function of the error $\bar{\delta}$ for this estimation can be obtained by

$$\begin{aligned} f(\bar{x}/\bar{z}) = f(\underline{\hat{x}}+\bar{\delta}/\bar{z}) &= K\exp\left[-\frac{1}{2}(\bar{x}-\underline{\hat{x}})^T V_{\tilde{x}}^{-1}(\bar{x}-\underline{\hat{x}})\right] \\ &= K\exp\left[-\frac{1}{2}\bar{\delta}^T V_{\tilde{x}}^{-1}\bar{\delta}\right] = f\left(\bar{\delta}/\bar{z}\right) \end{aligned} \tag{18}$$

where $\bar{\delta} = \begin{bmatrix}\delta_x \\ \delta_y\end{bmatrix}$; $\quad \delta_x = x-\hat{x}; \quad \delta_y = y-\hat{y}$

and the inverse of the $V_{\tilde{x}}$ - the variance of the error $\bar{\delta}$ - is given by

$$V_{\tilde{x}}^{-1} \triangleq V_x^{-1} + H^T V_v^{-1} H \tag{19}$$

Clearly, since the last relation (Eq. (19)) is independent of $\bar{z}$, we have

$$f(\bar{\delta}/\bar{z}) = f(\bar{\delta}) = K\exp\left[-\frac{1}{2}\bar{\delta}^T V_{\tilde{x}}^{-1}\bar{\delta}\right] \tag{20}$$

Using MAP estimation according to Eq. (10) requires an apriori knowledge of μ_x, V_x and V_v. However, if V_x is not known, it is possible to assume $V_x = \infty$. Thus,

$$\underline{\hat{x}}_{MAP} = (H^T V_v^{-1} H)^{-1} H^T V_v^{-1}\bar{z} \tag{21}$$

and

$$V_{\tilde{x}}^{-1} = H^T V_v^{-1} H = (H^T H)\sigma^{-2} \tag{22}$$

This assumption permits a simple evaluation of the Cramer - Rao bound for the error obtained during MAP estimation.

According to MAP or MV estimation, if the probability that the real point of measurement is located out of the circle with radius δ_x (error) around $\hat{x}$ is P_F $(|\hat{x}-x|>\delta_{x,min})\leq 0.0027$, then

$$\delta^2_{x,min} \geq 3^2 \cdot V_{\tilde{x}} \approx 9(H^T H)^{-1}\sigma^2 \tag{23}$$

It is possible to improve the correlation, using MAP algorithm, by selecting lines with larger slope H. Clearly, by enlarging H, through the selection of optimal lines, it is possible to reduce $V_{\tilde{x}}$.

The $\hat{\underline{x}}_{MAP}$ of Eq. (10) can be further developed for correlation of straight lines, where:

$$\hat{\underline{x}} = \left(H^T H\right)^{-1} H^T \bar{z} \tag{24}$$

Therefore,

$$\left(H^T H\right) = \begin{pmatrix} \sum^N b_n^2 & -\sum^N b_n \\ -\sum^N b_n & N \end{pmatrix} \tag{25}$$

$$\left(H^T H\right)^{-1} = \begin{pmatrix} N & \sum^N b_n \\ \sum^N b_n & \sum^N b_n^2 \end{pmatrix} \left(N \sum^N b_n^2 - \left(\sum b_n\right)^2 \right)^{-1} \tag{26}$$

$$H^T \bar{z} = \begin{pmatrix} \sum^N b_n z_n \\ -\sum^N z_n \end{pmatrix} \tag{27}$$

Thus,

$$\hat{\underline{x}} = \begin{bmatrix} \hat{x} \\ \hat{y} \end{bmatrix} = \begin{pmatrix} N \sum^N (b_n z_n) - \sum^N b_n \sum^N z_n \\ \sum^N b_n \sum^N (b_n z_n) - \sum^N b_n^2 \sum^N z_n \end{pmatrix} \left(N \sum^N b_n^2 - \left(\sum^N b_n \right)^2 \right)^{-1} \tag{28}$$

ANALYSIS OF THE MSD ALGORITHM RESULTS

The following will prove the identity between the result obtained by the above described MAP algorithm and that of the MSD (Mean Square Difference) algorithm.

Starting with Eq. (12) for the pattern

$$d_{i,n}(x_i, y_i) = b_n x_i + a_n - y_i \triangleq y_{i,n} - y_i$$

it is possible to define the performance index I as:

$$I = \sum_{n=1}^{N} (y - b_n x - a_n + z_n)^2 \tag{29}$$

The minimum for the criterion index is obtainable by differentiating I with respect to x and y and comparing to zero.

$$\frac{\partial I}{\partial x} = \sum^{N} (y - b_n x - a_n + z_n) b_n = 0 \tag{30}$$

$$\frac{\partial I}{\partial y} = \sum^{N} [y - b_n x - a_n + z_n] = 0 \tag{31}$$

Applying Eqs. (30) and (31) to measurements accompanied by errors, an estimation for location $(\hat{x}, \hat{y})$ is obtained. This estimation can be explicitly expressed by simple algebraic manipulations of the above equations:

$$\hat{x} = \frac{N\left[\sum^{N} (a_n b_n) - \sum^{N} (b_n z_n)\right] - \sum^{N} b_n \sum^{N} a_n + \sum^{N} b_n \sum^{N} z_n}{\left(\sum^{N} b_n\right)^2 - N \sum^{N} b_n^2} \tag{32}$$

$$\hat{y} = \frac{\hat{x} \sum^{N} b_n + \sum^{N} a_n - \sum^{N} z_n}{N} \tag{33}$$

These results are identical to the MAP estimator results, for which it has been assumed that $a_n = 0$ ($z_{MAP} = z_n - a_n$).

Equations (30) and (31) can be combined, based on distances between lines instead of the location of a single line, and the resulting equation permits direct evaluation of $\hat{x}$. Let D_{nj} be the distance between line n and line j of the pattern,

$$D_{nj} \triangleq d_n - d_j = (b_n - b_j)x + a_n - a_j \triangleq B_{nj}x + A_{nj} \tag{34}$$

$$B_{nj} \triangleq b_n - b_j, \qquad A_{nj} \triangleq a_n - a_j \tag{35}$$

and the measured distance be

$$Z_{nj} \triangleq z_n - z_j \tag{36}$$

The role of the correlation algorithm is to minimize the performance index:

$$I = \sum_{n,j}^{N} \left[D_{nj} - Z_{nj}\right]^2 = \sum_{n,j}^{N} \left[(b_n - b_j)x + (a_n - a_j) - z_n + z_j\right]^2 \tag{37}$$

where each possible pair n, j appears only once. The minimization depends on a single parameter (x) only, and therefore

$$\frac{\partial I}{\partial x} = 2\sum_{n,j}^{N} (b_n - b_j)[(b_n - b_j)x + a_n - a_j - z_n + z_j] = 0 \tag{38}$$

from which

$$\hat{x} = \left[\sum_{n,j}^{N} (B_{nj} z_{nj}) - \sum_{n,j}^{N} (B_{nj} A_{nj})\right] \Big/ \sum_{n,j}^{N} B_{nj}^2 \tag{39}$$

It will be proved now that $\hat{x}$ obtained from Eq. (39) is identical to $\hat{x}$ from Eq. (32) for the same set of measurements.

By comparing respective terms, the following relations are obtained:

a. $$\sum_{n,j}^{N} B_{nj}^2 = \sum_{n,j}^{N} (b_n - b_j)^2 = N\sum_{n=1}^{N} b_n^2 - \left(\sum_{n=1}^{N} b_n\right)^2$$

b. $$\sum_{n,j}^{N} (B_{nj} z_{nj}) = \sum_{n,j}^{N} (b_n - b_j)(z_i - z_j) = N\sum_{n=1}^{N} (b_n z_n) - \sum_{n=1}^{N} b_n \sum_{n=1}^{N} z_n$$

c. $$\sum_{n,j}^{N} (B_{nj} A_{nj}) = \sum_{n,j}^{N} (b_n - b_j)(a_n - a_j) = N\sum_{n=1}^{N} (b_n a_n) - \sum_{n=1}^{N} b_n \sum_{n=1}^{N} a_n \tag{40}$$

It is therefore possible to solve the correlation problem in two stages: First, $\hat{x}$ will be found, using distance differences and employing Eq. (39). Then, $\hat{y}$ is obtained from substitution of $\hat{x}$ in Eq. (33).

In order to reduce the computational effort, some degenerate versions of I can be derived and used instead of I itself. The terms in Eq. (37) can be grouped according to the distances between each pair of lines:

$$I = \sum_{n,j}^{N} \left[D_{nj} - Z_{nj}\right]^2 = \sum_{n,j=n+1}^{N-1} \left[D_{nj} - Z_{nj}\right]^2 + \sum_{n,j=n+2}^{N-2} \left[D_{nj} - Z_{nj}\right]^2 + \cdots + \left[D_{Nl} - Z_{Nl}\right]^2 \quad (41)$$

To reduce the computational load in minimizing I, it is sufficient to consider only the following pairs:

$$I_1 \triangleq \sum_{n,j=n+1}^{N-1} \left[D_{nj} - Z_{nj}\right]^2 \quad ; \quad n = 1,3 \ldots N-1 \ \text{(even N)} \quad (42)$$

These selections of pairs ensure independency between the errors of the measurement Z_{nj}, as each line d_n is considered in only one pair D_{nj}. The simulation results, presented below, demonstrate the advantages this selection has over possible computations based on the differences from one single line, such as:

$$I_2 \triangleq \sum_{j=2}^{N} \left[D_{1j} - Z_{1j}\right]^2 \quad (43)$$

This is because the correlation errors are identical, although the number of correlated points in I_2 is double that of I_1.

Another selection is based on pairs of adjacent lines, where each line is common to two neighboring pairs:

$$I_3 \triangleq \sum_{n,j=n+1}^{N} \left[D_{nj} - Z_{nj}\right]^2 + \left[D_{Nl} - Z_{Nl}\right]^2, \quad n = 1,2,,3 \ldots N-1 \quad (44)$$

Simulation shows (see section 9) that minimizing I_3 yields similar results to minimizing I, in spite of the dependency among the errors in measuring Z_{nj}. In particular, absolute identity between I and I_3 exists for the three-line case.

OBTAINING AN IMPROVED CRITERION BY MINIMIZING I_1

As defined by Eq. (34), D_{nj} is the difference between two pattern lines d_n and d_j, measured with a standard deviation of σ. It is possible to write

$$\sigma_1 = \sqrt{2}\,\sigma \quad (45)$$

where σ_1 is the standard deviation in measuring Z_{nj}.

When I_1 is used instead of I, it is still possible to improve the correlation results by selecting those pairs of lines that provide large changes in D for the same $\bar{\delta}$. The improved I_1 criterion, where the pairing is performed by selecting line j for line n according to

$$B_n \underline{\Delta} B_{nj} = \underset{j}{MAX}(B_n - B_j) \tag{46}$$

(where any particular line appears in one pair only), is denoted by I_4.
The constraints $\bar{\delta}$ obtained by the fast criterion of Eq. (23) will be the same if

$$\frac{[D(x)-D(x,\delta_x)]^T\,[D(x)-D(x,\delta_x)]}{\sigma_1^2} \cong \frac{\left[d(x,y)-d(x,y,\bar{\delta})\right]^T\left[d(x,y)-d(x,y,\bar{\delta})\right]}{\sigma^2} \tag{47}$$

This equality exists when

$$\begin{aligned}&[D(x)-D(x+\delta_x)]^T[D(x)-D(x+\delta_x)\\ &\cong 2[d(x,y)-d(x,y,\bar{\delta})]^T[d(x,y)-d(x,y,\bar{\delta})\end{aligned} \tag{48}$$

which is true in most of the cases where D_{nj} is selected according to Eq. (46). This assumption is justified by the simulation (see section 9), as well.

The largest error obtainable in the y-direction is estimated by substituting $\hat{x}$, which was estimated by minimization of I according to Eq. (31)

$$\sum^{N}\left[\hat{y} - b_n\hat{x} - a_n + z_n\right] = 0 \tag{49}$$

$$\text{Since } \sum^{N}\delta z_n \cong \sum w_n \cong 0, \text{ then: } \sum^{N}\left[\delta_y - b_n\delta_x\right] = 0 \tag{50}$$

where δ_x and δ_y are the errors in x and y respectively. As $0 \le |b_n| \le 1$ (it will later be shown that the selected pattern lines have maximal reclining angle of 45°)

$$\left(\delta_y\right)_{max} \approx \frac{(\delta_x)_{max}\sum |b_n|}{N} \cong 0.5(\delta_x)_{max} \tag{51}$$

This last result has been verified by the simulation as well.

It should be pointed out that it is not always necessary or possible to find the vector $\overline{H}$ along the error path. Therefore, it is not always possible to use the MAP algorithm (Eq. 10) or the minimization of the squared differences (Eq. 29) to find $\hat{x}$. The correlation between the measurement z and the recorded pattern d will then be performed according to the minimization of the squared differences between z and d (d is the pattern stored in the computer memory), as follows:

$$\text{MIN } I \underline{\Delta} \min\sum_{n=1}^{N}\left[(z_n - d_n)^T(z_n - d_n)\right] \tag{52}$$

EFFECT OF MEASUREMENT ERRORS ON THE CORRELATION ACCURACY

It will be shown that the uncertainty in yaw angle does not cause correlation errors in the y-axis nor in the yaw angle ψ.

The criterion used for the correlation is the minimization, according to the parameters x, y, ψ of the squared distances between the measured lines and their pattern, as stored in the memory map. First, the effect of the robot yaw angle ψ on its lateral location, in respect to the coordinates of the measuring system, is studied. According to Fig. 2, it is assumed that the reference system, x'y' (length and width respectively), which moves with the robot, moves with yaw angle ψ with respect to the fixed coordinates xy.

The pattern of the lines along the path is given and kept in the computer memory exploiting the xy system. For example, the n'th path line in the computer memory is given, in terms of the xy coordinates, as:

$$y_n = b_n x + a_n \tag{53}$$

The relations between the xy and x'y' systems are:

$$\begin{vmatrix} x' \\ y' \end{vmatrix} = \begin{vmatrix} \cos\psi & \sin\psi \\ -\sin\psi & \cos\psi \end{vmatrix} \begin{vmatrix} x \\ y \end{vmatrix} \tag{54}$$

or

$$\begin{vmatrix} x \\ y \end{vmatrix} = \begin{vmatrix} \cos\psi & -\sin\psi \\ \sin\psi & \cos\psi \end{vmatrix} \begin{vmatrix} x' \\ y' \end{vmatrix} \tag{55}$$

Expressed in the coordinates of the moving system, x'y', and assuming small ψ ($\sin\psi \approx \psi$, $\cos\psi \approx 1$), Eq. (53) is given by:

$$y_n = b_n(x' - y'_n\psi) + a_n = x'\psi + y'_n \tag{56}$$

from which it is possible to obtain the explicit form:

$$y'_n = \frac{b_n - \psi}{1 + b_n\psi}x' + \frac{a_n}{1 + b_n\psi} \tag{57}$$

Assuming $b_n\psi \approx 0$ leads to:

$$y'_n = (b_n - \psi)x' + a_n \tag{58}$$

or

$$y_n' + \psi x' = b_n x' + a_n \tag{59}$$

The assumption of small ψ is valid if there is no accumulation of large navigation errors. Taking $\psi b_n \approx 0$ is justified if the straight sectional lines, used for the navigation, are not too close to the normal to the robot line of motion (or that such lines are not used for the correlation). With these assumptions, it is impossible to differentiate between errors in y and in ψ. However, leaving out the assumption $b_n\psi \approx 0$ permitting the detection of the angle ψ by the correlation

$$y_n' + \frac{\psi x'}{1+b_n\psi} = \frac{b_n}{1+b_n\psi}x' + \frac{a_n}{1+b_n\psi} \tag{60}$$

This requires correlation according to three non-parallel pattern lines, as is proven in the following section.

FINDING THE YAW ANGLE BY CORRELATION

Fig. 3 shows why correlation according to only two pattern lines cannot yield a unique solution in the case of freedom in yaw angle. It is possible, however, to determine x, y and ψ uniquely if three (or more) non-parallel lines are used.

First, consider the possibility of finding ψ by minimizing the index I, as given by Eq. (29). This equation is rewritten as:

$$I \triangleq \sum^{N} (d_n - z_n)^2 = \sum^{N} [y_n(x,y) - y - z_n]^2 \tag{61}$$

$$d_n \triangleq y_n - y \tag{62}$$

where y_n is the lateral location of pattern line n in the fixed coordinate system xy. Similar to the relationships in Eq. (59), it can be assumed, for the fixed coordinate system as well as for the moving one, that if $\psi \approx 0$ or varies around zero, then

$$y_n(x, \psi) = (b_n - \psi)x + a_n \tag{63}$$

The optimal values of x, y, ψ are estimated by differentiating I according to these variables and comparing to zero.

$$\frac{\partial}{\partial y}\sum^{N} [-y + y_n(x,\psi) - z_n]^2 = \sum^{N} [y - (b_n - \psi)x - a_n + z_n] = 0 \tag{64}$$

$$\frac{\partial}{\partial x}\sum^{N} [y - y_n(x,\psi) + z_n]^2 = \sum^{N} [y - (b_n - \psi)x - a_n + z_n]b_n = 0 \tag{65}$$

$$\frac{\partial}{\partial \psi}\sum^{N} [-y + y_n(x,\psi) - z_n]^2 = \sum^{N} [y - (b_n - \psi)x - a_n + z_n]x = 0 \tag{66}$$

Eq. (66) is essentially similar to Eq. (64). Thus, as long as the assumption $\psi b_n \approx 0$ prevails, it is impossible to differentiate between y and ψx, and the estimate is for the combined variable $y - \psi x$. However, use of y_n, similar to $\dot{y}_n$ from Eq. (57) in Eq. (66), without neglecting $b_n\psi$, brings in a condition which is different from Eq. (64) due to the presence of $1 + b_n\psi$ in the denominator of the expression for y_n.

It is possible to prove algebraically that there exists a unique correlation state which maintains true distances among the lines, for more than three pattern lines. Therefore, minimizing the performance index I yields a zero only for the correct ψ. Referring to Fig. 4, the following set of eight equations with eight unknowns is obtained:

$$\begin{array}{ll} y_1 = b_1 x_1 & y_3 = b_3 x_3 + a_3 \\ y_1 = B x_1 + A & y_3 = B x_3 + A \\ y_2 = b_2 x_2 & D_{12} = \sqrt{(x_2 - x_1)^2 + (y_2 - y_1)^2} \\ y_2 = B\, x_2 + A & D_{23} = \sqrt{(x_3 - x_2)^2 + (y_3 - y_2)^2} \end{array} \tag{67}$$

where $x_1, y_1, x_2, y_2, x_3, y_3, A,$ and B are the eight unknowns. The use of more than three lines will contribute to reducing the correlation errors obtained from random measurement errors of all the lines.

CANCELLATION OF MEASUREMENT ERRORS WITH COMMON BIAS

Berman (1982) showed that the measurement errors with common bias for all the lines, without uncertainty in the yaw angle during correlation along the x-axis perpendicular to the measurement, are cancelled out as follows:

Assuming error d_n in the n-th measurement in which the value z_n has been obtained for the n-th line from location x_o, y_o according to

$$z_n = b_n x_o + a_n - y_o + d_n \tag{68}$$

Substituting z_n of Eq. (68) in Eqs. (64) and (65) and multiplying them by $\sum^N b_n$ and by N respectively, yields

$$x - x_o = \frac{+N\sum^N (b_n d_n) - \sum^N d_n \sum^N b_n}{N\sum b_n^2 - (\sum b_n)^2} \tag{69}$$

$$y - y_o = \frac{\sum^N b_n \cdot \sum^N (b_n d_n) - \sum^N b_n^2 \cdot \sum^N d_n}{N\sum b_n^2 - (\sum b_n)^2} \tag{70}$$

If the error d_n consists of a common part d for all lines and a non common part $\tilde{d}_n$ unique for each line, then:

$$x - x_0 = \frac{N\sum(b_n\tilde{d}_n) - \sum\tilde{d}_n\sum b_n}{N\sum b_n^2 - (\sum b_n)^2} \tag{71}$$

$$y - y_0 = \frac{\sum b_n\sum(b_n\tilde{d}_n) - \sum b_n^2\sum\tilde{d}_n}{N\sum b_n^2 - (\sum b_n)^2} - d \tag{72}$$

Thus, the common part d does not affect the $\Delta x(= x - x_o)$. In the y axis, however, the error d causes a deflection of the estimated measurement point by $-d$. Thus, the effect of the common part of the error is a virtual deflection of the robot from the reference line.

COMPUTER SIMULATION OF LINEAR ALGORITHMS FOR LINE CORRELATION

Each of the performance indices I, I_1, I_2, I_3 (Eqs. 41 - 44) and the improved index I_4 (Eq. 46) can be minimized by the maximum difference method. These minimization algorithms are $A_I, A_{I_1}, A_{I_2}, A_{I_3}$, and A_{I_4}, respectively. Due to the linear pattern it is possible to minimize I according to Equations (32) - (33) and the rest of the indices according to Eqs. (39) and (33).

To compare the algorithms, simulations of straight line correlations have been performed. The standard deviations σ_{v_x} and σ_{v_y} of the correlation errors according to the five algorithms were checked. The varying parameters were: the number N of reference pattern lines, the standard deviation σ of the measurement error, and the range of change of b_n - the slope of the pattern lines, d_n, according to Eq. (12). The standard-deviation calculation was performed on 500 samples for which the measurement noise was normally distributed. The parameters b_n were randomly determined, in uniform distribution between -0.5 and + 0.5 (Tables 1 and 2) and -1.0 to + 1.0 (Table 3).

These simulations provide justification of several equations used and assumptions made above.

a. From Tables 1 and 2, it is possible to see the absolute identity between the correlation errors of A_I and A_{I_3} for the three-line case.

b. According to Eq. (51)

$$\delta_{y_{max}} \leq \frac{\delta_{x_{max}}\sum|b_n|}{N} \tag{73}$$

$|b_n| \leq 0.5$ yields $\delta_y \leq \frac{1}{4}\delta_x$ and therefore $\sigma_{v_y}^2 \leq \frac{1}{16}\sigma_{v_x}^2$.

$|b_n| \leq 1$ yields $\delta_y < \frac{1}{2}\delta_x$ and therefore $\sigma^2_{v_y} \leq \frac{1}{4}\sigma^2_{v_x}$.

c. For large enough N it is possible to write:

$$\sum_n^N b_n^2 \approx N \cdot E(b_n^2) = \frac{N}{2|b|_{max}} \int_{-|b|_{max}}^{|b|_{max}} x^2 dx = \frac{N}{3} b^2{}_{max} \tag{74}$$

where $E(b_n^2)$ represents the expectation of b_n^2. Since the differences B_n have been formed according to conditions of Eq. (46), it is feasible to state that:

$$\sum_{n=1,3...}^{N-1} B_n^2 \simeq \frac{N}{2} E(B_n^2) = \frac{\frac{N}{2}}{4|b|max} \cdot \int_{-2|b|max}^{2|b|max} x^2 dx = \frac{2N}{3} b^2 max \tag{75}$$

Therefore, when $|b|_{max} = 0.5$, then $\sum_{n=1,3...}^{N-1} B_n^2 = \frac{N}{6}$, $n = 1,3...$, and when $|b|max = 1.0$, then $\sum B_n^2 = \frac{2N}{3}$. Eqs. (74) and (75) are used to verify the relations of Eq. (48)

$$\sum_n^{N-1} B_n^2 \approx 2 \sum_n^N b_n^2 \qquad n = 1,3.... \tag{76}$$

and from Eq. (76), as mentioned, the correlation errors obtained from the two algorithms I and I_4 are very similar. Since the time required to execute the correlation according to A_{I4} is half that of A_I, its superiority over all the other algorithms is obvious.

The simulation results prove the above assumptions (Eqs. 74, 75 and 76) as well.

d. According to Eq. (23)

$$V_{\tilde{x}} \triangleq \sigma^2_{MAP} = \left(H^T H\right)^{-1} \sigma^2_{I4} = 2(H^T H)^{-1} \sigma^2 \tag{77}$$

where,

$V_{\tilde{x}}, \sigma^2_{MAP}$ is the variance of the estiamtion error according to MAP,
σ_{I4} is $\sqrt{2}\sigma$, where σ is the measurement noise (Eq. 45), and
$H^T H$ is given by Eq. (25).

From Tables 1 and 2, it can be observed that $\sigma^2_{v_x}$, obtained according to A_{I4}, is very close to σ^2_{MAP}.

e. According to Eqs. (75), (77) and (25), it is possible to write

$$\sigma_x^2 \approx \frac{2\sigma^2}{\sum B^2} \approx \frac{3}{N}\frac{\sigma^2}{b_{max}^2} \tag{78}$$

Therefore, the estimation error decreases when a larger number of lines for the correlation is used, and increases if the measurement error is increased. Table 3 verifies the effect of changing b_{max} on the estimation error as well: doubling b_{max} (from 0.5 to 1) reduces the error in x four-fold (Eq. 78), while, according to Eq. (73), the error in y is not affected.

CONCLUSIONS

The novel correlation algorithm A_{I_4} (Eq. 46) described in this paper has near optimum performance. Its simplicity allows real-time estimation of the robot location, with minimum computational effort.

In future work, this algorithm will be integrated in a navigation system of an autonomous moving robot, based on a Kalman filter, in order to track a required path using feedback measurements.

REFERENCES

[1] Clary, J.B. and Russell, R.F. All-Digital Correlation for Missile Guidance, SPIE - Application of Digital Image Processing, Vol. 119, pp. 36-46, 1977.

[2] Reed, C.G. and Hogan, J.J. Range Correlation Guidance for Cruise Missiles, IEEE Trans. on Aerospace and Electronic Systems, Vol. AES-15, No. 4, pp. 547-554, 1979.

[3] Moskowitz, S.Terminal Guidance by Pattern Recognition - A New Approach, IEEE Trans. on Aerospace and Navigational Electronics, Vol. ANE-11, pp. 254-265, 1964.

[4] Boland, J.S., Pinson, I.J., Peters, E.G., Kane, G.R. and Malcolm, W.M. Design of a Correlator For Real-Time Video Comparisons, IEEE Trans. on Aerospace and Electronic Systems, Vol. AES-15, No. 1, pp. 11-19, 1979.

[5] Hinrichs, P. R. Advanced Terrain Correlation Techniques, IEEE Position, Location and Navigation Symposium, pp. 89-96, 1976.

[6] Wessely, H.W.Image Correlation, Part II (Theoretical Basis), Rand Corp. Santa Monica Calif., NTIS, AD-A036, 482, 1976.

[7] Young, T.Y. and Calvert, T.W. Classification, Estimation and Pattern Recognition, Elsevier, pp. 12-44, 1974.

[8] Sage, A.P. and Melsa, J.L. Estimation theory with Application to Communications and Control, McGraw-Hill, pp. 182-183, 1971.

[9] Berman, A. Terrain Contour Matching Navigation, Master thesis, Technion - Israel Inst. of Tech. October 1982.

ACKNOWLEDGEMENT

The paper is based on Mr. A. Berman's MS thesis, supervised by Dr. Z. Meiri whose contribution is greatly appreciated.

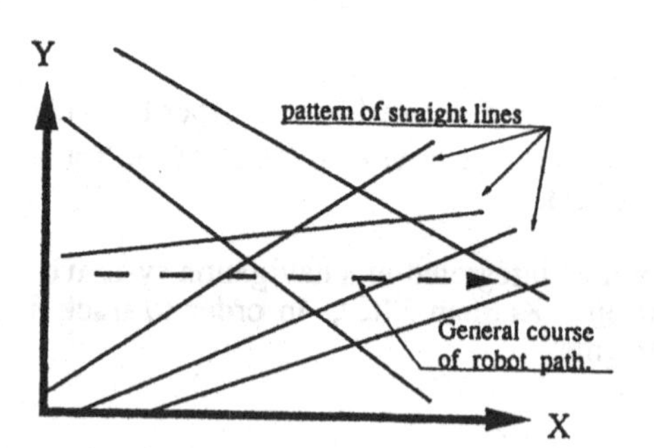

Fig. 1: Pattern of straight lines to guide the robot

Fig. 1: Pattern of straight lines to guide the robot.

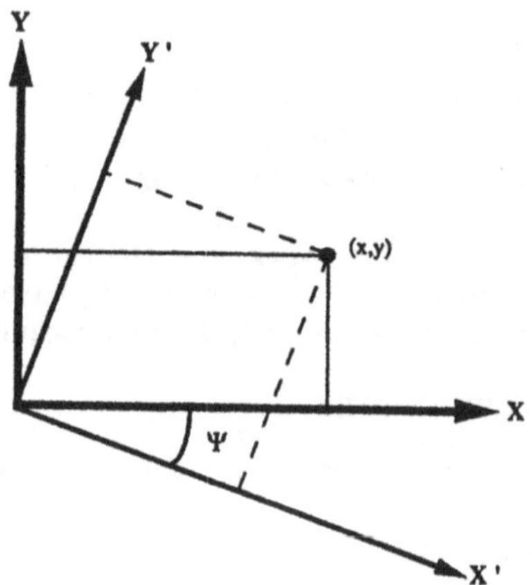

Fig. 2: Coordinate system rotation.

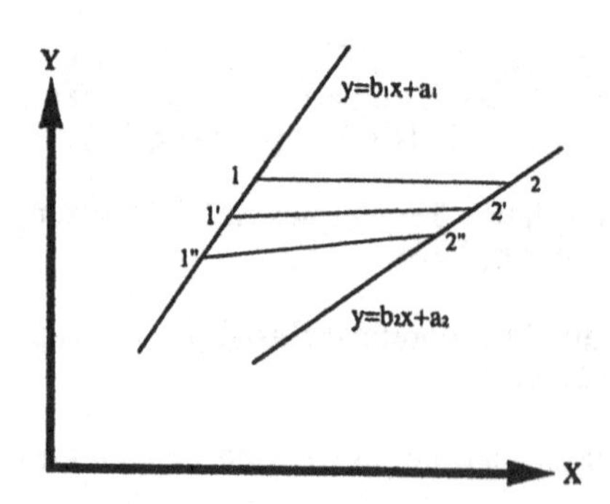

Fig. 3: Possible positions to measure the distance between points 1 and 2 of two lines, y_1 and y_2.

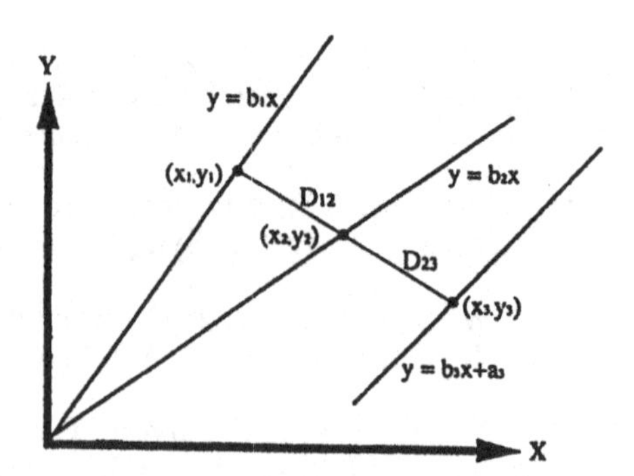

Fig. 4: Three-line correlation.

Table 1 - Correlation errors $\sigma^2_{v_x}$, $\sigma^2_{v_y}$ dependent on number of lines N for $|b| \leq 0.5$ (x factor) first distribution and measurement error $\sigma = 20$.

N	$\sigma^2_{v_x}$					$\sigma^2_{v_y}$						
	A_I	A_{I_1}	A_{I_2}	A_{I_3}	A_{I_4}	A_I	A_{I_1}	A_{I_2}	A_{I_3}	A_{I_4}	$\sum_{i=1,3}^{N-1} B_i^2$	$\sum_i^N b_i^2$
3	7469.1		7873.4	7469.1		209.9		214.4	209.9			
4	925.6	10615.9	1372.4	933.2	3862.1	105.9	101.3	105.9	106.0	95.7	.20	.14
5	904.7		1510.4	1039.1		76.0		77.4	75.7			
6	644.9	3347.0	1272.9	653.6	3563.7	64.7	75.4	68.9	64.8	76.0	.21	.14
7	1511.3		2685.5	2150.6		81.2		95.0	90.8			
8	723.2	4119.6	1473.2	878.9	749.9	63.6	81.7	89.8	68.7	50.7	1.12	.75
9	756.2		1311.5	882.9		47.6		51.1	48.0			
10	684.7	562.5	695.8	942.5	584.0	53.8	47.9	53.5	58.0	48.3	1.36	.85
11	562.0		946.8	669.0		59.3		80.5	62.7			
12	269.6	1007.5	927.4	402.0	382.3	33.8	33.4	33.9	33.8	31.7	1.95	1.11
13	272.5		1014.4	359.7		32.3		38.5	33.5			
14	441.5	513.1	1751.5	627.0	477.3	31.5	42.4	40.2	32.6	42.9	1.74	1.40
15	401.8		1108.9	504.1		27.9		32.5	28.6			
16	399.1	376.8	1475.2	661.8	275.3	23.2	25.7	25.8	23.8	25.7	3.06	1.72
17	322.8		1468.7	492.5		26.2		29.9	26.8			
18	321.8	1213.2	1311.6	457.6	612.2	23.4	24.4	25.7	23.6	23.8	1.28	.77
19	369.9		576.1	578.5		19.5		19.5	19.5			
20	206.5	849.3	825.7	245.6	431.4	19.3	19.8	19.3	19.4	19.8	1.88	1.21

A_{I_1}, A_I, A_{I_2}, A_{I_3}, A_{I_4} – Correlation algorithms.

Table 2 - Correlation errors $\sigma^2_{v_x}$, $\sigma^2_{v_y}$ dependent on number of lines N for b ≤ 0.5 (x factor) second distribution and measurement error $\sigma = 20$.

N	$\sigma^2_{v_x}$				$\sigma^2_{v_y}$				$\sum_{i=1,3}^{N-1} B_i^2$	$\sum_{i=1}^{N} b_i^2$
	A_I	A_{I2}	A_{I3}	A_{I4}	A_I	A_{I2}	A_{I3}	A_{I4}		
3	1714.5	1784.6	1714.5		129.9	130.4	129.9			
4	1054.0	1740.0	1147.0	1164.3	123.0	135.7	127.2	97.8	.69	.37
5	570.1	989.5	627.6		74.6	76.3	74.9			
6	1707.0	1789.0	1833.0	1795.2	82.2	83.4	83.3	102.0	.51	.45
7	1018.9	1258.5	1176.2		70.1	75.5	74.0			
8	577.8	1187.3	596.6	951.9	57.3	57.9	57.3	45.3	.82	.53
9	819.0	911.7	1016.0		41.1	41.0	41.1			
10	734.3	1139.1	877.0	1157.1	39.9	41.4	39.9	53.7	.70	.59
11	476.8	1422.7	524.6		34.8	36.2	34.8			
12	1061.7	3451.8	1524.9	466.5	107.1	259.9	135.9	33.1	1.62	.93
13	497.6	1914.2	615.9		35.7	53.4	37.5			
14	301.4	1226.9	460.0	717.8	27.4	27.9	27.4	28.7	1.14	.62
15	258.9	1032.9	306.5		27.5	28.5	27.5			
16	317.5	430.7	570.8	397.1	24.0	24.4	24.1	26.5	1.99	1.41
17	327.3	1659.8	403.1		28.2	40.0	29.0			
18	487.5	1857.7	657.8	343.1	36.5	73.3	40.4	22.6	2.17	
19	332.3	564.5	433.3		20.4	20.7	20.6			
20	295.6	1545.8	410.8	295.1	20.3	24.6	20.0	20.0	2.80	1.69

A_I, A_{I2}, A_{I3}, A_{I4} - Correlation algorithms

Table 3 - Correlation results of A_{I4} algorithm for different b (factors of x) and measurement noises dependent on number of lines N.

N	$\|b\| \leq 0.5$						$\|b\| \leq 1.0$			
	$\sigma = 10$		$\sigma = 20$		$\sigma = 40$		$\sigma = 20$		$\sigma = 40$	
	$\sigma^2_{v_x}$	$\sigma^2_{v_y}$	$\sigma^2_{v_x}$	$\sigma^2_{v_y}$	$\sigma^2_{v_x}$	$\sigma^2_{v_y}$	$\sigma^2_{v_x}$	$\sigma^2_{v_y}$	$\sigma^2_{v_x}$	$\sigma^2_{v_y}$
4	965.5	23.9	3862.1	95.7	15448.6	383.0	965.5	95.7	3862.1	383.0
6	890.9	19.0	3563.7	76.0	14254.9	304.3	890.9	76.0	3563.7	304.3
8	187.4	12.6	749.9	50.7	2999.8	203.0	187.4	50.7	749.9	203.0
10	146.0	12.0	584.0	48.3	2336.3	193.3	146.0	48.3	584.0	193.3
12	95.5	7.9	382.3	31.7	1529.2	127.0	95.5	31.7	382.3	127.0
14	119.3	10.7	477.3	42.9	1909.4	171.9	119.3	42.9	477.3	171.9
16	68.8	6.4	275.3	25.7	1101.3	103.0	68.8	25.7	275.3	103.0
18	153.0	5.9	612.2	23.8	2449.1	95.3	153.0	23.8	612.2	95.3
20	107.8	4.9	431.4	19.8	1725.7	79.3	107.8	19.8	413.4	79.3

A Multiple Views Robotic Stereo Method for 3-D Shape Perception

M.A. Arlotti, M.N. Granieri
IBM Italy Rome Scientific Center, via Giorgione 159, 00147 Roma, Italy

ABSTRACT

A 3-D machine perception technique is presented using stereo intensity images and an anthropomorphic robot. Multiple stereo views allow the perception of the third dimension by the solution of a correspondence problem defined in two stereo pairs. The zero crossing technique is used to detect edges on the images and to classify them in order to reduce the number of possible solutions. By the eye-in-hand configuration, several stereo pairs of a scene can be taken moving the robot arm. For a limited number of objects , stationary on the robot table, this technique allows the derivation of a set of three-dimensional sample points that correspond to the physical object edges. The technique mainly consist in a geometrical match, in the 3-D space, of some hypothesized solutions formulated on a couple of stereo pairs, taken from two points of view. Considering objects with straight edges and with no highly textured surfaces, some geometrical properties are statistically verified for the most solution points, so the correspondence problem can be robustly solved.

INTRODUCTION

One of the major problems for an intelligent robot is the ability to analyze the operational scenery and extract knowledge about it, in order to intelligently interact with it. In this context visual sensors play an essential role, since they can give the robot, despite other sensors the most complete information about the surrounding world. Of course, such an information should reproduce the 3-D world's structure and render the maximum amount of details about the objects contained in it.

One way to achieve these goals consists in exploiting the mobility of the robot to explore the scene: the robot itself handles its optical sensors (TV cameras), locating them in points of the space fro which the scene can be suitably seen. This helps the robot to acquire multiple 2-D images about the scene, which can be used to reconstruct its 3-D structure. Techniques of this kind, in analyzing a scene from different view points, are generally known as *multiple views* based techniques, and have been proposed with different approaches [1].

To perceive objects details with high accuracy, stereometric techniques can be used, combined with some other 3-D methods concerning, for instance, surfaces perception (textures, color, etc).
Many approaches to stereovision have been proposed, which can be classified into *lateral stereo* and *axial stereo* families. Most of the former are based on the Marr-Poggio human stereo vision theory [2], [3], while the latter (see [4],[5]) try to recover the scene's depth information, which is lost in the image formation process, moving the camera along its optical axis. All these systems must seek to extract a set of significant points from each image (edge points, for instance) and to establish an association between each point of either image, so that all coupled points are projections of the same physical point. This is the well known *correspondence problem.*

This paper describes a perception technique in which stereovision is combined with the multiple views technique, implemented using a robot manipulator, to solve, the correspondence problem, by a geometric-statistical approach. Using a TV camera pair attached to the robot arm, known as the *stereo-visor*, stereo image pairs of the world scene can be acquired and related to each other, and used to reconstruct scene's 3-D entity. The results of this perception technique are 3-D sample points, corresponding to physical objects edges or contours of objects' surface regions such as brightness changes, holes, labels etc. . This process is termed *spatial* or *3-D perception* and is part of an early vision system. The obtained 3-D points can then be used, in further processes, to localize the object as well as to recognize it among a collection of predefined ones.

IMAGE FORMATION AND TV CAMERA MODEL

A brief description of the image formation process is given in order to understand how it can affect the data detected and measured from the images, such those in our 3-D vision system.
Ideally, a camera can be modeled as a pinhole system (see fig.1), and the image formation can be modeled using the perspective projection. The pinhole corresponds to the lens center and the image plane to the optical sensor. The image plane is located at a distance equal to the focal length. This is an approximation and is not generally true in real devices: the distance of the image plane from the lens center depends on that between the object and the camera. Anyway, for large object distances, the image plane distance can be well approximated by the focal length. The resulting image is scaled version of the scene, in which the depth information has been lost (perspective projection).
This model does not take into account lens distortion which affects precision. This can be modeled in order to be compensated. In general two kinds of distortion can be considered, radial and tangential. Each one affects the error between real image coordinates and those computed using the perspective projection and can be expressed with numerical series. R. Tsai [6] has proposed to consider only radial distortion and to express this as a nonlinear function of the radius on the image plane.
The camera calibration concerns the construction of a global camera model and the estimation of its parameters [7]. In general two categories of parameters are to be estimated:

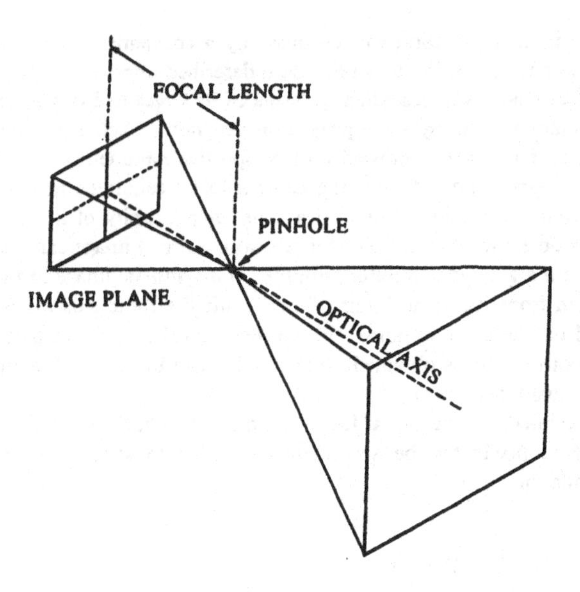

Figure 1. Image formation geometry. Camera model and its parameters

extrinsic parameters concerning camera spatial positioning

intrinsic parameters concerning image formation

The former (3 translations and 3 rotations) can be expressed by a unique 4x4 homogeneous coordinates transformation matrix, the latter are the focal length, the image center and the distortion function coefficients. Various procedures have been proposed, leading to different accuracy in parameter estimation (see Tsai [6] for instance) but requiring complex computations.

In the described vision system a pinhole camera model is assumed and an empirical camera calibration procedure is used. For seek of simplicity, no distortions of the lens are considered. This is because the objective focal length (16 mm.) is sufficiently long for such factors to be insignificant compared to other factors, such as the robot positioning precision and uncertainties in the parameters characterizing the TV camera pair mounting. The two cameras, CCD Hitachi KP 140 models, with a sensor resolution of 500 x 582 pixels, are equipped with two 16 mm Rainbow lens. The video signal is acquired by an IBM PS/2 equipped with an "AMS Photon" frame grabber, having a 720x590 geometrical resolution and an 8 bit per pixel brightness resolution. By a suitable calibration procedure the calibration parameters are estimated; these include the view frame of the stereo-visor and, for each camera, pinhole point position, rotation matrix, focal length and image center.

In particular, concerning the focal length, if the entire acquisition hardware is considered, the camera model and the system software require a global focal length, taking into account all the signal conversions after the image formation on the sensor. If the pixel width is assumed as the unit of measure, such a focal length, which is not equal to the optical focal length of the camera (physical), can be expressed in terms of these units. For example, our image acquisition hardware has a global focal length of 1510 pixels width . Its determination can be made by a comparison between a real image and a synthetic one, according to a procedure described afterwards. As well known, a simple object can be easily described in terms of its edges and displayed on a graphic computer monitor by a perspective projection program. Such a program, beside data to be displayed, requires the knowledge of perspective parameters, i.e. view frame and focal length. Suppose now, to put the camera to be calibrated in a known position and with a given orientation of its optical axis, in proximity of a scene containing a simple known object located at a known position. An real image can then be acquired and displayed on a graphic computer monitor. If a synthetic image of the same object, supposed taken from the same point of view with the same axis orientation, is then superimposed on the same monitor, the synthetic focal length can be tuned until a a matching of both images is obtained. Such a value can be assumed as the wanted estimate of the global focal length.

Once a good estimate of the global focal length is kwown, the direction of optical rays corresponding to pixels can be accurately computed in space: this is crucial for a correct triangulation.

STEREO VISOR GEOMETRY

The stereo-visor consists in two identical TV cameras mounted in a rigid body, connected to the robot flange (fig. 2a). The parallel axis arrangement requires that both cameras are rigidly mounted, with their optical axes on the same plane and parallel. Using this geometric configuration, the fixation point may be assumed to be at infinity and the horizontal scan lines of both cameras are parallel to the baseline: the disparity between image points is only "along the horizontal", i.e. along to the rows. Thus, given a generic point in one image, its corresponding point in the other image (any point is supposed to be visible on both) lies on the same scan line. This is the well known *epipolar constraint*. Figure 3 shows the entire stereo image formation process. This property reduces the research of the corresponding points to homologous rows.

In a typical robotic environment all the objects to be perceived lie in a limited workspace. This entails an admitted distance range or equivalently, disparity range on the stereo pair. In fact, setting an upper limit for the distance of the object from the stereo-visor results in a lower limit for the disparity. Conversely, an upper limit for the disparity implies a minimum admitted distance. The distance between the focal axes (*baseline* length) in the implemented stereo-visor arrangement is 80 mm and the mounting set can be adjusted, calibrating position and orientation of the cameras, by turning suitable screws.

The scene is illuminated by passive ambient light plus a neon circular lamp positioned around the visor, to obtain diffused illumination. The illumination arrangement is shown in fig.2b.

The robot used is an UNIMATION PUMA 562, an six degrees of freedom anthropomorphic manipulator. In order to know the robot flange frame and then the stereo-visor frame, with respect to a fixed reference frame connected to the robot table, the manipulator's direct kynematics is solved. This process is sufficiently accurate in comparison with the overall system's accuracy. Typical stereo image pair are shown in fig. 4.

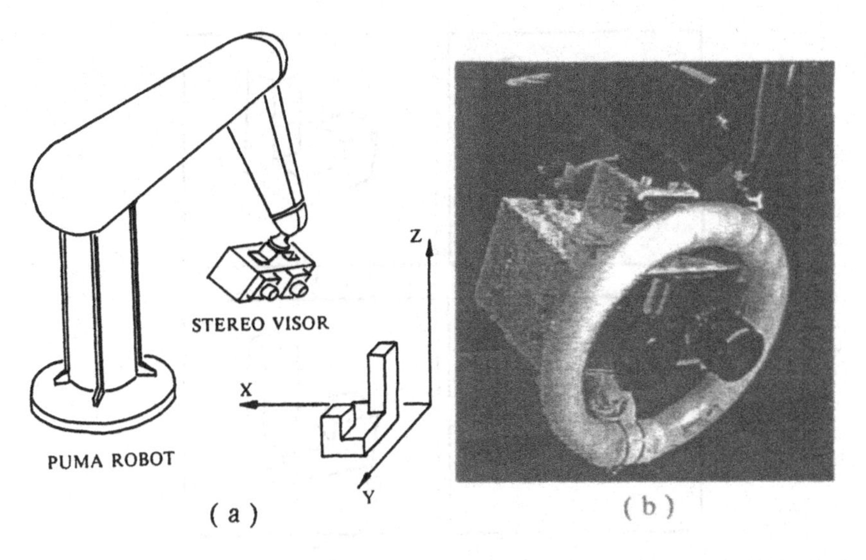

Figure 2. The acquisition arrangement. a) robot mounting of the stereo visor; b) detail of the visor and its illumination equipment.

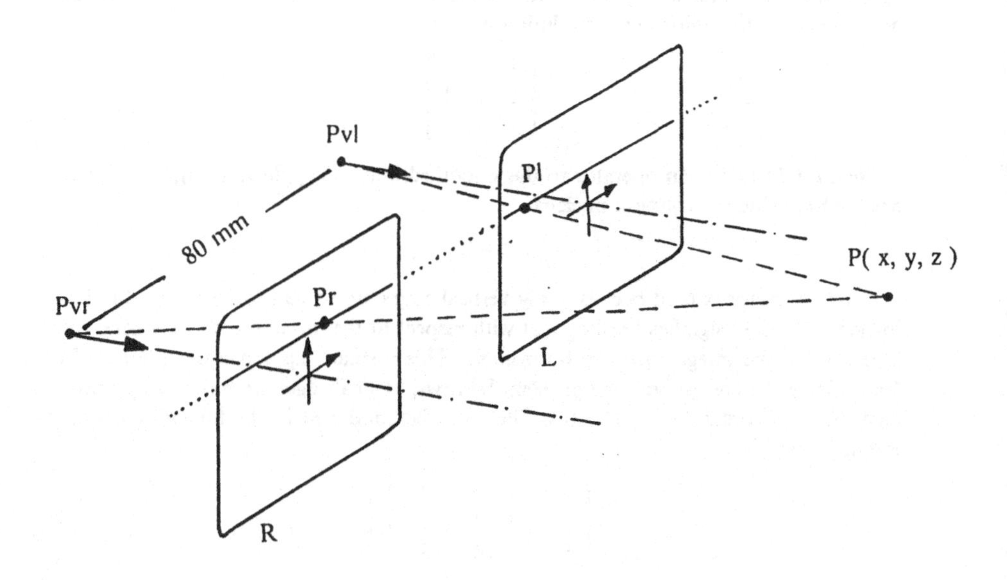

Figure 3. Stereo image formation

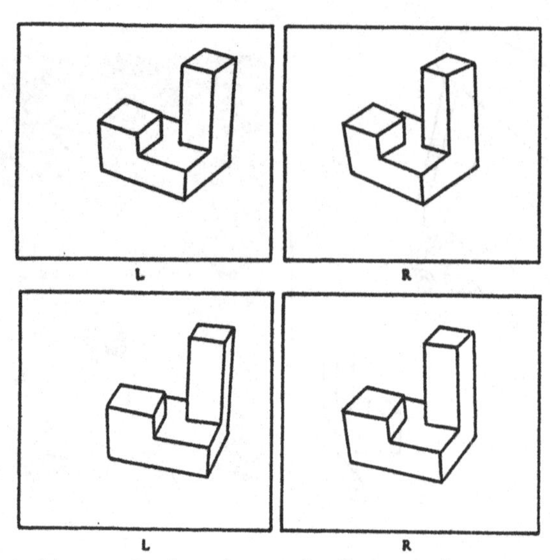

Figure 4. The graphic reproduction of a couple of stereo pairs.. (Acquired from two different view points)

IMAGE PROCESSING AND TRIANGULATION

To extract significant points from images a specific operator is needed. The chosen edge detector is a zero crossing operator [8], consisting of the following steps. First a simple low pass filter is applied to the acquired images in order to smooth out the noise. This has the following convolution nucleus:

$$\begin{bmatrix} 1 & 1 & 1 \\ 1 & 1 & 1 \\ 1 & 1 & 1 \end{bmatrix}$$

Then a 1-D Laplacian operator to the smoothed image is applied; its digital estimate has the following convolution nucleus:

$$[1 \quad -2 \quad 1]$$

A 1-D operator is used because only vertical edges are to be searched. fflIn the following, "vertical" signifies "orthogonal with respect to the image rows"; in general the baseline and the image rows are horizontal. This justifies the above definition". In fact, solving the correspondence problem by rows, vertical edges are more easily recovered than horizontal ones. The two filters can be combined in the following convolution 3x5 operator:

$$\begin{bmatrix} 1 & -1 & 0 & -1 & 1 \\ 1 & -1 & 0 & -1 & 1 \\ 1 & -1 & 0 & -1 & 1 \end{bmatrix}$$

which is applied directly to the raw images. Finally, the zero-crossing points, along each row of the filtered images, corresponding to edge points, are determined. Such

edge points can be easily classified as positive or negative, by testing the direction of the zero crossing, corresponding to the sign of the brightness discontinuity (*gap sign*).
Considering couples of homologous scan lines (rows), one for each image, each one will generally contain a different number of edge points, due to the different positioning of the cameras. This may cause a partial or total occlusion of edges in one image with respect to the other. Thus, points without any correspondence in the other image must be considered. In this case the correspondence problem cannot be solved using only one stereo pair. In fact, edge points are not characterized by suitable attributes, to simplify the process, other than the gap sign. The chosen approach exploits the mobility of the robot, using the following strategy; the stereo-visor is first positioned near the object, a stereo pair is then acquired, the robot arm is moved to reposition the visor and another stereo pair is acquired. The two visor points of view must not be far from one another. At each stereo pair acquisition, many hypothetical 3-D points are generated, by triangulating all possible couples of image points found at homologous rows of the two images. The number of 3-D points formed of course, are much more than those physically existing, being formed exhaustively, without solving the correspondence problem. A reduction is operated by checking the gap sign and the allowed distance range. In particular, suppose that the ith row of the images contains m and n edge points respectively, with m^+, n^+ positive and m^-, n^- negative edges ($m = m^+ + m^-, n = n^+ + n^-$). Thus, $k^+ = m^+ \times n^+$ plus $k^- = m^- \times n^-$ hypotheses of correspondence are generated, and then $k = k^+ + k^-$ 3D points.
By imposing an admitted distance range (e.g. fig.5, points must lie inside the limits "too near" and "too far"), some hypotheses are *a priori* discarded. Hence, a set of $h \leq k^+ + k^-$ 3-D points is produced. Considering all rows ($i = 1,..n$) , in all $H = \sum_i h_i$ 3-D points, are generated.
Repeating the process from the second point of view, another similar set of 3-D points is obtained. Both contain sample edge points plus many dummy points, arising from incorrect correspondence hypotheses.

MATCHING ALGORITHM

The matching algorithm essentially consists in the extraction of the intersection of the two sets of 3-D points, obtained by the procedure described above. To understand it, it is useful to decompose the problem into couples of homologous rows and project all generated 3-D points onto a 2-D plane. Suppose that the two data acquisitions are made holding the baseline on the same plane, such a plane could be used to project all spatial points. The situation is depicted in fig.5, where the projection plane is horizontal. In this figure, several spatial points are obtained by intersecting the rays corresponding to all couples of image points extracted at the first acquisition (for simplicity, positive edges are not distinguished from negative ones). In the same way, another set can be similarly obtained at a second data acquisition.
Each set contains several edge points, visible from both points of view, plus many incorrect hypotheses points, not corresponding to any physical entity. As can easily be shown, the former remain unchanged (confirmed) from one view to the other, while latter are not in general fixed. This is the main property underlying the solution of the correspondence problem. It may be that some incorrect hypotheses are confirmed. Incorrect points are statistically few compared with the correct ones and are usually called "phantom points". They are usually located without a clear spatial distribution.

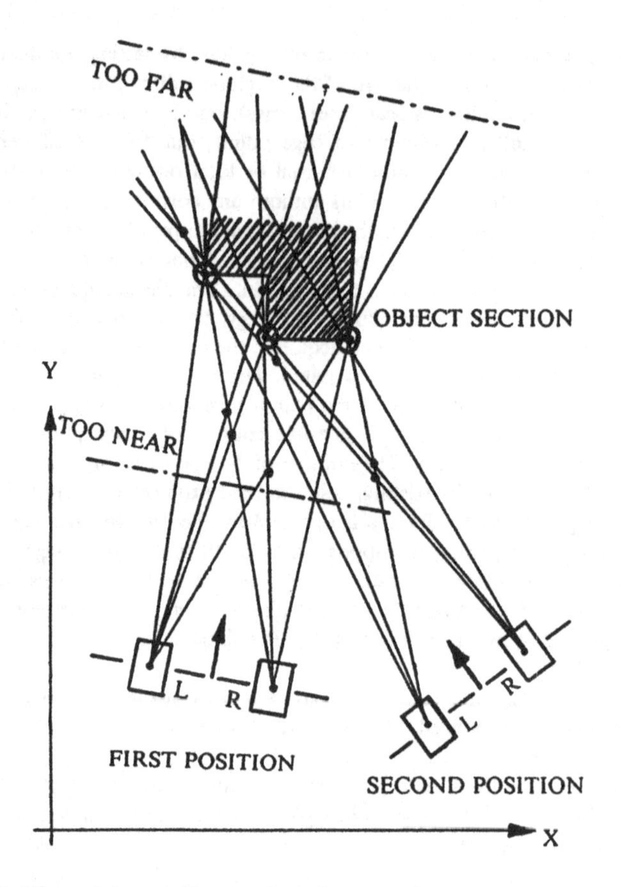

Figure 5. The points matching, projected onto a plane

Conversely, the edge points follow some known distribution rules. For example they are grouped into clouds seeming lines and curves. The analysis of spatial distribution of clusters of points can help to eliminate phantom points.

The scene illumination equipment generates some shadows. These are but easily filtered because, the lamp is rigidly moved with the visor. In this way all shadow lines, which could be interpreted as edges, by the system software, appear at different spatial position at each acquisition, so cannot be matched and then perceived.

RESULTS

The images acquired by an IBM PS/2 using an AMS frame grabber, are processed by an IBM RS/6000 running AIX, connected via Token Ring Local Area Network. Each image occupies about 300 kbytes and the network communicates at a mean transfer rate of 100 kbytes/sec., so it takes about 3 seconds to transfer a single image to the RS/6000. A C program on the RS/6000 analyzes a stereo pair and produces a set of 3-D points, taking about 2.5 sec. This process is repeated for each stereo pair. Another C program extracts the intersection of the previous sets, taking about 1 sec cpu time , for a mean scene complexity. Adding the above times to those needed to

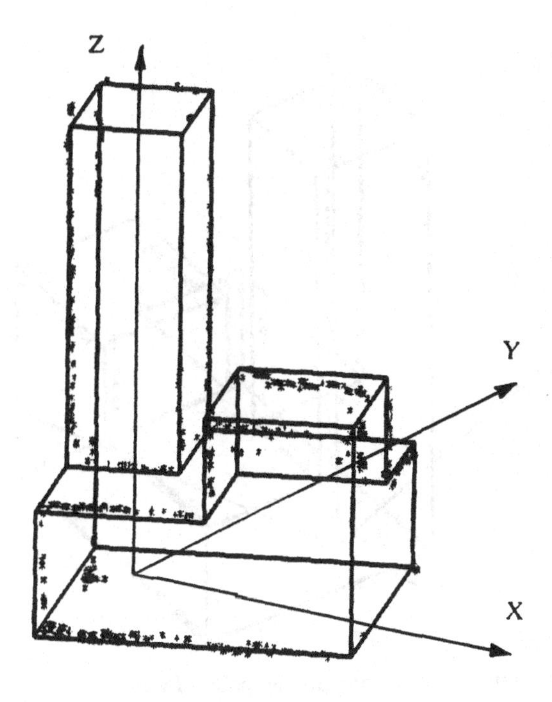

Figure 6. The result of the perception process.. Stars show the perceived edge points, lines the wire frame model

communicate with the robot and the duration of its movements , a total time of about 20 seconds is required to obtain 3-D data from two stereo pairs (base process).
Fig. 6 and 7 show the 3-D data extracted by this technique in comparison with a pre-stored 3-D wire frame model of the perceived object. In these figures, the data are obtained by four repetitions of the base perception process in order to perceive horizontal and vertical object edges, from front and back sides.

During calibration of the intrinsic and extrinsic parameters of the TV cameras and of the robot arm, we have reached a mean error for the sample points position, which is less than 1% the distance between the visor and the perceived object. As can be seen in figures 6 and 7, not every physical object edge is well sampled. This is because of limited contrast between them and the scene background. However, the figures show the combined result of four base perception process (double stereo acquisistion, e.g. fig.4). Each base process produces about 200 samples points, for an object complexity like that considered.

CONCLUSIONS

A simple multiple stereo perception technique has been presented, based on a robotic eye-in-hand configuration. The stereo visor position and orientation are known by solving robot arm direct kynematics.
The goal is to extract 3-D sample edge points from two stereo pairs, taken from different points of view. The image processing extracts 2-D edges points by a zero crossing technique. Due to the visor geometry the edge detector is specific for edges

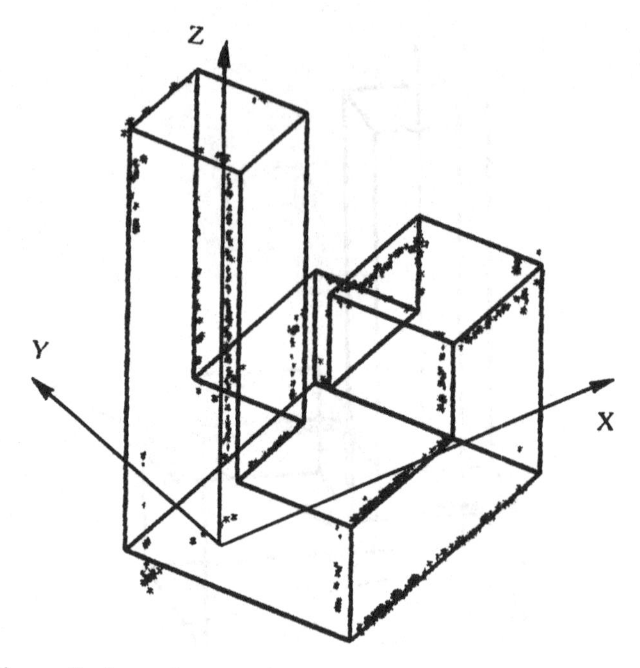

Figure 7. The results shown from another point of view

"along the image columns" and the correspondence between image points is solved row by row. The only considered feature for 2-D edge points, was the sign of the zero crossing. All the possible correspondence pairs are formulated, discarding couples outside an allowed distance range. These pairs yeld 3-D points by triangulation. The process is repeated for two visor positions, giving two sets of points. The intersection of such sets, matching points within a given tolerance, contains 3-D edges points plus some phantom points. These are typically a small number and can be discarded by checking for their irregular spatial density.

By a suitable calibration of the camera and the robot, the position of the 3-D points is estimated with a mean error less than 1% with the distance visor-object. The basic process can be iterated moving the robot arm around the object. At each step (two stereo acquisitions) the amount of 3-D data is increased.
Working on this set, many 3-D object features can be extracted. A curve growing algorithm allows to obtain a partial wire-frame of the object. This allows to estimate the dimensions, the volume and the location of the object. A recognition technique then allows the perceived objects to be recognized from a set of predefined objects. Conversely, for a known object, it is possible to localize it by a partial matching algorithm.

Future research issues concern a better characterization of the edge points, to reduce the possible incorrect correspondences, and an improvement in precision for calibrating cameras and robot arm. Color TV cameras and new image processing techniques should also allow us to obtain more information from images, and so to perceive more complex objects.

REFERENCES.

[1] V.Cappellini, R.Casini, M.T.Pareschi, C.Raspollini **From multiple view to Object Recognition** IEEE Trans.on Circuits and Systems, vol. cas-34, n.11, nov.1987

[2] D.Marr, T.Poggio **A theory of human stereo vision** M.I.T. Tech. Rep. A.P. 451, 1977

[3] D.Marr, T.Poggio **A computational theory of human vision** Proc.Roy.Soc. London, vol. B204, 1979

[4] H.Itoh, A.Miyauchi,S.Ozawa **Distance measuring method using only simple vision constructed for moving robots** IEEE Proc. 7th Conf. Pattern Rec., vol.1, Montreal, Canada, 1984

[5] N.Alvertos, D.Brzakovic, R.Gonzalez **Camera geometries for Image Matching in 3-D Machine Vision** IEEE Trans.on P.A.M.I., vol.11,n.9, September 1989

[6] R.Y.Tsai **A versatile camera calibration technique for high accuracy 3-D machine vision metrology using off-the-shelf tv cameras and lenses** IEEE J.Robotics and Automation,RA 3,n.4,Aug.1987

[7] O.D. Faugeras , G. Toscani **The calibration problem for stereo** Proc. CVPR'86 Miami Beach, Fl , pp. 15-20 , 1986.

[8] A.C. Kak **Depth perception for robots** TR-EE 83-44, Purdue University, 1983.

SECTION 7: KNOWLEDGE ELICITATION AND REPRESENTATION

Data Acquisition and Expert Knowledge Elicitation for Expert Systems in Construction

J. Christian

Department of Civil Engineering, University of New Brunswick, Fredericton, New Brunswick, Canada

ABSTRACT

The continuing work on the development of prototype expert systems applied to problems in the construction industry is described in this paper. Emphasis is placed on the acquisition of experts' knowledge and elicitation of experts' experiences as this process has been found to be a very important aspect in the use of artificial intelligence and development of prototype expert systems in an industry as diverse and fragmented as the construction industry.

Following development work in the creation of a prototype expert system using a shell program for predicting the cost-time profiles of various construction operations, the Construction Engineering Group at the University of New Brunswick in Canada were asked to investigate the feasibility of using artificial intelligence in the creation of an expert system to produce a pre-engineering estimate at a very early phase in the life of a project.

INTRODUCTION

The identification and acquisition of experts' knowledge and experiences has often been glossed over in the tremendous surge of research and development in expert systems in the last decade. This part of the development of a prototype expert system, however, often requires considerable effort, particularly in the construction industry which is characterized by being a very diverse and fragmented industry.

A knowledge based expert system was developed to determine the variation in cost-time profiles of construction activities. An example is shown in figure 1. A more detailed description of this work can be found in Reference [1].

The knowledge base contains a collection of facts, parameters, heuristics, assumptions and rules for the prediction of the cost-time profiles. A shell program (Personal Consultant Plus) uses the knowledge base to interact with the user and the database. The knowledge base was created from the knowledge acquired on twenty-eight building projects. In order to separate the rules into logical divisions, the knowledge base is divided into eleven rule groups. These groups are: knowledge-based

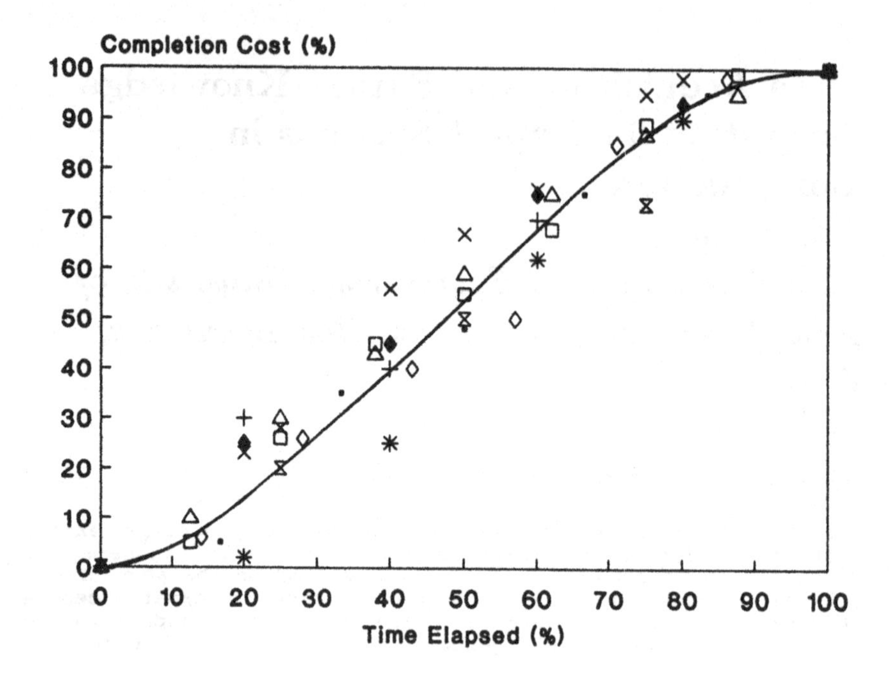

Figure 1 Cost-Time Profile (Masonry-Educational)

rules, site-work, concrete, masonry, steel, wood, thermal, finishes, door, mechanical and electrical rules. The operation of the expert system can be thought of as a dialogue between an expert and a prospective contractor, or owner, who is at the early phase in the development of a project.

A similar process of knowledge elicitation and data acquisition used in the development of the prototype system for the cost-time profiles was contemplated in continuing research on the application of expert systems for use in pre-engineering estimates.

During the construction estimating process four types of estimates may be used. In chronological order, the first estimate of a project may be a pre-engineering, or order of magnitude estimate. This estimate may be made when very little planning or design has been carried out. Following some initial planning and engineering or architectural design, a preliminary estimate is made which is often based on the area of the facility in square metres.

After a considerable amount of planning and design, a pre-tender estimate may be made and is often used for budgeting. Finally a unit price estimate, based on the exact quantities and dimensions of the completely designed facility is made by the contractors tendering for the construction project. The percentage error decreases, and the level of confidence increases, as each estimate is made in chronological order.

A distinguishing characteristic of a pre-engineering (order of magnitude) estimate is that it constitutes the first opinion on probable future costs. No matter how tentative the estimate is, and regardless of how thoroughly any subsequent estimate is prepared, the first estimate in many circumstances tends to have a high degree of influence on future, more accurate alterations and revisions to the cost estimate. It is for this reason that a more accurate forecasting system such as the one outlined in this paper is a welcome aid to the forecasting and decision making process in the initial conceptual stages of a construction project.

Some provincial governments in Canada have implemented a Capital Projects Approval Process which requires all government departments to initially identify their capital works requirements. Following this initial step in the process, a pre-engineering estimate is required for the proposed capital works. For example, a Department of Education may decide that an elementary school needs to be enlarged to accommodate an increase in enrolment. Another government department is then faced with the task of estimating the cost of this additional space, prior to the engineering or architectural design, with only the required area and historical architectural trends as a guide.

Several problems were identified after these Capital Projects Approval Processes were put into operation. The period of time between a department identifying its need and the request for funding the capital works was short. Also, historical data on the cost of construction was disjointed, vague, or was filed at various district offices, even within one province.

To overcome these problems the Construction Engineering Group at the University of New Brunswick was requested to investigate the feasibility of creating and utilizing an expert system to provide a quick and accurate pre-engineering estimate. It was realized that this process would also concurrently reveal the problems encountered in the data and knowledge elicitation of the various buildings studied.

PREVIOUS WORK ON PRE ENGINEERING ESTIMATES

Research and information on pre-engineering or architectural estimating is limited. A system by Brandon [2] requires the following user input:

1. Area required.
2. Number of storeys.
3. Substructure and structural details.
4. Quality of finishes required.
5. Site details.
6. Construction time constraints.

Kouskoulas and Koehn [3] found that the cost of a building could be predicted using six independent variables.

1. Location.
2. Price index.
3. Building type.
4. Height.
5. Quality.
6. Technology.

The price index predicts cost based on historical inflation data; quality is a measure or workmanship and materials, building use, design effort, and material type; and technology considers the cost resulting from special types of buildings or savings resulting from new technology. In all the models mentioned, some architectural design work had been carried out to establish the building layout, structural materials, and finishes. This however would not be the case in the research work described in this paper as the estimate was required before any architectural or engineering design had been undertaken.

DEVELOPMENT WORK

In the development work for the expert system creating the cost-time profiles, many contractors, consultants and owners were interviewed who had been concerned with the construction of a total of twenty-eight similar buildings in order to establish a meaningful knowledge base.

The following elicitation process was used in developing the knowledge base:
1) Analysis of published knowledge.
2) First unstructured interviews.
3) Second interviews and follow-up questionnaires for further elaboration to the knowledge base.

The data base was developed by using the following process:
1) Analysis of published data.
2) First unstructured interviews.
3) Analysis of company and owner cost records.
4) Follow up interviews for further elaboration.

These processes were relatively easy to follow as the interviewer was able to go to each particular office or department in town and eventually obtain the required knowledge and data.

De La Gorza and Ibbs [4] discuss knowledge elicitation for construction scheduling in a general way and also briefly consider total cumulative cash flow curves by cautiously suggesting measures for approximately determining the total cost-time profiles for midsize buildings in the context of front-end loading.

The following three lists were contained in the knowledge base:
1) list of hypotheses
2) list of observations
3) list of rules relating hypotheses and observations.

Although the shell program which was used is written in LISP, all rules were written in a language called Abbreviated Rule Language (ARL). This useful feature of the PC Plus shell program made the whole expert system development for the cost-time profiles easier, quicker and more efficient.

Additionally, the shell program permits the user to ask why a particular question is asked, and it then shows the user the logic being used. A review option is incorporated in every stage of the consultation to enable the user to modify the answers. Furthermore, at the end of the consultation, the user can print the consultation history, or save the responses in a file.

Following the development work on the cost-time profiles, research and development work proceeded with the investigation on the feasibility of creating and utilizing a prototype expert system to provide a quick and accurate pre-engineering estimating system for certain types of school buildings.

For this second phase of development work, similar processes for knowledge elicitation and data acquisition were used but more difficulties arose in spite of certain similarities between the estimating system and the cost-time profile system. Following the first interviews with a limited number of experts, it was found necessary to trace the files on thirteen recent school construction projects to develop the knowledge and data base before conducting follow up interviews.

After some considerable research and development work into the knowledge and cost data for school buildings in New Brunswick with similar construction methods and finishes, it was discovered that costs varied with floor area, the number of floors, the year of construction (i.e. construction price index), and the area allocated for each student.

The method used in developing the prototype estimating system consisted of two approaches. The first approach was the traditional expert system approach of knowledge acquisition from experts, and the second approach was a statistical analysis of historical school construction cost data.

KNOWLEDGE ELICITATION AND DATA ACQUISITION FOR ESTIMATE

By using the two approaches in the development work, it was necessary to elicitate the knowledge from the experts and also acquire and analyze historical school construction cost data. It was hoped that the two approaches would complement each other.

Data collection consumed a large proportion of time in the investigation mainly because a concise summary of all of the school construction projects had not been collated or recorded. Details of the construction of each school was filed individually, consisting of several file folders containing all the relevant correspondence and data.

Estimators were contacted and interviewed who worked, or had been working, for the government department. There was therefore considerable difficulty acquiring and collating the knowledge on conceptual estimating, compared to the cost profiles, as no individual person, or department, had previously made the entire conceptual estimate.

However, from the expert knowledge that was acquired, the estimators made the following observations:

1. Conceptual estimates are based on costs per square metre.
2. Larger schools have lower unit costs.
3. Senior high schools are more expensive than others.
4. Unit costs increase with the number of floors up to three storeys.
5. If municipal services are available then costs are lower.
6. A powerful school board can increase costs.
7. Political interference can increase costs.

Data on school construction costs was collected in the government offices. All drawings and files were obtained from the Data Centre which contained most of the drawings and job files for the entire Province of New Brunswick. The scope of data collection was restricted to elementary, junior high, or combined elementary/junior high schools and restricted to projects which were new construction or additions. Costs collected were based on the low bid and were divided into the following fourteen divisions:

1. Foundations and slab-on-grade.
2. Structure.
3. Exterior walls.
4. Roofing.
5. Interior finishes.
6. Mechanical plumbing.
7. Mechanical fire protection.
8. Mechanical HVAC.
9. Electrical.
10. Elevator.
11. Built-in-furniture
12. General conditions, overhead, profit, administration, and contingencies.
13. Exterior works.
14. Special construction.

Gross floor areas of each school were divided into:

1. Gymnasium.
2. Auditorium.
3. Cafeteria.
4. Kitchen.
5. Ground floor.
6. First floor.
7. Second floor.

Analysis of the data was performed using Statgraphics, a statistical software package site licensed by the University of New Brunswick, to analyze the information by regression.

This statistical technique determines the equation of the line or curve which minimizes the deviations between the observed data and the regression equation value.

It can be stated in numerical terms as follows:

Minimize $S = \Sigma(Y_i - \hat{Y}_i)^2$

where S = Sum of squares
Y_i = Observed values
$\hat{Y}_i$ = Values estimated by equation.

For each school record in the database the key items of data were the total gross floor area (in m^2), the number of floors, and the base construction cost corrected by using a construction cost index. The base construction cost is the total cost of construction less the costs of four of the fourteen divisions; these are the cost of the elevator (if any), the sprinkler system (if any), the exterior work (such as landscaping or paving, if required) and the installation of utilities in the building. These four divisions are subtracted because the costs varied randomly and depended upon code and budget requirements and the geographical location of the building.

When an estimate is required, certain questions are asked in the expert system procedure, then a spreadsheet performs a linear regression of the corrected base cost against the number of floors and the total gross floor area. The program will optionally allow outlying data points to be ignored and re-perform the regression. The results of the regression are then used to predict the base cost of the proposed construction from its floor area and number of storeys. The program then breaks down this base cost into each of the ten divisions, based on average percentages from the project database, and adds estimates of costs for each of the four divisions that are excluded from the base cost, when appropriate. Refer to Table 1.

Proposed Construction (m^2)		2,300		Cost (Can$)/			
Number of Floors..........		2		Sq. Metre			
				....			% OF
COST BREAKDOWN	LOW	AVERAGE	HIGH	LOW	AVG	HIGH	COST
Foundations	118,000	152,000	186,000	51	66	81	8.5
Structure	170,000	214,000	259,000	74	93	113	12.0
Exterior Walls	179,000	262,000	344,000	78	114	150	14.7
Roofing	63,000	79,000	95,000	27	34	41	4.4
Interior Finishes	271,000	311,000	351,000	118	135	153	17.4
Plumbing	85,000	128,000	172,000	37	56	75	7.2
Fire Protection	19,000	32,000	45,000	8	14	20	1.8
KVAC	116,000	148,000	180,000	50	64	78	8.3
Electrical Systems	192,000	211,000	229,000	83	92	100	11.8
Elevator	24,000	28,000	32,000	10	12	14	1.6
Built-in Furnishings	32,000	48,000	65,000	14	21	28	2.7
General Conditions	101,000	171,000	240,000	44	74	104	9.6
SUB-TOTAL		1,784,000			776		100
Exterior Work	90,000	124,000	159,000	39	54	69	7.0
Special Construction	16,000	34,000	51,000	7	15	22	1.9
GRAND TOTAL		1,942,000			844		108.9

Table 1 Estimated Price of School

CONCLUSION

The experts' knowledge and experience elicitation and acquisition process differed in the two studies although some knowledge and data overlapped. A comparison of these differences highlights the difficulties and variations in an industry as diverse and fragmented as the construction industry. In the creation of the prototype expert system for the cost-time profiles, experts that were interviewed and who answered questionnaires had fairly well documented data because fairly accurate progressive monthly records had been kept of the costs of the operations of various projects and their knowledge was relatively easy to acquire. In the creation of the prototype expert system for pre-engineering estimates, however, the knowledge and data was much more difficult to acquire, mainly because of the different projects and no one person or department had previously done the entire conceptual estimate. The knowledge and data was therefore scattered in various offices.

In the development of the prototype expert system for pre-engineering estimates two approaches evolved. For the sample taken in the study, ie. construction of thirteen schools, an expert system was used to initiate the process but then the other statistical approach combined with algorithmic programming was utilized and combined in the preparation of the estimate.

When the prototype is developed further, however, in order to consider other facilities such as health care facilities, offices, warehouses, and prisons, further knowledge identification and acquisition will be necessary. The inference mechanisms used in the fully developed system, will then require more procedural paths than the simpler algorithmic paths used in the feasibility study for school construction cost estimating and therefore an expert system will then be much more useful and meaningful.

REFERENCES

1. Christian, J., and Kallouris, G., Using Knowledge and Experiences for Planning the Cost-Time Profile of Construction Activities. AIENG Computational Mechanics Publication, Boston, Vol. 2, 1990.

2. Brandon, P.S., Basden, A., Stockley, J., and Hamilton, I., Expert Systems and Strategic Planning of Construction Projects, RICS.London. 1988.

3. Kouskoulas, V., and Koehn, E., Predesign Cost-Estimating Function for Buildings. Journal of Construction Engineering and Management, ASCE, New York, Vol. 100, No. C04, 1974.

4. De La Gorza, J., and Ibbs, C.W., Knowledge-Elicitation Study in Construction Scheduling Domain. Journal of Computing in Civil Engineering, ASCE, New York, Vol. 4 No. 2, 1990.

Representing the Engineering Description of Soils in Knowledge Based Systems

D.G. Toll, M. Moula, N. Vaptismas

School of Engineering and Applied Science, University of Durham, UK

ABSTRACT

A scheme has been devised to fully represent the engineering description of soils. The language of soil description can be rich, containing much information. Many of the descriptive terms convey quantitative information to the geotechnical expert. However, some of the more detailed parts of the description may play only a minor role in engineering design. A complete representation scheme could become cumbersome, if all the detailed information had to be represented. It is recommended that the complete soil description should be stored as a text field in geotechnical databases. This can be parsed by a Knowledge Based System to abstract the relevant information for internal representation, or an additional part of a database structure can store the description at a simplified level. In this way the loss of information due to simplification is prevented.

INTRODUCTION

Developing Knowledge Based Systems in Geotechnical Engineering involves representing the engineering descriptions of soils. A full engineering description of a soil is a carefully structured list of varying types of information. Although qualitative in form, many of the descriptive terms convey quantitative information to an engineer with expertise in geotechnics. Some of these qualitative-quantitative links have been made explicit in codes of practice and standards which set out ranges of numerical parameters for the descriptive terms.

Although full engineering soil descriptions will be an important part of the representation of soils, very often broader classifications of soils are needed, particularly at the early stages of a project when over-all feasibility is being considered, and a detailed investigation of the soils has not been carried out.

REPRESENTING A SOIL PROFILE

A PROFILE will be defined as the sequence of ground conditions with depth at a particular location (Toll [1]). The PROFILE can be subdivided into different GEOLOGICAL HORIZONs. A GEOLOGICAL HORIZON is a sequence of LAYERs having the same geological origin. Each LAYER may be described by a broad classification or by a full engineering description. A hierarchy based on the British Soil Classification System (British Standard 5930 [2]) is shown in Figure 1. This only shows the completed branch for non-organic soils.

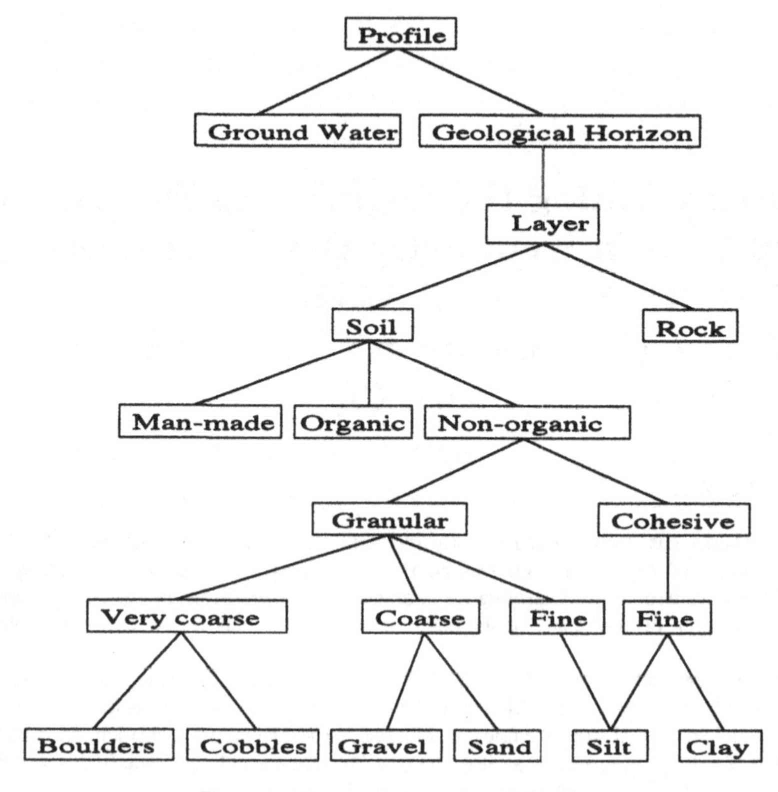

Figure 1 - The components of a soil profile

In the heirarchy shown in Figure 1 a soil is identified at the most detailed level by the dominant soil type (Clay, Silt, Sand, Gravel, Cobbles, Boulders). A full soil description contains considerable additional information.

REPRESENTING A SOIL DESCRIPTION

The required pieces of information and recommended order of an engineering soil description have been given in BS 5930. A slightly extended methodology for soil description is given by Burland [3] based on the work of Jennings and Brink [4]. The main components of a soil description can be set out as follows:-

M - Moisture condition
C - Consistency
C - Colour
S - Structure
S - Soil type
O - Other features
O - Origin
W - Ground water conditions.

The first six features are factual items which a field engineer can identify from examining a sample of soil. Origin requires interpretation based on a knowledge of the geology of the area in which the soil exists, and is a property of the GEOLOGICAL HORIZON. Ground water conditions apply to a complete PROFILE rather than being specific to a particular LAYER, as is shown in Figure 1.

A full soil description for a particular LAYER could be as complex as *'Moist stiff reddish brown closely fissured thinly bedded silty sandy CLAY with a little dark greenish grey sub-rounded fine gravel'.* This can be broken down into individual components for representation.

In some cases a LAYER may be made up of multiple soils, for example *'Interbedded SAND and SILT'* or *'CLAY with pockets of sandy silt'.* If so it is not possible to separate the soils into individual layers and they must be dealt with as different soils within a defined layer.

A SOIL is represented by MOISTURE, CONSISTENCY, STRUCTURE and SOIL TYPE as shown in Figure 2. STRUCTURE in this case simply applies to an individual SOIL. Where multiple soils exist within a layer an additional term LAYER_STRUCTURE is used at LAYER level to define the inter-relationships. An example of a full soil description and the associated representation is given in Figure 3. For many of these items a range can be specified so a LOW and HIGH value must be stored for each. This is not shown in the following for the sake of simplicity.

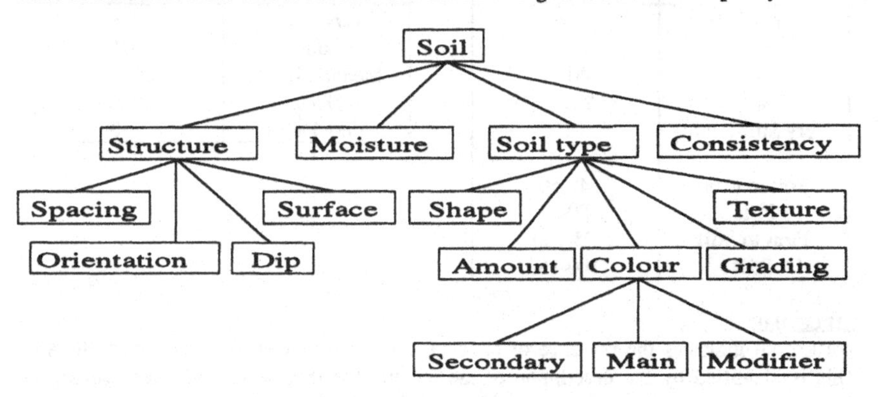

Figure 2 - The components of a soil description.

MOISTURE CONDITION

MOISTURE CONDITION is not included in BS 5930 and the terminology is not widely used in Britain. However, it is important, particularly in drier climates, hence it is included by Jennings and Brink [4]. It may often be given as a single term, although sometimes a range is specified eg *'Dry to moist'*. Possible descriptive terms are:-

Dry, Slightly moist, Moist, Very moist, Wet

There is no direct linkage between the descriptive terms for MOISTURE CONDITION and the quantitative measurement of Moisture Content, since the MOISTURE CONDITION depends on the soil type. For example a clay with a moisture content of 10% could be described as *Dry*, whereas a sand with the same moisture content could appear *Wet*.

CONSISTENCY

The descriptive terms for CONSISTENCY depend on soil type. A single term may be used, or a range specified e.g. *'Very Soft to Soft'*.

Very soft, Soft, Firm, Stiff, Very stiff (for cohesive soils)
Very loose, Loose, Medium dense, Dense, Very dense (for granular soils)
Firm, Spongy, Plastic (for peats).

Quantitative links with the descriptive terms are set out in BS 5930 for the Undrained Shear Strength of cohesive soils and N-values from the Standard Penetration Test, an empirical measure of the relative density of granular soils. There is possible confusion in the terms for cohesive soils since the terms *'Soft to Firm'* and *'Firm to Stiff'* are widely used as subdivisions of *'Firm'* and *'Stiff'* as shown in Table 1, rather than indicating a range of strengths.

Table 1
Definition of Terms for Cohesive Soil

Term	Undrained Shear Strength (kPa)
Very Soft	<20
Soft	20-40
Firm	40-75
Stiff	75-150
Very Stiff (Hard)	>150
Soft to Firm	40-50
Firm	50-75
Firm to Stiff	75-100
Stiff	100-150

Table 2
Definition of Terms for Granular Soil

Term	SPT N-values (blows/ 300mm penetration)
Very Loose	0-4
Loose	4-10
Medium Dense	10-30
Dense	30-50
Very Dense	>50

STRUCTURE

STRUCTURE indicates the presence of bedding, discontinuities or shearing within the SOIL. This is identified by the description of the feature, the spacing, dip and orientation, and details of the surface finish. Some possible terms for STRUCTURE are

Bedding, Lamination, Fissure, Joint, Fracture, Slip surface, Shear zone, Gouge, Intact

Spacing and dip can be defined by descriptive terms, or by quantitative observations. Some descriptive terms are

SPACING *Very Thick, Thick, Medium, Thin, Very Thin* (for bedding and lamination)
Very Wide, Wide, Medium, Close, Very Close, Extremely Close (for discontinuities)

DIP *Vertical, Sub-vertical, Sub-horizontal, Horizontal*

SURFACE *Polished, Striated, Slickensided, Grooved, Open, Closed, Tight*

<table>
<tr><td>MOISTURE</td><td>Moist</td><td></td><td></td><td></td><td></td></tr>
<tr><td>CONSISTENCY</td><td>Stiff</td><td></td><td></td><td></td><td></td></tr>
<tr><td>STRUCTURE</td><td>Bedding</td><td>SPACING
DIP
ORIENTATION
SURFACE</td><td>Thin</td><td></td><td></td></tr>
<tr><td>STRUCTURE</td><td>Fissure</td><td>SPACING
DIP
ORIENTATION
SURFACE</td><td>Close</td><td></td><td></td></tr>
<tr><td>SOIL TYPE</td><td>Clay</td><td>AMOUNT
COLOUR

GRADING
SHAPE
TEXTURE</td><td>Main</td><td>
MAIN
SECONDARY
MODIFIER</td><td>
Brown
Red</td></tr>
<tr><td>SOIL TYPE</td><td>Silt</td><td>AMOUNT
COLOUR

GRADING
SHAPE
TEXTURE</td><td>Secondary</td><td>
MAIN
SECONDARY
MODIFIER</td><td></td></tr>
<tr><td>SOIL TYPE</td><td>Sand</td><td>AMOUNT
COLOUR

GRADING
SHAPE
TEXTURE</td><td>Secondary</td><td>
MAIN
SECONDARY
MODIFIER</td><td></td></tr>
<tr><td>SOIL TYPE</td><td>Gravel</td><td>AMOUNT
COLOUR

GRADING
SHAPE
TEXTURE</td><td>Minor

Fine
Subrounded</td><td>
MAIN
SECONDARY
MODIFIER</td><td>
Grey
Green
Dark</td></tr>
</table>

Figure 3 - A representation for a soil layer: ***'Moist stiff reddish brown closely fissured thinly bedded silty sandy CLAY with a little dark greenish grey sub-rounded fine gravel'.***

SOIL TYPE

SOIL TYPE relates to the range of particle sizes of the different components of a soil. These might be as follows:-

Clay, Silt, Sand, Gravel, Cobbles, Boulders, Peat

Associated with each SOIL TYPE are AMOUNT, COLOUR, GRADING, SHAPE, and TEXTURE. These need to be associated with each SOIL TYPE, rather than the over-all SOIL since there may be individual colour, grading, shape or texture information given for each constituent soil type.

AMOUNT If the SOIL TYPE is the dominant soil type (normally given in capitals in a soil description) then AMOUNT is given as **Main**. For the terms *'Very silty', 'Very sandy'* etc the AMOUNT is given as **Major**. For *'Silty', 'Sandy'* etc AMOUNT is given as **Secondary**. **Minor** represents *'Slightly silty', 'Slightly sandy'* etc. Other descriptive forms which do not conform with BS 5930 (but are still widely used) can also be represented in this way such as *'with a little .....'* or *'with some ......'*. Some examples are given below.

'Very silty slightly sandy CLAY with a little gravel' would be represented as:-

SOIL TYPE - Clay	AMOUNT - Main
SOIL TYPE - Silt	AMOUNT - Major
SOIL TYPE - Sand	AMOUNT - Minor
SOIL TYPE - Gravel	AMOUNT - Minor

'Silty clayey SAND and GRAVEL' would be represented as:-

SOIL TYPE - Sand	AMOUNT - Main
SOIL TYPE - Gravel	AMOUNT - Main
SOIL TYPE - Silt	AMOUNT - Secondary
SOIL TYPE - Clay	AMOUNT - Secondary

COLOUR This needs to be defined by MAIN COLOUR, SECONDARY COLOUR and MODIFIER. Main and Secondary colour may need multiple entries. Some examples are:-

'Pale greeny reddish grey' can be represented as:-

MAIN COLOUR	- [Grey]
SECONDARY COLOUR	- [Green,Red]
MODIFIER	- Pale

'Mottled red-brown' can be represented by:-

MAIN COLOUR	- [Red, Brown]
SECONDARY COLOUR	- [ ]
MODIFIER	- Mottled

GRADING, SHAPE and TEXTURE These can often be given as single terms, although a range of values may be given eg *'fine and medium'* or *'sub-angular to rounded'*. For clays where true grading information is not valid, descriptions of plasticity can be included under GRADING. Typical descriptive terms for each heading are:-

GRADING — *Fine, Medium, Coarse, Well graded, Poorly graded, Gap graded, Uniform* (for granular soils)
Non plastic, Low plasticity, Intermediate plasticity, High plasticity, Very high plasticity, Extremely high plasticity, Lean, Fat, Heavy (for cohesive soils)

SHAPE — *Angular, Sub-angular, Sub-Rounded, Rounded, Flat, Elongate, Irregular*

TEXTURE — *Rough, Smooth, Polished* (for granular soils)
Remoulded, Reworked, Softened, Friable, Disturbed (for cohesive soils)
Fibrous, Amorphous (for peats)

IMPLEMENTATION

The first knowledge representation scheme used, based simply on the classification system shown in Figure 1, was implemented in Turbo Prolog on a PC. In this hierarchy a soil was identified at the most detailed level by the dominant soil type (Clay, Silt, Sand, Gravel, Cobbles, Boulders). Using a straightforward pair of (attribute, value), the properties of each class were initially identified and could be passed down the hierarchy to apply to lower classes by inheritance. So, at the level of COARSE in the structure, it was known to be GRANULAR, NON-ORGANIC and SOIL by inheritance from above. An alternative solution would be GRANULAR, ORGANIC and SOIL. It was also possible to generate all the members of a class by searching down the hierarchy i.e. possible members of COARSE are GRAVEL and SAND.

It was found to be necessary to have two types of attribute: (i) Attributes which are indentified once and are carried down the structure to the current level. These allow identification of the exact position within the structure. (ii) Attributes which appear at several levels within the structure and whose values change, becoming more specific going down the structure. These are illustrated in Figure 4 which shows a path through the structure, from LAYER to SAND.

The first type of attributes, GROUND TYPE, SOIL TYPE, SOIL CHARACTER etc are identified at one level. The second type, MIN GRAIN SIZE and MAX GRAIN SIZE are redefined at a number of levels.

This simple format of (attribute, value) pairs hides the restriction that an attribute like GRAIN SIZE had to be expressed as two attributes: MIN GRAIN SIZE and MAX GRAIN SIZE, since it could take a range of values, defined by a minimum and maximum numerical value. It was also found that the straightforward (attribute, value) pair could not easily handle a more detailed representation of the soil. To overcome these problems a more complicated format was introduced:

```
Attribute,      [ (Value1),      (Factor1)
                  (Value2),      (Factor2)
                     "              "
                     "              "    ]
```

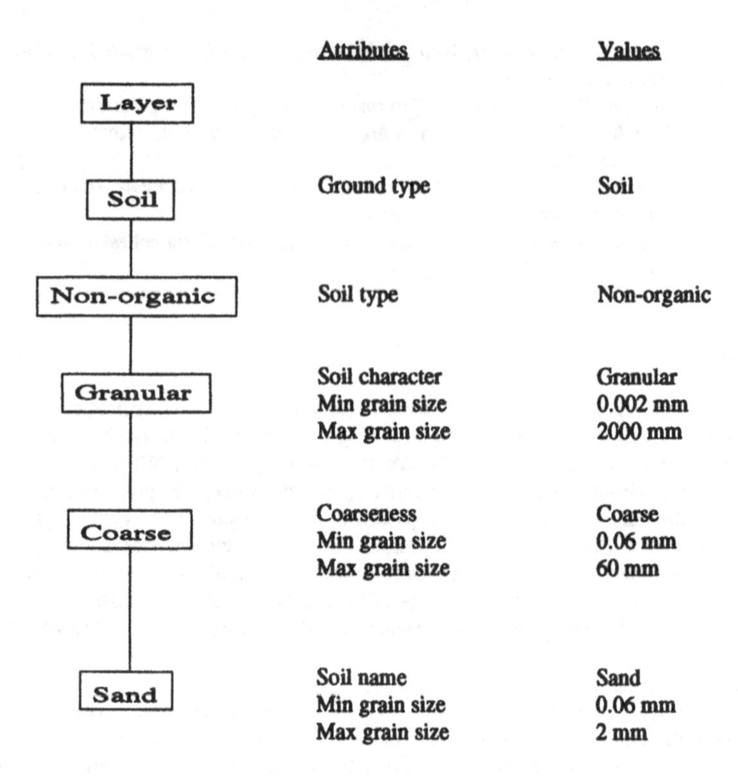

Figure 4 - A path through the soil classification hierarchy

In this format the attribute does not need to take a unique value, but can take a number of different values depending on an external factor (or factors). These values may be specified as a single value or a (minimum, maximum) pair. The Prolog clause for defining the class of SAND would be:

```
class ( sand, [ ],
          [att ( soil_name,
                 [ val ( [ sand ],          fact ( [ ]          ))]),
           att ( grain_size,
                 [ val ( [ 0.06, 2   ],     fact ( [ ]          )),
                   val ( [ 0.06, 0.2],      fact ( [ fine ]     )),
                   val ( [ 0.2,  0.6],      fact ( [ medium ] )),
                   val ( [ 0.6,  2   ],     fact ( [ coarse ]   ))])]).
```

The terms att( ), val( ) and fact( ) indicate the use of Turbo Prolog compound symbols which have been defined to identify attribute, value and factor.

The empty list, [ ], following the class name 'sand' indicates that the class of SAND is at the base of the structure. For higher classes the list would contain the members of that class e.g.

```
class ( coarse, [ sand, gravel ],
          [att ( coarseness,
                    [ val ( [coarse ],          fact ( [ ]          ))]),
           att ( grain_size,
                    [ val ( [0.06, 60 ],         fact ( [ ]          ))])]).
```

This representation is being used in the development of a system for selecting the most appropriate type of test (laboratory or insitu) for each soil type. It has been possible to represent both the soil information and the test information using the same structures. It has been found to be easy to add and delete information without changing the over-all structure.

The more complete representation of the engineering descriptions of soil described in this paper has been implemented using the relational database system INGRES on a Sun workstation. This allows full engineering soil descriptions to be stored in a structured database for access by a KBS through Structured Query Language (SQL). Work is progressing on the development of a parsing routine which enables a soils description to be broken down into its constituent parts for automated input to the database. The parser is built up of layered functions which are capable of identifying specific key words/phrases and allocating these keys to a set of variables.

DISCUSSION

A full representation scheme which takes into account all possible pieces of information which may be included in a soil description becomes rather cumbersome. Much of the detailed information will play an insignificant role in engineering design. Many branches in the decision making process will often be governed by crude classifications such as between *'Organic'* and *'Non-organic'* or *'Granular'* and *'Cohesive'* (See Figure 1). Nevertheless there are some decisions which do require a more detailed representation of the soil, for example when a decision is being made as to whether two soils have the same geological origin. Here, detailed colour or structure information may help to reinforce other comparisons on the basis of soil type or consistency. However, even here it is unlikely that details of the subsidiary soil types (eg colour, shape or texture) will be of major importance, even if they are for the main soil type. Some rationalisation can therefore be contemplated.

A problem for any representation scheme is that it imposes a restriction on the allowable format of the data. Soil descriptions currently have a richness both in form and vocabulary which will be lost by the adoption of rigid constraints. This process has already started by the introduction of standards such as BS 5930, and is certain to continue with the development of geotechnical databases and KBSs. However, even if the process does continue, KBSs will still have to deal with existing geotechnical information. Old borehole logs may well contain rich descriptions which will have to be simplified in order to represent them. While simplification will be essential in order for the KBS to assimulate the information, it is recommended that geotechnical databases store the soil descriptions as they were originally recorded by the site engineer. An additional representation at the appropriate level of detail will also be required, or the KBS can extract the description, parse it and internally represent it to suit the appropriate level of decision making. If this is not done, and only a simplified representation of the soil is stored, then information may well be lost which could be of value in a different context.

CONCLUSIONS

A soil description contains a large amount of varying types of information. Although qualitative in form, much of the information conveys quantitative ideas to a geotechnical expert. A scheme has been devised to fully represent a soil description, in all its richness, but it could be rather cumbersome, and much of the information would play only a minor role in engineering design.

It is recommended that the complete soil description (as originally recorded by the site engineer) should be stored as a text field in geotechnical databases. This can be parsed by a Knowledge Based System to abstract the relevant information for internal representation, or an additional part of a database structure can store the description at a simplified level. In this way the loss of information due to simplification is prevented.

ACKNOWLEDGEMENTS

The work of implementing the scheme in INGRES has been done by Andy Oliver who has also contributed to the ideas. Thanks to Prof. Peter Attewell for his contribution.

REFERENCES

1. Toll D.G. Representing the ground. Proc. NATO Advanced Study Institute: Optimisation and Decision Support Systems in Civil Engineering, Heriot-Watt University, Vol II, 1989.

2. British Standard 5930. Code of practice for site investigations. British Standards Institution, London, 1981.

3. Burland J.B. The teaching of soil mechanics - a personal view. The Nash lecture, Proc. 9th European Conf. Soil Mechanics and Foundation Engineering, Vol 3, pp 1427-1447, Balkema, Rotterdam, 1987.

4. Jennings J.E. and Brink A.B.A. Application of geotechnics to the solution of engineering problems. Proc. Instn. Civ. Engrs, Part 1, 64, pp 571-589, 1978.

SECTION 8: THEORY AND METHODS FOR SYSTEM DEVELOPMENT

Expert System Verification and Validation Part I: Defining the Concepts

T.W. Satre (*), J.G. Massey (**)

(*) Division of Psychology and Philosophy, Sam Houston State University, Huntsville, Texas 77341-2447, USA

(**) Department of Forest Science, Texas A&M University, College Station, Texas 77843-2135, USA

ABSTRACT

The goal of this paper is to distinguish between verification and validation as they relate to the total expert system evaluation process. Our thesis is that V&V are two distinct processes, which together completely encompass the needed elements of expert system evaluation. Our approach is to state and defend definitions of these terms, and then to demonstrate their application in practice. The definitions derive the meanings of the terms "verification" and "validation" from the literature and common usage. From the definitions, we demonstrate the application of V&V to a forest products facility expert system.

INTRODUCTION

Two general types of doubt occur in reading the literature concerning verification and validation of expert systems. First, because expert systems are uniquely different from other traditional types of software, there are doubts about whether the traditional methods of software verification and validation apply to expert systems. Second, as Culbert, et al.[3] have written, "One basic problem with V & V of expert systems has been the lack of consistent definitions for both validation and verification." Both types of doubt are addressed in this paper. First, we offer definitions of "verification" and "validation" which we argue give rigor to the use of the terms and will provide guidance in the task of actually verifying and validating expert systems. Second, we argue that the same concepts of verification and validation apply both within software engineering and the design of expert systems, even though the methodologies for carrying out verification and validation differ greatly.

Throughout this discussion we shall speak of the 'elements' of an expert system. By an element of a system we shall mean any part or process of the system which is to be evaluated. While these elements are described in different ways in the literature, Feigenbaum [4] basic statements which constitute the knowledge base of the system, and a set of inferential procedures or an inference engine. Ivan Bratko [1] adds user interface as a third distinct component. We shall claim that validation is the evaluation of the specific system as it has been installed, and we shall claim that verification is the evaluation of the various elements of the system and testing of very specific features of the system. A system as a whole is validated by a process of verifying that the system yields correct results. However, verification processes are processes for showing that the specific elements of the system have a a variety of properties which are important if the system is to meet the requirements for which it has been designed and installed.

VERIFICATION

A. The Concept of Verification

A study of the literature on verification of expert systems reveals that the term "verification" is used in various ways, and there is often a confusing of the definition of verification with discussion of helpful methods by which to verify that the elements of an expert system have the properties that the user and designers intend it to have. That verification differs from validation is suggested even by the ordinary uses of the terms. While one may have a validated parking permit, one would not be said to have verified the permit. Rather, one would verify the elements of the permit. For example an officer may verify each element such as driver's identification, address, date of birth, etc. If each element can be verified (proved true), then the license is valid. Elements are verified or falsified, and if all the relevant elements are verified, then the license has been validated or confirmed as valid. When we speak of the elements being verified, this is elliptical for saying that the statements about those elements are verified. If the elements of the driver's license include date of birth, then the verification consists of checking the truth of the statements about a person's date of birth. The crucial difference between verification and validation is that it is statements which are verified or falsified. The term "verification" finds its use in the testing of statements or hypotheses for truth, and this is the interpretation which is given by logicians and philosophers of science to the use of the term. (Copi [2]) The verification of expert systems occurs in the testing of statements about the various elements of the system.

B. Problems in the current technical use of the term "verification" in the literature of artificial intelligence

In the literature of expert systems, there appears to be much disagreement about what constitutes verification of software and whether the concept of software verification even applies to the verification of expert systems. It will be our claim here that this disagreement occurs because writers confuse the definition of verification, which is quite simple, with the methods and procedures for verifying different types of software. These methods and procedures are not simple and the choices among these methods are very much dictated by the types of software which is being tested.

The IEEE [7] defines verification as determining that the product created was what was agreed upon: "...the process of determining whether the products of a given phase of software development meet all the requirements established during the previous phase." One first notes that this definition would limit verification to products that had undergone phase by phase development. While the major methodologies of software engineering teach phase by phase development, the methodologies differ on what sorts of things go on in a given phase and what sorts of requirements are specified for the various phases. Further, such a definition not only builds into the definition of "verification" the structure of the method of verification, it is a structure which may not be applicable to the verification of expert systems. Still, this much is correctly suggested by the definition: one of the claims to be verified at each stage of the development of piece of software is that the software satisfies any requirements which might have been laid down at the previous stage. But verifying this claim is not the same as verification generally.

Another example of confusing the definition of verification with methods of verification is found in an article by Geissman and Schultz. [6] They state, "Verification is a determination that software has been developed in a formally correct manner in accordance with a specified software engineering methodology." A few sentences later they state that the process of verification is the process of answering two questions: "Was a sound methodology such as structured analysis or a design language with a preprocessor followed? Is the design reasonable and traceable?" [6] As in the definition given by the IEEE, there are two important claims given here which one might well wish to verify about the development of a particular expert system, namely, that the system was developed according to a sound methodology and that the design was reasonable and traceable. But these are just two of several claims that one might need to verify in the testing of software and the verification of these two is not the definition of verification. Geissman and Schultz then propose their six stage process of verification which is based on the assumption that the development stages begin with the creation of prototypes of the system which is eventually to be created. Their definition of verification is thus very much tied to their own theory or proposal for software development. For the types of software which are designed in this way, this procedure for testing is well conceived. The claim in this paper is simply that they have confused a procedure for testing with the definition of verification of software.

Other authors have approached testing of software with other methodologies which they describe as though they were giving the definition of what verification is for all cases of software verification. Suwa et al. [14] define expert system verification as the process of testing for two characteristics -- completeness and consistency. Nguyen, et al.[8] describe procedures as well as the underlying logical concepts for the understanding of consistency and completeness of expert systems. They too have discussed methods of testing which would be appropriate to verifying that a system was consistent and complete, in the senses of the terms given in that paper. Understanding and testing for consistency and completeness is important to the testing the adequacy of an expert system, but this does not give us a definition of what verification is.

C. A proposed definition of "verify" and "verification" for evaluation of expert systems

The definition of verification proposed here is this following:

Definition of "verify" and "verification"

> A statement or hypothesis is said to be verified if it has been shown to be true, or confirmed to a degree of strength which is reasonable for the acceptance of the statement. Verification is any process which has as the outcome the verifying of a statement. An expert system is said to be verified if evaluatory statements about its elements have been verified.

There are several reasons for conceptualizing verification of expert systems, and perhaps software generally, in the above manner which separates the definition of verification from the description of the particular methodologies for testing software. First, this account is consistent with the use of "verify" within epistemology and the philosophy of science. That literature does treat verification as justifying the belief of hypotheses by showing them to be true or at least confirming them to a degree which warrants us in believing them.

Second, this definition of "verification" is consistent with the fact that there are several types of claims which must be verified during the testing process including claims about the system in its operation, and claims about the type of design and methodology that went into the construction of the system. In each case claims must be tested in appropriate ways and the claims are typically about limited features of the elements of the system rather than the system as a whole. So, it is the elements of the system which are subject of the verification process.

Third, this definition enables one to understand how the apparently disagreements in the literature concerning software verification are just apparent disagreements. They are disagreements concerning what are the central claims about software to be verified and about the best methods for the design and testing of software, but they are not disagreements about the nature of verification. Verification is the justifying of claims about the elements of the system. Those justifications are stronger or weaker in part according to the reliability of the methods which are used in the process of justification. The importance of methodology in the construction and testing of expert and other software systems is that the use of reliable methodologies offers significant support to the claims which are to be verified during the testing process.

VALIDATION

A. Background

The topic of validity can be introduced by two analogies both of which will be useful to our discussion. The first is a comparison of validity of expert systems with validity of reasoning and argument. Within logic, validity is a property of an inference. The inference is valid when the conclusion set of statements follows from the premise set. To show that an inference is valid is just to confirm that its conclusion follows from the premises. A second analogy is that of validating parking permits. A parking permit is confirmed as valid if the local conditions or requirements have been met by the vehicle or the person holding the permit. The premise set of requirements having been met permits or authorizes acceptance of a conclusion that the vehicle may be parked in a location. The two analogies capture two important features of validation of expert systems. Validation is confirming that the system as a whole yields information by acceptable or valid reasoning and that the system as a whole authorizes the acceptance of conclusions on the basis of the information on which the system acts. The utility of the expert system is found both in its extending our knowledge by inference and by its authorizing our action by the type of advice or answers it gives to our questions. The analogies capture these two central features of the utility or workability of an expert system. Like human experts, expert systems both yield knowledge and authorize action at a particular time and place. Similarly, an expert system is evaluated for validity at a particular time in a particular setting where it is installed, and this will be captured by the definition which follow later in the paper.

B. Problems in the current technical use of the term "valid" in the literature of AI

The issue for this paper is not whether there is a common usage of "validate" but whether there is a clear technical usage of the term which enables us to distinguish whether a testing procedure or practice is one of validation or verification. A second issue is whether the activities of verification and validation together comprise the procedure necessary and sufficient for satisfactory evaluation. A brief look at statements in the literature does not give much evidence of a clear and consistent use of the term at all. For example, IEEE [7] defines "validation" as "...the process of evaluating software at the end of the development process to ensure compliance with software requirements." This definition can be criticized on two accounts. First, it is vague in so far as the expression "software requirements" is vague. Second, if the requirements set for the software are themselves inadequate, then a program may comply with those inadequate requirements even though one might hesitate there to say that the system had been validated because the system might not yield accurate results. Thus, the IEEE definition of validation may itself be open to refutation by counterexample.

A second definition of "validation" comes from Geissman and Schultz. [6] They define validation for expert systems as the "...ensuring [that] the expert system satisfies its users' needs." This definition suffers in ways like that of the definition from the IEEE. First, the phrase "users' needs" in the definition is both vague and broad, since the users' needs will include needs for such properties as efficiency of the the software as well. Second, if "users' needs" is taken to mean accurate or true claims by the system, then the definition comes closer to a clearer account suggested below. However, if "users' needs" is taken to mean "needs as described by the users", then the a system might meet users' needs where the users had given inadequate account of their needs and yet the system lack validity. In general, users have needs far exceeding the limits they themselves can identify.

A third account of validation suffers from

vagueness as well as confusion of concepts. O'Keefe [10] writes,"Validation is the process of checking that the model component of the system behaves like the 'real world'." First, this account is confused. What an expert system behaves like is another expert system, though those are part of the 'real world'. What a system models or represents is information and judgments about a part of the real world and its operation. Second, this account limits validation to the evaluation of a "model component" of an expert system. However, it is the system as a whole that is said to be valid, and it is the system as a whole which represents or even gives to the user judgments about the subject matter or processes which are described or modeled by the system. Third, satisfactory performance of any system does not depend on how the system arrives at an acceptable answer but, rather, that an acceptable answer is arrived at.

A fourth account of validation reflects still another difference in the way in which validation and verification are described. Finlay et al. [5], define verification as an area of validation with the distinction being that verification is concerned with whether the system operates correctly and validation whether the system is correct or appropriate for the problem to be solved. We believe that this use of the terms only indicates the need for the conceptual inquiry undertaken here, especially if the processes of verification and validation are indeed distinct, though related, processes as we believe them to be.

C. A proposed definition of "validity" and "validation" for evaluation of expert systems

There is a deeper problem with all of the above attempts to define validation. All miss what is central in the sense of "valid" as it is used in the empirical sciences, namely, that experimental methods measure within some limit of accuracy what they are intended to measure. There is a way in which this might be applied to expert systems so as to get a clearer account of validity. Let us call a statement which is supported or asserted a "claim". Experimental methods support claims. To say that the method is valid is to say that the claims which it supports or offers reasons to accept are true or at least highly probable. Expert systems also support claims. After all, they are used to support claims or judgments about the subject matter or process or system because they embody information and a method of reasoning. Let us say that an expert system makes a claim if that claim is the system's output to a question or if the system offers an affirmative answer when that claim is put to it as a question. Then a clear sense of the use of the terms "valid" and "validate" can be given as follows.

Definition of "validate" and "validation"

> A valid, or validated, expert system is one which yields correct claims for any given set of conditions, and validation consists of confirming that the expert system produces situationally relevant, correct answers for the location where it is installed. The expert system is validated if it has the capability to produce answers which are correct within the limits of accuracy which are reasonable for the particular subject matter.

Several features of this definition should be noted. First, it makes validity or being validated a property of the expert system itself based upon its performance. It is not a relationship between user needs and outcomes of a system, for example. Utilitarian attributes such as usefulness are important in the evaluation of an expert system, but those attributes are just different from that of being validated. O'Keefe et al. [9] uses the phrase "performance validation", and perhaps one should think of validation in this sense as a property of a installed system based upon that system's performance. In any case, this definition is consistent with that usage of the term "validation."

Second, this use of the term validation places validation procedures in the context of procedures and tests which justify the acceptance of the results of the expert system. There is a precedent for this use of the term "validation" in the literature of the philosophy of science where some have distinguished the context of validation (justification) from the context of discovery of hypotheses. The late Richard Rudner described this as follows:

> ..the context of validation is the context of our concern when, regardless of how we have come to discover or entertain a scientific hypothesis or theory, we raise questions about accepting or rejecting it. Rudner [12].

Similarly, Wesley C. Salmon [13] claims that we are concerned with the context of justification when we ask of a statement "What reasons do we have for accepting it is true?"

Third, the definition offered above for validation of expert systems is at least in the spirit of use of the term "validity" in the literature of experimental design where the validity of a method or procedure is its capability of accurately measuring or representing the attribute which it is intended and interpreted as measuring.

Finally, the definition above enables us to relate the use of the property of validity of expert systems to the property of logical validity. It is assumed that the inference engine permits or carries out inferences which are valid in the logician's sense. That is, the inference engine will not yield as a conclusion any statement which is false given that the premises or the knowledge statements to which the inferences apply are true. There is this connection between validity in the logician's sense and being validated as a property of an expert system. The operation of the expert system takes the form of inferring conclusions from premises. The inference or argument has the knowledge base as premises and the output answers to questions as the conclusion set. Validity of the system, as logical validity of an an argument, gives the user a basis for accepting the conclusions the system yields and so, a basis for accepting the system as a whole for the environment in which it is installed.

SOFTWARE ENGINEERING VS. EXPERT SYSTEM BUILDING

Thus far we have ducked the important issue of whether the concepts of verification and validation of expert systems are uniquely different from the concepts as they occur in the literature of software engineering. While it has not been the purpose of this paper to discuss whether the major methodologies of software engineering are applicable to the development of expert systems, it must be noted that there is disagreement among professionals concerning whether expert systems are so radically different from procedural software as to make the methodologies of software engineering inappropriate to the design of expert systems.

The claims of this paper are general. The concepts of verification and validation are the same whether one speaking of expert systems or of procedural software which is designed using methods from software engineering. That the concepts do not appear to be the same can perhaps be explained by noting that in software engineering methodologies, the processes of verification of the elements of the systems are, so far as possible, built into the steps of the methodology itself. One could interpret the history of software design from the development of structured programming through the current design methodologies such as transform analysis as attempts to insure that the systems will be verified by the fact that the entire development process has followed reliable procedures. These procedures range in type from the identification of user's needs to the writing of code.[11] The importance of reliability of the design methodology is that it increases the confidence that the elements of the system could be verified. If one knows that in general any piece of software designed by a certain methodology will have such virtues as portability, flexibility, and generality, then one does not have to verify separately that each system does indeed have those virtues.

In some of the literature of software engineering, the term "testing" is used in the manner in which we have used "validation". For example, Yourdon and Constantine [15] describe testing as "...the process of demonstrating that the system does what it is supposed to do...," which is very close to what we claim is the general concept of validation. The concern with testing, or as we say, validating, in software engineering rather than verification is due to the building of verification into the very design methodology. One can conjecture that a similar phenomenon is occurring in the design of expert systems. Use of verified shell systems in the development of an expert system makes at least some verification unnecessary for the system developer. This does not mean that the resulting system is not verified, but that the verification of some elements of the system is virtually insured by the reliability of the verified shell. This does not change the concept of verification, but rather it builds verification into the procedures of system development.

Thus, we conjecture that the concepts of verification and validation as they are defined here apply in the same sense whether they are applied to the testing of expert systems or of procedural software. It is the methodologies of verification and validation that differ, and this fact should not be surprising given the differences in these types of system.

SUMMARY AND CONCLUSION

The definitions of validation and verification together make it easier to understand the importance of the one process to the other, even though they are different. Among the claims to be verified about a system is the claim that the system is valid. That a system is valid is itself verified by verifying that the output claims of the system are accurate and perhaps also that the systems inference engine, in the case of expert systems, follows logically valid rules and reliable heuristics in its solving of problems. However, verification is not the same as validation, since there are other claims to be verified about a system which are not claims about validation but claims about such characteristics as efficiency, reliability, and capability of modification. Further, these definitions are consistent with the view that it is systems that are validated and, so certified, for use, but that it is particular claims that are said to be verified. This is parallel to the ordinary usage in which one can be said to validate a parking permit, but not said to verify it, and be said to verify the answers given on a questionnaire, but not said to validate them. While ordinary usage is not the final word on the technical use of terms, it is a first word and should not be ignored.

Based on our analysis of the concepts of verification and validation, we conclude:

* The current definitions that have evolved and are in use today contain ambiguity or are confusing.

* Verification seeks to confirm statements about an expert system.

* Validation seeks to confirm that the output is correct for the application of the expert system.

* Verification and validation are the necessary and sufficient conditions for having evaluated an expert system.

REFERENCES

[1] Bratko, Ivan. 1986. PROLOG: PROGRAGMMING FOR ARTIFICIAL INTELLIGENCE (Reading: Addison-Wesley, 1986).

[2] Copi, Irving M. 1972. INTRODUCTION TO LOGIC. New York: Macmillan Publishing Company, 1972.

[3] Culbert, Chris, Gary Riley, Robert T. Savsky. 1987. "Approaches to the Verification of Rule-based Expert Systems", National Technical Information Service, U. S. Dept. of Commerce, (October, 1987) p. 191.

[4] Feigenbaum, E. A. 1983. "Knowledge-based Reasoning Systems" in J. E. Hayes and D. Michie INTELLIGENT SYSTEMS. New York: John Wiley, 1983.

[5] Finlay, Paul N., Gareth J. Forsey, and John M. Wilson. 1988. "The Validation of Expert Systems -- Contrasts with Traditional Methods." JOURNAL OF THE OPERATION RESEARCH SOCIETY 39, (No. 10, 1988), pp. 933-938.

[6] Geissman, James R., and Roger D. Schultz. 1988. "Verification & Validation of Expert Systems," AI EXPERT 3 (February, 1988), 26-33.

[7] IEEE "Standard Glossary for Software Engineering Terminology." 1983. IEEE Std. 729-1983. Los Alimitos, California.

[8] Nguyen, Tin A., Walton A. Perkins, Thomas J. Laffey, and Deanne Pecora, "Knowledge Base Verification." AI MAGAZINE 8 (Summer 1987), 69-75.

[9] O'Keefe, Robert M., O. Balce, and E. P. Smith. 1988. "Validating Expert System Performance." AI APPLICATIONS 2 (No.'s 2-3, 1988), 35-43.

[10] O'Keefe, Robert M. 1989. "The Evaluation of Decision-Aiding Systems: Guidelines and Methods." INFORMATION & MANAGEMENT 17 (1989), 217-236.

[11] Powers, Michael J., David R. Adams, and Harlan D. Mills. COMPUTER INFORMATION SYSTEMS DEVELOPMENT: ANALYSIS AND DESIGN. CINCINNATI: South-Western Publishing Co., 1984.

[12] Rudner, Richard S. 1966. PHILOSOPHY OF SOCIAL SCIENCE. Englewood Cliffs: Prentice-Hall, 1966.

[13] Salmon, Wesley C. 1973. LOGIC, Second edition. Englewood Cliffs: Prentice-Hall, 1973.

[14] Suwa, M, A. C. Scott, and E. H. Shortliffe. 1982. "Completeness and Consistency in a Rule-Based System." AI MAGAZINE 3 (1982), 16-21.

[15] Yourdon, Edward, and Larry L. Constantine. STRUCTURED DESIGN. Englewood Cliffs: Prentice-Hall, 1978, p. 378.

Expert System Verification and Validation Part II: Implementing the Concepts

J.G Massey (*), T.W. Satre (**), C.D. Ray (***)

(*) Department of Forest Science, Texas A&M University, College Station, Texas 77843-2135, USA

(**) Division of Psychology and Philosophy, Sam Houston University, Huntsville, Texas 77341-2447, USA

(***) Temple-Inland Forest Products Company, Diboll, Texas 75941, USA

ABSTRACT

Verification and validation (V&V) have been defended as necessary and sufficient elements of the process for evaluating the expert system. An expert system is viewed as "verified" to the extent that its output statements describing the system are confirmed to a reasonable degree of strength. It is viewed as "validated" to the extent that it produces situationally relevant, correct answers at the location where it is installed. A gypsum wallboard expert system is evaluated in terms of these definitions. During the system's development, its elements and inferencing processes were verified through a process of having facility personnel confirm the system's elements and inferencing. On installation, the system was validated through an experiment that provided the basis for a statistical comparison of diagnoses by personnel and the system. The V&V procedure is offered as a means to improve the acceptability of expert systems to facilities where the system is to be installed.

INTRODUCTION

The goal of the expert system is to help the production process improve product quality or facility productivity (Massey et al. 1989). The underlying reason for installing such a system is that either there is more information than a person can analyze in the time available (cognitive overload), or existing expertise will not always be available when needed.

The two related procedures, verification and

validation, have been presented in Part I as needed elements to demonstrate that the system is internally correct and relevant to the process for which it was developed. In the development and implementation of a gypsum wallboard manufacturing expert system, the study team developed a process for verification and validation designed to provide an acceptable level of confidence in the system to the users as well as to the study team itself.

The goal of this paper is to present the verification and validation process for the development and installation of the gypsum wallboard manufacturing expert system (GQES). After a brief description of the wallboard manufacturing process itself and GQES, the authors describe the verification and validation processes and results of the V&V analysis.

O'Keefe et al. (1988) separate the evaluation process into five components: verification, validation, useability, efficiency and cost effectiveness. The first two components are the ones that provide the user with assurance of system correctness. Suwa et al. (1982) expressed verification as determining the completeness and consistency of the system. Does each input set result in an outcome? Does the same input set always result in the same outcome? Geissman and Schultz (1988) provide a very pragmatic test for validity of the system: Does it satisfy the users' needs? In Part I, we have argued rather that verification asks: Are both the rules and their associated reasoning processes correct? We distinguished between this and validation: Do sessions with the implemented expert system result in diagnoses and corrective action recommendations that are acceptable to the facility when the system is used? In distinguishing between the two V&V terms, the system developer makes certain that the system is internally correct and situationally useful.

THE MANUFACTURE OF GYPSUM WALLBOARD

The study mill was a large gypsum wallboard manufacturing facility located in West Memphis Arkansas. The manufacture of gypsum wallboard consists of a continuous process of converting gypsum rock to a gypsum/chemical mass sanwiched between paper coverings, machined to required dimensions (figure 1). The finished quantity is considered a commodity and is used almost exclusively in residential and commercial construction.

Figure 1. A schematic diagram of the manufacture of gypsum wallboard

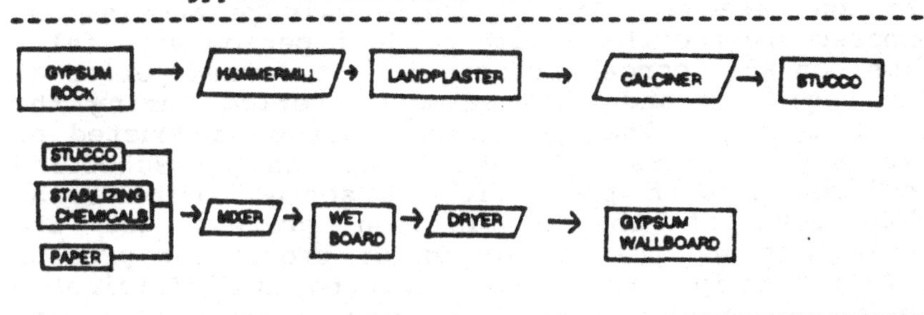

Gypsum rock (calcium sulfate dihydrate: $CaSO_4.H_2O$) is first hammermilled to landplaster (pebbles less than 3/4" in size). It is then pulverized and calcined into stucco (heated, converting the landplaster to calcium hemi-hydrate: $2CaSO_4.H_2O$). It is mixed into a slurry of 150% of the water needed to return it to its hydrated form and with stabilizing chemicals. The slurry is metered onto a base paper, covered, sideboarded with additional paper, and formed to desired width and thickness. The exothermic crystallization reaction of the hemi-hydrate and water is augmented with dryer heat, resulting in the finished wallboard that is cut to desired lengths.

THE GYPSUM WALLBOARD EXPERT SYSTEM

GQES was developed for the study mill from that mill's production expertise and installed in June, 1989 (Ray 1991). Written for the microcomputer in the Personal Consultant Plus (PC+) expert system shell, GQES diagnoses production and quality problems and then produces a prioritized list of corrective actions. Its inference process uses automated process control data, information in a knowledge base, and user-supplied information. The process control information is accessed from Allen-Bradley monitors, using that company's ADVISOR system. ADVISOR stores decision-tree knowledge in PC+'s internal format and stores general manufacturing information in DBaseIII+ files (directly accessible by the expert system). The system combines the information accessed from the user, data from process monitors, and information from its knowledge base as the basis for determining diagnoses and recommendations. In its first year of use, it was accessed 185 times. Mill management credited the system with a quality improvement that well exceeded the cost of the project to the facility.

The project duration was from January, 1988 through December, 1989. The knowledge acquisition process spanned the entire project period with full-time, on-site concentrations in a two-week period at the beginning and a three-month period during the first summer. The knowledge engineer conducted an extensive series of scheduled and unscheduled conferences with production personnel individuals. The focus of the acquisition process was the collection and transformation of production process information into rules that could be used efficiently by a computer system. It was a goal that the information not only be workable in terms of software, but that it retain the intuitive logic in the actual production process.

Verification and validation provided an assurance to the study team and facility management that the system was correct and useable. The approach of the study team was to maximize the involvement of the management in the process. Verification consisted of insuring that the elements of the system were correct from the points of view of each level of management in the facility. Correctness of inferencing was included. Validation consisted of an experiment and statistical anlaysis designed to provide the basis the for comparison of the system's performance with that of facility employees.

VERIFICATION OF THE KNOWLEDGE BASE

It is essential that correct knowledge be gathered and that knowledge retain its correctness in its final expert system rule form. Further, if mill personnel will be expected to alter the expert system later, it is essential that the final rules follow the same reasoning pattern that can be followed by the personnel. To this end, an iterative process was used for verification as the original idea was transformed through intermediate forms to the final rule. Figure 2 presents the evolution of a fact from its interview form to final rule in the knowledge base.

To evaluate correctness of the system, the study team established a "sign off" procedure that corresponded to the line of authority in the management of the facility where the system was

Figure 2. An example of knowledge verification from the initial interview to the final rule in the knowledge base

Issue raised in the initial conference:

Question: Discuss the problem of wallboard "wet peel."

Answer: Wet peel is a condition where the boards are coming out of the dryer wet and are therefore peeling. Partial bonding of the core is apparent, but because of the excess water, the bond crystals are not completely formed. Action:

.

.

4. Board weights. If the board weight is too light, the water/stucco ratio is too high. Heavy up on stucco or decrease water

Issue translated into a rule:

IF: problem is wet peel and wet weight is too low,
THEN: cause is too little stucco (cf 60) and too much main water (cf 20)

Issue as re-translated into final rule:

IF: problem is wet peel and wet weight is too low,
THEN: cause is too little stucco (cf 60) and too little main water (cf 20)

developed. The manager, the final authority in the chain of command at the mill, appointed a "point man" to be the primary contact and source of information. The knowledge engineer scheduled regular conferences with any of the production personnel, with the preferred source of information being this appointed point man. The information was then verified as correct first by the source of the knowledge, then approved first by the appointed "point man," and then approved finally by the facility manager. Thus, the information was verified through three distinct levels.

Level 1: Verifying the Knowledge Fact Itself

A database was set up in DBase III+ with the fact, a keyword for that fact, the date of the interview, and a verbal description of the fact. A database was chosen because it allowed the listing of issues in a more English-oriented framework, easier to be understood by the mill personnel. The field VALIDITY related directly to verification. It contained a number between 0 and 100 that reflected the degree to which the fact was acceptable for inclusion in the expert system. A value of 100 indicated that the fact was sufficiently verfied to be included in the knowledge base of the expert system. The value first assigned to a given fact depended on the objectivity of fact and the degree to which it would need to be verified before it could be used by the system. The knowledge engineer re-wrote each discernable "fact" from regularly scheduled interviews as a record in the database. That each fact was stored in the English form of a database allowed all involved personnel to evaluate it easily and have a permanent record of the contribution of the facility to the project. The first level of verification consisted of having the person who supplied a fact confirm that it was correct when reduced to a single fact and placed in a database. The field VALIDITY was subsequently raised to a number closer to 100. If the fact was acquired from someone beside the appointed point man, this point man was shown the fact for his approval.

Level 2: Verification of Database Facts by Management

Each week the facility manager was given a printout of the facts collected that week and verified by the source of expertise and the point man. The manager's responsibility was to confirm that knowledge elements were correct for inclusion in the expert system's knowledge base. He signed off on each element. After this final approval, the field VALIDITY in the fact's record was given a value of 100. At this point, it was considered verified to the extent that it could be transformed into rules and included in the system's knowledge base.

Level 3: Verification of the Rule by the Expert

After transformation in a rule, the expert was shown the rule and given an explanation of how the rule related to the original fact in the database. He was asked at this final level of verification to confirm that the rule in the knowledge base reflected correctly the original fact from the interview. In figure 2, the expert did not believe that a rule that contained the "and" could be accepted as true, since the rule as read was contrary to the intuitive logic. Therefore, the rule was rephrased to restore its intuitive logic while maintaining the concept.

To summarize this verification process, the facts extracted from interviews were placed in an easily read database. These facts were verified in accordance with the chain of command in the facility. This whole procedure satisfied the need for ensuring the correctness of the knowledge base. But, perhaps just as important, this procedure maintained the continuous involvement of facility personnel in the evolution of the expert system. The management was more apt to accept the system as correct since it had been involved in its development in carefully defined steps. In addition, the database gave a tangible measure of work being accomplished before the expert system was in place and working.

VALIDATING THE KNOWLEDGE BASE

It was essential to the ultimate success of the system that, after installation, the system's reasoning and recommendations be demonstrated as correct in the context of the facility where it was to be used. To this end, an experiment was designed to provide information for a statistical test to compare the system's performance with that of facility personnel. The system would be regarded as correct to the extent to which it could provide diagnoses for problems that actually occurred in the mill environment. Further, the system's effectiveness in choosing among possible corrective actions would be judged by comparing how well it solved actual problems when compared to how well experienced employees solved the same problems.

The experiment for validation was adapted from the Cohen (1960) for agreement between two raters. In his work, Cohen compared percentages of agreement observed with those expected by chance. Cohen's work needed to be adapted to allow for the typical expert system situation in which any number of diagnoses can be identified by the mill employee or expert system.

At the end of one year, nine production problems that the system had been used in diagnosing were selected as test problems. For each, data about the production process along with the actual cause of the problem were known. Six facility personnel and the "point man" were gathered for the experiment. They were instructed to list and rank all possible causes for the problems. For each problem they were given all the relevant production data collected at the time. Each was asked to rank all possible diagnoses for the problems. Figure 3 presents a sample diagnosis by the expert system and the seven persons. In the figure, the system ranks the known diagnosis second along with one of the persons, while two of the persons rank the known diagnosis as first. In such an environment, any high ranking needs to be given some form of credit since it would be expected that several corrections to the process might be tried before the actual cause of the problem was located. Thus, the statistical test was adapted to allow for credit to be given for high ranks.

Figure 3. The diagnosis of GQES and seven facility persons for the diagnosis known to be "starch too low."

Identified Diagnosis	GQES rank	Ranks given by employees #1	#2	#3	#4	#5	#6	Point man
Pulp too high	1							
Starch low	2	1		2			1	
Worn mixer	3							
Bad stucco	4		2		2	3	3	2
Bad dryer belt	5							
Bad motor	6							
Too little stucco	7							
Heavy smoother bars	8							
Too little sugar	9							
Too much heat	-	2	1		1	1	1	1

For the nine problems, similar to that presented in figure 3, a chart was constructed to reflect the percent of the time that the system and each person selected the known cause as his first, second, third, and fourth diagnosis (Figure 4). Thus, GQES selected the actual problem as its first diagnosis 22% of the time. Person #1 was a little worse, selecting the cause as his first choice only 11% of the time. But, more reflective of the effectiveness of the diagnostic process is a measure of what percentage of the time the diagnostician selected the known cause in his top choices. The top four choices were selected on the basis of how diagnoses were examined at this facility. For other applications, a different number might well be chosen. Based on this analysis, GQES did the best, identifying the actual cause of problems in its top four choices 55% of the time.

Figure 4. The percentage of the time that GQES and the seven persons selected the actual cause of the problem for the nine problems

Predictor	1st choice	2nd	3rd	4th	Total
GQES	0.22	0.11	0.11	0.11	0.55
Point man	0.00	0.00	0.11	0.11	0.22
Person #1	0.11	0.00	0.00	0.00	0.11
Person #2	0.00	0.33	0.11	0.00	0.44
Person #3	0.00	0.11	0.00	0.00	0.11
Person #4	0.00	0.22	0.33	0.11	0.66
Person #5	0.00	0.11	0.00	0.11	0.22
Person #6	0.22	0.11	0.11	0.00	0.44

CONCLUSIONS

Verification establishes the correctness of the knowledge base and the inferencing techniques used by the expert system to arrive at its diagnoses and corrective action recommendations. In this paper, the authors have presented a means by which information is verified at several distinct points as information is evaluated for correctness. Other authors have presented a case for completeness and consistency. We have found in our applications that the overriding demand by mill personnel is to have a system that is correct in the context of the line of authority of the facility.

Validation is a test of utility. We devised a test to comapre system performnace with that of seven mill employees. We would conclude that the expert system performed as well as did the seven facility personnel who were presented with the data. But, in fairness to the persons, when a person is removed from the normal stimulus associated with solving problems on the plant floor, he might well perform at a lower level. Also, the expert system itself, which performed at 55% accuracy, was based in large part on knowledge given or reviewed by the point man, who scored only 22%.

Finally, it is entirely possible that an expert system fully verified would not be useful to a given facility. Such would be the case if an expert system were designed around problems that no longer ever occurred in a given mill. But, validation ensures that the expert system can respond to a given problem, and that its response to that problem produces the same answer as does the mill's experts. Perhaps, more to the point, an expert system could be deemed valid provided that its responses were "acceptable" to mill personnel, even if mill experts might have done something different.

LITERATURE CITED

Cohen, J. 1960. A Coefficient of Agreement for Nominal Scales. Educational and Psychological measurement 20: 37-46.

Geissman, J.R. and R.D. Schultz. 1988. Verification and validation of expert systems. AI Expert. February: 26-33.

Massey, J.G., R.P. Thompson, and C.N. deHoop. 1989. The utility of expert systems to the forest products industry. Forest Products Journal. 38(11/12):37-40.

O'Keefe, R.M., O. Balci, and E.P. Smith. 1988. Validating expert system performance. AI Applications. 2(2&3):35-43.

Ray, C.F. 1991. A Prescriptive Methodology for Comprehensive Evaluation of Manufacturing Expert System Projects. Unpublished dissertation. Texas A&M University. 179 pp.

Satre, T.W. and J.G. Massey. 1991. Expert System Verification and Validation. Part I: Defining the Concepts. Sixth Annual Conference of the Applications of Artificial Intelligence to Engineering. Oxford. (In Press)

Suwa, M., A.C. Scott, and E.H. Shortliffe. 1982. Completeness and consistency in a rule-based system. AI Magazine. 3:16-21.

Development of a Systems Theory using Natural Language

J. Korn, F. Huss, J.D. Cumbers
Middlesex Polytechnic, Bounds Green Road, London N11 2NQ, UK

ABSTRACT

Attempts to predict the occurrence of events have long been a preoccupation of intellectual endeavour. Apart from inspired guesses, prediction requires a vehicle that can be manipulated and which enables the predictor to make inferences. Such a vehicle, whether it be the entrails of an animal, or the insight into human nature possessed by the oracle of Delphi, is called an inference machine. The mathematical theory based on a grouping of mathematical models in accordance with chosen principles, is such an inference machine. However, due to its very specific requirements, the construction of such a theory is limited to situations which can be modelled in quantitative terms.

The objective of this paper is to describe, mostly through examples, how natural language, perhaps the most general of symbolic models, suitably processed, can be used for the construction of an inference machine. Rich and complex declarative sentences are reduced to 'basic constituents' by linguistic analysis. The basic constituents are then represented as a semantic diagram which can be expressed in terms of propositional logic and Prolog. It is shown that a semantic diagram is built up of distinct types of causal chains of events which can be used to obtain expressions for the outcomes of a situation in logical terms.

An inference machine based on a linguistic model, is capable of predicting the existence of outcomes, if any, in the face of various disturbances which may arise as a result of removal of objects playing a part in a scenario, or changing the characteristics of selected objects and interactions. Quantitative aspects of a problem can also be considered by means of computations by subprograms.

INTRODUCTION

The unusually high level of inventions of devices and material produced during the last few decades, compared with that of the pre–war era, can be identified as a characteristic feature of a modern society. New systems using these devices and materials have enabled an unprecedented expansion of human activity to take place on a vast scale and at high intensity, involving complex, interacting organisations in industry, transport and other areas. Such activities imply considerable financial, environmental and social risks and can give rise to problems of a human and technical nature.

Risks may be reduced and problems alleviated if the consequences of human and technical activities can be predicted. Perhaps with this in mind, general systems theories and other approaches were attempted [1,2,3,4]. A feature of these methods is that they use 'variables' as the central theoretical construct, leading to the generation of mathematical theories as inference machines from which prediction can be made. The use of variables implies:

1. that only very specific aspects of a chosen part of the world or a 'real object', enter its model;
2. that these aspects of real objects recur with unfailing regularity or that a class of theoretical objects can be constructed; this is possible in the physical sciences and in many parts of technology, but not in questions involving human activities;
3. that the aspects modelled are quantifiable;
4. that the variables enter into relations to form a mathematical model which then stands for, or replaces, a theoretical object; for such a mathematical model to be of any generality, many different and diverse real objects must consistently exhibit the same aspects, a requirement easily met only in the physical sciences.

Problem–solving, or diagnostic–oriented hard and soft system methodologies have also been developed to alleviate the problems mentioned above [5,6,7]. Such methodologies do not lead to the construction of an inference machine; they are also difficult to apply, due to their imprecisely defined rules and theoretical constructs. Neither of the two theoretical developments, briefly described, appear to involve or relate to the existing and accepted branches of knowledge.

Linguistic modelling, however, considers human and technical activities, or situations, which exhibit regular features of a very general nature. The proposed modelling technique involves the use of natural language suitably processed so as to make possible the application of propositional logic and PROLOG, all of which then constitute the required inference machine. The method, thus, operates according to precise rules, and its theoretical constructs are derived from knowledge of linguistics and of the nature and operation of the human and technical constituents of situations.

The objective of this paper is to outline the steps necessary for evolving an inference machine which can then be used for predicting the existence of outcomes of situations subject to specified disturbances.

AN EXAMPLE

The following is the description of a situation which will be used as a source of illustrations: 'Shortly after the war, the central government with a large, parliamentary majority, urged farmers to increase their cereal output in order to make the country self-sufficient. The scientists of the day suggested to the farmers a wide use of new chemical pesticides, and the chemical industry advised the government to allow farmers to borrow on favourable terms the money needed for equipment. Although the farmers soon achieved vastly higher cereal output, dangerous chemicals became widely used and were contaminating much of the environment'.

BASIC CONSTITUENTS

A state and a change of state can be described by declarative statements consisting of a topic, or name of a perceived object, and a comment regarding it. This arrangement corresponds syntactically to the grammatical concept of Subject and Predicate. The verb phrase (VP) part of the predicate is stative when it expresses a state, and dynamic when it refers to a change of state. Stative and dynamic verbs attract a number of noun phrases (NP) when designating a state, or change of state. The basic constituents of a linguistic model of a situation are statements involving not more than two nouns. Each basic constituent may be represented by a semantic diagram. A number of such diagrams together form a visual expression of the whole situation. For example, the statement 'Shortly after the war, the central government urged farmers' diagrammed in Fig. 1, involves two nouns. Statement 'Dangerous chemical became widely used' involves only one noun. The diagrams also show the symbolic representation of objects, adjectives, adverbials and verbs expressing processes.

In Fig. 1, the dotted line between the 'blobs' denotes a change of state, and between a flag and a blob, an acquired property. A continuous line between blobs specifies a relation, and between a flag and a blob, an inherent and an initial property.

From the top diagram in Fig. 1 the following statements are made:

$$o^{c}_{c1} \supset p_{c}(1) \qquad 1.$$

$$b(1) \cdot r^{b}_{c}(1) \cdot s_{c}(1) \supset o^{c}_{c1}$$

from which

$$s_{b(1)} \cdot r^{b}_{c}(1) \cdot s_{c(1)} \supset p_{c(1)} \qquad 2.$$

Eq. 2 says that 'If the central government urged the farmers who are as yet undisturbed, then these farmers are in fact urged'.
The logical implications of eq. 1 can be expressed as the implication part of a full Prolog program.

```
farmers(urged) :- ao(farmers(undisturbed),bec,farmers(urged)).
ao(farmers(undisturbed),bec,farmers(urged)) :-
government(central),urgedshortlyafterthewar(government,farmers),
farmers(undisturbed).
```

In Fig. 1 the primitive form of an event is expressed as a change of state shown by the dotted lines and designated as 'ao', or 'alter–object', in the Prolog program leading to an acquired property.

In conclusion, one– and two–place statements have been identified as the basic constituents of which a semantic diagram, representing a situation can be built up.

LINGUISTIC ANALYSIS

In general, natural language is syntactically too complex, too rich in vocabulary and in the use of metaphors to be an effective means of creating an inference machine. Linguistic analysis identifies the structures suitable for the development of such a machine together with those which are too complex. By means of various techniques based on linguistics [8,9,10], it then operates on these complex linguistic structures so as to make them amenable for the creation of an inference machine.

The following is a summary of linguistic structures so far identified and capable of being converted into a combination of basic constituents.

Clause Structures: The predicate part of a simple sentence is divisible into its immediate constituents, the Auxiliary and the Verb Phrase. The former represents a complex of person, number, tense, aspect and mood features essential to the realization of the verb process. The latter represents any one of eight clause patterns analysable in terms of the process types expressed by the verbs they accommodate. These patterns are as follows: 1. NP + V, 2. NP^1 + V + Attribute, 3. NP^1 + V + Adv(Location), 4. NP^1 + V + Adv(Direction), 5. NP^1 + V + Adv(Source), 6. NP^1 + V + (Prep)NP^2, 7. NP^1 + V + NP^2 + Attribute, 8. NP^1 + V + NP^2 + Prep + NP^3. Although four main types of verb process have been identified [8], the semantic roles played by the nominal elements within the clause are best expressed in Case Grammar terms [9]. the roles played by these elements are seen as: agentive, instrumental, locative, dative, objective, and factive. Such analysis helps to identify the cause, the affected object and the means of achieving a change of state.

Passivization and Multiple–place Verbs: The diagramming convention requires that the arrow carried by a continuous line should point towards the affected object, i.e. the one to undergo a change of state. In most two–place goal–transitive statements, the role of affected object is played by the grammatical object, but when in certain cases the grammatical subject is the affected element, the statement is expressed in passive form. Thus, for example, 'John received a letter' becomes: 'A letter was received by John'.

The verbs in clause pattern 8 above, for example, involve more than two nouns, as in: 'John sold a car to Bill'. In order to fit into the diagramming convention, such structures are reduced to two, or more equivalent statements, so as to preserve the meaning of the original.

Outer Clause Relations: By outer clause relation is meant the way in which information of one clause is understood in the light of information contained in another. Each statement is meaningful on its own and their relation is indicated by a marker. Clause relations, such as 'cause' and 'means' may be signalled by the words 'because' (or 'since') and 'by...ing' respectively. These are then converted into a series of one– and two–place statements replacing the marker, so as to preserve the meaning of the original sentence. Such statements contain the verbs 'cause' and 'enable'.

Complex sentences: These are defined as deriving from two or more underlying simple sentences which have been put together in a variety of ways, involving the notions of: embedding, replacement, addition, complementation and nominalization. The analysis of such complex structures enables us to unravel the often hidden meaning of the constituent clauses, which leads on to the formulation of statements acceptable to the diagramming convention.

LOGIC IN LINGUISTIC MODELLING

The result of linguistic analysis is shown as a semantic diagram in Fig. 2. The diagram is expressed in terms of basic constituents forming 'causal chains' as indicated by the thick continuous and dotted lines. The thin continuous lines designate the interaction between the causal chains which 'prompt' further changes.

The causal chains in Fig. 2 can be constructed as:

1. $o_{h2}, o_{h1}, o_{h}, o_{g1}, o_{g}$

2. $o_{f2}, o_{f1}, o_{f}, o_{c1}, o_{c}, o_{b}$

3. $o_{c4}, o_{c3}, o_{c2}\ o_{c1}$

4. o_{b1}, o_{b}, o_{a}

5. o_{e1}, o_{e}, o_{d}

For each causal chain a series of logical implications, which indicate the propagation of changes of state, or events, can be written. For example, for causal chain 3.

$$o^{c3}_{c4} \supset p_{c4}$$

$$p_{c3} \cdot q^{c3}_{c3} \supset o^{c3}_{c4}$$

$$o^{c2}_{c3} \supset p_{c3} \qquad 3.$$

$$p_{c2} \cdot r^{b1}_{c2} \supset o^{c2}_{c3}$$

$$o^{c1}_{c2} \supset p_{c2}$$

$$p_{c1} \cdot r^{e1}_{c1} \supset o^{c1}_{c2}$$

From eq. 3

$$p_{c1} \cdot r^{e1}_{c1} \cdot r^{b}_{c2} \cdot q^{c3}_{c3} \supset p_{c4} \qquad 4.$$

which says: 'The farmers will have money if (they are urged) and (the use of pesticide was suggested to them) and (the government allowed them) (to borrow money for equipment on favourable terms).

Expressions similar to eq. 3 can be developed for the other causal chains so that 'shared' properties such as p_{ci}, can be eliminated, leading to an overall relation similar to eq. 4. Such a relation then indicates the conditions for the occurrence of an outcome, which could, in this case, be the property 'contrasted'. A full Prolog program can also be developed on the basis of expressions like eq. 3. The program then enables the possibilities of existence of outcomes to be investigated following selected disturbances; removal of particular objects and/or alteration of properties.

CONCLUSIONS

Devising methods for predicting the occurrence of future events has been a major preoccupation of human intellectual activity. Discarding guesses, such methods are based on the construction of some model, the manipulation of which can generate the substance of prediction. This paper has described an attempt at constructing a means of predicting the occurrence of future events not amenable to mathematical methods. Events arising from human activity situations serve as examples to which the method described, is applicable.

A framework of a predictive theory has been established using natural language as the symbolic model. Complex linguistic structures have been identified and it has been discussed how such structures can be transformed into statements which are then used for constructing a semantic diagram. A semantic diagram representing a situation, is seen to consist of inferential logic forms called causal chains exhibiting distinct patterns. These forms can then be translated into a Prolog program which is capable of working out whether an outcome of the situation under consideration, exists or not. The possibility of occurrence of an outcome may be varied by introducing disturbances in the form of deleting, or adding, objects, or changing the manner of interactions.

Work is in progress to extend the capability of Prolog to take into account properties expressible in numerical terms and to perform calculations the results of which can affect the chances of occurrence of an outcome. Methods developed in linguistic modelling are now applied to design procedures [11].

SELECTED REFERENCES

1. von Bertalanffy, L. General systems theory. General Systems, pp. 1–10, v1, 1956.
2. Klir, G.J. An approach to general systems theory. Van Nostrand Reinhold Co., 1969.
3. Chestnut, H. Systems engineering tools. J. Wiley, NY, 1966.
4. Churchman, C.W., Ackoff, R.L., Arnoff, E.L. Introduction to operational research. J. Wiley, NY, 1957.
5. Jenkins, G.M. The systems approach. Journal of Systems Engineering, n1, v1, 1969.
6. Checkland, P.B. Systems thinking, systems practice. J. Wiley & Sons, 1981.
7. Anon. Systems behaviour. The Open University Press, 1982.
8. Halliday, M.A.K., Fawsett, P., Robins, R. New developments in systemic linguistics. Batsford, 1982.
9. Filmore, C. The case for case. Universals in linguistic theory, ed. Bach and Harms, University of Texas, 1968.
10. Korn, J., Huss, F., Cumbers, J.D. Linguistic modelling of situations. UKSS Conference, University of Hull, 12–15 July, 1988.
11. Korn, J. An alternative approach to design. ICED 88 Conference, 23–25 August, 1988, Budapest, Hungary.

central $s_{b(1)}$ undisturbed $s_{c(1)}$ urged p_{c1}

government urged farmers farmers

o_b $r_c^b(1)$ o_c o_{c1}^c o_{c1}

Diagram of a two-place statement

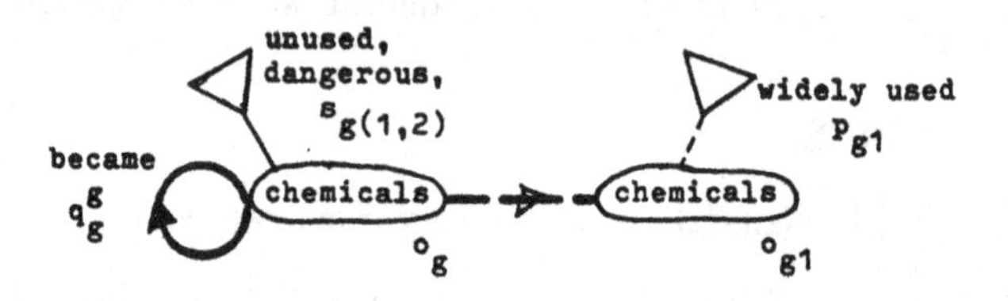

Fig.1: Diagram of a one-place statement.

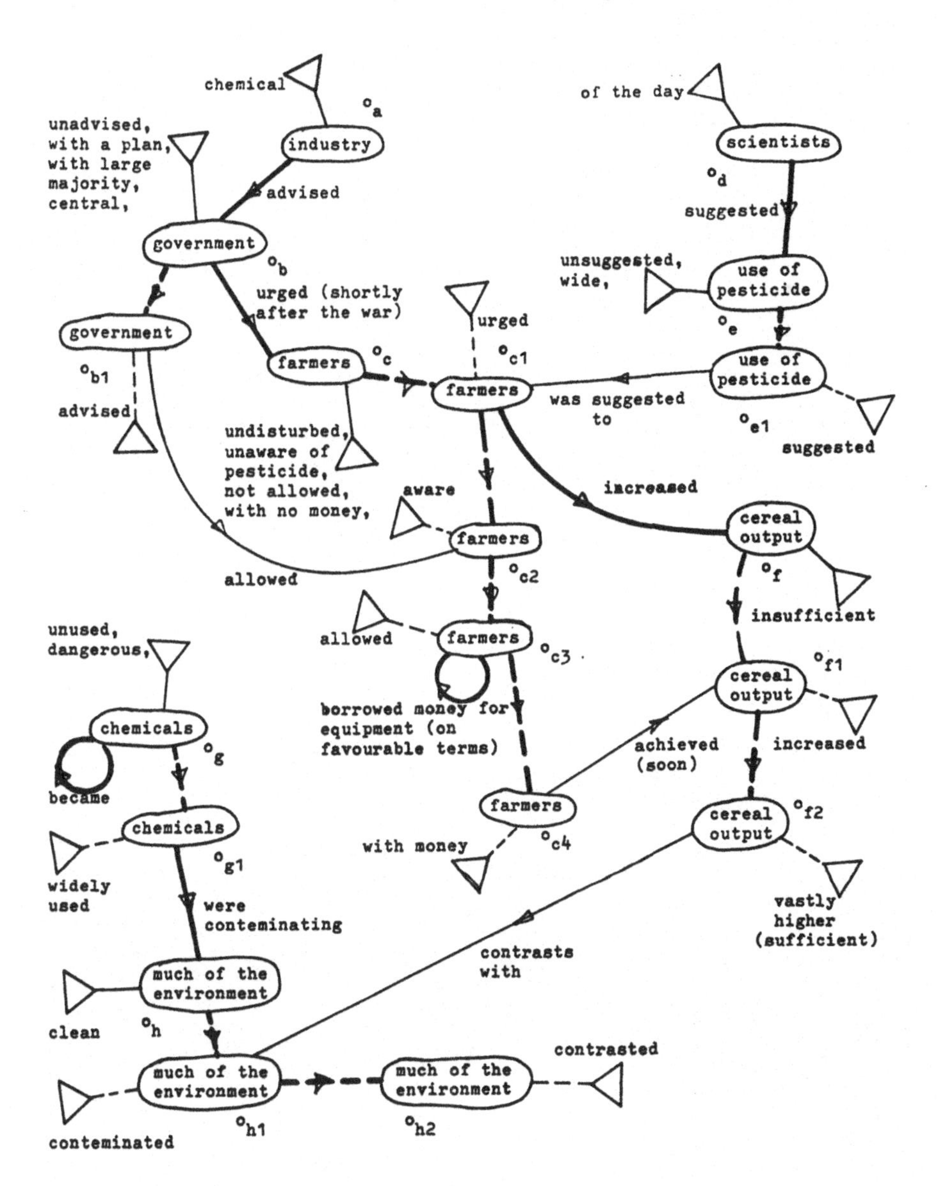

Fig.2: Semantic diagram representing example.

Computer Aided Validation Expert System

R.J. Smith

Sira Research and Development, South Hill, Chislehurst, Kent, BR7 5EH, UK

ABSTRACT

Computer Integrated Manufacturing (CIM) is seen as a key route to achieving future commercial success in industry. Its philosophy aims to achieve integration of all business functions from marketing through to support of the delivered product. One such area which will benefit from integration into the design of the product is that embracing the measurement, inspection and test activities necessary for validation of the delivered product.

This particular integration application is currently being tackled by a collaborative project which is supported by the Department of Trade and Industry. The project is aimed at developing techniques for analysing validation needs and for planning validation methods.

Within the project an experimental Computer Aided Validation Expert System (CAVES) is being constructed. This system will operate with a generalised model of the validation process and will help with all design stages from the specification of the product requirements through to the analysis of feedback data from the manufacturing process. The kernel of the system will be a knowledge base structure containing information related to the manufacturing process capabilities and the available inspection and test facilities.

It is planned to integrate the CAVES system into an advanced CIM facility for demonstration and evaluation.

INTRODUCTION

The ability to supply Quality products is critical to the success of any enterprise in todays highly competitive market. One of the keys to achieving this goal is the ability to build Quality into a product at the design stage. This means specifying a product that truly satisfies the real need, then designing an implementation that can be made efficiently and which can be proven to match the specification.

With the accelerating rate of technological innovation product life cycles are becoming shorter. As a result product development time is becoming increasingly critical. To reduce development times it is crucial to tighten the integration of all functions from specification through to validation of the product. This will ultimately lead to faster product development, with fewer iterations of design and trial production. It should also provide designs which are known to be within the capabilities of the manufacturing facilities and that can be verified by known procedures.

INTEGRATION

Computer Integrated Manufacturing (CIM) and Total Quality Management (TQM) are seen as the two primary paths to integration, but applied individually they apparently take different routes. The former is perceived to place emphasis on physical integration - interconnecting computer assisted functions by electronic communications paths. The latter places emphasis on functional integration - linking together functions, regardless of whether these are performed manually or with computer aid, so that they cooperate within total systems.

Whilst these two drives are converging, and must continue to do so, there are many as yet unsolved problems holding them apart. This is particularly true in relation to computer integration of the validation function.

Functional integration, in the form of multi-discipline teams as used in Simultaneous Engineering, goes a long way in solving these problems. The performance of such strategies though could, and should, be enhanced by the use of physical or computer integration.

This style of integration would require computer representation and manipulation of specifications of product requirements and of their relationships with design features, process parameters and measurement and test parameters.

At the present time, however, the links established in CIM systems are weak. They provide electronic communication of data rather than functional integration. Further more many of the links are for communication in one direction only; CAD or CAE sub-systems will for example typically pass data to a CAM subsystem but not use feedback of information from manufacturing.

The commonly implemented CIM elements that should ideally cooperate in establishing integration of design and inspection are the CAE and the horizontally integrated manufacturing control subsystem. However the currently available CAE systems, which have evolved from CAD systems, have serious limitations. Available systems integrate facilities for design analysis, such as solid modelling and finite element analysis, and they provide specialised word processing facilities for documenting engineering specifications; they do not however have capabilities to know about and reason about product characteristics and toleranced design features. Using the very latest and most comprehensive systems the designer still retains the bulk of the design in his/her head; it is only one narrow representation of what has been designed that is known to the CAE system. Integration of CAE and inspection (as part of the manufacturing system) extends little further than establishing communications links. Systems have been implemented that generate inspection data at the CAD station and pass this automatically through to inspection cells. However these systems do not have facilities for reasoning about the design (which might even defy validation by any practical means) and they have no capability to provide any justification for the selected inspections, or to provide any assurance that these are sufficient to prove the product.

It is now widely accepted that to overcome these limitations it will be necessary to exploit developments in object orientated programming and knowledge based systems. This is the approach being taken in developing the CAVES system.

THE VALIDATION PROCESS

Validation is an essential part of the total product creation process. It is a process in its own right which should be tackled at the start of any project to introduce a new product. True Quality will be difficult to achieve unless there is a strategy for proving that the design matches the specification and for proving that the manufactured products conform with the design.

The objective of the validation process is to provide confidence of product quality, ie knowledge that the product is consistent with its requirement. The process should be a combination of offensive measures - based upon proving the product design, proving of the production system and control of the production process - and defensive inspection and test measures. It should be designed and planned as a part of the product development plan showing the total proving process, and showing within this the roles of these offensive and defensive measures.

The validation process can be viewed as a succession of filters:

- firstly the product design is analysed to determine which product characteristics are inherently secured by the design and are independent of manufacturing method.

- secondly the manufacturing method is analysed to determine which features, and hence some of the remaining product characteristics, are inherently secured by the manufacturing process and are independent of manufacturing process variations.

- thirdly the manufacturing processes are analysed to determine which additional features, and hence some further product characteristics, are inherently secured by the repeatability of the proposed processes - with a given set of process controls.

- finally measurement, inspection and test is used to secure the product characteristics falling through the above offensive filters.

In practise these filters are iterative. At each stage the analysis may highlight needs or opportunities to modify the design or manufacturing method to reduce the need for inspection and testing.

This process should lead to designs well within the manufacturing process capabilities and consequently have intrinsic production repeatability. It should also lead to appropriate design of process controls to selectively improve process capabilities and further ensure production consistency. Finally it should lead to inspection and testing being used in the most effective manner, at the best points within the production system and to check the necessary and sufficient collection of specific features and functions of components, assemblies and product. The practise of attempting to "inspect quality into the product" at the production stage is avoided. Emphasis is placed upon the more effective goal of proving the design and the manufacturing process ahead of going into production. Whilst measurement, inspection and testing plays an important role in the pre-production proving its primary purpose once in production should be acquisition of knowledge that can be used for on-going improvement and new development.

This validation methodology removes the problematic questions surrounding complex products, where the particular collection of design tests, process controls, inspections and product tests is all too often assembled in an ad-hoc manner ie "How much confidence do the ad-hoc measures provide? Are some of the measures unnecessary? Are additional measures required? Could simple design changes make the validation process easier, or more thorough?"

CAVES

The objectives in developing the CAVES system are to provide a means for functionally and physically integrating design and validation. The system will be designed to assist product development teams in:

- defining the desired functions and characteristics of appearance, reliability etc, in an unambiguous and explicit (computer understandable) form.

- analysing the way in which a proposed design achieves these requirements, to determine the criticality (tolerances) of the various "features" in the design.

- analysing the way in which, and the consistency with which, a proposed production method will produce the "features", to determine the needs for "feature" verification during production.

- generating an inspection and test plan (identifying the inspection and test stages within production, the frequency, the methods and the procedures), that uses defined inspection and test facilities to achieve the verification needs in a cost effective manner.

A knowledge base of manufacturing methods and measuring, inspection and testing methods is used to perform these functions and feedback of inspection and test results is used by the system to update this knowledge base.

The scope of the CAVES system and its method of operation is as follows.

<u>Product specification</u>
This function is essential for ultimate "Design to Product" CIM systems and should be regarded as an input to the CAVES system rather than a system function. It would ideally be integrated into the CAE element. However existing CAE systems operate at a lower level knowing only about highly abstracted and very specific representations of what has been designed - not why or how this design has evolved. To ensure product requirements are totally and unambiguously specified (which is essential for the validation process and manufacturing planning activities) it is necessary to include within the CAVES system its own MMI (Man Machine Interface) to assist in generating the product specification.

This interface elicits statements, in an unambiguous and explicit computer understandable form, about the attributes of the product (such as the required functions and characteristics of appearance, reliability etc) and for each of these the criticality, the describing parameters and the permissible variations in the parameter values. By adopting a knowledge based approach it is possible where necessary for parameters values to be expressed qualitatively.

The interface uses an interactive natural language dialogue which is directed by using a knowledge base of attributes relevant to classes of products and parameters relevant to classes of attributes.

<u>Product design analysis</u>
This function should also ideally be integrated with the CAE element rather than being included in the CAVES system. However whilst existing CAE systems have capabilities to support analysis of certain very specific product-attribute / design-feature dependencies (using for example Finite Element Analysis) the major part of the reasoning remains in the designers head. For this reason it has been necessary to include within CAVES a higher level function related to product design analysis. This does not attempt to embrace the individual specific analysis tools or at this stage to have access to them. It does provide a facility for drawing together information on the analyses performed and the results obtained in terms of proving of product attributes.

CAVES interacts with the designer, and ideally integrates with developing tools such as feature based CAD, to elicit information on the design "features" that combine to provide the product attributes, the relevant parameters of each feature, and the dependencies of the product attribute parameters upon each design "feature" parameter. From this the system can reason about the criticality and permissible tolerance of each feature parameter value.

To support the interaction with the designer and the reasoning CAVES has within its knowledge base knowledge of features used to implement product attributes, attributes relevant to classes of feature and parameters relevant to classes of attributes.

<u>Manufacturing design analysis</u>
This function is beyond the scope of any of the commonly implemented CIM elements, except perhaps the latest generative CAPP (Computer Aided Process Planning) systems. It is a key design to inspection link, integrating design to manufacture via production engineering.

CAVES interacts with the production engineer to define the manufacturing process that will be used to replicate the product. It analyses the proposed manufacturing design to determine the dependencies of the feature parameters upon the manufacturing process parameters and to identify the feature parameters not assured by the proposed manufacturing method. To achieve this it has, within its knowledge base, knowledge of the processes used to implement features, the manufacturing facilities available and the process capabilities of these facilities.

<u>Verification planning and analysis</u>
The functions of verification planning and verification analysis must clearly be tightly integrated and have links with the manufacturing design analysis function and with the product design analysis function.

The verification planning function is beyond the scope of any of the commonly implemented CIM elements. It can be argued that it is a part of the manufacturing design function, following the philosophy of regarding measurement, inspection and test as an integral part of the manufacturing system. In this respect CAPP systems probably comes closest to addressing this opportunity. There are however no commonly implemented CIM sub-systems for this level of manufacturing planning.

CAVES assists in planning measurement, inspection and test (identifying the stages within production and the methods and frequencies) by using a model of the validation process described above. From the analysis of the design and the manufacturing method CAVES identifies the feature parameters and/or the product attributes not sufficiently assured. It then interactively builds into the manufacturing plan a measurement, inspection and test plan. Within this process it analyses the plan to determine the extent to which it assures the necessary feature parameters or product attributes.

To perform these functions CAVES uses knowledge of the processes used for measurement, inspection and test, the facilities available and their capabilities.

To link the verification planning and analysis function to the measurement, inspection and test facilities a further function is required: generation of verification procedures and data. In practice this function, which determines the implementation of the measurements, inspections and tests, is extremely dependent upon the details of the specific measurement, inspection or test facility used. For this reason it is ideally integrated into the measurement, inspection and test cells. The function includes translation between the general (non cell specific) language used to define what is to be measured , inspected or tested, and the cell specific control and reference data. Future instrument systems used for measurement inspection and test should accept and respond with information in feature property terms rather than individual items of data. For this reason the generation of cell specific procedures and data will not be included in the CAVES system.

Feedback analysis
This function is closely related to many of the above functions but would not logically be integrated with any particular one. In practise it will usually be distributed as a collection of functions associated with each of the functions activated to formulate the corrective action.

For this reason the role of CAVES in the feedback analysis is restricted to the high level integrating function. It integrates the distributed functions by collecting feedback data and using this to compare product attribute and feature parameter achievement with that predicted in the design and development phase. It updates the data/knowledge base forming the basis for those predictions, highlights discrepancies and alerts attention to needs for re-entry to the design and development phase, triggering consideration of possible changes in the product, manufacturing, or even the verification designs.

THE CAVES DEMONSTRATOR

The CAVES development project is due to finish in May 1992. In keeping with projects of this nature the early work has revealed areas presenting severe challenges. There have been, as anticipated at the outset, several major hurdles in developing the system.

The first hurdle was finding an unambiguous and explicit (computer understandable) means for defining the requirement characteristics. This has been achieved by the development of a generalised knowledge structure.

Approaching the problem in this way also provided a high level solution to how "features" could be related to product characteristics.

The knowledge-based system being developed will model the validation process knowing about features used in designs, about processes used in production, and about facilities used for inspection. It will have the ability to reason about product features defined by the designer, to check to what extent these features will be affected by manufacturing tolerances and to decide which features will need to be inspected. It will use this knowledge to help decide how to carry out the necessary inspection, acting as a "designers assistant", providing advice concerning the practical inspection implications of the features in the evolving design.

Whilst all of the issues discussed above are being considered the programme is concentrating upon the inspection aspects of validation, and on developments that will assist rather than totally automate.

REFERENCES

1. Pugh, S. Total Design, Addison-Wesley Publishing Company, 1990.

2. Nevins, J.L. and Whitney, D.E. Concurrent Design of Products & Processes, McGraw-Hill Publishing Company, 1989.

3. Pugh, S. Knowledge-based systems in the design activity, Design Studies, October 1989, pp. 219-227.

4. Tannock, J., and Maull, R. The integrated quality system in CIM, Computer-Integrated Manufacturing Systems. Vol.1, No.4, pp. 228-234, 1988.

5. Shaw, N.K., Bloor, M.S. and de Pennington, A. Product Data Models, Research in Engineering Design, Vol.1, pp. 43-50, 1989.

6. Brown, C.D. and Chandrasekaran, B. Design Problem Solving, Knowledge Structures and Control Strategies, Pitman, London, 1989.

Fuzzy BOXES as an Alternative to Neural Networks for Difficult Control Problems

N. Woodcock, N.J. Hallam, P.D. Picton

Faculty of Technology, The Open University, Milton Keynes, MK7 6AA, UK

ABSTRACT

Three benchmarks of highly non-linear, difficult control problems are used to test the claim that "fuzzifying" thresholds in state space can improve the performance of a controller which has been trained using the BOXES algorithm. Experiments are reported which suggest that the fuzzy controllers produce more erratic performance than their non-fuzzy counterparts.

INTRODUCTION

Control of a dynamic system has traditionally required detailed knowledge of the system dynamics and an expression of the system's desired behaviour. Then, given the state of the system at a particular time, a control signal can be determined that will drive the system towards a desired state. However, there exist many difficult control problems for which traditional methods of control are inadequate, for a mathematical model may be unavailable, incomplete or highly non-linear; or the desired behaviour of the controller at a given time may be unknown.

In the present paper we describe experiments performed on several difficult benchmark control problems. For the purposes of this paper we restrict ourselves to cases where the mathematical model of the system is not used in the design of the controller and the desired behaviour of the controller cannot be specified in advance, either analytically or by example. The only information available to the controller, besides the current state, is a reinforcement signal, indicating when the system has reached a desirable or undesirable state. Over a series of training runs the controller must learn an appropriate control strategy. This can be achieved by "learning with a critic" [20], whereby a controller has an on-line critic that monitors its performance, facilitating sensible parameter changes to improve performance. The critic is usually some performance function, applied to important characteristics of the system such as time to failure. The idea of "learning with a critic" can be viewed as an intermediate learning regime lying between supervised learning (i.e. "learning with a teacher") and unsupervised learning: the performance function provides a rough indication of a network's output error, without directly comparing actual output with desired output.

It has been claimed [5] that the actions of a controller may be improved by the use of "fuzzy" techniques, which can be used to provide a smooth transition between control actions learnt for different regions of the system state space. The experiments described below constitute an investigation into the use of such techniques with Michie and Chambers' BOXES learning algorithm [12] in the class of difficult control problems outlined above. The experiments described in this paper show that fuzzy techniques are not guaranteed to produced improvements in performance.

The paper is set out in the following manner. First we briefly describe the background to our work, including neural network techniques for control and the BOXES algorithm. We then describe three benchmark non-linear control problems - the cart-and-pole, the bioreactor and the tractor-trailer - and present the results of our work on computer simulations of these problems. Finally we discuss conclusions that can be drawn from our results.

BACKGROUND

Neural networks have been widely recognised as a useful tool for tackling difficult control problems [1], [4], [13], [15]. Networks of sufficient size are, in theory, capable of approximating to any desired accuracy highly non-linear multidimensional mappings [14]. The domain of the mapping could be the state space of a system and the range could be the the space of possible control actions. The difficulty lies in obtaining suitable values for weights in the network, which determine the mapping.

Simple feedforward networks can be trained using the backpropagation learning algorithm [15] to determine network weights. However, such supervised learning algorithms require knowledge of the desired state or control action at time $t = 0$, which is not always known if the goal is simply to drive the system to a desired state at time $t = K$ but the route to that goal is unspecified (fig. 1). Such knowledge can be obtained by monitoring a human controller [6], although this is not always satisfactory, particularly in systems that are too complex for a human to control well. An alternative is to use a network with feedback trained by recurrent backpropagation through time [9], [13]. Such a network learns in two stages: firstly it learns to emulate the system dynamics from runs of the uncontrolled system; then it learns to drive the system towards the desired state by making use of its internal emulation. So this type of network makes use of a model of the system dynamics but the model is learnt during training and not explicitly used for the design of the controller.

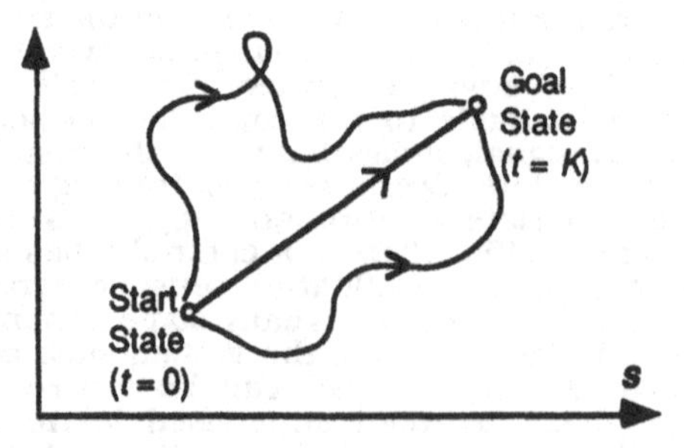

Figure 1: Routes from a start state to a goal state.

Although backpropagation is the most well-known learning algorithm, other algorithms have benefits in control applications. For example, the ACE/ASE architecture of Barto, Sutton and Anderson [4] contains evaluator and controller elements which use the strategies of temporal difference [17] and reinforcement learning [22] respectively to determine connection weights. These strategies allow simultaneous adaptation of the evaluator and controller and allow training to continue throughout a run, not just on receipt of a failure signal. Temporal difference learning is used to train the evaluator by adjusting weights on the basis of changes in predictions of the evaluator; meanwhile reinforcement learning is used to set the controller's weights on the basis of a performance measure provided by the evaluator. Another promising approach is that taken by Wieland [21], who used genetic algorithms [10] to search the weight space for a suitable set of weights.

All the aforementioned strategies for obtaining connection weights require the careful setting of various parameters (learning rate, momentum, number of processing elements, number of hidden layers, etc.). So we are, in effect, substituting one non-linear optimization problem (i.e determining suitable control actions) for another (i.e. setting network parameters at a "meta-level"). Harp [7] suggested a novel approach to setting parameters using genetic algorithms. Unfortunately genetic algorithms themselves require the careful setting of various parameters, such as mutation and crossover rates. In an attempt to find a control strategy less dependent on parameters we returned to basics and began experimenting with Michie and Chambers's BOXES algorithm [12], which was the starting point for Barto, Sutton and Anderson's work on the ACE/ASE architecture. However, rather than take their neural network approach to improving the BOXES algorithm, we used fuzzy sets [23] to smooth the passage between different regions of the state space. This is similar to using a trained CMAC network as a fuzzy look-up table, as described by Barto [3]. We shall now briefly describe the BOXES algorithm and our attempts to improve it.

THE BOXES ALGORITHM

Imagine some system whose state can be described, without loss of generality, by just two variables, say x and y. We wish to build a controller that associates with each state vector (x, y) some control action $F(x, y)$ in such a way that the system will be driven towards some desired state and kept out of illegal states. We assume no knowledge of the system dynamics and further assume that the only information available to the controller is the current state and a failure signal indicating when an illegal region of the state space has been entered. Assume, again without loss of generality, that there are only two possible control actions, represented by +1 and -1, providing "bang-bang" control. Then the BOXES algorithm provides a means for the controller to learn appropriate control actions, in the following way.

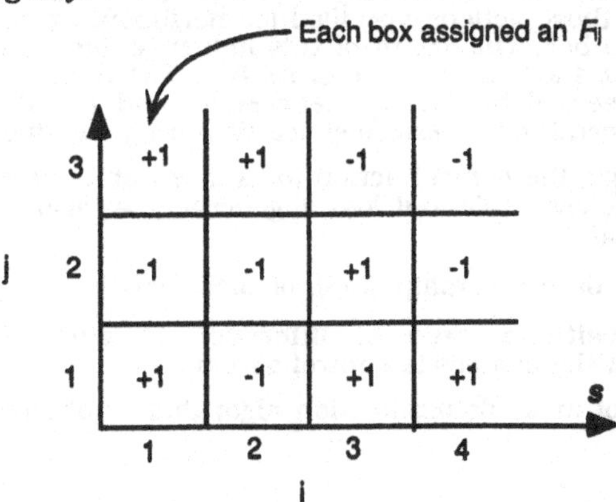

Figure 2: Division of the state space into non-fuzzy "boxes".

Firstly we must divide the state space into separate regions (or "boxes"), each of which will be assigned a control action $F_{ij} \in \{+1, -1\}$, where i and j are used to index the boxes, as shown in fig. 2. This assignment is made at random initially. We then allow the system to run, applying the control action specified by the current box, at given time intervals. The controller learns to improve its performance by adjusting the boxes' control actions. Each box keeps a record of average time to failure following a +1 or -1 action. When a box is entered the following occurs:

(1) The box's values of average lifetime expectancy following +1 and -1 decision are compared. Whichever is greater determines the current action (F_{ij}) for that box.

(2) A counter records time to failure.

(3) On receipt of a failure signal the average lifetime associated with the current F_{ij} is updated.

Note that the records of average times to failure could be replaced by a more sophisticated performance measure if time to failure were not the most important criterion. This is illustrated in the tractor-trailer example later.

FUZZY BOXES

We stated above that a motivation for using the BOXES algorithm was a desire for a less parametric-dependent control algorithm. In our experiments the box boundaries were set by hand, in apparent conflict with our initial criterion. Other researchers have investigated the possibility of setting these boundaries automatically [4], [6] - we tried to make the choice less critical by using fuzzy sets. Techniques taken from Zadeh's fuzzy set theory [23] and fuzzy logic [24] have been used for over a decade in the domain of intelligent control. (Mamdani and Assilian [11] devised a "fuzzy linguistic controller", essentially a rule based controller where the control rule for any given state vector may be modified by rules pertaining to nearby states. Control rules may be elicited directly from a human expert, or indirectly during a training period, as described by Takagi and Sugeno [18].) In our approach only the thresholds of the boxes need be provided beforehand. These are automatically "fuzzified" using triangular fuzzy sets (fig. 3). Control actions for each box are learnt using the usual non-fuzzy BOXES algorithm. Then, after training, the compositional rule of inference [8], [24] is used to determine a composite control action for a given state vector. By this means the actual action taken is influenced by those actions specified for neighbouring boxes, as well as the vector's own box. The extent of this influence depends on the proximity of the current state to the edge of its box. The final composite control action is represented by a value between -1 and +1, allowing a smooth transition between the two extremes learnt during training.

In terms of fuzzy logic the control action for a given state vector $\mathbf{s} = (s_1, s_2)$ is evaluated as follows. (Without loss of generality we again assume that $\mathbf{s}$ is two-dimensional.)

(1) Calculate the degree of membership, $\mu_{ij}(\mathbf{s})$, of each fuzzy box.

(2) Apply the compositional rule of inference to determine the membership function of the composite control action.

(3) Apply the "centre of area" defuzzification algorithm to obtain a non-fuzzy control action.

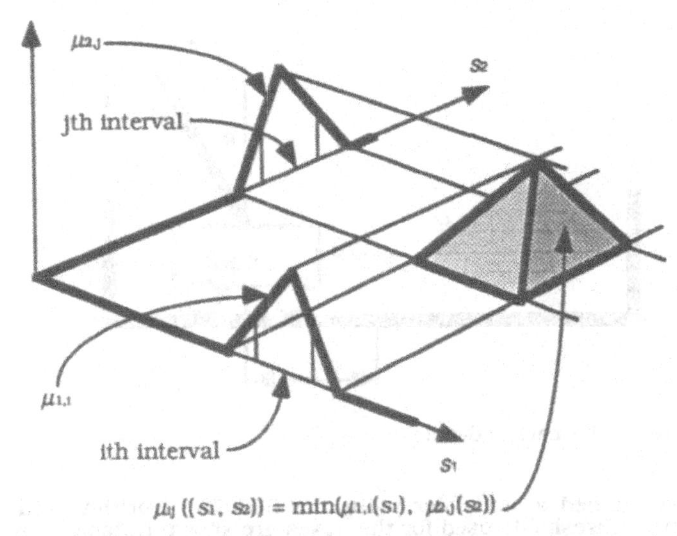

Figure 3: Using triangular fuzzy sets to represent degrees of box membership.

Stages (2) and (3) can be condensed into the following formula, which evaluates a composite non-fuzzy control action F for non-fuzzy $\mathbf{s}$:

$$F = \frac{\sum_i \sum_j \mu_{ij}(\mathbf{s}) F_{ij}}{\sum_i \sum_j \mu_{ij}(\mathbf{s})},$$

where $\mu_{ij}(\mathbf{s}) = \min\{\mu_{1,i}(s_1), \mu_{2,j}(s_2)\}$, $\mu_{1,i}$ and $\mu_{2,j}$ being the membership functions for the ith and jth intervals on s_1 and s_2 respectively.

The results of fuzzifying various controllers trained using the BOXES algorithm are presented below. The experiments were performed on three benchmark problems, namely the cart-and-pole, the bioreactor and the tractor-trailer.

THE CART AND POLE

The cart-and-pole problem has been used extensively as a benchmark for intelligent control techniques [1], [6], [16], [21]. The system is illustrated in fig. 4. A pole is attached by a pivot to a cart which is allowed to move along a track of finite length. Movement of the cart and the pole is restricted to two dimensions. The state of the system is given by four parameters: the cart's position, x, and velocity, $\dot{x}$, and the pole's angle, θ, and angular velocity, $\dot{\theta}$. The equations of the model used to construct the simulation can be found in the appendix. Note that the model was only used in designing the simulation and not in any way by the controller.

Failure of the system occurs when the pole falls past a certain angle, 12 degrees in our case, or when the cart hits the end of the track. Control of the system can be achieved by the application of a sequence of left and right forces to the trolley.

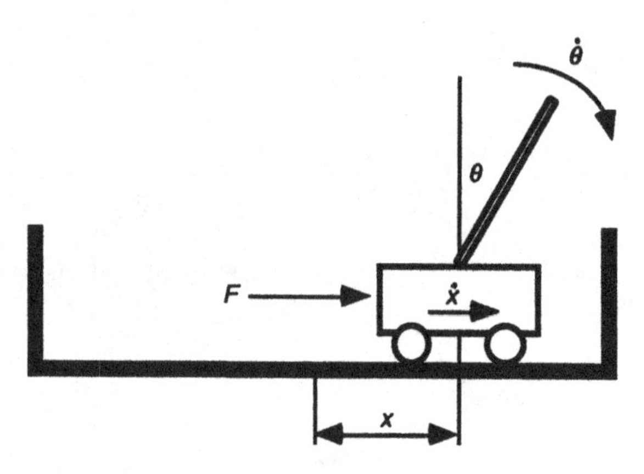

Figure 4: The cart-and-pole

We trained a controller using the BOXES algorithm and bang-bang control. Thresholds used for the boxes are shown in table 1. After training we "fuzzified" the boxes and compared the performance of the controller with that achieved using non-fuzzy boxes. The results of three separate training runs are shown in fig. 5. The graphs show the performance of the fuzzy and non-fuzzy controllers at stages during learning. At each stage learning was halted temporarily and the average time to failure over 10 runs was recorded for each controller. Plots (a) and (b) are typical runs. Plot (a) shows the fuzzy controller consistently outperforming the non-fuzzy controller, while plot (b) shows no single controller having the best performance throughout a run. Occasionally the fuzzy controller would achieve extremely long runs (over an hour in one case), which had to be terminated manually, as occurred in the run illustrated by plot (c). The non-fuzzy controller never achieved such outstanding runs.

x (m):	[-1.5, -0.8, 0.8, 1.5]
x (m/s):	($-\infty$, -0.5, 0.5, ∞)
θ (degrees):	[-12, -6, -1 0, 1 6, 12]
$\dot{\theta}$ (degrees/s):	($-\infty$, -50, 50, ∞)

Table 1: Thresholds for the cart-and-pole boxes

The reasons for the erratic behaviour of the fuzzy controller are unclear at present: future work will involve an investigation into this. However, we postulate the following. The effectiveness of the fuzzy controller depends on the nature of the control actions learnt during training for different regions of the state space. For example, when the pole is near the vertical, which is a point of unstable equilibrium, the actions in neighbouring boxes will often be different, in order to keep the pole oscillating about that point. Then the effect of fuzzifying the box thresholds is to reduce the magnitude of the actual force applied. Since, in this region of the state space, fine control is required, the fuzzy controller should then perform better in this region than the non-fuzzy one. However, elsewhere in the state space more vigorous control is required to keep the system away from failure. Here a reduction in the magnitude of the applied force may be undesirable and so the fuzzy controller performs badly. Further experiments are required to confirm this hypothesis.

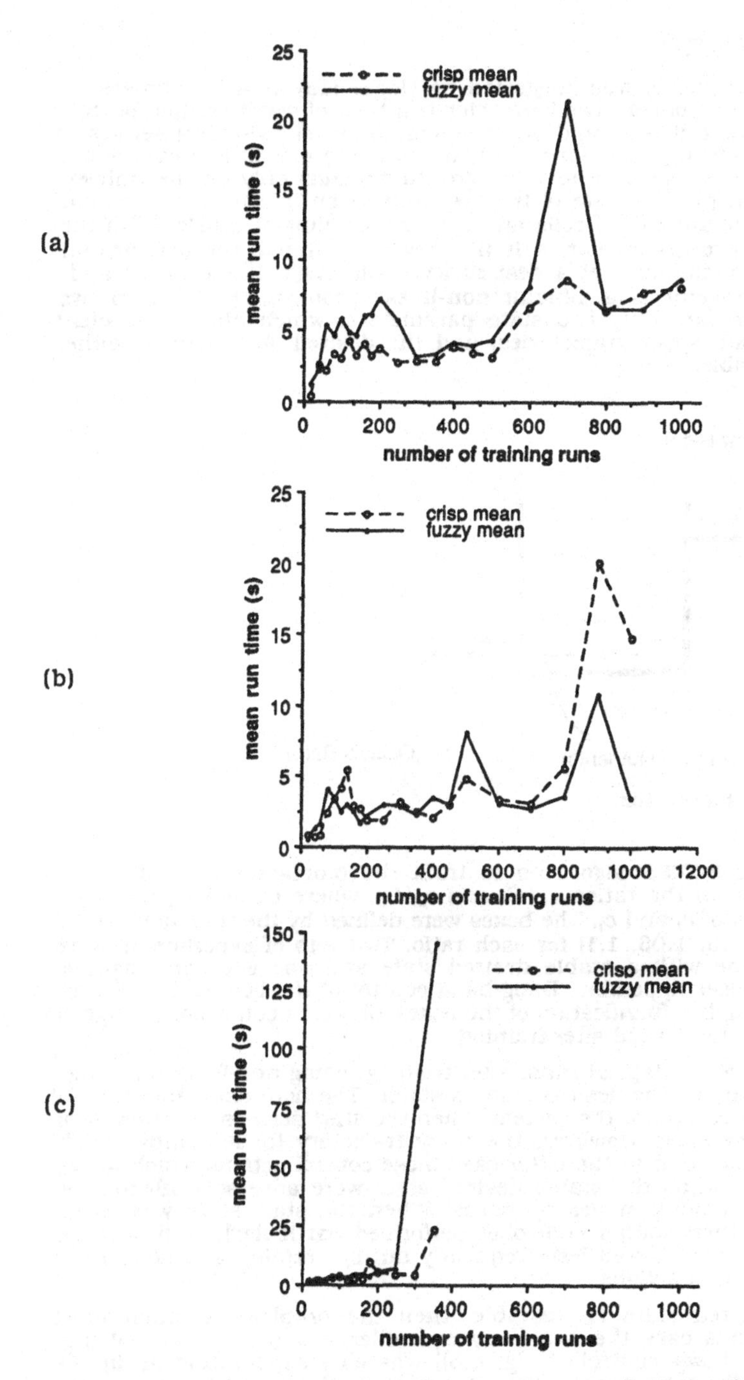

Figure 5: Performance of the cart-and-pole controller over three separate training runs

THE BIOREACTOR

The bioreactor (fig. 6) was suggested by Ungar [19] as a benchmark for adaptive process control. The bioreactor is a tank of water containing cells which consume nutrients and produce more cells. The simplest version of the problem (see appendix for the equations of the model) is a continuous flow stirred tank reactor where cell growth depends only on the nutrient fed to the system. The state of the system is given by the vector (c_1, c_2), where c_1 is the amount of cells and c_2 is the amount of nutrients in the tank. The control parameter, r, is the flow rate through the reactor and the aim is to maintain c_1 at a desired level. Although, like the cart-and-pole, the bioreactor is a difficult non-linear problem, it differs in two respects: there are only two state parameters, which allows the clear display of state space trajectories, and the desired state can be either stable or unstable.

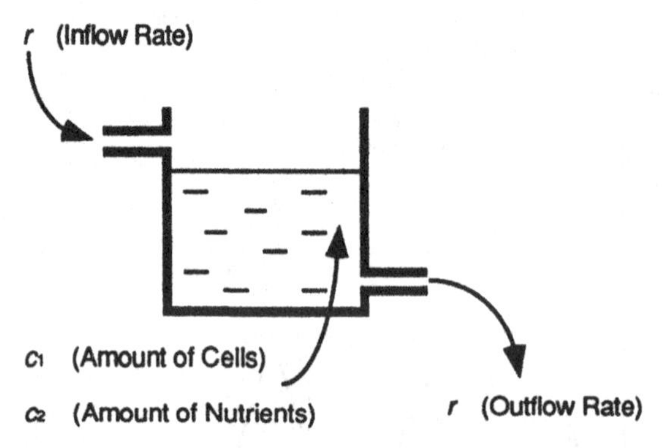

Figure 6: The bioreactor

To use the BOXES algorithm to train the bioreactor controller, we defined boxes on the ratios c_1/C_1 and c_2/C_2, where C_1 and C_2 are the desired values of c_1 and c_2. The boxes were defined by the thresholds (0.9, 0.94, 0.97, 1.03, 1.06, 1.1) for each ratio. Two sets of experiments were performed, one with a stable desired state and one with an unstable desired state (see appendix). Bang-bang control of the flow rate was used during training but fuzzification of the boxes allowed a continuous range of control values to be used after training.

Fig. 7 shows two typical runs, after training, using non-fuzzy and fuzzy boxes respectively. The desired state is stable. The non-fuzzy trajectory (a) displays good control of the system, when the start state is relatively near to the desired state. However, the fuzzy trajectory (b) is comparatively poor. We believe that in the latter case those control actions which forced the system towards the stable desired state were adversely affected by poor control actions in nearby boxes. When the start state was in an outlying box then neither controller performed particularly well because those boxes were entered less frequently during training and so did not learn good control actions.

If the desired state is unstable, then the problem is much more difficult. In this case the non-fuzzy controller still performed well (fig. 8(a)) but the fuzzy controller (fig. 8(b)) was as disappointing as in the stable case. The experiments were also performed with 10% noise on the controller inputs, with no significant change in performance.

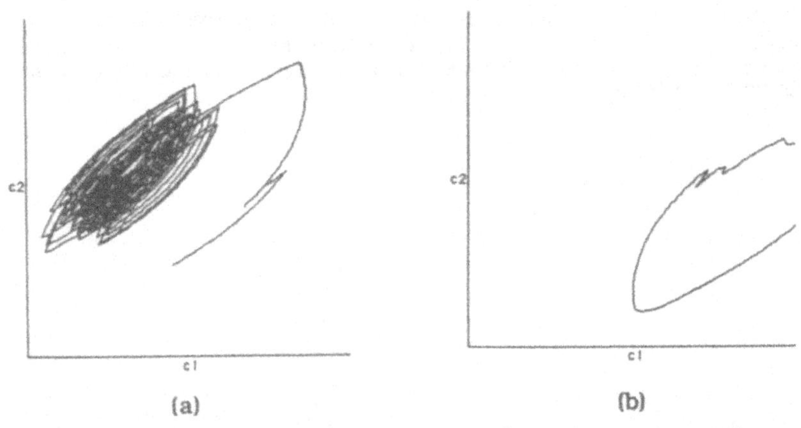

Figure 7: Typical runs, after training, of the bioreactor with a stable desired state. (a) Non-fuzzy control. (b) Fuzzy control.

Figure 8: Typical runs, after training, of the bioreactor with an unstable desired state. (a) Non-fuzzy control. (b) Fuzzy control.

THE TRACTOR-TRAILER

The two problems described so far simply required their controllers to keep the system out of illegal regions of the state space. Each failure was an equal "distance" from the desirable state as any other. The tractor-trailer problem described below is slightly different in that certain failures are "further" (in some sense) from the desired state than others. Therefore, during training, it is desirable to punish the controller more the further the failure from the desired state. How this can be achieved is described below.

The tractor-trailer problem was first used by Nguyen and Widrow [13] to test recurrent backpropagation networks. The equations of the model can be found in the appendix and the scenario is illustrated in fig. 9. It consists of a tractor and trailer which must be reversed to a point O on a loading bay so that the trailer is perpendicular to the loading bay. Control is achieved by steering the wheels of the tractor, which move a fixed distance between each time step. The state parameters are the distance,

r, and angle, θ_r, of the centre of the rear of the trailer from the point O; the angle, θ_s, of the trailer with the horizontal; and the angle, θ_c, of the tractor with the horizontal. The control parameter, u, is the angle of the wheels with the tractor.

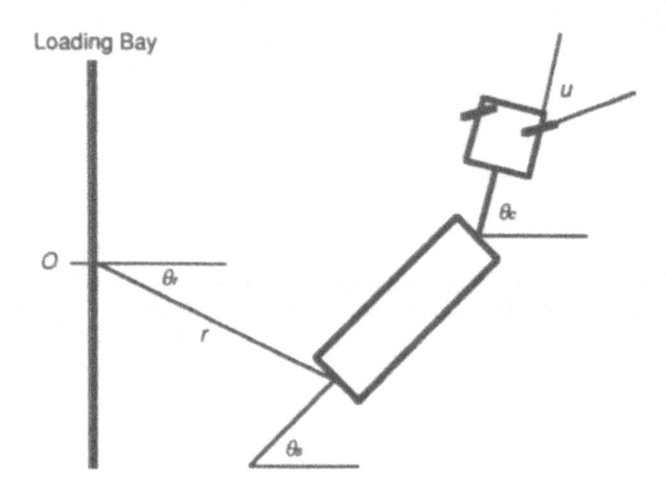

Figure 9: The tractor-trailer

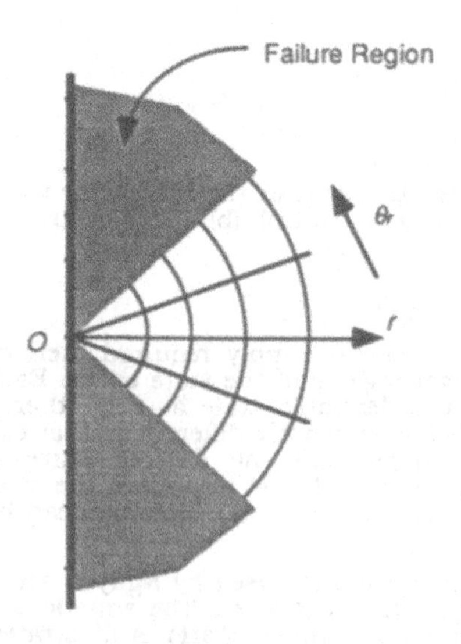

Figure 10: Boxes for r and θ_r

In Nguyen and Widrow's original model failure occurs if $|\theta_s - \theta_c|$ exceeds 90 degrees or if the trailer hits the loading bay away from point *O*. They used cartesian coordinates to measure distance from *O* but we used polar coordinates and set the failure criteria to be whenever $|\theta_r|$ exceeded 30 degrees, $|\theta_s - \theta_c|$ exceeded 90 degrees or *r* exceeded 50 m. Use of polar coordinates is sensible in this case with the BOXES algorithm, since it means that the boxes radiate out from the desired state *O*, as shown in fig. 10. Table 2 shows the thresholds used to define the boxes' boundaries. This structure of the boxes allow a sensible measure of "distance" from *O* to be used, instead of time to failure, as a performance measure in the BOXES algorithm. Fig. 11 shows the function used. The effect of this function is to punish the controller more for failures at high values of $|\theta_s|$ and reward it less for failures at large values of *r*. This persuades the controller to try to steer the tractor-trailer towards *O* along a trajectory perpendicular to the loading bay. During training bang-bang control was used so that steering was only either hard left or hard right, i.e. $|u| = 70$ degrees. Fuzzifying the boxes after training allowed finer steering of the tractor.

r (m):	[0, 15, 25, 35, 50]
θ_r (degrees):	[-30, -15, 0, 15, 30]
θ_s (degrees):	[-180, -15, 0, 15, 180)
$(\theta_s - \theta_c)$ (degrees):	[-90, -45, 0, 45, 90]

Table 2: Thresholds for the tractor-trailer boxes

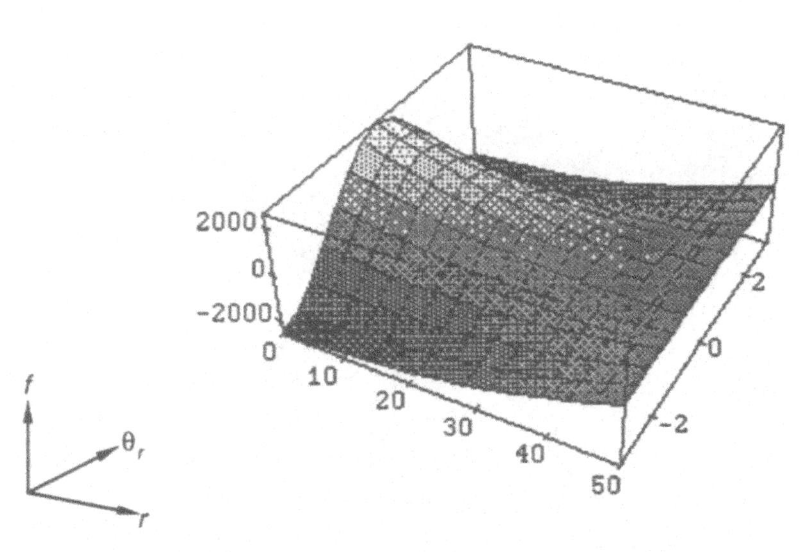

Fig. 11: The performance function $f(r, \theta_s) = (50 - r)(50\cos\theta - r)$

The trajectories of a fuzzy and non-fuzzy run from the same start state are shown in fig. 12. They show that the fine steering provided by the fuzzy controller has produced a smoother trajectory than the non-fuzzy one. In certain regions of the state space this is an undesirable effect. For example, if the tractor-trailer is in a position almost parallel to the loading bay then it requires "full-lock" on the steering in order to turn to meet the bay at the appropriate position. Fuzzifying the boxes prevents full-lock being applied.

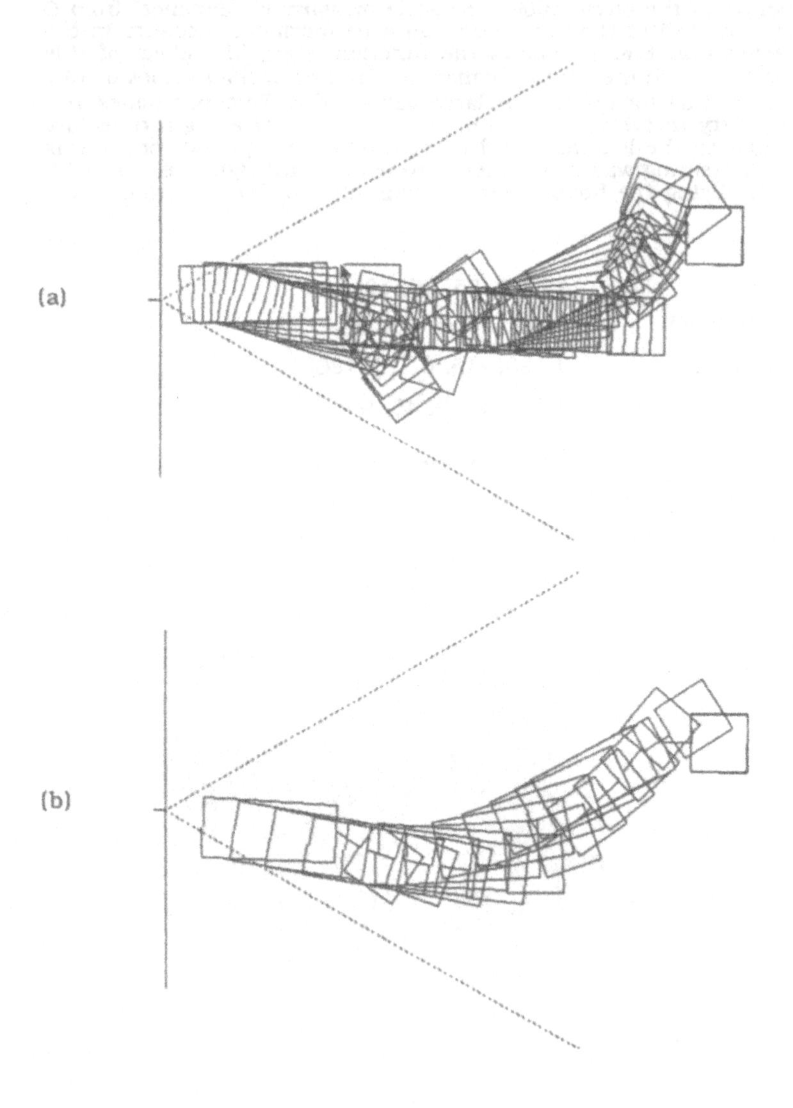

Figure 12: Trajectories of the tractor-trailer. (a) Non-Fuzzy control. (b) Fuzzy control.

CONCLUSIONS

The problems described in this paper are benchmarks of difficult non-linear control problems. Our experiments compared non-fuzzy controllers, whose actions had been learnt using the BOXES algorithm, with their fuzzy counterparts, with the aim of testing claims that fuzzifying threshold values in the state space can improve performance of the controller in this type of problem.

Initial results show that fuzzifying state space thresholds does not automatically lead to improved performance. Results from the cart-and-pole and truck-and-trailer problems suggest that certain regions of the state space may benefit from fuzzy thresholds while others may not. The bioreactor problem shows a consistent drop in performance when fuzzy thresholds are used.

Future work involves performing more comprehensive experiments to test the relative merits of the fuzzy and non-fuzzy controllers described in this paper. In particular automatic setting of the boxes' thresholds will be investigated and the use of fuzzy and non-fuzzy techniques in different regions of the state space will be explored. Although our experiments used computer simulations of real-world problems, the results are still useful as an initial comparison of techniques. To test for robustness, mechanical apparatus must be used: this is a long term goal.

REFERENCES

[1] C. W. Anderson (1989): Learning to Control an Inverted Pendulum Using Neural Networks, *IEEE Control Systems Magazine*, April 1989, pp 31 - 37.

[2] C. W. Anderson & W. T. Miller (1990): Challenging Control Problems, *Neural Networks for Control* (eds. W. T. Miller, R. S. Sutton & P. J. Werbos), MIT Press.

[3] A. G. Barto (1990): Connectionist Learning for Control, *Neural Networks for Control* (eds. W. T. Miller, R. S. Sutton & P. J. Werbos), MIT Press.

[4] A. G. Barto, R. S. Sutton & C. W. Anderson (1983): Neuron-like adaptive elements that can solve difficult learning control problems, *IEEE Transactions on Systems, Man, and Cybernetics*, vol. SMC-13, no. 5, pp 834 - 846.

[5] J. A. Bernard (1988): Use of a Rule-Based System for Process Control, *IEEE Control Systems Magazine*, October 1988, pp 3 - 13.

[6] E. Grant & B. Zhang (1989): A Neural-Net Approach to Supervised Learning of Pole Balancing, *Proceedings IEEE International Symposium on Intelligent Control 1989*, pp 123 - 129.

[7] S. A. Harp (1991): Genetic Synthesis of Neural Network Architecture, *Handbook of Genetic Algorithms* (ed. L. Davis), Van Nostrand Reinhold.

[8] C. J. Harris & C. J. Moore (1989): Intelligent identification and control for autonomous guided vehicles using adaptive fuzzy-based algorithms, *Engineering Applications of AI*, vol. 2, December 1989, pp 267 - 285.

[9] R. Hecht-Nielsen (1990): *Neurocomputing*, Addison Wesley, ch. 6.1.

[10] J. H. Holland (1975): *Adaptation in Natural and Artificial Systems*, The University of Michigan Press.

[11] E. H. Mamdani & S. Assilian (1975): An experiment in linguistic synthesis with a fuzzy logic controller, *International Journal of Man-Machine Studies*, vol. 7, no. 1, pp 1 - 13.

[12] D. Michie & R. A. Chambers (1968): BOXES: An experiment in Adaptive Control, *Machine Intelligence 2*, Oliver & Boyd, pp 137 - 152.

[13] D. H. Nguyen & B. Widrow (1990): Neural Networks for Self-Learning Control Systems, *IEEE Control Systems Magazine*, April 1990, pp 19 - 23.

[14] P. D. Picton, N. J. Hallam & N. Woodcock (1991): Are Three Layers Enough? A Review of Kolmogorov's Existence Theorem, *Technical Report for the Open University*.

[15] D. Psaltis, A. Sideris & A. A. Yamamura (1988): A Multilayered Neural Network Controller, *IEEE Control Systems Magazine*, April 1988, pp 17 - 21.

[16] D. W. Russell (1990): The Trolley and Pole Revisited: Further Studies in AI Control of a Mechanically Unstable System, *Applications of Artificial Intelligence in Engineering V, Vol. 2: Manufacture and Planning*, Computational Mechanics Publications / Springer-Verlag, pp 307 - 322.

[17] R. S. Sutton (1988): Learning to Predict by the Methods of Temporal Differences, *Machine Learning*, vol. 3, pp 9 - 44.

[18] T. Takagi & M. Sugeno (1985): Fuzzy Identification of Systems and Its Applications to Modeling and Control, *IEEE Transactions on Systems, Man, and Cybernetics*, vol. SMC-15, no. 1, pp 116 - 132.

[19] L. H. Ungar (1990): A Bioreactor Benchmark for Adaptive Network-based Process Control, *Neural Networks for Control* (eds. W. T. Miller, R. S. Sutton & P. J. Werbos), MIT Press.

[20] B. Widrow, N. K. Gupta & S. Maitra (1973): Punish/Reward: Learning with a Critic in Adaptive Threshold Systems, *IEEE Transactions on Systems, Man, and Cybernetics*, vol. SMC-3, no. 5, pp 455 - 465.

[21] A. P. Wieland (1990): Evolving Controls for Unstable Systems, *Connectionist Models: Proceedings 1990 Summer School*, (eds. D. S. Touretsky, J. L. Elman, T. J. Sejnowski, G. E. Hinton), Morgan Kaufmann.

[22] R. J. Williams (1988): On the use of backpropagation in associative reinforcement learning, *Proceedings IEEE International Conference on Neural Networks, San Diego*, pp 263 - 270.

[23] L. A. Zadeh (1965): Fuzzy Sets, *Information and Control*, vol. 8, pp 338 - 353.

[24] L. A. Zadeh (1975): Fuzzy Logic and Approximate Reasoning, *Synthese*, 30, pp 407 - 428.

APPENDIX

The mathematical models for the three benchmark problems described in this paper were adapted from those described by Anderson and Miller [2]. They are reproduced below.

The Cart-and-Pole

State

θ	Angle of pole from upright (degrees)
$\dot{\theta}$	Angular velocity of pole (degrees/s)
x	Horizontal position of cart's centre (m)
$\dot{x}$	Velocity of cart (m/s)

Control

F Force on cart (N)

Constraints

-12 degrees $< \theta <$ 12 degrees; -1.5 m $< x <$ 1.5 m; -10 N $< F <$ 10 N

Initial Conditions

$\theta[0]$	A random variable from uniform distribution from -1 degree to 1 degree
$\dot{\theta}[0]$	A random variable from uniform distribution from -30 degrees/s to 30 degrees/s
$x[0]$	A random variable from uniform distribution from -0.04 m to 0.04 m
$\dot{x}[0]$	A random variable from uniform distribution from -0.25 m/s to 0.25 m/s

Equations of Motion

$$\theta[t+1] = \theta[t] + \Delta\dot{\theta}[t]$$

$$\dot{\theta}[t+1] = \dot{\theta}[t] + \Delta\frac{Mg\sin\theta[t] - \cos\theta[t]\,(F[t] + ml(\dot{\theta}[t]\,\pi/180)^2\sin\theta[t])}{(4/3)Ml - ml\cos^2\theta[t]}$$

$$x[t+1] = x[t] + \Delta\dot{x}[t]$$

$$\dot{x}[t+1] = \dot{x}[t] + \Delta\frac{F[t] + ml((\dot{\theta}[t]\,\pi/180)^2\sin\theta[t] - \ddot{\theta}[t]\,\pi/180\cos\theta[t])}{M}$$

Parameters

g	9.8	Acceleration due to gravity
M	1.1	Combined mass of cart and pole
m	0.1	Mass of pole
l	0.5	Distance from pivot to pole's centre of mass
Δ	0.02	Sampling interval

Control Interval

Δ

Controller Input

$\theta[t]$, $\theta[t]$, $x[t]$, $x[t]$

Controller Output

$F[t]$

The Bioreactor

State

c_1 Amount of cells

c_2 Amount of nutrients

Control

r Flow rate

Constraints

$0 \leq r \leq 2; 0 \leq c_1, c_2 \leq 1$

Initial Conditions

$c_1[0]$ A random variable from uniform distribution on the interval $(0.9C_1, 1.1C_1)$

$c_2[0]$ A random variable from uniform distribution on the interval $(0.9C_2, 1.1C_2)$

$r[0]$ A random variable from uniform distribution on the interval $(0.9R, 1.1R)$

C_1, C_2 and R are desired values of c_1; c_2 and r respectively. For a stable state, $C_1 = 0.1207$, $C_2 = 0.8801$ and $R = 0.75$. For an unstable state, $C_1 = 0.2107$, $C_2 = 0.7226$ and $R = 1.25$.

Equations of Motion

(These are based on Ungar's equations [19], which differ from Anderson and Miller.)

$$c_1[t+1] = c_1[t] + \Delta(-c_1[t]r[t] + c_1[t]\,(1 - c_2[t])\,\exp(c_2[t]/\gamma))$$

$$c_2[t+1] = c_2[t] - \Delta c_2[t]r[t] + \Delta c_1[t]\,(1 - c_2[t])\,\exp(c_2[t]/\gamma)\ \frac{(1+\beta)}{(1+\beta - c_2[t])}$$

Parameters

β 0.02 Growth rate parameter

γ 0.48 Nutrient inhibition parameter

Δ 0.01 Sampling interval

Control Interval

50Δ

Controller Input

$c_1[t]$, $c_2[t]$, for $t = 0, 50, 100, \ldots$

Controller Output

$r[t]$ for $t = 0, 50, 100, \ldots$

$r[t] = r[t-1]$ for all other values of t

The Tractor-Trailer

(Note: in these equations the position of the centre of the rear of the trailer is in *cartesian* coordinates.)

State

x, y Cartesian coordinates of centre of rear of trailer (m)

θ_s Angle of trailer, measured from positive x with anticlockwise being positive (degrees)

θ_c Angle of cab, measured from positive x with anticlockwise being positive (degrees)

Control

u Steering angle of front wheels relative to cab orientation, with anticlockwise being positive (degrees)

Constraints

$0 < x$; $|\theta_s - \theta_c| \leq 90$ degrees; $|u| \leq 70$ degrees

Initial Conditions

$x[0]$ A random variable from uniform distribution from 0 m to 100 m

$y[0]$ A random variable from uniform distribution from -50 m to 50 m

$\theta_s[0]$ A random variable from uniform distribution from -90 degrees to 90 degrees

$\theta_c[0]$ A random variable from uniform distribution from $\theta_s[0]$ - 10 degrees to $\theta_s[0]$ + 10 degrees

Equations of Motion

$$A = p \cos u[t]$$

$$B = A \cos(\theta_c[t] - \theta_s[t])$$

$$C = A \sin(\theta_c[t] - \theta_s[t])$$

$$x[t+1] = x[t] - B \cos \theta_s[t]$$

$$y[t+1] = y[t] - B \sin \theta_s[t]$$

$$\theta_c[t+1] = \arctan \frac{(d_c \sin \theta_c[t] - p \cos \theta_c[t] \sin u[t])}{(d_c \cos \theta_c[t] + p \sin \theta_c[t] \sin u[t])}$$

$$\theta_s[t+1] = \arctan \frac{(d_s \sin \theta_s[t] - C \cos \theta_s[t])}{(d_s \cos \theta_s[t] + C \sin \theta_s[t])}$$

where arctan is from -180 degrees to 180 degrees

Parameters

p 0.2 Distance front tyres move in one time step (m)

d_c 6.0 Length of cab, from pivot to front axle (m)

d_s 14.0 Length of trailer, from pivot to ftrailer rear (m)

Controller Input

$x[t]$, $y[t]$, $\theta_s[t]$, $\theta_c[t]$

Controller Output

$u[t]$

Optimal Solutions for Deep and Shallow Engineering Expert Systems

K. Preiss (*)(**), O. Shai (*)

(*) Department of Mechanical Engineering,
Ben Gurion University, Beer Sheva, Israel
(**) The Sir Leon Bagrit Professor of CADCAM
E-mail: preiss@bengus.bgu.ac.il

(**) also, the Pearlstone Center for Aeronautical Engineering Studies.

ABSTRACT

The paper shows how application of modern graph and network theory to problems of artificial intelligence in engineering can lead to solutions which use fundamental or deep knowledge to provide optimal or near-optimal solutions to significant problems in reasonable computing time. The paper shows implementation in the Prolog language.

INTRODUCTION

From the beginning, work in artificial intelligence represented a problem as a graph, and searched for a solution within that graph. In traversing the graph to reach a goal state from the intial state, it was realised that the number of intermediate states in real problems is large, so that special techniques had to be used to search for a goal state from the initial state in reasonable time. Nilsson's seminal book [1], which summed up the state of the artificial intelligence art in the 1970's provides a good summary of that approach. The techniques used in much of today's work on artificial intelligence and expert systems is based on those methods. As an example, we see in Nilsson's book [1] emphasis placed on heuristics and the A* algorithm. The word "heuristic" is often used in the general sense as meaning a helpful rule of thumb. However, strictly speaking, a heuristic rule is a rule which satisfies a mathematical condition, for instance the A* condition. If a heuristic rule conforms to the criterion of A*, then the solution of a problem which would otherwise take much search time, can be completed in reasonable computation time, and the solution is guaranteed optimal in the sense that the number of moves (transformations, or steps of search) is minimal.

It should be noted that it is rare to find a practical problem for which a heuristic rule can be found which satisfies the A* criterion, hence guaranteeing optimality, and which also allows the computation to complete in a reasonable time, (Preiss and Shai [2]). As a result, heuristic rules usually do not guarantee optimality. The solution may be turn out to be optimal, but that is not guaranteed by the heuristic rule.

A common approach in current work in expert systems is to ask an expert what the facts and rules are in her/his field, then to write a list of these facts and rules into a program, and have the computer program search for a solution from those. The search can be carried out from initial state to goal state, or in the reverse direction, or in both directions meeting in the middle, and the search can be guided by all kinds of rules. The system is then used to find solutions for many test problems, and will give some answers which the expert finds to be wrong. If a wrong solution is found the expert adds more rules and facts, until the program is deemed to be satisfactory. A problem with this approach is that because the facts, transformation rules, and heuristics are input by one or more human experts, by the time one has introduced thousands of such items in the program, the system as a whole exhibits "brittleness" and logical inconsistency. In a "brittle" system, a combination of facts and rules can generate an internal state, from which the computer can navigate no further, nor can backtrack. The only way out is to have an expert edit the facts and rules, changing, retracting or adding facts and rules. A logically inconsistent system arises because different search techniques and different heuristic rules will use the facts and transformation rules in a different order, and then in one case a particular state may be generated, but for another heuristic rule applied to the same problem, the negation of that same state can be generated. These are serious problems, at the foundation of the methodology of expert systems and artificial intelligence.

One can therefore understand the reasons for Parnas' well-known and severe criticism [3] of artificial intelligence work. "I find the approaches taken in AI to be dangerous and much of the work misleading. The rules that one obtains by studying people turn out to be inconsistent, incomplete, and inaccurate. Heuristic programs are developed by a trial and error process in which a new rule is added whenever one finds a case not handled by the old rules. This approach usually yields a program whose behaviour is poorly understood and hard to predict. AI researchers accept this evolutionary approach to programming as normal and proper. I trust such programs even less than I trust unstructured conventional programs. One never knows when the program will fail.

On occasion I have had to examine closely the claims of a worker in AI I have always been disappointed. On close examination the heuristics turned out to handle a small number of obvious cases but failed to work in general. The author was able to demonstrate spectacular behavior on the cases that the program handled correctly. He marked other cases as extensions for future researchers. In fact, the techniques being used often do not generalize and the improved program never appears".

Rules and facts generated by a human lead to systems called "shallow". The system is not subject to overall supervision of a proven mathematical theory to control the consistency. A "deep" system is a system based on a proven, consistent, mathematical formulation, so that conclusions and solutions to problems have provable properties of correctness, consistency and hopefully, optimality. Results of research in the algorithms of graph theory over the last decade or so, allow us to go further than before in the representation of an artificial intelligence problem as a graph, using proven methods of graph theory to give a deep solution.

An approach often adopted to the problem of applying deep knowledge in an artificial intelligence problem is to postulate a state-space graph, where all states can be generated, then to develop a set of facts and rules derived from fundemental considerations. The deep knowledge is then encoded in those facts and rules. Our approach is different. We encode the knowledge in the graph representation, then solve the graph using known theorems from graph theory. The deep knowledge is then encoded into the graph representation itself.

This paper shows how representation of an engineering

problem for solution by methods of artificial intelligence is facilitated by graph theory algorithms, especially for a type of graph called a "network flow graph". These algorithms give solutions which are optimal or close to optimal, in reasonable computing time. In order to take advantage of them, the problem has to be represented as a suitable graph. Once the correct graph representation is established, finding the solution reduces to the problem of applying a known algorithm to the graph. The key to the approach is in defining the graph representation. Once that is done, the solution will have proven mathematical properties of consistency and optimality, because the solution algorithm has such proven properties. Such an approach hence can be called "deep". It is based on the proven mathematical properties of a graph, not on a collection of man-made rules and facts.

GRAPH THEORY AND NETWORKS FOR ENGINEERING PROBLEMS

Terminology

A graph is composed of an ordered pair of two sets (V,E), where V is the vertex set with n vertices and E is the edge set of the graph with m edges. The fact that each edge joins a pair of vertices is written $E \subseteq V \times V$, and there is function Γ, such that for all $e \in E$, $\Gamma(e) = (v_i, v_j)$, where v_i and v_j are end points of edge e.

A path is a sequence of edges $P = (e_1, e_2, , , e_n)$ such that each pair of successive edges has a common end point, where e_1 is the start edge and e_n is the end edge of the path.

A graph is said to be connected, if there exists a path from every vertex to every other vertex. A circuit is a path for with the same start and end vertex; it ends at the start vertex.

Complexity and algorithms in graphs

Complexity of an algorithm is the function T(N) which gives the running time and/or storage space requirement of the algorithm in terms of the size N of the input data.

In order to estimate the running time of an algorithm, one usually analyses the limiting worst case of the performance of the algorithm by the big O notation, defined as follows: The function $T(N)$ is said to be order of $O(f(N))$ if there exists a constant C and an integer number K such that $T(N) \leq C * f(N)$ for all $N > K$.

It should be noted that for most problems with a small number of input data elements N, the running time will be reasonably small, whatever the solution method. However, for practical engineering problems N is often in the range of many hundreds or thousands. In that case, there is a fundemental difference if T(N) is an exponential or factorial function, which will give values so high as to be impractical, or polynomial or logarithmic, which give practical solution times.

The field of artificial intelligence tackled problems of high complexity, known as NP complete. The heuristics and other methods, mentioned above, are used to provide solutions in practical running time to those problems. The theme of this paper is that many engineering problems (but of course, not all) can be tractably solved using new algorithms in graph theory, giving solutions preferable to classical AI search.

Network Flow Graphs

A network, (Chen [4]) is a finite directed graph $G(V,E)$ with two vertices s and t, s the source vertex which has no incoming edges, and t the sink vertex, a vertex which has no outgoing edges. With every edge is associated a non-negative number, called the capacity of the edge. The capacity of edge (i,j) is denoted by $c(i,j)$. A cut $(V1,V2)$ of a flow network is a partition of V into V1 and V2, while $V1=V-V2$ is such that $s \in V1$ and $t \in V2$.

A flow in G is a real valued function $f: V \times V \to R$ which satisfies the following properties:

Capacity constraint: For all $(i,j) \in E$, $f(i,j) \leq c(i,j)$

Flow conservation: For all $v \in V - \{s,t\}$ and for all $u \in V$, the conservation condition is $\Sigma f(v,u)=0$

The value of a flow f is defined as:
for all $v \in V$, $|f| = \Sigma f(s,v) = \Sigma f(v,t)$

From these properties another useful and important property which can be deduced is:
The maximum flow is equal to the minimum cut in the network.

New algorithms for finding the maximum flow in

networks with reasonable complexity have been developed over the last decade (Goldberg and Tarjan [5]). These algorithms enable us to find optimal solutions for engineering problems, when the problem can be represented as a problem in network flow.

The next section will be devoted to examples of problems which can usefully be represented as graphs. Two examples are represented as maximum flow problems (and hence minimal cut problems) in a flow graph; this is the same as the problem of matching in bipartite graphs, and two are problems of finding the shortest path in a graph.

It is helpful to note that there are many different engineering problems which have the same representation in graph theory. The same graph algorithm can then be used and analogous conclusions deduced automatically when solving these problems. The two examples below, the secondary operation planning problem and the problem of production planning, have the same graph representation so the same algorithm can solve both problems.

REPRESENTATION OF PROBLEMS

After representing a given problem as a problem in graph theory much knowledge of the solution is encoded in the computer, without the user having to write explicit rules to find the solution, because the algorithm used to solve the problem on the graph uses known and proven properties. The question then to face is how to represent the problem; currently this requires knowledge of graph theory and algorithms, and creative thinking, but there is hope that in the future this process will be automated in the computer. Building the graph will be demonstrated in the following problems.

Matching and maximal flow problems

Example 1. Planning Secondary Operations

In this problem four main operations are to be carried out on a product, together with four secondary operations. A secondary operation is carried out between main operations such that between two successive main operations there will be one and only one secondary operation. Every secondary operation can be done before or after a main operation, arbitrarily, as shown in Table 1.

secondary operation	can be done after or before the following main operations
s1	m1,m2,m3,m4
s2	m1,m2,m3,m4
s3	m1,m4
s4	m1,m4

Table 1 Information for planning the secondary operations

According to the constraints in Table 1, the computer has to find a solution (and if there is no solution to write 'fail'), such that main operations will be done in the order m1, m2, m3, m4, m1 with the correct secondary operations between them. If one uses one of the conventional methods of artificial intelligence such as depth first search or breadth first search, without using deep algorithmic understanding of the problem, the complexity of the solution will be O(n!), where n is the number of the secondary operations. The time for solution will go up as the order of n!, which increases rapidly to an impractically long time.

However, if one thinks about the problem by using the terminology of graph theory, the problem becomes easily tractable. Restating the problem, the solution is a traverse of the series
m1, X, m2, Y, m3, Z, m4, W, m1
where the big letters represent the wanted secondary operations. In graph theory this is a requirement to find a circuit where the vertices represent the operations, and the edges (mi, R) represent the fact that secondary operation R will be done before or after main operation i. This circuit must have an even number of vertices, so the algorithm can automatically infer the fact that it can be represented as bipartite graph, then use that fact for the solution. Because we want to match a secondary operation to every two successive main operations, the problem becomes that of matching vertices in a bipartite graph, where the vertices in one side are the successive main operations and the vertices in the other side are the secondary vertices. The edge ({ mi, mj }, sk) exists in the graph if secondary operation sk can be done before or after main operations mi and mj, and can hence come between them as shown in Fig. 1.

hence come between them as shown in Fig. 1.

When solving this problem by classical search methods, to ensure optimality one has to use depth first or breadth first search, of complexity $O(n!)$ and hence an impossibly long solution time; alternatively, the use of heuristics leaves one unsure about the optimality of the solution. On the other hand, having stated the problem as finding a matching in a graph, much knowledge is known about the solution.

Once the problem is identified by graph theory as a problem of matching in bipartite graph, we know from graph theory that this matching problem can be solved as a network flow algorithm (Even [6]). To do this requires definition of a dummy source and sink, and construction of extra edges. The new graph is now not a one to one match of the physical problem, but has edges added which do not appear in the original problem statement. Instead of using classical search to satisfy the constraints we use the known algorithm of maximum network flow which has a complexity of $O(n\, m \log(n^2/m))$, much better than a complexity of $O(n!)$.

We build a network G', such that $V' = V \cup \{s,t\}$, $V = V1 \cup V2$, $E' = \{(s,v) \mid v \in V1\} \cup E \cup \{(v,t) \mid v \in V2\}$ and for all $v \in V1$ $c(s,v) = 1$ and for all $u \in V2$ $c(u,t) = 1$, as shown in Fig. 2.

Example 2. Production Planning

In this problem, there are 4 machines, and 3 products which have to be manufactured by the machines. Table 2 is a list of information from which the computer has to decide the maximum amount of products which can be produced, and by which machines to do the production, given the following constraints; the manufacturing capability of each machine; the types of products which each machine can manufacture; and the maximum amount of each product that it is worthwhile to manufacture.

Type of machine	Machine's production capability	Types of products
m1	c(m1)	p1,p2
m2	c(m2)	p3
m3	c(m3)	p1,p2
m4	c(m4)	p2,p3

Product type	Quantity required
p1	c(p1)
p2	c(p2)
p3	c(p3)

Table 2 Information for the Production Planning example

This is a problem of matching, but here the edges have capacities according to the constraints. We build a network G, as shown in Fig. 3, such that V1 = {s, m1, m2, m3, m4}, V2 = {p1, p2, p3, t}, and the edges have capacity according to the information of Table 2. Using the network flow algorithm, an optimal solution is obtained.

There are more types of graphs which can represent groups of engineering problems from many various subjects.

Shortest Path Problem in Graphs

The following two problems are solved as shortest path problem in a graph, even though they are quite different engineering problems.

Example 3. Optimal Process Planning

Given two production lines for two products, with machines in common. The common machines can work on one product at a time, and production is limited by the capacity of those machines.

Table 3 shows the time that is needed for each station in each production line and also the constraints deriving from the fact that the machine common to both lines can be devoted to work on one production line only, at any given moment.

Station	1	2	3	4	5
production line 1	2	3 A	2	2 B	4 C
production line 2	2	2 A	3 B	2	4 C

Table 3 Time for each station in each production line. The letters A, B, C indicate machines A, B, C, used in both production lines. All other stations use different machines.

Fig.4 shows a graph for that problem, in which the optimal solution minimizes the waiting time, in other words a minimal path in the graph.

Example 4. Deciding a Policy of Machine Investment

Given data, for each pair of years i and j the value c(i,j) which is the sum of buying a new machine at year i, plus the cost of maintenace during the years i to j, minus the salvage value of the machine in year j.

Now, for a machine which was bought in year s, and will be needed until year k, one wants to decide at any intermediate year whether it is more economic to keep the machine or to buy a new one. The desired solution is the shortest path from year s to year k in the graph as shown in Fig. 5. (Phillips and Garcia-Diaz [7]).

IMPLEMENTATION OF THE TECHNIQUES IN THE COMPUTER

A computer programming language suitable to implement these techniques is Prolog. This is because Prolog is both a declarative and a procedural language as can be seen by different interpretations of a clause:

```
Goal :- Subgoal_1, Subgoal_2, . . .  Subgoal_n
```

Declarative Interpretation: Goal is true if Subgoal_1 is true and Subgoal_2 is true and . . . Subgoal_n is true.

Procedural Interpretation: To prove Goal prove

Subgoal_1 and then prove subgoal_2 and then . . Subgoal_i, and at last prove Subgoal_n.

There are two types of operations which can be done by the computer:

Type 1: Declarative approach: Concluding information from the representation and from data which is typical to the problem.

Type 2: Procedural Approach: Invoking an algorithm which solves the known problem from graph theory.

Type 1: This consists of inference rules which enable the computer to represent the problem as a problem in graph theory, and than to decide how to solve the graph theory problem.

For example, a rule of inference would be:

Graph G is bipartite <u>if</u> all the circuits C in the graph have an even number of edges.

This rule written in Prolog, is as follows:

```
bipartite(G) :-        /* Instantiate variable G as a
                                bipartite graph */
  find_all(G,C), /* IF in variable C will be data of
                               all the circuits */
  all_circuit_even(C).  /* AND all the circuits in C
                               have even edges. */
```

all_circuit_even(C) is evaluated as follows.

```
all_circuit_even([]).     /* termination condition */
all_even_circuit([H|T]) :- /* check the list items */
  even_circuit(H), all_even_circuit(T).
```

Type 2: This consists of graph theory algorithms, with low complexity. For example:

```
match(G,Match) :-  /* In order that the matching of
                        graph G will be Match do */
  change(G,Network),   /* First: change the graph G
                       to a network 'Network' and */
  algorithm_goldberg(Network,Mathc). /* Second use
     the algorithm written by Goldberg which finds
                  the maximum flow in a network. */
```

OPTIMALITY AND COMPLEXITY

An important consideration when solving engineering problems by computer is to obtain the optimal solution, or if the problem is NP-complete to get a solution which is near the optimal solution, and to achieve that with low complexity. When the problem is represented as a graph, there is a possibility of using algorithms, many published in the last decade, developed after considerable effort in graph theory, algorithms and computer science.

For instance, in example 1, Planning Secondary Operations if one does not use deep knowledge, but encodes an algorithm intended to solve the problem by conventional methods of search, the complexity is $O(n!)$, which is intractable. By representing the problem as problem in graph theory and using network flow algorithms, the optimal or near optimal solution is found in complexity of $O(m\, n \log(n^2/m))$, which is a polynomial time algorithm.

Occasionally an opportunity presents itself to use knowledge from graph theory to find an easy solution to an engineering problem. For instance, suppose that one wants to find the maximum matching in example 1 and the complexity of running time of the algorithm is not important. Here one can use the Konig theorem (Bondy and Murty [8]), which states that in a bipartite graph, the size of maximum matching is equal the size of the minimum vertex cover. VC is a vertex cover if for each edge in the graph at least one of its end points belongs to the vertex cover set. Because of that, when the program finds a matching set M and a vertex cover set VC such that $|VC| = |M|$, the program terminates with the knowledge that optimal solution has been found. This can be written as follows:

```
max_matching(G, M):- /* M is the maximum matching in
                                               graph G */
  match(G, M),   /* IF there exists some matching M */
  v_cover(G, VC), /* AND there exists a
                                      vertex cover VC */
  |M|=| VC|.      /* AND the size of the matching M
       is equal to the size of the vertex cover VC. */
```

CONCLUSIONS

Much effort is devoted to writing expert systems which use knowledge of experts, and to building a program which will trace his or her thinking. The authors suggest that a fruitful line of effort in the future will be to represent engineering problems in the mathematical language of graph theory, especially network flow theory. By doing so, deep or fundemental knowledge is implicitly encoded in the graph representation itself. Proven theorems of graph theory can then be applied, and there is no need to try to figure out a collection of facts and rules which encode the deep knowledge for the problem being considered.

It is important to note that when approaching the problems in this way, the same graph representation can be found to apply to very different engineering problems. This shows up interesting analogies which can be helpful in solving a problem, and in understanding it.

Encoding the deep knowledge into the graph representation itself helps in two fundemental difficulties with current applications of AI techniques in engineering. The first is that the rules and facts as put into the computer over a period of time, can lead to a logical contradiction. In other words, a solution can be found to the problem, for which some rules or facts are neglected, and if these had been used, then a different solution would have been found. The other problem is that the heuristics used to speed up the solution can lead to a situation where a solution, although it exists, is not found. For the first problem, when the representation of the problem is in terms of graph theory, and the inference rules which are used to implement knowledge are based on proven theorems of graph theory, a contradiction will not occur. Regarding the second problem, using algorithms (many of them new), with low complexity of running time, there will be less need to use heuristic techniques; deep knowledge will then prevent situations which were mentioned above.

Implementation of the results is conveniently done in Prolog, using it both as a procedural and declarative language, rather than a procedural language only as is customary.

ACKNOWLEDGEMENT

This research is partially supported by the Paul Ivanier Center for Robotics and Production Management, Ben Gurion University of the Negev.

REFERENCES

1. Nilsson N. J. "Problem-Solving Methods in Artificial Intelligence", McGraw Hill, 1971.

2. Preiss, K. and Shai O. "Process Planning by Logic Programming", Robotics and Computer-Integrated Manufacturing, Vol. 5, No. 1, pp 1 - 10, 1989

3. Parnas D. L. "Software Aspects of Strategic Defense Studies", Communications of the ACM, Vol 28, No 12, pp 1326 - 1335, 1985.

4. Chen W. K. "Theory of Nets: Flows in Networks", John Wiley & Sons, 1990.

5. Goldberg A.V. and Tarjan R.E, "A New Approach to the Maximum Flow Problem", Journal of the Association for Computing Machinery, Vol. 35, No. 4, pp 921 - 940, October 1988.

6. Even S. "Graph Algorithms", Computer Science Press, 1979.

7. Phillips, D.T. and Garcia-Diaz, A. "Fundamentals of Network Analysis", Prentice Hall, 1981.

8. Bondy, J. A. and Murty, U. S. R., "Graph Theory with Applications", Macmillan Press, 1976.

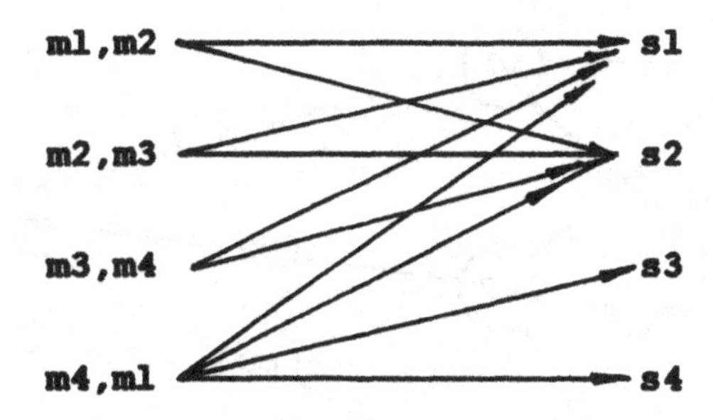

Fig 1 GRAPH REPRESENTING THE RELATIONS BETWEEN MAIN AND SECONDARY OPERATIONS

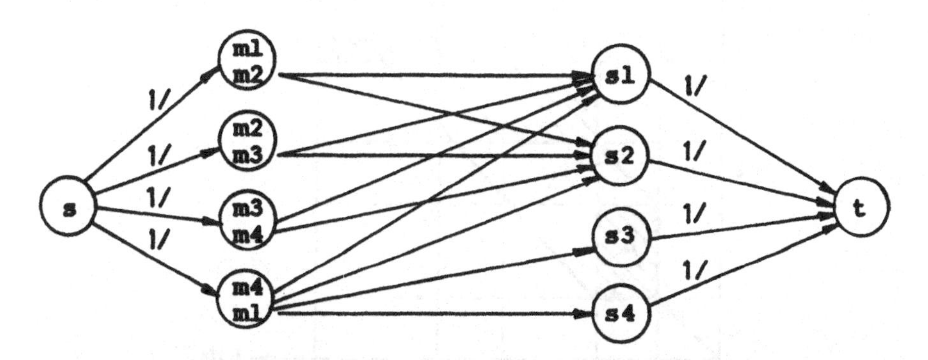

Fig 2 PLANNING SECONDARY OPERATIONS AS A NETWORK FLOW PROBLEM

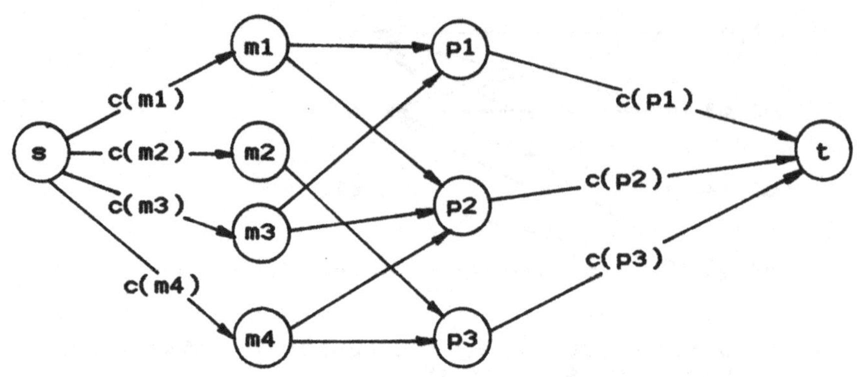

Fig 3 GRAPH REPRESENTATION OF THE PRODUCTION PLANNING PROBLEM

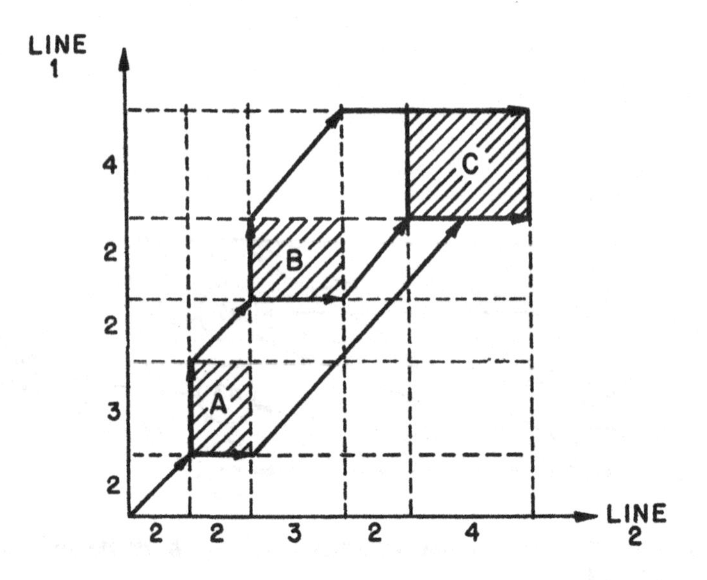

Fig 4 OPTIMAL PROCESS PLANNING GRAPH

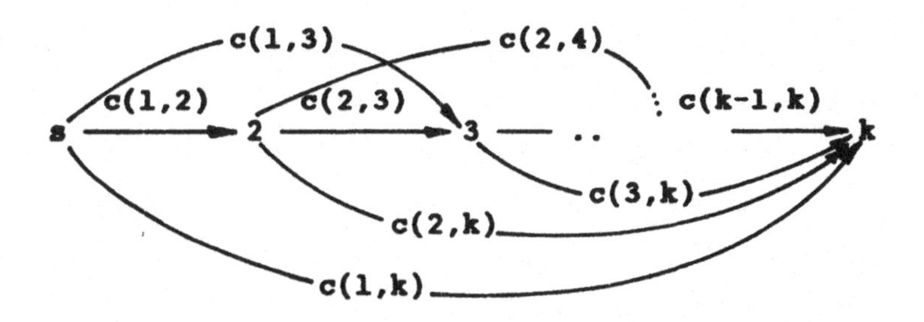

Fig 5 REPRESENTATION OF THE MACHINE INVESTMENT PROBLEM AS A SHORTEST PATH PROBLEM

Optimisation of Neural-Network Structure using Genetic Techniques

N. Dodd

Research Initiative in Pattern Recognition, Royal Signals and Radar Establishment, Malvern, WR14 3PS, UK

E-mail: nigel@uk.mod.hermes

Abstract

We present arguments that it is necessary to use *structured* neural networks for the solution of certain problem types. Structure is imposed on connectivity, activation functions and other parameters of the network to simultaneously optimise generalisation ability and compactness of the network. An analogy is made between the development of biological nervous systems from their genetic coding and the generation of artificial neural networks from a parametric description. Experiments are described which use genetic techniques to optimise network structure for a specific class of problem. Results are given which demonstrate the effectiveness of genetic optimisation of network specifications in comparison with other optimisation techniques. The parallel asynchronous implementation of genetic algorithms on a network of Sun's is briefly described.

1 Introduction

Biological neural networks are rich in structure. Connectivity, transmission mechanisms and activation properties have evolved to perform computation which confers an advantage in natural selection. By comparison, the science of artificial neural networks has been confined to an impoverished diversity of network architectures, stereotyped by the fully connected layered network.

"Designer" networks have been used for some application areas where certain properties of the data are well understood. These properties might, for instance, be transformations of the input to which the output of the network should be invariant. For example, if it is known that a handwritten character should be similarly classified wherever it might occur on the retina, we can construct a network of shared weights that replicates over the retina in such a way that recognition is translation invariant.

The properties of the data that need to be addressed by the network for successful classification, however, are not always known. Furthermore, even when they are known, it is not always obvious how to build the network appropriately. This has led to a number of research projects [4, 7] directed towards formulating a method by which networks can be optimally structured for specific problem classes.

An important property of the multi-layer perceptron (MLP) is *generalisation*. Without this ability an MLP may be considered to be simply a look-up table. Many examples from the literature demonstrate that, when the structure of a network addresses the underlying symmetries and invariances of the data classification problem, the network is able to generalise well, learn quickly and is often of a compact form. These examples include

Parity A conventional fully connected, layered MLP with logistic activation functions fails to learn parity of more than 4 bits. When the network is structured appropriately, the solution for an arbitrary number of bits is always found quickly using error backpropagation and the network is extremely compact [2].

Sequential Context Much work, [6], has been done using structured MLPs for speech recognition. [3] describes networks with delay-line and recurrent sub-structures which perform to the Baum-Welch limit in state-sequence classification.

Transform Invariance Recognition of handwritten characters [1] requires invariance to translation. Networks predisposed to these and other invariances can be built and are termed "spread networks".

For problem types where the mechanism of data production is well understood, human engineering of a suitable network structure is often possible. Where we do not have this knowledge it is desirable to make use of an automatic method for network optimisation. The formulation of such a method is the aim of this research.

The manner in which the genes map onto a network architecture is of crucial importance. If the mapping is such that a gene controls the presence or absence of a node or of a link, then it is possible for the generation process to create networks without a path existing between input and output or networks with unreachable subnetworks or other congenital defects [4]. Ideally we would like a genetic space of network structure which does not contain any unviable network genotypes, but which spans the space of all potentially useful network genotypes. A parametric description of a specific network type is such a space, and we used this level of genetic description for our experiments, choosing two examples whose invariance required a "spread" network. Results are presented for automatic generation of network structure for two example problems, one toy (with 100 connections) consisting of a line orientation problem, and one real (with 5000 connections) consisting of the identification of dolphin sounds. Both problems exhibit spatial invariance along one or two axes and so are amenable to the general class of spread networks. Optimisation is done over the height and width of the replicated weight pattern, the horizontal and vertical increment of this sub-structure and learning rate. Fitness is taken to be inversely proportional to residual error and network complexity.

```
for(number of populations)
for(number in population)
 {
 select parents from population with
 probability proportional to fitness

 mate
 create 1 offspring
 calculate fitness
 }
```

Figure 1: Pseudo-code for Crossover

```
select a random genome G1
evaluate fitness, f1m, of G1
for(number of evaluations)
    {
    mutate G1, call the mutation G2
    evaluate fitness, f2, of G2
    f2m = mean(any previous
     fitness valuations of G2)
    if G1 = G2 then reevaluate f1m,
     the new mean
    if f2m > f1m then
     G1 = G2, f1m = f2m
    }
```

Figure 2: Pseudo-code for PGD

2 Criteria for Choice of Optimisation Technique

In general the parameters of a network specification are qualitatively different from one another. It is desirable to use an optimisation method that is insensitive to correlations between the network parameters. Different network types will have quite different descriptive parameters and a general optimisation technique will cater for all network types.

It is desirable to explore several different regions of search in parallel since the parameters may have no correlation and the search space may be irregular. It may also be productive to swap the values of the parameters between the various search points.

Runs of the same network with different starting weights can yield quite different figures of performance. The optimisation technique must be tolerant of such noise present on estimates of fitness.

These and other requirements lead to the use of genetic algorithms [5] to find near optimal combinations of the network generation parameters. Two, more conventional, optimisation techniques were implemented for comparison.

3 Detail of Optimisation Techniques

Pure crossover was implemented as follows. A starting generation was produced from a flat distribution over the genetic space. Parents F and M were selected from the starting generation with a probability equal to their normalised fitness. From two parents an offspring genotype was created by randomly selecting a splice point within or at the extremes of the chromosome. To the left of the splice point genes of parent F were inherited, and to the right of the splice point genes of parent M were inherited. A population equal in number to the starting population was built up in this way, and the fitness of each member of the population was evaluated. Subsequent populations were similarly generated. Pseudo-code is given in figure 1

The two conventional optimisation techniques tested on the "lines" data were firstly a random search, and secondly a pseudo-gradient descent (PGD). The crossover technique finished with a generation of fit genotypes. A generation size of 100 individuals was selected as a satisfactory compromise. The comparison techniques were also implemented to finish with a set of 100 fit individuals. Estimates of fitness are taken as the mean over each set of 100 individuals.

1. For the random search, a number of evaluations equal to the total number of evaluations made in the genetic crossover approach were made. The values of the genes were uniformly distributed over the allowable genetic range.

2. The PGD pseudo-code is listed in figure 1. It is only possible to evaluate exactly a phenotype. The fitness of a genotype, of which the phenotype is an expression, is plagued with noise. By repeated evaluations of genotypes near the optima in gene space, the PGD method reduces the noise of genotype evaluation in the interesting regions of gene space.

The three algorithms described in this paper were all capable of parallel implementation and many of the experimental results were obtained using an asynchronous parallel implementation on a network of Sun 3's. A semaphore protocol was employed to avoid file access contention.

4 Comparison of the Optimisation Techniques

A comparison of the three optimisation techniques is shown in the following table.

	random	PGD	Crossover
optimum	10.2	19.3	17.0
re-evalutaion	6.1	8.5	17.0

It can be seen that the pseudo-gradient descent algorithm finds the fittest phenotypes, whereas crossover finds the fittest genotypes. We are interested in finding the fit genotypes since we want a network description which, given any weight start, will perform well. The phenotype in this context is the network plus a set of random weight starts which the PGD finds effectively. These apparently fit individuals, however, do not perform well when re-evaluated with another set of random weight starts. The crossover technique is, by comparison, very effective in avoiding individuals which evaluate well with a special starting weight set. As expected for the crossover algorithm, the mean fitness of a re-evaluation of a generation is approximately identical to the original evaluation of fitness of the generation.

As a demonstration of the increased fitness of successive generations, histograms are given in figure 3. This shows the generations of dolphin sound recognition networks increasing by a factor of three, or so, over 30 generations. Even with the toy problem, it is found that the genetically derived optimal networks differ from, and are more efficient than, the networks engineered by humans.

5 Conclusion

For many classification and recognition problems, structured neural networks perform better than standard fully connected layered networks. Designing optimally

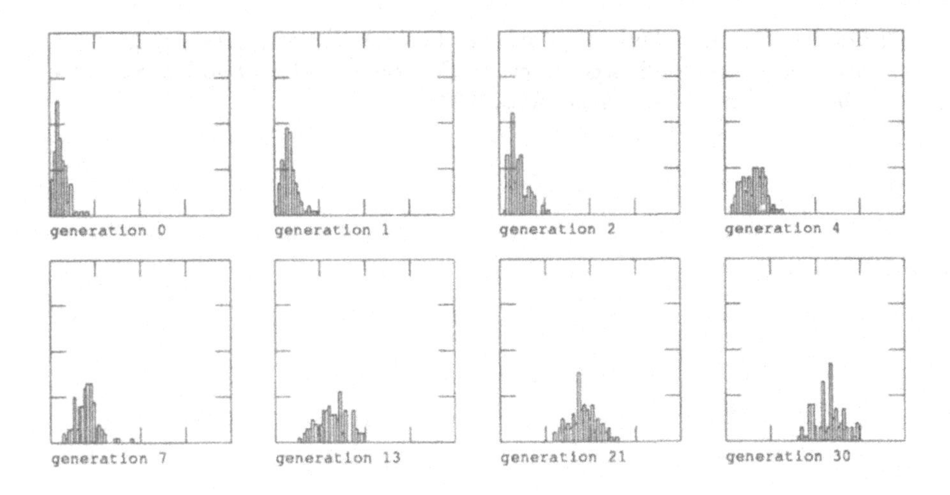

Figure 3: Frequency histograms (frequency of occurrence vs. fitness) for the "dolphin" data. The last number in the label corresponds to the generation number.

structured networks by hand is frequently not possible. Genetic optimisation has been shown, in this paper, to be more effective than other conventional techniques for optimising network structure.

References

[1] Y. Le Cun et al. Hand written digit recognition with a back-propagation network. In *Proceedings of Neural Information Processing Systems—Natural and Synthetic*, 1989.

[2] Nigel Dodd. The gmlp program used as a state sequence classifier. Technical Report RIPR/1000/42/89, Research Initiative in Pattern Recognition, 1989.

[3] Nigel Dodd. Optimisation of network structure using genetic techniques. Technical Report RIPRREP/1000/63/89, Research Initiative in Pattern Recognition, 1989.

[4] Steven Alex Harp, Tariq Samad, and Aloke Guha. Towards the genetic synthesis of neural networks. In *Proceedings of the Third International Conference on Genetic Algorithms*, 1989.

[5] John H Holland. *Adaptation in Natural and Artificial Systems.* Ann Arbor: University of Michigan Press, 1975.

[6] Richard P. Lippmann. Review of neural networks for speech recognition. *Neural Computation*, 1(1):1–38, 1989.

[7] Geoffrey F. Miller, Peter M. Todd, and Shailesh U. Hedge. Designing neural networks using genetic algorithms. In *Proceedings of the Third International Conference on Genetic Algorithms*, 1989.

Interval Algebras and Order of Magnitude Reasoning

S. Parsons
Department of Electronic Engineering, Queen Mary and Westfield College, Mile End Road, London E1 4NS, UK

ABSTRACT

Qualitative reasoning has been extended by several authors to encompass reasoning about the order of magnitude of quantities. This paper discusses how interval algebras based on interval arithmetic [9] may form a basis for modelling three schemes for order of magnitude reasoning.

1.0 INTRODUCTION

Qualitative reasoning [1] was introduced as a formalism for reasoning about physical systems that captures many of the features of human reasoning. Qualitative reasoning reduces the quantitative precision of the behavioural descriptions of such systems whilst retaining crucial distinctions. Real valued variables are replaced with qualitative variables which can adopt only a small number of values, usually +, 0 and -. The behaviour is described in terms of changes in the qualitative value of a number of state variables and their first and second derivatives, and these values are related by means of qualitative differential equations, often called confluences. For example, we write:

$$\begin{aligned} dP_{IN,OUT} + dQ_{*1(V,V)} &= 0 \\ dP_{OUT,SMP} + dF_{*1(S)} + dF_{A(M)} &= 0 \\ dP_{IN,SMP} &= + \end{aligned}$$

to qualitatively describe part of the behaviour of a pressure regulator, where dP is shorthand for dP/dt, the first time derivative of P. In theory there is no reason to limit the information used to just the first two derivatives, but in practice it is extremely difficult to obtain higher order relations. All time derivatives are continuous, so that no variable may jump from one qualitative state to another without passing through any intervening states, and variables are combined by means of combinator tables giving the result of every possible combination of inputs, so for qualitative addition $\oplus$ we have :

⊕	+	0	-
+	+	+	?
0	+	0	-
-	?	-	-

Table 1.

The basic model has been extended in a number of directions in recent years, generally in an attempt to defeat its tendency to over abstract, obscuring important detail and rendering some simple problems insoluble. Notable attempts at providing more precise systems include Raiman's [12] FOG, Mavrovouniotis and Stephanopoulos' [8] O[M], and Dubois and Prade's [5] system based on fuzzy arithmetic. These systems are often collectively known as order of magnitude systems since they make explicit use of the relative size between the quantities with which they deal.

Other important work has been carried out by Travé-Massuyès and Piera [14] who have provided a mathematical characterisation of some of the properties of order of magnitude models, and give a set of axioms for the algebras that they use. Struss [13] has provided an analysis of the soundness and completeness of qualitative models and has demonstrated that their difficulties in this respect are not alleviated by recasting them as algebras of more restricted intervals.

In this paper we briefly review the work of Raiman, Mavrovouniotis and Stephanopoulos, and Dubois and Prade. We introduce semiqualitative models as a generalisation of the basic qualitative calculus, and sketch their relationship with interval arithmetic [9]. Interval algebras are discussed as an extension of these ideas, and their relationship with Travé-Massuyès and Piera's qualitative algebras outlined. The main part of the paper analyses the ways in which interval and qualitative algebras may form a simple basis for order of magnitude reasoning.

2.0 AN OVERVIEW OF ORDER OF MAGNITUDE TECHNIQUES

There seem to be three distinct approaches to the formalisation of human order of magnitude reasoning; the process by which the relations between the magnitude of quantities is used to simplify the qualitative models used to predict the behaviour of devices and systems. Dubois and Prade [5] showed that fuzzy set theory can provide a basis for an absolute order of magnitude system by dividing the real numbers into an arbitrarily large set of fuzzy intervals. Raiman [12] proposed a system FOG that uses relational operators to denote the size relation between quantities. Finally, Mavrovouniotis and Stephanopoulos [7] introduced a system O[M] based on the relative magnitude of quantities that overcame what they saw as serious flaws in FOG.

In the remainder of this section we will briefly discuss the properties of these three approaches as a necessary precursor to demonstrating that interval arithmetic is capable of providing an underlying formal basis for the techniques.

2.1 Absolute order of magnitude and fuzzy intervals

Dubois and Prade [5] propose that order of magnitude reasoning can be performed by increasing the precision of qualitative reasoning, splitting the quantity space {-, 0, +} [1] into smaller fuzzy intervals; negative large (NL), negative medium (NM), negative small (NS), zero (0), positive small (PS), positive medium (PM), positive large (PL). These may be viewed as a partitioning of the real numbers:

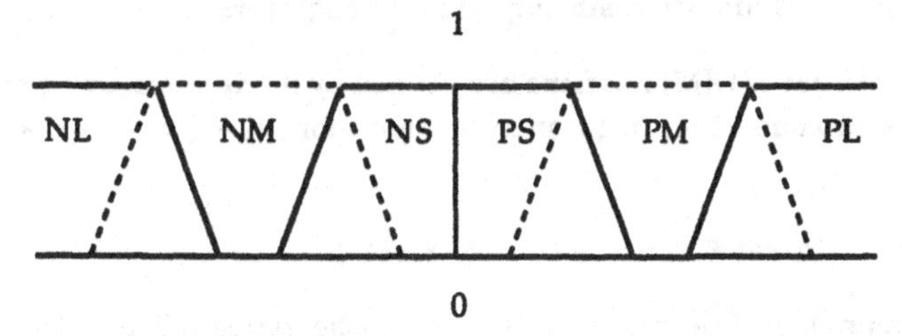

It is suggested that a meaningful calculus should conform to the following conditions:

C1: The combinator tables which define the arithmetic operations of qualitative reasoning should be maintained for efficient order of magnitude reasoning.

C2: The calculus should be consistent with the real values that it approximates, and the operations over them.

Dubois and Prade show that the closure of the combinator tables for the fuzzy interval system can be obtained by defining the operations over all possible combinations of adjacent intervals in addition to the intervals themselves. Thus we have:

$$PS \oplus PS = +$$
$$PM \oplus PM = PM+$$
$$PM \oplus PL = PL$$

where ≈ is the addition operator for the calculus, and PM+ = [PM, PL], + = [PS, PL] for $[s_i, s_j] = \{s_k \mid s_i \le s_k \le s_j\}$ with $s_i, s_j, s_k \in$ {NL, NM, NS, 0, PS, PM, PL}.

2.2 Relative order of magnitude and O[M]

Mavrovouniotis and Stephanopoulos [7] constructed a system of order of magnitude reasoning based on seven primitive relations:

$A \ll B$	A is much smaller than B
$A -< B$	A is moderately smaller than B
$A \sim< B$	A is slightly smaller than B
$A == B$	A is exactly equal to B

A >~ B	A is slightly larger than B
A >- B	A is moderately larger than B
A >> B	A is much larger than B

Further relations are constructed as disjunctions of two or more consecutive relations, so that 'A is less than B' is denoted 'A << ... ~< B'. A total of 21 compound relations may be formed including all those that Mavrovouniotis and Stephanopoulos claim are commonly used by engineers.

The semantics of O[M] relations are defined by fixing the allowed range of the ratio of the two quantities in question. We have for rel ∈ {<<, -<, ~<, ==, >~, >-, >>}:

$$\text{A rel B} \quad \equiv \quad \text{A/B rel 1}$$

and the bounds on the intervals into which the ratios fall are defined by the values e_1, e_2, e_3, and e_4 (see Figure 2(a)). A strict interpretation requires:

$$\begin{aligned} e_3 &= e_1 + 1 \\ e_4 &= 1/e_1 \\ e_2 &= 1/e_3 \end{aligned} \qquad (1)$$

allowing the intervals to be determined by a single application specific *accuracy parameter* e (Figure 2 (b)):

$$e_1 = e \qquad (2)$$

This interpretation was felt to be too error prone, so the set of intervals was replaced with a pair of sets of overlapping (Figure 2 (c)) and non-exhaustive (Figure 2(d)) intervals, creating a so-called heuristic interpretation. Inferences proceed from antecedents whose values are restricted to the non-exhaustive intervals to create consequents that have overlapping interval values.

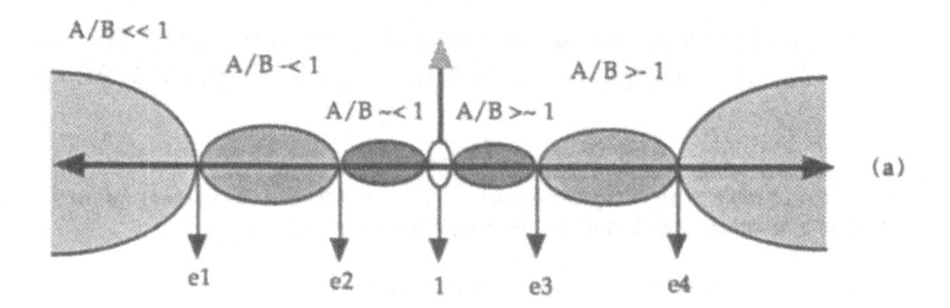

Figure 2.

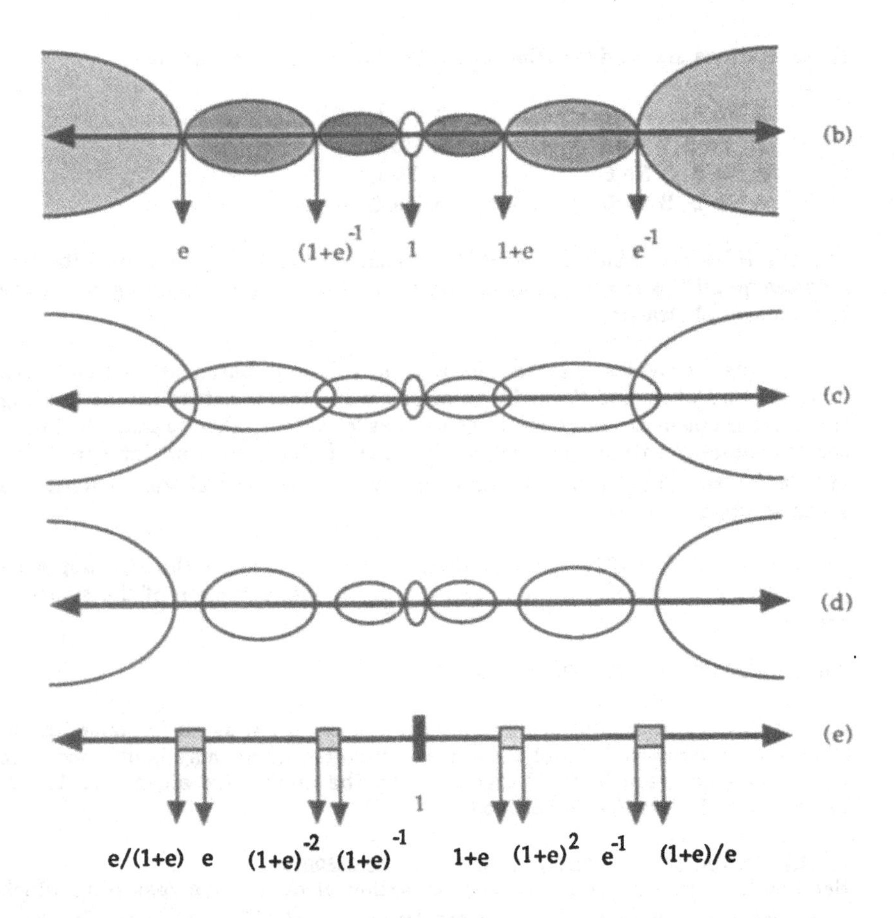

Figure 2.

These consequent values are shrunk to non-exhaustive values when used as the basis for further reasoning. This behaviour mirrors the human tendency to chain approximate inferences regardless of the inherent uncertainty, reasoning that:

$$A \succ\!\sim B,\ B \succ\!\sim C \quad \rightarrow \quad A \succ\!\sim C$$
$$A \succ\!\sim C,\ C \succ\!\sim D \quad \rightarrow \quad A \succ\!\sim D$$

The boundaries are those of Figure 2(e).

2.3 FOG and Symbolic relations

Raiman [12] introduced the FOG system based on three operators, Ne, Vo, Co representing what he claimed were intuitive order of magnitude concepts:

A Ne B :	**A is negligible with respect to B**
A Vo B :	**A is close to B so that A - B is negligible with respect to B**
A Co B :	**A is comparable to B so that A has the same sign and order of magnitude as B and if B Ne C, then A Ne C.**

these relations are used to define a set of 31 inference rules such as:

A Vo B	→	B Vo A
A Vo B, B Vo C	→	A Vo C
A Ne B, B Ne C	→	A Ne C
A Vo B, B Ne C	→	A Ne C

and the rules are then used with that axiom A Vo A to propagate relations between quantities to solve problems that defeat qualitative reasoning due to the latter's over abstraction.

Raiman provides a semantics for the relations based on non-standard analysis. As Dubois and Prade [4] point out, while this is natural in the limiting case, it is problematic when relations such as 'close' are taken to mean 'within a certain interval' rather than 'infinitely close'. Indeed, such an interpretation makes the transitivity of Vo unacceptable, and forces Raiman to limit the repeated use of such rules.

Dubois and Prade [5] suggest using fuzzy relations to model the FOG approach to order of magnitude reasoning, giving a fuzzy interpretation of the symbolic relations.

3.0 A FAMILY OF INTERVAL ALGEBRAS

In this section we discuss semiqualitative reasoning [3] as an implementation of interval arithmetic [9] and show how interval algebras may be defined. The relationship between interval algebras and the qualitative algebra of Travé-Massuyès and Piera [14] is outlined.

3.1 Interval arithmetic and semiqualitative reasoning

Semiqualitative reasoning is a generalisation of qualitative reasoning which increases precision by splitting the quantity space into $(2k + 1)$ intervals whose boundaries are the ordered set of values:

$$\{ I_{(-k)}, I_{(-(k-1))}, \dots, I_{(-1)}, I_{(0)}, I_{(1)}, \dots, I_{(k-1)}, I_{(k)} \}$$

where $I_{(+j)}$ is the upper boundary of the jth positive interval, $I_{(-j)}$ is the lower boundary of the jth negative interval and $I_{(0)}$ is the zero interval. Clearly the standard quantity space $\{+, 0, 1\}$ is that obtained for $k = 1$, $I_{(-k)} = -\infty$ and $I_{(k)} = \infty$.

The size of the intervals is tailored to suit specific problems, and the sequence of boundaries may be in an arithmetic or geometric progression, or chosen to cover values of interest [3]. The arithmetic properties of semiqualitative systems are defined by Moore's interval arithmetic, and are summarised by:

$$[a, b] \otimes_j [c, d] = [\min(a \otimes_j c, a \otimes_j d, b \otimes_j c, b \otimes_j d), \max(a \otimes_j c, a \otimes_j d, b \otimes_j c, b \otimes_j d)] \quad (3)$$

where $\otimes_j \in \{+, -, \times, \div\}$ and $\otimes_i$ is its interval equivalent. Division by intervals containing 0 is not defined. Note that operations over degenerate intervals [a, b] where a = b, are equal to the equivalent operations over the respective reals:

$$[a, b] \otimes_i [c, d] \quad = a \otimes_j c = a \otimes_j d = b \otimes_j c = b \otimes_j d$$

for a = b and c = d. Thus interval arithmetic may be considered to be a generalisation of real arithmetic. In practice the results of arithmetic operations are specified by means of combinator tables. Consider a system using the ordered set of intervals { [a, b], [b, c], [c, d] } with equal spacing so that (b - a) = (c - b) = (d - c) and d = 3b. The tendency of interval arithmetic to extend the interval bounds means that the full set of operands include compound intervals, giving the following operator table for addition ⊕:

⊕	[a, b]	[a, c]	[a, d]	[b, c]	[b, d]	[c, d]
[a, b]	[a, c]	[a, d]	[a, d]	[b, d]	[b, d]	[c, d]
[a, c]	[a, d]	[a, d]	[a, d]	[b, d]	[b, d]	[c, d]
[a, d]	[a, d]	[a, d]	[a, d]	[b, d]	[b, d]	[c, d]
[b, c]	[b, d]	[b, d]	[b, d]	[b, d]	[b, d]	[c, d]
[b, d]	[b, d]	[b, d]	[b, d]	[b, d]	[b, d]	[c, d]
[c, d]	[c, d]	[b, d]	[c, d]	[c, d]	[c, d]	[c, d]

The closure of the operations is clearly a problem for arbitrarily specified intervals. A suitable approximation may be, as above, to allow the upper and lower intervals to absorb all values that exceed their bounds. This is unlikely to appeal to purists, and an alternative approach is to consider new maximum and minimum intervals extending from the largest positive and negative values to ± ∞. Despite such theoretical problems, semiqualitative techniques have been successfully applied to number of engineering problems which are not soluble by pure qualitative reasoning. Such problems include simulations of chemical reactions [6] and bioengineering processes [2]

3.2 Interval and Qualitative Algebras

A particular interval algebra may be defined from a ordered set of values $\mathcal{V} = \{v_1, \ldots, v_n\}$. The set of intervals over which the operators of the algebra are closed is the set of all intervals $[v_i, v_j]$ such that $v_i < v_j$ and $v_i, v_j \in \mathcal{V}$. We can define an order $<_{Q3}$ [11] over the interval such that $[v_i, v_j] < [v_k, v_l]$ iff $(v_i - v_j) < (v_k - v_l)$.

Travé-Massuyès and Piera [14] present a mathematical framework to support order of magnitude reasoning that explicitly distinguishes between different

levels of description. Given a set S and an order $\lesssim$ defined over S, qualitative equality $\approx$ is defined as:

$$a \approx b \text{ if there exists } x \in S \text{ such that } x \lesssim a \text{ and } x \lesssim b.$$

A qualitative algebra is a pair $(S, \approx)$ provided with operations $\oplus$ and $\otimes$, which are:

(i) qualitatively associative: $a \otimes (b \otimes c) \approx (a \otimes b) \otimes c$ and $a \oplus (b \oplus c) \approx (a \oplus b) \oplus c$
(ii) qualitatively commutative: $a \otimes b \approx b \otimes a$ and $a \oplus b \approx b \oplus a$
(iii) $\otimes$ is qualitatively distributive with respect to $\oplus$: $a \otimes (b \oplus c) \approx (a \otimes b) \oplus (a \otimes c)$

Travé-Massuyès and Piera prove that a qualitative algebra $(S, \approx, \oplus, \otimes)$ and a subalgebra $(T, \approx, \oplus, \otimes)$ where $T \subset S$ and $T \neq \emptyset$ are embedded in one another, and that it is possible to dynamically refine a model during processing by switching from T to S.

It is possible to show [10] that interval algebras defined as above are qualitative algebras $(S_i, \approx, \oplus, \otimes)$ for interval addition $\oplus$, interval multiplication $\otimes$, and Q-equality defined by the following:

$[v_i, v_j] = [v_k, v_l]$ iff $v_i = v_k$ and $v_j = v_l$.
$[v_i, v_j] \lesssim_{IA} [v_k, v_l]$ if $[v_l - v_j] <_{Q3} [v_k - v_i]$ or $[v_i, v_j] = [v_k, v_l]$
$[v_l - v_j] \approx [v_k - v_i]$ if there exists $[v_a, v_b]$ such that $[v_a, v_b] \lesssim [v_i, v_j]$ and $[v_a, v_b] \lesssim [v_k, v_l]$.

4.0 ORDER OF MAGNITUDE REASONING BY INTERVALS

Having discussed several proposals for order of magnitude reasoning in some detail, we now demonstrate how they may be interpreted in terms of interval algebras. Whilst interval methods may be said to subsume FOG and O[M] in some sense, it makes little sense to claim that it subsumes the more general fuzzy approach. Instead we can merely claim that the full mechanism of fuzzy mathematics is not required for the comparatively simple domain discussed by Dubois and Prade.

4.1 Absolute order of magnitude

As suggested by Dubois and Prade, their absolute order of magnitude scheme may be modelled by adjacent intervals. In keeping with the underlying interval arithmetic, we will reference the intervals by their endpoints so that NL = $[\infty, nm]$, NM = $[nm, ns]$, NS = $[ns, 0]$, PS = $[0, ps]$, PM = $[ps, pm]$ and PL = $[pm, \infty]$. This gives:

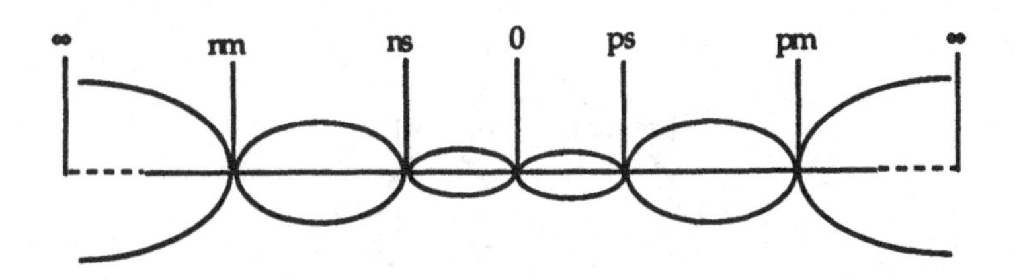

Defining the intervals by means of their boundaries gives us an easy means of obtaining closure, since the operations over two intervals are defined by equation (3). The additional intervals are formed automatically, and follow directly from the set of landmark values. The combinator table for the addition operation $\oplus$ includes Table 3 for the addition of positive valued intervals.

$\oplus$	[0, ps]	[0, pm]	[0, ∞]	[ps, pm]	[ps, ∞]	[pm, ∞]
[0, ps]	[0, ∞]	[0, ∞]	[0, ∞]	[ps, ∞]	[ps, ∞]	[pm, ∞]
[0, pm]	[0, ∞]	[0, ∞]	[0, ∞]	[ps, ∞]	[ps, ∞]	[pm, ∞]
[0, ∞]	[0, ∞]	[0, ∞]	[0, ∞]	[ps, ∞]	[ps, ∞]	[pm, ∞]
[ps, pm]	[ps, ∞]	[ps, ∞]	[ps, ∞]	[ps, ∞]	[ps, ∞]	[pm, ∞]
[ps, ∞]	[ps, ∞]	[ps, ∞]	[ps, ∞]	[ps, ∞]	[ps, ∞]	[pm, ∞]
[pm, ∞]	[pm, ∞]	[pm, ∞]	[pm, ∞]	[pm, ∞]	[pm, ∞]	[pm, ∞]

Table 3.

With the full $\oplus$ table, and a combinator table for interval multiplication $\otimes$ as defined by (3) we can specify a family $\mathfrak{S}$ of algebras over the basic set of intervals { [-∞, nm], [nm, ns], [ns, 0], [0, ps], [ps, pm], [pm, ∞]} such that $\langle \mathcal{S}_i, \pm, \approx, \otimes \rangle \in \mathfrak{S}$ and $\mathcal{S}_i \subseteq$ { {[-∞, nm], [nm, ns], [ns, 0], [0, ps], [ps, pm], [pm, ∞]}, {[-∞, ns], [nm, 0], [ns, ps], [pm, 0], [ps, ∞]}, {[-∞, 0], [nm, ps], [ns, pm], [0, ∞]}, {[-∞, ps], [nm, pm], [ns, ∞]} {[-∞, pm], [nm, ∞]}, {[-∞, ∞]} }. Note that the intervals [0, ∞] and [-∞, 0] correspond to the qualitative intervals + and - respectively, and that [-∞, ∞] is equivalent to the qualitative value ? introduced to maintain closure [1]. There is an implied order on the $\mathcal{S}_i$ that is summarised by Figure 4.

Since the family $\mathfrak{S}$ are interval algebras and thus obey Travé-Massuyès and Piera's axioms, we can switch between levels of granularity at will.

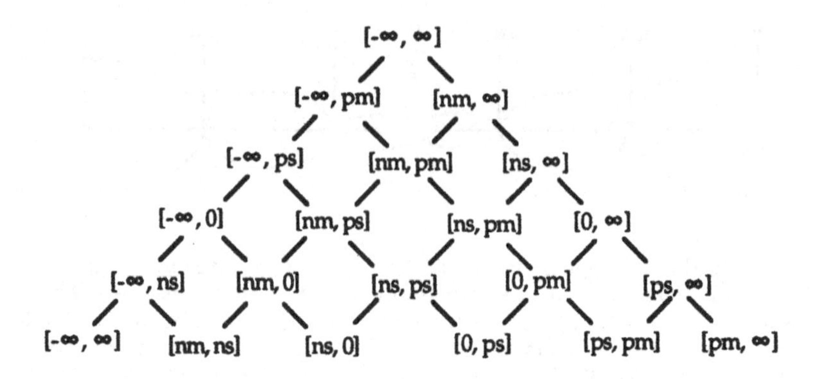

Figure 4.

It is also possible to consider the fuzzy intervals as defining overlapping intervals on the real axis. This interpretation may also be modelled using adjacent intervals by distinguishing between those regions that are, say, purely in PM, and those that are in PS and PM or PL and PM. Such an interpretation is given in Figure 5. Here NL = [∞, nm-], NM = [nm+, ns-], NS = [ns+, 0], PS = [0, ps+], PM = [ps-, pm+] and PL = [pm-, ∞]. A further family of interval algebras may be based on this set of values.

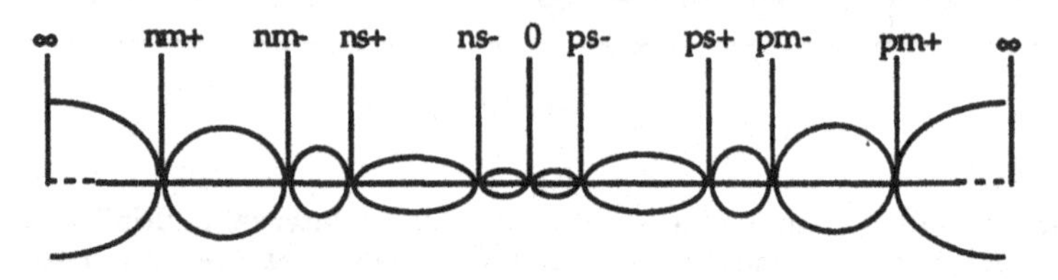

Figure 5.

<u>4.2 Relative order of magnitude models</u>
Here we discuss how we can interpret quantities in terms of the O[M] model. We can both propagate interval values, and deal with order of magnitude relations, which are essentially interval based, between relations.

<u>4.2.1 The strict interpretation</u> O[M] relations are defined by means of the interval containing the result of dividing one quantity by another. Thus the representation has an natural interval arithmetic interpretation. We first define a mapping [[ ]] between the ratio between the quantities and their O[M] relation:

$$\text{For any } A = [a_1, a_2]\ \ B = [b_1, b_2],\ \ [[\,A\ B\,]] = \begin{cases} A \ll B & \text{if } e_1 > [A/B \\ A \mathrel{-\!<} B & \text{if } e_1 < [A/B] < e_2 \\ A \mathrel{\sim\!<} B & \text{if } e_2 < [A/B] < 1 \\ A == B & \text{if } 1 = [A/B] \\ A \mathrel{>\!\sim} B & \text{if } 1 < [A/B] < e_3 \\ A \mathrel{>\!-} B & \text{if } e_3 < [A/B] < e_4 \\ A \gg B & \text{if } e_4 < [A/B] \end{cases}$$

where $[A/B] = [a_1/b_2, a_2/b_1]$ is the interval valued result of dividing A by B.

The mapping loses no generality by being described as operating over intervals (see Section 3.1) and the interval definition will be required to establish the result of order of magnitude inference. The constants e_1, e_2, e_3 and e_4 are related to the accuracy parameter e by (1) and (2). A reverse mapping $[[\ ]]'$ generates the permitted interval value of A/B from the relation between them. For $rel \in \{\ll, \mathrel{-\!<}, \mathrel{\sim\!<}, ==, \mathrel{>\!\sim}, \mathrel{>\!-}, \gg\}$, $A = [a_1, a_2]$ $B = [b_1, b_2]$, and $[A/B] = [a_1/b_2, a_2/b_1]$:

$$\text{For any } A, B, rel,\ [[\,A\ rel\ B\,]]' = \begin{cases} e_1 > a_2/b_1 & \text{if } rel = \ll \\ e_1 < [A/B] < e_2 & \text{if } rel = \mathrel{-\!<} \\ e_2 < [A/B] < 1 & \text{if } rel = \mathrel{\sim\!<} \\ 1 = [A/B] & \text{if } rel = \,== \\ 1 < [A/B] < e_3 & \text{if } rel = \mathrel{>\!\sim} \\ e_3 < [A/B] < e_4 & \text{if } rel = \mathrel{>\!-} \\ e_4 < [A/B] & \text{if } rel = \gg \end{cases}$$

Initial relations between quantities may established either by $[[\ ]]$ or by definition. Propagation is achieved by using $[[\ ]]'$ to get the bounds on the ratios between quantities which are then manipulated using interval arithmetic. $[[\ ]]$ is then applied to obtain the final relations. Thus given $A \mathrel{-\!<} B$ and $B \mathrel{-\!<} C$ we can use $[[\ ]]'$ to obtain $e_1 < [A/B] < e_2$ and $e_1 < [B/C] < e_2$ from which it is trivial to establish $e_1^2 < [A/C] < e_2^2$. This interval ratio can be converted by $[[\ ]]$ into the relationship $A \ll ... \mathrel{-\!<} C$. The extension of the interval mirrors the cautious, and arguably correct, tendency of interval arithmetic to accentuate the underlying uncertainty. Given the bounds on $\mathrel{-\!<}$ it is possible that the value of A/C is less than the lower limit on $\mathrel{-\!<}$.

<u>4.2.2 The heuristic interpretation</u> If such cautious inferences are to be avoided we must support the heuristic interpretation, which requires another pair of mappings; $[[\ ]]_h$ to map onto the extended intervals, and $[[\ ]]_h'$ to retrieve the non-exhaustive intervals. So for $A = [a_1, a_2]$ $B = [b_1, b_2]$ and $[A/B] = [a_1/b_2, a_2/b_1]$:

$$\text{For any } A, B,\ [[\ A\ B\ ]]_h = \begin{cases} A \ll B \text{ if } e_1 > [A/B] \\ A -\!< B \text{ if } e_5 < [A/B] < e_2 \\ A \sim\!< B \text{ if } e_6 < [A/B] < 1 \\ A == B \text{ if } 1 = [A/B] \\ A >\!\sim B \text{ if } 1 < [A/B] < e_7 \\ A \gg B \text{ if } e_4 < [A/B] \end{cases}$$

where e_1, e_2, e_3 and e_4 are related to e by (1) and (2), and $e_5 = e/(1+e)$, $e_6 = (1 + e)^{-2}$, $e_7 = (1 + e)^2$, and $e_8 = (1 + e)/e$ are the boundaries of the interval overlaps. We also have:

$$\text{For any } A, B\ , rel,\ [[\ A\ rel\ B]]_h^- = \begin{cases} e_5 > [A/B] & \text{if } rel = \ll \\ e_1 < [A/B] < e_6 & \text{if } rel = -\!< \\ e_2 < [A/B] < 1 & \text{if } rel = \sim\!< \\ 1 = [A/B] & \text{if } rel = == \\ 1 < [A/B] < e_3 & \text{if } rel = >\!\sim \\ e_7 < [A/B] < e_4 & \text{if } rel = >\!- \\ e_8 < [A/B] & \text{if } rel = \gg \end{cases}$$

using the same procedure as before we can take $A \sim\!< B$ and $B \sim\!< C$, obtain $(1 + e)^{-1} < [A/B] < 1$ and $(1 + e)^{-1} < [B/C] < 1$ from which we deduce $(1 + e)^{-2} < [A/C] < 1$ which enables us to conclude that $A \sim\!< C$.

Note that while the mappings given here cover just the basic O[M] relations, there is no reason why they cannot be extended to deal with all 21 compound relations. For instance we can define $[[\ A\ B\ ]] = A -\!<\ldots \sim\!< B$ if $e_1 < [A/B] < 1$, $[[A -\!<..>\!- B]]' = e_1 < [A/B] < e_4$, $[[\ A\ B\ ]]_h = A -\!<\ldots \sim\!< B$ if $e_5 < [A/B] < 1$ and $[[A -\!<..>\!- B]]_h' = e_1 < [A/B] < e_4$.

<u>4.2.3 An example</u> To demonstrate that our approach captures the essence of O[M], we solve the heat exchanger example introduced in Mavrovouniotis and Stephanopoulos [7]. We have a countercurrent heat exchanger as pictured in Figure 6, with a hot flow that is cooled and a cold flow that is heated. The important parameters are the molar heat capacities of the hot and cold streams, K_h and K_c and the molar flows F_h and F_c. The following temperature differences may be defined:

$$\Delta T_h = T_{h1} - T_{h2} \qquad \Delta T_c = T_{c1} - T_{c2}$$
$$\Delta T_1 = T_{h1} - T_{c1} \qquad \Delta T_2 = T_{h2} - T_{c2} \qquad (4)$$

so that ΔT_h is the drop in temperature of the hot stream, ΔT_c the rise in temperature of the cold stream, and ΔT_1 and ΔT_2 are the differences in temperature at either end of the exchanger.

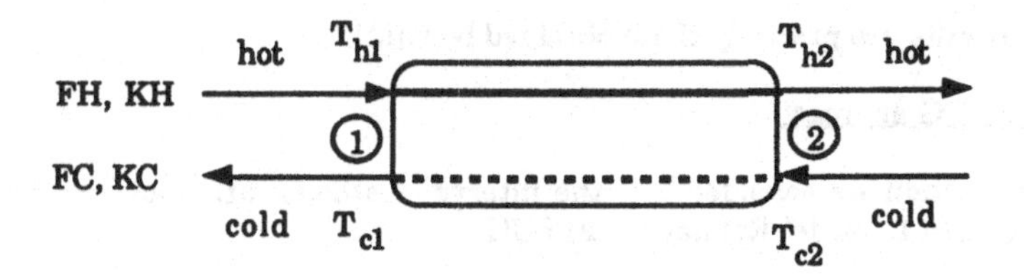

Figure 6.

The definitions (4) constrain the values of the parameters:

$$\Delta T_h - \Delta T_1 - \Delta T_c + \Delta T_2 = 0 \qquad (5)$$

and since all the energy entering the cold stream must come from the hot stream:

$$\Delta T_h \cdot K_h \cdot F_h = \Delta T_c \cdot F_c \cdot K_c \qquad (6)$$

Following Mavrovouniotis and Stephanopoulos we will assume:

$$\Delta T_2 \;\; -\!< \;\; \Delta T_1 \qquad (7)$$
$$\Delta T_1 \;\; << \;\; \Delta T_h \qquad (8)$$
$$K_h \;\; >\!\sim \;\; K_c \qquad (9)$$

(a) Applying [[ ]]$_h$' to (7) and (8) we get $\Delta T_2/\Delta T_1 = [e_1, e_6]$ and $\Delta T_1/\Delta T_h < e_5$. This simply yields $\Delta T_2/\Delta T_h < e_1 e_5 < e_1$, and applying [[ ]]$_h$ gives

$$\Delta T_2 \;\; << \;\; \Delta T_h \qquad (10)$$

(b) From (5) we know that $\frac{\Delta T_c}{\Delta T_h} = 1 - \frac{\Delta T_1}{\Delta T_h} + \frac{\Delta T_2}{\Delta T_h}$. Using [[ ]]$_h$' on (8) and (10) we can rewrite this as $\Delta T_c / \Delta T_h = [1 - e/(1+e),\ 1 + e/(1+e)]$. Clearly, since $e < 1$, $\Delta T_c / \Delta T_h \subseteq [e_2, e_3]$ so [[ ]]$_h$ gives us :

$$\Delta T_c \;\; \sim\!<\!\ldots\!>\!\sim \;\; \Delta T_h \qquad (11)$$

(c) Applying [[ ]]$_h$' to (8) and (11) we get $\Delta T_c/\Delta T_h = [e_2, e_3]$ and $\Delta T_1/\Delta T_h < e_5$. This gives us $\Delta T_1/\Delta T_c < e_5\,(1+e) < e_1$, and applying [[ ]]$_h$ gives

$$\Delta T_1 \;\; << \;\; \Delta T_c \qquad (12)$$

(d) Rewriting (6) as $\frac{\Delta T_c}{\Delta T_h} = 1 - \frac{\Delta T_1}{\Delta T_h} + \frac{\Delta T_2}{\Delta T_h}$ and substituting in the results of applying [[ ]]$_h$' to (9) and (11), we get $F_c/F_h = [e_2, e_3{}^2]$ which by [[ ]]$_h$ gives us:

$$F_c \;\; \sim\!<\!\ldots\!>\!\sim \;\; F_h \qquad (12)$$

These results are precisely those obtained by O[M].

4.3 The FOG approach

In this section we consider how the interval methods introduced in Sections 3 may be used to model Raiman's [12] FOG.

4.3.1 An interval interpretation of FOG Raiman's approach may modelled by interpreting the relations Vo, Co and Ne in terms of the basic relations supplied by O[M]. Mavrovouniotis and Stephanopoulos [8] suggest the equivalences:

$$\begin{array}{lll} A \text{ Ne } B & \equiv & A \ll B \\ A \text{ Vo } B & \equiv & A \sim\!\langle ... \rangle\!\sim B \\ A \text{ Co } B & \equiv & A -\!\langle ... \rangle\!- B \end{array}$$

in which case the technique we have discussed for O[M] relations may be applied. Alternatively we can adopt Raiman's definitions and obtain Vo and Co in terms of Ne:

$$\begin{array}{lll} A \text{ Vo } B & \text{iff} & (A - B) \text{ Ne } B \\ A \text{ Co } B & \text{iff} & \forall C \;\; A \text{ Ne } C \Leftrightarrow B \text{ Ne } C \end{array} \qquad (14)$$

then borrowing from O[M] and defining Ne by the ratio between $A = [a_1, a_2]$ and $B = [b_1, b_2]$, we have, say:

$$A \text{ Ne } B \quad \text{iff} \quad [A/B] < e_5 \qquad (15)$$

since $A - B = [a_1 - b_2, a_2 - b_1]$, (14) and A Vo B → B Vo A give us:

$$A \text{ Vo } B \quad \text{iff} \quad \frac{a_2 - b_1}{b_1} < e_5 \text{ and } \frac{b_2 - a_1}{a_1} < e_5 \qquad (16)$$

As far as Co is concerned, we know from (14) that for A Co B then for $C = [c_1, c_2]$ such that $a_2/c_1 < e_5$ then $b_2/c_1 < e_5$. This condition alone does not preclude A Ne B, so we need to encode the intuitive notion that $A/B = [1 - k, 1 + k] \approx 1$. We will conjecture these limits as $A/B = [1 - e_2, 1 + e_2]$ so that:

$$A \text{ Co } B \quad \text{iff} \quad A = [(1 - e_2)b_2, (1 + e_2)b_1]$$

Since the intervals are bounded by real parameters derived from e which describes the granularity of the reasoning, both these approaches allow us to perform inferences that are not possible in FOG such as:

$$A \text{ Vo } 0.1, B \text{ Vo } 1000 \quad \rightarrow \quad A \text{ Ne } B$$

4.3.2 A further example Raiman provides a simple mechanics problem that defeats simple qualitative techniques as motivation for the development of FOG. Here we show how it may be solved using the interval schemes described above. Two masses M and m (Figure 7), where m is negligible in comparison with M, and roughly equivalent speeds V_i and v_i, are moving in opposite directions in one dimension. What happens after impact?

M ■ → ← ■ m

V_i v_i

Figure 7.

We have:

$$m \quad Ne \quad M \tag{17}$$

$$v_i \quad Vo \quad -V_i \tag{18}$$

The conservation of energy and momentum give us, respectively:

$$MV_i^2 + mv_i^2 \quad Vo \quad MV_f^2 + mv_f^2 \tag{19}$$

$$MV_i - mv_i \quad Vo \quad MV_f + mv_f \tag{20}$$

Using the O[M] interpretations of Ne and Vo suggested by Mavrovouniotis and Stephanopoulos, and applying $[[\ ]]_h'$, we can rewrite (19) and (20) as:

$$\frac{V_i\left\{1 - \frac{mv_i^2}{MV_i^2}\right\}}{V_f\left\{1 + \frac{mv_f^2}{MV_f^2}\right\}} = [e_2, e_3] \tag{21}$$

$$\frac{V_i\left\{1 - \frac{mv_i}{MV_i}\right\}}{V_f\left\{1 + \frac{mv_f}{MV_f}\right\}} = [e_2, e_3] \tag{22}$$

Now, (17) and (18) give us the values $m/M \in e_5$ and $v_i/V_i \in [e_2, e_3]$, so, taking the accuracy parameter e to be 0.1, we can solve (21) and (22) and calculate $V_f = [0.84, 1.1]\, V_i$, ignoring the physically impossible solution, so that:

$$V_f \quad Vo \quad V_i$$

Applying [[ ]]$_h$' gives us V_f/V_i = $[e_2, e_3]$ which, along with (22) allows us to deduce:

$$V_f \quad Vo \quad V_f$$

Comparing this with Raiman's results suggests that this interpretation of Vo is perhaps closer to his idea of Co than Vo. Solving the problem with the second interpretation of the symbolic relations, we can rewrite (16) as:

$$A\ Vo\ B \qquad \text{iff} \qquad [A/B] = [\frac{1}{1+e_5}, 1 + e_5]$$

With $m/M < e_5$ as before, we can solve (21) and (22) to get $V_f = [0.84, 1.1]\, V_i$, so that:

$$V_f \quad Vo \quad V_i \qquad \text{and} \qquad v_f \quad Vo \quad V_f$$

<u>4.3.3 Related work</u> In a recent paper Dubois and Prade [4] suggest another means of dealing with Raiman's relations, providing a new semantics based on crisp intervals. Whereas our approach makes Raiman's inference rules redundant to some extent by propagating interval values and mapping between intervals and relations, Dubois and Prade augment the rules by parameterising the symbol relations with tolerance intervals into which the ratios of the related quantities fall:

$$A\ Vo(M)\ B \Leftrightarrow 1 - e \leq \min(A/B, B/A) \leq 1 \text{ where } M = [1 - e, 1] \text{ and } 0 \leq e \leq 1$$

As inference proceeds, the tolerance interval evolves, its growth depending on which rules are fired. The size of the interval can be used to terminate inference when a given limit is exceeded, rather than after an arbitrary number of steps.

5.0 CONCLUSION

We have demonstrated that despite its simplicity, interval analysis, and the interval algebras that may be built using the techniques of interval analysis, are subtle and powerful enough to provide an underlying computational basis for several different types of system of order of magnitude reasoning. We can build a set of absolute intervals, defined by means of intuitive landmark values, that fall into a natural hierarchy of intervals of varying discrimination. These can form the basis of a reasoning system whose precision can be altered as desired. Such a system may be used for simple order of magnitude computations of the form $100 + 0.5 \approx 100$ in the manner of that of Dubois and Prade [5], or as the basis for a scheme for reasoning about variables and constants which is essentially an extension of Raiman's [12] system that incorporates numerical information. An alternative scheme can be constructed across a ratio scale so that a given interval represents the range of values of the ratio between two quantities. This set of values can be used to reason about the relationship between quantities, forming the underlying mechanism of another type of order of magnitude reasoning. This allows the modelling of the approach adopted by Mavrovouniotis and Stephanopoulos [7], as well as another means of modelling Raiman's treatment.

ACKNOWLEDGEMENTS

This work was supported by a grant from ESPRIT Basic Research Action 3085 DRUMS, Defeasible Reasoning and Uncertainty Management Systems. Thanks, as ever, to all my colleagues on the project for continual help and encouragement. Particular thanks to Mirko Dohnal for suggesting semiqualitative reasoning, Henri Prade for timely advice, Louise Travé-Massuyès for supplying papers, and Didier Dubois for helping me to beat the deadline.

REFERENCES

1. Bobrow, D., G. Qualitative Reasoning about Physical Systems, Elsevier Science Publishers B. V., Amsterdam, 1984.
2. Dohnal, M. Naive Models as Active Expert Systems in Bioengineering and Chemical Engineering, Coll. Czech. Chem. Comm., vol.53,1476-1499, 1988.
3. Dohnal, M., Krause, P., and Parsons, S. Some considerations on semiqualitative reasoning. DRUMS RP4 Workshop, Palma da Mallorca, 1990.
4. Dubois, D., and Prade, H. Semantic considerations on order of magnitude reasoning, Proc. IMACS workshop on Qualitative Reasoning and Decision Support Systems, Toulouse, March 1991.
5. Dubois, D., and Prade, H. Fuzzy arithmetic in qualitative reasoning. In: Modelling and Control or Systems in Engineering, Quantum Mechanics, Economics and Biosciences (Proc. Bellman Continuum Workshop 1988 Sophia Antipolis, France) (A. Blaquière, ed.) Springer Verlag, Berlin, 457-467. 1989.
6. Koivisto, R., Dohnal,M., Likitalo, A. Deep and shallow knowledge integration: a case study of an AI diagnosis of a chemical reactor, The Second Scandinavian Conference on Artificial Intelligence, Tampere,1989.
7. Mavrovouniotis, M., L., and Stephanopoulos, G. Order-of-magnitude reasoning with O[M], Artificial Intelligence in Engineering, 4, pp106-114, 1989.
8. Mavrovouniotis, M., L., and Stephanopoulos, G. Reasoning with orders of magnitude and approximate relations, Proc. 6th National Conf. on Artificial Intelligence (AAAI 1987), Seattle, WA, pp 626-630, 1987.
9. Moore, R., E., Interval Analysis. Prentice-Hall, Inc. Englewood Cliffs, N.J. 1966.
10. Parsons, S. Qualitative, semiqualitative and interval algebras. Technical Report, Dept Elec. Eng. Queen Mary and Westfield College, 1991.
11. Parsons, S. On using qualitative algebras in place of metatheories for reasoning under uncertainty: a preliminary report. Technical Report, Dept Elec. Eng. Queen Mary and Westfield College, 1990.
12. Raiman, O. Order of magnitude reasoning, Proc. 5th National Conf. on Artificial Intelligence (AAAI 1986), Philadelphia, PA, pp 100-104, 1986.
13. Struss, P. Mathematical aspects of qualitative reasoning, AI in Engineering, 3, pp156-169. 1988.
14. Travé-Massuyès, L., and Piera, N. The orders of magnitude models as qualitative algebras. Proc. 11th Inter. Joint Conf. on Artificial Intelligence (IJCAI 1989), Detroit Michigan, pp 1261-1266, 1989.

ACKNOWLEDGEMENTS

[illegible]

[illegible]

An Algorithm for the Automated Generation of Rheological Models

A.C. Capelo, L. Ironi, S. Tentoni
Istituto di Analisi Numerica - C.N.R.,
C.so C. Alberto 5, 27100 Pavia, Italy

ABSTRACT

This paper presents a computational approach to the generation of rheological models. A rheological structure can be analogically described as a set of basic components connected in series or in parallel. The models are automatically created in two different forms: a symbolic qualitative relation and a mathematical equation. In both cases, the designed algorithm uses the same knowledge representation scheme which is based on rooted binary tree-like graphs. The mathematical model of a rheological structure, made up of *n* basic components, is built from the basic models of each component by exploiting connection laws. This work is part of a more ambitious project aiming at carrying out a system for automated reasoning about rheological systems. Beside a model-building task, such a system should perform a simulation and diagnostic task. Therefore it should provide methods for the simulation, both qualitative and quantitative, of the behavior of a rheological structure and methods for the identification of a model of an actual material, given its behavior.

1. INTRODUCTION

Automated analysis of physical systems has received more and more attention in the last few years (eg. Bobrow [3]; Weld and De Kleer [16]). Most proposed approaches focus on methodologies to create a qualitative representation of a physical system and algorithms to infer all of its possible behaviors (eg. De Kleer and Brown [6]; Forbus [10]; Kuipers [12]; Williams [17]). As far as the formulation of the model is concerned these qualitative reasoning systems rely on the user. More recent work pointed out the need to develop reasoning systems capable of creating automatically the model and integrating the qualitative knowledge with the quantitative one (eg. Aubin [1]; Berleant and Kuipers [2]; De Kleer [5]; Dormoy and Raiman

[7]; Falkenhaier and Forbus [9]; Forbus and Falkenhaier [11]; Macfarlane [13]).

This paper deals with an on–going program of research aiming at carrying out a system for the automated generation and analysis of rheological models. The system should provide an environment in which an engineer, having a library of models at his disposal, can use different methods (both qualitative and quantitative simulation, paramenters identification and so on) to analyze and classify materials on the basis of their behaviors.

Rheology has been defined as the science of the deformation and flow of matter but a much narrower meaning of the term is usually adopted (eg. Capelo [4]; Reiner [14]). More precisely, in this paper rheology means a set of methods for building models able to predict the behavior of materials when they are subject to external forces, that is, methods for establishing the constitutive laws of materials. Besides modeling a broad spectrum of physical situations, the choice of rheology as application domain is mainly motivated by the actual difficulties in building by hand rheological models even when the number of basic components is low. A rheological model can be associated with an analogical model in which mechanical components, reproducing the fundamental rheological responses of materials (elasticity, viscosity and plasticity), are connected in series or in parallel. In this paper models of visco-elastic materials are considered.

The approach presented in this paper is based on a component–connection paradigm and it is domain dependent, although ideas and techniques can be applied to other physical domains, such as electronics.

A graphical knowledge representation scheme, based on rooted binary tree-like graphs, provides the underlying data structures for the rheological models and allows the system to map a rheological model directly to a symbolic qualitative description, and to a mathematical model. This capability allows for a powerful approach to integrate various symbolic reasoning methods with standard numerical methods.

The basic steps in the design and implementation of the system may be identified as follows:

1. *Enumerate the rheological models, given n components.* At this step a model consists of a symbolic qualitative description of the structure, that is of the rheological components and types of the connections between them. The symbolic expression is strictly analog to an arithmetic one and can be graphically shown as a circuit-like configuration.
2. *Generate the corresponding mathematical model, given a model built in step 1.* The followed method to derive the rheological equation uses the internal state variables whose evolution is described by ordinary differential equations. In this way a library of rheological models of ideal materials is built.

3. ***Simulate, both qualitatively and quantitatively, the dynamic behavior of a model.*** Of course, the quantitative simulation uses the traditional numerical methods for differential equations. The algorithm for the qualitative simulation is designed to operate at the lowest level of description, that is on the symbolic qualitative model and it is mainly based on the response of each component. The response of the model as a whole is assembled taking into account how the action of the external forces is propagated and distributed on different components.

4. ***Derive a model of an actual material, given its behavior.*** First, the model of an ideal material that qualitatively best reproduces the behavior of the real material is looked for in the library of models built in step 1 and 2. Then, through techniques of fitting of experimental data, values of the parameters in the equation can be obtained and therefore a model for the real material can be identified.

The steps 1, 2 are fully designed and their implementation is in progress. The other two steps are sketched out. This paper discusses the methods and algorithms concerning the generation of the symbolic qualitative and mathematical models. First, the domain-specific knowledge for building and analyzing rheological models is discussed. This knowledge is split into two parts: modeling assumptions and basic rheological definitions, basic component models and connection laws. The component models are described by the constitutive laws of materials behaving in a specific rheological way (in our case, elastic or viscous) via internal variables, while the connection laws provide the interaction rules between the different rheological components. Then the algorithm for the automated creation of all rheological models made up of n elements is given in detail. This algorithm generates, according to a suitable definition of equivalence, only all non–equivalent models.

2. THE BASICS OF RHEOLOGY

The most usual classification of materials uses the state of molecular aggregation and divides them into solids, liquids and gases. In engineering this distinction is of little practical value, while it is more important and useful to know how materials react to the action of external forces. As a first classification, materials are defined as elastic, plastic and viscous, according to their behavior. Beside elastic, plastic and viscous materials, whose properties are well defined and fundamental, and which can be considered as ideal materials, there are many other materials which can not be classified, not even approximatively, in these archetypes. But it is worth while noticing that the behavior of almost all materials is characterized by the combination, at various degrees, of these three fundamental properties.

The main goal of rheology is precisely to build models of materials, with composite rheological properties, explaining how the different fundamental rheological properties are combined.

2.1. Modeling Assumptions and Definitions

In this paper a material is assumed to be a continuous and homogeneous medium. Our materials are also assumed to be isotropic, but, here, only one-dimensional external forces and responses are actually considered.

The properties of the materials we study and the external forces acting on them, are mechanical ones. Since unstable materials are not considered, the change over time of the geometrical configuration of a body, that is a deformation, solely occurs when energy is supplied or has been supplied. In our case, such energy is mechanical and is provided through surface forces, such as tractions, or body forces, such as gravity.

When mechanical energy, a *load*, acts on a body, a deformation usually occurs since the material points, making up the body, change their relative positions from the initial state to the final one. How the deformation is carried out, how it depends on the load, and what happens when the load action stops, is generically called the *response* of the material to the action of the load.

The measure of the deformation of a continuous medium is essentially a geometrical problem. A second order tensor is usually used to describe the deformation, but in the one-dimensional case a scalar can be sufficient. Let us consider, just to focus the attention, a very thin cylindrical bar (with the cross-section much shorter than the longitudinal one). A traction or a compression can act along the longitudinal axis of this bar. Let us denote by l_0 the original length of the bar at time $t = t_0$, and by $l = l(t)$ the length of the bar at time t. A measure of the deformation of the bar is given by the ratio $e(t) = (l(t) - l_0)/l_0$, called ***strain***.

Beside the deformation, an internal tension state is concomitantly created in the continuous medium. It is generally described by a second order tensor, the stress tensor, but in the one-dimensional case by a scalar. A measure of the internal tension state of the bar, called ***stress***, is given by the ratio $s(t) = F(t)/A_0$, where A_0 is the area of the cross–section of the bar at time t_0 when the external force is not yet acting, and $F(t)$ is the modulus of the applied external force (this force being considered positive, if it is a traction, or negative, if it is a compression).

The response of the material to the load can be expressed by a functional relation $R(s(t), e(t)) = 0$, where the meaning of the symbol R depends on the approach used to model the material (integral equation, ordinary differential equation, qualitative differential equation).

2.2. Basic component models

The response of an actual material to an external force usually depends on the history of the material itself and on the current value of the applied force. The method we adopted for describing the response of a material is based on the *internal state variables*. Therefore the history of a material is supposed to be expressed only through the current values of a set of variables, the internal state variables, whose time evolution is described by opportune functional relationships. In this approach $R(s, e) = 0$, the *rheological equation*, is a differential equation where the time derivatives of $s(t)$ and $e(t)$ and some parameters (modulus and coefficients of the material) appear.

These rheological equations model ideal complex materials which can be considered the new archetypes. In order to map each actual material to its respective archetype, the responses of actual materials to some typical rheological experiments have to be matched against those of ideal materials.

In this paper, for the sake of simplicity, materials, whose composite responses are the results of a suitable combination of elastic and viscous responses, are considered. But this limitation does not jeopardize the general validity of the used approach, which indeed allows an extension to responses resulting from the combination of the previously mentioned fundamental rheological responses and the plastic ones as well. Moreover, visco–elastic responses describe the behavior of a wide set of actual materials.

Each fundamental response corresponds to an ideal material which can be represented by a mechanical analogous device. More precisely, the purely elastic response is associated to a material H analogically represented by a spring. Similarly, the purely viscous response corresponds to a material N analogically represented by a piston. It is worth noticing that such components can not be located in the actual structure of materials, although it is possible to build mechanical devices made up of springs and pistons and simulating the response of actual materials. But our interest is not focused on this matter: in our view an analogical model is nothing but a diagram where components, graphically represented by springs and pistons, are connected in series or in parallel.

Let us denote by $\mathcal{C} = \{H, N\}$ the set of the basic components of a rheological model and by $\mathcal{R}$ the set of ideal mechanical devices built by connecting in series or in parallel any number of elements of $\mathcal{C}$. Each complex response is assumed to be associated with one or more elements of $\mathcal{R}$.

The response of the elastic component H is described by the rheological equation:

$$s = Ee, \tag{1}$$

generally known as Hooke's law, where E, called Young's modulus, is a constant.

The purely viscous response of the component N is given by the equation (Newton's law):

$$s = \eta \dot{e}, \tag{2}$$

where η is the viscosity coefficient.

In equations (1) and (2) we assume a linear dependence of s on e and $\dot{e}$; in some circumstances it could be useful to consider non-linear relationships: the only difficulty is in the construction of operator R and, from the point of view of parameter identification, in a greater number of introduced parameters.

The simple mathematical models (1), (2) could directly be reformulated in qualitative terms, for examples by confluences or qualitative differential equations, and connection rules could be accordingly given in order to build qualitative models. But for reasoning qualitatively on rheological models, a description of the structure at a lower level is enough. More precisely, information on the modality of interaction between H and N basic components are sufficient to assemble the qualitative response from the basic ones. For this reason qualitative models, in the usual meaning of the term, are not considered in this paper.

2.3. Connection laws

Let $O = \{-, |\}$ be the set of basic connections between elements: in series ($-$) and in parallel ($|$) connections.

Let us consider a system with two components C_1 and C_2 and assume a longitudinal external force (a traction, for example) is acting on the system. The whole system is described, at any time t, by the stress $s(t)$ and the strain $e(t)$, which both are observable. On the contrary, the stresses and the strains, characterizing each component C_1 $(s_1(t), e_1(t))$ and C_2 $(s_2(t), e_2(t))$, need not be observed and are the internal variables. By analyzing from the mechanical point of view the two different kinds of connection, rules for combining the basic models of the components C_1 and C_2 can be provided. These rules allow for the generation of the model of the material as a whole.

Let us examine first the system S_s where C_1 and C_2 are connected in series (Figure 1a). At the lowest level of description, the system can be described by the symbolic qualitative equation $S_s = C_1 - C_2$. At any time, the same force is applied on the whole system and on each component; therefore the stress is the same in each element. The strain of each component is independent from the other and the total strain is obviously given by their sum.

A system S_p, where C_1 and C_2 are connected in parallel (Figure 1b), can be described, at the lowest symbolic level, by the equation $S_p = C_1|C_2$. When the elements are connected in parallel, the total stress on the material results from the sum of the stresses on each component. Moreover the components are simultaneously and equally deformated, so the deformation is the same in each element.

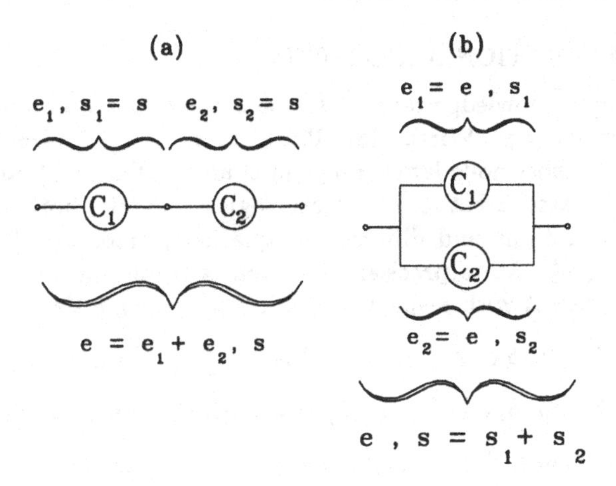

Figure 1. (a) Components C_1 and C_2 connected in series; (b) components C_1 and C_2 connected in parallel. The strains e_1, e_2 and the stresses s_1, s_2 are internal variables while e and s are respectively the observable strain and stress.

The total stress and strain of the systems can be summarized by the following rules:

$$\begin{cases} s = s_1 = s_2 \\ e = e_1 + e_2 \end{cases} \qquad (C_1, C_2 \text{ connected in series}) \tag{3}$$

$$\begin{cases} s = s_1 + s_2 \\ e = e_1 = e_2 \end{cases} \qquad (C_1, C_2 \text{ connected in parallel}) \tag{4}$$

Since the components C_1 and C_2 have not necessarily to be basic components, but could be themselves composed by connected components, the equations (3), (4) give the general rules for building the mathematical models associated with any rheological structure.

In order to create models of materials which exhibit actually different behaviors, filtering rules based on the commutative and associative laws of the compositional rules (3), (4) are exploited. Besides commutativity and associativity, the "behavioral equivalence" of two connected equal components to a single component, that is two series or parallel–connected springs (or pistons) behave like a single spring (respectively piston), is taken into account. In the following, this latter law is denoted by the term *absorption*. Finally, behavioral equivalence based on equation structure similarities will be taken into account while building the differential equations associated to the models.

3. THE GENERATION ALGORITHM

Using a unique knowledge representation scheme, based on rooted binary tree-like graphs (eg. Wirth [18]; Reingold et al. [15]; Dougherty and Giardina [8]) whose node labels range in $\mathcal{O}$ and $\mathcal{C}$ (Figure 2), an algorithm for the automated creation of all rheological models, both as symbolic qualitative equations and differential equations, made up of n elements can be designed. More precisely, our aim is to obtain only the models which are different with respect to the following properties:

$(P1)$ $\forall o \in \mathcal{O}$ $\forall a, b \in \mathcal{R}$ $aob = boa$ (commutativity)

$(P2)$ $\forall o \in \mathcal{O}$ $\forall a, b, c \in \mathcal{R}$ $(aob)oc = ao(boc)$ (associativity)

$(P3)$ $\forall o \in \mathcal{O}$ $\forall a \in \mathcal{C}$ $aoa = a$ (absorption)

A binary tree structure, when representing a rheological model with n elements, has internal node labels ranging in $\mathcal{O}$ (operator identifiers) and leaf labels in $\mathcal{C}$ (component identifiers). A correct balance of operators and operands is guaranteed when internal nodes have exactly two children subtrees and amount to $n-1$, whereas the number of leaves is n; such trees are called *admissible*.

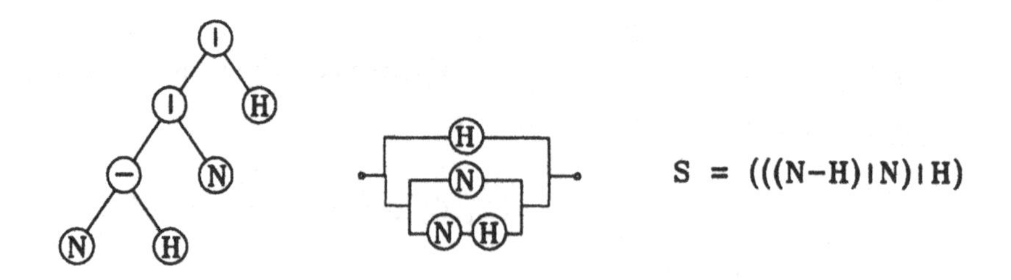

Figure 2. Different representations of the same rheological structure: rooted-binary tree, series-parallel circuit, symbolic qualitative expression.

The fundamental and subsequent steps of the designed algorithm are:

- the automated generation of admissible unlabeled binary tree structures;

- the automated generation of symbolic qualitative equations by labeling the tree structures and filtering out the equivalent ones;

- the automated creation of differential equations.

A description of the algorithm is given in the next subsections. The first two steps are fully implemented, while the implementation of the latter one is still in progress.

3.1. Automated generation of admissible unlabeled binary tree structures

In this subsection we deal with unlabeled trees: no node labels (operator/element identifiers) are specified, and trees are regarded as structured sets of unassigned variables. Our attention is now focused only on the node hierarchy, i.e. on the tree shape.

A reduced set of admissible unlabeled binary trees is generated by grouping structures with respect to an equivalence relation based on property $(P1)$ and by selecting only one representative for each equivalence class. This will allow for a reduction of the combinatorial explosion occurring when trees are labeled.

Let $\mathcal{T}_n$ be the set of all binary n-leaf tree structures which underlie n-component rheological models. Let us denote by $\mathcal{T} = \bigcup_n \mathcal{T}_n$ and, $\forall t \in \mathcal{T}$, by $T_{left}(t)$ (respectively $T_{right}(t)$) the left (respectively right) subtrees of the root of t and define the following recursive relation χ in $\mathcal{T} \times \mathcal{T}$:

Definition 1: $t_1 \chi t_2 \Leftrightarrow$ either t_1 and t_2 have both just one node
or $T_{left}(t_1) \chi T_{left}(t_2)$ and $T_{right}(t_1) \chi T_{right}(t_2)$
or $T_{left}(t_1) \chi T_{right}(t_2)$ and $T_{right}(t_1) \chi T_{left}(t_2)$.

Remark 1: χ is an equivalence relation in $\mathcal{T} \times \mathcal{T}$. In fact it is obviously reflexive and symmetric. By induction let us assume that transitivity holds for trees with depths $\leq m-1$. Let t_1, t_2, t_3 be trees such that $t_1 \chi t_2$, $t_2 \chi t_3$ and with depths $\leq m$. From χ definition: $T_{left}(t_1) \chi T_{left}(t_2)$ and $T_{right}(t_1) \chi T_{right}(t_2)$ or $T_{left}(t_1) \chi T_{right}(t_2)$ and $T_{right}(t_1) \chi T_{left}(t_2)$, and similarly for t_2 and t_3. From the induction hypothesis it is deduced that $T_{left}(t_1) \chi T_{left}(t_3)$ and $T_{right}(t_1) \chi T_{right}(t_3)$ or $T_{left}(t_1) \chi T_{right}(t_3)$ and $T_{right}(t_1) \chi T_{left}(t_3)$, which is the same as saying that $t_1 \chi t_3$.

By means of the χ relation, it is basically stated that trees are equal apart from left/right subtree permutations, according to property $(P1)$. Therefore $\mathcal{T}$ can be replaced by a smaller set $\mathcal{T}^* \subset \mathcal{T}$ gathering the representatives of χ-equivalence classes; the following recursive algorithm for creating such representatives is defined:

$$(\mathcal{A})\begin{cases} \mathcal{T}^* = \bigcup_n \mathcal{T}_n^*, \text{ where } \mathcal{T}_n^* \text{ are recursively generated :} \\ \bullet\ \mathcal{T}_1^* = \mathcal{T}_1 \\ \bullet \text{ Given } \mathcal{T}_{n-1}^*, \mathcal{T}_n^* \text{ is generated as follows :} \\ \quad \forall t \in \mathcal{T}_{n-1}^* \text{ leaves are numbered by post}-\text{order traversal of } t \text{ (Figure 3);} \\ \quad\quad \forall \text{ leaf } p \text{ of } t \text{ (following the above leaf}-\text{ordering)} \\ \quad\quad\quad \text{Let } t' \text{ be the } n-\text{leaf tree obtained by appending the left and right direct descendents of } p. \\ \quad\quad\quad \text{If } t'\chi t^* \text{ is false } \forall t^* \in \mathcal{T}_n^* \text{ then } \mathcal{T}_n^* := \mathcal{T}_n^* \cup \{t'\}; \end{cases}$$

(for the definitions of post-order traversal and other classical tree traversal schemes as well as for the usual tree terminology, see eg. Wirth [18]).

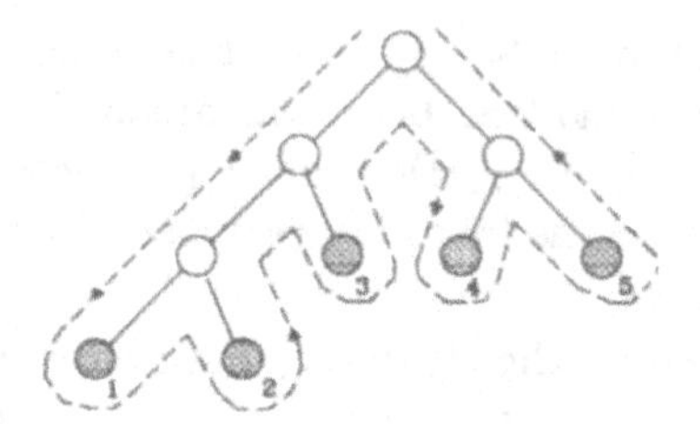

Figure 3. Leaf numbering scheme induced by post-order traversal.

Figure 4 shows all the tree structures generated for n=4 (Figure 4a) and n=5 (Figure 4b).

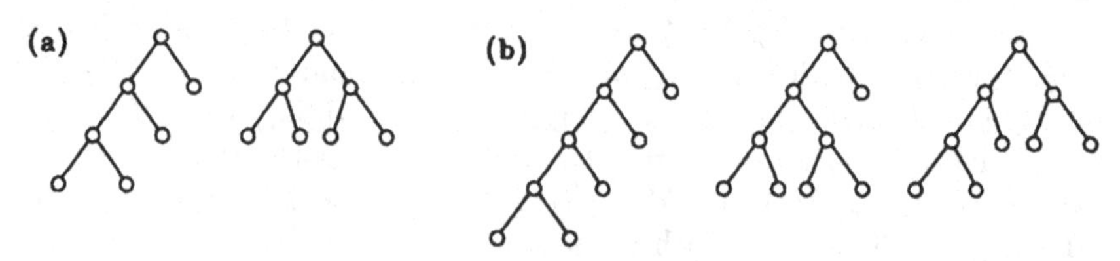

Figure 4. Unlabeled binary trees: (a) $\mathcal{T}_4^*$; (b) $\mathcal{T}_5^*$.

Remark 2: Only admissible $n-$leaf trees are generated (all internal nodes have two children subtrees) and only one representative for each χ-equivalence class is produced. Since the leaf numbering follows a left-to-right scheme, the algorithm $\mathcal{A}$ generates "left-biased" trees, i.e. structures whose right subtrees are not deeper and do not have more nodes than the left ones. Conversely, trees with such property are unique in each χ-class. This holds a characterization of the elements of $\mathcal{T}^*$.

3.2. Automated generation of symbolic qualitative equations

Unlabeled trees, viewed as structured sets of variables, are given values in $\mathcal{O}^{n-1} \times \mathcal{C}^n$, i.e. $(2n-1)$-tuples of identifiers are provided to label the tree nodes. Consequently the elements of the set $\mathcal{M}_n = \mathcal{T}_n^* \times \mathcal{O}^{n-1} \times \mathcal{C}^n$ can be identified with labeled trees, representing n-component rheological models, i.e. $\mathcal{M}_n$ and $\mathcal{M} = \bigcup_n \mathcal{M}_n$ are identified with subsets of $\mathcal{R}$. Adequate filter procedures based on properties $(P1)-(P3)$ are defined so that the set $\mathcal{M}_n$ can be replaced by the set of all distinct n-component rheological models.

In terms of trees, associativity means that trees like t_1 and t_2 of Figure 5 are equivalent. Let's notice that if $t_1 \in \mathcal{M}_n$ then (see Remark 2) $t_2 \in \mathcal{M}_n$ too. Indeed, if $t_1 \in \mathcal{M}_n$ then the subtree **a** is "heavier and deeper" than **b**σ**c** and then **a**σ**b** is "heavier and deeper" than **c**.

We can define the following relation α in $\mathcal{M}_n \times \mathcal{M}_n$:

Definition 2: $t_1 \alpha t_2 \Leftrightarrow t_2$ can be obtained from t_1 by a finite number of applications of $(P2)$.

It can be easily proved that α is an equivalence relation.

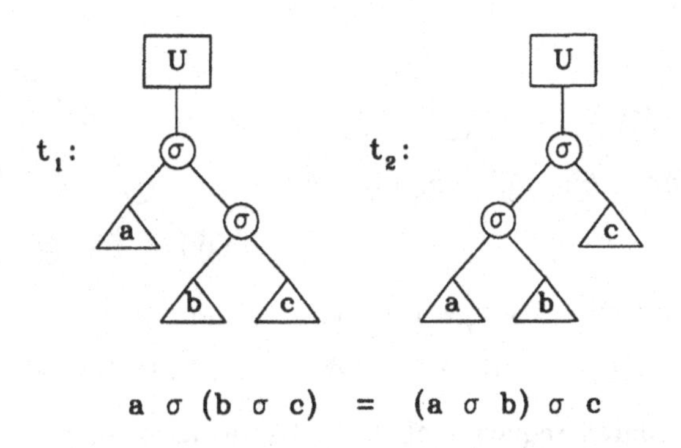

Figure 5. Meaning of associativity: trees t_1 and t_2 are equivalent. **a**, **b**, **c** denote subtrees, and **U** denotes any upper tree structure.

Since equal operator sequences can be evaluated in any order, a selection criterion for α-class representatives can be given by fixing a left-to-right evaluation order. This corresponds to discard all trees containing any right-branched sequence of equal operators, i.e. trees containing an operator node with equal valued right descendant.

Let's denote by $\mathcal{M}_n^* \subset \mathcal{M}_n$ the resulting set and $\mathcal{M}^* = \bigcup_n \mathcal{M}_n^*$. A criterion for taking into account the arbitrary disposition of operands related to the same operator (commutativity) has now to be considered. To this end, an ordering in $\mathcal{M}^*$ is introduced.

Let's define first the mappings:

Definition 3: $\forall p \in \mathcal{M}_n^*$ let
$\pi(p) = a_1 a_2 \ldots a_{2n-1}$ be the $(2n-1)$-tuple of identifiers obtained by pre-order traversal of p

and

$$\rho(p) = b_1 b_2 \ldots b_{2n-1}, \quad b_i = \begin{cases} 0 & \text{if } a_i \in C \\ 1 & \text{if } a_i \in O \end{cases}$$

be the *binary mask* of $\pi(p)$.

Assuming that an ordering for node identifiers is fixed (for example $H < N < - < |$), the following relation can be defined for $p,\ q \in \mathcal{M}^*$:

Definition 4: $p < q \Leftrightarrow$ either $\rho(p) < \rho(q)$
or $\rho(p) = \rho(q)$ and $\pi(p) < \pi(q)$

where binary masks are compared as binary integers and identifier strings are in lexicographic order.

Figure 6 shows an ordered sequence of labeled trees according to definition 4.

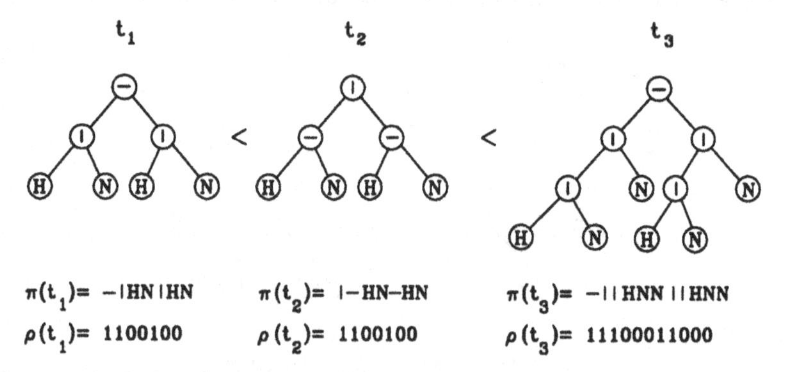

Figure 6. Ordered sequence of labeled trees according to definition 4. Unlabeled trees are first compared through ρ. In case of match labeled trees are compared through π.

It holds that relation $<$ is a total order in $\mathcal{M}^*$. Let's also define for $t_1, t_2 \in \mathcal{M}_n^*$ the following equivalence relation:

Definition 5: $t_1 \gamma t_2 \Leftrightarrow t_2$ can be obtained from t_1 by using property $(P1)$.

The set $\mathcal{M}_n^*$ can now be filtered out with respect to commutativity by requiring that the arguments of any sequence of equal operators are given left-to-right in descending order, and discarding labeled trees that do not fulfill such a condition. Figure 7 shows labeled trees which are equivalent with respect to commutativity.

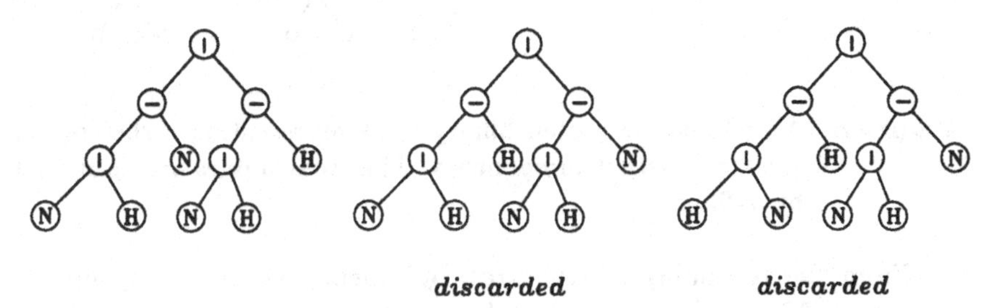

Figure 7. Labeled trees equivalent with respect to commutativity.

Absorption ($P3$) can be taken into account too, by additionally requiring that no couple of equal leaves appear as argument to any operation. When this occurs, the rheological model is actually equivalent to another one having a lower number of basic components. This filtering rule corresponds again to select representatives of the classes induced by an equivalence relation:

Definition 6: $t_1 \beta t_2 \Leftrightarrow t_2$ can be obtained by t_1 by using ($P3$).

Let $\mathcal{R}^*$ be the resulting set after filtering $\mathcal{M}^*$ with respect to relations γ and β.

Symbolic qualitative equations are directly obtained from labeled trees in $\mathcal{R}^*$ by in-order traversing and placing the parentheses in order to keep all hierarchical information, as it is usually done for arithmetic expressions represented by binary trees.

These symbolic expressions give a description of a rheological structure at the lowest level. Nevertheless they can be usefully exploited to simulate qualitatively the behavior of their corresponding ideal materials in response to simple load signals. The qualitative comparison of the simulated behaviors against the responses of actual materials to the same load signals allows to associate any material with a model.

3.2.1. Example In order to clarify how the previously described filter procedures work, let's consider the case n=4. The χ-filter leads to discard 3 structures out of 5 (Figure 8).

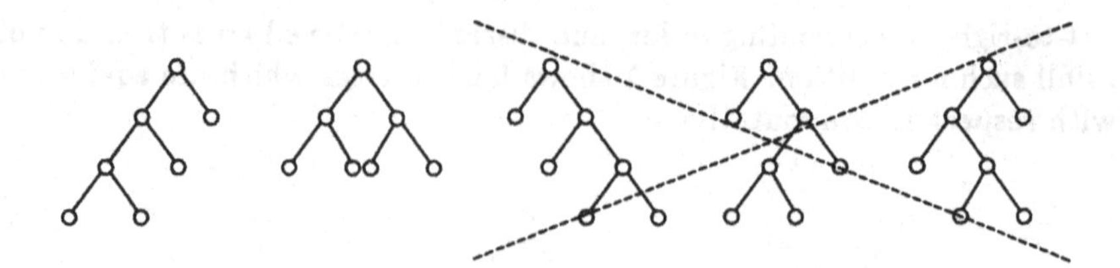

Figure 8. All admissible rooted binary trees representing a rheological structure with 4 components. The crossed ones are discarded by χ-filter.

From the remaining 2 structures, by labeling nodes in all possible ways, $2 * 2^{2n-1} = 2^8$ labeled trees (i.e. qualitative symbolic equations) are then considered. Most of them (244) are discarded by absorption or by the associativity, commutativity arguments. For example the following symbolic qualitative equations are all considered but only the first one is held while the others are discarded:

$$S = (((N - H)|N)|H)$$

$$\left.\begin{array}{ll} S = (((H - N)|N)|H); & S = (((N - H)|H)|N) \\ S = (((H - N)|H)|N) & \end{array}\right\} \quad \text{commutativity}$$

$$\left.\begin{array}{ll} S = ((N - H)|(N|H)); & S = ((N - H)|(H|N)) \\ S = ((H - N)|(H|N)); & S = ((H - N)|(N|H)) \end{array}\right\} \quad \text{associativity}$$

$$\left.\begin{array}{ll} S = ((N|H)|(N - H)); & S = ((N|H)|(H - N)) \\ S = ((H|N)|(H - N)); & S = ((H|N)|(N - H)) \end{array}\right\} \quad \begin{array}{l}\text{associativity +} \\ \text{commutativity}\end{array}$$

As a result, the set of models with 4-basic components consists of the following 12 different symbolic qualitative equations:

$$\begin{array}{ll} S = (((N - H)|H) - H); & S = (((N|H) - N) - H) \\ S = (((N - H)|N) - N); & S = (((N|H) - H)|H) \\ S = (((N - H)|N) - H); & S = (((N|H) - N)|H) \\ S = (((N - H)|H) - N); & S = (((N|H) - H)|N) \\ S = (((N - H)|N)|H); & S = (((N|H) - N)|N) \\ S = ((N - H)|(N - H)); & S = ((N|H) - (N|H)) \end{array}$$

3.3 Automated creation of differential equations

In order to automatically derive the differential equations, symbolic algebra procedures, tailored specifically to the problem, are used by exploiting the hierarchical information captured by the labeled trees, the related internal variables, the differential equations modeling the basic components, and the connection laws (3), (4).

In order to give an idea how mathematical models can be derived, let us consider, as an example, a system where $C_1 = H$ and $C_2 = N$. In this case, for the absorption and commutativity laws, there are only two rheological structures with behavioral differences: $S_s = H-N$, $S_p = H|N$. The components H and N are respectively described, in terms of internal variables, by $s_1 = Ee_1$ and $s_2 = \eta\dot{e}_2$. Let us consider first the material S_s. For the equation (3), it is $s = Ee_1$ and $s = \eta\dot{e}_2$. The addition of e_1 and e_2 cannot be directly performed since these internal variables appear with different derivative orders in the fundamental equations. Therefore the rule $e = e_1 + e_2$ can more conveniently be substituted by the expression of its first derivative. Finally the rheological differential equation $s + \frac{\eta}{E}\dot{s} = \eta\dot{e}$, describing the response of the composite material $S_s = H - N$, is obtained.

The mathematical model of the material $S_p = H|N$ is easily generated by using the rules (4): $s = s_1 + s_2 = Ee_1 + \eta\dot{e}_2 = Ee + \eta\dot{e}$.

For complex cases, that is when the structure is made up of more than two elements, the differential equation is built in an analogous way by combining the equations of two basic or composite components. Of course, some difficulties, due to the symbolic manipulation, arise when complex expressions must be linked together.

The set $\mathcal{R}^*$ can still be filtered out by using an equivalence relation based on the differential equations. More precisely two models are equivalent if the derivatives of e and s appear at the same order in the corresponding differential equations. In the example described in 3.2.1, this filter allows to further reduce the set of 12 elements to 4 elements (actually, two of them are equivalent to structures with 3 components).

4. CONCLUSIONS

We have proposed a systematic approach to building rheological models. The algorithm that we have designed and implemented automatically generates a library of models of ideal materials.

All of the parts of the generation algorithm have been implemented and tested except for the automated derivation of differential equations. Our program is written in C. The implementation of the derivation of differential equations is in progress and done in MATHEMATICA, a symbolic algebra package providing high level symbolic algebra functions for manipulating mathematical structures.

An important practical advantage of this algorithm is that also those complex situations which are quite hard to handle by hand can be modeled.

In fact, it is difficult, even for $n = 4$ to derive by hand both all the symbolic qualitative equations and differential equations which describe the different behaviors of a 4–components rheological structure. But the key contribution of this work consists in its capability to represent a model using two different formalisms in the same environment. This allows to carry out a versatile tool providing the user with a rich set of methods, ranging from the qualitative to the numerical ones, for the automated analysis of rheological systems.

The work is on-going in two directions. The first is to complete the implementation of the algorithm concerning the derivation of differential equations. The second, to design and implement simulation and parameters identification algorithms. Qualitative simulation algorithms will be given top priority.

One limitation in this implementation is that it does not take into account the plastic case. A further extension of the system will also concern with this case: the knowledge representation scheme and the set O are the same, and C is $\{H, N, S\}$, where S is the basic component representing the plastic material. The generation algorithm will always exploit commutative and associative filters but need some modifications mainly as far as the derivation of differential equations is concerned.

REFERENCES

1. Aubin, J.-P. Qualitative Simulation of Differential Equations, J. Differential and Integral Equations, Vol. 2, pp. 183–192, 1989.
2. Berleant, D. and Kuipers, B.J. Combined Qualitative and Numerical Simulation with $Q3^*$, Proceedings of the 4th Int. Work. on Qualitative Physics, pp. 140–152, Lugano, 1990.
3. Bobrow, D.G. (Ed.). Qualitative Reasoning about Physical Systems, North Holland, Amsterdam, 1984.
4. Capelo, A.C. Introduzione allo Studio dei Modelli Reologici, Technical Report, University of Pavia, 1983.
5. De Kleer, J. Compiling Devices and Processes, Proceedings of the 4th Int. Work. on Qualitative Physics, pp. 169–182, Lugano, 1990.
6. De Kleer, J. and Brown, J.S. A qualitative physics based on confluences, Artificial Intelligence, Vol. 24, pp. 7–83, 1984; also in: Bobrow, D.G. (Ed.). Qualitative Reasoning about Physical Systems, North Holland, Amsterdam, 1984.
7. Dormoy, J.-L. and Raiman, O. Assembling a Device, Proceedings of the Nat. Conf. on Artificial Intelligence (AAAI–88), Morgan Kaufmann, Los Altos, CA., pp. 330–335, 1988.

8. Dougherty, E.R. and Giardina, C.R. Mathematical Methods for Artificial Intelligence and Autonomous Systems, Prentice-Hall, Englewood Cliffs, 1988.
9. Falkenhaier, B. and Forbus, K.D. Compositional Modeling of Physical Systems, Proceedings of the 4th Int. Work. on Qualitative Physics, pp. 1–15, Lugano, 1990.
10. Forbus, K.D. Qualitative process theory, Artificial Intelligence, Vol. 24, pp. 85–168, 1984; also in: Bobrow, D.G. (Ed.). Qualitative Reasoning about Physical Systems, North Holland, Amsterdam, 1984.
11. Forbus, K.D. and Falkenhaier, B. Self-Explanatory Simulations: An integration of qualitative and quantitative knowledge, Proceedings of the 4th Int. Work. on Qualitative Physics, pp. 97-110, Lugano, 1990.
12. Kuipers, B.J. Qualitative Simulation, Artificial Intelligence, Vol. 29, pp. 289–338, 1986.
13. Macfarlane, J.F. Qualitative and Symbolic Analysis of Dynamic Physical Systems, Phd. Thesis Dissertation, University of Minnesota, 1989.
14. Reiner, M. Rheology, in: Encyclopedia of Physics (Ed. Flugg, S.), Vol. 6, pp. 434–550, Springer Verlag, Berlin, 1958.
15. Reingold, E.M., Nievergelt, J. and Deo, N. Combinatorial Algorithms, Prentice–Hall, Englewood Cliffs, 1977.
16. Weld, D.S. and De Kleer, J. Readings in Qualitative Reasoning About Physical Systems, Morgan Kaufman, Los Altos, CA., 1990.
17. Williams, B.C. Qualitative Analysis of MOS Circuits, Artificial Intelligence, Vol. 24, pp. 281–346, 1984; also in: Bobrow, D.G. (Ed.). Qualitative Reasoning about Physical Systems, North Holland, Amsterdam, 1984.
18. Wirth, N. Algorithms + Data Structures = Programs, Prentice–Hall, Englewood Cliffs, 1976.

Expert's Interface in a Generator of Process Supervision Systems: Relational Management of an Object-Oriented Knowledge

J. Pastor (*), P. Grivart, H. Grandjean

INSERM CJF 89-08, CHU La Grave, 31052 Toulouse Cedex, France

(*) Currently at Texas A&M University, Department of Forest Science, College Station, Texas 77843-2135, USA

ABSTRACT

When supervising a process, an operator uses two kinds of reasoning strategies, task-dependent (domain-free) or domain-dependent. The RONSART extended object-oriented model, the core of a generator of process supervision expert systems, aims at modeling the domain-free reasoning involved in supervision. Generating such a system using this specific generator consists essentially of acquiring the process-dependent knowledge, according to the model provided by RONSART. An interface, for the expert to use in building a system with the generator, was developed. Its goal was to provide a friendly environment for knowledge acquisition and mechanisms for managing the knowledge base and checking for its consistency with the RONSART model. This interface was implemented with a relational data base management system. The design of the relational knowledge base has used the object-oriented model as the information's conceptual schema. The attributes of the specific object features, such as inheritance, and the coherence of the knowledge base with the RONSART model have been implemented through dynamic integrity constraints.

INTRODUCTION

Currently, many industrial processes are automatically controlled, while their supervision is under an operator's responsibility. Supervision is defined here as the act of detecting failures, diagnosing their causes, predicting their consequences and determining corrective actions. Cognitive overload has resulted in many processes where the automated process controllers have generated more information that the operator can synthesize and interpret. A process supervision expert system is needed to give the operator guidance for obtaining and analyzing the pertinent information and making decisions. But, for the system to be accepted by the operator, its reasoning strategies must parallel the operator's.

Ergonomic studies (e.g. Alengry [1], Landeweerd et al. [2], Lejoly et al. [3], Sébillotte-Garnier [4]) have pointed out two kinds of reasoning, depending upon whether the situation the operator has to face is "routine" or "new". They have noted that, in "routine" tasks, the operator approaches each with a kind of task-specific, domain-free "mind set" and that, in "new" situations, he makes use of the more fundamental principles governing the process. This last type of reasoning can typically be simulated by a system using a deep model of the process, that is its structural, behavioral and functional description.

The supervision-specific, domain-free concepts and reasoning are the core of RONSART (*Representing with Objects the Notion of Supervision through Anomalies Related to Time*) (e.g. Pastor [5]), a generator of expert systems that are designed to help the operator supervise processes in "routine" situations. Generating a specific expert system in RONSART consists of four steps: 1) modeling the knowledge relative to the process, according to the frame provided by the supervision concepts, 2) acquiring the knowledge through the expert's interface, 3) storing the knowledge, for convenient management, in a relational data base and 4) translating the relational tables into RONSART object classes.

In this paper, we will present briefly the interface design and implementation, fully described by Grivart [6]. We will describe the context of this implementation, that is the RONSART model. We will also justify further our selection of a relational data base management system for knowledge acquisition and manipulation. We will then show how the translation between the two models has been solved and finally describe the interface's realization.

THE RONSART MODEL

The RONSART model (e.g. Pastor [7]) proposes an "extended" object-oriented representation of the concepts manipulated during the supervision of any process. In this model (figure 1), supervision concepts such as DEVICE, PARAMETER... are generic classes, organized around the central notion of ANOMALY by inter-class links. Each class is structured by intra-class links and has intrinsinc properties which are links between a class or instance and a value (string or number). The domain entities are expressed as subclasses of the generic classes (figure 2). Facts related to the functioning of a specific process are represented by classes' instances.

The ANOMALY class has a double organization: hierarchical, through KIND_OF, and causal, through CAUSE_OF. The hierarchical organization allows the system to reason, in a certain context, at the most general efficient level possible and thus, to limit the numbers of PARAMETER's measured. The CAUSE_OF relation supports the system's main reasoning and is explored in two directions. When an ANOMALY is detected, a diagnostic search is initiated to look for all the ANOMALIES that could be its direct causes. Prediction of the evolution of the process, under the detected ANOMALY, consists of verifying all the ANOMALYs that could be caused by the detected one.

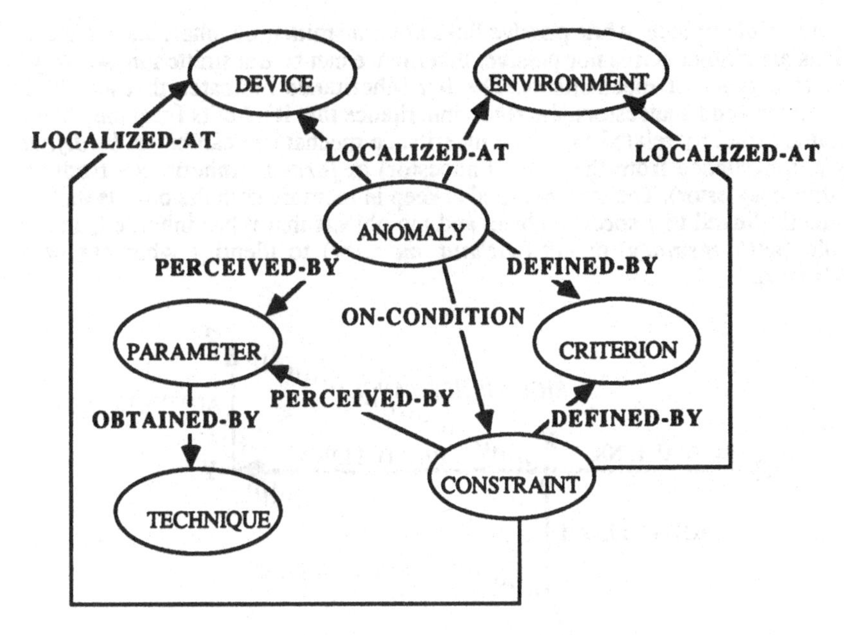

Figure 1. RONSART model

oil_leak	<u>KIND-OF</u>	ANOMALY
	<u>CAUSE-OF</u>	engine_lubrication_failure *(ANOMALY)*
	TYPE	temporary *(string values)*
	VALUE	(TRUE, FALSE)
	LOCALIZED_AT	engine *(DEVICE)*
	PERCEIVED_BY	oil_spot_existence *(PARAMETER)*
	DEFINED_BY	yes_answer* *(CRITERION)*

inter-class link
<u>intra-class link</u>
intrinsic property
* yes_answer returns a TRUE VALUE for the ANOMALY if the PARAMETER's VALUE is yes

Figure 2. A domain entity example

In order to deal with the complexity of the reasoning, an "extended" object-oriented model has been defined (e.g. Pastor et al. [8]). In this model, the inter and intra-class links, as well as the intrinsic properties, are defined as objects and are thus manipulated through their properties. One of the most important and complex properties is **inheritance** (figure 3) which becomes a local concept (i.e. can be different in each class) and can be parametrized. The links are classified according to their behavior towards inheritance. *Active* links convey inheritance, while *passive* links are transmitted by inheritance. *Neutral* links are neither active nor passive. Inheritance can be transmitted in two ways, by four types of mechanisms. The *left* inheritance indicates that an object inherits from its ancestors, the *right* inheritance that it inherits from the objects that are directly related to it. The inheritance mechanism can be set *by default* (i.e. inheritance from the nearest ancestor) or *forced* (inheritance from the farthest ancestor). The system can also keep in memory both the objects that are directly linked to a specific object and the objects that it has inherited, and be able (*with memory*) or not (*without memory*) to identify what has been inherited.

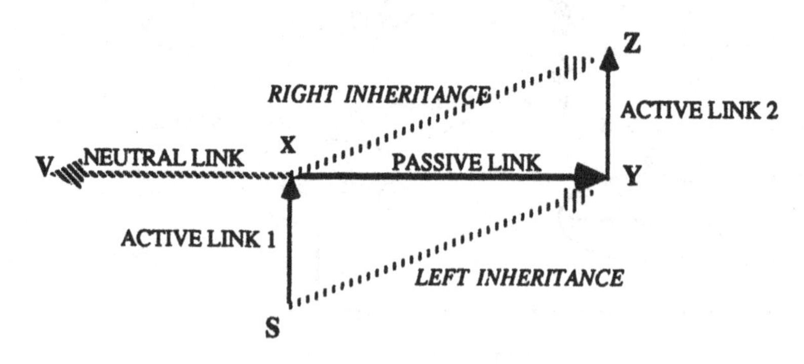

Figure 3. Inheritance mechanisms

The extended object-oriented model and the temporal and causal reasonings are complex features that have made the use of GENESIA-II *(Software designed by E.D.F. and commercialized by STERIA)* necessary. This software has a powerful second-order logic language, originated from the BOOJUM language, designed by Dormoy [9]. RONSART's specific knowledge is expressed two ways. The knowledge structure, that is the classes and links, is described by sets of GENESIA-II facts, i.e. by sets of triplets. The semantics of the knowledge are implemented through GENESIA-II rules.

Generating the expert system linked to the supervision of a specific type of process consists of structuring the domain entities according to the supervision concepts, that is, expressing the knowledge of the process as subclasses of the RONSART classes, and then writing them as GENESIA-II triplets. However, GENESIA-II does not offer any knowledge acquisition tool and manipulating triplets through a word processor has many drawbacks, including lack of readability and lack of management of the information's integrity or security. In fact, the knowledge acquisition must be in the expert's hands. He thus needs a friendly interface that can guide him in data acquisition by providing the frame of RONSART's concepts and that can manage the knowledge by checking the information's integrity and coherence.

INTERFACE DESIGN: THE CHOICES

The choice of a Data Base Management System (DBMS) for acquiring and manipulating the domain knowledge is straightforward. Such a system can manage large volumes of information and respect its integrity and security. Moreover, most available DBMS's provide tools for designing input screens, facilities for updating information and a high level language for accessing it.

A Relational DBMS (RDBMS) is the only realistic choice, for at least two reasons:
1) Object-oriented DBMS's are still prototypes and are more persistent object manipulators than full-fledged DBMS, as outlined by Bancilhon [10]; they still do not solve such important problems as the integrity and security of the DB or the management of concurrent accesses. Moreover the models provided by these systems are "weaker" than RONSART, especially in inheritance mechanisms.
2) Extended RDBMS, such as SABRINA *(INFOSYS, France)* or INGRES *(RELATIONAL TECHNOLOGY CORPORATION, U.S.A.)*, which have an object-oriented layer operating on relational data bases, cannot be used. Their object model is too limited compared to the RONSART model and their inferential capabilities are weak.

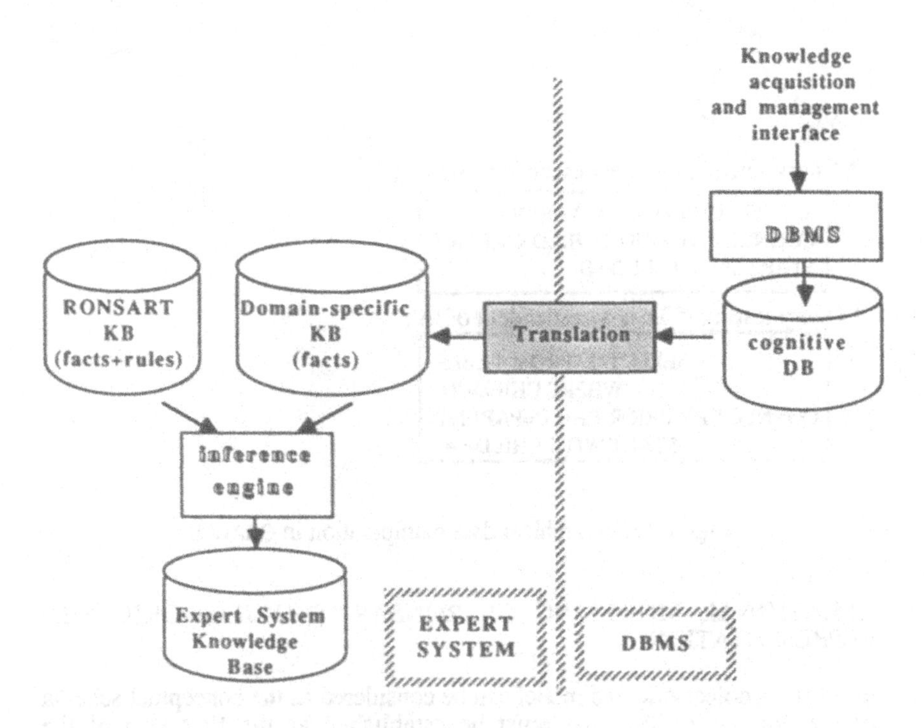

Figure 4. Generation of an expert system

The linkage between RONSART and the RDBMS consists of a translation (figure 4) of the relational knowledge base into its object-oriented counterpart, as the base is being created or updated. It has been chosen not to implement a strong coupling of the expert system and the RDBMS for several reasons, including the following. First, it can be hazardous for the expert system to access a knowledge base that could be updated by the expert while the system is used. Second, a real-time access of the relational knowledge base by the expert system would be less efficient than its use of a compiled knowledge base. By compiled, it is meant that the factual knowledge base is structured so that its access by the rules that manipulate it is optimized for a faster inference.

The RDBMS ORACLE *(ORACLE Corporation, U.S.A.)* has been chosen primarly because its extensions of the SQL language allow a manipulation of hierarchical data (figure 5). Its generator of transactional applications, SQL-FORMS, is a 4th generation language which can be used for quickly realizing input sheets and implementing simple integrity constraints. In addition, ORACLE is available on many different types of computers.

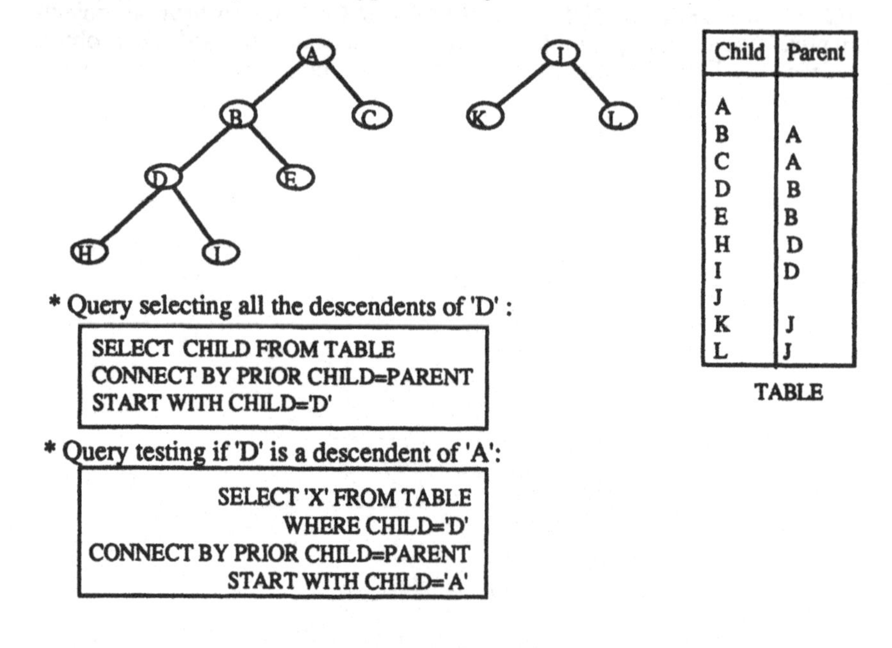
A
B
C
D
E
H
I
J
K
L
* Query selecting all the descendents of 'D' :
SELECT CHILD FROM TABLE
CONNECT BY PRIOR CHILD=PARENT
START WITH CHILD='D'
* Query testing if 'D' is a descendent of 'A':
SELECT 'X' FROM TABLE
WHERE CHILD='D'
CONNECT BY PRIOR CHILD=PARENT
START WITH CHILD='A'

Child	Parent
A	
B	A
C	A
D	B
E	B
H	D
I	D
J	
K	J
L	J

TABLE

Figure 5. Hierarchical data manipulation in ORACLE

RELATIONAL MODELING OF RONSART'S OBJECT-ORIENTED REPRESENTATION

RONSART's object-oriented model can be considered as the conceptual schema of the domain entities that must be established as the first step of the implementation of an information management system. Going from this conceptual schema to a relational schema follows two guidelines. The logical guideline consists of normalizing relations. The empirical guideline aims at retaining the entities and associations that have been found in the first step and at taking into account the integrity constraints as well as the specific problems linked to the implementation of the application. RONSART classes and links have a straightforward translation, according to Chen's entity-association model (figure 6).

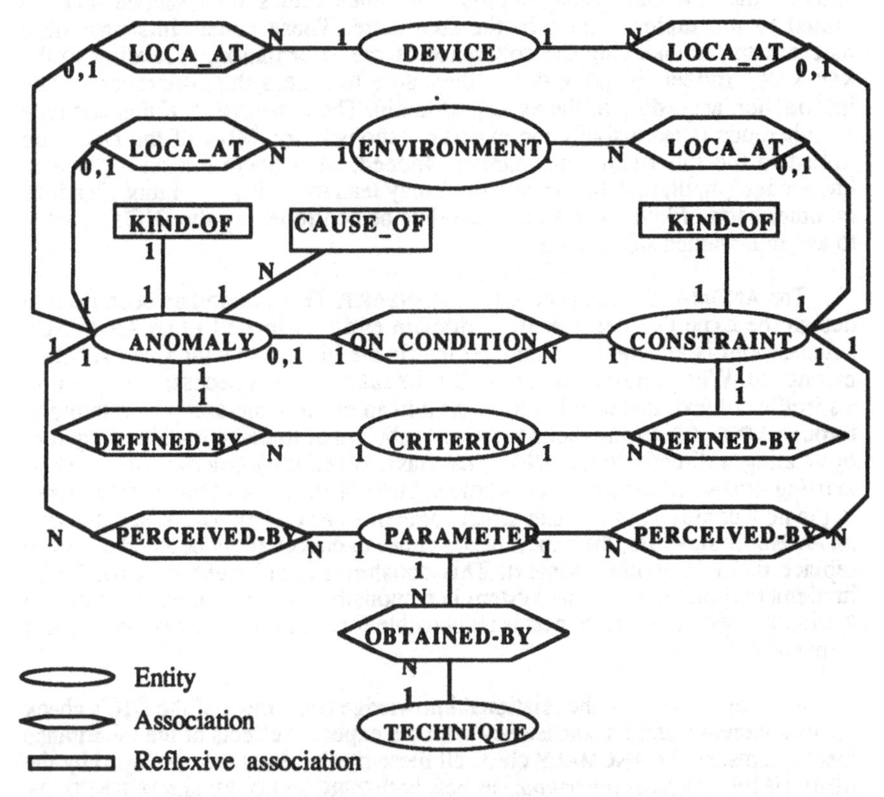

Figure 6. Entity-association model for RONSART

The relational schema that corresponds directly to this entity-association model is in the third normal form. In this schema, each RONSART class is represented by a relation, the attributes of which are the class' intrinsic properties. Each link, whether it is intra or inter, is a table too, whose attributes are the names of the "input" class and "output" class for the link. Since RONSART represents an accurate conceptual model of the information, the object-oriented representation, the entity-association schema and the relational model are very close to each other. This is convenient for designing a knowledge acquisition interface that is very close to the knowledge organization in the expert system.

The semantics of RONSART classes and links have been partially translated by Dynamic Integrity Constraints (DIC). DIC's are sets of logical conditions which must be verified so that, each time a coherent state of the data base is updated, the new state is coherent too. The tuples' values in the second state are linked to the tuples' values in the first state. These constraints provide a mechanism for checking the coherence of the information, according to the RONSART model. In no way are they able to assess the coherence of the information according to the expert's domain. The assessment of this last type of coherence is exclusively the expert's responsibility. Most of the DIC's are linked to the inheritance mechanism. Indeed, each operation on a piece of knowledge (input, update, cancellation) may lead to unsuspected modifications on other parts of knowledge and must thus be controlled by the DIC's in order to avoid unwanted side-effects.

The ANOMALY class is central to RONSART. This notion has been used to design the expert's interface. Each domain entity is related to an ANOMALY subclass and is attached to a context that is the hierarchy which this subclass belongs to. When *creating a new ANOMALY subclass,* it is necessary to put it in a specific context, that is either to insert it in an existing hierarchy by relating it, through KIND_OF, to an other ANOMALY subclass or to make it a hierarchy root by relating it directly to the ANOMALY class. When *updating the context of an existing ANOMALY subclass,* the whole subtree of its descendents is transferred in the new context, but this can affect some links as CAUSE_OF. When *deleting an ANOMALY subclass,* the expert needs either to cancel all its descendents or to replace them in another context. This constraint is coherent with the DIC's implementation strategy: the system is responsible for the coherence with the RONSART model, but the expert is responsible for the knowledge taxonomy and semantics.

After operations on the relational knowledge base, most of the DIC's check for its coherence and its consistency with the expected effects of the inheritance mechanisms. In the ANOMALY class, all these mechanisms are conveyed by the KIND_OF link. *Default inheritance* affects both PERCEIVED_BY and DEFINED_BY links. These links are thus transmitted from the parent class to the child class, when they are not directly defined at the level of the child class. In the relational model, this is translated by the fact that a tuple's value may depend on another tuple's value. *Forced inheritance*, which affects the LOCALIZED_AT link, consists of forcing each member of an ANOMALY's subtree to inherit the ANOMALY's localization (figure 7). In the relational knowledge base, the information on the forced inherited links is duplicated at each level of the hierarchy. This increases the volume of stored information, but improves the efficiency of the access to this information. The expert can get it directly instead of retrieving it at the root of the hierarchy.

The *inheritance without memory*, can be illustrated by the fact that each ANOMALY subclass exists ON_CONDITION that a certain CONSTRAINT is verified **and** ON_CONDITION that all the CONSTRAINTs linked to its ancestors are verified. In the relational knowledge base, only the direct ON_CONDITION link is stored. The drawback is that is the indirect conditions are not readily available for the expert. However, this strategy has many advantages: no redundant information is stored and no modification of the ANOMALY context can affect its ON_CONDITION link. The *inheritance with memory,* which concerns the CAUSE_OF link, is both a *right inheritance* and a *left inheritance* (figure 8).

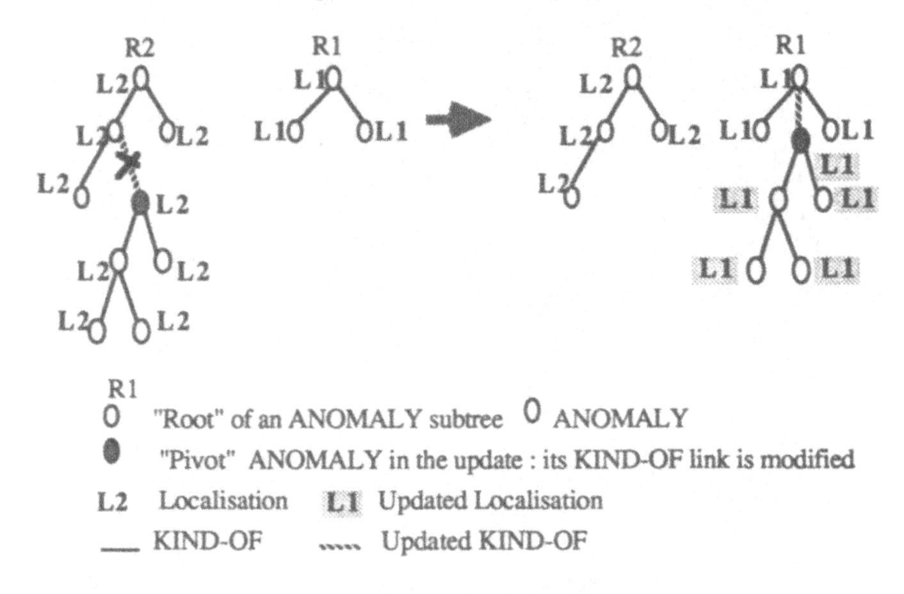

Figure 7. Forced inheritance

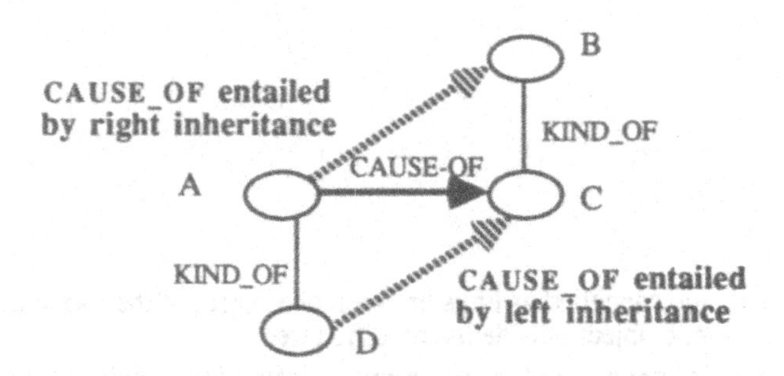

Figure 8. Right and left inheritance for CAUSE_OF

For example, the ANOMALY A is CAUSE_OF C (direct link), and also of B (right inheritance); D is CAUSE_OF C (left inheritance). The inheritance is with memory because the expert system recognizes that C is direct and B is inherited, and treats them differently. In the relational data base, only the direct link is kept. This strategy allows the minimum amount of information to be stored. Moreover, a CAUSE_OF link can be cancelled without any consequence on the relational data base, or an ANOMALY's context can be modified without updating the CAUSE_OF links that are inside its subtree. However, any update of an ANOMALY's context has an influence on the CAUSE_OF links starting from or terminating at its subtree (figure 9).

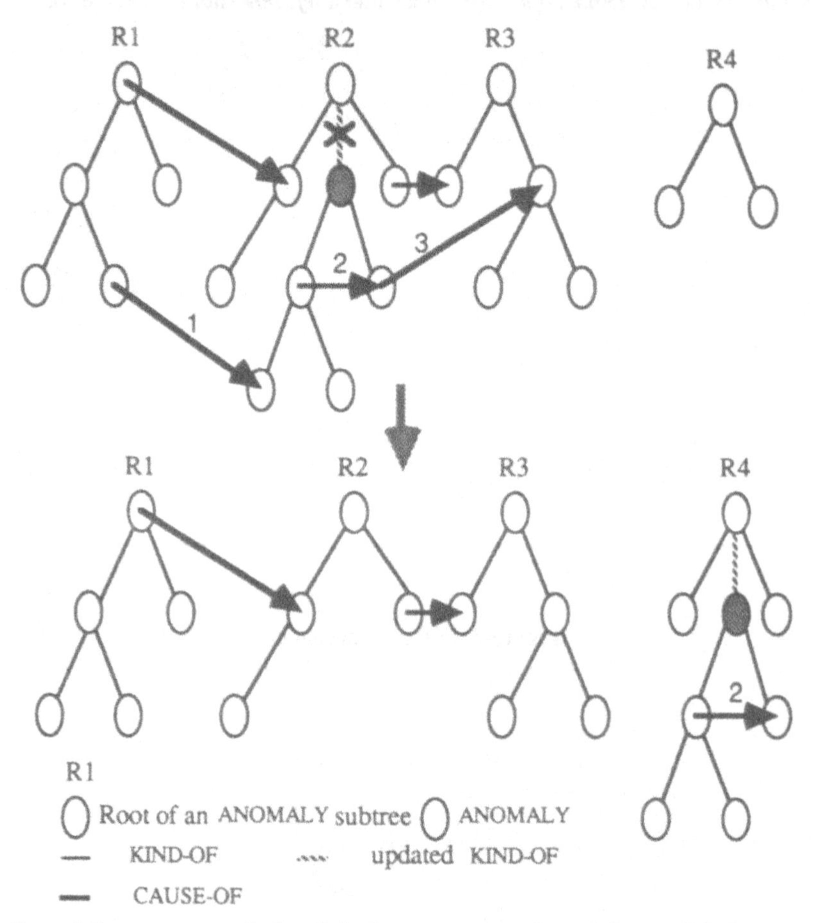

1 and 3 : are cancelled as links between one object of the modified tree and one object outside the modified tree

2 : is kept as a link between two objects of the modified tree

Figure 9. Maintenance of the CAUSE_OF link

The CAUSE_OF links between two ANOMALYs that belong to the subtree are kept, as their context does not change. The links between a subtree's ANOMALY and an ANOMALY outside the subtree are cancelled. Keeping these last links and updating the causal network accordingly would be hazardous, because it could change the semantics of the knowledge.

When a new direct CAUSE_OF link has to be created, it is necessary to ascertain whether or not it can be deduced from already existing links or if it can conflict with them. Figure 10 describes the interactions between a new CAUSE_OF link L and pre-existing links. It points out the fact that L cannot be deduced from links 1, 2 or 3 but that 2 and 3 can be entailed by L. In the relational knowledge base, however, these two links are not cancelled even if redundant. On one hand, this is neither very much useful nor easy and, on the other hand, this would cause the loss of the links if L was deleted. L can be deduced from 4 or 5, according to the left inheritance, but not from 6. In fact, the only links which allow the deduction of L and thus prevent its creation, are the links between any of its causal ANOMALY's ancestors and any of its caused ANOMALY's descendents. Although this corresponds to a rather complex dynamic integrity constraint, it can be implemented by a single SQL query.

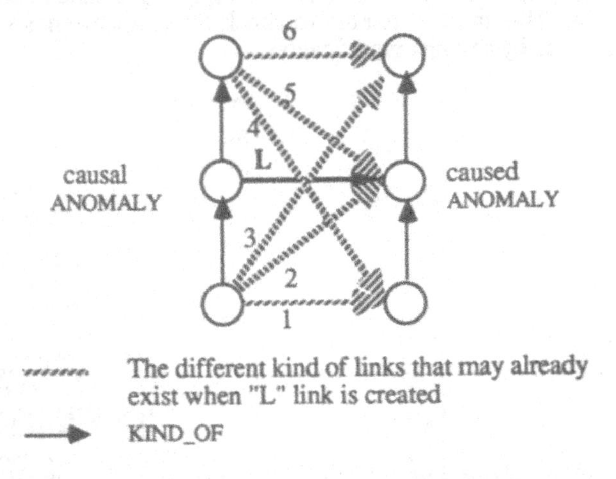

Figure 10. Interactions between new and pre-existing CAUSE_OF links

The above examples outline the real advantages of ORACLE's extended SQL for building dynamic integrity constraints applied to hierarchical information. However, it is not intended to make any deduction or inference or to implement any inheritance mechanism. The rich semantics of an object-oriented model can hardly be represented by the relative bareness of the relational model. The interface's goal is thus to build a "good" domain knowledge base, that is a knowledge base which is coherent with the RONSART's representation and which is easily accessible. In order to achieve this last point, it is needed to keep a good balance between the redundancy, which makes the information readily available, and the queries' complexity, which appears when the redundancy decreases.

IMPLEMENTATION

The knowledge acquisition and updating has been implemented with the fourth generation language SQL-FORMS, which allows a quick construction of input screens and an easy implementation of simple static integrity constraints. This language uses an event-based logic (the programs can be seen like daemons, which are executed each time a certain event appears). This gives flexibility to the tool, but the lack of structured programming can make the implementation of dynamic integrity constraints very difficult.

The interface structure follows RONSART's logic: its central point is the creation or the updating of an ANOMALY. Each subclass of a RONSART class is acquired or updated through an input screen which asks for its name, the names of the subclasses which are related to it either by an inter or an intra-class link, and the values of its properties (figure 11). A module, that is a specific piece of software written in SQL-FORMS, is associated to each screen. Each RONSART link is represented by a table in the relational knowledge base. Updating a link is done through a specific screen. This allows a specific module to be associated with each link. This module is able to check for inconsistencies or lack of coherence entailed by any link modification.

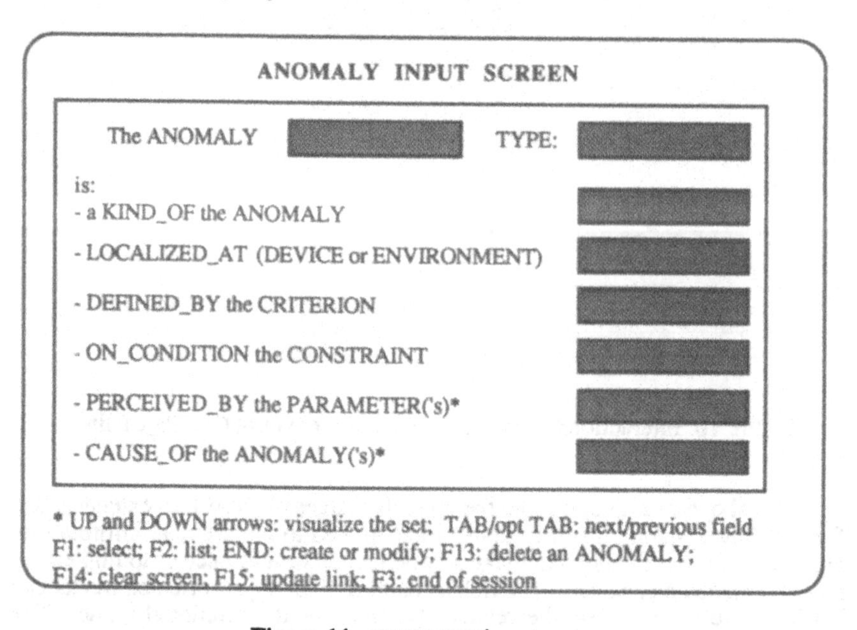

Figure 11. ANOMALY input screen

A specific expert system, capable of providing advice on the supervision of a given process, is generated as soon as its knowledge base is created, according to the frame given by RONSART, and is translated in GENESIA-II's triplet format. The knowledge base encompasses both the supervision-specific concepts and the process-specific information. This information is acquired, by the expert, through the interface and stored in the relational domain knowledge base. In order to generate objects described by sets of triplets, the base's relations and tuples are translated, according to the following rules:

- each relation LINK($OBJECT_1$, $OBJECT_2$) has tuples (O_1,O_2) which are translated by (O_1, LINK, O_2) triplets,

- each relation CLASS(ID, $PROP_1$,...,$PROP_N$) has tuples (I, P_1,..., P_N) that are translated by N triplets (I, $PROP_1$, P_1) ... (I, $PROP_N$, P_N) .

It can be noticed that distinct links and/or classes in the RONSART model correspond to distinct tables and input screens in the relational data base. Any addition or update of the object model would be thus very quickly propagated to the data base.

CONCLUSION

According to the RONSART model, supervision concepts and domain entities are represented by object classes and subclasses, while the causal and temporal reasoning is expressed by second-order logic rules. As no domain knowledge is expressed by rules, generating a specific process-supervision expert system consists mainly of adding the domain-specific subclasses to the knowledge base.

The process knowledge is acquired and managed through an expert's interface. This interface is domain-free and guides the expert in the process-specific knowledge acquisition, with the help of the frames provided by the supervision concepts. It also checks the data base to assure that it is in a coherent state and is consistent with the RONSART model. The verifications are implemented through dynamic integrity constraints. As the interface is independent from any specific process, the knowledge validation is strictly syntactic (the knowledge must fit the model). The semantic validation is exclusively done by the expert. It could only be automated, by coupling the system with a process' deep knowledge model.

In the generator of expert systems which is presented here, each programming tool is used for what it can do the best: the relational DBMS ORACLE, for the knowledge acquisition and management, and the second-order logic language GENESIA-II, for the implementation of a complex deductive mechanism. The cooperation between the two systems is realized through the translation between an object-oriented representation and a relational model. It is looser than a coupling between an expert system and a data base (e.g. Ballou et al. [11]) and stronger than the simple use of the object-oriented model as a tool for designing a relational data base schema (e.g. Vernadat [12]).

The goal of this work has been essentially to build an efficient tool, able to solve a specific problem: the generation of a process-supervision expert system. This implies that the knowledge must be acquired according to the model which is the best suited for the specific task (and for the deductive mechanism). This point of view is more pragmatic than the one of the general purpose knowledge acquisition systems (e.g. Aussenac et al. [13], Motta et al. [14]).

REFERENCES

1. Alengry, P. Evaluation d'un Dispositif d'Assistance à l'Opérateur dans le Secteur Industriel: Ergonomie du Dialogue, Processus de Traitement d'Information, Organisation Socio-Technique. Rapport de Recherche n°437, I.N.R.I.A.-Rocquencourt, France, 1985.
2. Landeweerd, J.A., Seegers, H.J.J.L. and Praagman, J. Effects of Instruction, Visual Imagery and Educational Background on Process Control Performance. Ergonomics, Vol. 24, n°2, pp. 133-141, 1981.
3. Lejoly, S. and Housiaux, A. La Nature des Mécanismes Cognitifs chez l'Ingénieur et chez l'Opérateur dans une Situation de Conduite d'une Unité de Production d'Energie Electrique. In Proceedings of the "Septième Conférence Annuelle Européenne: Prise de Décision et Contrôle Manuel", pp. 175-182, Paris, 1988.
4. Sébillotte-Garnier, S. Les processus de diagnostic au cours du déroulement de la grossesse. Thèse de Doctorat de Troisième Cycle en Psychologie Appliquée, Université de Paris V, France, 1982.
5. Pastor, J. RONSART: Représentation Objet du Raisonnement dans un Système d'Aide à la Surveillance de Processus. Thèse de Doctorat de l'Université de Toulouse III, France, 1990.
6. Grivart, P. Couplage d'un Système Expert de Suivi de la Grossesse avec une Base de Données: Liens entre une Représentation Objet de la Connaissance et un Schéma Relationnel de Données Factuelles. Mémoire du Diplôme d'Ingénieur en Informatique, Conservatoire National des Arts et Métiers, Toulouse, France, 1990.
7. Pastor, J. Object Representation in Computer Aided Process Surveillance. In Proceedings of "Applications of Artificial Intelligence in Engineering V, Vol.2: Manufacture and Planning" (Ed. Rzveski, G.), pp. 179-192, Computational Mechanics Publications, Springer-Verlag, Berlin, 1990.
8. Pastor, J., Fournié, A. and Grandjean H. RONSART: a Knowledge Representation Model in Computer Aided Proces Surveillance. In Proceedings of COGNITIVA 90, pp. 77-84, Madrid, Spain, 1990.
9. Dormoy, J.L. Résolution Qualitative: Complétude, Interprétation Physique et Contrôle; Mise en œuvre dans le langage BOOJUM. Thèse de Doctorat de l'Université de Paris VI, France, 1987.
10. Bancilhon, F. Système de Gestion de Bases de Donnés Orienté Objet. Ecole d'Eté d'Informatique E.D.F./C.E.A./I.N.R.I.A., Bréau-Sans-Nappe, France, 1989.
11. Ballou, N., Chou, H-T., Garza, J.F., Kim, W., Petrie, C., Russinof, D., Steiner, D. and Woelk, D. Coupling an Expert System Shell with an Object-Oriented Database System. JOOP, June/July, pp. 12-21, 1988.
12. Vernadat, F. Représentation des Connaissances et Bases de Données Relationnelles. In Proceedings of "Des Bases de Données aux Bases de Connaissances" (Ed. Miranda, S.), pp. 83-103, Editests, Paris, 1987.
13. Aussenac, N., Frontin, J., Rivière, M-H. and Soubie, J-L. A Mediating Representation to Assist Knowledge Acquisition with MACAO. In Proceedings of the "Third European Workshop on Knowledge Acquisition" (Ed. Boose, J., Gaines, B. and Ganascia, J-G.), pp 516-529, Paris, France, 1989.

14. Motta, E., Eisenstadt, M., Pitman, K. and West, M. Support for Knowledge Acquisition in the Knowledge Engineer's Assistant (KEATS). Expert Systems, Vol. 5, n°1, pp. 6-28, 1988.

KEY WORDS

Expert System Generation, Expert's Interface, Knowledge Acquisition, Object Representation, Relational Model, Process Supervision

ACKNOWLEDGEMENTS

The authors would like to thank Dr. Deborah F. Cook and Pr. Joseph G. Massey, from the Dept. of Forest Science, Texas A&M University, for their fruitful comments and suggestions and for having thus greatly contributed to improve the quality of this paper.

Default Reasoning in DIPSY-E System

M. Kantardžić, V. Okanović, A. Filipović, H. Glavić
Faculty of Electrical Engineering, LERT,
Toplička bb, 71000 Sarajevo, Yugoslavia

ABSTRACT:

Incomplete information in reasoning process is one of the problems in expert systems. Reasoning based on **default assumptions**, called **default reasoning**, is a possible solution to that problem. **Default reasoning** in DIPSY-E system is based on three types of **default assumptions**, as well as on introduction of additional data structures and management upon them, used in situations of **default reasoning** process activation.

1. INTRODUCTION

Incomplete and unaccessible information in a reasoning process is one of crucial problems for expert systems. One of the possible approaches to solution of that problem can be reasoning based on default assumptions, called default reasoning. A common opinion is that the default reasoning offers several advantages [5, 6]:

- it allows reasoning process to generate conclusion when there is no complete information.

- it provides a way to shorten the reasoning process which could become time over-consuming if total belief into conclusions should be sought.

- it enables use of common-sense knowledge.

Default reasoning is often treated from theoretic point of view with few practical implementations [4, 6, 8, 9]. An approach to introduction, as well as hints on implementation of default reasoning in relation to a particular development tool, DIPSY-E, are presented in this paper.

DIPSY-E is a software environment for development of expert systems [1, 2, 3]. The system have been developed in LERT laboratory. It is oriented towards knowledge based applications in real-time process control [1].

Default reasoning in DIPSY-E system is based on implementation of so called default assumptions. This paper discusses default assumptions definition manner and gives algorithms describing the reasoning process based on these assumptions. Data structures use in situations of default reasoning activation in DIPSY-E system, as well as management of these structures are described. Illustrative examples the process of reasoning with default assumptions in DIPSY-E system are given.

2. DEFAULT ASSUMPTIONS

A knowledge base in DIPSY-E system is composed of two basic parts:

- declarative knowledge base
- procedural knowledge base

Declarative knowledge is the knowledge on a data level. Data (facts) are represented in DIPSY-E system in a form of an object-attribute-value triple.

Procedural knowledge represents a level of operators upon the data. In DIPSY-E system, it is represented in a form of IF-THEN rules. By default assumptions, we mean assumptions defined thus they represent some common-sense knowledge that can be used when there is no complete information, needed in the reasoning process. There are three types of default assumptions in DIPSY-E system:

1. **Default reasoning on an attribute value level**: if a particular object attribute value is not known, a common-sense value is assumed. This value we call **default attribute value.** Collection of all such values represent **default assumptions base**. The facts in the **default assumptions base** are described by the following syntax:

<defaults> :: **default** <facts>
<facts> ::= <object> { <object> }
<object> ::= <object_name> (<attribute_value> { <attribute_value> })
<attribute_value> ::= <attribute_name> <const_value>

2. **Default reasoning on a rule condition level**: if there are no facts in the base, needed for a particular condition analysis, an assumption is introduced that the condition is satisfied. This condition we call **default condition. Default condition** is labeled by an additional word **'default'** after the text of condition thus differing from non **default condition.**

3. **Default reasoning on a rule conclusion level**: there are common-sense knowledge rules in a reasoning process, able to generate data. Such a datum we call **default conclusion**. **Default conclusion** is labeled by an additional word '**default**' after the text of conclusion, thus differing from non **default conclusion**. The **default conclusion** becomes an element of the **default assumptions base**. It can be used in the reasoning process as a **default attribute value**. This type of **default assumptions** is, in fact, the most commonly found with other authors.

The definition of the **default assumptions** is performed trough a knowledge base editor available in DIPSY-E system. During the edition of knowledge base, **default attribute values** are defined in the **default assumptions base**, while **default condition** and **default conclusion** are defined in rules.

3. THE REASONING PROCESS BASED ON DEFAULT ASSUMPTIONS

3.1. *Notation and data structures*

In order to describe the reasoning process based on **default assumptions**, the following notation is introduced:

K_b - knowledge base represented as:

$K_b = \{ R_b, F_b, D_b \}$

R_b - rule base

F_b - base of facts

D_b - default attribute value assumptions base

$R_b = \{ R_i;\ i = 1..t \}$

$R_i - \{ U_{ij}, k_{ie};\ j = 0..m,\ e = 1..l \}$

U_{ij} - j-th condition of the rule R_i

k_{ie} - e-th conclusion of the rule R_i

R_i - i-th rule in the base

$F_b = \{ f_i,\ i = 1..n_1 \}$

$f_i = (o_i, a_i, v_i)$

o_i - object in f_i

a_i - attribute in f_i

v_i - value in f_i

f_i - i-th fact in F_b

$D_b = \{ f_{di};\ i = 1..n_d \}$

$f_{di} = (o_{di}, a_{di}, v_{di})$

o_{di} - object in f_{di}
a_{di} - attribute in f_{di}
v_{di} - default value in f_{di}
f_{di} - i-th fact in D_b

Besides the existing data structures in DIPSY-E, some additional structures, in a form of lists are introduced for the purpose of the reasoning process using default assumptions. A root of a list is a default assumption used at the beginning of a default reasoning process. Elements of the list are conclusions (facts) generated by the assumption. This list we call **a list of defult facts,** L_d. In any particular moment of the reasoning process an arbitrary number of such lists can exist, depending on a number of assumptions. In order to describe **default reasoning** in DIPSY-E, some additional notation is introduced:

$L_d = \{h_d, l_{di}, i = 1..n_1\}$
$h_d = \{ v_{di} \vee U_d \vee k_d \}$
U_d - default condition
k_d - default conclusion
h_d - root of L_d
$l_{di} = (o_{1i}, a_{1i}, v_{1i})$
l_{di} - i-th element of L_d
L_d - list of dependencies of default facts
$f_d = (o_d, a_d, v_d)$
f_d - fact currently concluded in reasoning process

Flags with the following meaning are used in reasoning process of DIPSY-E system:

- *inference engine flag* e_f: Activation of default reasoning is followed by the activation of the *inference engine flag* e_f.
- *rule flag* r_f: Activation of a default assumption in a rule is followed by the activation of the *rule flag* r_f.
- *flag of a rule with respect to a defult assumption* r_d. Forbids the use of the particular default assumption in the given rule.

Description of a **default reasoning** process in DIPSY-E system is described through algorithms used in accordance to a type of a **default assumption** used in the reasoning process. Yet, there are two algorithms used in **default reasoning** independently of the type of the assumption. These are algorithm1 for creation and algorithm2 for modification of a list L_d.

Algorithm1 updates the list L_d, adding elements l_{di} which represent conclusions f_d generated by default assumptions f_{di}. *Algorithm1* is activated in the moment of generation of conclusion when the rule flag r_f is set.

ALGORITHM1:

Any condition f_d generated by default assumption f_{di} is added to the list L_d. Formal notation of this algorithm is:

```
IF r_f THEN
      ADD(L_d, f_d)
```

Algorithm2 performs modification of reasoning when either contradiction or confirmation of an element of a list L_d arises. *Algorithm2* is activated after the *algorithm1*, if the rule flag r_f is set, or after conclusion if r_f is not set, but e_f is set.

Algorithm2:

```
IF ( r_f and (f_d = f_j)) or
   ( e_f and not r_f and (f_d = l_di)
THEN
 TRANSFER(L_d, f_d, F_b);
 DELETE(L_d, f_d);
ELSE IF ( r_f and (o_d, a_d) = (o_j, a_j) and (v_d <> v_j)) or
        (e_f and not r_f and (o_d, a_d) = (o_1i, a_1i) and (v_d <> v_1i))
    THEN
        DELETE(L_d, h_d);
        RESET(r_d);
```

If either some fact f_d, equal to some fact f_j from the knowledge base, is generated in a reasoning process, using a **default assumption**, or an element l_{di} of a list L_d is confirmed during the reasoning process, where by confirmation is meant repeated conclusion of the same fact, but with no assumptions taken, the confirmed fact, along with all the consequent facts, no more depend on the introduced assumption h_d. These facts starting with the assumption f_d are entered into the base of facts F_b, and removed from L_d. On the other hand, if some element l_{di} causes a contradiction in regard to some fact f_i (for the same object-attribute have different values), all the conclusions, being consequences of **default assumption** h_d are invalid and the whole list L_d is removed. At this point the flag r_d is reset thus forbidding they use of the assumption h_d in given rule.

3.2. *Default reasoning based on an default attribute value*

Default reasoning process based on a default attribute value is an extension to forward chaining inference engine in DIPSY-E system, using algorithms 1 and 2 described in the previous chapter, as well as an algorithm3 used for creation of a root of a list L_d.

The algorithm3 is activated during the analysis of a condition U_{ij} if a fact $f_k(o_k, a_k, v_k)$, not existing in a base of facts K_b, is used in the condition U_{ij}. The algorithm3 consist of the following steps:

ALGORITHM3:

```
IF ((f_k∈U_ij) and (f_k∉F_b) and (f_dl∈D_b) and (o_k, a_k) = (o_dl, a_dl))
THEN
     IF U_ij(f_dl) THEN
      h_d := f_dl;
      set(r_f);
      set(e_f);
```

If there is a fact f_{dl}, from a default assumptions base D_b, that matches the fact f_k used in the condition U_{ij} (the facts match to each other if they refer to the same object and attribute), then if the fact f_{dl} satisfies the condition U_{ij} the assumption f_{dl} becomes a root h_d of some list dependencies L_d. Flags r_f and e_f are set thus indicating that default reasoning is active.

3.3. *Default reasoning based on default conditions*

Default reasoning process based on a default condition is an extension to forward chaining inference engine in DIPSY-E system, using algorithm 1 and 2 described in a chapter 3.1., as well as an algorithm4 used for creation of a root of a list L_d.

The algorithm4 is activated during the analysis of a condition U_d, if it uses a fact f_k which does not exist in the base of facts F_b. The algorithm4 consists of the following steps:

ALGORITHM4:

IF $(f_k \in U_{ij})$ and $(f_k \notin F_b)$ and $(U_{ij}=U_d)$ THEN
    $h_d := U_d$;
    set(r_f);
    set(e_f);

The default condition U_d becomes a root h_d of some list of dependencies L_d. Flags r_f and e_f are set thus indicating that default reasoning is active.

3.4. *Default reasoning based on default conditions*

Default reasoning process based on a default conclusion is an extension to forward chaining inference engine in DIPSY-E system, using algorithm 1 and 2 described in a chapter 3.1., as well as an algorithm5 used for creation of a root of a list L_d. The algorithm5 is activated if there is no fact $f_k(o_k, a_k, v_k)$, from the base of facts F_b, which have the same object (o_k) and attribute (a_k) as the concluded fact k_d. The algorithm5 consists of the following steps:

Algorithm5:

IF $(f_d=k_d)$ THEN
    $h_d := k_d$;
    set(e_f);

The default conclusion k_d becomes a root h_d of some list of dependencies L_d. Flag e_f is set thus indicating that default reasoning is active.

3.5. *Final actions in* **default reasoning** *process*

If there are no more applicable rules in the knowledge base, i.e. no additional conclusions can be generated, the reasoning process is finished. If some lists L_d exists in that moment, all the **default assumptions** taken and not rejected during the reasoning process, the roots of the lists L_d, as well as conclusions generated with regard to these assumptions (the elements of the lists L_d), are reported.

4. ILLUSTRATION OF DEFAULT REASONING PROCESS IN DIPSY-E

Scenario1:

Let there be an engine on a test in a factory. If an environment temperature of the testing place is unknown, we will assume that it has a value of 20 (centigrade degree). If environmental agents are represented by an object 'environment' then this assumption, written in DIPSY-E language, looks like:

default environment (temperature 20)

Let this be an element of a default attribute value assumptions base. Let there be the following two rules in the knowledge base (in the rules all the conditions and conclusions irrelevant to the example, i.e. with no interference to default reasoning process, are omitted):

```
RULE1: IF

          environment (temperature >15)

       THEN

          engine (temperature 80)
```

```
RULE2: IF

          engine (temperature >70)

       THEN

          engine (oil_pressure 5)
```

Let the rule1 be analyzed in a reasoning process. Let there be no environmental temperature entry in the base of facts. In this case the activation conditions of the algorithm3 are satisfied, and a list L_d is created with the assumption

environment (temperature 20)

as a root. Thus we have:

$L_d = \{ h_d \}$

h_d = (environment (temperature 20))

Besides, the flags r_f and e_f are set. Assuming that all the other conditions of the rule1 are satisfied, the fact

engine (temperature 80)

will be concluded. This fact is valid only if the assumption h_d is valid. Thus the concluded fact differs from the facts generated without default assumptions. In order to denote this difference we include the fact

engine (temperature 80)

as the second node, i.e. element l_{d1} of the list L_d. Thus we have:

$L_d = \{ h_d, l_{d1} \}$

l_{d1} = (environment (temperature 80))

Now let the rule2 be analyzed in the reasoning process. The condition is obviously satisfied. Suppose that all the other conditions of the rule2 are satisfied. Then the fact

engine (oil_pressure 5)

is concluded.

Since this conclusion depends on the fact l_{d1}

engine (temperature 80)

which also depends on the introduced assumption h_d

environment (temperature 20)

it appears that the last conclusion also depends, thou indirectly, on the assumption h_d. Thus this fact becomes the next element l_{d2} of the list L_d. Now we have:

$$L_d = \{ h_d, l_{d1}, l_{d2} \}$$
$$l_{d2} = (\text{engine (oil\_pressure 5)})$$

Let us analyze a different case. Let the condition of the rule1 now have the form

environment (temperature 15)

If all the other assumptions from the previous example are unchanged when this condition is analyzed the algorithm3 is activated again, but now the test

IF $U(f_{di})$ THEN

fails, since the assumption

environment (temperature 20)

does not satisfy the condition

environment (temperature 15)

The reasoning process continues as if there were no default assumptions, i.e. the list L_d is not created at all.

Let us analyze the following example. Besides the previous two, let there be the rule:

```
RULE3: IF
          cooling_liquid (temperature [x])
       THEN
          engine (temperature [x])
```

in the knowledge base, and let there be a fact

cooling_liquid (temperature 80)

in the base of facts. Analyzing the rule3 the reasoning process concludes the fact

engine (temperature 80)

The same fact is concluded after the rule1, but in that case, it have depended on the default assumption. Now this dependency disappears. Simultaneously, the dependence of the fact

l_{d2}= (engine (oil_pressure 5)),

generated by the rule2, disappears too.

Actually, when the fact

engine (temperature 80)

is concluded by the rule3, because of the e_f is set, the algorithm2 is activated. It transforms the list L_d from the form:

$$L_d = \{ h_d, l_{d1}, l_{d2} \}$$

into $L_d = \{ h_d \}$,
transferring the elements l_{d1} and l_{d2} from the list into the base of facts.
If the fact

cooling_liquid (temperature 80)

in the base of facts is replaced by the fact

cooling_liquid (temperature 50)

using the rule3 the reasoning process would conclude the fact

cooling_liquid (temperature 50)

Using the algorithm2 a contradiction between this conclusion and the fact l_{d1}

cooling_liquid (temperature 80)

would be discovered. According to algorithm2 the list L_d would be removed and the flag r_d would be reset.

Scenario2:

Let the rule1 from the scenario1, have the following form:

RULE1: IF

environment (temperature >15) **default**

THEN

engine (temperature 80)

The condition of the rule becomes default condition, because the word '**default**' is specified behind the condition. Let the reasoning process analyze the rule1. The condition is analyzed according to algorithm4. A list L_d is created with the assumption

environment (temperature >15) default

as a root. Thus we have:

$L_d = \{ h_d \}$

h_d = (environment (temperature >15))

Besides, the flags r_f and e_f are set. Assuming that all the other conditions of the rule1 are satisfied, the fact

engine (temperature 80)

will be concluded. This fact is valid only if the assumption h_d is valid. Thus the concluded fact differs from the facts generated without default assumptions. In order to denote this difference we include the fact

engine (temperature 80)

as the second node, i.e. element l_{d1} of the list L_d. Thus we have:

L_d = { h_d, l_{d1}}

l_{d1} = (environment (temperature 80))

Using the knowledge base from the scenario1, the reasoning process is continued identically to the reasoning process described in scenario1 from the point where the rule2 have been analyzed.

Scenario3:

Let us introduce a rule0 as a common-sense rule meaning: "If, among the weather conditions, we do not no anything about the temperature, but we know that the season is a spring, we shall assume that the temperature is 20". If the weather conditions are represented by the object 'environment', we can describe this assumption, using DIPSY-E language, as:

```
RULE0: IF
           global_situation ( season spring )
       THEN
           environment (temperature 20) default
```

A conclusion of a rule becomes default conclusion specifying the word '**default**' behind the conclusion. Reasoning process using a default conclusion can be described on the example of the knowledge base from scenario1 as follows:

Let the reasoning process analyzes the rule0 and let the condition of the rule0 is satisfied. According to algorithm5 a list L_d is created, having the concluded fact as a root:

L_d = { h_d }

h_d = (environment (temperature 20))

and the flag e_f is set.

The reasoning process is continued identically to the reasoning process described in scenario1 from the point where the list L_d have been created.

5. CONCLUSION

The possibilities of implementation of default reasoning in expert systems are considered in this paper. An approach, based on introduction of three types of **default assumptions**, as well as implementation of **default reasoning** in DIPSY-E system is given. Data structures and algorithms used in default reasoning in DIPSY-E are described. Illustrative examples of the reasoning process with **default assumptions** are also given in this paper.

REFERENCE:

1. M. Kantardžić, U. Jusupović, A. Filipović, N. Gujić, K. Delić, H. Glavić, 'Software Environment for Real-Time Expert System', Proceedings of The International Conference on Systems Management, Hong Kong, June, 1990.
2. M. Kantardžić, M. Jeftović, H. Glavić, A. Filipović, D. Gajić, N. Miličić, 'A Comprehensive Environment for Computer Aid in Tehnical Security System Design', Proceedings of The International Conference "DEXA '90", Springer-Verlag, Vienna, Austria, August, 1990.
3. M. Kantardžić, N. Gujić, H. Glavić, A. Filipović, 'Knowledge Refinement in DIPSY-E system', IITT Conference "EXPERT SYSTEMS APPLICATION", Los Angeles, USA, November, 1990.
4. D. Poole, 'A Logical Framework for Default Reasoning ', Artificial Intelligence 36, Elsevier Science Publishers B.V. North-Holand, 1988.
5. Marie-Odile Cordier, ' Sherlock : Hypothetical Reasoning in an Expert System Shell ', Proceedings "ECAI '88", Minhen, August, 1988.
6. B. Selman, H. A. Kautz , ' Model-Preference Default Theories', Artificial Intelligence Vol 45, Elsevier Science Publishers B.V. North-Holand,, October, 1990.
7. H. Freitag, M. Reinfrank, ' A Non-Monotonic Deduction System Based on (A)TMS', Proceedings "ECAI '88", Minhen, August, 1988.
8. G. A. Cleveland, R. H. Brown, 'Mutations and Their Consequences; A Study of Non-Monotonic Behavior', IEEE Expert, 1985.

9. D. McDermott, ' Nonmonotonic Logic II: Nonmonotonic Modal Theories', Journal of the Association for Computing Machinery, Vol 29, No 1, January, 1982.

Knowledge-Based Modelling Systems for Research of Engineering Objects

V. Vittikh

Kuibyshev Branch of Institute for the Study of Machines, USSR Academy of Sciences, 1 Pervomayskaya St., 443100 Kuibyshev, USSR

ABSTRACT

New opportunities for the improvement of research process are opened by the development and application of modelling systems based on hypothetic-deductive method and software tools for computer knowledge representation and processing. In the researcher-computer dialogue methodology of constructing computer models of objects provides for the periodically repeated sequential fulfillment of the following interconnected procedures: experiments -- inductive generalization -- computer knowledge representation -- model designing (deduction) -- model investigation and comparison with test data -- new experiments. To estimate the advantages of knowledge-based modelling system application for scientific research some results in this field obtained in Kuibyshev Branch of Institute for the Study of Machines are considered.

INTRODUCTION

The products of the scientific research process fixed in the brain or outside it on a suitable data carrier is a model providing gradually specified representation about object investigation which is realized by means of advancing and verifying hypotheses. Knowledge contained in the constructed model can be used as a base for getting new knowledge, which determines specific character of research: relative conditionality of the research level and obligatory capability of the model for expansion [1].

Hypothesis in the empirical (natural and engineering) sciences is an inductive generalization of experimental data expressed by some language that is to provide natural reflection of real world objects in language constructions conformed to perception inherent to a man [2]. Determining the statement truth is either an immediate result of the experiment or a result of using hypothetic-deductive method [3], consisting in advancing hypotheses and verifying conclusions drawn by means of deductions. It turns out possible to judge about correctness of hypotheses by confirming or disproving conclusions. In all cases experience or in a broader sense practice is the final source of determining truth [2]. This method is so inherent to natural sciences that natural scientific method is rather often identified with hypothetic-deductive one [3].

In those cases when a person deals with a lot of hypotheses he suffers from natural difficulties in obtaining and verifying consequences as his capabilities are limited. That is why it is reasonable to give the researcher certain tools. As far back as the 17-th century logic-constructive type including not only expressive and communicative abilities but also tools was formulated [4]. G. Leibnitz's ``universal symbolics'' (or ``universal characteristics'') is the most interesting and well-known [5]. The essence of his idea consisted in systematizing knowledge, revealing main concepts and expressing all others through them. After such analysis being complete definite signs (symbols) should be given to every concept. But the most important is that G. Leibnitz set the task of creating tool enabling ``finding in all the cases consequence from the basic truth or facts as though by computation no less accurate and simple than arithmetical or algebraical computation'' [5]. If it ``were introduced into practice people who learnt it and were trained in it would surpass all others in everything else their equals to such an extent as a versed man surpasses an ignored one, a learned man surpasses ignoramus, perfect specialist in geometry -- a schoolboy, brilliant specialist in algebra -- usual ledger-clerk. ...And, at last, if the invention of telescopes and microscopes has been so useful for studying nature, it is easy to imagine how much more useful this new organon should be, by which, as far as it is in the power of man, the very intellectual vision will be equipped'' [5].

Provided the scientist had similar (certainly not so universal as G. Leibnitz had it in mind) tools, he would obtain qualitativity new possibilities for improvement of research process. First of all it is the possibility to increase the degree of models adequacy because number of concepts and relations among them (hypotheses) with which he can never operate while designing a model with the help of traditional data representation tools. It is well-known, for instance, that when constructing a mathematical model forced decrease of dimensions might make the essence of the task vapid. No less important is time saving for hypotheses verification as with the aid of such a tool a scientist could satisfy his curiosity rapidly (at the rate of thinking), getting answers to a chain of interconnected questions of the ``What would it be if ... ?''-type. Lastly, it is the possibility to integrate heterogeneous knowledge within one model: both the results of the experiments given in the form of tables, empirical dependences and fundamental laws expressed in mathematical equations and separate logical statements. The above mentioned possibilities (the list of them could be broader) can be realized by using new information technology - methods and software tools of artificial intelligence [6]. Researcher-computer system based on the hypothetic-deductive method and software tools for knowledge representation and processing and intended for the construction and research of the computer models of objects where cognition results are fixed can be called knowledge-based modelling system.

At Kuibyshev Branch of Institute for the Study of Machines of the USSR Academy of Sciences methodology, software tools and applied knowledge-based modelling systems which could be used in research of machinery, technological processes, materials and complex systems of the ``machine-man-environment''-type are being developed. Below tasks of research are formulated and a brief review of results obtained is given.

METHODOLOGY OF COMPUTER MODELS CONSTRUCTION

Computer models of objects (CMO) construction begins from the registration of measurement products characterizing qualitative and quantitative features of objects obtained by observation and experiments. For connecting and regulating the facts concepts are introduced which are coded by means of signs. Then with the help of inductive generalization hypotheses in the form of relations among concepts are advanced. This first stage of model constructing includes rather considerate extent of heuristics and formalizes poorly. That is why at present formulation of concepts and relations among them is done by specialists in corresponding objects domains. However, this fact does not mean that the process of inductive generalization can not be automated as there are research works proving the possibility of algorithming these procedures for the subsequent application of computer.

To such researches the work [8] may be referred, where the inductive concept formation is based on the comparison of descriptions of similar objects, assigned by the whole set of measurement products and on the selection of the most typical fragments of these descriptions. Examples of using algorithms of generalization by measurement products for scientific research in chemistry, metallurgy and economics are also given. Modelling of experimenter's inductive statements is possible by means of JSM-method developed by V.K. Finn [9], based on fundamental research of induction by J.S. Mill [10]. JSM-method makes automatical formations of hypothesis-regularity possible, giving an opportunity of automatically completing knowledge base by experimental data processing. Computer data representation implies their systematization that could be done in many ways. Considerating the specific character of scientific research it is suitable to classify them according to the extent of hypotheses substantiation [11].

Roughly, three classes of knowledge can be singled-out: isolated inductive generalization, empirical and fundamental laws. Isolated inductive generalization is hypotheses connecting facts between each other at the basis of private observation; they can be obtained by direct generalization of experimental data. Extent of trust to them is not great as the researcher's intuition is not supported by sufficient experimental material. Empirical laws present ``uniformities which existence is indicated by observation or experiment but which the researchers do not dare transfer to the cases more or less considerably different from really observed ones -- do not dare because they do not see the reason for the existence of such a law'' [10]. Fundamental laws display interconnections among real life phenomena which truth is universally accepted (at least at a given stage of science development).

During research with the use of modelling systems knowledge base should enable easily and flexibly modify its contents because a part of primary inductive generalizations can be disproved, and some of them, vice versa, can be proved and transfer to a class of empirical laws. Besides, some new hypotheses can be advanced that should complete knowledge base. One can define parameters of empirical laws more exactly and, if necessary, broaden a range of fundamental laws.

A whole set of separate logical statements, tables, empirical dependences, mathematical equations forms a system of basic models, i.e. knowledge components that should be presented to computer in a form suitable for formalized conclusions. Thus, further integration of heterogeneous knowledge within unified computer model of object obtained on the foundation of basic models by deductive conclusions is provided.

In outline, construction of design and research of computer models comes to the following [12]. When object description in object domain language is input to the computer the program is automatically synthesized and necessary for modelling knowledge components (basic models) selected from the knowledge base unite into a computational model. By means of a computer model so designed a researcher can according to his wish vary model parameters in a close to natural objects domain language, change model structure flexibility, follow the computer experiment, visualize and document the modelling results, compare them to the data of natural experiments, make decisions on conducting new experiments.

In accordance with the described methodology of designing computer models of objects (Fig. 1) the following functional subsystems can be classified in knowledge-based modelling systems:

- automation of experiments;
- automation of inductive generalizations (hypotheses advance);
- computer knowledge representation (in a form that can be used for conclusions);
- automation of deductive conclusions (models construction);
- dialogue communication with the user (for investigating models and comparing results of modelling with experimental data).

SOFTWARE TOOL ``RESOURCE''

To implement the described methodology of CMO construction special software tools are necessary making automation of developing functional subsystems given above possible. Such software tools are being created in Kuibyshev Branch of Institute for the Study of Machines. Software tool RESOURCE developed by Tsybatov V.A. [13] is meant for automation of construction and research of CMO. Its application when constructing general resource model of main production at the bearing plant [13], several computer models of technological processes and dynamic systems enables to see the advantages of the suggested approach compared to the traditional ones. To illustrate these advantages consider the construction of fuel-speed computer model of automobile done together with experts from Technical-Scientific Centre of Volga Automobile Associated Works (AVTOVAZ, Togliatti).

System of the basic concepts necessary for the construction of fuel-speed model includes such concepts as speed, acceleration, fuel consumption, power loss, etc., as well as a number of constants used in computations. Relations among concepts are given in various forms. For instance, acceleration and speed are connected by the sign of derivative, power at the engine shaft and its rotation speed -- by means of a table, dependence of the drag resistance on the speed is determined by empirical formula, and the number of gear is connected with the engine revolutions during automobile acceleration by logical statement (it revolutions have come to the value when engine power is maximum put in the next gear). Tables of concepts, relations among them and initial conditions are input into computer (in our case into IBM PC/AT).

Software tool RESOURCE automatically constructs a computer model in the form of computational network consisting of interconnected functional modules [13]. However, it can fail to do so at first time (and so it happens more than often) as a set of basic models (relations among concepts) can be incomplete for a simple reason that experts missed some initial data needed during model construction. Then a programmer (knowledge engineer, to be exact) asks the experts to supply certain missing data and after that he inputs them to the computer. The system RESOURCE is run again and such iteration may be repeated several times. At last, computer model is constructed. This means that the system of basic models is complete, and the sequence characterizing concepts interconnection has no breaks. In other words, collection of causal relationship forms a close cycle.

However, the construction of the model is not finished yet. It is necessary to have an adjustment stage comparing the results of modelling with natural experiments data (such an opportunity was given due to the availability of test stands and software tools for experiments automation at the Technical-Scientific Centre of Volga Automobile Associated Works (AVTOVAZ, Togliatti). It is hypothetic-deductive method that we use in this case giving the statement about correctness of hypothesis (basic models) confirming or disproving consequences (modelling product).

Constructed with the help of RESOURCE system program enables observe and fix values of any parameter, which can compare them with the results of experiments rather simply. If there are divergences (provided the experiments data are true) it is necessary to check the correctness of initial hypotheses, i.e. basic models. Having made the information more exact, it is possible to compare it with the experiment once again, and up to the time when one is convinced in the model adequacy. It was in this very way that the fuel-speed computer model of automobile VAZ-2108 was adjusted. So we may say that the constructed computer model is metrologically certified.

After that the model was passed to Technical-Scientific Centre of AVTOVAZ and was used while solving applied tasks. For example, it helped to estimate fuel consumption (for 100 km) for this model of ``Lada'' with greater reliability than the data obtained during natural automobile tests. The reason for this is that our estimation was given using modelling system based on hypothetic-deductive method and computer knowledge representation and processing software got from experts.

The extent of trust to the results of modelling is higher because the estimation of fuel consumption in this particular case characterizes a whole class of automobiles but not a separate automobile or a group of them chosen for test arbitrary.

It turned out possible to demonstrate the advantages of the constructed computer model more clearly because there was a fuel-speed model constructed with the use of traditional modelling methodology and programming techniques in Technical-Scientific Centre of AVTOVAZ. Its main drawback is inaccessibility to the end user, who is forced to solve the problems of modelling with the help of a intermediary who is a program developer. Only he can, having spent some time, introduce changes into initial data, display necessary set of parameters, etc. That is why user tries to apply to him as seldom as possible. The situation becomes a deadlock when the program developer leaves his job. As the program is developed for VAX computer it can't be applied to personal computers. Based on the suggested methodology system RESOURSE computer model of automobile is free from the described drawbacks. The user interacts with computer without intermediary in the language of his own object domain. He can change model structure flexibly, display the parameters which he is interested in at a given moment. Structuring of the basic knowledge system enabling to speak about their completeness and consistency is extremely important. And, lastly, fuel-speed model is implemented at the IBM PC/AT, which makes it available to a wide range of users.

KNOWLEDGE-BASED MODELLING SYSTEM FOR KINEMATIC ANALYSIS OF PLANAR MECHANISMS

System ANMEC designed for kinematic analysis of planar mechanisms is developed by Budyatchevsky I.A. in Kuibyshev Branch of Institute for the Study of Machines for IBM PC/AT using principles of models construction automation on the basis of knowledge representation systems [12, 14]. Traditional approach to the mechanism research supposes that construction of mathematical model is fulfilled by a user on the foundation of given kinematic scheme. There are special techniques for this purpose which application is completed with composition of the systems of equations, non-linear (transcendental) as a rule, then solved in a computer.

As a world of mechanisms is rather various a lot of concepts and classifications is used to describe them in the theory of machines and mechanisms. Simultaneously, any kinematic scheme can be described by means of the simplest concepts: a point, its coordinates, a straight line section, an angle, etc. Relations between the concepts are determined with the help of elementary geometric and trigonometric rules studied at secondary school (Pythagorean proposition, theorem of sines, equation of point of two straight lines intersection, equality of the angles sum in a triangle to 180°, etc.) However, while constructing a mathematical model this simple description is translated into a complex language of formalized models (non-linear equations). The question arises whether the kinematic analysis of mechanisms can be done in terms of the initial language of object domain. If a traditional technology of computer application is used it is not possible, as the theory of machines and mechanisms operates with a set of derived-concepts and implies input of formal schemes to the computer, constructed with the use of them. At the same time, if it were possible to construct a software system which, using the supplied kinematic scheme of mechanism, could construct a computer model at once, the answer would be positive. ANMEC system does fulfil this function.

Availability of a tool automating the processes of computer models design and research gives the opportunity to relate the kinematic section in the theory of machines and mechanisms in a new way. The perfection of the relation can be achieved by reducing the number of the applied concepts and ties among them, by excluding formalisms which were to be used in traditional technology and by transferring the centre of gravity to the analysis of the results of modelling and the synthesis of mechanisms. Could the system ANMEC be widely used it would not be necessary to have various reference manuals in kinematic analysis of mechanisms, as having prescribed scheme of mechanisms all its characteristics could be got immediately (the same refers to the data bases). And a wide range of system application can be guaranteed because it is intended for an ordinary user but not for an expert in the field of theory of machines and mechanisms. An obvious advantage of the ANMEC system -- a considerable time saving for analysis -- is accompanied with a simultaneous improvement in quality of mechanisms research and design, as the user can make computer experiments with models many times, flexibly changing the mechanism structure and parameters in the researcher-computer dialogue. Moreover, many errors in the designer's decisions can be found at the designing stage.

And, finally, the application of ANMEC system in education opens new possibilities for intensifying educational process and improving its quality [15].

KNOWLEDGE-BASED MODELLING SYSTEM FOR MATERIAL RESEARCH

Methodology described above is used when developing knowledge-based modelling system for material research.

Account of material structure under external influences makes the task of constructing their mathematical models that could be efficiently used during research extremely complex, and even more often unsolved. The search for new approaches to this problem solution led to the necessity of developing methods and tools for automating construction and research of computer models of structural materials.

Knowledge structuring that rests on the basic physical concepts and their relations is applied to the definite physical processes so far. For example, for the solidification (crystallization) of metal from melt process concepts of a cell, a phase transition temperature, a grains boundary and their space orientation, etc. are used, and relations among them are expressed as rules of cell transition from liquid to solid state and other physical regularities. Several hypotheses can be formulated about the kinetics of crystallization processes. That is why design and application of software for computer models construction (in other words, tools for drawing conclusions from hypotheses automatically) enables to implement here the major advantage of hypothetic-deductive method: comparing results of computer modelling with the data obtained during physical experiments one can draw a conclusion about hypotheses correctness. Possibility of multiple conduction of the cycle ``hypotheses advance -- consequence verification'' in the knowledge-based modelling system can help to understand the mechanism of solidification process better and increase the adequacy extent of the obtained computer models.

In Kuibyshev Branch of Institute for the Study of Machines knowledge-based modelling system for research of solidification processes for IBM PC/AT is created by Kalinin B.V. Relations describing intensity of the process of nucleation centre (seed) formation, temperature boundary conditions local rules of change of phase state of the melt microvolumes (cells) are input to the computer. On the basis of this knowledge simulation of solidification process is fulfilled, and user can watch this process on display screen till polycrystal structure is formed. The image obtained can be compared to the polycrystal structures got during natural experiment under the same physical solidification conditions as in a computer experiment. Thus, the adjustment of the model can be made.

CONCLUSION

The first results obtained while constructing knowledge-based modelling systems and their functional subsystems make it possible to conclude that such systems application gives new capabilities for developing research process, first of all by improving the adequacy extent of the models of objects and intensifying research and education. These capabilities are provided with methodology of computer model construction, which is based on using both hypothetic-deductive method and software tools of knowledge representation and processing. The process of sequential refinement of model relies on multiple repetition and comparison of the results of computer and natural experiments, and this makes possible to speak about getting metrologically certified computer models.

Suggested approach to the construction of knowledge-based modelling systems implies conducting research in the following fields:

- automation of inductive generalization (hypotheses advance);
- computer knowledge representation taking into account specific character of scientific research;
- automation of deductive conclusions (models construction).

Without diminishing the importance of investigations in every separate trend one should have in mind that modelling system can give considerable effect only in case that the network of interconnected procedures is closed: experiment -- inductive generalization -- computer knowledge representation -- model construction -- model investigation and comparison with experimental data -- new experiments. Then complex solution to the key problems of computer modelling can be found: securing model adequacy, diminishing time for its construction, accessibility to the end user and saving of computational resources.

REFERENCES

1. Peshel, M. Modelirovanie signalov i sistem. p.13, Mir, Moskva, 1981.

2. Rubashkin, V.Sh. Predstavlenie i analiz smysla v intellektualnykh informatsionnykh sistemakh. pp. 6-16, Nauka, Moskva, 1989.

3. Rusavin, G.I. Gipotetiko-deductivnyi metod.In the book: Logika i empiricheskoe poznanie. pp. 86-113, Nauka, Moskva, 1972.

4. Kuzicheva, Z.A. Yazyki nayki, yazyki logiki, yestestvennye yazyki. In the book: Logika nauchnogo poznaniya. pp. 57-73, Nauka, Moskva, 1987.

5. Leibnitz, G. Sochineniya (in four volumes).Vol.3, pp. 501-502, Mysl, Moskva, 1984.

6. Pospelov, G.S. Iskusstvennyi intellekt -- osnova novoi informatsionnoi tekhnologii, Nauka, Moskva, 1988.

7. Vittikh, V.A. Sistemy modelirovaniya baziruyutsh-iesya na znaniyakh. GF IMASH, preprint No.34, Gorky, 1990.

8. Vagin, V.G. Deductsiya i obobtshenenie v sis-temakh prinyatiya reshenii. pp. 235-273, Nauka, Moskva, 1988.

9. Predstavlenie znanii v cheloveko-machinnykh i roboto-tekhnicheskikh sistemakh. In: Fundamen-talnye issledovaniya v oblasti predstavleniya znanii. p.262, VTS AN SSSR, VINITI, Moskva, 1984.

10. Mill, J.S. Sistema logiki (sillogicheskoi i in-ductivnoi). Leman, Moskva, 1914

11. Karnap, R. Philosofskie osnovaniya phiziki. pp. 59-71, Progress, Moskva, 1971.

12. Vittich, V.A., Budyachevsky, I.A. Avtomatizatsiya sinteza modelei obyektov mashinostroeniya na osnove sistem predstavleniya znanii. Mashinostroeniye, No.1, pp. 5-10, Moskva, 1989.

13. Vittikh, V.A., Tsybatov, V.A. Obobshyonnye resursnye modeli sistem mashina-chelovek-sreda. Problemy mashinostroyeniya i nadyozhnosti mashin, No.1, Moskva,1990.

14. Budyachevsky, I.A., Vittikh, V.A. A knowledge-based system for interactive computer-aided kinematik analysis of planar mechanisms, pp. 355-358, Proceedings of 5th Int. Conf. on Artificial Intelligence and Information-Control Systems of Robots, Strbske Pleso, Czechoslovakia, 6-10 November, 1989.

15. Vittikh, V.A., Peregudov, F.I. and Petrov, O.M. Kompyuternaya tekhnologiya posnaniya. In: Primenenie vychislitelnoi tekhniki v fizicheskom eksperimente, pp. 7-15, IPF AN SSSR, Gorki, 1987.

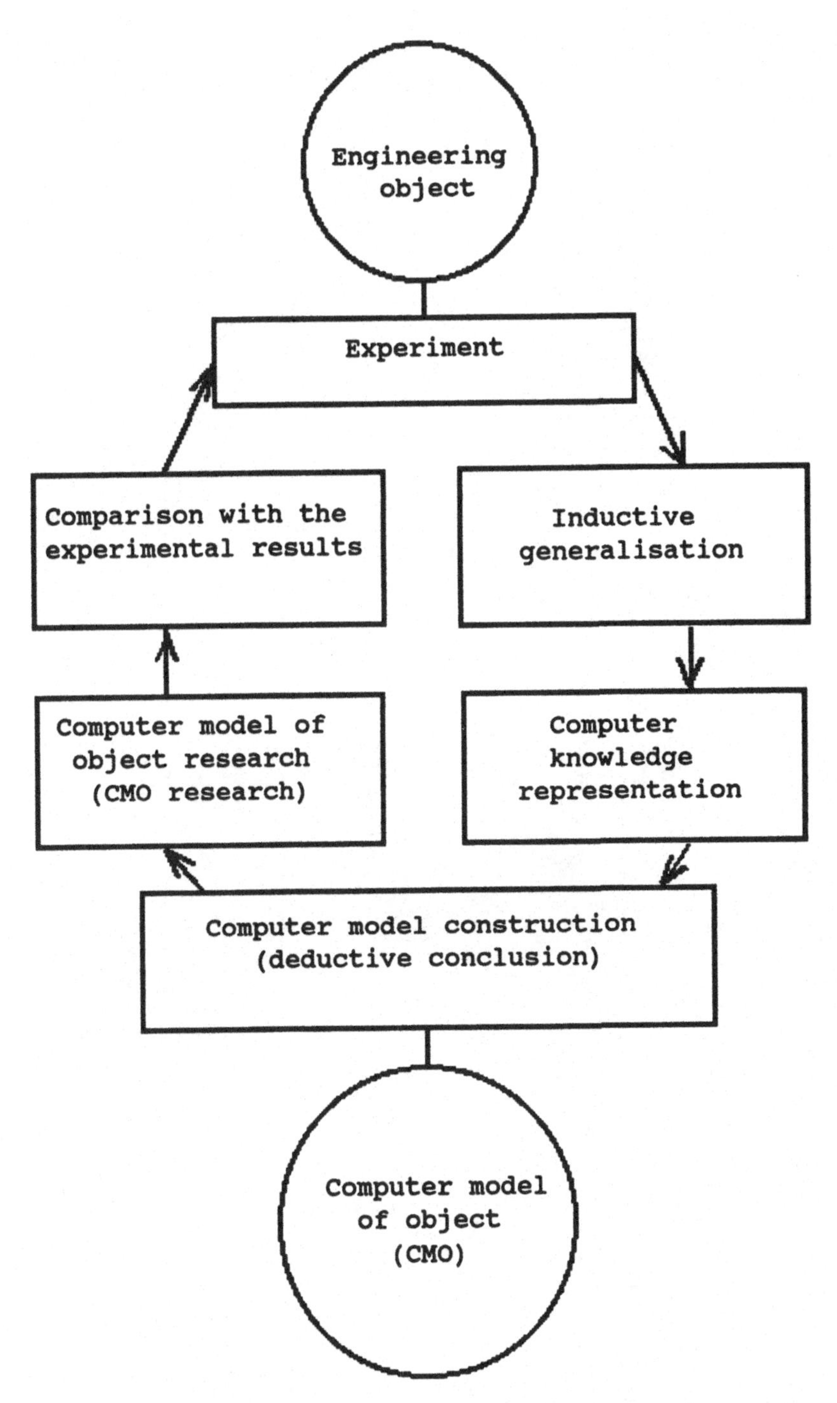

Figure 1. Construction and research methodology of computer models of object.

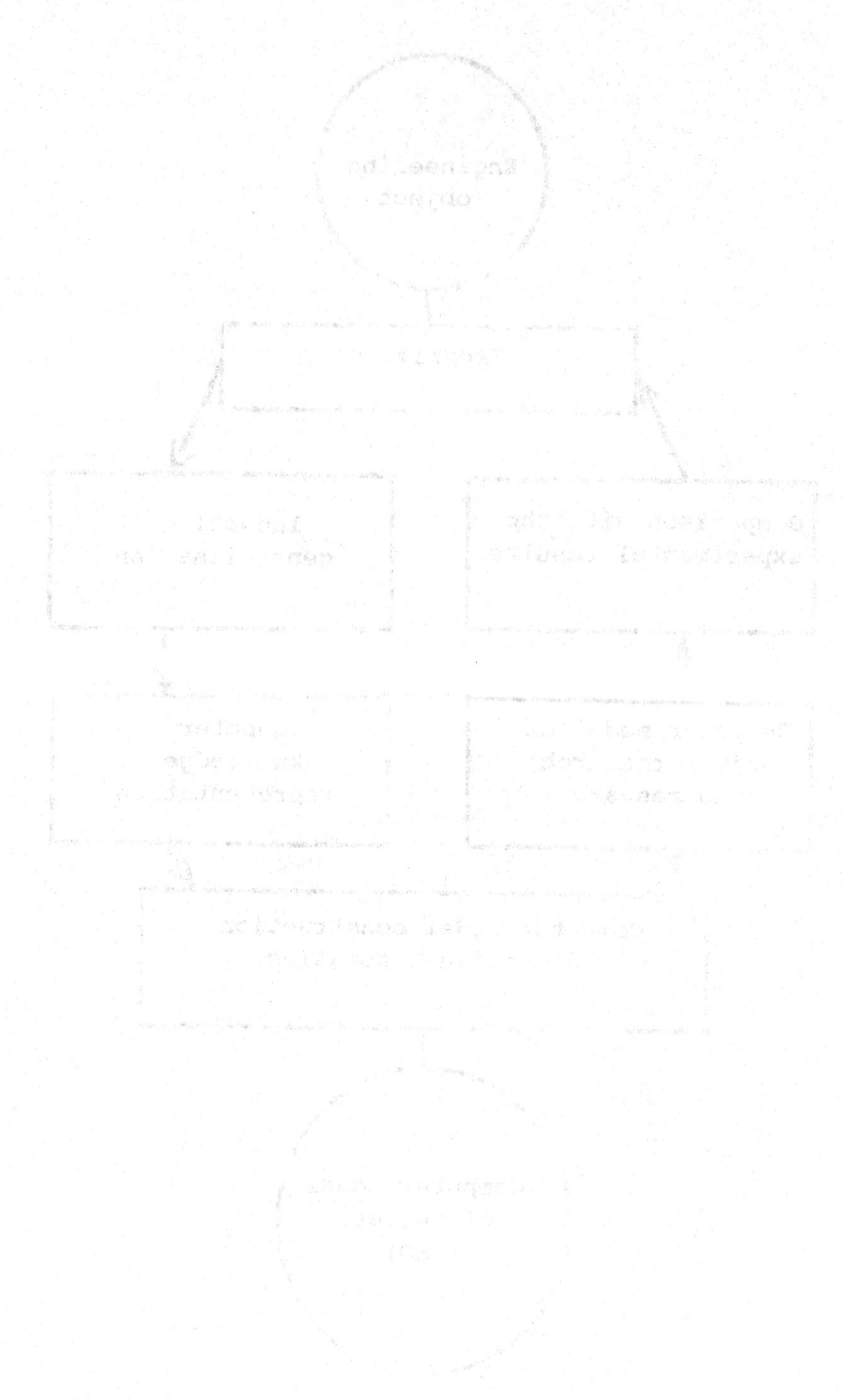

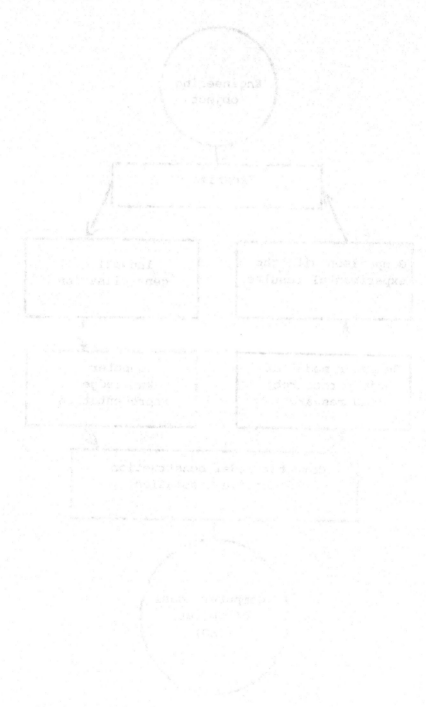

Integration of Natural Language in Developing the Inference Mechanism Provides Powerful Benefits for Engineering Applications

E.L. Parkinson (*), A.K. Sunol (**)

(*) Department of Industrial Engineering and Management Systems, University of Central Florida, Orlando, Florida, 32816, USA

(**) Department of Chemical Engineering, University of South Florida, Tampa, Florida, 33620, USA

ABSTRACT

This article discusses reasons for using a human model and natural language processing in developing the inference mechanism of expert systems. It brings out the importance for developing this technology with engineering applications. A particular focus is on evolving this systems architecture with attention to specific computer programming design features.

INTRODUCTION

There is a tremendous market for expert systems and a need to have expert systems that support real-world applications. In the design architecture of any system, one considers the intended operational use of the system, and alternatives for interfaces and system hardware and software. Appropriate consideration for these features can bring about improvement in system performance. Due to the time and costs for development, research is being done to develop generic design architectures that provide more reliability, flexibility and robustness for expert systems applications.

Strides are being made for expert system architectural development and tools for real-world markets. However, it is important to recognize that as the song states, "We've Only Just Begun." One might compare where we are in expert system architectural development with the development of capability to transport a human from place-to-place. In this comparison, expert system architecture is perhaps in the horse and buggy stage versus the automobile or airplane. This need for architectural features affords opportunity for basic research. It indicates a need for further development and evolution of the technology base. This article discusses a feature of architectural development and integration -- the use of natural language as an integral part of the inference mechanism.

Expert systems are typically considered to be composed of a knowledge base and an inference engine or inference mechanism (Figure 1). In this context, there are two basic functional requirements of the inference mechanism. First, the architecture includes a method to accept input from a user. Second, the architecture includes an approach to access knowledge to support advancing to a conclusion or goal-state response.

From classical artificial intelligence objectives, an expert system can be viewed as a communication device just as a human expert is a communication device. The human uses characteristics of natural language to provide success in performing inferencing functions. Thus, a basic premise for this research is that there is power, in particular, long-term

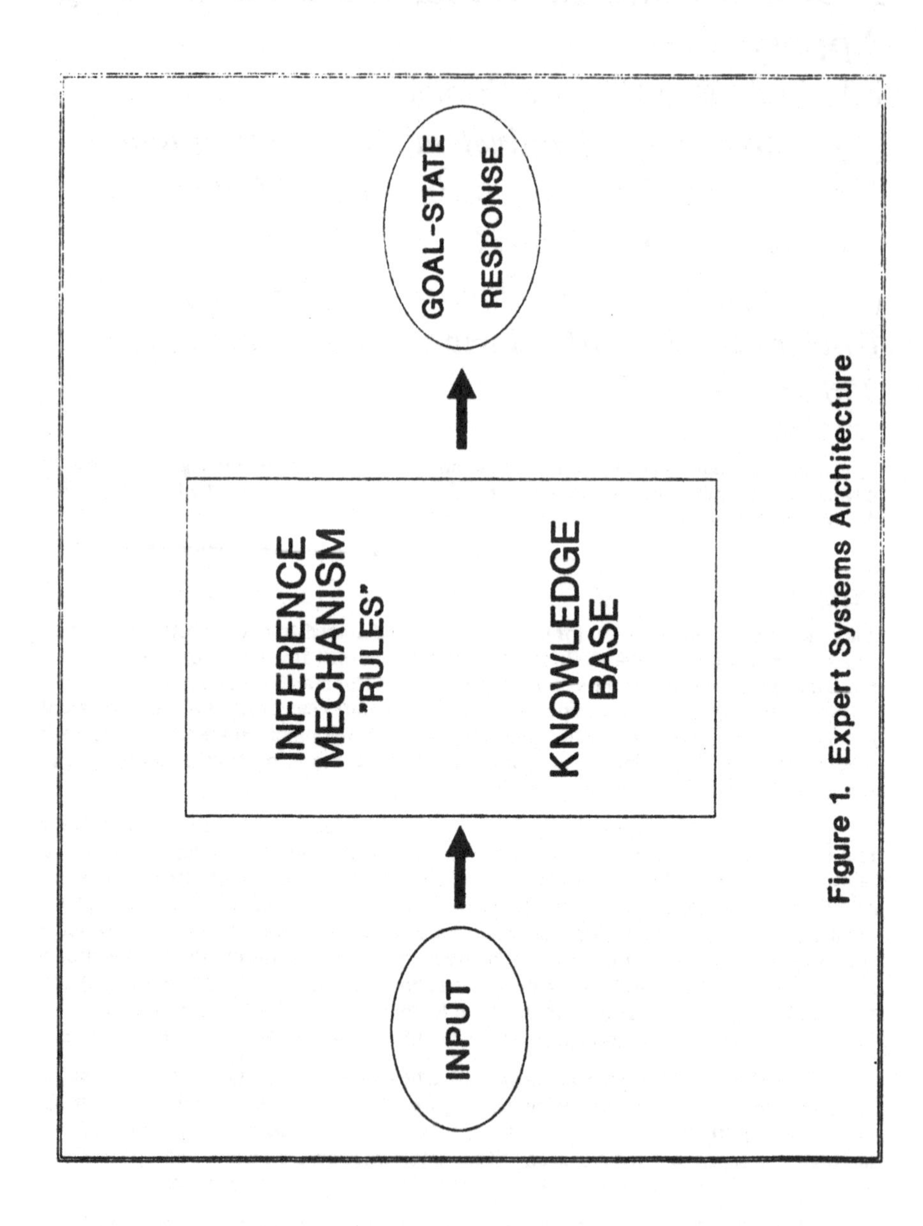

Figure 1. Expert Systems Architecture

advantage in evolving human characteristics of natural language in the architecture for expert system development. This article focuses on top-level characteristics of human inferencing in natural language. Features of the use of natural language are discussed from a perspective that can be helpful in the integration and synthesis with software programming and for further development of reasoning and response capability. This article shows how engineering applications are a natural test-bed for development of the systems architecture for natural language. These points are used as a basis to defend the need for the engineering profession to increase emphasis in the incorporation of natural language in basic architecture for expert system development.

TECHNICAL

In many corners of the engineering profession, particularly in academia, use of the terms natural language, parsing, human cognition and human modelling are avoided. The reasons for avoiding these topics are not part of this discussion; however, the technical basis for this article uses these topics. Helping to break through these apparent barriers for the engineering profession are an important secondary objective of this article. The primary thrust is to discuss some interesting and powerful reasons for developing the inference mechanism of expert systems using a human model for natural language processing.

Expert System Architecture Development to meet Categories of Application Areas

By definition, expert systems simulate to some degree the capabilities that a human contributes in a particular area of expertise. There are many ways to consider categorizing expert systems (Figure 2). One might consider functional application categories; such as, safety, construction, environmental, specification development, network design, cost estimation, risk analysis, system reliability analysis, test development, training systems, configuration management and logistics support analysis. One might consider developing systems within academic areas; such as, areas of mathematics, physics, chemistry, engineering, psychology, medical, education and business. One might wish to make distinctions based on complexity or type of reasoning. Distinctions may be made for expert systems that are essentially a compilation of more than one human expert or a human expert that integrates knowledge of many experts. In some applications, the expert system may perform reasoning functions of a single human expert, a group of experts or integrate expert knowledge with data from other sources. In these applications, the expert system may provide further support by performing tests or other calculations and assessments needed to obtain results. This could be a part or combination of any of the categories.

Upon further investigation of expert systems developed to meet these applications, one may notice that architecture of the inferencing mechanism is probably significantly different for most of the applications. In fact, it may be difficult to identify a generic architecture, below the point of agreeing that they have an inference mechanism, knowledge base and rules.

The importance for considering how one categorizes expert systems is that it can have a direct bearing on the generic architectural development of the inference mechanism. A researcher typically develops an inference mechanism that supports a part of one of these categories of expert systems or perhaps a part of more than one category, as in the case when using expert system development tools. The inference mechanism might be embedded as an integral part of a specialized computer program or part of an expert system shell. However, in most cases, by developing the inference mechanism in this focussed manner, essentially a new development effort is needed to expand the inferencing capability within the particular category and especially to support another category.

FUNCTIONAL AREAS	ACADEMIC SUBJECTS	COMPLEXITY	INTEGRATED KNOWLEDGE
SAFETY CONSTRUCTION DESIGN SPECIFICATION DEVELOPMENT COST ESTIMATION RELIABILITY ANALYSIS	ENGINEERING MATHEMATICS PHYSICS CHEMISTRY MEDICAL PSYCHOLOGY BUSINESS	TYPES OF REASONING	MANY EXPERTS

Figure 2. Categories For Expert System Development

This is not meant to infer that it is poor or inadequate to have specialized expert system developments or tools allowing support for setting up the inference mechanism. The technical base for inference mechanism development is an evolving process as it should be.

A Natural Building Block for Expert Systems Technology Based on Human Modelling

What is suggested is that further emphasis and examination be given in evolving generic architectures for inferencing using the human model for natural language processing. Let us begin to see how this framework will allow for expanding support within a category and for other categories of expert systems. Consider using a building block for the inference mechanism, based on "modelling of the human and how his mind operates." Lets consider what we get and what we do not get. As with classical approaches to programming this architecture uses rules. However, the suggested framework focusses on the use of rules for natural language processing. Thus, ultimately we can get a "generic" systems architecture that will have the strengths and limitations of the human. However, before the ultimate, if we develop the architecture in a focussed and systematic manner, what can have a natural evolution of the technology and program for different generic application areas. This plan for technical base evolution is a major step for consideration.

At the highest level, this inferencing capability can allow an expert system to do some combination of three things (Figure 3):

(1). answer questions,
(2). respond to requests, and
(3). learn.

The human element portion in most of the categories of real-world expert systems applications (Figure 2), is supported through his mind processes. This architecture allows him to provide the three capabilities listed above. He uses his inference mechanism to understand the type of communication. This feature is accomplished through processing or communication in natural language. If he recognizes the communication as a question, then he tries to answer the question. In the human model, having the first words of natural language input, such as "can", "will" and "could" trigger that a question is being asked.

If the communication is in the form of a request, then he tries to respond to the request. Engineering applications are a natural for evolving this technology in natural language. This is because there are many engineering applications where the degree of flexibility, reliability and robustness can be relatively easy to control or limit to the domain of application. The technology can begin by limiting the natural language input to the domain, such as "I would", "I want" or "get".

When given declarative knowledge input, or when his reasoning capabilities generate new knowledge, the human model tries to learn this in a manner such that he can use it later. A natural language structure lends itself to relative ease for knowledge acquisition. The learning of new procedures can be partitioned in a manner for easy expansion as the technology base evolves. It is a matter of focussing on the objective to develop generic routines that integrate the use of natural language within all components of the system. One example of an efficient design structure and approach for parsing and integration with the elements of an expert system is outlined in the Surface Knowledge Communication (SK-COMM) System, Parkinson[1,2].

To carry the theme further, in showing the power of using natural language in the inferencing mechanism, let us look at how this systems infrastructure flows through the categories of expert systems shown in Figure 2. Using this concept in development of the inference mechanism, allows support through all categories of expert system application areas (Figure 4). Perhaps the biggest problem with this architecture is in the level of reasoning needed for many application areas. This is brought out, particularly in robust

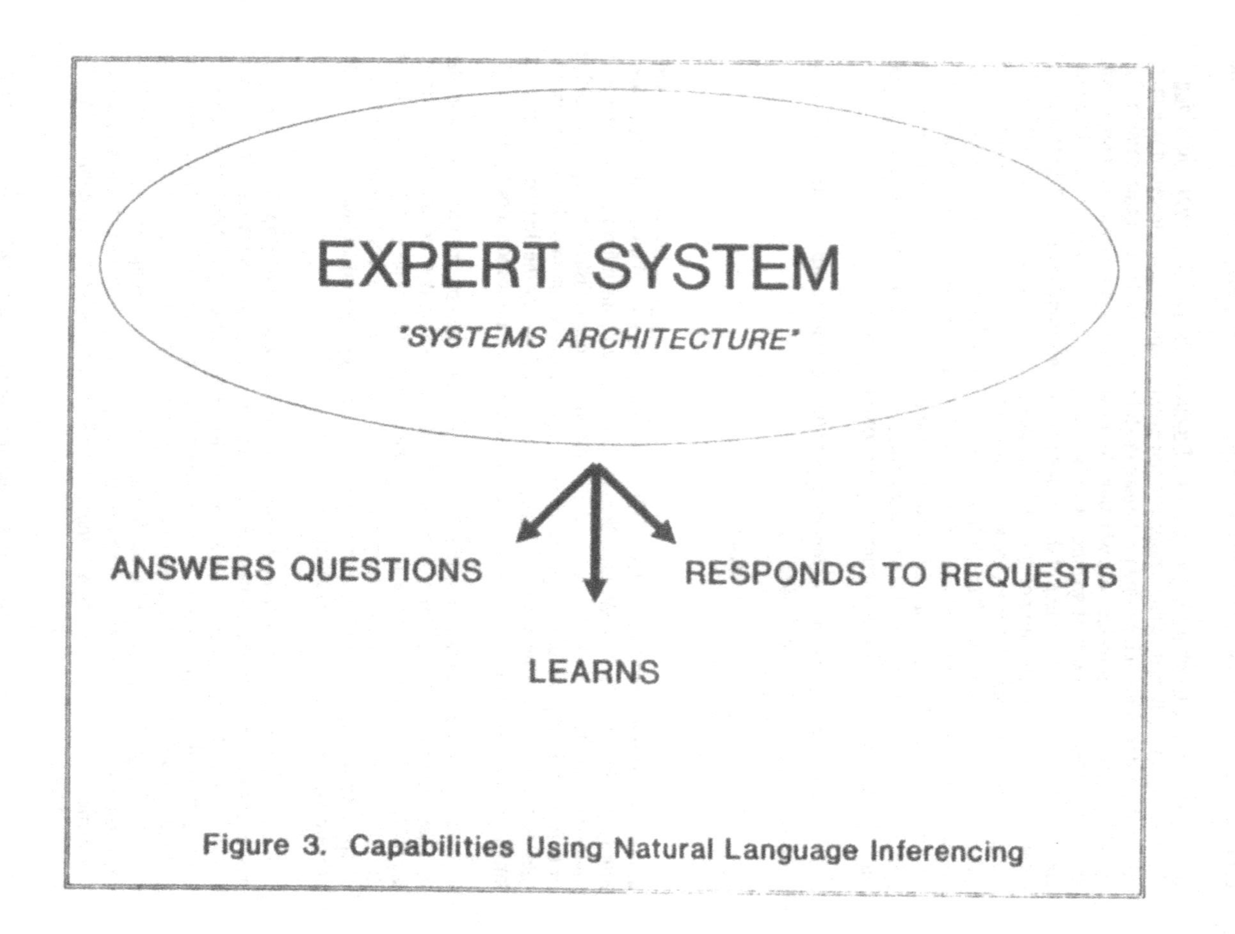

Figure 3. Capabilities Using Natural Language Inferencing

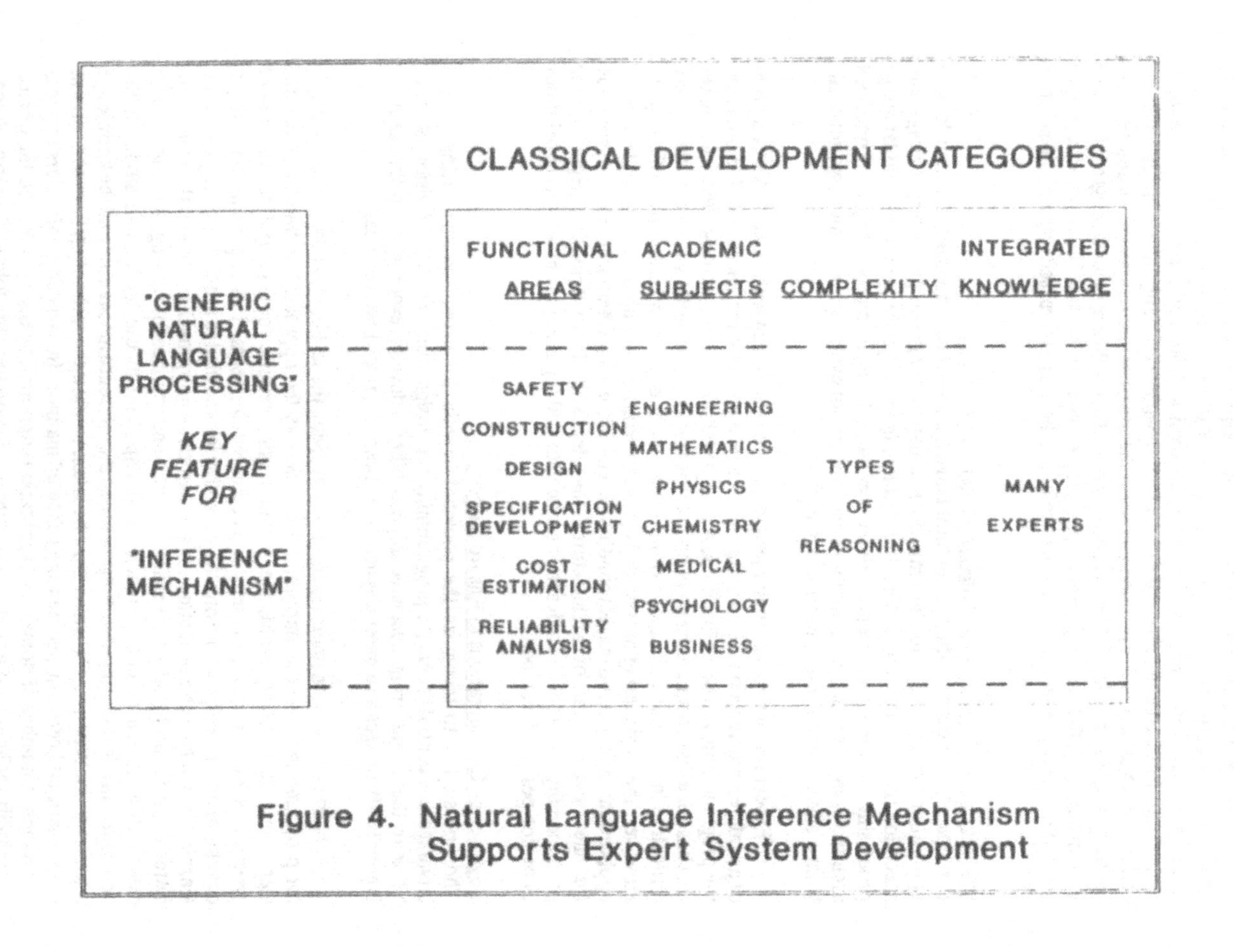

Figure 4. Natural Language Inference Mechanism Supports Expert System Development

systems desiring significant flexibility for natural language input of domain knowledge and for applications requiring deep reasoning through many levels of knowledge, Anderson[3]. However, for engineering applications an interesting feature of this architecture is that the market for a particular expert system may not require a large degree of reasoning. In addition, many industrial and organizational applications can easily limit the vocabulary used for domain knowledge and for communication with the expert system for assistance. This allows for marketing a standard architecture while expanding and evolving technical capability. Thus, the problem is actually not in the use of natural language. In addition, the problem gets less and less as the technology is evolved -- thus a natural progression. The benefits of using this developmental approach, as with the human, is that when a deeper level of reasoning capability has been developed, there are increased opportunities for its use.

Do Not Forget that Expert Systems Are Computer Programs

Perhaps many research efforts in natural language have been criticized for their lack of application potential due to the limitations in natural language processing, storage and execution time requirements. Was the problem in the objective to use natural language processing, or was the problem in either how natural language processing was conceived architecturally, how the programming was done or not evolving the technology, generally from a small capability to greater capability?

Expert systems are computer programs. It is one challenge to make a block diagram of an inferencing function. It is another challenge to program the function and to integrate the programming with other functions. A design architecture may be presented in a manner that has benefits in communicatinga point for educationalor scholarly purposes, that is very difficult to implement and expand or improve once it is programmed as part of a system. However, the engineering profession is primarily concerned with development leading to applications in real-world market conditions. Therefore, added emphasis needs to continue for the development of architectures that lends itself to smooth integration for programming. This also includes the necessity for making changes and adding capabilities as prototypes are developed.

Engineering Considerations to Ensure Success

The first step is to recognize the potential benefits for expert systems, using natural language as a central theme for programming and design architecture. The vision needs to be clear, that engineering must have this capability to develop expert systems that gain the inherent advantages and meet classical objectives of artificial intelligence.

Next comes an engineering thrust for evolving the technology base. The approach for programming is just as important for success as the approach for human modelling. Care must be taken to evolve the program using generic routines and partitioning of procedures for responding to query and action requirements. It is important to identify a development strategy, for evolving the capability for knowledge acquisition and procedural learning, by studying the programming and partitioning approach in the initial layout of the system. Care must be taken to ensure that rules and routines for parsing simulate to some extent, to a human model approach. It is important to use the human model in the development of smooth interworkings between the inference mechanism and the knowledge representation scheme. Care also must be taken to allow the technology base to evolve in a systematic and carefully planned and validated manner. In particular, this refers to levels of reasoning, number of domains of application supported by one system, and the degree of flexibility for handling natural language input. If these steps are followed, then prototype systems can be evaluated and improved incrementally. Operational limitations can be studied and fixed without compromising downstream improvements. When good design concepts for architectural development are followed, improvements in the system capability have a chance of being integrated without major program changes.

"Basic Research" for Expert Systems Architectural Development is an Engineering Challenge

For this discussion let us categorize research as; basic and applied, Blanchard[4]. Problems exist by embarking on applied research leading to product application when the technical base for basic research has not been adequately evolved. This is the main point when considering the product life-cycle for expert system development. Engineering, as it should, puts emphasis in applied research opportunities for real-world application areas. There are many real-world examples where engineering has backed-up and performed degrees of basic research. However, typically engineering does the design and integration, using the tools developed through basic research and other application efforts. When these tools are not adequate to meet the real-world operational application needs, problems arise. In other words, inadequate tools leads to cost, schedule and performance problems.

In the development of expert systems, the engineer uses the computer, the language and in some cases a "shell" for prototyping. The basic research going into the "shell" may or may not be adequately evolved for meeting real-world application areas. When this occurs, in most cases, the result is that a new system has to be developed -- the product life-cycle essentially starts over. Based on lessons learned, is improvement possible by the development of basic research of the inference mechanism, using the human model for natural language processing? Should this be done within the engineering community or should engineering wait for the research to be developed elsewhere? The human expert has proven to work satisfactory in real life and the model is readily available for researchers to simulate. As with other system development efforts, when basic research is needed for engineering applications, then engineering charges ahead. For the most part, this technology could be considered a primary engineering design responsibility. The reason is that when it is developed, for the most part, the applications will be merely a packaging and training challenge versus conceptual through full-scale development.

CONCLUSIONS

This article has attempted to shed light on why development of the architecture for inference mechanism should be evolved using the human model for natural language processing. The design of expert systems has to take into account the design of the computer program. Important design considerations, from lessons learned, were provided to aid in planning of developments. For many engineering applications, it is relatively easy to limit the domain of natural language. Therefore, it should be perceived, as natural, that the engineering profession takes the lead in evolving of technology.

REFERENCES

1. Parkinson, E.L., *Surface Knowledge Communication (SK-COMM) System*, Ph.D. Dissertation; University of South Florida, 1990.

2. Parkinson, E.L., "Surface Knowledge Communication (SK-COMM) System - Natural Language Processing Module", *Applications in Artificial Intelligence in Engineering V, Volume 2, Manufacturing and Planning*, ed. Rzevski, G., pp 411-425, Computational Mechanics Publications, Springer-Verlag, 1990.

3. Anderson, J.R., *The Architecture of Cognition*, University Press, Cambridge, Massachusetts, 1983.

4. Blanchard, B., and Fabrycky, W., *Systems Engineering and Analysis*, Prentice-Hall Inc., Englewood Cliffs, New Jersey, 1990.

Organizational System as Hierarchy of Information Processes

S. Gudas

Department of Management Systems, Kaunas University of Technology, Lithuania

ABSTRACT

An attempt to explain a basic nature of the information processing hierarchy in Organizational Systems is made. A framework for intelligent information processing at managing is presented. Organizational System (Enterprise) as the framework of so called elementary management cycles is decomposed. The structure of elementary management cycle (EMC) is presented and its properties discussed. A taxonomy of information processing hierarchies is based and presented. Such considerations seems to be important for the design of intelligent management information systems.

1. INTRODUCTION

Although the term "hierarchy" is widely used in the area of information processing in the Organizational Systems it is rarely given a systemic analysis of that (Elzas[1], Mesarovic[2], Ackoff[3], Gudas[4]). We shall offer an attempt at systemic investigation of information processes hierarchy in Organizational Systems which is oriented toward a pragmatic incorporating of intelligence into management information systems. Thus, the structural model of information units and their interactions at managing of Enterprise is needed. Such framework should include the hierarchical system of interactions between data, knowledge and objective items as well as between technological objects. It seems that investigation of the hierarchy of information processes in the Organizational System should have influence for the problem of intelligence of information systems as well as for the problem of artificial intelligence on the whole.

2.PECULIARITIES OF ORGANIZATIONAL SYSTEM MODELLING

The analysis of information flow in Organizational Systems is complicated due to such features of that systems :
a)two interrelated technologies - technology of materials and energy processing (TMP), and technology of information processing (TIP) - are interacted;
b)both technologies can be conceptualized adequatly only as multilevel hierarchical systems;
c)it seems that hierarchies of several types (for example,layers,stratas and echelons as presented in (Mesarovic[2]) may be singled out;
d)information is considered to be a complex item which consists of data,knowledge and objective components;
e)information manipulations at managing of Enterprise are intelligent because includes syntactic as well as semantical and pragmatical aspects of information.

Recently the requirements for the intelligence level of information systems are going up.Thus ,a comprehensive frameworks for modelling of information processing in Organizational Systems are necessary.A systemic view to the management of Organizational System as to the hierarchical structure of information processing seems to be the one of possible points of view. In order to make adequate decomposition of information processing in Enterprise the concept of Space of Processes (SP) was introduced in (Gudas[5]). The choice of three-dimentional SP was predetermined as a mean for representation of data,knowledge and objective items in all three aspects of information manipulations (syntactic,semantical,pragmatical) at managing of Organizational System .Some results of investigation based on such approach are presented in this paper. As the basic unit of the management model presented in this paper an elementary management cycle (EMC) is introduced.The EMC is considered as the "systemic factor" in this approach,and may be considered as the "gene" of properties of the whole Organizational System.The EMC is suppossed to be the smallest part in partitioning the body of management process (a unit of management).A hierarchy of information processing in Enterprise as the structure of EMC is presented.Departments of Enterprise are considered to be "devices" for information processing, and administrative hierarchy (hierarchy of echelons in (Mesarovic[2])) as the sequel of information processing hierarchy is suppossed.

3. ELEMENTARY MANAGEMENT CYCLE AS A UNIT OF HIERARCHICAL STRUCTURE

The elementary management cycle (EMC) is concidered to be the basic structural unit in the framework of information processing at managing of Enterprise. Formally the structure of EMC can be defined as diagram in notations of the theory of categories (Gudas[8]) and is presented in figure 1.

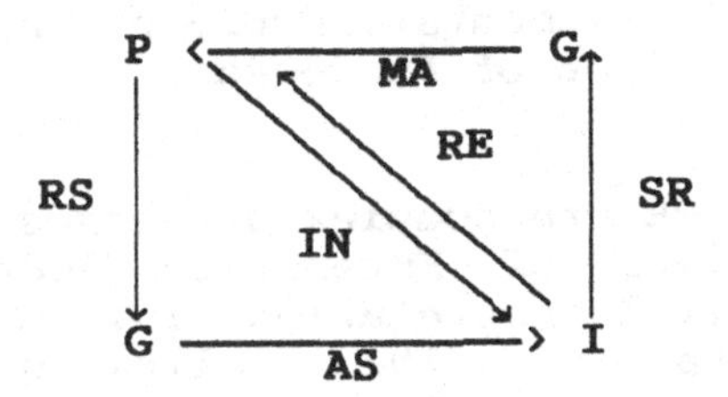

Fig.1 The diagram of elementary management cycle

The elementary management cycle consists of following elements: model of material-energy processing TMP (category P); model of information processing TIP (category I); structure of goals (objectives) of Enterprise (category G); process of restructuring of TMP in accordance with definite goal (morphism RS); process of aspectisation of TMP, i.e. semantisation of TIP in accordance with definite goal (morphism AS); process of interpretation as composition of RS and AS (morphism IN); process of decision structuring (morphism SR) and implementation of it (morphism MA) in accordance with definite goal; process of realisation (morphism RE) as composition of SR and MA.

Suppose that all units of EMC are necessary - the managing of Enterprise doesn't fit within the quality if the lack of any unit of any EMC occur. Decomposition of the EMC in the so called Space of Processes (Gudas[5]) makes the data, knowledge and objectives interactions at managing evident (figure 2). The technology of material-energy processing (TMP) is related with technology of information processing (TIP) by feedback loop created by the interpretation (IN) and realisation (RE).

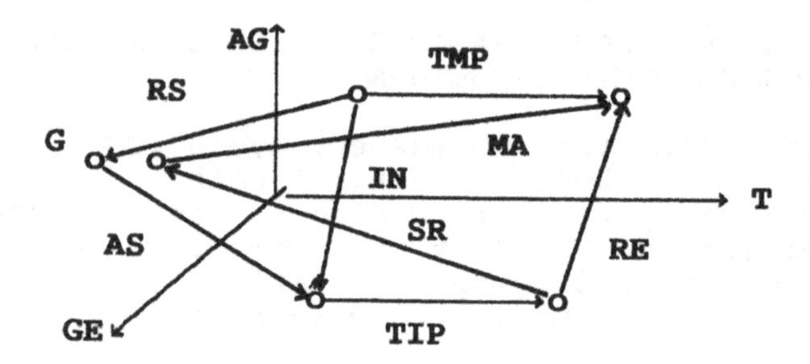

Fig.2. Elementary cycle of management in the Space of Processes

IN and RE are complex processes closely related with objectives of Enterprise located on the plane (AG,GE). Interpretation (IN) of situation in TMP consists of the following two stages : restructuring (RS) of TMP (real world) to the set of equally important entities making no structure; aspectisation (AS), i.e. transformation of a set of entities into a certain semantic frame in accordance with a definite goal of investigation. Realisation (RE) of some managing solution consists of the following two stages : structuring (SR) of the managing solution in terms ("language") of some adequate subgoal ,and materialisation (MA), i.e. changing the state of TMP.

The principal feature of the Organizational System - the intelligence of information processes at management - is emphasized by this decomposition. This is confirmed by the character of such information manipulations as interpretation IN and realisation RE. Namely the essence of interpretation is the acquisition of data and knowledge from the real world (TMP) when semantic model of reality is established adequately to definite subgoal of managing.The realisation of some managing solution is considered as the manipulation when semantic of solution by the way of its materialisation is transformed adequately to the fixed goal (subgoal) of Enterprise (or department) and influence to the state of TMP is made.So all three aspects of information - syntactic, semantical and pragmatical - are included in information processing at management.

4.ORGANIZATIONAL SYSTEM AS THE HIERARCHY OF ELEMENTARY MANAGEMENT CYCLES

The processes of aggregation (AG) and generalization (GE) divide the set of the EMC into levels and in this way two different types of EMC hierarchies arise in the Space of Processes : hierarchy of aggregation and hierarchy of generalization of EMC. Formally in such case each EMC is identified by two indexes : i - the number of aggregation level, and j - the number of generalization level. Naturally, the third index r for the type of activity of Organizational System must be included.The graphical representations of EMC hierarchies are presented in figures 3 - 5. In the figure 3 all activities r,r+1,..., r+n, ...,r+m are on the same level i of aggregation. Every mark j,j+1,... on the axis GE notes the level of generalization of EMC.Each single activity r may be managed by different number of EMC of various levels of generalization. Besides the generalization level of the same activity may be different on different levels of aggregation as presented in figure 4.

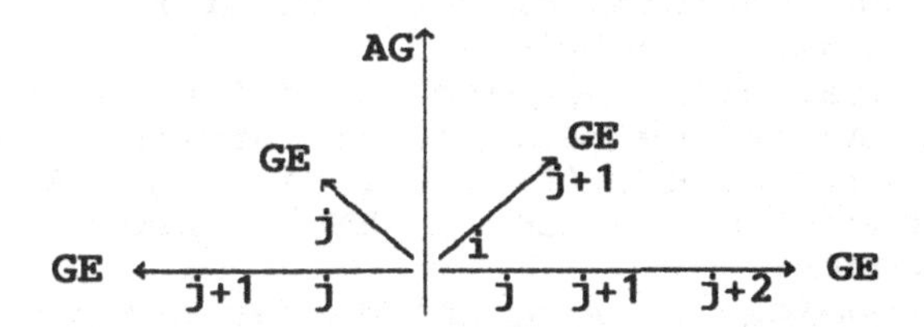

Fig.3.The set of EMC of the same level i of aggregation.

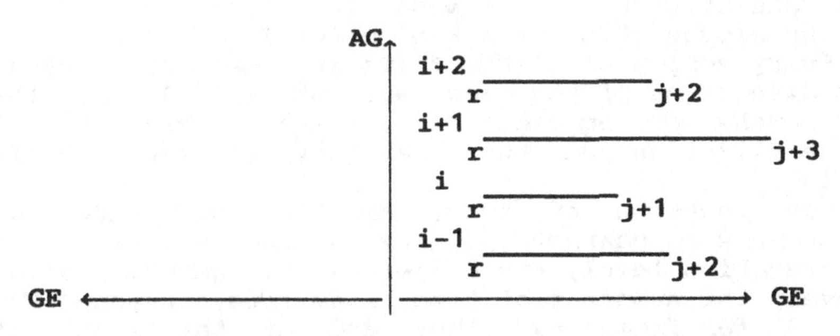

Fig.4.The different level of generalization of the same activity r in different levels of aggregation.

And ,finally the Organizational System as a system of EMC hierarchies is presented in figure 5.This is the graphical model of information processing hierarchy in Organizational Systems decomposed in the Space of Processes. Several

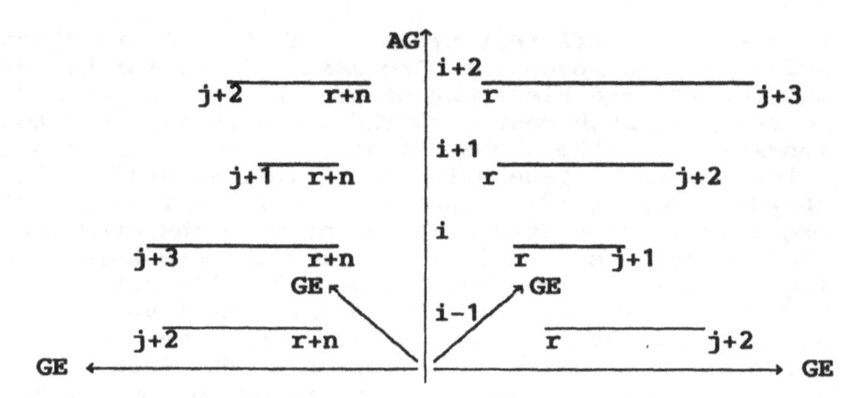

Fig.5.The hierarchy of EMC as the model of management of Organizational System.

important features of the such hierarchical system of EMC must be outlined :

1.An interaction between two different EMC in this hierarchy may be realised as interface between any components of EMC (fig.1) Thus, one EMC can make influence to another EMC through such its components:

a) technology of material-energy processing P;
b) interpretation IN;
c) data processing DP;
d) decision making DM;
e) realisation RE;
f) structure of goals G;
g) data structures D of model I;
h) knowledge structures K of model I.

2.Every component of EMC forms its own hierarchies in directions of the axes AG, GE and T, so the hierarchy of aggregation , the hierarchy of generalization and the hierarchy of time levels arise.

3.The location of every EMC in the Space of Processes is characterized by : the number i of aggregation level; the number j of generalization level; the number t of time item.Thus, anyone EMC may differ from any other EMC by the level of aggregation, level of generalization and number of time period.

4.A set of EMC is implemented in Organizational System by a set of departments (or other units of administrative system).In general the correlation between the EMC and departments may be as follows:

a) a single EMC is carried out by one department;
b) a part of the EMC is carried out by one department,i.e. a set of departments implements one

EMC; c) some EMC are implemented by one department.
5. Information which is used for management of Organizational System can be devided into three diverse components : data, knowledge and objectives(goals) .Every component has its own place in the Space of Processes: data and knowledge structures are clossely related and are placed (fig.2) on the plane (AG,T) and the plane (GE,T).The structure of objectives of Enterprise is placed on the plane (AG,GE) as separate structure, which is related with TMP and TIP by the processes of interpretation IN and realisation RE.

5.TAXONOMY OF HIERARCHIES

The set of hierarchical relation types which take place in model of information processing shown in figure 5 is presented in figure 6.

	The units of E M C										
	IN		I		G	DP	DM	RE		P	
	RS	AS	D	K				SR	MA	TMP	TIP
AG	H^{AG}_{IN}		H^{AG}_{I}		H^{AG}_{G}	H^{AG}_{DP}	H^{AG}_{DM}	H^{AG}_{RE}		H^{AG}_{P}	
	H^{AG}_{RS}	H^{AG}_{AS}	H^{AG}_{D}	H^{AG}_{K}				H^{AG}_{SR}	H^{AG}_{MA}	H^{AG}_{TMP}	H^{AG}_{TIP}
GE	H^{GE}_{IN}		H^{GE}_{I}		H^{GE}_{G}	H^{GE}_{DP}	H^{GE}_{DM}	H^{GE}_{RE}		-	
	H^{GE}_{RS}	H^{GE}_{AS}	H^{GE}_{D}	H^{GE}_{K}				H^{GE}_{SR}	H^{GE}_{MA}	-	-

Fig.6 Taxonomy of hierarchies of Organizational System

Consiquently, hierarchical structures of the following types may arise in the direction of aggregation process (fig.2):
- aggregation hierarchy of interpretation processes H^{AG}_{IN};
- aggregation hierarchy of data processing H^{AG}_{DP};
- aggregation hierarchy of decision making processes H^{AG}_{DM};
- aggregation hierarchy of realisation processes H^{AG}_{RE};
- aggregation hierarchy of material units of technologies (products technology TMP and information technology TIP) H^{AG}_{P};
- aggregation hierarchy H^{AG}_{I} of information elements of model I, which can be decomposed into two separate hierarchies of data items H^{AG}_{D} and knowledge

items H_K^{AG}.

Theoretically several EMC may be located on the same aggregation level i for managing of the same object. In this way the generalization hierarchy for every stage of EMC arises. That may occur then several methods of different levels of abstraction are used to solve the same problem. Thus, the following taxonomy of generalization hierarchies of information units at management arise in Enterprise:

- generalization hierarchy of interpretation H_{IN}^{GE} which defines the abstraction levels of interpretation rules and procedures;
- generalization hierarchy of data processing H_{DP}^{GE} which defines the levels of abstraction of data structures and procedures;
- generalization hierarchy of decision making H_{DM}^{GE} which includes data, knowledge structures and procedures;
- generalization hierarchy of decision realisation processes H_{RE}^{GE};
- generalization hierarchy H_I^{GE} of information elements of model I, which can be decomposed into seperate hierarchies of data items D and knowledge items K : H_D^{GE} and H_K^{GE}.
- generalization hierarchy of goals H_G^{GE} of Organizational System ;

Complex processes IN and RE are compositions of RS, AS and SR, MA adequatly (fig.1,2), and so hierarchies H_{RS}^{AG}, H_{AS}^{AG}, H_{RS}^{GE}, H_{AS}^{GE} and H_{RS}^{AG}, H_{AS}^{AG}, H_{RS}^{GE}, H_{AS}^{GE} are included in taxonomy (fig.6).

The taxonomy of hierarchies given above can be compared with that in (Mesarovic[2]). For example, the hierarchy of layers can be expressed as both the aggregation hierarchy H_{DM}^{AG} and generalization hierarchy H_{DM}^{GE} of decision making; the strata's hierarchy in (Mesarovic[2]) is adequate for the aggregation hierarchy H_P^{AG} of technologies units. Other types of EMC hierarchies would be named too in order to outline the matter of them. For example, H_{IN}^{AG} may be named "the hierarchy of competence"; H_{DP}^{AG} - "the hierarchy of decision support"; H_{RE}^{AG} -"the hierarchy of decision executives".

6. HIERARCHICAL RELATIONS BETWEEN DEPARTMENTS

A set of EMC is implemented in Organizational System by a set of departments (or other units) of administrative system. The correlation between the EMC and departments can be various : a single EMC is carried out by one department; a set of departments implements one EMC; some EMC are implemented by one department.So ,a set of the different types of information relations arises between departments. For example, a set of types of relations between two departments when each of them implements all steps of EMC is presented in figure 7.

N.	The basic types of hierarchical relations													
	H^{AG}_{IN}	H^{AG}_{I}	H^{AG}_{G}	H^{AG}_{DP}	H^{AG}_{DM}	H^{AG}_{RE}	H^{AG}_{TMP}	H^{AG}_{TIP}	H^{GE}_{IN}	H^{GE}_{I}	H^{GE}_{G}	H^{GE}_{DP}	H^{GE}_{DM}	H^{GE}_{RE}
1	+	-	-	-	-	-	-	-	-	-	-	-	-	-
2	-	+	-	-	-	-	-	-	-	-	-	-	-	-
...														
14	-	-	-	-	-	-	-	-	-	-	-	-	-	+
15	+	+	-	-	-	-	-	-	-	-	-	-	-	-
...														
16382	+	+	+	+	+	+	+	+	+	+	+	+	+	-
16383	+	+	+	+	+	+	+	+	+	+	+	+	+	+

Fig.7 The set of relation types between two departments

The total number of different types of relations between two departments is very high : $2^{14} - 1 = 16383$.This is the best illustration of complexity of information processes at management.

7.TAXONOMY OF THE CO-ORDINATION PROCESSES

The hierarchical system of information units and relations discussed above is ruther complicated. The managing of Organizational System is interpreted as a process of interactions of EMC in that hierarchical system.Interactions between departments of different levels of hierarchy in (Mesarovic[2]) are named "co-ordination".The taxonomy of hierarchies presented above makes a basis for investigation of co-ordination process.

It seems the classification of hierarchical interactions between departments (or other administrative units) of Enterprise may be useful for successfull investigations of intelligent information processing.The classes of interfaces between departments of Enterprise can be separated in accordance with mutual disposition of EMC in the Space of Processes.From what has been stated above it can be concluded that such situations are

reasonable:
1.The stages of the single EMC are executed by different departments.So all that departments are located on the same level of aggregation and on the same level of generalization hierarchies. Thus, the relations between different stages of EMC are not hierarchical.They are related in time as the steps of the one single managing process.That can't be classified as co-ordination because interactions among agents of the same EMC are not hierarchical.
2.Two or more EMC of different level of generalization are used to manage the same object on the one level of aggregation.Thus, several solutions of the problem will be generated, but only one of them can be realised.In such case the concretisation of the solution must be done.The co-ordination of this type can be named "horizontal concretisation".
3.Two EMC of the same object are located on the different levels of aggregation hierarchy, but the level of generalization is the same. In this case the elements of the lower EMC became the units of the structure for the stages of the higher level EMC. Consequently, the EMC of higher level can co-ordinate the lower EMC by acting on stages (one or some) of them. So, the vertical co-ordination of the following types arises - co-ordination by interface with processes of : a)interpretation, b)data processing, c)decision making, d)realisation of solution, and co-ordination by interface with structures of : e)data, f)knowledge, g)goals in both model I and model G.
The main feature of such vertical co-ordination is the influence over information on the lower levels of aggregation hierarchy (detailized information).This type of co-ordination can be named "detailizing co-ordination".
4.Two EMC are on the different levels of both - aggregation and generalization -hierarchies. In this case two-dimentional interfaces between EMC and their elements arise : along the AG axis and along the GE axis. Co-ordination of such type may be named "two-dimentional co-ordination".
The classification of co-ordination is continued now taking into account the technological relations between the objects of TMP, i.e. the third axis T (Time) of Space of Processes.
5.Two or more EMC are located on the same levels of aggregation and generalization, but the objects of managing are different. Objects are in special technological interdependence as elements of technology (TMP) .Consequently, the additional EMC of the higher level of aggregation is necessary.The object of managing of that EMC includes (as the

elements of its structure) the objects of the lower EMC.So ,the "vertical technological co-ordination" can be separated.

6.Two EMC are located on the same level of aggregation and on different levels of generalization hierarchies ,the objects of managing are interrelated. The actions of departments which implement these different EMC must be co-ordinate in time and in level of solution concretisation (generalization).Thus a supplementary EMC (department) for the co-ordination of such type (concretizing- technological) is necessary.

7.Two EMC are on the same level of generalization and on different levels of aggregation hierarchies, the objects of managing are interrelated.In this case co-ordination of activities in time and in level of solution detalisation is necessary. Thus the supplementary EMC (department) for co-ordination of such type (detalizing-technological) must be included.

8.The levels of aggregation and generalization of two EMC are different, objects of managing are technologicaly related.This is the general case of mutual disposition of different EMC , and so, the separate type of co-ordination (three-dimentional) must be implemented.

EMC has the structure (fig.1,2) , so one can conclude that each type of co-ordination is the interactions between EMC elements IN, G, I, RE ,P. Thus in this way all types of co-ordination mentioned above can be decomposed into set of subtypes.

8.RELATION BETWEEN THE HIERARCHY OF DEPARTMENTS AND GENERALIZATION LEVEL OF INFORMATION

A set of departments of administrative system of organization usually is presented as equilateral triangle and the number of departments decreases as the level of their hierarchy (the level of aggregation) increases.What can be stated about the regularity of information generalization in different levels of administration? It seems it would be complicated to define the precise dependence, but some characteristic cases can be picked out (figure 7).

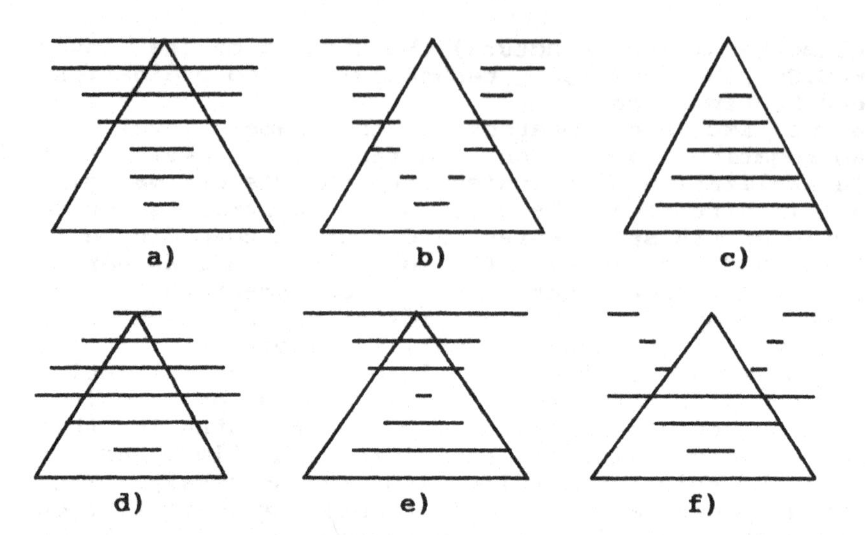

Fig.8.The relation between the administrative hierarchy and the generalization level of information.

The level of generalization of information in figure 8a goes up together with the aggregation level of departments.It seems to be the most natural case in managing of Organizational System. It can be supposed that lower levels of generalization of information in the departments of the higher levels are not important and so , can be regarded (figure 8b). In the figure 8c the level of "competence" decreases as the aggregation level of departments increases (in organizations with the distinct bureaucratic character).The competence of the departments in middle levels of organization is the reason for efficiency (figure 8d) and the obstacle for improvement (figure 8e) of the situation.
The figure 8f picks up the difference between information used in functional (specialized) departments on lower levels and in administrative departments on the higher levels of aggregation hierarchy.

CONCLUSIONS

This is the attempt at a theoretical foundation of framework for information processing in Organizational Systems which is oriented to incorporating intelligence into management information systems (MIS).The decomposition of information processes of management in the Space of Processes picks up that the interfaces of different information structures in management includes all three aspects of information : sintactic , semantical and pragmatical.It seems that this feature of information processing is nearly the basic for the intelligent processes.Thus, the framework discussed above emphasizes the main conceptual elements of intelligent information systems and interfaces between them.

The basic results of the approach to the problem of hierarchy of information processes in Organizational Systems presented in this paper are as follows:

1. The concept "hierarchy" has a sense only if the concrete space is considered. For example, the hierarchies of aggregation, of generalization and of time units are separated in the Space of Processes.
2. The reason for hierarchy in Organizational Systems are the processes of aggregation and generalization of information - data,.knowledge and goal items.The administrative departments are suppossed to be the bearers ("hardware") for these information processes ("software").
3. The elementary management cycle (EMC) is considered to be the basic unit of information processing hierarchy in Organizational Systems. Thus the Organizational System is suppossed to be the hierarchy of EMC for various activities. All departments which implements the steps of one EMC are on the same levels of both hierarchies - the hierarchy of aggregation and the hierarchy of generalization.
4. The co-ordination of EMC located on the different levels of aggregation and generalization hierarchies are necessary to obtain the global goal of Organizational System. The co-ordination can be implemented by influence to everyone element of EMC in two different directions: along the axis of aggregation (AG) and along the axis of generalization (GE).
5. The framework of information processing hierarchy in Organizational Systems given above seems to be the structure suitable for the development of intelligent information systems.

REFERENCES

1. Elzas,M.S.The Kinship between Artificial Intelligence, Modelling and Simulation: an appraisal. Modelling and Simulation Methodology in the Artificial Intelligence Era, (Ed.M.S.Elzas, T.I.Oren, B.P. Zeigler), pp.3-13,North-Holland, 1986.

2. Mesarovic,M.D., Mako,D., Takahara,Y. Theory of Hierarchical, Multilevel Systems. Academic Press, N.Y., London, 1970.

3. Ackoff,R. Towards a System of Systems Concepts, Management Science, V.77, N.11, pp.661-671, 1971.

4. Gudas,S. Organizational System as a type of Systems,Scientific Works of Lithuania Higher Schools, V.19, Vilnius, pp.26-34, 1989.

5. Gudas,S. A Framework for research of Information processing in Management, Proceedings of Fifth International Conference "Applications of AI in Engineering", 17-20 July, 1990, Boston.

6. Gudas,S. Formalization of Management Process Unit for Organizational Systems,Scientific Works of Lithuania Higher Schools, V.20, Vilnius, pp.28-41, 1989 .

AUTHORS' INDEX

Printed in the United States
By Bookmasters